LIBRARY-LRC
TEXAS HEART INSTITUTE

Stem Cell Biology and Regenerative Medicine

Series Editor
Kursad Turksen, Ph.D.
kturksen@ohri.ca

For other titles published in this series, go to
www.springer.com/series/7896

Ira S. Cohen • Glenn R. Gaudette
Editors

Regenerating the Heart

Stem Cells and the Cardiovascular System

Editors
Ira S. Cohen
Department of Physiology and Biophysics
Stony Brook University
Stony Brook, NY
USA
ira.cohen@stonybrook.edu

Glenn R. Gaudette
Department of Biomedical Engineering
Worcester Polytechnic Institute
Worcester, MA
USA
gaudette@wpi.edu

ISBN 978-1-61779-020-1 e-ISBN 978-1-61779-021-8
DOI 10.1007/978-1-61779-021-8
Springer New York Dordrecht Heidelberg London

Library of Congress Control Number: 2011922496

© Springer Science+Business Media, LLC 2011
All rights reserved. This work may not be translated or copied in whole or in part without the written permission of the publisher (Humana Press, c/o Springer Science+Business Media, LLC, 233 Spring Street, New York, NY 10013, USA), except for brief excerpts in connection with reviews or scholarly analysis. Use in connection with any form of information storage and retrieval, electronic adaptation, computer software, or by similar or dissimilar methodology now known or hereafter developed is forbidden.
The use in this publication of trade names, trademarks, service marks, and similar terms, even if they are not identified as such, is not to be taken as an expression of opinion as to whether or not they are subject to proprietary rights.
While the advice and information in this book are believed to be true and accurate at the date of going to press, neither the authors nor the editors nor the publisher can accept any legal responsibility for any errors or omissions that may be made. The publisher makes no warranty, express or implied, with respect to the material contained herein.

Printed on acid-free paper

Humana Press is part of Springer Science+Business Media (www.springer.com)

Contents

Contributors

Francesco Angelini
Department of Experimental Medicine,
Cenci-Bolognetti Foundation, University "Sapienza" of Rome, Rome, Italy

Priya R. Baraniak
The Wallace H. Coulter Department of Biomedical Engineering at Georgia
Institute of Technology and Emory University, Atlanta, GA, USA

Lucio Barile
Department of Biotechnology and Biosciences,
University "Bicocca" of Milan, Milan, Italy

Alexander M. Becker
The Heart Center Göttingen,
George-August University, Göttingen, Germany

Antonio Paolo Beltrami
Centro Interdipartimentale di Medicina Rigenerativa (CIME),
Università degli Studi di Udine, Udine, Italy

Carlo Alberto Beltrami
Centro Interdipartimentale di Medicina Rigenerativa (CIME),
Università degli Studi di Udine, Udine, Italy

Ofer Binah
The Sohnis Family Stem Cells Center, The Rappaport Family Institute for
Research in the Medical Sciences, The Department of Physiology, Ruth and Bruce
Rappaport Faculty of Medicine, Technion – Israel Institute of Technology,
Haifa, Israel

Muath Bishawi B.S
Department of Surgery, Stony Brook University Medical Center,
Stony Brook, NY, USA

Peter R. Brink
Department of Physiology and Biophysics, Stony Brook University, Stony Brook, NY, USA

Daniela Cesselli
Centro Interdipartimentale di Medicina Rigenerativa (CIME), Università degli Studi di Udine, Udine, Italy

Isotta Chimenti
Department of Experimental Medicine, Cenci-Bolognetti Foundation, University "Sapienza" of Rome, Rome, Italy

Ira S. Cohen
Department of Physiology & Biophysics, Stony Brook University, Stony Brook, NY, USA

Kenneth Day
Hudson Alpha Institute for Biotechnology, Huntsville, AL, USA

Tanja Dominko
Department of Biology and Biotechnology, Bioengineering Institute, Worcester Polytechnic Institute, Worcester, MA, USA
and
Cellthera, Inc., Southbridge, MA, USA

J. Kevin Donahue
Heart and Vascular Research Center, MetroHealth Hospital, Case Western Reserve University, Cleveland, OH, USA

Sergey Doronin
Department of Physiology and Biophysics, Stony Brook University, Stony Brook, NY, USA

Yi Duan
Department of Biomedical Engineering, Columbia University, New York, NY, USA

Sarah Fernandes
Center for Cardiovascular Biology,University of Washington Medicine at South Lake Union, Seattle, WA, USA

Loren J. Field
The Riley Heart Research Center, Herman B Wells Center for Pediatric Research, Indiana University School of Medicine, Indianapolis, IN, USA

Elvira Forte
Department of Experimental Medicine, Cenci-Bolognetti Foundation, University "Sapienza" of Rome, Rome, Italy

Yingli Fu
Russell H. Morgan Department of Radiology
and Radiological Science School of Medicine,
The Johns Hopkins University, Baltimore, MD, USA

Keiichi Fukuda
Department of Cardiology, KEIO University School of Medicine, Tokyo, Japan

Roberto Gaetani
Department of Experimental Medicine, Cenci-Bolognetti Foundation,
University "Sapienza" of Rome, Rome, Italy

Glenn R. Gaudette
Department of Biomedical Engineering, Worcester Polytechnic Institute,
Worcester, MA, USA

Lior Gepstein
The Sohnis Family Research Laboratory for Cardiac Electrophysiology
and Regenerative Medicine and the Rappaport Family Institute
for Research in the Medical Sciences, The Bruce Rappaport Faculty
of Medicine, Technion – Israel Institute of Technology, Haifa, Israel

Alessandro Giacomello
Department of Molecular Medicine and Pathology,
Cenci-Bolognetti Foundation, University "Sapienza" of Rome,
Rome, Italy

Amandine F. G. Godier-Furnémont
Department of Biomedical Engineering,
Columbia University, New York, NY, USA

Jacques P. Guyette
Department of Biomedical Engineering, Worcester Polytechnic Institute,
Worcester, MA, USA

Tracy A. Gwyther
Department of Biomedical Engineering, Worcester Polytechnic Institute,
Worcester, MA, USA

Simon P. Hoerstrup
Swiss Center for Regenerative Medicine, Department of Surgical Research
and Center for Clinical Research, University of Hospital of Zurich,
Switzerland

Vittoria Ionta
Department of Experimental Medicine, Cenci-Bolognetti Foundation,
University "Sapienza" of Rome, Rome, Italy

Joseph Itskovitz-Eldor
Chairman, Department of Ob-Gyn, Rambam Health Care Campus Head, Stem Cell Center, Technion – Israel Institute of Technology, PoB 9602, Haifa 31096, Israel

André G. Kléber
Department of Physiology, University of Bern, Bühlplatz 5, 3012 Bern, Switzerland

Dara L. Kraitchman
Russell H. Morgan Department of Radiology and Radiological Science, School of Medicine, The Johns Hopkins University, Baltimore, MD, USA

Kenneth R. Laurita
Heart and Vascular Research Center, MetroHealth Hospital, Case Western Reserve University, Cleveland, OH, USA

Robert Maidhof
Department of Biomedical Engineering, Columbia University, New York, NY, USA

Shinji Makino
Center for Integrated Medical Research, and Department of Cardiology, Therapeutics, KEIO University School of Medicine, Tokyo, Japan

Christopher Malcuit
Bioengineering Institution, Worcester Polytechnic Institute, Worcester, MA, USA
and
Cellthera, Inc., Southbridge, MA, USA

Richard T. Mathias
Department of Physiology and Biophysics, Stony Brook University, Stony Brook, NY, USA

Todd C. McDevitt
The Wallace H. Coulter Department of Biomedical Engineering at Georgia Institute of Technology and Emory University, Petit Institute for Bioengineering and Bioscience, Atlanta, GA, USA

Elisa Messina
Department of Experimental Medicine, Cenci-Bolognetti Foundation, University "Sapienza" of Rome, Rome, Italy

Timothy J. Nelson
General Internal Medicine and Transplant Center Mayo Clinic, Rochester, MN, USA

Raymond L. Page
Department of Biomedical Engineering,
Department of Biology and Biotechnology, Bioengineering Institute,
Worcester Polytechnic Institute, Worcester MA, USA;
Cellthera, Inc., Southbridge, MA, USA

Hans Reinecke
Center for Cardiovascular Biology, University of Washington
Medicine at South Lake Union, Seattle, WA, USA

Richard B. Robinson
Department of Pharmacology and Center for Molecular Therapeutics,
Columbia University College of Physicians and Surgeons,
New York, NY 10032, USA

Marsha W. Rolle
Department of Biomedical Engineering, Worcester Polytechnic Institute,
Worcester, MA, USA

Michael R. Rosen
Department of Pharmacology, Department of Pediatrics, Center for Molecular
Therapeutics, Columbia University, New York, NY, USA

Todd K. Rosengart
Department of Surgery, Stony Brook University Medical Center,
Stony Brook, NY, USA

Michael Rubart
The Riley Heart Research Center, Herman B Wells Center for Pediatric Research,
Indiana University School of Medicine, Indianapolis, IN, USA

Oshra Sedan
The Sohnis Family Stem Cells Center, The Rappaport Family Institute for
Research in the Medical Sciences, The Department of Physiology, Ruth &
Bruce Rappaport Faculty of Medicine, Technion – Israel Institute of Technology,
Haifa, Israel

Andre Terzic
Marriott Heart Disease Research Program, Division of Cardiovascular Diseases,
Department of Medicine, Molecular Pharmacology
and Experimental Therapeutics and Medical Genetics,
Mayo Clinic, Rochester, MN, USA

Gordana Vunjak-Novakovic
Department of Biomedical Engineering, Columbia University,
New York, NY, USA

Benedikt Weber
Swiss Center for Regenerative Medicine,
Department of Surgical Research and Center for Clinical Research,
University of Hospital of Zurich, Zurich, Switzerland

Michal Weiler-Sagie
The Sohnis Family Research Laboratory for Cardiac Electrophysiology
and Regenerative Medicine, The Rappaport Family Institute
for Research in the Medical Sciences, The Bruce Rappaport Faculty of Medicine,
Technion – Israel Institute of Technology, Haifa, Israel

Kai C. Wollert
Hans-Borst Center for Heart and Stem Cell Research,
Department of Cardiology and Angiology, Hannover Medical School,
Hanover, Germany

Zipora Yablonka-Reuveni
Department of Biological Structure, University of Washington
School of Medicine, Seattle, WA, USA

Naama Zeevi-Levin
Sohnis and Forman Families Center for Stem Cell and Tissue Regeneration
Research, Ruth & Bruce Rappaport, Faculty of Medicine, Technion, Haifa, Israel

Introduction

Ira S. Cohen and Glenn R. Gaudette

The twentieth century witnessed many positive changes for heart disease. In particular, our ability to treat myocardial ischemia was greatly improved. Developments from coronary artery bypass surgery to drug-eluding stents restored blood flow and helped keep coronary blood vessels open, thereby extending patients lives. Electrophysiology also made great strides, with implantable electronic pacemakers and automatic internal defibrillators providing patients with electrical rhythm control. Public awareness of heart disease also increased over the past 100 years. Multiple groups have formed to fight heart disease and educate the general public with regard to this debilitating disease. Automatic external defibrillators are showing up in public locations, demonstrating public awareness of the seriousness of heart disease. As we look forward to the twenty-first century and consider the next great challenges, we see the potential of cell therapy to address many cardiovascular diseases. From heart failure to atrioventricular nodal dysfunction, the young but promising field of cell therapy is likely to play a significant role in developing a cure during this century.

Both of us entered the stem cell field less than a decade ago; one of us an electrophysiologist, the other as a mechanical engineer. Like others we were attracted by the opportunity for a real breakthrough. Arrhythmias and heart failure had one thing in common: neither pharmacology nor devices were a panacea. Instead, therapies represented the best that modern medicine had to offer, but certainly were far short of a cure. With our backgrounds, we both faced the same problem: How do we accumulate sufficient knowledge in this burgeoning field to think creatively? Together we attended meetings sponsored by the National Heart, Lung, and Blood Institute, and by the American Heart Association and found them helpful, but ultimately we continuously found ourselves at a disadvantage, almost as if we were entering in the middle of a long conversation without a good source to quickly catch up on the field. A book to document where the field has been, where it is, and where

I.S. Cohen (✉)
Department of Physiology and Biophysics,
Stony Brook University, SUNY, Stony Brook, NY, USA
e-mail: ira.cohen@sunysb.edu

I.S. Cohen and G.R. Gaudette (eds.), *Regenerating the Heart*, Stem Cell Biology and Regenerative Medicine, DOI 10.1007/978-1-61779-021-8_1,
© Springer Science+Business Media, LLC 2011

it is heading would be helpful. To our surprise, we were recently invited to write a book on stem cells and the cardiovascular system. Fortunately, we had sufficient self-insight to "know what we know and also know what we don't know" and declined the offer. However, a renewed offer to edit such a book held much more interest. First knowing what we did not know, this text offered us the opportunity to learn from the experts we invited. Second, we could organize this missing knowledge in a manner that would afford others like us an opportunity to learn as well.

Our major challenge for this book was to organize the field into a tractable body of knowledge. We decided to organize the text into four major sections. The first section considers mechanical regeneration. When the heart fails as a pump, the major cause is the loss of contractile elements (possibly more than a billion myocytes). In this first section of the book, we consider approaches to mechanical regeneration of cardiac function. Despite the large number of patients suffering from myocardial infarction, currently, the only clinical method available to add contractile myocytes is whole heart transplantation. However, the demand for hearts for transplant far exceeds the supply. Here, the potential for cell therapy is large. Multiple stem cells have demonstrated cardiogenic potential and so these cell types had to be reviewed individually. Embryonic stem cells, bone-marrow-derived cells, cardiac stem cells, and induced pluripotent stem cells are all considered. Both the basic properties of these cells and the methods to drive them toward cardiac lineages are considered. Skeletal myocytes may not be cardiogenic but do contract and are also included in this section because of their early role in clinical attempts at cardiac cellular myoplasty. However, not all cardiac regeneration by these cells occurs through cardiac differentiation and thus other regenerative mechanisms are considered. Further, the ability to differentiate stem cells into myocytes in vitro is necessary but not sufficient to achieve regeneration of mechanical function in vivo and so translational efforts in both animal and human trials are reviewed.

We next consider electrical regeneration. The mechanical function of the heart is triggered by the orderly electrical activation of each of its myocytes through a predefined electrical pathway. Each myocyte is electrically connected to all others, creating a functional electrical syncytium. Here the problem is somewhat different. It is not the massive loss of myocytes that creates the problem, but the punctuate loss of electrical connectivity or decreased excitability that is at fault. To consider the therapeutic potential of stem cells it is necessary to understand the genesis of arrhythmias and the basis of electrical connectivity in biological systems. Each of these topics is considered in individual chapters. Arrhythmias are classified into two types: bradyarrhythmias due to excessive slowing of heart rate, and tacchyarrhythmias due to excessive speeding of heart rate. Stem cell approaches to each of these common problems are discussed. Finally, it is worth considering what future stem cells have in the panoply of alternative therapies for electrical dysfunction and a chapter looking to the future concludes this section.

The heart is a complex tissue that subserves its mechanical function with various tissue types. The third section of our book considers cardiac tissues. These include heart valves which separate the upper and lower chambers as well as the lower chambers and the systemic and pulmonic circulations. One chapter reviews attempts to create biologic solutions to their replacement. Vessels which carry the blood throughout the systemic and pulmonic circulations frequently fail and exciting new approaches to vessel replacement are also considered. Finally, it is a dream of all cardiac researchers to replace not only individual myocytes or blood vessels but also complete cardiac tissue, and approaches to engineer such tissue are also considered.

Finally, this text would not be complete without considering the approaches employed to evaluate stem cell therapies in vivo. In the last section, there are chapters which consider methods for stem cell delivery to the myocardium, methods to track the delivered stem cells, and finally methods for how to assess their contributions to mechanical function.

We have learned greatly from the preparation of this text. We thank the authors for their fine contributions and hope that it contributes to the education of newly committed and veteran stem cell researchers alike.

Part I
Stem Cells for Regeneration of Mechanical Function

Inducing Embryonic Stem Cells to Become Cardiomyocytes

Alexander M. Becker, Michael Rubart, and Loren J. Field

Abstract Many forms of heart disease are associated with a decrease in the number of functional cardiomyocytes. These include congenital defects (e.g. hypoplastic and noncompaction syndromes) as well as acquired injuries (e.g. exposure to cardiotoxic agents or injuries resulting from coronary artery disease, hypertension, or surgical interventions). Although the adult mammalian heart retains some capacity for cardiomyocyte renewal (resulting from cardiomyocyte proliferation and/or cardiomyogenic stem cell activity), the magnitude of this regenerative process is insufficient to effect repair of substantively damaged hearts. It has become clear that exogenous cardiomyocytes transplanted into adult hearts are able to structurally and functionally integrate. It has also become clear that embryonic stem cells (ESCs), as well as induced progenitors with ESC-like characteristics, are able to generate bona fide cardiomyocytes in vitro and in vivo. These cells thus constitute a potential source of donor cardiomyocytes for therapeutic interventions in damaged hearts. This chapter reviews spontaneous cardiomyogenic differentiation in ESCs, methods used to generate enriched populations of ESC-derived cardiomyocytes, and current results obtained after engraftment of ESC-derived cardiomyocytes or cardiomyogenic precursors.

Keywords Cardiac differentiation • Intracardiac engraftment • Cell therapy

1 Introduction

The structure and cellular composition of the adult mammalian heart are complex; consequently, myocardial disease can manifest itself at many different levels, and can impact multiple structures and cell types (valves, coronary arteries, capillaries,

L.J. Field (✉)
The Riley Heart Research Center, Herman B Wells Center for Pediatric Research,
Indiana University School of Medicine,
Indianapolis, Indiana, USA
e-mail: ljfield@iupui.edu

I.S. Cohen and G.R. Gaudette (eds.), *Regenerating the Heart*, Stem Cell Biology and Regenerative Medicine, DOI 10.1007/978-1-61779-021-8_2,

© Springer Science+Business Media, LLC 2011

endothelial cells, veins, interstitial fibroblasts, nodal cells, conduction system cells, working cardiomyocytes, etc.). Advances in surgical and pharmacologic interventions, as well as the development of electrophysiologic and mechanical devices, have steadily advanced and currently provide a wide variety of viable treatments for many forms of heart disease. The elucidation of the molecular underpinnings of cell lineage commitment and morphogenesis provide additional avenues of treatment, particularly in the area of angiogenesis. Unfortunately, the ability to promote widespread replacement of lost contractile units (i.e. cardiomyocyte replacement) has remained elusive.

Developmental and molecular studies have identified progenitor cells which give rise to cardiomyocytes in the developing heart. Proliferation of immature but contracting cardiomyocytes is a major contributor to the increase in cardiac mass observed during fetal development. The proliferative capacity of cardiomyocytes decreases markedly in postnatal life. Additionally, several progenitor cell populations with cardiomyogenic activity identified during development are depleted or have lost their ability to form new cardiomyocytes in neonatal life. Nonetheless, evidence for cardiomyocyte proliferation and/or apparent cardiomyogenic stem cell activity has been reported in the adult heart. For example, quantitation of radioisotope incorporation into cardiomyocyte nuclei of individuals alive during atmospheric atomic bomb detonations suggested an annual cardiomyocyte renewal rate of approximately 1% in young adults [1], a value remarkably similar to that extrapolated from shorter pulse/chase tritiated thymidine incorporation studies in mice [2]. Although the findings of these studies collectively are more consistent with the notion of cardiomyocyte renewal via proliferation, they do not rule out potential contributions from cardiomyogenic stem cells. Indeed, studies employing an elegant conditional reporter transgene system suggested stem-cell-based regeneration following injury in adult mice [3]. The notion of cardiomyocyte renewal in the adult heart has been with us for a long time – at issue is the magnitude of the regenerative response, a point which is the subject of intense research and debate among cardiomyocyte aficionados. What is clear is that the adult heart lacks the ability to reverse damage following the loss of large numbers of cardiomyocytes.

Studies from the 1990s demonstrated that donor cells could successfully engraft the hearts of recipient animals. Proof-of-concept experiments showed that cardiomyocytes from enzymatically dispersed fetal mouse hearts structurally integrated into the hearts of adult recipients following direct intracardiac injection [4, 5]. Subsequent analyses demonstrated that the donor cardiomyocytes formed a functional syncytium with the host myocardium, using the presence of intracellular calcium transients as a surrogate marker for contractile activity [6]. Although promising, it was soon apparent that only a small fraction of the injected cardiomyocytes survived and engrafted [7], a problem which remains a major obstacle for clinical efficacy of this approach. Nonetheless, several studies have reported that intracardiac injection of fetal cardiomyocytes could preserve cardiac function following experimental injury in rodents [8–10].

In light of these observations, considerable effort has been invested to identify potential sources of donor cardiomyocytes, or alternatively progenitor cells with cardiomyogenic activity. Toward that end, many cells with apparent cardiomyogenic activity have been reported in the recent literature, a remarkable observation given that the intrinsic regenerative rate in the adult myocardium is quite low. Many factors likely contribute to this phenomenon. For example, the presence of multiple markers, or alternatively the transient expression of different markers, could result in an individual cell or cell lineage being categorized as multiple cells/lineages. The ability of some cell types to fuse with cardiomyocytes [11] could result in their false identification as cardiomyogenic stem cells. The relative rigor of the assays employed to detect cardiomyogenic activity could also contribute to the identification of false positives. It is also possible that reprogramming during in vitro propagation unmasked or enhanced cardiomyogenic potential. Given the intense activity in the field, it is likely that the true in vitro and in vivo cardiomyogenic activity of the various progenitor cells identified to date will rapidly be either validated or repudiated.

It is well established that embryonic stem cells (ESCs) are able to generate bona fide cardiomyocytes [12]. ESCs are derived from the inner cell mass (ICM) of preimplantation embryos [13, 14]. ESCs can be propagated in vitro in an undifferentiated state, and when allowed to differentiate can form endodermal, ectodermal, and mesodermal derivatives in vitro and in vivo. ESCs thus constitute a potential source of donor cardiomyocytes (or alternatively, donor cardiomyogenic progenitors) for therapeutic interventions targeting diseased hearts. In this chapter we review the spontaneous cardiomyogenic differentiation in ESCs, the various methods which have been developed to generate enriched populations of ESC-derived cardiomyocytes, and the current status of preclinical studies aimed at regenerating myocardial tissue via engraftment of ESC-derived cardiomyocytes or cardiomyogenic precursors. We then consider the challenges which must be overcome for successful translation to the clinic.

2 ESCs and Spontaneous Caridomyogenic Differentiation

After fertilization, initial growth of the preimplantation mammalian embryo is characterized by rapid cell division. Cells within the embryo begin to differentiate at the 16-cell stage (morula). As development proceeds, cells on the periphery of the morula give rise to trophoblasts (which, together with maternal endometrium, form the placenta) and cells in the center of the morula give rise to the ICM (which forms the embryo). The resulting blastocyst remains surrounded by the zona pellucida. Blastocysts can be cultured on feeder layers of mitomycin-treated mouse embryonic fibroblasts (MEFs). In the example shown in Fig. 1a, the MEFs were derived from transgenic mice carrying a transgene encoding leukemia inhibitory factor (LIF). LIF activates the Janus kinase/signal transducer and activator of transcription (JAK/STAT) and mitogen-activated protein kinase pathways and suppresses differentiation in mouse ESCs (but is not required for generating human ESCs).

After several days of culture, the zona pellucida of the preimplantation embryo will rupture, allowing the outgrowth of both trophoblasts and ICM cells (Fig. 1b–d). The two cell types were readily distinguished by phase-contrast microscopy, with the

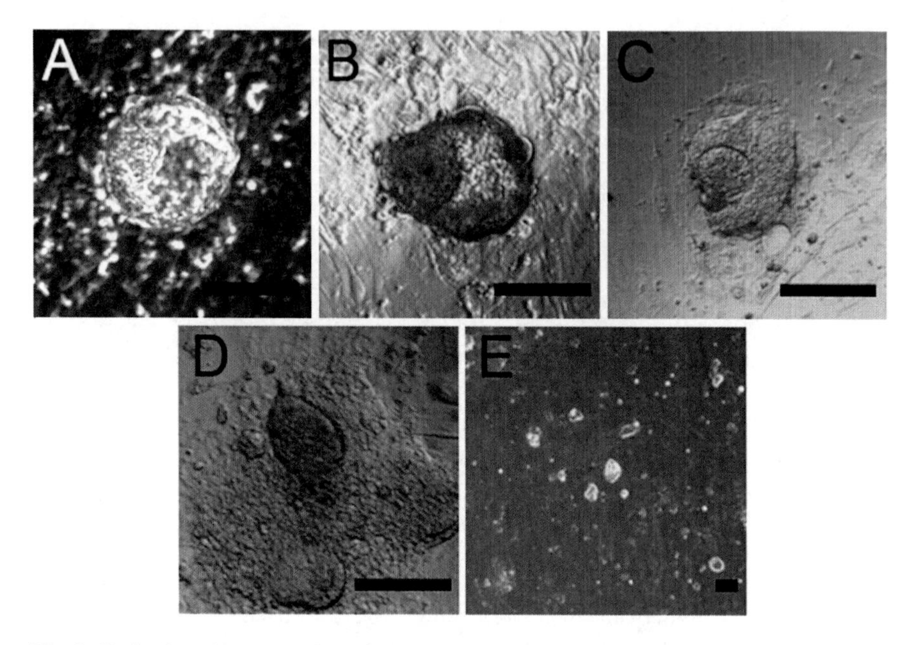

Fig. 1 Derivation of mouse embryonic stem cell (ESC) lines. (**a**) Blastocyst isolated at 3.5 days post coitus and culture for 3 days on a mouse embryonic fibroblast (MEF) feeder layer. Note the presence of the zona pellucida (a refractile ring surrounding the embryo) and the inner cell mass. (**b–d**) Blastocysts after 6, 7, and 8 days of culture on the MEF feeder layer. Note that the zona pellucida ruptures with time, releasing inner cell mass and trophoblast cells. (**e**) ESC lines after three passages. *Bar* 100 μm

ICM derivatives exhibiting a very dense, refractile morphology. Clusters of refractile cells were then physically isolated, dispersed, and replated onto MEF feeder layers. This process was repeated until clonal ESC lines were established (Fig. 1e). Mouse ESC lines can be propagated extensively in an undifferentiated state as long as care is exercised to maintain high levels of LIF and to limit colony size.

Early studies demonstrated that, when cultivated in suspension, ESCs form multicellular aggregates which have been termed "embryoid bodies" (EBs) [14]. Stochastic signaling between different cell types within the EBs mimics in vivo developmental induction cues, and upon further differentiation (either in suspension or adherent culture) the EBs give rise to ecto-, endo-, and mesodermal derivatives. Wobus and colleagues [15] developed a very useful technique to generate EBs with reproducible ESC content (which in turn resulted in more reproducible patters of differentiation). This entailed placing microdrops of medium seeded with a fixed number of undifferentiated ESCs on the inner surface of a tissue culture dish lid. The lid was then gently inverted so as to prevent mixing of the microdrops, and was placed on a tissue culture dish containing medium. The resulting "hanging drops" provide an ideal environment for the ESCs to coalesce and form EBs in a highly reproducible manner. Subsequent studies by Zweigerdt and colleagues demonstrated that EBs with reproducible ESC content could be generated in bulk in tissue culture dishes on rotating devices [16] or in stirred bioreactors [17].

To document cardiomyogenic differentiation using the hanging drop approach, ESCs were generated using blastocysts derived from myosin heavy chain (MHC)-enhanced green fluorescent protein (EGFP) transgenic mice. These mice carry a transgene comprising the cardiomyocyte-restricted α-MHC promoter and an EGFP reporter. The transgene targets EGFP expression in cardiomyocytes [6], and thus provides a convenient reporter to trace cardiomyogenic activity in differentiating ESC cultures, as illustrated in Fig. 2 (Fig. 2a–c shows phase-contrast images of the EBs and adherent cultures, and Fig. 2d–f shows epifluorescence images of the same field). Individual dispersed ESCs were plated in hanging drops; after several days in culture, the ESCs formed EBs which continued to grow and differentiate. No EGFP epifluorescence was apparent, consistent with the absence of cardiomyogenic differentiation at this stage (Fig. 2a). The EBs were transferred from

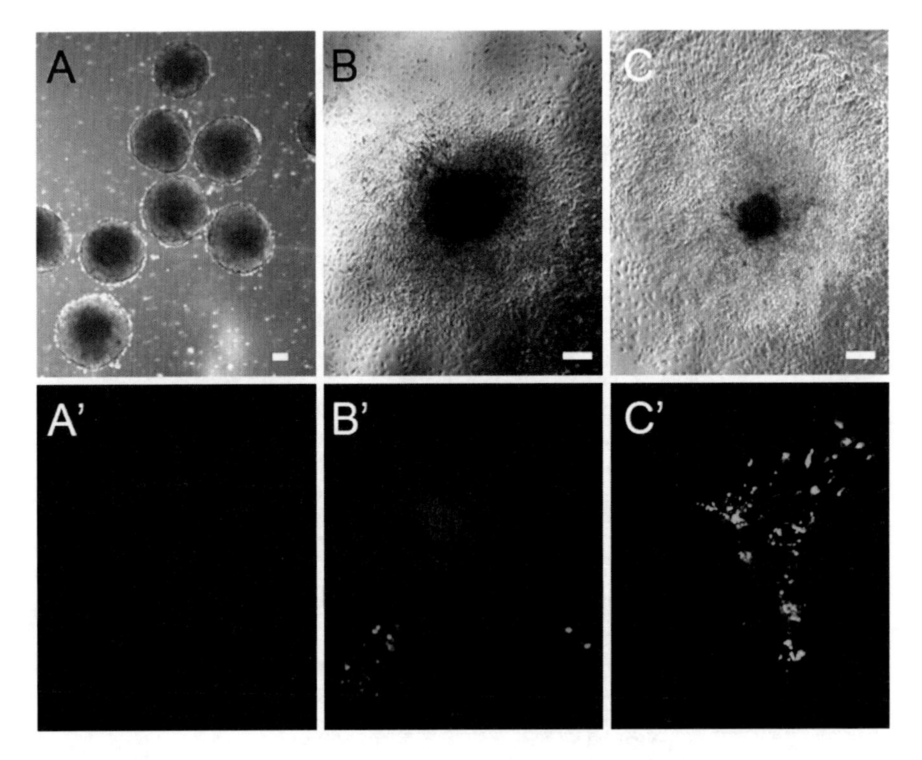

Fig. 2 Timeline of cardiomyogenic differentiation in mouse ESCs carrying a myosin heavy chain (MHC)-enhanced green fluorescent protein (EGFP) reporter transgene. (**a**, **a′**) Phase-contrast and epifluorescence images, respectively, of embryoid bodies (EBs) generated by the hanging drop procedure after 4 days of suspension culture. The absence of EGFP epifluorescence indicates that cardiomyocyte differentiation has not yet occurred. (**b**, **b′**) Phase-contrast and epifluorescence images, respectively, of an EB after 5 days of suspension culture and 2 days of adherent culture. A few scattered cells with EGFP epifluorescence indicates the initial onset of cardiomyocyte differentiation. (**c**, **c′**) Phase-contrast and epifluorescence images, respectively, of an EB after 5 days of suspension culture and 5 days of adherent culture. Most cardiomyogenic differentiation has occurred by this time. *Bar* 100 μm

suspension culture to adherent culture after 5 days of differentiation. Expression of the cardiomyocyte-restricted reporter transgene was first detected after 7 days of differentiation (i.e. 5 days of suspension culture and 2 days of adherent culture; Fig. 2b); however, contractile activity was not apparent until 3 days of differentiation. This reflected the time differential between the induction of myofiber structural protein gene expression (and, consequently, activation of the reporter transgene expression) and the assembly of functional myofibers and the requisite intracellular machinery for the generation and propagation of action potentials and calcium transients. It was also apparent that cardiomyocytes constituted only a small fraction of the total cell population during spontaneous ESC differentiation (Fig. 2c). With the development of human ESC lines [18], in vitro cardiomyocyte differentiation was rapidly observed and characterized [19].

3 Inducing ESCs to Produce Cardiomyocytes

Numerous approaches have been developed to generate enriched cultures of ESC-derived cardiomyocytes (Table 1). Perhaps the most obvious approach entails the identification of growth factors which enhance cardiomyocyte differentiation. Indeed,

Table 1 Approaches to enhance cardiomyocyte yield from embryonic stem cells (*ESCs*)

Approach	Comments	References
Growth factors	Retinoic acid enhanced cardiomyocyte differentiation in mouse ESCs	[20]
	Exogenous glucose, amino acids, vitamins, and selenium enhanced cardiomyocyte differentiation in mouse ESCs	[21]
	LIF enhances and inhibits cardiomyocyte commitment and proliferation in mouse ESCs in a developmental stage-dependent manner	[22]
	Reactive oxygen species enhanced cardiomyocyte differentiation in mouse ESCs	[23, 24, 25, 26]
	Endoderm enhanced cardiomyocyte differentiation in mouse ESCs	[27, 28]
	A TGF/BMP paracrine pathway enhanced cardiomyocyte differentiation in mouse ESCs	[29]
	Activation of the MEK/ERK pathway enhanced cardiomyocyte differentiation in mouse ESCs	[30]
	Verapamil and cyclosporine enhanced cardiomyocyte differentiation in mouse ESCs	[31]
	5-Aza-2′-deoxycytidine enhanced cardiomyocyte differentiation in human ESCs	[32]
	Endoderm cell lines enhanced cardiomyocyte differentiation in human ESCs	[33, 34]
	Ascorbic acid enhanced cardiomyocyte differentiation in human ESCs	[35]
	Directed differentiation with activin A and BMP4 in monolayers of human ESC	[36]

(continued)

Table 1 (continued)

Approach	Comments	References
Genetic engineering	Lineage-restricted drug resistance gene resulted in highly purified cardiomyocyte cultures from mouse ESCs	[37, 17, 38]
	Highly purified cardiomyocyte cultures generated by FACS of mouse ESCs expressing a lineage-restricted EGFP reporter	[39]
	Targeted expression of α-1,3-fucosyltransferase enhanced cardiomyocyte differentiation in mouse ESCs	[40]
	Coexpression of EA1, dominant negative p53, and dominant negative CUL7 enhanced cell cycle in mouse ESC-derived cardiomyocytes	[41]
	Expression of SV40 T antigen enhanced cell cycle in mouse ESC-derived cardiomyocytes	[42]
	Antagonization of Wnt/β-catenin enhanced cardiomyocyte differentiation in mouse ESCs	[43]
	Lineage-restricted drug resistance gene resulted in highly purified cardiomyocyte cultures from human ESCs	[44, 45]
Miscellaneous	A single 90-s electrical pulse applied to day 4 EBs increased cardiomyocyte differentiation in mouse ESCs	[46]
	Application of mechanical loading enhanced cardiomyocyte differentiation in mouse ESCs	[47, 48]
	FACS for transient Flk-1 isolated cardiomyogenic progenitors from mouse ESCs	[49]
	Cardiomyocyte enrichment using density centrifugation and cultures of cell aggregates in human ESCs	[50]
	Activin A, BMP4, bFGF, VEGF, and DKK1 treatment, followed by KDR$^+$/c-kit$^-$ FACS, identified cardiovascular progenitor cells in human ESCs	[51]

LIF leukemia inhibitory factor, *TGF* transforming growth factor, *BMP* bone morphogenetic protein, *MEK* mitogen-activated protein kinase, *ERK* extracellular-signal-regulated kinase, *FACS* fluorescence-activated cell sorting, *EGFP* enhanced green fluorescent protein, *EBs* embryoid bodies, *bFGF* basic fibroblast growth factor, *VEGF* vascular endothelial growth factor

many studies have reported modest to moderate increases in cardiomyocyte yield in differentiating ESC cultures. Perhaps the most impressive work was from Murry and colleagues [36], who demonstrated that treatment of monolayers of human ESCs with a combination of activin A and bone morphogenetic protein 4, followed by gradient centrifugation, resulted in an average final cardiomyocyte content of 82%. The degree to which this approach can be scaled up for the production of large numbers of donor cardiomyocytes (and, in particular, if directed differentiation is effective in suspension as opposed to in monolayer cultures) remains to be determined.

One of the earliest approaches to enhance cardiomyocyte yield entailed introduction of a lineage-restricted selectable marker. In one example, the cardiomyocyte-restricted MHC promoter was used to target expression of aminoglycoside phosphotransferase (MHC-neo transgene). After spontaneous differentiation, cultures were enriched for cardiomyocytes by simple treatment with G418 [37]. Cultures with more than 99% cardiomyocyte content can routinely be obtained.

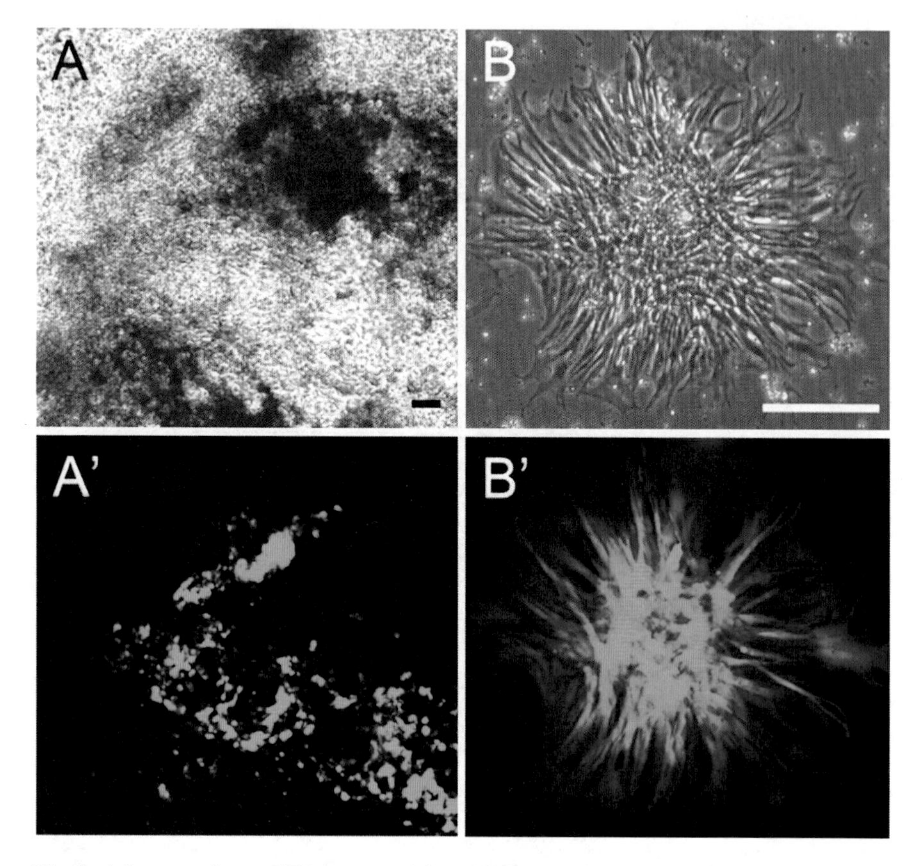

Fig. 3 Adherent culture of EBs generated from ESCs carrying the MHC-EGFP and MHC-neo transgenes after a total of 23 days of differentiation in the absence (**a**, **a′**) or presence (**b**, **b′**) of 11 days of G418 selection. (**a**, **b**) Phase-contrast images; (**a′**, **b′**) epifluorescence images. Note the marked cardiomyocyte enrichment in the G418-treated sample (**b**, **b′**). *Bar* 100 μm

To illustrate this approach, ESCs carrying the MHC-EGFP reporter transgene described earlier as well as the MHC-neo transgene were generated. The ESCs were allowed to differentiate spontaneously, and were then cultured in the absence or presence of G418. In the absence of G418, cardiomyocyte constituted only a small portion of the cultures, in agreement with the data presented above (Fig. 3a). In contrast, G418 treatment effectively eliminated the noncardiomyocytes, resulting in highly enriched cultures (Fig. 3b). This selection approach was readily scalable to bioreactors [52], and could yield more than 10^9 cardiomyocytes per 2-L reaction vessel in preparations seeded with dispersed ESC cultures [17]. Similarly, lineage-restricted expression of an EGFP reporter has been employed in conjunction with fluorescence-activated cell sorting (FACS) to generate highly enriched cardiomyocyte cultures [39]. Importantly, both the selection-based and the FACS-based approaches were readily used for the generation of human ESC-derived cardiomyocytes [44, 45, 53].

4 Intracardiac Transplantation of ESCs or ESC-derived Cardiomyocytes

Given that the ability to form teratomas in syngeneic or immune-compromised hosts is a major criterion for ESC identification, one would a priori expect that delivery of undifferentiated ESCs into the heart would also give rise to teratomas. Indeed, teratomas were reported following ESC injection into normal [37] or infarcted [54] myocardium. Nonetheless a number of studies have delivered undifferentiated ESCs and failed to report teratoma formation (Table 2). This could reflect compromised differentiation capacity in the ESCs being tested, or alternatively the insufficient

Table 2 Intracardiac transplantation of ESCs or ESC-derived cardiomyocytes

Donor/host species	Comments	References
Mouse/mouse	Genetically selected cardiomyocytes engrafted in normal myocardium	[37]
Mouse/mouse	In vivo cardiomyocyte differentiation of ESCs required TGF and BMP2	[29]
Mouse/mouse	Intravenous ESC delivery improved cardiac function during viral myocarditis	[55]
Mouse/mouse	Cardiomyocyte-enriched cells plus VEGF enhanced postinfarct function	[56]
Mouse/mouse	Growth factors enhanced ESC engraftment in infarcted hearts	[57, 58]
Mouse/mouse	ESC-seeded synthetic scaffolds improved postinfarct function	[59]
Mouse/mouse	Allogenic ESCs evoked an immune response following heart transplant	[60, 61]
Mouse/mouse	Matrigel enhanced ESC seeding in infarcted hearts	[62]
Mouse/mouse	Genetically selected cardiomyocytes improved postinfarct function	[63]
Mouse/mouse	ESCs improved function in infarcted hearts	[64, 65]
Mouse/mouse	TNF enhanced cardiomyocyte differentiation and lessened teratoma potential	[66]
Mouse/mouse	Cardiomyocytes improved postinfarct function via paracrine mechanisms	[67]
Mouse/mouse	In vivo MR imaging of transplanted cardiomyocytes in infarcted hearts	[68]
Mouse/mouse	Allogenic ESCs formed teratomas when transplanted into infarcts	[54]
Mouse/mouse	Cardiomyocyte engraftment blocked adverse post-MI remodeling	[69]
Mouse/rat	ESC transplantation improved function following myocardial infarction	[70]
Mouse/rat	Density-gradient-enriched cardiomyocytes improve postinfarct function	[71]
Mouse/rat	Differentiated ES cultures survived in immune-suppressed normal heart	[72]
Mouse/rat	ESC-seeded synthetic scaffolds improved postinfarct function	[73]
Mouse/rat	ESCs improved cardiac function in aging hearts	[74]

(continued)

Table 2 (continued)

Donor/host species	Comments	References
Mouse/rat	GCSF enhanced cardiomyocyte engraftment in infarcted hearts	[75]
Mouse/rat	Intravenously delivered ESCs homed to infarcted myocardium	[76]
Mouse/rat	ESCs formed teratomas when transplanted into infarcted hearts	[77]
Mouse/rat	Chitosan hydrogel enhanced ESC seeding and postinfarct function	[78]
Mouse/sheep	Enriched cardiomyocytes improved postinfarct function	[79, 80]
Human/mouse	Allopurinol/uricase/ibuprofen increased postinfarct cardiomyocyte survival	[81]
Human/mouse	Cardiomyocyte impact on adverse post-MI remodeling is transient	[82, 83]
Human/mouse	KDR progenitors for 3 lineages in vivo improved post-MI function	[51]
Human/rat	In vivo MR imaging of transplanted ESCs	[84]
Human/rat	Microdissected cardiomyocytes improved function in infarcted hearts	[85]
Human/rat	Cardiomyocytes engrafted athymic hearts after ischemia/reperfusion	[86]
Human/rat	Cardiomyocyte engraftment blocked adverse post-MI remodeling	[36, 87]
Human/rat	Cardiomyocytes from BMP2 treatment engrafted infarcted hearts	[88]
Human/rat	ESCs do not form teratomas when engrafted into infarcted hearts	[89]
Human/rat	Physically enriched cardiomyocytes engrafted normal athymic rat heart	[90]
Human/guinea pig	Mixed SAN and cardiomyocyte transplants provided pacemaker activity	[91]
Human/pig	Mixed SAN and cardiomyocyte transplants provided pacemaker activity	[92]

MR magnetic resonance, *MI* myocardial infarction, *GCSF* granulocyte colony stimulating factor, *SAN* sinoatrial node

histologic analyses of the engrafted hearts. It has also been suggested that the milieu of the normal or infarcted heart may be sufficient to drive lineage-restricted differentiation of progenitor cells. Nonetheless, the bulk of available data suggest that this is not the case for transplanted ESCs.

Since the initial observation that ESC-derived cardiomyocytes could successfully engraft recipient hearts [37], a large number of experiments have been performed to examine the impact of injecting ESCs or ESC-derived cardiomyocytes into normal or injured hearts (Table 2). Of note, many of these studies indicated that animals receiving ESCs or ESC-derived cardiomyocytes following experimental injury exhibited superior cardiac function as compared with those which did not receive cells. In almost all instances, cardiac function was not improved in the engrafted hearts. Rather, the process of engraftment appeared to attenuate the deleterious postinjury ventricular remodeling and concomitant decreases in cardiac function. Similar results have been reported with a number of donor cell types. In particular, work by Dzau and colleagues using mesenchymal stem cells strongly suggests that the benefit of cell transplantation in their studies likely reflects the secretion of proangiogenic and antiapoptotoic factors from donor cells [93–97]. Such a mechanism would readily explain how engraftment of a relatively small number of ESC-derived cells could impact function in injured hearts. Unfortunately, studies from the Mummery laboratory suggest that this improvement in postinjury remodeling may be transient in nature [82, 83].

Although there are direct data at the cellular level supporting the functional engraftment of fetal cardiomyocytes in recipient hearts [6], the current data available with ESC-derived cardiomyocytes are more circumstantial in nature. Gepstein and colleagues [92] demonstrated that ectopic pacemaker activity originated at the site of engraftment of human ESC-derived cells following atrioventricular node blockade in swine, consistent with the notion that the donor cells were functionally integrated. Similar results were obtained with guinea pig [91]. Despite these promising observations, it would be prudent to directly assess at the cellular level the ability of ESC-derived cardiomyocytes to functionally integrate following engraftment, as formation of a functional syncytium is an absolute requirement for regenerative repair. This is particularly important for studies wherein human ESC-derived cells promoted better function when engrafted into rodent hearts, as it is not at all clear that human cells can sustain rapid rates for extended periods of time. Indeed, rapid pacing is often used to induce heart failure in larger experimental animals [98].

5 Future Challenges

The discussions herein suggest that donor cardiomyocytes likely functionally integrate following transplantation into recipient hearts, that methods are available to eliminate the risk of teratoma formation following transplantation of ESC-derived cardiomyocytes, and that approaches to the large scale generation of ESC-derived cells are in hand. Perhaps the greatest challenge facing the use of ESC-derived cardiomyocytes for myocardial regeneration is the limitation in graft size using current approaches. Arguably the best study to date, by Murry and colleagues [36], utilized a combination of materials to enhance survival of donor cells after engraftment. This intervention permitted on average 4% replacement of an infarct which constituted 10% of the left ventricle (which correlates to only 0.4% of the ventricular mass). Thus, we have a long way to go before we will be able to replace transmural myocardial defects.

A number of approaches can be explored to attempt to enhance graft size. For example, many cardiomyocyte prosurvival pathways have been identified [99]. Targeting these pathways in donor ESC-derived cardiomyocytes, either by genetic intervention prior to cardiomyocgenic differentiation or via pharmacologic interventions, may facilitate enhancement of donor cell survival, as exemplified by the work of Murry and colleagues [36]. Similarly, many pathways which impact cardiomyocyte cell cycle activity have been identified [100, 101]. Once again, genetic modification of the ESCs prior to cardiomyogenic differentiation might permit enhanced growth of the cardiomyocyte grafts. Several recent studies suggest that pharmacologic interventions might also be exploited to enhance cardiomyocyte proliferation, particularly if the engrafted cardiomyocytes are not yet terminally differentiated [102–104]. Tissue engineering approaches may also enhance donor cell engraftment. For example, comparably large myocardial replacement was achieved by surgical attachment of collagen-based casts seeded with neonatal rat cardiomyocytes [105, 106].

An alternative strategy to enhance graft size is to transplant ESC-derived committed cardiomyogenic progenitors as opposed to differentiated ESC-derived cardiomyocytes. This approached is based on the notion that progenitor population may exhibit better survival characteristics, and may also be able to undergo postengraftment expansion, thereby resulting in enhanced graft size. Cardiomyogenic progenitors have been identified in differentiating ESC cultures based on transient expression of the vascular endothelial growth factor 2 receptor [49, 51], Nkx2.5 [107, 108], Isl-1 [109, 110], MESP1 [111], or a combination of OCT4, SSEA-1, and MESP1 [112]. Most of the progenitors have been shown to give rise to endothelial and/or smooth muscle cells, in addition to cardiomyocytes. The presence of vascular progenitors might enhance postengraftment donor cell survival and facilitate graft growth. Indeed, enhanced graft size was noted using progenitors isolated by virtue of Isl-1 expression [113]. Transplantation of ESC-derived progenitors into nonhuman primates [112] has also recently been reported.

Clinical use of established ESC lines will likely require some level of immune suppression. The development of immune suppression protocols used for allogenic cadaveric β-cell transplantation [114] would likely be directly transferable to the transplantation of ESC derivatives. The ability to generate autologous ESCs or ESC-like cells would circumvent the need for immune suppression. Several approaches have been developed to accomplish this, including nuclear transfer [115] as well as the generation of ESC-like cells from spermatogonial [116] and mesenchymal [117] stem cells. The ability to generate induced pluripotent stem (iPS) cells developed by Yamanaka and colleagues [118–121] provides a very powerful approach for the generation of autologous donor ESC-derived cells. Importantly, iPS cells exhibit robust and bona fide cardiomyogenic differentiation [122, 123], and many of the interventions and results described above will likely be directly transferrable to iPS derivatives. The main limitation that will likely affect the clinical use of reprogrammed cells is the time requirements for the reprogramming event(s) and implementation of quality control measures to ensure that the individual lines are competent for differentiation and nontumorigenic.

Collectively the studies reviewed herein raise the hope that ESC-derived cells might be useful for the treatment of heart disease, and specifically for the replacement of lost cardiomyocytes. The field has advanced remarkably since the initial report of successful cardiomyocyte engraftment 16 years ago. Given the influx of talented basic and clinical researchers in the field, it is hopeful that the challenges limiting the clinical use of ESC-derived cells will be overcome, and that ESC-derived cardiomyocyte (or cardiomyocyte precursor) transplantation will become a viable option for individuals with end-stage heart failure.

References

1. Bergmann, O., Bhardwaj, R.D., Bernard, S., et al. (2009) Evidence for cardiomyocyte renewal in humans. *Science*, 324, 98–102
2. Soonpaa, M.H. and Field, L.J. (1997) Assessment of cardiomyocyte DNA synthesis in normal and injured adult mouse hearts. *Am J Physiol*, 272, H220–6
3. Hsieh, P.C., Segers, V.F., Davis, M.E., et al. (2007) Evidence from a genetic fate-mapping study that stem cells refresh adult mammalian cardiomyocytes after injury. *Nat Med*, 13, 970–74
4. Koh, G.Y., Soonpaa, M.H., Klug, M.G., et al. (1995) Stable fetal cardiomyocyte grafts in the hearts of dystrophic mice and dogs. *J Clin Invest*, 96, 2034–42
5. Soonpaa, M.H., Koh, G.Y., Klug, M.G., et al. (1994) Formation of nascent intercalated disks between grafted fetal cardiomyocytes and host myocardium. *Science*, 264, 98–101
6. Rubart, M., Pasumarthi, K.B., Nakajima, H., et al. (2003) Physiological coupling of donor and host cardiomyocytes after cellular transplantation. *Circ Res*, 92, 1217–24
7. Muller-Ehmsen, J., Whittaker, P., Kloner, R.A., et al. (2002b) Survival and development of neonatal rat cardiomyocytes transplanted into adult myocardium. *J Mol Cell Cardiol*, 34, 107–16
8. Li, R.K., Jia, Z.Q., Weisel, R.D., et al. (1996) Cardiomyocyte transplantation improves heart function. *Ann Thorac Surg*, 62, 654–60; discussion 660–1
9. Muller-Ehmsen, J., Peterson, K.L., Kedes, L., et al. (2002a) Rebuilding a damaged heart: long-term survival of transplanted neonatal rat cardiomyocytes after myocardial infarction and effect on cardiac function. *Circulation*, 105, 1720–6
10. Sakai, T., Li, R.K., Weisel, R.D., et al. (1999) Fetal cell transplantation: a comparison of three cell types. *J Thorac Cardiovasc Surg*, 118, 715–24
11. Alvarez-Dolado, M., Pardal, R., Garcia-Verdugo, J.M., et al. (2003) Fusion of bone-marrow-derived cells with Purkinje neurons, cardiomyocytes and hepatocytes. *Nature*, 425, 968–73
12. Doetschman, T.C., Eistetter, H., Katz, M., et al. (1985) The in vitro development of blastocyst-derived embryonic stem cell lines: formation of visceral yolk sac, blood islands and myocardium. *J Embryol Exp Morphol*, 87, 27–45
13. Evans, M.J. and Kaufman, M.H. (1981) Establishment in culture of pluripotential cells from mouse embryos. *Nature*, 292, 154–56
14. Martin, G.R. (1981) Isolation of a pluripotent cell line from early mouse embryos cultured in medium conditioned by teratocarcinoma stem cells. *Proc Natl Acad Sci U S A*, 78, 7634–8
15. Maltsev, V.A., Rohwedel, J., Hescheler, J., et al. (1993) Embryonic stem cells differentiate in vitro into cardiomyocytes representing sinusnodal, atrial and ventricular cell types. *Mech Dev*, 44, 41–50
16. Zweigerdt, R., Burg, M., Willbold, E., et al. (2003) Generation of confluent cardiomyocyte monolayers derived from embryonic stem cells in suspension: a cell source for new therapies and screening strategies. *Cytotherapy*, 5, 399–413
17. Schroeder, M., Niebruegge, S., Werner, A., et al. (2005) Differentiation and lineage selection of mouse embryonic stem cells in a stirred bench scale bioreactor with automated process control. *Biotechnol Bioeng*, 92, 920–33
18. Thomson, J.A., Itskovitz-Eldor, J., Shapiro, S.S., et al. (1998) Embryonic stem cell lines derived from human blastocysts. *Science*, 282, 1145–47
19. Kehat, I., Kenyagin-Karsenti, D., Snir, M., et al. (2001) Human embryonic stem cells can differentiate into myocytes with structural and functional properties of cardiomyocytes. *J Clin Invest*, 108, 407–14

20. Wobus, A.M., Kaomei, G., Shan, J., et al. (1997) Retinoic acid accelerates embryonic stem cell-derived cardiac differentiation and enhances development of ventricular cardiomyocytes. *J Mol Cell Cardiol*, 29, 1525–39
21. Guan, K., Furst, D.O. and Wobus, A.M. (1999) Modulation of sarcomere organization during embryonic stem cell-derived cardiomyocyte differentiation. *Eur J Cell Biol*, 78, 813–23
22. Bader, A., Al-Dubai, H. and Weitzer, G. (2000) Leukemia inhibitory factor modulates cardiogenesis in embryoid bodies in opposite fashions. *Circ Res*, 86, 787–94
23. Buggisch, M., Ateghang, B., Ruhe, C., et al. (2007) Stimulation of ES-cell-derived cardiomyogenesis and neonatal cardiac cell proliferation by reactive oxygen species and NADPH oxidase. *J Cell Sci*, 120, 885–94
24. Sauer, H., Rahimi, G., Hescheler, J., et al. (2000) Role of reactive oxygen species and phosphatidylinositol 3-kinase in cardiomyocyte differentiation of embryonic stem cells. *FEBS Lett*, 476, 218–23
25. Sharifpanah, F., Wartenberg, M., Hannig, M., et al. (2008) Peroxisome proliferator-activated receptor alpha agonists enhance cardiomyogenesis of mouse ES cells by utilization of a reactive oxygen species-dependent mechanism. *Stem Cells*, 26, 64–71
26. Wo, Y.B., Zhu, D.Y., Hu, Y., et al. (2008) Reactive oxygen species involved in prenylflavonoids, icariin and icaritin, initiating cardiac differentiation of mouse embryonic stem cells. *J Cell Biochem*, 103, 1536–50
27. Bader, A., Gruss, A., Hollrigl, A., et al. (2001) Paracrine promotion of cardiomyogenesis in embryoid bodies by LIF modulated endoderm. *Differentiation*, 68, 31–43
28. Rudy-Reil, D. and Lough, J. (2004) Avian precardiac endoderm/mesoderm induces cardiac myocyte differentiation in murine embryonic stem cells. *Circ Res*, 94, e107–16
29. Behfar, A., Zingman, L.V., Hodgson, D.M., et al. (2002) Stem cell differentiation requires a paracrine pathway in the heart. *Faseb J*, 16, 1558–66
30. Kim, H.S., Cho, J.W., Hidaka, K., et al. (2007) Activation of MEK-ERK by heregulin-beta1 promotes the development of cardiomyocytes derived from ES cells. *Biochem Biophys Res Commun*, 361, 732–38
31. Sachinidis, A., Schwengberg, S., Hippler-Altenburg, R., et al. (2006) Identification of small signalling molecules promoting cardiac-specific differentiation of mouse embryonic stem cells. *Cell Physiol Biochem*, 18, 303–14
32. Xu, C., Police, S., Rao, N., et al. (2002) Characterization and enrichment of cardiomyocytes derived from human embryonic stem cells. *Circ Res*, 91, 501–508
33. Mummery, C., Ward-van Oostwaard, D., Doevendans, P., et al. (2003) Differentiation of human embryonic stem cells to cardiomyocytes: role of coculture with visceral endoderm-like cells. *Circulation*, 107, 2733–40
34. Mummery, C.L., Ward, D. and Passier, R. (2007) Differentiation of human embryonic stem cells to cardiomyocytes by coculture with endoderm in serum-free medium. *Curr Protoc Stem Cell Biol*, Chapter 1, Unit 1F 2
35. Takahashi, T., Lord, B., Schulze, P.C., et al. (2003) Ascorbic acid enhances differentiation of embryonic stem cells into cardiac myocytes. *Circulation*, 107, 1912–16
36. Laflamme, M.A., Chen, K.Y., Naumova, A.V., et al. (2007) Cardiomyocytes derived from human embryonic stem cells in pro-survival factors enhance function of infarcted rat hearts. *Nat Biotechnol*, 25, 1015–24
37. Klug, M.G., Soonpaa, M.H., Koh, G.Y., et al. (1996) Genetically selected cardiomyocytes from differentiating embronic stem cells form stable intracardiac grafts. *J Clin Invest*, 98, 216–24
38. Zandstra, P.W., Bauwens, C., Yin, T., et al. (2003a) Scalable production of embryonic stem cell-derived cardiomyocytes. *Tissue Eng*, 9:4
39. Muller, M., Fleischmann, B.K., Selbert, S., et al. (2000) Selection of ventricular-like cardiomyocytes from ES cells in vitro. *Faseb J*, 14, 2540–8
40. Sudou, A., Muramatsu, H., Kaname, T., et al. (1997) Le(X) structure enhances myocardial differentiation from embryonic stem cells. *Cell Struct Funct*, 22, 247–51
41. Pasumarthi, K.B., Tsai, S.C. and Field, L.J. (2001) Coexpression of mutant p53 and p193 renders embryonic stem cell-derived cardiomyocytes responsive to the growth-promoting activities of adenoviral E1A. *Circ Res*, 88, 1004–11

42. Huh, N.E., Pasumarthi, K.B., Soonpaa, M.H., et al. (2001) Functional abrogation of p53 is required for T-Ag induced proliferation in cardiomyocytes. *J Mol Cell Cardiol*, 33, 1405–19
43. Singh, A.M., Li, F.Q., Hamazaki, T., et al. (2007) Chibby, an antagonist of the Wnt/beta-catenin pathway, facilitates cardiomyocyte differentiation of murine embryonic stem cells. *Circulation*, 115, 617–26
44. Anderson, D., Self, T., Mellor, I.R., et al. (2007) Transgenic enrichment of cardiomyocytes from human embryonic stem cells. *Mol Ther*, 15, 2027–36
45. Xu, X.Q., Zweigerdt, R., Soo, S.Y., et al. (2008) Highly enriched cardiomyocytes from human embryonic stem cells. *Cytotherapy*, 10, 376–89
46. Sauer, H., Rahimi, G., Hescheler, J., et al. (1999) Effects of electrical fields on cardiomyocyte differentiation of embryonic stem cells. *J Cell Biochem*, 75, 710–23
47. Gwak, S.J., Bhang, S.H., Kim, I.K., et al. (2008) The effect of cyclic strain on embryonic stem cell-derived cardiomyocytes. *Biomaterials*, 29, 844–56
48. Shimko, V.F. and Claycomb, W.C. (2008) Effect of mechanical loading on three-dimensional cultures of embryonic stem cell-derived cardiomyocytes. *Tissue Eng Part A*, 14, 49–58
49. Kattman, S.J., Huber, T.L. and Keller, G.M. (2006) Multipotent flk-1+ cardiovascular progenitor cells give rise to the cardiomyocyte, endothelial, and vascular smooth muscle lineages. *Dev Cell*, 11, 723–32
50. Xu, C., Police, S., Hassanipour, M., et al. (2006) Cardiac bodies: a novel culture method for enrichment of cardiomyocytes derived from human embryonic stem cells. *Stem Cells Dev*, 15, 631–9
51. Yang, L., Soonpaa, M.H., Adler, E.D., et al. (2008) Human cardiovascular progenitor cells develop from a KDR+ embryonic-stem-cell-derived population. *Nature*, 453, 524–28
52. Zandstra, P.W., Bauwens, C., Yin, T., et al. (2003b) Scalable production of embryonic stem cell-derived cardiomyocytes. *Tissue Eng*, 9, 767–78
53. Huber, I., Itzhaki, I., Caspi, O., et al. (2007) Identification and selection of cardiomyocytes during human embryonic stem cell differentiation. *Faseb J*, 21, 2551–63
54. Nussbaum, J., Minami, E., Laflamme, M.A., et al. (2007) Transplantation of undifferentiated murine embryonic stem cells in the heart: teratoma formation and immune response. *Faseb J*, 21, 1345–57
55. Wang, J.F., Yang, Y., Wang, G., et al. (2002) Embryonic stem cells attenuate viral myocarditis in murine model. *Cell Transplant*, 11, 753–58
56. Yang, Y., Min, J.Y., Rana, J.S., et al. (2002) VEGF enhances functional improvement of postinfarcted hearts by transplantation of ESC-differentiated cells. *J Appl Physiol*, 93, 1140–51
57. Kofidis, T., de Bruin, J.L., Yamane, T., et al. (2004) Insulin-like growth factor promotes engraftment, differentiation, and functional improvement after transfer of embryonic stem cells for myocardial restoration. *Stem Cells*, 22, 1239–45
58. Kofidis, T., de Bruin, J.L., Yamane, T., et al. (2005b) Stimulation of paracrine pathways with growth factors enhances embryonic stem cell engraftment and host-specific differentiation in the heart after ischemic myocardial injury. *Circulation*, 111, 2486–93
59. Ke, Q., Yang, Y., Rana, J.S., et al. (2005) Embryonic stem cells cultured in biodegradable scaffold repair infarcted myocardium in mice. *Sheng Li Xue Bao*, 57, 673–81
60. Kofidis, T., deBruin, J.L., Tanaka, M., et al. (2005c) They are not stealthy in the heart: embryonic stem cells trigger cell infiltration, humoral and T-lymphocyte-based host immune response. *Eur J Cardiothorac Surg*, 28, 461–66
61. Swijnenburg, R.J., Tanaka, M., Vogel, H., et al. (2005) Embryonic stem cell immunogenicity increases upon differentiation after transplantation into ischemic myocardium. *Circulation*, 112, I166–72
62. Kofidis, T., Lebl, D.R., Martinez, E.C., et al. (2005d) Novel injectable bioartificial tissue facilitates targeted, less invasive, large-scale tissue restoration on the beating heart after myocardial injury. *Circulation*, 112, I173–77
63. Kolossov, E., Bostani, T., Roell, W., et al. (2006) Engraftment of engineered ES cell-derived cardiomyocytes but not BM cells restores contractile function to the infarcted myocardium. *J Exp Med*, 203, 2315–27
64. Nelson, T.J., Ge, Z.D., Van Orman, J., et al. (2006) Improved cardiac function in infarcted mice after treatment with pluripotent embryonic stem cells. *Anat Rec A Discov Mol Cell Evol Biol*, 288, 1216–24

65. Singla, D.K., Hacker, T.A., Ma, L., et al. (2006) Transplantation of embryonic stem cells into the infarcted mouse heart: formation of multiple cell types. *J Mol Cell Cardiol*, 40, 195–200
66. Behfar, A., Perez-Terzic, C., Faustino, R.S., et al. (2007) Cardiopoietic programming of embryonic stem cells for tumor-free heart repair. *J Exp Med*, 204, 405–20
67. Ebelt, H., Jungblut, M., Zhang, Y., et al. (2007) Cellular cardiomyoplasty: improvement of left ventricular function correlates with the release of cardioactive cytokines. *Stem Cells*, 25, 236–44
68. Ebert, S.N., Taylor, D.G., Nguyen, H.L., et al. (2007) Noninvasive tracking of cardiac embryonic stem cells in vivo using magnetic resonance imaging techniques. *Stem Cells*, 25, 2936–44
69. Singla, D.K., Lyons, G.E. and Kamp, T.J. (2007) Transplanted embryonic stem cells following mouse myocardial infarction inhibit apoptosis and cardiac remodeling. *Am J Physiol Heart Circ Physiol*, 293, H1308–14
70. Min, J.Y., Yang, Y., Converso, K.L., et al. (2002) Transplantation of embryonic stem cells improves cardiac function in postinfarcted rats. *J Appl Physiol*, 92, 288–96
71. Hodgson, D.M., Behfar, A., Zingman, L.V., et al. (2004) Stable benefit of embryonic stem cell therapy in myocardial infarction. *Am J Physiol Heart Circ Physiol*, 287, H471–9
72. Naito, H., Nishizaki, K., Yoshikawa, M., et al. (2004) Xenogeneic embryonic stem cell-derived cardiomyocyte transplantation. *Transplant Proc*, 36, 2507–508
73. Kofidis, T., de Bruin, J.L., Hoyt, G., et al. (2005a) Myocardial restoration with embryonic stem cell bioartificial tissue transplantation. *J Heart Lung Transplant*, 24, 737–44
74. Min, J.Y., Chen, Y., Malek, S., et al. (2005) Stem cell therapy in the aging hearts of Fisher 344 rats: synergistic effects on myogenesis and angiogenesis. *J Thorac Cardiovasc Surg*, 130, 547–53
75. Cho, S.W., Gwak, S.J., Kim, I.K., et al. (2006) Granulocyte colony-stimulating factor treatment enhances the efficacy of cellular cardiomyoplasty with transplantation of embryonic stem cell-derived cardiomyocytes in infarcted myocardium. *Biochem Biophys Res Commun*, 340, 573–82
76. Min, J.Y., Huang, X., Xiang, M., et al. (2006) Homing of intravenously infused embryonic stem cell-derived cells to injured hearts after myocardial infarction. *J Thorac Cardiovasc Surg*, 131, 889–97
77. He, Q., Trindade, P.T., Stumm, M., et al. (2008) Fate of undifferentiated mouse embryonic stem cells within the rat heart: role of myocardial infarction and immune suppression. *J Cell Mol Med*, 13(1), 188–201
78. Lu, W.N., Lu, S.H., Wang, H.B., et al. (2008) Functional improvement of infarcted heart by co-injection of embryonic stem cells with temperature-responsive chitosan hydrogel. *Tissue Eng Part A*, 15, 1437–47
79. Cai, J., Yi, F.F., Yang, X.C., et al. (2007) Transplantation of embryonic stem cell-derived cardiomyocytes improves cardiac function in infarcted rat hearts. *Cytotherapy*, 9, 283–91
80. Menard, C., Hagege, A.A., Agbulut, O., et al. (2005) Transplantation of cardiac-committed mouse embryonic stem cells to infarcted sheep myocardium: a preclinical study. *Lancet*, 366, 1005–12
81. Kofidis, T., Lebl, D.R., Swijnenburg, R.J., et al. (2006) Allopurinol/uricase and ibuprofen enhance engraftment of cardiomyocyte-enriched human embryonic stem cells and improve cardiac function following myocardial injury. *Eur J Cardiothorac Surg*, 29, 50–55
82. van Laake, L.W., Passier, R., Doevendans, P.A., et al. (2008) Human embryonic stem cell-derived cardiomyocytes and cardiac repair in rodents. *Circ Res*, 102, 1008–10
83. van Laake, L.W., Passier, R., Monshouwer-Kloots, J., et al. (2007) Human embryonic stem cell-derived cardiomyocytes survive and mature in the mouse heart and transiently improve function after myocardial infarction. *Stem Cell Res*, 1, 9–24
84. Tallheden, T., Nannmark, U., Lorentzon, M., et al. (2006) In vivo MR imaging of magnetically labeled human embryonic stem cells. *Life Sci*, 79, 999–1006
85. Caspi, O., Huber, I., Kehat, I., et al. (2007) Transplantation of human embryonic stem cell-derived cardiomyocytes improves myocardial performance in infarcted rat hearts. *J Am Coll Cardiol*, 50, 1884–93

86. Dai, W., Field, L.J., Rubart, M., et al. (2007) Survival and maturation of human embryonic stem cell-derived cardiomyocytes in rat hearts. *J Mol Cell Cardiol*, 43, 504–16
87. Leor, J., Gerecht, S., Cohen, S., et al. (2007) Human embryonic stem cell transplantation to repair the infarcted myocardium. *Heart*, 93, 1278–84
88. Tomescot, A., Leschik, J., Bellamy, V., et al. (2007) Differentiation in vivo of cardiac committed human embryonic stem cells in postmyocardial infarcted rats. *Stem Cells*, 25, 2200–5
89. Xie, C.Q., Zhang, J., Xiao, Y., et al. (2007) Transplantation of human undifferentiated embryonic stem cells into a myocardial infarction rat model. *Stem Cells Dev*, 16, 25–29
90. Laflamme, M.A., Gold, J., Xu, C., et al. (2005) Formation of human myocardium in the rat heart from human embryonic stem cells. *Am J Pathol*, 167, 663–71
91. Xue, T., Cho, H.C., Akar, F.G., et al. (2005) Functional integration of electrically active cardiac derivatives from genetically engineered human embryonic stem cells with quiescent recipient ventricular cardiomyocytes: insights into the development of cell-based pacemakers. *Circulation*, 111, 11–20
92. Kehat, I., Khimovich, L., Caspi, O., et al. (2004) Electromechanical integration of cardiomyocytes derived from human embryonic stem cells. *Nat Biotechnol*, 22, 1282–89
93. Gnecchi, M., He, H., Liang, O.D., et al. (2005) Paracrine action accounts for marked protection of ischemic heart by Akt-modified mesenchymal stem cells. *Nat Med*, 11, 367–68
94. Gnecchi, M., He, H., Noiseux, N., et al. (2006) Evidence supporting paracrine hypothesis for Akt-modified mesenchymal stem cell-mediated cardiac protection and functional improvement. *Faseb J*, 20, 661–69
95. Mangi, A.A., Noiseux, N., Kong, D., et al. (2003) Mesenchymal stem cells modified with Akt prevent remodeling and restore performance of infarcted hearts. *Nat Med*, 9, 1195–201
96. Mirotsou, M., Zhang, Z., Deb, A., et al. (2007) Secreted frizzled related protein 2 (Sfrp2) is the key Akt-mesenchymal stem cell-released paracrine factor mediating myocardial survival and repair. *Proc Natl Acad Sci U S A*, 104, 1643–48
97. Noiseux, N., Gnecchi, M., Lopez-Ilasaca, M., et al. (2006) Mesenchymal stem cells overexpressing Akt dramatically repair infarcted myocardium and improve cardiac function despite infrequent cellular fusion or differentiation. *Mol Ther*, 14(6), 840–50
98. Kaab, S., Nuss, H.B., Chiamvimonvat, N., et al. (1996) Ionic mechanism of action potential prolongation in ventricular myocytes from dogs with pacing-induced heart failure. *Circ Res*, 78, 262–73
99. Kang, P.M. and Izumo, S. (2003) Apoptosis in heart: basic mechanisms and implications in cardiovascular diseases. *Trends Mol Med*, 9, 177–82
100. Lafontant, P.J., Field, L.J. (2007) "Myocardial regeneration via cell cycle activation", in Rebuilding the infarcted heart, K.C. Wollert and L.J. Field, Eds., Informa Healthcare, London, UK., 41–54
101. Pasumarthi, K.B. and Field, L.J. (2002) Cardiomyocyte cell cycle regulation. *Circ Res*, 90, 1044–54
102. Bersell, K., Arab, S., Haring, B., et al. (2009) Neuregulin1/ErbB4 signaling induces cardiomyocyte proliferation and repair of heart injury. *Cell*, 138, 257–70
103. Engel, F.B., Schebesta, M., Duong, M.T., et al. (2005) p38 MAP kinase inhibition enables proliferation of adult mammalian cardiomyocytes. *Genes Dev*, 19, 1175–87
104. Tseng, A.S., Engel, F.B. and Keating, M.T. (2006) The GSK-3 inhibitor BIO promotes proliferation in mammalian cardiomyocytes. *Chem Biol*, 13, 957–63
105. Yildirim, Y., Naito, H., Didie, M., et al. (2007) Development of a biological ventricular assist device: preliminary data from a small animal model. *Circulation*, 116, I16–23
106. Zimmermann, W.H., Melnychenko, I., Wasmeier, G., et al. (2006) Engineered heart tissue grafts improve systolic and diastolic function in infarcted rat hearts. *Nat Med*, 12, 452–58
107. Christoforou, N., Miller, R.A., Hill, C.M., et al. (2008) Mouse ES cell-derived cardiac precursor cells are multipotent and facilitate identification of novel cardiac genes. *J Clin Invest*, 118, 894–903
108. Wu, S.M., Fujiwara, Y., Cibulsky, S.M., et al. (2006) Developmental origin of a bipotential myocardial and smooth muscle cell precursor in the mammalian heart. *Cell*, 127, 1137–50

109. Bu, L., Jiang, X., Martin-Puig, S., et al. (2009) Human ISL1 heart progenitors generate diverse multipotent cardiovascular cell lineages. *Nature*, 460, 113–17
110. Moretti, A., Caron, L., Nakano, A., et al. (2006) Multipotent embryonic isl1+ progenitor cells lead to cardiac, smooth muscle, and endothelial cell diversification. *Cell*, 127, 1151–65
111. David, R., Brenner, C., Stieber, J., et al. (2008) MesP1 drives vertebrate cardiovascular differentiation through Dkk-1-mediated blockade of Wnt-signalling. *Nat Cell Biol*, 10, 338–45
112. Blin, G., Nury, D., Stefanovic, S., et al. (2010) A purified population of multipotent cardiovascular progenitors derived from primate pluripotent stem cells engrafts in postmyocardial infarcted nonhuman primates. *J Clin Invest*, 120, 1125–39
113. Moretti, A., Bellin, M., Jung, C.B., et al. (2010) Mouse and human induced pluripotent stem cells as a source for multipotent Isl1+ cardiovascular progenitors. *FASEB J*, 24, 700–11
114. Shapiro, A.M., Lakey, J.R., Ryan, E.A., et al. (2000) Islet transplantation in seven patients with type 1 diabetes mellitus using a glucocorticoid-free immunosuppressive regimen. *N Engl J Med*, 343, 230–38
115. Lanza, R.P., Cibelli, J.B. and West, M.D. (1999) Human therapeutic cloning. *Nat Med*, 5, 975–77
116. Guan, K., Nayernia, K., Maier, L.S., et al. (2006) Pluripotency of spermatogonial stem cells from adult mouse testis. *Nature*, 440, 1199–203
117. Jiang, Y., Jahagirdar, B.N., Reinhardt, R.L., et al. (2002) Pluripotency of mesenchymal stem cells derived from adult marrow. *Nature*, 418, 41–49
118. Lewitzky, M. and Yamanaka, S. (2007) Reprogramming somatic cells towards pluripotency by defined factors. *Curr Opin Biotechnol*, 18, 467–73
119. Nakagawa, M., Koyanagi, M., Tanabe, K., et al. (2008) Generation of induced pluripotent stem cells without Myc from mouse and human fibroblasts. *Nat Biotechnol*, 26, 101–106
120. Okita, K., Ichisaka, T. and Yamanaka, S. (2007) Generation of germline-competent induced pluripotent stem cells. *Nature*, 448, 313–37
121. Takahashi, K., Tanabe, K., Ohnuki, M., et al. (2007) Induction of pluripotent stem cells from adult human fibroblasts by defined factors. *Cell*, 131, 861–72
122. Mauritz, C., Schwanke, K., Reppel, M., et al. (2008) Generation of functional murine cardiac myocytes from induced pluripotent stem cells. *Circulation*, 118, 507–17
123. Zovoilis, A., Nolte, J., Drusenheimer, N., et al. (2008) Multipotent adult germline stem cells and embryonic stem cells have similar microRNA profiles. *Mol Hum Reprod* 14(9), 521–29

Regenerating Function In Vivo with Myocytes Derived from Embryonic Stem Cells

Priya R. Baraniak and Todd C. McDevitt

Abstract Myocardial infarction, or heart attack, is a principal cause of congestive heart failure and adult morbidity and mortality in the western world. Since the adult myocardium lacks the inherent ability to repair itself following ischemic injury, a number of exogenous cell sources with differing cardiomyogenic potential have been investigated for the restoration of infarcted myocardium. To this end, the ability of embryonic-stem-cell-derived cardiomyocytes (ESC-CMs) to successfully engraft within host myocardium, fully differentiate to a mature cardiomyogenic phenotype, and electromechanically couple with host cardiomyocytes upon transplantation has been a subject of much inquiry in recent years. Overall, these studies demonstrate that the use of ESC-CMs alone or in conjunction with a biodegradable scaffold serves as a novel route to restore cardiomyocytes to the heart and thereby facilitate myocardial repair and functional regeneration.

Keywords Myocardial infarction • Congestive heart failure • Cellular cardiomyoplasty • Cell transplantation • Cell sourcing • Embryonic stem cells • Cardiomyocytes • Embryonic-stem-cell-derived cardiomyocytes • Biomaterials • Polymer scaffolds • Tissue engineered constructs • Paracrine mechanisms

1 Clinical Significance

Cardiovascular disease has been the leading cause of death in the USA since 1900, accounting for more American deaths each year than the next four leading causes of death combined [1]. It is estimated that over 850,000 people in the USA suffer from a myocardial infarction (MI; or heart attack) every year [1]. When one or more

T.C. McDevitt (✉)
The Wallace H. Coulter Department of Biomedical Engineering at Georgia Institute of Technology and Emory University, Petit Institute for Bioengineering and Bioscience, Atlanta, GA, USA
e-mail: Todd.McDevitt@bme.gatech.edu

I.S. Cohen and G.R. Gaudette (eds.), *Regenerating the Heart*, Stem Cell Biology and Regenerative Medicine, DOI 10.1007/978-1-61779-021-8_3,
© Springer Science+Business Media, LLC 2011

of the coronary arteries supplying blood to the heart are blocked, resulting in MI, oxygen and nutrient deprivation leads to ischemic injury and death of muscle cells in the infarct zone. Since the mammalian heart cannot adequately regenerate after injury, progressive muscle atrophy (due to injury or death of cardiomyocytes) and the formation of noncontractile scar tissue result in cardiac cachexia [2]. Furthermore, intrinsic remodeling of the left ventricle (LV) following MI results in LV dilation, wall thinning, and sphericity. Although allowing the heart to maintain cardiac output by increasing stroke volume, in the long-term, this remodeling leads to maladaptive changes in myocardial structure and function resulting in congestive heart failure (CHF) [3]. MI is a principal cause of CHF, with approximately 22% of men and 46% of women being disabled with CHF within 6 years of their first MI [1, 4]. CHF affects five million Americans and 22 million people worldwide and is a leading cause of adult hospitalizations, resulting in an estimated healthcare cost of $29.6 billion in the USA in 2006 alone [1]. As such, medical interventions for MI and CHF are in dire need.

The most effective treatment for CHF, to date, is heart transplantation; however, in the last 13 years, although the number of organ donors has tripled, the overall number of transplants has remained constant and is currently limited to fewer than 3,000 patients worldwide each year owing to severe organ shortage. Furthermore, organ transplantation requires long-term immunosuppression in patients, resulting in substantial postoperative risk [5]. Thus, alternatives to heart transplantation are needed. Mechanical circulatory assist devices (i.e., LV assist devices) provide a "bridge to transplant" for many patients awaiting organ donation, but a substantial number of patients still die on the waiting list, whereas others are subject to the same immunosuppression regimens and subsequent perioperative risks as transplant patients [6]. Pharmaceutical strategies to antagonize LV remodeling (such as treatment with beta-blockers and angiotensin-converting enzyme inhibitors) have proven limited in their ability to stem the progression from MI to CHF, and surgical interventions for advanced CHF have been limited to patients with severe dilation and dysfunction (as with the Dor and Batista procedures) or have resulted in complications arising from the intervention (e.g., foreign body encapsulation of the epicardium following epicardial restraint therapy) [3].

In light of these limitations, in recent years, regenerative medicine approaches including cell transplantation and the use of bioactive biomaterials for myocardial repair and regeneration have emerged as promising alternatives to heart transplantation. The challenge of regenerating functional myocardial tissue is multifaceted and complex. Cardiac muscle is highly organized and vascularized, containing a high density of metabolically active, contractile cells (cardiomyocytes) in a complex three-dimensional assembly of endothelial and smooth muscle cells and collagen fibers in parallel. Cardiomyocytes contract synchronously with electrical stimulation, and myocardial fibers are therefore both electrically and mechanically anisotropic. For this reason, cells implanted into infarcted myocardium must not only survive and proliferate, but must also become electrically coupled with the surrounding native myocardium and be able to withstand the mechanical loads imposed by contractile myocardium.

2 Cell Therapy for Cardiac Repair and Regeneration

2.1 *Cellular Cardiomyoplasty*

Cellular cardiomyoplasty, also known as cellular cardiomyogenesis, is the transplantation of cells into injured myocardium to restore blood flow and contractility to infarcted, scarred, or dysfunctional heart tissue, thereby reversing heart failure [7]. To this end, a number of cell types have been tested in the laboratory using experimental models of heart failure, and have led to multiple human clinical trials [8–14]. Although cell transplantation has resulted in improvements in cardiac function, this approach has several limitations, including poor cell engraftment and survival, cell differentiation, and function after transplantation, and cell sourcing issues. The choice of optimal cell type for cellular cardiomyoplasty may depend on the type of injury and the clinical objective for the patient in question. However, at a fundamental level, any cell type used to repair and regenerate the myocardium must engraft and survive in the heart, differentiate into cardiomyocytes and vasculature, and integrate both mechanically and electrically with native cardiomyocytes.

2.2 *Modes of Cell Delivery*

Several methods exist to deliver cells to infarcted myocardium, and it has been demonstrated that the success of cell transplantation therapies depends not only on the cell population used for treatment, but also on the mode of cell delivery [15]. Intramyocardial (IM) injection (the injection of cells into the infarcted region of the myocardium, usually at the epicardial surface) enables cell delivery directly to the site of injury. However, this method requires invasive, open-heart surgery and can be limited by cell washout from the area of delivery, inhomogeneous distribution of cells, and the potential for arrhythmia formation owing to the presence of transplanted cells within the myocardium [15]. In contrast, intracoronary (IC) transplantation is a catheter-based technique for direct cell delivery to the infarcted myocardium via the coronary circulation. Although much less invasive than IM injection, IC transplantation has been associated with poor cell engraftment rates and requires access to the heart through the femoral artery in the leg, thereby carrying some associated risk [16]. Finally, endocardial injection of cells (via a catheter as with IC delivery) is also much less invasive than IM delivery; however, electromechanical mapping is necessary to guide this procedure [16]. Owing to the limitations associated with cell injection strategies, researchers have more recently begun to examine the use of cell-seeded natural and synthetic scaffolds implanted at the site of injury as "cardiac patches" to restore cellularity to infarcted myocardium [17–19]. The evaluation of these and other cell delivery strategies in animal models is currently ongoing and may provide valuable insights for the successful clinical application of cell transplantation in the future.

2.3 In Vitro Cardiomyogenic Differentiation of Embryonic Stem Cells

Embryonic stem cells (ESCs) are pluripotent cells derived from the inner cell mass of preimplantation blastocysts. ESCs can be readily expanded in culture while maintaining the potential to differentiate to cells from all three germ lineages (endoderm, ectoderm, and mesoderm) and can be differentiated to all three major cardiac cell types in vitro [20, 21], making them an attractive cell source for cardiac repair and regeneration. In particular, ESCs remain one of the only types of stem cells definitively proven to differentiate into cardiomyocytes. However, the efficiency of ESC commitment to the cardiomyocyte lineage in vitro has been limited [22], spurring experimentation with a number of growth factors, cytokines, genetic manipulations, and cell culture systems attempting to enhance cardiomyogenic differentiation of these cells.

The differentiation of ESCs in vitro is commonly induced through the formation of three-dimensional cell aggregates known as embryoid bodies (EBs). Spontaneously contracting cells are often visible in EBs after several days of culture under appropriate cell culture conditions, and ESC-derived cardiomyocytes (ESC-CMs) have been shown to resemble native cardiac muscle cells on the basis of cardiac-specific gene expression profiles and the appearance of sarcomeric contractile filaments [23–26]. However, spontaneously contracting ESCs typically account for a very small percentage (generally less than 1%) of the overall population of differentiated cells within EBs [22, 24]. As such, methods to augment the in vitro cardiac differentiation of ESCs and to purify populations of ESC-CMs have been employed to enhance the number of cells needed for cell transplantation studies.

The addition of soluble factors such as dimethyl sulfoxide [27, 28], oxytocin [28, 29], retinoic acid [27, 30, 31], hepatocyte growth factor [32], neuregulin-1 [33], basic fibroblast growth factor [34], transforming growth factor-β_2 [35], bone morphogenic protein (BMP)-2 and BMP-4 [36, 37], leukemia inhibitory factor (LIF) [37], and platelet-derived growth factor [38] to cell culture media has been reported to increase the cardiomyogenic differentiation of ESCs in vitro. Additionally, signals emanating from the visceral endoderm (likely comprising several of the aforementioned soluble factors) have been implicated in stimulating cardiac morphogenesis during development. Accordingly, the co-culture of ESCs with endoderm-like cells and conditioned media from endoderm-like cells has been shown to increase the in vitro cardiomyogenic commitment of ESCs [39–41]. Such media manipulations generally result in an enhanced cardiomyocyte population comprising 10–60% of differentiated cells [26, 36, 42].

Owing to the potential for undifferentiated ESCs to form teratomas upon transplantation in vivo [21, 26], obtaining pure populations of differentiated cells has been deemed critical to the ultimate safety and success of ESC-derived cell therapies. Selection of ESC-CMs has been accomplished through genetic manipulation using promoters for cardiac-specific markers such as α-myosin heavy chain (α-MHC) [43–46], ventricular myosin light chain-2 (MLC-2v) [47], and islet-1 (isl-1) [48]. Additionally, the combinatorial expression of isl-1, Nkx2.5, and flk1 has been used to differentially select for ESC-derived cardiac progenitors capable of differentiating into mature cardiomyocytes and pacemaker cells, and to distinguish the cardiomyogenic population from cells giving rise to endothelial and smooth muscle cells [48, 49]. Enrichment of ESC-CMs has also been accomplished through the use of physical separation methods such as Percoll density gradient centrifugation (yielding populations that are up to 90% pure) [47, 50, 51] and cell surface marker selection methods such as fluorescence-activated cell sorting [47] and through the use of dynamic suspension culture systems [24, 31]. Although ESC-CMs can be generated in vitro using the aforementioned protocols, the ability of these cells to successfully engraft within the host myocardium, fully differentiate to a mature cardiomyogenic phenotype, and electrically couple with host cardiomyocytes upon transplantation has been a subject of much debate and inquiry in recent years.

2.4 In Vivo Transplantation of ESC-CMs

2.4.1 Cell Transplantation

As early as 1996, purified murine ESC-CMs (mESC-CMs) were demonstrated to engraft and survive within the myocardium. A highly purified population (more than 99%) of mESC-CMs, obtained using an α-MHC promoter construct coupled to a neomycin resistance gene and transplanted via IM injection into the hearts of adult dystrophic mice, was shown to engraft and survive for up to 7 weeks after transplantation [44]. Transplanted ESC-CMs aligned with host cardiomyocytes and exhibited normal myocardial topography, without tumor formation or eliciting a significant host immune response. Spontaneously beating mESC-CMs (30,000 cells, dissected from EBs) transplanted into infarcted, adult Wistar rat myocardium were also shown to engraft and differentiate into mature cardiomyocytes (assessed by their rod shape and the appearance of striations typical of cardiomyocytes) 6 weeks after transplantation [52]. IM ESC-CM transplantation into rat infarcts decreased infarct size, attenuated LV hypertrophy, and improved isometric contractility. These beneficial effects were attributed to cardiogenesis, with approximately 7% of LV cardiomyocytes in the infarcted myocardium originating from donor cells. More recently, the IM injection of 50,000 mESC-CMs, obtained by differentiating ESCs in the presence of BMP-2 and LIF, in a murine model of acute MI resulted in significantly improved LV diastolic dimensions compared with control animals and those receiving grafts of undifferentiated ESCs 4 weeks after transplantation [37]. In addition, treatment with either undifferentiated ESCs or ESC-CMs improved fractional shortening in hearts after MI compared with untreated controls, although no significant difference in the level of improvement was seen between groups treated with ESCs and ESC-CMs. Finally, in a clinically relevant, large-animal ovine model of MI, 30 million IM injected mESC-CMs (differentiated using LIF and BMP-2) engrafted within infarcted myocardium, differentiated into mature cardiomyocytes expressing connexin 43, and improved LV ejection fraction compared with controls 1 month after transplantation [53]. Beneficial functional effects were observed in both immunosuppressed and immunocompetent sheep, suggesting that ESC-CMs are potentially immune-tolerant.

Similar to the initial results with mESC-CMs, transplantation of human ESC-CMs (hESC-CMs) into ischemic myocardium has been reported to exert various functional benefits. When 500,000 to ten million hESC-CMs (purified using Percoll gradient centrifugation) were transplanted via IM injection into athymic rat infarcted myocardium, cells engrafted and expressed cardiomyocyte-specific markers (including β-myosin heavy chain, MLC-2v, and atrial natriuretic factor) by 4 weeks after transplantation [54]. Interestingly, hESC-CMs engrafted in greater numbers (i.e., formed larger grafts) when heat-shocked prior to transplantation, and engrafted cells proliferated (assessed by Ki-67 and 5-bromo-2′-deoxyuridine incorporation) and induced angiogenesis over a 4-week period in vivo. Subsequent work by the same group found that despite the persistence of some transplanted cells, most hESC-CMs die after transplantation into infarcted myocardium [36]. However, hESC-CMs delivered with a prosurvival cocktail (consisting of Matrigel to prevent anoikis, Bcl-XL to block mitochondrial cell death, pinacidil to mimic ischemic preconditioning, cyclosporine A to attenuate cyclophilin D dependent mitochondrial pathways, insulin-like growth factor-1 to activate Akt pathways, and the caspase inhibitor ZVAD-fmk) were found to survive in vivo and improve LV wall thickness and cardiac function 4 weeks after transplantation [36]. Another study aiming to circumvent cell death upon transplantation administered cytoprotective agents (allopurinol and uricase) and an anti-inflammatory agent (ibuprofen) to mice before and after the IM transplantation of one million hESC-CMs to infarcted myocardium [55]. hESC-CMs engrafted, differentiated to mature cardiomyocytes, and improved LV ejection fraction in mice receiving cytoprotective and anti-inflammatory agents compared with infarcted controls. Others have also demonstrated that transplantation of hESC-CMs to uninjured and infarcted myocardium in immunocompetent and immunosuppressed rats results in cell engraftment, proliferation, alignment, and the formation of gap junctions between grafted and host cells 4 and 8 weeks after transplantation [21, 26]. Eight weeks after transplantation, hESC-CM-treated infarcted hearts demonstrated attenuated adverse LV remodeling in terms of fractional shortening, wall motion, and LV diastolic dimensions [21]. Longer-term (12-week) studies of the effects of hESC-CM IM transplantation into infarcted rat and mice hearts confirmed previous findings of cell survival and improved cardiac function at 4 weeks after transplantation; however, these beneficial effects were not sustained at 12 weeks after transplantation, despite graft survival, suggesting that the results of shorter-term studies should be interpreted with caution [56, 57]. Improvement in cardiac function at 4 weeks was attributed to increased vascularity within infarcted myocardium independent of graft size [58], and the cotransplantation of endothelial cells with hESC-CMs enhanced the formation of functional capillaries, thereby increasing survival of grafted hESC-CMs for up to 24 weeks in vivo [59]. Finally, hESC-CMs (differentiated using activin A and BMP-4) aggregated to form macroscopic disc-shaped patches (ranging from approximately 2 to 11 mm in diameter depending on the initial cell seeding density) of electrically coupled, beating cardiomyocytes when subjected to rotary suspension culture in vitro [60]. However, these patches failed to survive and form significant grafts upon implantation in vivo, likely due to the lack of vascular and stromal elements within patches [61], warranting further examination of such constructs.

Additional studies examining the prolonged effects (i.e., 3 months or greater) of ESC-CM transplantation into infarcted myocardium and in clinically relevant, large-animal models are necessary before this cell therapy can become a clinically viable option. Although short-term preclinical studies have demonstrated improved cardiac function following ESC-CM transplantation, the potential for teratoma formation and the plausible in vivo differentiation of ESC-derived cells to unintended cell phenotypes must be addressed prior to their clinical use [62]. Additionally, the high heart rates of rodents and other small animals can mask arrhythmias that may occur in slower-paced human hearts, thereby warranting studies in large animals with heart rates resembling those of humans. Large-animal studies are also needed to assess the engraftment, prolonged survival, and electromechanical coupling of significantly larger numbers (approximately 10^9) of cells within infarcts that are closer in size to those found in human hearts and to elucidate and successfully modulate host immune response upon cell transplantation.

2.4.2 Cell-Seeded Biomaterials

Although the aforementioned preclinical ESC-CM transplantation studies have resulted in improvements in cardiac function, issues of poor cell engraftment and survival within the host myocardium still need to be resolved. Therefore, to increase cell viability and the efficiency of cell retention within infarcted myocardium, researchers have turned to tissue-engineered constructs as cell delivery vehicles. The use of these biodegradable scaffolds provides cells with temporary mechanical support while allowing them time to deposit their own extracellular matrix and integrate within the host myocardium. Furthermore, the shape, size, and mechanical properties of these constructs can be precisely controlled, and scaffolds can be loaded with a number of biomolecules, thereby modulating cell behavior in vitro and in vivo. Suitable biomaterials for myocardial repair must be able to meet the difficult mechanical demands of cardiac tissue without failing mechanically, and must also support angiogenesis. To this end, a number of natural and synthetic ESC-CM-seeded biomaterial scaffolds have been investigated for myocardial repair following MI.

Biodegradable polyurethanes (PUs) have been shown to possess appropriate mechanical properties for myocardial tissue repair and to support cardiomyocyte gap junction formation in vitro [3, 63]. As such, mESC-CMs have been shown to proliferate and contract on collagen- and laminin-coated PU films for up to 30 days in vitro [17]. In addition, ESC-CMs could be successfully cultured in vitro on three-dimensional PU constructs fabricated via either electrospinning or a thermally induced phase separation (TIPS) method [64]. Despite morphological differences, cells on fibrous electrospun scaffolds exhibited an elongated morphology typical of more mature cardiomyocytes, whereas cells on porous TIPS scaffolds were more rounded like less mature cardiomyocytes, yet hESC-CMs on both types of scaffolds contracted and expressed cardiac markers (sarcomeric myosin heavy chain and connexin 43). ESC-CMs have also been used to construct engineered cardiac tissue in vitro. Percoll-gradient-enriched mESC-CMs cultured on collagen supplemented with Matrigel could be stretched in vitro for 7 days, resulting in synchronous beating and responsiveness to pharmacological stimuli [65]. Further analyses revealed that constructs resembled native neonatal cardiomyocytes in the appearance of aligned sarcomeres and cell–cell junctions. More recently, ESC-CMs cultured on a PEGylated fibrinogen hydrogel matured (on the basis of striated patterns of α-sarcomeric actin and gap junction expression) and remained responsive to pharmacological stimuli for up to a 2-week period in vitro [66].

Although in vitro studies with ESC-CM-seeded polymer constructs are promising, they are few in number, and those assessing such tissue-engineered scaffolds in an in vivo model of MI are even fewer. However, it has been demonstrated that a scaffold consisting of growth-factor-free Matrigel seeded with ESC-CMs implanted into infarcted mouse myocardium engrafted and restored LV wall dimension and function 2 weeks after implantation, and did so significantly better than Matrigel or ESC-CM transplantation alone [67]. Additionally, polyglycolic acid meshes seeded with ESC-CMs restored LV function and decreased mortality in mice compared with untreated animals and those receiving a cell-free patch 8 weeks after implantation [68]. However, it should be noted that a cell-only control group was not included, and therefore, conclusions as to the advantages of the use of this cell-seeded scaffold over direct cell transplantation cannot be drawn from this study. Finally, Matrigel-supplemented collagen constructs seeded with ESC-CMs were implanted subcutaneously in nude mice and assessed for survival and tumorigenesis at 4 weeks [65]. Cell-seeded constructs were highly vascularized, contained cells that were positive for cardiac troponin T and troponin I, and did not result in teratoma formation. The efficacy of these constructs was not, however, examined in an animal model of MI. Although these preliminary results demonstrate potential for ESC-CM-seeded constructs in myocardial repair, additional studies on the in vivo biocompatibility, integration, and vascularization of these cardiac patches are necessary. Furthermore, many of the same concerns surrounding cell transplantation must be addressed with cell-seeded constructs before their viable clinical use.

3 Conclusions and Future Directions

Most in vivo experiments on ESC transplantation into the heart (either uninjured or following ischemic injury) have been conducted in small-animal models (primarily mouse and rat), and the subset of studies focused on ESC-CM transplantation into infarcted myocardium is even more limited in number and the preclinical animal models used. Therefore, before ESC-derived cell therapies can be considered for clinical use, large-scale, clinically relevant, large-animal studies are needed. Existing small-animal studies have, however, provided enhanced cardiac function after cell transplantation. Consequently, the use of ESC-CMs alone or in conjunction with a biodegradable scaffold holds much promise for myocardial repair and regeneration.

Positive functional outcomes after cell transplantation into infarcted myocardium have often been observed despite poor cell engraftment, survival, and commitment to a cardiomyogenic or vascular phenotype. However, the precise mechanisms behind these functional improvements remain elusive. To this end, the investigation of the paracrine actions of various stem cell and stem-cell-derived cell populations is an active field of research, and it has been demonstrated that these cells produce and secrete a number of growth factors, chemokines, cytokines, and immunosuppressive molecules that play a role in regenerative events [69]. As such, the use of ESC-CM-derived biomolecules and acellular matrices for myocardial repair and regeneration may provide alternative therapies that overcome current cell sourcing issues.

References

1. Heart Disease and Stroke Statistics – 2005 Update. 2005, American Heart Association: Dallas, TX.
2. Rosenthal, N. and A. Musaro, *Gene therapy for cardiac cachexia?* Int J Cardiol, 2002. **85**(1): p. 185–91.
3. Fujimoto, K.L., et al., *An elastic, biodegradable cardiac patch induces contractile smooth muscle and improves cardiac remodeling and function in subacute myocardial infarction.* J Am Coll Cardiol, 2007. **49**(23): p. 2292–300.
4. Gnecchi, M., et al., *Paracrine action accounts for marked protection of ischemic heart by Akt-modified mesenchymal stem cells.* Nat Med, 2005. **11**(4): p. 367–8.
5. Ott, H.C., J. McCue, and D.A. Taylor, *Cell-based cardiovascular repair – the hurdles and the opportunities.* Basic Res Cardiol, 2005. **100**(6): p. 504–17.
6. Lietz, K. and L.W. Miller, *Will left-ventricular assist device therapy replace heart transplantation in the foreseeable future?* Curr Opin Cardiol, 2005. **20**(2): p. 132–7.
7. Taylor, D.A., *Cell-based myocardial repair: how should we proceed?* Int J Cardiol, 2004. **95 Suppl 1**: p. S8–12.
8. Chachques, J.C., et al., *Myocardial assistance by grafting a new bioartificial upgraded myocardium (MAGNUM trial): clinical feasibility study.* Ann Thorac Surg, 2008. **85**(3): p. 901–8.
9. de la Fuente, L.M., et al., *Transendocardial autologous bone marrow in chronic myocardial infarction using a helical needle catheter: 1-year follow-up in an open-label, nonrandomized, single-center pilot study (the TABMMI study).* Am Heart J, 2007. **154**(1): p. 79.e1–7.
10. Lunde, K., et al., *Autologous stem cell transplantation in acute myocardial infarction: the ASTAMI randomized controlled trial. Intracoronary transplantation of autologous mononuclear bone marrow cells, study design and safety aspects.* Scand Cardiovasc J, 2005. **39**(3): p. 150–8.
11. Lunde, K., et al., *Exercise capacity and quality of life after intracoronary injection of autologous mononuclear bone marrow cells in acute myocardial infarction: results from the autologous stem cell transplantation in acute myocardial infarction (ASTAMI) randomized controlled trial.* Am Heart J, 2007. **154**(4): p. 710.e1–8.
12. Ripa, R.S., et al., *Bone marrow derived mesenchymal cell mobilization by granulocyte-colony stimulaWting factor after acute myocardial infarction: results from the Stem Cells in Myocardial Infarction (STEMMI) trial.* Circulation, 2007. **116**(11 Suppl): p. I24–30.

13. Schachinger, V., et al., *Transplantation of progenitor cells and regeneration enhancement in acute myocardial infarction: final one-year results of the TOPCARE-AMI trial.* J Am Coll Cardiol, 2004. **44**(8): p. 1690–9.
14. Wollert, K.C., et al., *Intracoronary autologous bone-marrow cell transfer after myocardial infarction: the BOOST randomised controlled clinical trial.* Lancet, 2004. **364**(9429): p. 141–8.
15. Hamdi, H., et al., *Cell delivery: intramyocardial injections or epicardial deposition? A head-to-head comparison.* Ann Thorac Surg, 2009. **87**(4): p. 1196–203.
16. Moscoso, I., et al., *Analysis of different routes of administration of heterologous 5-azacytidine-treated mesenchymal stem cells in a porcine model of myocardial infarction.* Transplant Proc, 2009. **41**(6): p. 2273–5.
17. Alperin, C., P.W. Zandstra, and K.A. Woodhouse, *Polyurethane films seeded with embryonic stem cell-derived cardiomyocytes for use in cardiac tissue engineering applications.* Biomaterials, 2005. **26**(35): p. 7377–86.
18. Shimko, V.F. and W.C. Claycomb, *Effect of mechanical loading on three-dimensional cultures of embryonic stem cell-derived cardiomyocytes.* Tissue Eng Part A, 2008. **14**(1): p. 49–58.
19. Wang, X., et al., *Scalable producing embryoid bodies by rotary cell culture system and constructing engineered cardiac tissue with ES-derived cardiomyocytes in vitro.* Biotechnol Prog, 2006. **22**(3): p. 811–8.
20. Singla, D.K. and B.E. Sobel, *Enhancement by growth factors of cardiac myocyte differentiation from embryonic stem cells: a promising foundation for cardiac regeneration.* Biochem Biophys Res Commun, 2005. **335**(3): p. 637–42.
21. Caspi, O., et al., *Transplantation of human embryonic stem cell-derived cardiomyocytes improves myocardial performance in infarcted rat hearts.* J Am Coll Cardiol, 2007. **50**(19): p. 1884–93.
22. Singla, D.K., et al., *Transplantation of embryonic stem cells into the infarcted mouse heart: formation of multiple cell types.* J Mol Cell Cardiol, 2006. **40**(1): p. 195–200.
23. Kehat, I., et al., *Human embryonic stem cells can differentiate into myocytes with structural and functional properties of cardiomyocytes.* J Clin Invest, 2001. **108**(3): p. 407–14.
24. Sargent, C., G. Berguig, and T. McDevitt, *Cardiomyogenic differentiation of embryoid bodies is promoted by rotary orbital suspension culture.* Tissue Engineering Part A, 2009. **15**(2): p. 331–32.
25. Wei, H., et al., *Embryonic stem cells and cardiomyocyte differentiation: phenotypic and molecular analyses.* J Cell Mol Med, 2005. **9**(4): p. 804–17.
26. Dai, W., et al., *Survival and maturation of human embryonic stem cell-derived cardiomyocytes in rat hearts.* J Mol Cell Cardiol, 2007. **43**(4): p. 504–16.
27. Chen, Y., et al., *Cyclic adenosine 3′,5′-monophosphate induces differentiation of mouse embryonic stem cells into cardiomyocytes.* Cell Biol Int, 2006. **30**(4): p. 301–7.
28. Paquin, J., et al., *Oxytocin induces differentiation of P19 embryonic stem cells to cardiomyocytes.* Proc Natl Acad Sci USA, 2002. **99**(14): p. 9550–5.
29. Hatami, L., M.R. Valojerdi, and S.J. Mowla, *Effects of oxytocin on cardiomyocyte differentiation from mouse embryonic stem cells.* Int J Cardiol, 2007. **117**(1): p. 80–9.
30. Bugorsky, R., J.C. Perriard, and G. Vassalli, *N-cadherin is essential for retinoic acid-mediated cardiomyogenic differentiation in mouse embryonic stem cells.* Eur J Histochem, 2007. **51**(3): p. 181–92.
31. Niebruegge, S., et al., *Cardiomyocyte production in mass suspension culture: embryonic stem cells as a source for great amounts of functional cardiomyocytes.* Tissue Eng Part A, 2008. **14**(10): p. 1591–601.
32. Roggia, C., et al., *Hepatocyte growth factor (HGF) enhances cardiac commitment of differentiating embryonic stem cells by activating PI3 kinase.* Exp Cell Res, 2007. **313**(5): p. 921–30.
33. Wang, Z., et al., *Neuregulin-1 enhances differentiation of cardiomyocytes from embryonic stem cells.* Med Biol Eng Comput, 2009. **47**(1): p. 41–8.

34. Khezri, S., et al., *Effect of basic fibroblast growth factor on cardiomyocyte differentiation from mouse embryonic stem cells.* Saudi Med J, 2007. **28**(2): p. 181–6.
35. Singla, D.K. and B. Sun, *Transforming growth factor-beta2 enhances differentiation of cardiac myocytes from embryonic stem cells.* Biochem Biophys Res Commun, 2005. **332**(1): p. 135–41.
36. Laflamme, M.A., et al., *Cardiomyocytes derived from human embryonic stem cells in pro-survival factors enhance function of infarcted rat hearts.* Nat Biotechnol, 2007. **25**(9): p. 1015–24.
37. Rajasingh, J., et al., *STAT3-dependent mouse embryonic stem cell differentiation into cardiomyocytes: analysis of molecular signaling and therapeutic efficacy of cardiomyocyte precommitted mES transplantation in a mouse model of myocardial infarction.* Circ Res, 2007. **101**(9): p. 910–8.
38. Sachinidis, A., et al., *Identification of platelet-derived growth factor-BB as cardiogenesis-inducing factor in mouse embryonic stem cells under serum-free conditions.* Cell Physiol Biochem, 2003. **13**(6): p. 423–9.
39. Graichen, R., et al., *Enhanced cardiomyogenesis of human embryonic stem cells by a small molecular inhibitor of p38 MAPK.* Differentiation, 2008. **76**(4): p. 357–70.
40. Mummery, C., et al., *Differentiation of human embryonic stem cells to cardiomyocytes: role of coculture with visceral endoderm-like cells.* Circulation, 2003. **107**(21): p. 2733–40.
41. Xu, X.Q., et al., *Chemically defined medium supporting cardiomyocyte differentiation of human embryonic stem cells.* Differentiation, 2008. **76**(9): p. 958–70.
42. Singla, D.K., *Embryonic stem cells in cardiac repair and regeneration.* Antioxid Redox Signal, 2009. **11**(8): p. 1857–63.
43. Doss, M.X., et al., *Global transcriptome analysis of murine embryonic stem cell-derived cardiomyocytes.* Genome Biol, 2007. **8**(4): p. R56.
44. Klug, M.G., et al., *Genetically selected cardiomyocytes from differentiating embryonic stem cells form stable intracardiac grafts.* J Clin Invest, 1996. **98**(1): p. 216–24.
45. Bugorsky, R., J.C. Perriard, and G. Vassalli, *Genetic selection system allowing monitoring of myofibrillogenesis in living cardiomyocytes derived from mouse embryonic stem cells.* Eur J Histochem, 2008. **52**(1): p. 1–10.
46. Zandstra, P.W., et al., *Scalable production of embryonic stem cell-derived cardiomyocytes.* Tissue Eng, 2003. **9**(4): p. 767–78.
47. Muller, M., et al., *Selection of ventricular-like cardiomyocytes from ES cells in vitro.* FASEB J, 2000. **14**(15): p. 2540–8.
48. Moretti, A., et al., *Multipotent embryonic isl1+ progenitor cells lead to cardiac, smooth muscle, and endothelial cell diversification.* Cell, 2006. **127**(6): p. 1151–65.
49. Bu, L., et al., *Human ISL1 heart progenitors generate diverse multipotent cardiovascular cell lineages.* Nature, 2009. **460**(7251): p. 113–7.
50. Xu, C., et al., *Characterization and enrichment of cardiomyocytes derived from human embryonic stem cells.* Circ Res, 2002. **91**(6): p. 501–8.
51. Xu, C., et al., *Cardiac bodies: a novel culture method for enrichment of cardiomyocytes derived from human embryonic stem cells.* Stem Cells Dev, 2006. **15**(5): p. 631–9.
52. Min, J.Y., et al., *Transplantation of embryonic stem cells improves cardiac function in postinfarcted rats.* J Appl Physiol, 2002. **92**(1): p. 288–96.
53. Menard, C., et al., *Transplantation of cardiac-committed mouse embryonic stem cells to infarcted sheep myocardium: a preclinical study.* Lancet, 2005. **366**(9490): p. 1005–12.
54. Laflamme, M.A., et al., *Formation of human myocardium in the rat heart from human embryonic stem cells.* Am J Pathol, 2005. **167**(3): p. 663–71.
55. Kofidis, T., et al., *Allopurinol/uricase and ibuprofen enhance engraftment of cardiomyocyte-enriched human embryonic stem cells and improve cardiac function following myocardial injury.* Eur J Cardiothorac Surg, 2006. **29**(1): p. 50–5.
56. van Laake, L.W., et al., *Human embryonic stem cell-derived cardiomyocytes and cardiac repair in rodents.* Circ Res, 2008. **102**(9): p. 1008–10.
57. van Laake, L.W., et al., *Human embryonic stem cell-derived cardiomyocytes survive and mature in the mouse heart and transiently improve function after myocardial infarction.* Stem Cell Res, 2007. **1**(1): p. 9–24.

58. van Laake, L.W., et al., *Improvement of mouse cardiac function by hESC-derived cardiomyocytes correlates with vascularity but not graft size.* Stem Cell Res, 2009. **3**(2–3): p. 106–12.
59. van Laake, L.W., et al., *Extracellular matrix formation after transplantation of human embryonic stem cell-derived cardiomyocytes.* Cell Mol Life Sci, 2010. **67**(2): p. 277–90
60. Stevens, K.R., et al., *Scaffold-free human cardiac tissue patch created from embryonic stem cells.* Tissue Eng Part A, 2009. **15**(6): p. 1211–22.
61. Stevens, K.R., et al., *Physiological function and transplantation of scaffold-free and vascularized human cardiac muscle tissue.* Proc Natl Acad Sci USA, 2009. **106**(39): p. 16568–73.
62. Nussbaum, J., et al., *Transplantation of undifferentiated murine embryonic stem cells in the heart: teratoma formation and immune response.* FASEB J, 2007. **21**(7): p. 1345–57.
63. McDevitt, T.C., et al., *Spatially organized layers of cardiomyocytes on biodegradable polyurethane films for myocardial repair.* J Biomed Mater Res A, 2003. **66**(3): p. 586–95.
64. Fromstein, J.D., et al., *Seeding bioreactor-produced embryonic stem cell-derived cardiomyocytes on different porous, degradable, polyurethane scaffolds reveals the effect of scaffold architecture on cell morphology.* Tissue Eng Part A, 2008. **14**(3): p. 369–78.
65. Guo, X.M., et al., *Creation of engineered cardiac tissue in vitro from mouse embryonic stem cells.* Circulation, 2006. **113**(18): p. 2229–37.
66. Shapira-Schweitzer, K., et al., *A photopolymerizable hydrogel for 3-D culture of human embryonic stem cell-derived cardiomyocytes and rat neonatal cardiac cells.* J Mol Cell Cardiol, 2009. **46**(2): p. 213–24.
67. Kofidis, T., et al., *Novel injectable bioartificial tissue facilitates targeted, less invasive, large-scale tissue restoration on the beating heart after myocardial injury.* Circulation, 2005. **112**(9 Suppl): p. I173–7.
68. Ke, Q., et al., *Embryonic stem cells cultured in biodegradable scaffold repair infarcted myocardium in mice.* Sheng Li Xue Bao, 2005. **57**(6): p. 673–81.
69. Baraniak, P.R. and T.C. McDevitt, *Stem cell paracrine actions and tissue regeneration.* Regen Med, 2010. **5**(1): p. 121–43.

Excitation–Contraction Coupling, Functional Properties, and Autonomic and Hormonal Regulation in Human Embryonic Stem Cell Derived Cardiomyocytes

Oshra Sedan and Ofer Binah

Abstract Recent years have witnessed numerous publications concerning the therapeutic efficacy of myocardial cell therapy utilizing cardiomyocytes from different sources, including human embryonic stem cell derived cardiomyocytes (hESC-CMs). Since for a favorable outcome of cell therapy (e.g. following myocardial infarction) the transplanted tissue must integrate with the host myocardium, it is conceivable that functional compatibility between the tissues is likely to improve the prospects of cardiac cell therapy. In view of the therapeutic potential of hESC-CMs in myocardial regeneration, in recent years several groups, including our own, have investigated the functional properties as well as the autonomic (e.g. β-adrenergic stimulation) and hormonal regulation of hESC-CMs. Hence, our goal in this chapter is to share with the readers some of these studies, specifically those focusing on intracellular Ca^{2+} handling and mechanical function.

Keywords Embryonic stem cells • Calcium handling • Cardiac myocytes • Excitation–contraction coupling • Calcium channels • Calcium currents

O. Binah (✉)
The Sohnis Family Stem Cells Center, The Rappaport Family Institute for Research in the Medical Sciences, The Department of Physiology, Ruth and Bruce Rappaport Faculty of Medicine, Technion – Israel Institute of Technology, Haifa, Israel
e-mail: binah@tx.technion.ac.il

I.S. Cohen and G.R. Gaudette (eds.), *Regenerating the Heart*, Stem Cell Biology and Regenerative Medicine, DOI 10.1007/978-1-61779-021-8_4,
© Springer Science+Business Media, LLC 2011

1 The Excitation–Contraction Coupling in Human Embryonic Stem Cell Derived Cardiomyocytes

1.1 Basic Functional Properties of Human Embryonic Stem Cell Derived Cardiomyocytes

The excitation–contraction (E-C) coupling of the mature myocardium depends on a sequence of events which include electrical activation and Ca^{2+} influx via the L-type Ca^{2+} channels which trigger Ca^{2+}-induced Ca^{2+} release from sarcoplasmic reticulum (SR). These Ca^{2+} ions, which exit the SR via the ryanodine receptor (RyR) channels, are the major Ca^{2+} source for the contractile machinery. In general, in adult cardiomyocyte, Ca^{2+} ions originating from the SR contribute approximately 70% of the Ca^{2+} required for contraction, whereas the residual 30% is derived from the extracelluar pool. As will be discussed below, in human embryonic stem cell derived cardiomyocytes (hESC-CMs) these key processes and the functionality of various components of the E-C coupling machinery differ from those in the adult myocardium.

Rather than describing the individual components of the E-C coupling in the order they proceed, we will first focus on the contractile performance of hESC-CMs and on the differences from the mature myocardium, and only thereafter address upstream elements (e.g. the intracellular Ca^{2+} handling machinery); the rationale for doing so will be established as we progress. The fundamental properties of any cardiac cell include its ability to generate a propagated action potential, to increase diastolic intracellular Ca^{2+} concentration, and to contract. Indeed, activation and propagation can be readily recorded from cardiomyocyte networks by means of the microelectrode array data acquisition setup, which provides the means to determine conduction velocity and activation patterns as well as testing the efficacy and toxicity of new drugs (e.g. Q-T prolongation) (Fig. 1). Further, by means of Fura 2 fluorescence and a video edge detector, Ca^{2+} transients and contractions can be recorded from hESC-CMs (Fig. 1); these signals represent the seemingly normal functionality of the E-C coupling machinery in hESC-CMs.

1.2 Force–Frequency Relations and Postrest Potentiation

Although the basic features of the Ca^{2+} transients and contractions are similar to those of adult cardiomyocytes, the following contractile features (determined in the clones studied in our laboratory) appear to differ from those of the adult myocardium: force–frequency relations and postrest potentiation. Since both phenomena are associated with augmented SR Ca^{2+} release, their adult-dissimilar characteristics suggest that SR function in hESC-CMs differs from that in the adult myocardium.

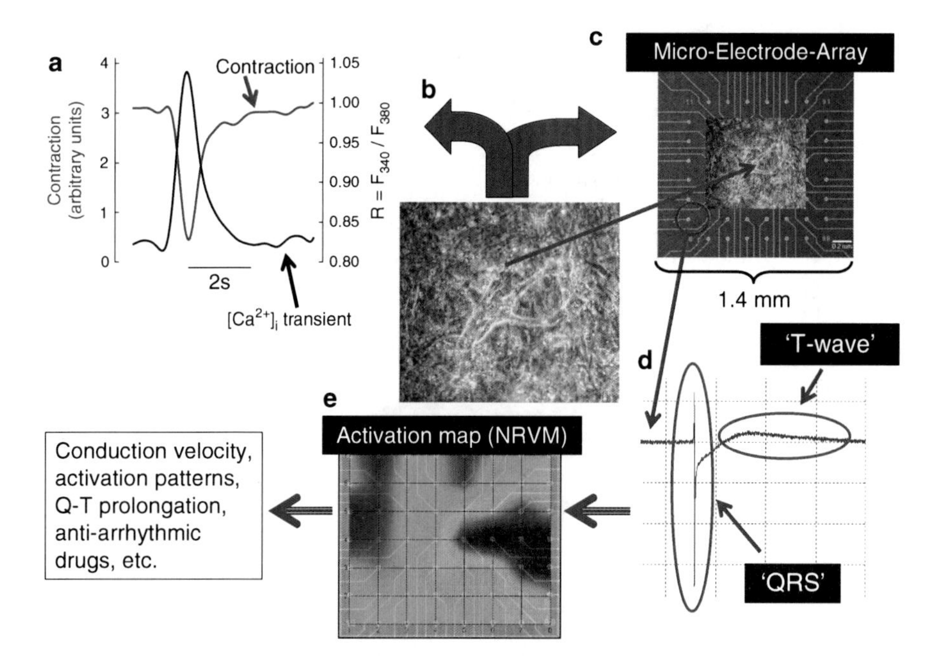

Fig. 1 Recordings of Ca^{2+} transients, contractions, and extracellular electrograms obtained from human embryonic stem cell derived cardiomyocytes (hESC-CMs). Simultaneously recorded traces of Ca^{2+} transients and contractions from a spontaneously contracting embryoid body (EB) (**a**). An example of an EB used in these experiments (**b**). The microelectrode array on which an EB is "mounted" for illustration purposes (**c**). A representative electrogram recorded from an EB (**d**). An example of an "activation map" generated from a neonatal rat ventricular myocyte (*NRVM*) culture (**e**)

1.2.1 Force–Frequency Relations

A basic attribute of the human myocardium is its ability to augment contractile force by increasing heart rate. This phenomenon utilized by the heart to increase cardiac output to meet higher metabolic demands during stress conditions or exercise is termed "positive force–frequency relations." In contrast with the adult heart, hESC-CMs exhibit *negative force–frequency relations*; namely, stimulating hESC-CMs at 0.5, 1.0, 1.5, 2.0, and 2.5 Hz causes a progressive decrease in the Ca^{2+} transient and contraction amplitude (Fig. 2). That intracellular Ca^{2+} handling is immature in hESC-CMs is suggested not only by the negative force–frequency relations, but also by the observation that increasing the stimulation rate elevates diastolic intracellular Ca^{2+} concentration as well resting tension (shown by the arrows in Fig. 2); these rate-dependent changes suggest that the mechanisms responsible for intracellular Ca^{2+} removal (e.g. sarco/endoplasmic reticulum Ca^{2+}-ATPase, SERCA) are not sufficiently effective. A variety of mechanisms (yet to be investigated in hESC-CMs) such as Ca^{2+}-dependent inactivation of the L-type

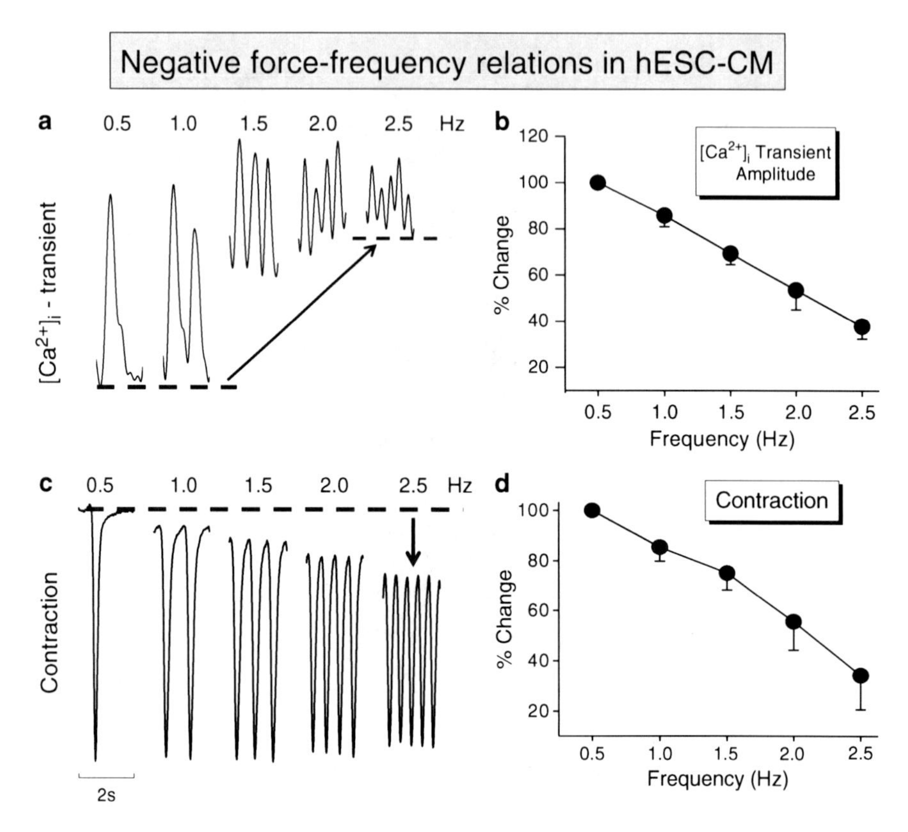

Fig. 2 Force–frequency relations in hESC-CMs from clone H9.2. To generate the force–frequency relations, contracting EBs were stimulated at 0.5, 1, 1.5, 2, and 2.5 Hz. (**a**, **c**) Representative recordings of Ca^{2+} transients and contractions, respectively, in a 55-day-old contracting EB from clone H9.2, stimulated at different rates. The *arrows* indicate the increased diastolic intracellular Ca^{2+} concentration (**a**) and resting tension (**c**). (**b**, **d**) Summary ($n=5$ EBs) of the effects of stimulation rate changes on the Ca^{2+} transient and contraction amplitude in 47–55-day-old EBs from clone H9.2: (**b**) Ca^{2+} transient amplitude; (**d**) contraction amplitude. The results are expressed as the percent change from the values at 0.5 Hz (From [5])

Ca^{2+} current ($I_{Ca,L}$) can account for negative force–frequency relations in cardiomyocytes. Among the possible explanations for the negative force–frequency relations in hESC-CMs, Dolnikov et al. [5] tested the hypothesis that the contractile machinery is unresponsive to increased intracellular Ca^{2+} concentration. However, because elevating extracellular Ca^{2+} concentration from 2 to 4 and 6 mM increased both diastolic and systolic intracellular Ca^{2+} concentration as well as resting and active force (Fig. 3), the authors proposed that the negative force–frequency relations in hESC-CMs do not result from the inability to respond to elevated intracellular Ca^{2+} concentration.

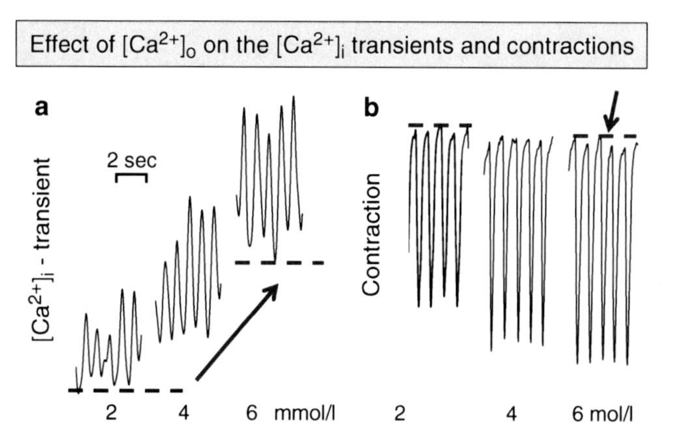

Fig. 3 The effects of different extracellular Ca^{2+} concentrations on the Ca^{2+} transients and contractions in hESC-CMs. (**a**, **b**) Representative Ca^{2+} transients and contractions recorded from a contracting EB exposed to 2, 4, and 6 mM extracellular Ca^{2+}. The EBs were stimulated at 1 Hz (From [5])

1.2.2 Postrest Potentiation

An additional mechanical feature related to a functional SR is postrest potentiation, expressed as the occurrence of a stronger contraction following a pause in the regular beating sequence. In principle, a stronger postrest contraction may result from a larger SR Ca^{2+} filling during the rest period, and a concomitant increased SR Ca^{2+} release after the rest period [2, 6]. Experimentally, the magnitude of postrest potentiation is determined by interrupting the regular stimulation with pauses of various lengths, followed by resumption of the regular stimulation protocol. Indeed, an indication that SR function in hESC-CMs is immature emerges from the findings that whereas adult mouse ventricular cardiomyocytes exhibit prominent postrest potentiation, this mechanical capacity is absent in hESC-CMs (Fig. 4). Collectively, the negative force–frequency relations and lack of postrest potentiation are both suggestive of at least a partially dysfunctional SR Ca^{2+} release, which inspired us to further study intracellular Ca^{2+} handling in hESC-CMs.

1.3 The L-Type Ca^{2+} Channels and E-C Coupling in hESC-CMs

The first step of the E-C coupling process is opening of voltage-dependent L-type Ca^{2+} channels and influx of Ca^{2+} ions, which in turn activate the contractile machinery directly, as well as indirectly by releasing additional Ca^{2+} from the SR via the RyR. Hence, in this section we will provide experimental support for the molecular and electrophysiological expression as well as for the function of the

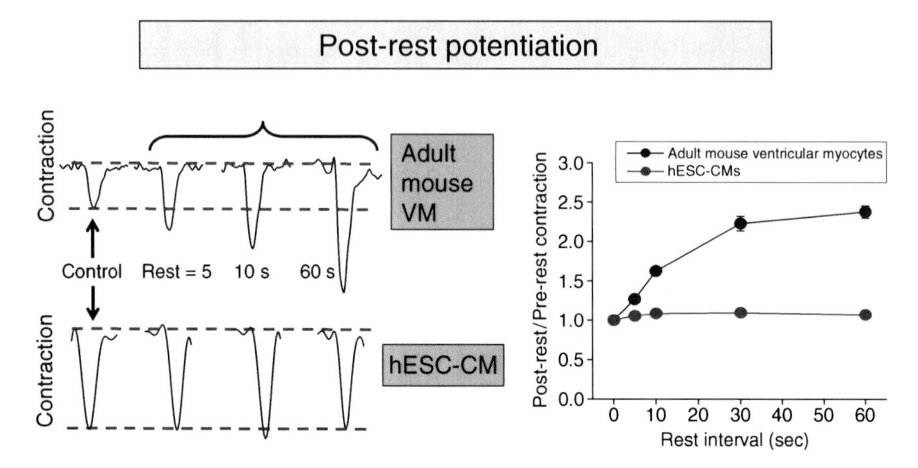

Fig. 4 Postrest potentiation in adult mouse ventricular myocytes and hESC-CMs. *Left*: Representative contraction tracings recorded from an adult mouse ventricular myocyte and from a contracting EB, depicting the control contractions recorded at 1 Hz and the first postrest contractions after rest periods of 5, 10, and 60 s. Note the prominent postrest potentiation in the mouse ventricular myocyte and its absence in the hESC-CM. *Right*: Average percent change in postrest contraction amplitude/prerest contraction amplitude versus the rest length in adult mouse ventricular myocytes ($n=3$) and in contracting EBs ($n=3$). *VM* ventricular myocyte (From [5])

L-type Ca^{2+} channel in hESC-CMs. The molecular expression of the L-type Ca^{2+} channel α-subunit (α1c) was demonstrated by several groups. For example, both Sartiani et al. [26] and Mummery et al. [20] showed prominent messenger RNA (mRNA) expression of the channel's α1c subunit in hESC-CMs (Fig. 5a). Next, the electrophysiological expression of the Ca^{2+} channel was nicely illustrated by Sartiani et al. by recording from isolated cardiomyocytes rapidly activating, slowly inactivating L-type Ca^{2+} currents ($I_{Ca,L}$), resembling the $I_{Ca,L}$ properties in adult cardiomyocytes (Fig. 5b, c). Finally, the function of $I_{Ca,L}$ in hESC-CMs was demonstrated by several groups. For example, Xu et al. [32] and Mummery et al. [20] showed that the Ca^{2+} channel blockers diltiazem and verapamil, respectively, attenuated the spontaneous beating rate of hESC-CMs. In contrast, Satin et al. [27] reported that neither diltiazem nor nifedipine affected automaticity of hESC-CMs from line H9.2 (Fig. 5e), but at the same time demonstrated that nifedipine markedly shortened action potential duration, supporting the contribution of $I_{Ca,L}$ to the hESC-CM action potential. However, although several studies indeed demonstrated that Ca^{2+} channel blockers affect the spontaneous activity and action potential configuration, these findings do not necessarily mean that $I_{Ca,L}$ contributes to the E-C coupling process. A more direct support for the contribution of $I_{Ca,L}$ to the contraction was obtained by our group [5] by showing that verapamil completely blocked the contractions in hESC-CMs (line H9.2) *stimulated* (thus eliminating the effect on automaticity) at 1.0 Hz (Fig. 5d).

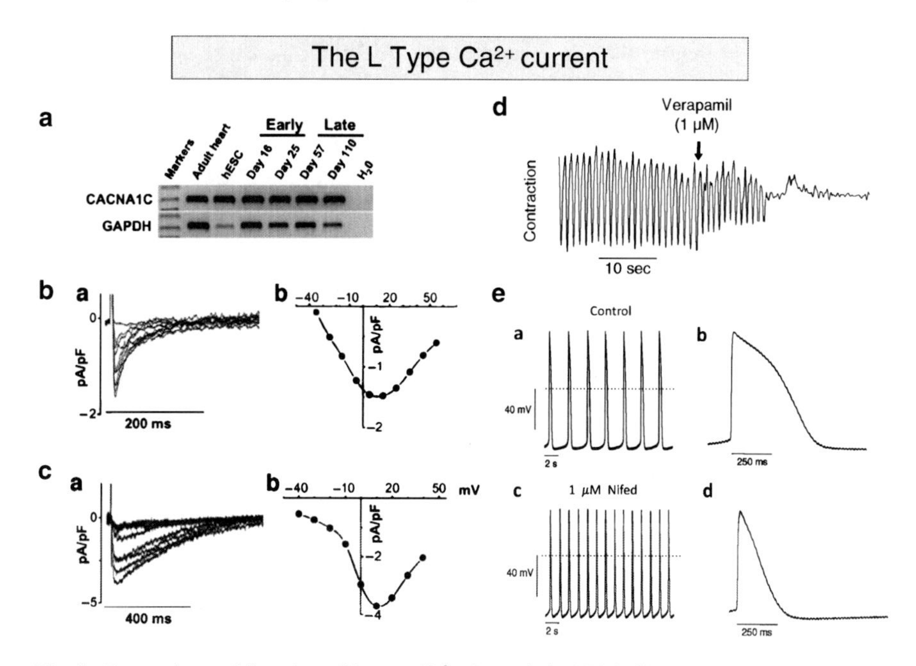

Fig. 5 Expression and function of L-type Ca^{2+} channels in hESC-CMs. (**a**) Messenger RNA of the active subunit CACNA1C is present in undifferentiated human embryonic stem cells (hESCs) and at all stages of differentiation. $I_{Ca,L}$ was measured both in hESC (**b**, plot *a*) and in cardiomyocytes (**c**, plot *a*, 57 days). Activation curve of $I_{Ca,L}$ density versus step potentials in hESCs (**b**, plot *b*) and cardiomyocytes (**c**, plot *b*). (**d**) Verapamil (1.0 μM) blocks the contraction of hESC-CMs stimulated at 1.0 Hz. Verapamil was included in the Tyrode's solution superfusing the recording bath. (**e**) Nifedipine increases automaticity and shortens action potential duration in hESC-CMs from clone H9.2. In (**e**), plots *a* and *c* are representative recordings of spontaneous action potentials recorded from a cell cluster in the absence and presence of 1 μM nifedipine. In (**e**), plots *b* and *d* are action potential recordings in the absence and presence of nifedipine. These representative recording show that nifedipine caused a positive chronotropic effect and markedly shortened action potential duration. *GAPDH* glyceraldehyde 3-phosphate dehydrogenase, *pA* picoampere, *pF* picofarad. [(**a**–**c**) From [26]; (**d**) from [5]; (**e**) from [27]]

1.4 Intracellular Ca^{2+} Handling in hESC-CMs; Comparison with the Adult Heart

Having demonstrated that basic contractile features differ between hESC-CMs and adult myocardium, we will now focus on key elements of the intracellular Ca^{2+} handling machinery, attempting to highlight the disparities from the mature heart.

1.4.1 The Expression and Function of RyRs in hESC-CMs

Over the past few years, several groups have investigated the mRNA and protein expression of RyRs in hESC-CMs. Using the real-time PCR technique, Satin et al. [28] demonstrated that in hESC-CMs, the mRNA levels of RyR_2 are approximately

1,000-fold lower than in the adult human heart. Although the expression of RyRs in hESC-CMs was demonstrated by immunofluorescence staining [13, 20, 28], both Liu et al. [13] and Satin et al. [28] indicated that the regularly spaced expression pattern of RyRs reported in adult ventricular cardiomyocytes was missing in the stained hESC-CMs; RyRs expression was throughout the cytosol, with intense staining in the perinuclear region. These pieces of evidence indicate that although RyRs are expressed in hESC-CMs, this does not necessarily imply that they are fully functional. And indeed, as discussed herein, there are inconsistent reports regarding the degree of compatibility/incompatibility of the hESC-CMs RyRs with the adult myocardium. Complete refractoriness or only partial SR responsiveness to pharmacological agents that affect SR Ca^{2+} release were reported in recent years. In clone H9.2 of hESC-CMs, Binah's group found that the Ca^{2+} transients and contractions are insensitive to ryanodine, thapsigargin, and caffeine (Fig. 6a), suggesting that SR Ca^{2+} release does not contribute to contraction [5]. In partial agreement with these findings, Liu et al. [13] showed that in hESC-CMs clones H1 and HES2 (different from those reported by Dolnikov et al.) only 35–40% of the cells were responsive to caffeine. Further, only the caffeine-responsive cells also responded to ryanodine and thapsigargin (the latter is a SERCA2a inhibitor) (Fig. 6b–d). Collectively, these results indicate that the E-C coupling of hESC-CMs is different from that of the adult myocardium.

1.4.2 The Expression and Function of Key Ca^{2+} Handling Proteins in hESC-CMs

Although the mechanisms underlying the absent/immature SR function in hESC-CMs are not fully understood, some clues have emerged from studies investigating the expression of key proteins of the Ca^{2+} handling machinery [5, 13, 27]. Western blot analysis has shown that hESC-CMs express SERCA2 and the Na^+/Ca^{2+} exchanger, but neither the SR Ca^{2+} handling protein calsequestrin nor the SERCA regulator phospholamban (Fig. 7). In contrast, both calsequestrin and phospholamban are expressed in specimens of adult human atria and ventricles (Fig. 7). Recently, Liu et al. [14] hypothesized that gene transfer of calsequestrin into hESC-CMs suffices to induce SR functional improvement. Their results show that the calsequestrin-transduced hESC-CMs exhibited greater SR Ca^{2+} content as indicated by a larger response to caffeine compared with green fluorescent protein transduced hESC-CMs. Additionally, the calsequestrin-transduced hESC-CMs exhibited increased Ca^{2+} transient amplitude as well as increased upstroke and decay velocities compared with controls. An important Ca^{2+}-binding protein located in the endoplasmic reticulum (ER) which releases Ca^{2+} resulting from activation of inositol 1,4,5-trisphosphate receptors (IP_3R) is calreticulin [4, 8]. Recently, Sedan et al. [30] and Satin et al. [28] have shown by means of western blot analysis and immunofluorescence techniques, respectively, that hESC-CMs express the ER

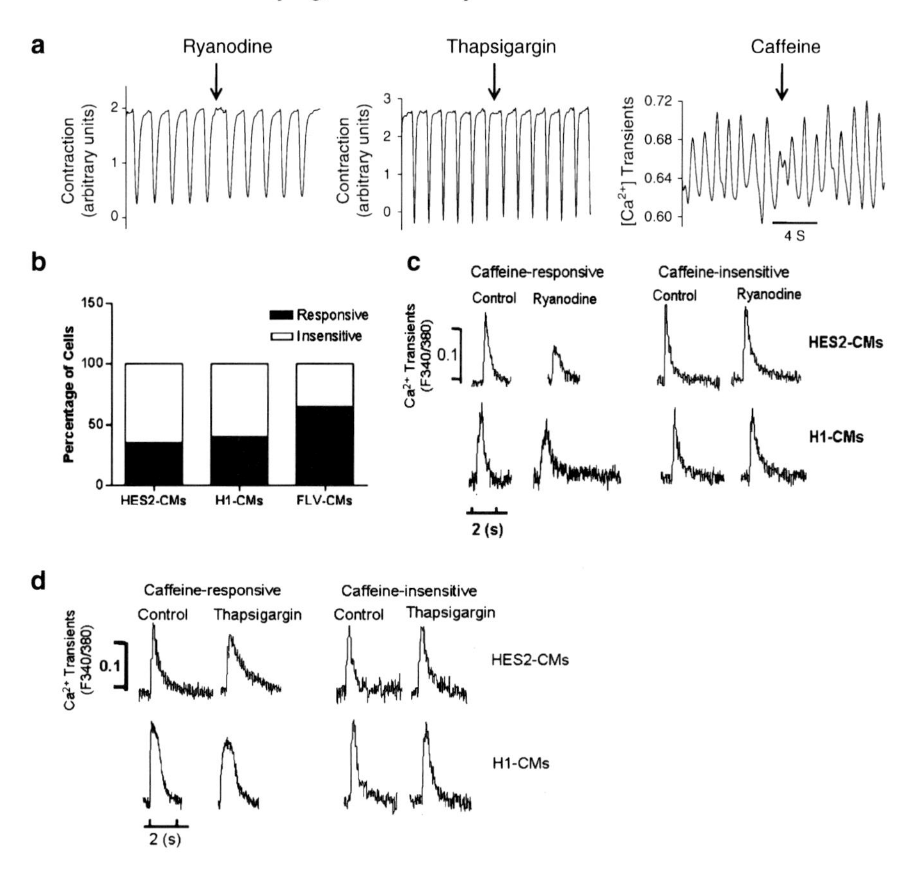

Fig. 6 The responsiveness of hESC-CMs to pharmacological agents which affect sarcoplasmic reticulum (SR) Ca^{2+} handling. (**a**) Representative contractions recorded from hESC-CMs (clone H9.2) stimulated at 1.0 Hz, before and 25 min after superfusion with ryanodine (10 μM) dissolved in Tyrode's solution (*left*); representative contractions (conditions as in the *left panel*) before and 25 min after superfusion with thapsigargin (100 nM) (*middle*); representative intracellular Ca^{2+} concentration transients recorded from hESC-CMs, before and after brief application of caffeine (10 mM) (*right*). (**b**) The percentages of caffeine-responsive and caffeine-insensitive cardiac myocytes. Total cell numbers were 20, 20, and 17 for HES2 cardiomyocytes (*HES2-CMs*), H1 cardiomyocytes (*H1-CMs*), and cardiomyocytes fetal left ventricular cardiomyocytes (*FLV-CMs*), respectively. (**c**) Representative tracings of Ca^{2+} transients in HES2-CMs and H1-CMs before and after incubation with ryanodine for 30 min. (**d**) Representative tracings of Ca^{2+} transients in HES2-CMs and H1-CMs before and after incubation with thapsigargin for 15 min. [(**a**) From [5]; (**b–d**) From [13]]

Ca^{2+} handling calreticulin. Interestingly, several studies have demonstrated that in embryonic cardiomyocytes calreticulin-ER-mediated, inositol 1,4,5-trisphosphate (1,4,5-IP_3)-triggered Ca^{2+} release plays an important role in excitability and pacemaker function [15, 16, 18, 19]. Hence, the precise role of ER Ca^{2+} release in hESC-CMs still needs to be determined.

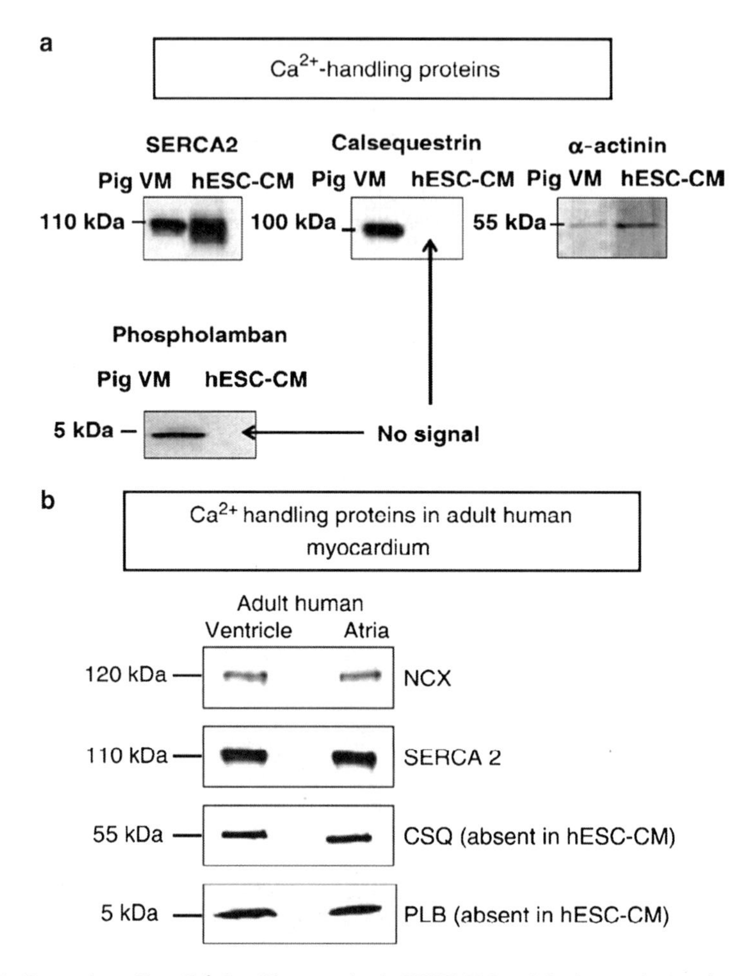

Fig. 7 Expression of key Ca^{2+} handling proteins in hESC-CMs, adult pig ventricular myocardium, and human atrial and ventricular specimens. (**a**) Representative western blots of the three SR proteins sarco/endoplasmic reticulum Ca^{2+}-ATPase 2 (*SERCA2*), calsequestrin (*CSQ*) and phospholamban (*PLB*), and α-actinin in pig myocardium and in hESC-CMs. These results were obtained in three different experiments, each analyzing five or six carefully excised spontaneously contracting areas of embryoid bodies. (**b**) Representative western blots of Na^{+}/Ca^{2+} exchanger (*NCX*), SERCA2, CSQ, and PLB in human atrial and ventricular specimens removed during open-heart operations [3]

2 Autonomic and Hormonal Regulation of the E-C Coupling in hESC-CMs

Because hESC-CMs are potential candidates for cardiac regeneration, it is important to determine their compatibility with the humoral environment of the host myocardium. Specifically, this section will focus on the autonomic and hormonal regulation of key E-C coupling elements. Regarding the autonomic modulation of pacemaker activity (not directly related to E-C coupling), the first demonstration (later supported by [23]) that hESC-CMs are responsive to autonomic agonists was provided by Kehat et al. [10]. In brief, the authors reported that positive and negative chronotropic responses were induced by the β-adrenergic agonist isoproterenol and by the muscarinic agonist carbamylcholine, respectively. Isoproterenol (10^{-6} M) increased the spontaneous contraction rate by 46% and the muscarinic agonist carbamylcholine (10^{-6} M) decreased the rate by 22%. In support of the functionality of the β-adrenergic stimulatory pathway, the direct adenylate cyclase activator forskolin and the phosphodiesterase inhibitor 3-isobutyl-1-methylxanthine increased the spontaneous rate by 82 and 52%, respectively. In agreement with the positive chronotropic effect, recent studies reported that isoproterenol also caused prominent positive inotropic and lusitropic effects (Fig. 8) [7, 29].

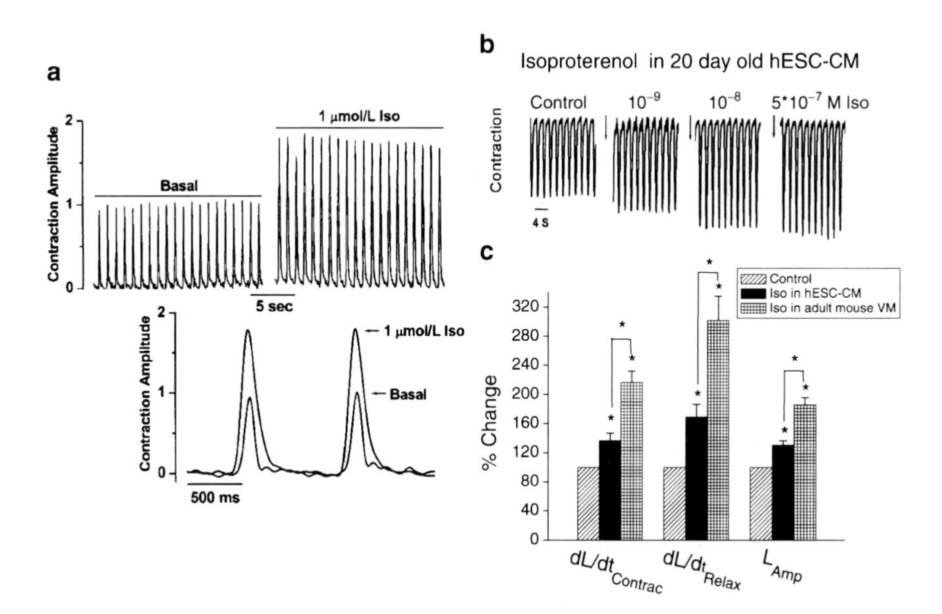

Fig. 8 The effects of β-adrenergic stimulation with isoproterenol on the contraction of hESC-CMs. (**a**) An EB (the specific clone was not indicated by the authors) was stimulated by an electric filed at 1 Hz and contractions were measured using video edge detection before and after application of 1 μM isoproterenol. (**b**) A representative experiment illustrating an increase in contraction amplitude in response to increasing concentrations of isoproterenol (the preparation was paced at 0.5 Hz). All six EBs (hESC-CM clone H9.2) studied responded similarly to isoproterenol. (**c**) Summary of the effects of isoproterenol on the contraction parameters in hESC-CMs and in adult mouse ventricular myocytes ($n=6$ and $n=5$, respectively). The plot illustrates the maximal effects of isoproterenol obtained at the concentration range studied, presented as percent change of the respective controls. *$P<0.05$. *Iso* isoproterenol, *VM* ventricular myocytes, $dL/dt_{Contrac}$ maximal rate of contraction, dL/dt_{Relax} maximal rate of relaxation, L_{Amp} contraction amplitude [(**a**) From [7]; (**b**) From [29]]

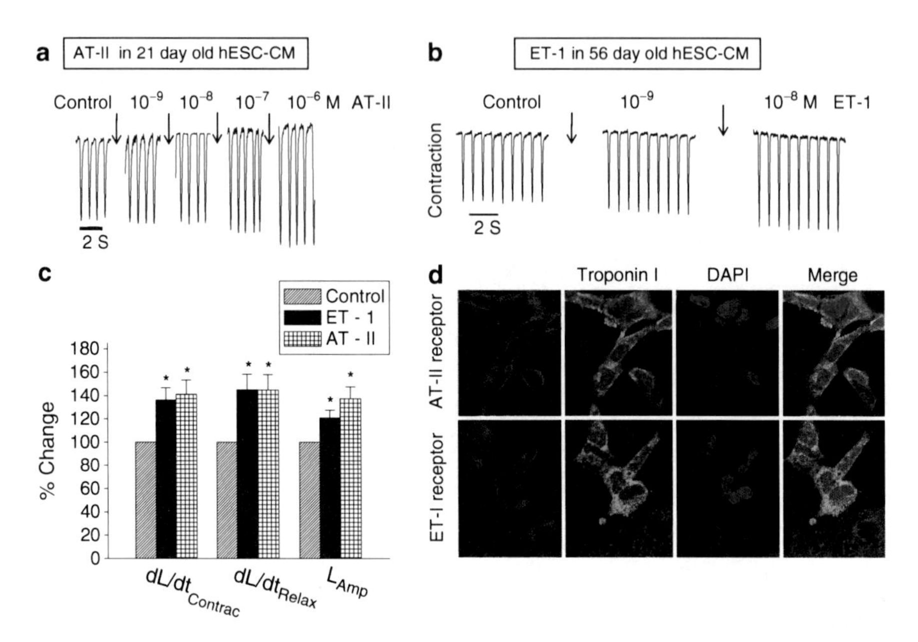

Fig. 9 The effects of angiotensin II (*AT-II*) and endothelin-1 (*ET-1*) on the contraction of hESC-CMs (clone H9.2) and immunofluorescence of the corresponding receptors. (**a**) A representative experiment illustrating the positive inotropic effect of increasing concentrations of AT-II. The preparation was paced at 1 Hz. (**b**) A representative experiment illustrating the positive inotropic effect of ET-1 (10^{-9} and 10^{-8} M). The preparation was paced at 0.3 Hz. (**c**) A comparison of the effects of AT-II (n=11–12 EBs) and ET-1 (n=4–5 EBs) on the contraction parameters. Abbreviations are as in the legend to Fig. 8. *P<0.05. (**d**) Immunofluorescence staining of AT-II and ET-1 receptors in hESC-CMs. AT-II staining: a 29-day old EB, magnification 40×4. ET-1 staining: a 31-day-old EB, magnification 40×2. *DAPI* 4′,6-diamidino-2-phenylindole (From [29])

In addition to the adult-like autonomic responsiveness, hESC-CMs also respond to the key hormones angiotensin II (AT-II) and endothelin-1 (ET-1), which induce their effects on contraction by activating the 1,4,5-IP_3 pathway [12, 21]. In support of a functional 1,4,5-IP_3 pathway, we [29] demonstrated that AT-II and ET-1 increase the contraction amplitude, and the rates of maximal contraction and relaxation, Accordingly, hESC-CMs express both the AT-II and the ET-1 receptors (Fig. 9). The contribution of 1,4,5-IP_3-dependent intracellular Ca^{2+} release to the inotropic effect of AT-II was further established by the following findings:

1. The effects of AT-II were blocked by 2 μM 2-aminoethoxyphenyl borate (2-APB; an IP_3R blocker) and by U73122 (a phospholipase C blocker) (Fig. 10a, b) [29]. Accordingly, Satin et al. [28] found that 2 μM 2-APB attenuated the rate of spontaneous beating of hESC-CMs (Fig. 10d, e).
2. hESC-CMs express types 1 and 2 of the 1,4,5-IP_3 receptor, as determined by immunofluorescence staining (Fig. 10c) [28, 29].

In summary, these findings demonstrate that hESC-CMs exhibit functional AT-II and ET-1 signaling pathways, as well as 1,4,5-IP_3-operated releasable Ca^{2+} stores.

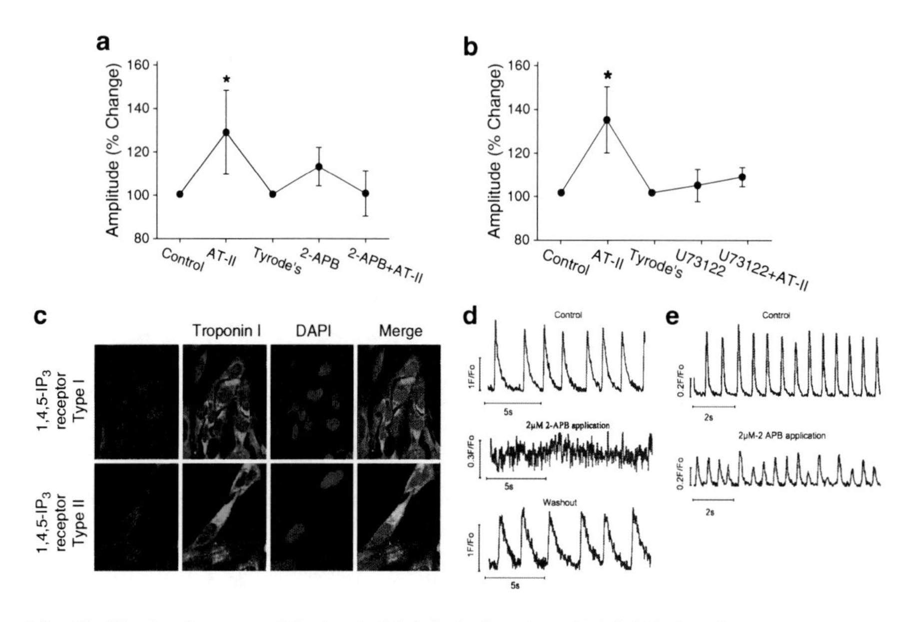

Fig. 10 The involvement of the inositol 1,4,5-trisphosphate (*1,4,5-IP$_3$*) signaling pathway in the inotropic response to AT-II: 2-aminoethoxyphenyl borate (*2-APB*) and U73122 block the positive inotropic effect of AT-II, the expression of 1,4,5-IP$_3$ receptors, and the involvement of 1,4,5-IP$_3$ receptors in the automaticity of hESC-CM clone H9.2. (**a**) Summary of four experiments in which the ability of 2-APB to block the positive inotropic effect of AT-II was investigated ($^*P<0.05$). (**b**) Summary of five experiments in which the ability of U73122 to block the inotropic effect of AT-II was investigated ($^*P<0.05$). The 2-APB and the U73122 experiments were performed as follows. Firstly, the EB was exposed to AT-II for 5–7 min, and contraction was recorded. Thereafter, the preparation was superfused with drug-free Tyrode's solution for 10 min, followed by exposure for 5 min to 2-APB (2 μM) or U73122 (2 μM), followed by 2-APB + AT-II or U73122 + AT-II. (**c**) The expression of 1,4,5-IP$_3$ type 1 and 2 receptors in hESC-CMs. 1,4,5-IP$_3$ type 1 positive staining (*upper row*) in a 35-day-old EB, magnification 63 × 1, $n=3$. 1,4,5-IP$_3$ type 2 positive staining (*lower row*) in a 45-day-old EB, magnification 40 × 1.9, $n=3$. (**d**) An example of a cell in which application of 2-APB (2 μM) completely and reversibly blocked all Ca^{2+} transients. (**e**) An example of a cell in which application of 2-APB decreased action-potential-initiated Ca^{2+} transients. [(**a**, **b**) From [29]; (**c**) from [30]; (**d**, **e**) from [28]]

2.1 *The Role of the 1,4,5-IP$_3$ Pathway During Cardiac Development*

Owing to the importance of the 1,4,5-IP$_3$ pathway in developing cardiomyocytes, we will present a short overview of IP$_3$Rs and their respective function. In contrast to the uncertain role of IP$_3$Rs in adult cardiomyocytes [1, 11, 17, 33], these receptors seem to have an important function during heart development and embryogenesis. As soon as the heart tube is created, spontaneous IP$_3$R-related (rather than RyR-related) Ca^{2+} oscillations were found to be essential for contraction, gene transcription, and structural arrangement of the developing cardiomyocytes [11, 22].

Specifically, IP_3R mRNA was detected at the earliest stages of the developing heart tube (5.5 days after egg fertilization in the mouse), whereas at a later development stage (8.5 days after egg fertilization), RyR_2 mRNA was only beginning to be detected [25]. Shortly after birth, when the mitotic potential of cardiomyocytes is lost, IP_3R expression declines and RyR_2 expression is on the rise, suggesting that IP_3Rs are required at least initially for the proliferation and differentiation of developing cardiomyocytes ([24]; reviewed in [31]). In addition, several groups demonstrated in mouse ESC-CMs that the IP_3R-mediated Ca^{2+} signaling pathway underlies the initiation of cardiac pacemaking, and thus constitutes a critical step in cardiogenesis [9, 18, 24]. In a recent review, Bootman's group [11] reported that IP_3Rs are important in establishing the cytoplasmic Ca^{2+} oscillations in the first few days of embryonic heart development. Shortly thereafter, the IP_3Rs give way to more mature-like contraction components such as RyRs, voltage-activated Ca^{2+} channels, and hyperpolarization cyclic-nucleotide-gated channels, the later responsible for pacemaker currents (I_f) in the adult myocardium.

3 Summary

During recent years it has become increasingly recognized that hESC-CMs constitute an important potential source for cardiac therapy and regeneration. In addition, inducible pluripotent stem cell derived cardiomyocytes (described in Chaps. 12–14) are also emerging as a novel source for cardiac cell therapy. However, although in many aspects hESC-CMs resemble the host myocardium, extensive basic research, including developing means for selection and enrichment of cardiomyocytes, is still required before cell transplantation can be successful implemented.

Acknowledgements This work was supported by the Israel Science Foundation, Israeli Ministry of Science and Technology, the Russell Barrie Nanotechnology Institute (RBNI) at Technion, and the Rappaport Family Institute for Research in the Medical Sciences.

References

1. Bers DM (2002) Cardiac excitation-contraction coupling. Nature 415:198–205
2. Bers DM, Bassani RA, Bassani JW et al. (1993) Paradoxical twitch potentiation after rest in cardiac muscle: increased fractional release of SR calcium. J Mol Cell Cardiol 25:1047–1057
3. Binah O, Dolnikov K, Sedan O, Shilkrut M, Zeevi-Levin N, Amit M, Danon A, Itskovitz-Eldor J (2007) Functional properties of human embryonic stem cells-derived cardiomyocytes. J Electrocardiology 40(6 suppl):5192–5196
4. Burns K and Michalak M (1993) Interactions of calreticulin with proteins of the endoplasmic and sarcoplasmic reticulum membranes. FEBS 318:181–185
5. Dolnikov K, Shilkrut M, Zeevi-Levin N et al. (2006) Functional properties of human embryonic stem cell-derived cardiomyocytes: intracellular Ca^{2+} handling and the role of sarcoplasmic reticulum in the contraction. Stem Cells 24:236–245

6. Ferraz SA, Bassani JW, Bassani RA (2001) Rest-dependence of twitch amplitude and sarcoplasmic reticulum calcium content in the developing rat myocardium. J Mol Cell Cardiol 33:711–722.
7. He JQ, Ma Y, Lee Y et al. (2003) Human embryonic stem cells develop into multiple types of cardiac myocytes. Action potential characterization. Circ Res 93:32–39
8. Imanaka-Yoshida K, Amitani A, Ioshii SO et al. (1996) Alterations of expression and distribution of the Ca(2+)-storing proteins in endo/sarcoplasmic reticulum during differentiation of rat cardiomyocytes. J Mol Cell Cardiol 28:553–562
9. Kapur N and Banach K (2007) Inositol-1,4,5-trisphosphate-mediated spontaneous activity in mouse embryonic stem cell-mediated cardiomyocytes. J Physiol 581:1113–1127
10. Kehat I, Kenyagin-Karsenti D, Snir M et al. (2001) Human embryonic stem cells can differentiate into myocytes with structural and functional properties of cardiomyocytes. J Clin Invest 108:407-414
11. Kockskämper, Zima AV, Roderick HL et al. (2008) Emerging roles of inositol 1,4,5-trisphosphate signaling in cardiac myocytes. J Mol Cell Cardiol 45:128–147
12. Li X, Zima AV, Sheikh F et al. (2005) Endothelin-1-induced arrhythmogenic Ca^{2+} signaling is abolished in atrial myocytes of inositol-1,4,5-trisphosphate(IP_3)-receptor type 2-deficient mice. Circ Res 96:1274–1281
13. Liu J, Fu JD, Siu CW et al. (2007) Functional sarcoplasmic reticulum for calcium handling of human embryonic stem cell-derived cardiomyocytes: insights for driven maturation. Stem Cells 25:3038–3044
14. Liu J, Lieu DK, Siu CW, Fu JD, Li R (2009) Facilitated maturation of Ca^{2+} handling properties of human embryonic stem cell-derived cardiomyocytes by calsequestrin expression. Am J Physiol Cell Physiol 297:C152–159
15. Lozyk MD, Papp S, Zhang X et al. (2006) Ultrastructural analysis of development of myocardium in calreticulin-deficient mice. BMC Dev Biol 6:54.
16. Lynch JM, Chilibeck K, Qui Y et al. (2006) Assembling pieces of the cardiac puzzle; calreticulin and calcium-dependent pathways in cardiac development, health, and disease. Trends Cardiovasc Med 16:65–69
17. Marks AR (2000) Cardiac intracellular calcium release channels: role in heart failure. Circ Res 87:8–11
18. Méry A, Aimond F, Ménard C et al. (2005) Initiation of embryonic cardiac pacemaker activity by inositol 1,4,5-trisphosphate-dependent calcium signaling. Mol Biol Cell 16:2414–2423
19. Michalak M, Lynch J, Groenendyk J et al. (2002) Calreticulin in cardiac development and pathology. Biochim Biophys Acta 1600:32–37
20. Mummery C, Ward-van Oosterwaard, Doevendans P et al. (2003) Differentiation of human embryonic stem cells to cardiomyocytes. Role of coculture with visceral endoderm-like cells. Circulation 107:2733–2740
21. Proven A, Roderick HL, Conway SJ et al. (2006) Inositol 1,4,5-trisphosphate supports the arrhythmogenic action of endothelin-1 on ventricular cardiac myocytes. J Cell Sci 119:3363–3375
22. Pucéat M and Jaconi M (2005) Ca^{2+} signalling in cardiogenesis. Cell Calcium 38:383–389
23. Reppel M, Boettinger C, Hescheler J (2004) Beta-adrenergic and muscarinic modulation of human embryonic stem cell-derived cardiomyocytes. Cell Physiol Biochem 14:187–196
24. Roderick HL and Bootman MD (2007) Pacemaking, arrhythmias, inotropy and hypertrophy: the many possible facets of IP_3 signaling in cardiac myocytes. J Physiol 581:883–884
25. Rosemblit N, Moschella MC, Ondriasá E et al. (1999) Intracellular calcium release channel expression during embryogenesis. Dev Biol 206:163–177
26. Sartiani L, Bettiol E, Stillitano F et al. (2007) Developmental changes in cardiomyocytes differentiated from human embryonic stem cells: a molecular and electrophysiological approach. Stem Cells 25:1136–1144
27. Satin J, Kehat I, Caspi O et al. (2004) Mechanism of spontaneous excitability in human embryonic stem cell derived cardiomyocytes. J Physiol 559:479–496

28. Satin J, Itzhaki I, Rapoport S et al. (2008) Calcium handling in human embryonic stem cell-derived cardiomyocytes. Stem Cells 26:1961–1972
29. Sedan O, Dolnikov K, Zeevi-Levin N et al. (2008) 1,4,5-Inositol trisphosphate-operated intracellular Ca(2+) stores and angiotensin-II/endothelin-1 signaling pathway are functional in human embryonic stem cell-derived cardiomyocytes. Stem Cells 26:3130–3138
30. Sedan O, Dolnikov K, Zeevi-Levin N, Fleishmann N, Spiegel I, Berdichevski S, Amit M, Itskovitz-Eldor J, Binah O. (2010) Human embryonic stem cell-derived cardiomyocytes can mobilize 1,4,5-inositol trisphosphate operated $[Ca^{2+}]_i$ stores; the functionality of angiotensin-II/endothelin-1 signaling pathways. Ann N Y Acad Sci 1188:68–77
31. Woodcock EA and Matkovitch SJ (2005) Ins(1,4,5)P_3 receptors and inositol phosphates in the heart: Evolutionary artefacts or active signal transducers? Pharmacol Ther 107:240–251
32. Xu C, Police S, Rao N et al. (2002) Characterization and enrichment of cardiomyocytes derived from human embryonic stem cells. Circ Res 91:501–508
33. Zima AV and Blatter LA (2004) Inositol-1,4,5-trisphosphate-dependant Ca^{2+} signaling in cat atrial excitation-contraction coupling and arrhythmias. J Physiol 555:607–615

Embryonic Stem Cell Derivatives for Cardiac Therapy: Advantages, Limitations, and Long-Term Prospects

Michal Weiler-Sagie and Lior Gepstein

Abstract The ultimate goal of cardiovascular regenerative medicine is to generate functional myocardial tissue that can engraft, survive, mature, and become well integrated with host myocardium leading to restoration of the myocardial electromechanical properties. Human embryonic stem cells (hESCs) (Science 282:1145–7, 1998) offer a number of theoretical advantages to achieve this objective because of their availability and expandability, their diverse differentiation capacity, and their proven capabilities to robustly differentiate into cardiovascular lineages yielding cells with the appropriate molecular, structural, and functional properties. Despite these potential advantages, there are still important limitations that need to be addressed before clinical cell therapy with hESCs can be realized. These include ethical and immunological issues, technical issues for efficient production and delivery of cardiomyocytes, and safety measures to prevent possible misbehavior of the transplanted cells. The recent introduction of the groundbreaking induced pluripotent stem cell technology, allowing the reprogramming of somatic cells into pluripotent lines resembling hESCs, has provided another boost to the field and further highlighted the potential use of cardiomyocytes derived from these pluripotent stem cells in several cardiovascular basic and translational research areas.

Keywords Human embryonic stem cell derived cardiac myocytes • Induced pluripotent stem cells • Immunogenicity • Arrhythmogenesis • Regulations

---

L. Gepstein (✉)
The Sohnis Family Research Laboratory for Cardiac Electrophysiology and Regenerative Medicine and the Rappaport Family Institute for Research in the Medical Sciences, The Bruce Rappaport Faculty of Medicine, Technion – Israel Institute of Technology, Haifa, Israel
e-mail: mdlior@tx.technion

I.S. Cohen and G.R. Gaudette (eds.), *Regenerating the Heart*, Stem Cell Biology and Regenerative Medicine, DOI 10.1007/978-1-61779-021-8_5,
© Springer Science+Business Media, LLC 2011

1 Potential Advantages of Human Embryonic Stem Cell Derived Cardiomyocytes

Cell therapy and tissue engineering are emerging as novel therapeutic paradigms for myocardial repair/regeneration. As reviewed recently [1–3], there is extensive animal experimental data suggesting that grafting of a wide array of different cell types can improve ventricular function following myocardial infarction. Although they were originally proposed to regenerate the heart, it seems that several of these cell types probably do not transform to generate significant amounts of new myocardium. This indicates that noncontractile mechanisms such as alteration of the infarct remodeling process, enhancement of angiogenesis, and augmentation of an endogenous repair mechanism probably underlie most of the benefit seen to date with cell transplantation [1–3]. It is therefore likely that further functional benefit could be achieved with a cell population that is characterized by typical cardiomyocyte properties and can contribute directly to contractility. Presently, the most obvious candidate cell type that can give rise to human cardiomyocytes with the appropriate phenotypic properties is human pluripotent stem cell lines such as human embryonic stem cells (hESCs) [4] and the recently described human induced pluripotent stem (hiPS) cells [5].

As discussed in previous chapters, hESCs are pluripotent stem cell lines that are derived from the inner cell mass of human blastocytes. These lines can be propagated in the undifferentiated state under special conditions and coaxed to differentiate into cell derivatives of all three germ layers, including bona fide cardiomyocytes [6–9]. Because of their inherent cardiomyocyte properties, hESC-derived cardiomyocytes (hESC-CMs) offer several advantages for potential myocardial regeneration strategies.

1.1 Quantity and Availability

It is estimated that a large myocardial infarction that results in heart failure is associated with the loss of several hundred million cardiomyocytes. Moreover, given the large number of donor cells that may be lost during the process of cell transplantation into the hostile ischemic myocardium [10], even a greater number of cells may be required for transplantation. Additionally, it may be better to transplant cells early after the insult before remodeling ensues.

The typical method for obtaining hESC-CMs is to form embryoid bodies (EBs) and then harvest the resultant spontaneously contractile cardiomyocytes by mechanical dissection [7], enzymatic methods, or on the basis of their physical properties (using Percoll gradient centrifugation) [9]. Only a small fraction of approximately 1–5% of the cells composing the EB cell population are indeed cardiomyocytes using this spontaneous differentiation system. Efforts to improve the efficiency of cardiogenesis in hESCs have used factors known to be released by endoderm during development and members of the transforming growth factor B superfamily. Co-culture with END-2 cells, a murine endoderm-like cell line, together with serum depletion resulted in enhanced efficiency of hESC-CM generation to 5–20% of the cell population [11]. Differentiation protocols using activin A and bone morphogenetic protein 4 in serum-free medium generate cultures with 30% hESC-CMs [12]. There have been reports of a threefold to tenfold rise in efficiency with the application of bone morphogenetic protein 2 and SU5402, a fibroblast growth factor (FGF) receptor inhibitor [13], and a fivefold increase when activin A and FGF-2 were added [14]. Thus, substantial progress has been made in our ability to efficiently direct hESC cardiogenesis.

Today, more than 400 hESC lines are available; some of them being well characterized and organized in international stem cell banks, for example, the European Human Embryonic Stem Cell Registry – hESCreg (http://www.hescreg.eu/) and the National Stem Cell Bank (http://www.nationalstemcellbank.org). The progress achieved in hESC-directed cardiogenesis implies that hESCs can provide, ex vivo, potentially an unlimited number of human cardiomyocytes for transplantation, especially if scaling-up methods are used during the entire process of differentiation such as by using bioreactors and related technologies [15]. On demand, cells can be thawed, cultured, and directed toward myocardial differentiation.

1.2 Proven Differentiation into Functioning Cardiomyocytes

hESCs have the capacity to undergo directed differentiation into genuine cardiomyocytes in vitro, making them more likely to achieve functional connection with host myocardium. Several lines of evidence confirmed the molecular, ultrastructural, and functional cardiomyocyte nature of the cells within the beating EBs [6–9]. Reverse transcription PCR analysis demonstrated the expression of cardiac-specific genes and transcription factors in the appropriate developmental temporal pattern. Transmission electron microscopy of hESC-CMs at various developmental stages showed the progressive ultrastructural maturation from an irregular myofilament distribution to a more mature sarcomeric organization (containing well-defined sarcomeres with recognizable A, I, and Z bands) in late-stage EBs [16]. Interestingly, simultaneously with this ultrastructural maturation, the hESC-CMs gradually withdrew from the cell cycle [16]. The hESC-CMs also portrayed functional properties of early-stage human heart cells. Detailed electrophysiological analyses revealed that hESC-CMs display action potentials with a variety of cardiac-like morphologies (atrial-, ventricular-, and pacemaker/conduction-system-like) [6] and that they express many of the same ion channels as mature cells [17, 18]. hESC-CMs express a large Na^+ current density and the action potential duration is shortened by L-type Ca^{2+} channel blockade. Interestingly, although only a paucity of different potassium channels were expressed in early-stage hESC-CMs, the expression of a more diverse set of potassium channels was found in late-differentiating cells [17]. This in vitro maturation process is significantly slower than that reported in the well-characterized mouse embryonic stem cell (ESC) model [19]; this is not surprising given the significant differences in the length of the gestation time between the two species.

Beyond the presence of typical extracellular and intracellular electrical activity, hESC-CMs also display intracellular calcium transients [7, 20] and an appropriate chronotropic response to adrenergic and cholinergic agents [7]. A recent study suggests that hESC-CMs contain functional ryanodine receptor regulated intracellular calcium stores and contraction is not only dependent on transsarcolemmal calcium influx [20]. Thus, all the components of normal cardiac excitation–contraction coupling were demonstrated to be present within these cells. Microelectrode array studies demonstrated the development of a functional cardiac syncytium in beating EBs with spontaneous pacemaker activity and action potential propagation and highlighted the potential of this system for cardiac electrophysiological drug screening [21, 22].

1.3 Potential to Differentiate into a Variety of Cardiac-Related Cell Types or Multipotent Cardiovascular Progenitor Cells

The hESC lines are pluripotent and therefore can be directed to differentiate into a variety of cell types that may have relevance for cardiac repair, including cardiomyocytes [6–9] that have been characterized extensively as well as vascular precursors [23] for angiogenesis. Moreover, hESCs were demonstrated to give rise to specialized cardiomyocyte subtypes (pacemaking, atrial, and ventricular cells) that may enable tailoring for specific applications [6]. The differentiation process is accomplished ex vivo and therefore cells can be selected for transplantation at different stages of differentiation. Cells can be selected for transplantation when they are fully differentiated or at the cardiac progenitor cell stage. In addition, owing to their clonal origin, hESC-CMs could lend themselves to extensive characterization and genetic manipulation to promote desirable characteristics such as resistance to ischemia and apoptosis, improved contractile function, and specific electrophysiological properties.

Lessons from embryonic developmental models have directed and inspired hESC researchers. This developmental biology approach has made it possible to recapitulate in ESC cultures the key events that regulate early-lineage commitment in the embryo, resulting in the generation of highly enriched differentiated cardiovascular cell-lineage populations [24].

Formation of the heart requires the coordinated functions of cardiac myocytes, smooth muscle cells, endothelial cells, and connective tissue elements. A number of studies revealed that these different cell types may arise from common progenitors.

Cardiac progenitor cells have been identified using ESC lines in which reporter genes were expressed under the control of early cardiac or mesodermal genes (brachyury, Isl1, and Nkx2.5) [25–27] or by fluorescence-activated cell sorting (FACS) selection using antibodies against specific cell-surface markers (cKit and

Flk-1 or the human equivalent KDR) [25, 27, 28]. The identified multipotent cardiovascular progenitors were shown to be able to give rise to cardiomyocytes, smooth muscle cells, and endothelial cells in vitro [24]. The transplantation of hESC-derived multipotent cardiovascular progenitor cells can potentially result in in vivo differentiation to cardiomyocytes as well as supporting cells (endothelial and smooth muscle cells), possibly improving engraftment and survival of the cells.

2 Potential Disadvantages and Challenges

Despite the important progress that has been made in the decade that has passed since the first hESC lines were established, many conceptual, ethical, scientific, technical, and regulatory obstacles remain on the road to clinical utilization of hESC-CMs for myocardial repair. These include the following: (1) strategies need to be developed further to allow efficient hESC-directed cardiomyocyte differentiation and enrichment; (2) strategies need to be developed for optimizing hESC culturing techniques and for scaling up the procedure to derive clinically relevant numbers of cardiomyocytes; (3) the transplantation technique should be developed to enable proper alignment of the graft tissue, to minimize long-term donor cell loss, and to minimize damage to host tissue; (4) ethical issues need to be addressed; (5) strategies aimed at preventing immune rejection should be established; and (6) safety issues including potential arrhythmogenesis, oncogenic transformation, and genetic manipulation need to be dealt with. The issues of directed cardiomyocyte differentiation, enrichment, and scaling up were discussed in the preceding sections, so we will mainly focus in the next sections on some of the remaining issues.

2.1 Regulatory, Intellectual Property, and Ethical Issues

In general, the fundamental regulatory and ethical requirements that are used in drug and other clinical trials apply equally to cell therapy [29]. However, cell therapy trials introduce new ethical issues applicable to all stem cell sources, including those of ownership of cell lines, intellectual property, and patents [30–33] as well as those related to the specific use of hESCs – the debate over the use of embryonic material for research. Such debates are influenced by attitudes of governmental administrations and different approaches to funding in different countries. This barrier has limited hESC research in the USA for over a decade and has been lifted only recently. The regulatory issues for new cardiovascular therapies, and in particular procedures in which cells are manipulated and engineered, are complex and evolving [34–37]. The lack of precedent in this area of science necessitates close interaction among regulators, scientists, clinicians, and the public because the potential for misunderstanding on all sides is substantial.

2.2 Immunogenicity

A major obstacle for the utilization of hESC derivatives for the regeneration of different organs is the expected immune rejection. The first question to be contended with is precisely how immunogenic are tissues derived from hESCs. Initial characterization of the immunogenicity of hESCs showed that they express HLA class-I molecules but not HLA class-II molecules [38]. Moreover, with use of the human/mouse trimera model (enabling the assessment of the direct alloimmune response of the human immune system to transplanted human cells), it was demonstrated that transplanted hESCs elicit only a relatively small immune response [39]. Studies in murine models have shown that the immunogenicity of ESCs increases with differentiation; and thus cardiac transplants of differentiated ESC derivatives are recognized and rejected by allogeneic recipients [40]. These results imply that immunosuppressive regimes for future hESC-based therapeutics may be reduced in comparison with the conventional organ transplantation situation.

The need for immunosuppression by no means precludes use of hESC-based cell therapies – heart transplant recipients receive immunosuppressive therapy because the benefits to the patient with end-stage heart disease outweigh the intrinsic risks of immunosuppression. A similar risk–benefit analysis would be required in this case if hESC-CM transplants were proven efficacious in human patients.

Nonetheless, since it is expected that hESC-CMs will be rejected following transplantation into the immunocompetent heart [40]; strategies to overcome this immune rejection process are being developed. Although a detailed discussion of this issue is beyond the scope of this chapter and this issue has been discussed in detail in a number of excellent reviews [41, 42], we will briefly discuss some of the suggested strategies. One approach to overcoming the immunological barriers to stem cell transplantation is to minimize the alloantigenic differences between the donor and the recipient. This may be achieved by establishing "banks" of major histocompatibility complex (MHC) antigen-typed hESCs and then matching donors and recipients, as is carried out routinely for tissue and organ transplantation. An alternative solution may be the generation of a universal donor hESC line. This could be achieved by silencing genes associated with the assembly or transcriptional regulation of the MHCs or by inserting or deleting other genes that can modulate the immune response. Another attractive strategy for inducing tolerance is by hematopoietic chimerism during which hESC-derived lymphohematopoietic stem cells are initially transplanted and this is followed by the engraftment of cell derivatives generated from the same hESC line.

Another possibility is to create isogenic stem cell lines that are not immunogenic. One such technology is somatic nuclear transfer technology (SCNT), also known as genomic replacement or therapeutic cloning [42]. The process of nuclear-transfer-derived ESC production involves transferring the nucleus of a mature somatic cell into the cytoplasm of an oocyte from which the original nucleus has been removed. Once a blastocyst is formed, cells from the inner cell mass can be utilized to produce pluripotent ESCs that are genetically identical to the nucleus donor's cells, with the exception of the mitochondrial DNA. Nevertheless, several conceptual, technical, and ethical barriers will probably significantly limit this approach in the near future.

An exciting alternative for the generation of isogenic lines that may bypass the significant technical and ethical hurdles associated with SCNT is direct cell reprogramming by pluripotency-related transcription factors. In 2006, Takahashi and Yamanaka [43] were able to reprogram mouse adult and embryonic fibroblasts by ectopic expression of just four transcription factors, Oct4, Sox2, c-Myc, and Klf4. The generated induced pluripotent stem (iPS) cells were then selected on the basis of the endogenous activity of the Oct4 or Nanog genes, yielding cells that display characteristics similar to those of ESCs [43, 44]. The mouse iPS cells were demonstrated to be able to differentiate into derivatives of all three germ layers, to form teratomas, and to give rise to chimeric embryos after blastocyst injection [43, 44]. In 2007, the first hiPS cell lines were described independently by Yamanka's [5] and Thomson's [45] groups. More recently, the ability of hiPS cells to form cardiomyocytes was also described [46].

2.3 Genetic Manipulation

The potential of cell therapy with hESCs has long been recognized but at the same time regarded as science fiction. It is much easier to embrace stem cell technologies as in vitro platforms for drug discovery and to avoid confrontation with the risk involved in in vivo studies in real patients. One of the major issues hindering the acceptance of hESCs as a therapeutic tool is the fear of transplanting genetically modified cells. It is clear that genetic manipulation of cells enhances our ability to select a homogeneous population of the required cells (e.g. by promoter-based antibiotic-resistance selection [47]) and to do so more efficiently – a basic requirement. But these genetic manipulations harbor the danger of insertional mutagenesis by integrating viral vectors, inappropriate expression of the transgene, and immune activation against the viral vector, the gene-engineered cells, or the transgene product [48–50]. The consequences of these events range from the failure to establish stable transgene expression, to elimination of the gene-modified cells because of immune reactions, to even transformed cell growth and oncogenesis [51].

Advancement in genetic manipulation of ESCs was triggered by the pivotal studies introducing reprogramming and creation of mouse iPS and hiPS cells. These cells are pluripotent as are hESCs but evade the ethical and immunological barriers that impede cell therapy with hESCs. iPS cells are created by genetic manipulation – by insertion of the reprogramming factors Oct4, sox3, klf4, and c-myc. Their conception is a consequence of genetic manipulation. The prospect of restoring a patient's heart function with iPS-cell-derived cardiomyocytes generated from cells biopsied from the patient's skin is fueling the search for a safe method of genetic manipulation.

Several groups have shown that it may be feasible to induce iPS cells without viral integration. Stadtfeld et al. [52] generated iPS cells from mouse hepatocytes using adenoviruses carrying the four reprogramming factors. Okita et al. [53] generated iPS cells from mouse embryonic fibroblasts using plasmids. Many of these plasmid-generated iPS cells did not show integration into the host genome by either PCR or Southern blotting. Several recent studies have reported viral-vector-free integration of reprogramming genes utilizing transposons, followed by their removal [54–56]. The most important and unique feature of this approach is that reintroduction of the *PB* transposase by transient transfection resulted in the traceless excision of the reprogramming cassette from the iPS cell [56]. This obviates most of the aforementioned concerns or limitations associated with the use of retroviral, lentiviral, or nonintegrating vectors, while maintaining a relatively robust reprogramming efficiency. As an alternative to transposase-mediated excision, it was also possible to excise the reprogramming cassette flanked by *loxP* sites after transient expression of the Cre recombinase [55]. However, in that case some residual trace elements outside the *loxP* sites, including the transposon repeats, could not be excised from the iPS cell genome. More recently, proof-of-concept studies for the ability to generate iPS cells by protein-based reprogramming have been reported.

2.4 Optimizing In Vivo Cell Delivery and Tissue Engineering

Recent studies by a number of groups have assessed the feasibility of hESC-CM grafting in the infarcted rodent heart [13, 57–60]. These studies demonstrated the ability of hESC-CMs to survive within the infarcted area and to favorably affect cardiac function. Despite these encouraging results, the underlying mechanism explaining the observed functional improvement and issues pertaining to its long-term persistence [60] remain a matter of debate. Moreover, clinical translation of the cell therapy approaches has been hampered by the large cell loss during the injection procedure and by the significant cell death following grafting into the hostile ischemic myocardium [10]. Hence, initially techniques should be developed to increase cell retention within the transplanted area. This may be achieved by improving the delivery techniques, by using improved injection media, and by developing different routs of administration. Similarly, strategies should be developed to increase the size of the in vivo cell grafts. This may be accomplished by establishing strategies to reduce the degree of cell death following transplantation (e.g. by overexpression of antiapoptotic proteins or by inducing angiogenesis) or alternatively by augmenting the in vivo ability of the hESC-CMs to proliferate.

One potential strategy to overcome the aforementioned obstacles lies in the emerging field of cardiac tissue engineering [61]. This multidisciplinary field applies the principles of engineering and life sciences to the development of biological substitutes aiming to restore, maintain, or improve tissue function. The use of cell-seeded scaffolds to promote tissue development is the hallmark of tissue engineering.

The scaffold serves many purposes, including the delivery of biological signals to control and enhance tissue formation, provision of adequate biomechanical support for the cell graft, control of graft shape and size, promotion of angiogenesis, and protection from physical damage. Several tissue-engineering approaches are currently being explored for cardiac repair. The most common approach is to seed cardiomyocytes onto biodegradable porous scaffolds that can be engrafted to the injured heart. This approach is based on the concept that the biodegradable scaffolds act as a useful initial alternative for the extracellular matrix and that seeded cells will eventually reform their native structure following scaffold biodegradation.

In vivo transplantation of degradable biopolymers into the injured heart of rats has yielded promising results. Several cell-seeded biopolymers have been transplanted in the infracted heart of rats, including alginate, fibrin glue, and polyurethane scaffolds, all of which resulted in a decrease in infarct size, induction of neovascularization, and attenuation of the remodeling process.

Insufficient graft vascularization is considered one of the main factors responsible for the limited graft survival and the thickness of the engineered transplanted tissue. Increase of the degree of graft vascularization may significantly improve the survival of the transplanted myocytes. Recent work from our group [62] described the generation of a three-dimensional engineered, human vascularized, cardiac tissue. In this work, poly-L-lactic acid (PLLA)/polylactic-glycolic acid (PLGA) biodegradable scaffolds were seeded with hESC-CMs, endothelial cells, and embryonic fibroblasts. Ultrastructural analysis, immunostaining, reverse transcription PCR, pharmacological, and confocal laser calcium imaging studies confirmed the presence of a functional cardiac tissue coupled with a developing microvasculature system. More recently, we performed engraftment studies of the human vascularized cardiac muscle construct in the rat heart [63]. These experiments showed that the preexisting vessels within the scaffold augmented its total vascularization. Moreover, these human vessels became functional, anastomosed with host rat vasculature, and perfused the biograft.

2.5 *Arrhythmogenesis*

To date, no arrhythmia or increased mortality has been reported in preclinical studies using hESC-CMs, but nearly all these studies have been performed in rodent models. The considerably higher heart rates in rodents as compared with humans could conceal arrhythmias generated by pacemaker activity or reentrant circuits that may occur in species with a lower heart rate. Nonetheless, the potential for increased arrhythmogenesis exists and has been observed in studies using skeletal myoblast transplantation. Important factors may be the efficiency of cell-to-cell coupling, electrical heterogeneity due to incomplete differentiation of cells, the distribution of action potentials at the cell–residual myocardial interface, and gap junction remodeling, in addition to the potentially antiarrhythmic effects of paracrine factors. hESC-CMs are heterogeneous and include myocytes with ventricular-like but also nodal- and atrial-like action potential properties. Even the apparently ventricular-like hESC-CMs show immature electrophysiologic properties and so, relative to adult ventricular myocytes, have a comparatively depolarized maximum diastolic potential, a slower action potential upstroke, and a much smaller cell size. It is not clear whether these differences will disappear with maturation. If the grafted hESC-CMs are not well coupled, they could promote arrhythmogenesis by introducing points of electrical heterogeneity into the myocardium. Preclinical safety studies in a large-animal model will be required to address whether this issue is a real or only a theoretical concern.

2.6 *Oncogenic Transformation*

The pluripotency of hESCs is a double-edged sword conveying the ability to provide quantities of specific differentiated cells together with the ability to form teratomas and the potential to undergo oncologic transformation [64–66]. The assay used to assess plutipotency in hESCs and hiPS cells is teratoma formation, thereby providing abundant experimental evidence of the fate of transplanted undifferentiated cells. Currently, there is no method for selecting a pure population of differentiated cells for transplantation. The danger of undifferentiated cells contaminating the selected population exists. In preclinical studies using animal models of myocardial ischemia to evaluate cell therapy with cardiomyocytes derived from hESCs, teratoma formation was not reported. This may be a result of the low number of cells surviving after transplantation and teratoma formation may become relevant as increasing numbers of cells survive. The potential risk to human patients from both teratoma and malignant tumors is quite real, yet remains difficult to estimate as no human trials of hESCs or iPS cells have been conducted.

There are several approaches to making stem-cell-based regenerative medicine safer (reviewed in [67]) by devising methods to eliminate undifferentiated cells before transplantation, during the selection process in vitro, or after transplantation. In vitro elimination of undifferentiated cells is usually achieved by selection of the differentiated cells. This process, termed "enrichment," is achieved, for example, by genetically manipulating the cells to express antibiotic resistance genes or fluorescent markers under tissue-specific promoters or by FACS or magnetic-based sorting using antibodies targeting cell-surface markers unique to cardiac progenitor cells or more mature cardiomyocytes. Recent work has shown that tumorigenic risk is abrogated through guided lineage specification or by selection of early progenitors [68].

Examples of in vivo selection strategies include suicide genes and directed antibodies. The transfer of a gene encoding a susceptibility factor can make a cell specifically sensitive to a drug. After genomic integration, such a potentially destructive gene is endogenously expressed and has therefore been coined a "suicide gene." The stable genetic introduction of a suicide gene such as thymidine kinase into stem cells has been reported to be effective in combination with ganciclovir treatment [69]. In this study, the suicide gene was expressed in cells at all stages of differentiation and demonstrated in vivo elimination of all transplanted cells at will. Expression of thymidine kinase driven by the Oct4 promoter enables selective elimination of only those ESCs that have escaped differentiation since this is a marker of undifferentiated ESCs [70].

Alternatively, killer antibodies directed against antigens present on the surface of undifferentiated hESCs such as SSEA-4 or a member of the TRA family can be used. New hESC surface antigens are currently being discovered and tested, such as podocalyxin-like protein-1, which appears quite promising as a cytotoxic agent specifically against undifferentiated hESCs [71, 72]. A major concern with this approach is how specific these methods would be at targeting stem cells as opposed to other cell types and the possible side effects.

3 Summary

The development of hESC lines and their ability to differentiate into cardiomyocyte tissue holds great promise for several cardiovascular research and clinical areas. Most importantly, the ability to generate, for the first time, human cardiac tissue provides an exciting and promising cell source for the emerging discipline of regenerative medicine, tissue engineering, and myocardial repair. Nevertheless, as described herein, several milestones have to be achieved to fully harness the enormous research and clinical potential of this unique technology.

References

1. Dimmeler S, Zeiher AM, Schneider MD. Unchain my heart: the scientific foundations of cardiac repair. J Clin Invest. 2005;115:572–83.
2. Laflamme MA, Murry CE. Regenerating the heart. Nat Biotechnol. 2005;23:845–56.
3. Segers VF, Lee RT. Stem-cell therapy for cardiac disease. Nature. 2008;451:937–42.
4. Thomson JA, Itskovitz-Eldor J, Shapiro SS, Waknitz MA, Swiergiel JJ, Marshall VS, et al. Embryonic stem cell lines derived from human blastocysts. Science. 1998;282:1145–7.
5. Takahashi K, Tanabe K, Ohnuki M, Narita M, Ichisaka T, Tomoda K, et al. Induction of pluripotent stem cells from adult human fibroblasts by defined factors. Cell. 2007;131:861–72.
6. He JQ, Ma Y, Lee Y, Thomson JA, Kamp TJ. Human embryonic stem cells develop into multiple types of cardiac myocytes: action potential characterization. Circ Res. 2003;93:32–9.
7. Kehat I, Kenyagin-Karsenti D, Snir M, Segev H, Amit M, Gepstein A, et al. Human embryonic stem cells can differentiate into myocytes with structural and functional properties of cardiomyocytes. J Clin Invest. 2001;108:407–14.
8. Mummery C, Ward-van Oostwaard D, Doevendans P, Spijker R, van den Brink S, Hassink R, et al. Differentiation of human embryonic stem cells to cardiomyocytes: role of coculture with visceral endoderm-like cells. Circulation. 2003;107:2733–40.
9. Xu C, Police S, Rao N, Carpenter MK. Characterization and enrichment of cardiomyocytes derived from human embryonic stem cells. Circ Res. 2002;91:501–8.
10. Muller-Ehmsen J, Peterson KL, Kedes L, Whittaker P, Dow JS, Long TI, et al. Rebuilding a damaged heart: long-term survival of transplanted neonatal rat cardiomyocytes after myocardial infarction and effect on cardiac function. Circulation. 2002;105:1720–6.

11. Passier R, Oostwaard DW, Snapper J, Kloots J, Hassink RJ, Kuijk E, et al. Increased cardiomyocyte differentiation from human embryonic stem cells in serum-free cultures. Stem Cells. 2005;23:772–80. doi:23/6/772 [pii] 10.1634/stemcells.2004-0184.
12. Xu C, Inokuma MS, Denham J, Golds K, Kundu P, Gold JD, et al. Feeder-free growth of undifferentiated human embryonic stem cells. Nat Biotechnol. 2001;19:971-4.
13. Tomescot A, Leschik J, Bellamy V, Dubois G, Messas E, Bruneval P, et al. Differentiation in vivo of cardiac committed human embryonic stem cells in postmyocardial infarcted rats. Stem Cells. 2007;25:2200–5.
14. Burridge PW, Anderson D, Priddle H, Barbadillo Munoz MD, Chamberlain S, Allegrucci C, et al. Improved human embryonic stem cell embryoid body homogeneity and cardiomyocyte differentiation from a novel V-96 plate aggregation system highlights interline variability. Stem Cells. 2007;25:929–38.
15. Zandstra PW, Bauwens C, Yin T, Liu Q, Schiller H, Zweigerdt R, et al. Scalable production of embryonic stem cell-derived cardiomyocytes. Tissue Eng. 2003;9:767–78.
16. Snir M, Kehat I, Gepstein A, Coleman R, Itskovitz-Eldor J, Livne E, et al. Assessment of the ultrastructural and proliferative properties of human embryonic stem cell-derived cardiomyocytes. Am J Physiol Heart Circ Physiol. 2003;285:H2355–63.
17. Sartiani L, Bettiol E, Stillitano F, Mugelli A, Cerbai E, Jaconi ME. Developmental changes in cardiomyocytes differentiated from human embryonic stem cells: a molecular and electrophysiological approach. Stem Cells. 2007;25:1136–44.
18. Satin J, Kehat I, Caspi O, Huber I, Arbel G, Itzhaki I, et al. Mechanism of spontaneous excitability in human embryonic stem cell derived cardiomyocytes. J Physiol. 2004;559:479–96.
19. Hescheler J, Fleischmann BK, Lentini S, Maltsev VA, Rohwedel J, Wobus AM, et al. Embryonic stem cells: a model to study structural and functional properties in cardiomyogenesis. Cardiovasc Res. 1997;36:149–62.
20. Satin J, Itzhaki I, Rapoport S, Schroder EA, Izu L, Arbel G, et al. Calcium handling in human embryonic stem cell-derived cardiomyocytes. Stem Cells. 2008;26:1961–72.
21. Caspi O, Itzhaki I, Arbel G, Kehat I, Gepstien A, Huber I, et al. In vitro electrophysiological drug testing using human embryonic stem cell derived cardiomyocytes. Stem Cells Dev. 2008;18(1):161–72.
22. Kehat I, Gepstein A, Spira A, Itskovitz-Eldor J, Gepstein L. High-resolution electrophysiological assessment of human embryonic stem cell-derived cardiomyocytes: a novel in vitro model for the study of conduction. Circ Res. 2002;91:659–61.
23. Levenberg S, Golub JS, Amit M, Itskovitz-Eldor J, Langer R. Endothelial cells derived from human embryonic stem cells. Proc Natl Acad Sci U S A. 2002;99:4391–6.
24. Murry CE, Keller G. Differentiation of embryonic stem cells to clinically relevant populations: lessons from embryonic development. Cell. 2008;132:661–80.
25. Kattman SJ, Huber TL, Keller GM. Multipotent flk-1+ cardiovascular progenitor cells give rise to the cardiomyocyte, endothelial, and vascular smooth muscle lineages. Dev Cell. 2006;11:723–32.
26. Moretti A, Caron L, Nakano A, Lam JT, Bernshausen A, Chen Y, et al. Multipotent embryonic isl1+ progenitor cells lead to cardiac, smooth muscle, and endothelial cell diversification. Cell. 2006;127:1151–65.
27. Wu SM, Fujiwara Y, Cibulsky SM, Clapham DE, Lien CL, Schultheiss TM, et al. Developmental origin of a bipotential myocardial and smooth muscle cell precursor in the mammalian heart. Cell. 2006;127:1137–50.
28. Yang L, Soonpaa MH, Adler ED, Roepke TK, Kattman SJ, Kennedy M, et al. Human cardiovascular progenitor cells develop from a KDR+ embryonic-stem-cell-derived population. Nature. 2008;453:524–8.
29. Lewis RM, Gordon DJ, Poole-Wilson PA, Borer JS, Zannad F. Similarities and differences in design considerations for cell therapy and pharmacologic cardiovascular clinical trials. Cardiology. 2008;110:73–80.
30. Zarzeczny A, Caulfield T. Emerging ethical, legal and social issues associated with stem cell research & and the current role of the moral status of the embryo. Stem Cell Rev. 2009; 5:96–101.
31. Jung KW. Perspectives on human stem cell research. J Cell Physiol. 2009;220:535–7.

32. Chapman AR. The ethics of patenting human embryonic stem cells. Kennedy Inst Ethics J. 2009;19:261–88.
33. Sugarman J. Human stem cell ethics: beyond the embryo. Cell Stem Cell. 2008;2:529–33.
34. Gersh BJ, Simari RD, Behfar A, Terzic CM, Terzic A. Cardiac cell repair therapy: a clinical perspective. Mayo Clin Proc. 2009;84:876–92.
35. Lo B, Kriegstein A, Grady D. Clinical trials in stem cell transplantation: guidelines for scientific and ethical review. Clin Trials. 2008;5:517–22.
36. Bartunek J, Dimmeler S, Drexler H, Fernandez-Aviles F, Galinanes M, Janssens S, et al. The consensus of the task force of the European Society of Cardiology concerning the clinical investigation of the use of autologous adult stem cells for repair of the heart. Eur Heart J. 2006;27:1338–40.
37. Martin JF. Stem cells and the heart: ethics, organization and funding. Nat Clin Pract Cardiovasc Med. 2006;3 Suppl 1:S136–7.
38. Drukker M, Katz G, Urbach A, Schuldiner M, Markel G, Itskovitz-Eldor J, et al. Characterization of the expression of MHC proteins in human embryonic stem cells. Proc Natl Acad Sci U S A. 2002;99:9864–9.
39. Drukker M, Katchman H, Katz G, Even-Tov Friedman S, Shezen E, Hornstein E, et al. Human embryonic stem cells and their differentiated derivatives are less susceptible to immune rejection than adult cells. Stem Cells. 2006;24:221–9.
40. Nussbaum J, Minami E, Laflamme MA, Virag JA, Ware CB, Masino A, et al. Transplantation of undifferentiated murine embryonic stem cells in the heart: teratoma formation and immune response. Faseb J. 2007;21:1345–57.
41. Drukker M. Recent advancements towards the derivation of immune-compatible patient-specific human embryonic stem cell lines. Semin Immunol. 2008;20:123–9.
42. Chidgey AP, Layton D, Trounson A, Boyd RL. Tolerance strategies for stem-cell-based therapies. Nature. 2008;453:330–7.
43. Takahashi K, Yamanaka S. Induction of pluripotent stem cells from mouse embryonic and adult fibroblast cultures by defined factors. Cell. 2006;126:663–76.
44. Okita K, Ichisaka T, Yamanaka S. Generation of germline-competent induced pluripotent stem cells. Nature. 2007;448:313–7.
45. Yu J, Vodyanik MA, Smuga-Otto K, Antosiewicz-Bourget J, Frane JL, Tian S, et al. Induced pluripotent stem cell lines derived from human somatic cells. Science. 2007;318:1917–20. doi:1151526 [pii] 10.1126/science.1151526.
46. Zwi L, Caspi O, Arbel G, Huber I, Gepstein A, Park IH, et al. Cardiomyocyte differentiation of human induced pluripotent stem cells. Circulation. 2009;120:1513-23.
47. Klug MG, Soonpaa MH, Koh GY, Field LJ. Genetically selected cardiomyocytes from differentiating embryonic stem cells form stable intracardiac grafts. J Clin Invest. 1996;98:216–24.
48. Hacein-Bey-Abina S, Garrigue A, Wang GP, Soulier J, Lim A, Morillon E, et al. Insertional oncogenesis in 4 patients after retrovirus-mediated gene therapy of SCID-X1. J Clin Invest. 2008;118:3132–42.
49. Manno CS, Pierce GF, Arruda VR, Glader B, Ragni M, Rasko JJ, et al. Successful transduction of liver in hemophilia by AAV-Factor IX and limitations imposed by the host immune response. Nat Med. 2006;12:342–7.
50. Raper SE, Yudkoff M, Chirmule N, Gao GP, Nunes F, Haskal ZJ, et al. A pilot study of in vivo liver-directed gene transfer with an adenoviral vector in partial ornithine transcarbamylase deficiency. Hum Gene Ther. 2002;13:163–75.
51. Hacein-Bey-Abina S, Von Kalle C, Schmidt M, McCormack MP, Wulffraat N, Leboulch P, et al. LMO2-associated clonal T cell proliferation in two patients after gene therapy for SCID-X1. Science. 2003;302:415–9.
52. Stadtfeld M, Nagaya M, Utikal J, Weir G, Hochedlinger K. Induced pluripotent stem cells generated without viral integration. Science. 2008;322:945–9.
53. Okita K, Nakagawa M, Hyenjong H, Ichisaka T, Yamanaka S. Generation of mouse induced pluripotent stem cells without viral vectors. Science. 2008;322:949–53.
54. Kaji K, Norrby K, Paca A, Mileikovsky M, Mohseni P, Woltjen K. Virus-free induction of pluripotency and subsequent excision of reprogramming factors. Nature. 2009;458:771–5.

55. Woltjen K, Michael IP, Mohseni P, Desai R, Mileikovsky M, Hamalainen R, et al. piggyBac transposition reprograms fibroblasts to induced pluripotent stem cells. Nature. 2009;458:766–70.
56. Yusa K, Rad R, Takeda J, Bradley A. Generation of transgene-free induced pluripotent mouse stem cells by the piggyBac transposon. Nat Methods. 2009;6:363–9.
57. Caspi O, Huber I, Kehat I, Habib M, Arbel G, Gepstein A, et al. Transplantation of human embryonic stem cell-derived cardiomyocytes improves myocardial performance in infarcted rat hearts. J Am Coll Cardiol. 2007;50:1884–93.
58. Laflamme MA, Chen KY, Naumova AV, Muskheli V, Fugate JA, Dupras SK, et al. Cardiomyocytes derived from human embryonic stem cells in pro-survival factors enhance function of infarcted rat hearts. Nat Biotechnol. 2007;25:1015–24.
59. Leor J, Gerecht S, Cohen S, Miller L, Holbova R, Ziskind A, et al. Human embryonic stem cell transplantation to repair the infarcted myocardium. Heart. 2007;93:1278–84.
60. van Laake LW, Passier R, Doevendans PA, Mummery CL. Human embryonic stem cell-derived cardiomyocytes and cardiac repair in rodents. Circ Res. 2008;102:1008–10.
61. Zimmermann WH, Didie M, Doker S, Melnychenko I, Naito H, Rogge C, et al. Heart muscle engineering: an update on cardiac muscle replacement therapy. Cardiovasc Res. 2006;71:419–29.
62. Caspi O, Lesman A, Basevitch Y, Gepstein A, Arbel G, Habib IH, et al. Tissue engineering of vascularized cardiac muscle from human embryonic stem cells. Circ Res. 2007;100:263–72.
63. Lesman A, Habib M, Caspi O, Gepstein A, Arbel G, Levenberg S, et al. Transplantation of a tissue-engineered human vascularized cardiac muscle. Tissue Eng Part A. 2009. doi:10.1089/ten.TEA.2009.0130.
64. Wong DJ, Liu H, Ridky TW, Cassarino D, Segal E, Chang HY. Module map of stem cell genes guides creation of epithelial cancer stem cells. Cell Stem Cell. 2008;2:333–44.
65. Singh SK, Clarke ID, Terasaki M, Bonn VE, Hawkins C, Squire J, et al. Identification of a cancer stem cell in human brain tumors. Cancer Res. 2003;63:5821–8.
66. Shih CC, Forman SJ, Chu P, Slovak M. Human embryonic stem cells are prone to generate primitive, undifferentiated tumors in engrafted human fetal tissues in severe combined immunodeficient mice. Stem Cells Dev. 2007;16:893–902.
67. Knoepfler PS. Deconstructing stem cell tumorigenicity: a roadmap to safe regenerative medicine. Stem Cells. 2009;27:1050–6. doi:10.1002/stem.37.
68. Behfar A, Perez-Terzic C, Faustino RS, Arrell DK, Hodgson DM, Yamada S, et al. Cardiopoietic programming of embryonic stem cells for tumor-free heart repair. J Exp Med. 2007;204:405–20.
69. Schuldiner M, Itskovitz-Eldor J, Benvenisty N. Selective ablation of human embryonic stem cells expressing a "suicide" gene. Stem Cells. 2003;21:257–65.
70. Hara A, Aoki H, Taguchi A, Niwa M, Yamada Y, Kunisada T, et al. Neuron-like differentiation and selective ablation of undifferentiated embryonic stem cells containing suicide gene with Oct-4 promoter. Stem Cells Dev. 2008;17:619–27. doi:10.1089/scd.2007.0235.
71. Choo AB, Tan HL, Ang SN, Fong WJ, Chin A, Lo J, et al. Selection against undifferentiated human embryonic stem cells by a cytotoxic antibody recognizing podocalyxin-like protein-1. Stem Cells. 2008;26:1454–63. doi:2007-0576 [pii] 10.1634/stemcells.2007-0576.
72. Tan HL, Fong WJ, Lee EH, Yap M, Choo A. mAb 84, a cytotoxic antibody that kills undifferentiated human embryonic stem cells via oncosis. Stem Cells. 2009;27:1792–801.

Methods for Differentiation of Bone-Marrow-Derived Stem Cells into Myocytes

Shinji Makino and Keiichi Fukuda

Abstract Although heart transplantation is the ultimate therapy for severe heart failure, it is not widely used owing to the inadequate supply of donor hearts. Therefore, cell-based therapies for the prevention or treatment of cardiac dysfunction have attracted significant interest. Since we first reported (in 1999) that bone marrow (BM) mesenchymal stem cells (MSCs) could differentiate into cardiomyocytes in vitro [1], research on regenerative medicine has advanced dramatically [2, 3]. In addition to BM MSCs, embryonic stem cells, cardiac tissue stem cells, adipose tissue stem cells, and induced pluripotent stem cells undergo myocardial differentiation; additional cell types may also prove to have cardiac cell differentiation abilities. An early-phase clinical trial involving the direct infusion of BM mononuclear cells and peripheral blood mononuclear cells into coronary arteries and the myocardium has been undertaken. However, there is a vast gap between demonstrating that a cell type can differentiate into myocardium and translating this result into clinical practice. The major challenges for the therapeutic use of stem cells include the effective harvesting and in vitro expansion of cells to ensure sufficient numbers and purity of the cells. This chapter focuses on methods for the differentiation of BM-derived stem cells into myocytes.

Keywords Bone marrow stem cells • Mesenchymal stem cells • Cardiac stem cells • Hematopoietic stem cells • Endothelial progenitor cells • Cell transplant • Myocardial infarction • Myocytes

---

S. Makino (✉)
Center for Integrated Medical Research,
and
Department of Cardiology,
KEIO University School of Medicine, Tokyo, Japan
e-mail: koshinji@sc.itc.keio.ac.jp

I.S. Cohen and G.R. Gaudette (eds.), *Regenerating the Heart*, Stem Cell Biology and Regenerative Medicine, DOI 10.1007/978-1-61779-021-8_6,
© Springer Science+Business Media, LLC 2011

1 Bone-Marrow-Derived Stem Cells

Stem cells are clonogenic cells that are capable of both self-renewal and differentiation into more specialized progeny. Traditionally, stem cells have been divided into two broad categories: adult stem cells and embryonic stem cells. Adult stem cells are derived from postnatal somatic tissues and are considered to be multipotent, meaning they can give rise to multiple differentiated cell types. Embryonic stem cells, which are derived from the inner cell mass of blastocyst-stage embryos, are pluripotent, meaning they can give rise to all the differentiated cell types of the postnatal organism. Differentiated somatic cell types can also be reprogrammed into a pluripotent state similar to that of embryonic stem cells via the forced expression of stem-cell-related genes, which represents the basis for a recent report on induced pluripotent stem cells [4].

Approximately one decade ago, several studies challenged the long-held view that adult stem cells give rise to only a restricted set of differentiated cell types. These reports described "transdifferentiation" events, whereby adult stem cells differentiated into unexpected cell types, and even across embryonic germ layer boundaries. Cardiac differentiation has been reported for a variety of expected and unexpected stem cell types. These manifestations of transdifferentiation continue to be sources of controversy.

The bone marrow (BM) is a very heterogeneous compartment that contains multiple stem cell populations with putative cardiac potential, e.g., hematopoietic stem cells (HSCs) [5], mesenchymal stem cells (MSCs) [1, 6–11], very small embryonic-like stem cells [12], and multipotent adult progenitor cells (MAPCs) [13]. In this chapter, we focus on these BM-derived progenitors, which have attracted considerable attention.

1.1 Mesenchymal Stem Cells

Friedenstein et al. first reported the existence of MSCs in the BM in 1966, terming them "bone formation progenitors" [14]. Subsequently, MSCs were reported to constitute 0.001–0.01% of the total nucleated cell population in the BM, which is far lower than the content of HSCs in the BM [15, 16]. BM-MSCs were initially believed to be the stem cells that gave rise to osteoblasts, chondroblasts, adipocytes, and connective tissues [17, 18]. Recent studies have demonstrated that BM-MSCs can also differentiate into neurons [19], skeletal muscle cells [20], and cardiomyocytes [1, 21, 22], both in vitro and in vivo. BM-MSCs are found in the stromal cell fraction, which can be easily separated from hematocytes in culture. These stem cells were initially isolated from the BM stromal cells on the basis of their characteristic proliferative activities and multipotencies. Cell-surface markers that can be used to isolate MSCs have yet to be determined. CD29, CD44, CD105, and Sca-1 (only in the mouse) are widely accepted cell-surface markers for MSCs, whereas the value of other markers is debated among researchers. In 1999, we [1, 22] observed that the exposure of immortalized murine MSCs to 5-azacytidine (5-AzaC), which demethylates methylcytosine and induces the transcription of critical transcription factors by demethylating the CpG islands in the promoter regions, resulted in the appearance of spontaneously beating foci. We have termed these cell lines "CMG" (*c*ardio*m*yo*g*enic), as they are from adult BM stromal cells. Through repeated limiting dilutions, we isolated hundreds of clones, and we identified several clones that could differentiate into cardiomyocytes that exhibited spontaneous beating. These experiments were repeatable and reproducible, although the percentages of cardiomyocyte differentiation varied among these clones. Phase-contrast photography revealed that the CMG cells had a fibroblast-like morphology before 5-AzaC treatment (week 0), and this phenotype was retained through repeated subcultures under nonstimulating conditions. After 5-AzaC treatment, the morphology of the cells gradually changed. Approximately 30% of the CMG cells increased gradually in size, attaining a ball-like appearance or lengthening in one direction, and showed a sticklike morphology after 1 week. These cells connected with adjoining cells after 2 weeks, and formed myotube-like structures at 3 weeks (Fig. 1). The differentiated CMG myotubes retained the cardiomyocyte phenotype and beat vigorously for at least 8 weeks after the final 5-AzaC treatment.

The cardiac phenotype of the treated cells was confirmed by a variety of techniques, including reverse transcription PCR (for the markers of atrial natriuretic peptide, myosin light chain 2a and myosin light chain 2v, GATA4, and Nkx2.5), immunocytochemistry (for the markers of sarcomeric myosin heavy chain (MHC) and α-actinin), and electron microscopy. An electrophysiology study was performed on the differentiated CMG cells 2–5 weeks after 5-AzaC treatment. Two types of morphologic action potentials were distinguishable: sinus-node-like potentials (Fig. 2a); and ventricular-myocyte-like potentials (Fig. 2b). All the action potentials recorded for the CMG cells until 3 weeks of 5-AzaC treatment were sinus-node-like action potentials. Ventricular-myocyte-like action potentials were recorded after 4 weeks, and the percentage of these action potentials gradually increased thereafter.

This outcome was surprising because at the time BM cells were thought to form only blood cell lineages or bone cells. This finding was followed up using a variety of approaches, revealing the potential of BM cells to differentiate into a variety of tissues, including cardiomyocytes. Although similar findings with 5-AzaC have been reported by others [6], some investigators have suggested that this type of cardiac induction requires "immortalized" MSCs [23]. Currently, less is known about methods for the specific induction of differentiation than is known about embryonic stem cells.

Shim et al. [7] isolated MSCs from the BM of human patients who were undergoing coronary artery bypass surgery, and treated the cells with insulin, dexamethasone, and ascorbic acid. The authors reported that the treated cells immunostained positively for α-MHC, β-MHC, and GATA4, but not for skeletal muscle markers, such as skeletal MHC and MyoD. However, the efficiency of cardiogenesis achieved using this approach appeared to be poor. The resultant "cardiomyocyte-like" cell cultures lacked appreciable spontaneous contractile activity, and only a

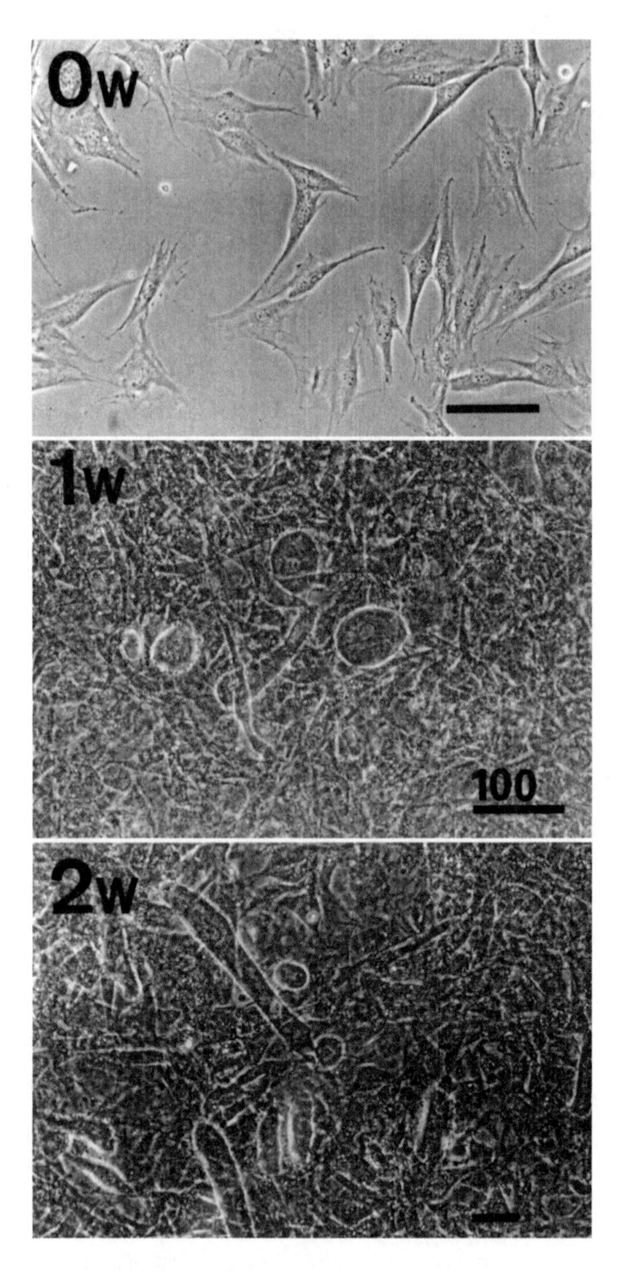

Fig. 1 Phase-contrast micrographs of CMG cells before and after 5-azacytidine (5-AzaC) treatment *Top*: CMG cells show fibroblast-like morphology before 5-AzaC treatment (week 0). *Middle*: CMG cells 1 week after treatment. Some of the cells have increased in size, assuming a ball-like or sticklike appearance. These cells began beating spontaneously thereafter. *Bottom*: CMG cells 2 weeks after treatment with 5-AzaC. Ball-like or sticklike cells are connected to adjoining cells, and are beginning to form myotube-like structures. *Bars* 100 μm

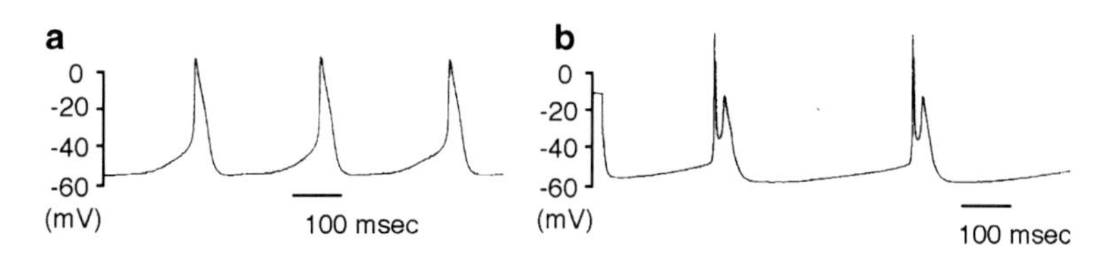

Fig. 2 Representative tracing of the action potential of CMG myotubes Action potential recordings were obtained for the spontaneously beating cells on day 28 after 5-AzaC treatment using a conventional microelectrode. These action potentials are categorized as a sinus-node-like action potential (**a**) and a ventricular-cardiomyocyte-like action potential (**b**)

small subset of the cells exhibited α-actinin-positive cross-striations. More recently, Shiota et al. have reported the cardiac induction of MSC-like progenitors derived using a complex culturing protocol that involves the formation of spheres by BM-derived adherent cells [24]. After treatment with 5-AzaC, the spheres showed spontaneous beating activity, as well as immunoreactivity for cardiac markers, including Nkx2.5 and myosin light chain 2v. The authors tested the capacity of these preparations to mediate cardiac repair in a murine infarct model. They reported functional improvements following the transplantation of green fluorescent protein (GFP)-tagged, sphere-derived cells, although the degree of remuscularization was extremely low. The latter study is one of many preclinical studies that assert beneficial effects for contractile function following the transplantation of MSCs in models of cardiac injury. Some [6, 8, 11], but not all [25], of these studies conclude that MSCs transdifferentiate into cardiomyocytes in vivo. In general, reports favoring myocardial repopulation by MSCs have shown only rare clusters of cells that lack the typical cardiomyocyte morphology but that immunostain positively for one or more cardiac markers.

In 2001, Beltrami et al. observed cardiomyocyte mitotic figures in human hearts after myocardial infarction (MI) [26]. In 2009, Bergmann et al. reported that cardiomyocytes undergo renewal, with a gradual decrease in annual turnover from 1% at 25 years of age to 0.45% at 75 years of age, according to carbon-14 measurements. Fewer than 50% of cardiomyocytes are exchanged during a normal life span [27]. Their report, which sparked controversy regarding cardiomyocyte induction, investigated the following possibilities: (1) whether the cells, which were thought to be terminally differentiated, had acquired the ability to proliferate; (2) whether immature cardiomyocytes differentiated from stem cells into cardiomyocytes and then began to proliferate; or (3) whether mature cardiomyocytes acquired the ability to proliferate by fusing with cells that had retained the proliferative capabilities of stem cells.

1.2 Hematopoietic Stem Cells

Recent advances in fluorescently activated cell sorting (FACS) techniques have enabled the prospective isolation of HSCs on the basis of their cell-surface antigen expression patterns and fluorescent dye efflux characteristics [28–30].

The FACS-derived $CD34^-$ c-kit^+ Sca-1^+ Lin^- tip side population (SP) cell fractions contained the HSC population in mice; [30] c-kit is a stem cell factor receptor, Sca-1 is a stem cell antigen that is specifically expressed in various stem cells (only in the mouse), and Lin is a mixture of antibodies against lineage markers for hematocytes (in mouse, Gra-1, Mac-1, B220, CD3, and Ter119; in human, CD3, CD4, CD8, CD19, CD33, and glycophylin A). In 2001, Orlic et al. reported cardiomyocyte differentiation following the transplantation of c-kit^+ Lin^- BM cells into peri-infarct tissue after MI [26]. They demonstrated directly that BM cells become cardiomyocytes in vivo. However, c-kit^+ Lin^- BM cells are predominantly HSCs, and even if BM cells differentiate into a variety of cells, including cardiomyocytes, controversy persists regarding, for example, whether HSCs transdifferentiate or MSCs differentiate. Moreover, in 2002, fluorescent in situ hybridization analysis revealed the presence of numerous cardiomyocytes that seemed to be recipient-derived after human heart transplantation [31]. In contrast, in 2003, numerous BM-derived cardiomyocytes were shown to be present in the recipient heart after BM transplantation [32]. In an experiment that gave very different results, Wagers et al. examined a variety of organs after transplanting GFP-labeled single HSCs (c-kit^+, Lin^-, Sca-1^+) into irradiated mice, and they concluded that if HSC transdifferentiation does occur, it is extremely rare, and that cardiomyocyte differentiation does not occur as a result of MI or induced injury [33]. Goodell et al. transplanted highly enriched HSCs into lethally irradiated mice, which were subsequently rendered ischemic by coronary artery occlusion for 60 min, followed by reperfusion; they reported that the transplanted BM cells differentiated into cardiomyocytes in the peri-infarct region at a prevalence of 0.02% [34]. In 2004, Balsam et al. investigated whether the c-kit^+ HSCs in BM are capable of differentiating into cardiomyocytes [35], by directly injecting BM cells into myocardial tissue instead of transplanting BM cells after irradiation as other groups had done. Importantly, they conducted their study to exclude irradiation, given the possibility that invasive treatment, including irradiation, contributes to a fusion phenomenon. They concluded that c-kit^+ HSCs do not include cells that are capable of differentiating into cardiomyocytes. Murry et al. investigated this differentiation ability in a similar manner, by directly infusing c-kit^+ Lin^- HSCs into the heart [36] and, as expected, they found that the HSCs were unable to differentiate into cardiomyocytes. In the same year, we examined the differentiation capabilities of HSCs using a c-kit^+ Sca-1^+ $CD34^-$ Lin^- SP ($CD34^-$KSL-SP) of HSCs [37]. When we transplanted whole BM cell populations, which included both HSCs and MSCs, from GFP-transgenic mice into lethally irradiated mice and subsequently induced MI, we found very few GFP^+ (BM-derived) cardiomyocytes. Interestingly, granulocyte colony stimulating factor (G-CSF) increased the number of GFP^+ cardiomyocytes and nonmyocytes in the infarcted or border zone area. In contrast, when we performed HSC transplantation followed by induction of MI and administration of G-CSF, cardiomyocytes were rarely found in the group that was transplanted with HSCs alone, although fibroblast-like cells were observed, and G-CSF increased their number. Moreover, we confirmed the predominance of MSC-derived GFP^+ cardiomyocytes in the group that was transplanted with cardiomyogenic cells, i.e., purified MSCs. It should be emphasized that in this type of BM transplantation experiment the dosage of radiation must be carefully determined, as the sensitivity to radiation of MSCs is much higher than that of HSCs. We propose that the differentiation by whole BM cells into organs (cells) other than hematopoietic populations is attributable to MSCs rather than HSCs, and that MSCs are mobilized from the BM into the bloodstream, in similarity to HSCs.

Nonetheless, the cardiac potential of HSCs remains controversial. The authors of the original study by Orlic et al. recently revisited this issue, and they concluded once again that in a mouse infarct model, the c-Kit$^+$ BM cells transdifferentiated following transplantation and formed extensive replacement myocardium [38].

1.3 BM-Derived Endothelial Progenitor Cells

Endothelial progenitor cells (EPCs) should be viewed as both circulating and BM stem cell types, since they are known to reside in both compartments. In 1997, Asahara and colleagues described the phenotype of EPCs, which proliferate in response to tissue ischemia, home to areas of injury, and either incorporate within or otherwise promote neovascularization [39, 40]. EPCs express the markers of Flk-1, CD34, and CD133, and can differentiate into definitive endothelial cells [39, 41–43]. Initial interest in the application of EPCs to cardiac repair was naturally focused on their angiogenic properties. The capacity of EPCs to transdifferentiate into cardiomyocytes was first reported by Dimmeler and colleagues in 2003 [44]. In that study, CD34$^+$ human EPCs were obtained from peripheral blood mononuclear cells of healthy adults or from patients with coronary artery disease. After coculture with neonatal rat cardiomyocytes, EPCs were reported to transdifferentiate into cardiomyocytes on the basis of morphology, α-sarcomeric actinin immunoreactivity (as assessed by flow cytometry), and the expression of other cardiac markers (as assessed by immunostaining or reverse transcription PCR with species-specific probes). Furthermore, the EPCs showed calcium transients that synchronized with adjacent rat cardiomyocytes, suggesting communication with the host myocardium through gap junctions. Coculturing experiments with paraformaldehyde-fixed cardiomyocytes revealed that cell fusion was not required for EPCs to acquire the cardiac phenotype [44–47]. However, the efficiency of cardiac induction by EPCs was very low; even after enhancement through inhibition of Notch signaling, less than 1% of the EPCs expressed α-sarcomeric actinin [47]. Asahara and colleagues reported even lower rates of cardiac transdifferentiation in vitro following coculturing of EPCs with the rat heart-derived H9C2 cell line [48]. The latter authors also reported the in vivo cardiac differentiation of a related preparation of human circulating cells following transplantation into a rodent infarct model. However, this conclusion is complicated by the definitive demonstration of cell fusion between host myocytes and graft cells, using species-specific fluorescent in situ hybridization probes [49]. Moreover, Gruh et al. were unable to confirm the in vitro cardiac differentiation of EPCs following coculturing with primary myocytes [50]. These authors found no expression of human cardiac transcripts, and they concluded that the rare, ostensibly transdifferentiated EPCs observed by FACS or epifluorescence microscopy were artifacts that resulted from overlying cells and/or autofluorescence. Thus, although the cardiac potential of EPCs remains a source of controversy, the report of Gruh et al. underscores the challenges inherent to interpreting coculture experiments.

1.4 Very Small Embryonic-Like Stem Cells

In 2006, employing multiparameter sorting, Kucia and colleagues identified in murine BM populations a homogenous population of rare (approximately 0.02% of BM mononuclear cells) Sca-1$^+$ Lin$^-$ CD45$^-$ cells that express SSEA-1, Oct-4, Nanog, and Rex-1 [51]. These cells are very small and display several features that are typical of primary embryonic stem cells. In vitro cultures of these cells are able to differentiate into all three germ layer lineages, including cardiomyocytes. For cardiac differentiation, GFP$^+$ Sca-1$^+$ Lin$^-$ CD45$^-$ or Sca-1$^+$ Lin$^-$ CD45$^+$ cells together with unpurified GFP$^-$ BM cells were plated in Dulbecco's modified Eagle's medium that was supplemented with 10% fetal bovine serum, 10 ng/ml basic fibroblast growth factor, 10 ng/ml vascular endothelial growth factor, and 10 ng/ml transforming growth factor β_1. Growth factors were added every 24 h, and the medium was replaced every 2–3 days.

Dawn et al. have reported that the transplantation of a relatively low number of very small embryonic-like stem cells is sufficient to improve left ventricular function and to alleviate myocyte hypertrophy after MI [12]. In that report, 10,000 very small embryonic-like stem cells in a 50-μl volume were injected intramyocardially using a 30-gauge needle.

1.5 Multipotent Adult Progenitor Cells

In 2002, Jiang et al. reported on pluripotent BM-derived cells, which they referred to as multipotent adult progenitor cells (MAPCs) [52]. When transplanted into blastocysts, MAPCs had the potential to differentiate into the three germ layers both in vitro and in vivo. These MAPCs were maintained using a low-density culture method, making independent corroboration of the findings by other laboratories rather difficult. In 2006, Zeng et al. showed that MAPCs could be derived from both postnatal and fetal swine BM. Swine MAPCs are negative for CD44, CD45, and major histocompatibility complex classes I and II, express octamer-binding transcription factor 3a messenger RNA and protein at levels close to those seen in human embryonic stem cells, and have telomerase activity, which prevents telomere shortening.

Transplantation of MAPCs (injected directly into heart at ten million cells per location diluted in 400 μl of saline) at the time of coronary artery ligation resulted in improved infarct zone contractile function and prevented peri-infarct border zone bioenergetic deterioration [13]. The left ventricular chamber response to cell transplantation resulted from the beneficial effects of sparing myocytes and increasing revascularization in both the infarct zone and the peri-infarct border zone. A direct structural contribution of the engrafted cells to cardiomyocyte regeneration appears to be unlikely.

2 Other BM-Derived Cells and Cell Fusion

In 2005, Yoon et al. identified a subpopulation of human BM stem cells (hBMSCs) that did not belong to the previously described class of BM-derived stem cells [53]. These cells were CD29$^-$, CD44$^-$, CD73$^-$, demonstrating minimal expression of CD90, CD105, and CD117, and could differentiate into the three germ layers. Intramyocardial transplantation of hBMSCs after MI resulted in robust engraftment of transplanted cells, which exhibited smooth muscle cell identity and colocalization with markers of cardiomyocytes and endothelial cells, which is consistent with the differentiation of hBMSCs into multiple lineages in vivo. Coculturing of hBMSCs with cardiomyocytes revealed that phenotypic changes in the hBMSCs result from both differentiation and fusion. Other laboratories have identified additional multipotent, CD45$^-$, nonhematopoietic BM-derived cells [40, 54, 55]. In some cases, it is likely that similar or overlapping populations of primitive stem cells in the BM detected using various experimental strategies have been assigned different names. The relationships among the BM-derived stem cells reported from different laboratories need to be clarified.

In 2002, Terada et al. suggested that a cell fusion phenomenon had to be considered with regard to the plasticity of the BM cells reported thus far [56]. Their coculture of adult animal BM cells with embryonic stem cells induced cell fusion naturally in the presence of interleukin-3, and although the karyotype was tetraploid, the cells acquired pluripotency and proliferative ability. More recently, the transplantation of whole BM cells into lethally irradiated mice resulted in fused cardiomyocytes but no transdifferentiation [57]. In addition, the same study aimed to identify the cell lineages in whole BM populations that are responsible for cell fusion, by transplanting CD45-Cre mouse BM into R26R mice. Fused cardiomyocytes were observed in this experimental system, and BM-derived leukocyte lineage cells were found to be responsible for the fusion. The lack of a clear definition for cell plasticity has led to confusion, with several reports failing to demonstrate that a single cell can indeed differentiate into multiple lineages at significant levels.

Studies using the Cre-lox recombination system revealed only rare MSC-derived cardiomyocytes, nearly all of which resulted from cell fusion [58].

3 Specific Culture Method for Cardiac Differentiation and Cell Fusion

Another obstacle to cell therapy is that specific culture methods for differentiating BM cells are only available for some target cells. Specific differentiation is achievable for osteoblasts, chondroblasts, and adipocytes. The use of 5-AzaC is effective for cardiomyocyte differentiation but it is clinically toxic. For cardiomyocytes, no methods have been established that use physiologic growth factors, cytokines, or nontoxic chemical compounds. Perhaps the most studied strategy to date with adult stem cells is the effect of 5-AzaC, a DNA demethylation reagent, on cardiac protein expression in MSCs [1, 59]. Several studies have demonstrated an increase in cardiac protein expression after treatment of MSCs with 5-AzaC [1]. Importantly, studies have consistently demonstrated improvement in cardiac function after the transplantation of 5-AzaC-treated MSCs, as compared with the transplantation of control MSCs [59–61]. As we begin to define the pathways, we can attempt to optimize further cardiac differentiation and functional effects [61]. For the further development of this field, it is necessary to find the small molecule and to elucidate the epigenetic status that can enhance cardiac differentiation from these stem cells [62].

Recently, Ge et al. reported the cardiomyocyte differentiation of rat BM-MSCs by treating the stem cells under conditions similar to those seen during MI [63]. The extract from the infarcted rat myocardium contained the same biochemical factors that arise after MI. Ge et al. found that the extract of infarcted myocardium could induce cardiomyocyte differentiation of BM-MSCs, as shown by the expression of cardiomyocyte-specific genes, including those for α-actin, connexin 43, Nkx2.5, MEF2c, GATA4, α-MHC, and troponin I. This approach could represent an alternative means of inducing cardiomyogenic differentiation in that it does not rely on gene demethylation or the use of viral vectors. The findings of that study appear to support the use of autologous extracts for the induction of stem cell differentiation and may have clinical implications for cardiac cell therapy.

Significant work has been performed to further understand the regulatory pathways involved in embryonic stem cell differentiation to cardiac myocytes [64–66]. These studies have suggested potential pathways that could be activated in adult stem cells so as to induce them to take on a cardiac phenotype [64, 66, 67].

Another approach that is being developed to direct the cardiac differentiation of adult stem cells is the delivery of chimeric proteins that encode cell-penetrating peptides (CPPs) and cardiac-specific transcription factors [68, 69] CPPs cause nonsecreted proteins to be secreted and to be internalized by surrounding cells. Bian et al. have demonstrated that the transplantation into the myocardium of cells that are genetically enhanced to express a CPP-GFP protein results in GFP expression in native cardiac myocytes [69]. To deliver functional transcription factors to the myocardium, Bian et al. developed a CPP-GATA4 construct and transplanted cardiac fibroblasts that were stably transfected with the CPP-GATA4 construct, 1 month after MI in the Lewis rat. The infarct border zones of the animals that received CPP-GATA4 demonstrated increased expression of cardiac myosin and Bcl-2 [69]. The modulation of GATA4-responsive gene expression led to hypertrophy of the cardiac myocytes at the infarct border zone and a global improvement in cardiac function [69]. These findings suggest that combining genetic enhancement of stem cells to deliver CPP–transcription factor chimeric proteins together with either stem cell homing agents or additional stem cells could lead to an increase in cardiac protein expression in the stem cells, cardiac myocyte regeneration, and further improvements in cardiac function.

4 Cardiospheres and Cardiac Extracts for Cardiomyogenesis

In 2004, Messina et al. described a novel technique for isolating resident cardiac progenitors from murine hearts, as well as subcultures of human atrial or ventricular specimens [70]. Mild enzymatic digestion of the tissue specimens yielded small, round, phase-bright cells that clustered together in suspension. These sphere-generating cells were allowed to adhere to poly(L-lysine)-coated plates, and were cultured in a medium that was supplemented with cytokines (epidermal growth factor, basic fibroblast growth factor, cardiotrophin-1, and thrombin). These "cardiosphere"-derived cells were self-renewing, clonogenic, and expressed both endothelial markers (KDR in human, flk-1 in mouse cells, and CD31) and stem cell markers (CD34, c-Kit, and Sca-1). Murine cardiosphere-derived cells showed spontaneous contractile activity, whereas human cardiosphere cells did so only after 24 h of coculturing with postnatal rat cardiomyocytes. The cardiosphere-derived cells from both human and mouse demonstrated trilineage differentiation into cardiomyocytes and endothelial and smooth muscle cells. However, quantitative data on the frequencies of these events were not reported. Cardiosphere-derived cardiomyocytes express cardiac markers, including cardiac troponin I, atrial natriuretic peptide, and cardiac MHC. In vivo, cardiosphere-derived cells have been reported to regenerate the infarcted mouse heart [70]. Subsequently, Smith et al. expanded on these findings by isolating cardiosphere-forming cells from human biopsy specimens [71]. These human cardiospheres, which were successfully isolated from 69 of the 70 biopsies tested, consistently expressed c-Kit but not the multidrug resistance gene MDR1, indicating that these cells were phenotypically distinct from the resident cardiac progenitors previously identified in situ (c-Kit$^+$, MDR1$^+$) [31, 72]. Consistent with the findings of Messina et al. [70] human cardiosphere-derived cells did not spontaneously contract, whereas coculturing with neonatal rat cardiomyocytes evoked calcium transients in synchrony with neighboring cardiomyocytes, action potentials, and fast inward sodium currents. Smith et al. also injected lentivirally transduced LacZ$^+$ human cardiosphere-derived cells into the border zones of infarcted SCID beige mice [71]. Twenty days later, the cardiosphere-derived cells were detected throughout the border regions of the mouse hearts, and occasional donor cells were immunostained for α-sarcomeric actin and von Willebrand factor. Echocardiography showed improvements in global left ventricular function, although given the apparently limited cardiomyocyte repopulation by LacZ$^+$ cells, these functional effects were attributed to a combination of regeneration and paracrine effects. On the basis of these studies, explant-derived cardiospheres appear to have cardiomyogenic potential and considerable promise for cardiac repair.

The differentiation of human adipose tissue stem cells to take on cardiomyocyte properties occurs following transient exposure to a rat cardiomyocyte extract [73–75]. Adult cardiomyocytes retain the capacity to induce cardiomyogenic differentiation of adult human MSCs. This approach could represent an alternative strategy to induce cardiomyogenic differentiation that does not rely on gene demethylation or the use of viral vectors.

5 Conclusions

Advances in stem cell and developmental biology have resulted in the identification of numerous candidate stem cell types with putative cardiogenic potential. The ideal cell type remains to be confirmed, despite all claims to the contrary. The cardiogenic potentials of BM-derived and circulating stem cells appear limited, whereas other candidates, including pluripotent stem cells, are clearly capable of more efficient cardiogenesis. We are optimistic that research into cell-based cardiac repair will eventually yield effective myogenic therapies, although success in this area will require rigorous cardiac phenotyping, cell fate mapping, and preclinical and clinical testing.

References

1. Makino S, Fukuda K, Miyoshi S, et al. Cardiomyocytes can be generated from marrow stromal cells in vitro. *J Clin Invest.* 1999;103(5):697–705.
2. Dimmeler S, Zeiher AM, Schneider MD. Unchain my heart: the scientific foundations of cardiac repair. *J Clin Invest.* 2005;115(3):572–583.
3. Laflamme MA, Murry CE. Regenerating the heart. *Nat Biotechnol.* 2005;23(7):845–856.
4. Takahashi K, Yamanaka S. Induction of pluripotent stem cells from mouse embryonic and adult fibroblast cultures by defined factors. *Cell.* 2006;126(4):663–676.
5. Orlic D, Kajstura J, Chimenti S, et al. Bone marrow cells regenerate infarcted myocardium. *Nature.* 2001;410(6829):701–705.
6. Tomita S, Li RK, Weisel RD, et al. Autologous transplantation of bone marrow cells improves damaged heart function. *Circulation.* 1999;100(19 Suppl):II247–II256.
7. Shim WS, Jiang S, Wong P, et al. Ex vivo differentiation of human adult bone marrow stem cells into cardiomyocyte-like cells. *Biochem Biophys Res Commun.* 2004;324(2):481–488.
8. Piao H, Youn TJ, Kwon JS, et al. Effects of bone marrow derived mesenchymal stem cells transplantation in acutely infarcting myocardium. *Eur J Heart Fail.* 2005;7(5):730–738.
9. Nagaya N, Kangawa K, Itoh T, et al. Transplantation of mesenchymal stem cells improves cardiac function in a rat model of dilated cardiomyopathy. *Circulation.* 2005;112(8):1128–1135.
10. Fazel S, Chen L, Weisel RD, et al. Cell transplantation preserves cardiac function after infarction by infarct stabilization: augmentation by stem cell factor. *J Thorac Cardiovasc Surg.* 2005;130(5):1310.
11. Dai W, Hale SL, Martin BJ, et al. Allogeneic mesenchymal stem cell transplantation in postinfarcted rat myocardium: short- and long-term effects. *Circulation.* 2005;112(2):214–223.
12. Dawn B, Tiwari S, Kucia MJ, et al. Transplantation of bone marrow-derived very small embryonic-like stem cells attenuates left ventricular dysfunction and remodeling after myocardial infarction. *Stem Cells.* 2008;26(6):1646–1655.
13. Zeng L, Hu Q, Wang X, et al. Bioenergetic and functional consequences of bone marrow-derived multipotent progenitor cell transplantation in hearts with postinfarction left ventricular remodeling. *Circulation.* 2007;115(14):1866–1875.
14. Friedenstein AJ, Petrakova KV, Kurolesova AI, et al. Heterotopic of bone marrow. Analysis of precursor cells for osteogenic and hematopoietic tissues. *Transplantation.* 1968;6(2):230–247.
15. Pittenger MF, Mackay AM, Beck SC, et al. Multilineage potential of adult human mesenchymal stem cells. *Science.* 1999;284(5411):143–147.
16. Le Blanc K, Pittenger M. Mesenchymal stem cells: progress toward promise. *Cytotherapy.* 2005;7(1):36–45.

17. Prockop DJ. Marrow stromal cells as stem cells for nonhematopoietic tissues. *Science*. 1997;276(5309):71–74.
18. Rickard DJ, Sullivan TA, Shenker BJ, et al. Induction of rapid osteoblast differentiation in rat bone marrow stromal cell cultures by dexamethasone and BMP-2. *Dev Biol.* 1994;161(1):218–228.
19. Kohyama J, Abe H, Shimazaki T, et al. Brain from bone: efficient "meta-differentiation" of marrow stroma-derived mature osteoblasts to neurons with Noggin or a demethylating agent. *Differentiation*. 2001;68(4–5):235–244.
20. Berghella L, De Angelis L, Coletta M, et al. Reversible immortalization of human myogenic cells by site-specific excision of a retrovirally transferred oncogene. *Hum Gene Ther*. 1999;10(10):1607–1617.
21. Fukuda K. Development of regenerative cardiomyocytes from mesenchymal stem cells for cardiovascular tissue engineering. *Artif Organs*. 2001;25(3):187–193.
22. Hakuno D, Fukuda K, Makino S, et al. Bone marrow-derived regenerated cardiomyocytes (CMG Cells) express functional adrenergic and muscarinic receptors. *Circulation*. 2002;105(3):380–386.
23. Liu Y, Song J, Liu W, et al. Growth and differentiation of rat bone marrow stromal cells: does 5-azacytidine trigger their cardiomyogenic differentiation? *Cardiovasc Res*. 2003;58(2):460–468.
24. Shiota M, Heike T, Haruyama M, et al. Isolation and characterization of bone marrow-derived mesenchymal progenitor cells with myogenic and neuronal properties. *Exp Cell Res*. 2007;313(5):1008–1023.
25. Silva GV, Litovsky S, Assad JA, et al. Mesenchymal stem cells differentiate into an endothelial phenotype, enhance vascular density, and improve heart function in a canine chronic ischemia model. *Circulation*. 2005;111(2):150–156.
26. Beltrami AP, Urbanek K, Kajstura J, et al. Evidence that human cardiac myocytes divide after myocardial infarction. *N Engl J Med*. 2001;344(23):1750–1757.
27. Bergmann O, Bhardwaj RD, Bernard S, et al. Evidence for cardiomyocyte renewal in humans. *Science*. 2009;324(5923):98–102.
28. Osawa M, Hanada K, Hamada H, et al. Long-term lymphohematopoietic reconstitution by a single CD34-low/negative hematopoietic stem cell. *Science*. 1996;273(5272):242–245.
29. Goodell MA, Rosenzweig M, Kim H, et al. Dye efflux studies suggest that hematopoietic stem cells expressing low or undetectable levels of CD34 antigen exist in multiple species. *Nat Med.* 1997;3(12):1337–1345.
30. Matsuzaki Y, Kinjo K, Mulligan RC, et al. Unexpectedly efficient homing capacity of purified murine hematopoietic stem cells. *Immunity.* 2004;20(1):87–93.
31. Quaini F, Urbanek K, Beltrami AP, et al. Chimerism of the transplanted heart. *N Engl J Med*. 2002;346(1):5–15.
32. Deb A, Wang S, Skelding KA, et al. Bone marrow-derived cardiomyocytes are present in adult human heart: A study of gender-mismatched bone marrow transplantation patients. *Circulation*. 2003;107(9):1247–1249.
33. Wagers AJ, Sherwood RI, Christensen JL, et al. Little evidence for developmental plasticity of adult hematopoietic stem cells. *Science*. 2002;297(5590):2256–2259.
34. Jackson KA, Majka SM, Wang H, et al. Regeneration of ischemic cardiac muscle and vascular endothelium by adult stem cells. *J Clin Invest*. 2001;107(11):1395–1402.
35. Balsam LB, Wagers AJ, Christensen JL, et al. Haematopoietic stem cells adopt mature haematopoietic fates in ischaemic myocardium. *Nature*. 2004;428(6983):668–673.
36. Murry CE, Soonpaa MH, Reinecke H, et al. Haematopoietic stem cells do not transdifferentiate into cardiac myocytes in myocardial infarcts. *Nature*. 2004;428(6983):664–668.
37. Kawada H, Fujita J, Kinjo K, et al. Nonhematopoietic mesenchymal stem cells can be mobilized and differentiate into cardiomyocytes after myocardial infarction. *Blood*. 2004; 104(12):3581–3587.
38. Rota M, Kajstura J, Hosoda T, et al. Bone marrow cells adopt the cardiomyogenic fate in vivo. *Proc Natl Acad Sci USA*. 2007;104(45):17783–17788.
39. Asahara T, Murohara T, Sullivan A, et al. Isolation of putative progenitor endothelial cells for angiogenesis. *Science*. 1997;275(5302):964–967.

40. Takahashi T, Kalka C, Masuda H, et al. Ischemia- and cytokine-induced mobilization of bone marrow-derived endothelial progenitor cells for neovascularization. *Nat Med.* 1999;5(4):434–438.
41. Shi Q, Rafii S, Wu MH, et al. Evidence for circulating bone marrow-derived endothelial cells. *Blood.* 1998;92(2):362–367.
42. Murohara T, Ikeda H, Duan J, et al. Transplanted cord blood-derived endothelial precursor cells augment postnatal neovascularization. *J Clin Invest.* 2000;105(11):1527–1536.
43. Rafii S. Circulating endothelial precursors: mystery, reality, and promise. *J Clin Invest.* 2000;105(1):17–19.
44. Badorff C, Brandes RP, Popp R, et al. Transdifferentiation of blood-derived human adult endothelial progenitor cells into functionally active cardiomyocytes. *Circulation.* 2003; 107(7):1024–1032.
45. Koyanagi M, Urbich C, Chavakis E, et al. Differentiation of circulating endothelial progenitor cells to a cardiomyogenic phenotype depends on E-cadherin. *FEBS Lett.* 2005;579(27):6060–6066.
46. Rupp S, Koyanagi M, Iwasaki M, et al. Genetic proof-of-concept for cardiac gene expression in human circulating blood-derived progenitor cells. *J Am Coll Cardiol.* 2008;51(23):2289–2290.
47. Koyanagi M, Bushoven P, Iwasaki M, et al. Notch signaling contributes to the expression of cardiac markers in human circulating progenitor cells. *Circ Res.* 2007;101(11):1139–1145.
48. Murasawa S, Kawamoto A, Horii M, et al. Niche-dependent translineage commitment of endothelial progenitor cells, not cell fusion in general, into myocardial lineage cells. *Arterioscler Thromb Vasc Biol.* 2005;25(7):1388–1394.
49. Iwasaki H, Kawamoto A, Ishikawa M, et al. Dose-dependent contribution of CD34-positive cell transplantation to concurrent vasculogenesis and cardiomyogenesis for functional regenerative recovery after myocardial infarction. *Circulation.* 2006;113(10):1311–1325.
50. Gruh I, Beilner J, Blomer U, et al. No evidence of transdifferentiation of human endothelial progenitor cells into cardiomyocytes after coculture with neonatal rat cardiomyocytes. *Circulation.* 2006;113(10):1326–1334.
51. Kucia M, Reca R, Campbell FR, et al. A population of very small embryonic-like (VSEL) CXCR4(+)SSEA-1(+)Oct-4+ stem cells identified in adult bone marrow. *Leukemia.* 2006;20(5):857–869.
52. Jiang Y, Jahagirdar BN, Reinhardt RL, et al. Pluripotency of mesenchymal stem cells derived from adult marrow. *Nature.* 2002;418(6893):41–49.
53. Yoon YS, Wecker A, Heyd L, et al. Clonally expanded novel multipotent stem cells from human bone marrow regenerate myocardium after myocardial infarction. *J Clin Invest.* 2005;115(2):326–338.
54. Ratajczak MZ, Kucia M, Reca R, et al. Stem cell plasticity revisited: CXCR4-positive cells expressing mRNA for early muscle, liver and neural cells 'hide out' in the bone marrow. *Leukemia.* 2004;18(1):29–40.
55. Kogler G, Sensken S, Airey JA, et al. A new human somatic stem cell from placental cord blood with intrinsic pluripotent differentiation potential. *J Exp Med.* 2004;200(2):123–135.
56. Terada N, Hamazaki T, Oka M, et al. Bone marrow cells adopt the phenotype of other cells by spontaneous cell fusion. *Nature.* 2002;416(6880):542–545.
57. Alvarez-Dolado M, Pardal R, Garcia-Verdugo JM, et al. Fusion of bone-marrow-derived cells with Purkinje neurons, cardiomyocytes and hepatocytes. *Nature.* 2003;425(6961):968–973.
58. Noiseux N, Gnecchi M, Lopez-Ilasaca M, et al. Mesenchymal stem cells overexpressing Akt dramatically repair infarcted myocardium and improve cardiac function despite infrequent cellular fusion or differentiation. *Mol Ther.* 2006;14(6):840–850.
59. Ye NS, Chen J, Luo GA, et al. Proteomic profiling of rat bone marrow mesenchymal stem cells induced by 5-azacytidine. *Stem Cells Dev.* 2006;15(5):665–676.
60. Burlacu A, Rosca AM, Maniu H, et al. Promoting effect of 5-azacytidine on the myogenic differentiation of bone marrow stromal cells. *Eur J Cell Biol.* 2008;87(3):173–184.
61. Yoon J, Min BG, Kim YH, et al. Differentiation, engraftment and functional effects of pre-treated mesenchymal stem cells in a rat myocardial infarct model. *Acta Cardiol.* 005;60(3):277–284.

62. Takeuchi JK, Bruneau BG. Directed transdifferentiation of mouse mesoderm to heart tissue by defined factors. *Nature*. 2009;459(7247):708–711.
63. Ge D, Liu X, Li L, et al. Chemical and physical stimuli induce cardiomyocyte differentiation from stem cells. *Biochem Biophys Res Commun*. 2009;381(3):317–321.
64. Lim JY, Kim WH, Kim J, et al. Involvement of TGF-beta1 signaling in cardiomyocyte differentiation from P19CL6 cells. *Mol Cells*. 2007;24(3):431–436.
65. Behfar A, Zingman LV, Hodgson DM, et al. Stem cell differentiation requires a paracrine pathway in the heart. *FASEB J*. 2002;16(12):1558–1566.
66. Arrell DK, Niederlander NJ, Faustino RS, et al. Cardioinductive network guiding stem cell differentiation revealed by proteomic cartography of tumor necrosis factor alpha-primed endodermal secretome. *Stem Cells*. 2008;26(2):387–400.
67. Kofidis T, de Bruin JL, Yamane T, et al. Stimulation of paracrine pathways with growth factors enhances embryonic stem cell engraftment and host-specific differentiation in the heart after ischemic myocardial injury. *Circulation*. 2005;111(19):2486–2493.
68. Popovic ZB, Benejam C, Bian J, et al. Speckle-tracking echocardiography correctly identifies segmental left ventricular dysfunction induced by scarring in a rat model of myocardial infarction. *Am J Physiol Heart Circ Physiol*. 2007;292(6):H2809–H2816.
69. Bian J, Kiedrowski M, Mal N, et al. Engineered cell therapy for sustained local myocardial delivery of nonsecreted proteins. *Cell Transplant*. 2006;15(1):67–74.
70. Messina E, De Angelis L, Frati G, et al. Isolation and expansion of adult cardiac stem cells from human and murine heart. *Circ Res*. 2004;95(9):911–921.
71. Smith RR, Barile L, Cho HC, et al. Regenerative potential of cardiosphere-derived cells expanded from percutaneous endomyocardial biopsy specimens. *Circulation*. 2007;115(7):896–908.
72. Urbanek K, Quaini F, Tasca G, et al. Intense myocyte formation from cardiac stem cells in human cardiac hypertrophy. *Proc Natl Acad Sci USA*. 2003;100(18):10440–10445.
73. Gaustad KG, Boquest AC, Anderson BE, et al. Differentiation of human adipose tissue stem cells using extracts of rat cardiomyocytes. *Biochem Biophys Res Commun*. 2004;314(2):420–427.
74. Rangappa S, Entwistle JW, Wechsler AS, et al. Cardiomyocyte-mediated contact programs human mesenchymal stem cells to express cardiogenic phenotype. *J Thorac Cardiovasc Surg*. 2003;126(1):124–132.
75. Schimrosczyk K, Song YH, Vykoukal J, et al. Liposome-mediated transfection with extract from neonatal rat cardiomyocytes induces transdifferentiation of human adipose-derived stem cells into cardiomyocytes. *Scand J Clin Lab Invest*. 2008;68(6):464–472.

Homing, Survival, and Paracrine Effects of Human Mesenchymal Stem Cells

Sergey Doronin

Abstract In this chapter we describe a snapshot of the ever-changing landscape for the use of mesenchymal stem cells (MSCs) as a therapeutic agent. To be used for cell therapy, they must reach their target and have a therapeutic effect. Determinants of MSC homing are considered in the first part of the chapter, with special emphasis on the dynamic nature of these cells reflected in changes of their chemokine receptor profile with the associated implication of these changes for homing to sites of injury. In the second part of the chapter we consider therapeutic effects of MSCs at their target sites mediated by the paracrine factors they produce.

Keywords Mesenchymal stem cells • Homing • Chemokine receptors • Paracrine factors • Angiogenesis

Mesenchymal stem cells (MSCs) were first described as fibroblast-like cells from the bone marrow that firmly adhere to plastic surfaces [1] and differentiate into osteoblasts and chondrocytes [2–5]. Given their ability to differentiate, MSCs are increasingly used as a therapeutic tool for the treatment of bone and cartilage disorders [6–8]. Besides being used for tissue replacement therapy, MSCs may play a role in supporting hematopoietic stem cell homing to the bone marrow [9–13] and improve functions of ischemic brain [14–19] and myocardium [20–23]. Whether systemically delivered or mobilized from the bone marrow, MSCs target tissues affected by radiation, infarction, or other kinds of trauma [24–35]. Most systemically delivered MSCs accumulate in the lungs, spleen, and liver [36–38]. After homing to injured tissue, MSCs release paracrine factors that reprogram the tissue's microenvironment, promoting cell survival and angiogenesis [39–41].

S. Doronin (✉)
Department of Physiology and Biophysics, Stony Brook University, Stony Brook, NY, USA
e-mail: sergey.doronin@stonybrook.edu

I.S. Cohen and G.R. Gaudette (eds.), *Regenerating the Heart*, Stem Cell Biology and Regenerative Medicine, DOI 10.1007/978-1-61779-021-8_7,
© Springer Science+Business Media, LLC 2011

Homing of cells circulating in the blood requires their adhesion to an endothelial monolayer which separates all tissues of the body from the bloodstream. The state of the endothelium at the site of injury can be affected by hypoxia, inflammation, and cell death. A unified model of cell homing to the injured tissue which takes into account all variations in the state of endothelium remains to be developed. In the absence of a common theory, researchers draw parallels between the homing of MSCs and leukocyte homing to tissues affected by inflammation. The mechanism of leukocyte homing postulates at least three distinguished but yet highly coordinated steps which include a recruitment stage, followed by firm adhesion, and then transmigration through the endothelial barrier [42].

At the recruitment stage, leukocytes are loosely attached and roll on the endothelial surface (Fig. 1a). The recruitment of the cells to the endothelial monolayer is regulated by the selectins (P, E, and L) and the type 1 transmembrane glycoproteins that bind to fucosylated and sialylated hydrocarbons present on their ligands. E-selectin and P-selectin are expressed on endothelial cells activated by proinflammatory stimuli. L-selectin is constitutively expressed by most leukocytes and is capable of interactions with E- and P-selectins of endothelial cells. Glycoprotein ligand-1 (PSGL1) protein is one of the major ligands of all three selectins. The binding of PSGL1 to

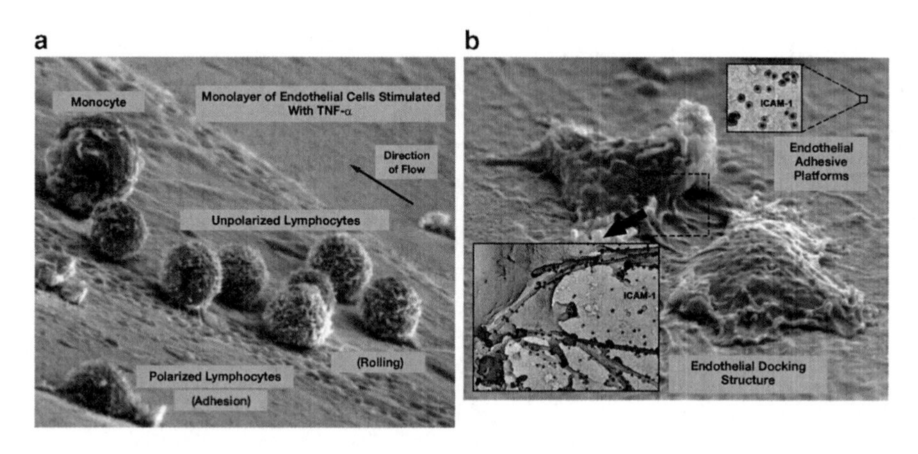

Fig. 1 The adhesion cascade. (**a**) A scanning electron microscope image showing a human endothelial monolayer treated with proinflammatory stimuli and perfused with human peripheral blood lymphocytes and monocytes at physiological flow (1.8 dyn/cm^2). Several unpolarized leukocytes have come into contact with the endothelium and have been captured during the rolling process. Also shown is a lymphocyte that has managed to firmly adhere to the endothelium and drastically changed its morphology from rounded to polarized. (**b**) Active role of the endothelium during extravasation. The scanning electron microscope image shows the organization of endothelial adhesion receptors in nanoclusters on the apical membrane [endothelial adhesive platforms; staining corresponds to intercellular adhesion molecule 1 (ICAM-1) using antibody coupled to colloidal gold]. When a leukocyte establishes contact with the endothelium, the endothelial adhesion receptors are concentrated in the endothelial docking structure, which keeps the leukocyte firmly adhered and prevents it from becoming separated owing to the force of the flow that it has to withstand (from [43])

P-selectin and E-selectin promotes capture of leukocytes by endothelium, whereas the binding of PSGL1 to L-selectin enables leukocyte–leukocyte interactions facilitating the capture of other leukocytes [44]. In addition to PSGL1, selectins may bind other glycoproteins such as CD44 and E-selectin ligand-1 (ESL1) [45]. The transient contact of leukocytes with endothelium is further stabilized by the interactions of $\alpha_4\beta_1$ and $\alpha_4\beta_7$ integrins with their ligands, such as vascular cell adhesion molecule 1 (VCAM-1), intercellular adhesion molecule 1 (ICAM-1), and mucosal addressing cell adhesion molecule 1 (MAdCAM-1) [46–48]. Cumulatively, the actions of selectins and integrins reduce the rolling velocity of leukocytes and set the stage for a second step in leukocyte homing – the firm adhesion.

The firm adhesion of leukocytes depends on availability of adhesion molecules (primarily VCAM-1 and ICAM-1) on the surface of endothelial cells and coordinated activation of integrins in adhering cells. Resting endothelial cells are considered to be nonadhesive owing to lack of or very low expression levels of adhesion molecules on their surface. During roughly 30 years of research it has been established that the activation of endothelial cells with inflammatory factors results in de novo synthesis and delivery of adhesion molecules to the surface of endothelial cells. Underlying molecular mechanisms of endothelial activation are well established [49]. Table 1 and Fig. 2 illustrate the typical outcome of endothelial cell activation with the inflammatory factor (IL-1β), resulting in the upregulation of a number of genes related to cell adhesion, where E-selectin was upregulated 34-fold, VCAM-1 was upregulated 30-fold and ICAM-1 was upregulated 22-fold (Table 1). Upregulation of messenger RNAs (mRNAs) for E-selectin, VCAM-1, and ICAM-1 was inhibited by actinomycin D, the inhibitor of RNA synthesis, which demonstrates that the upregulation of mRNAs for adhesion molecules is the result of de novo transcription of corresponding genes [49]. Upregulation of gene transcription translates into the upregulation of adhesion molecules on the surface of endothelial cells. The upregulation of adhesion molecules on the surface of endothelial cells can be eliminated by cycloheximide, an inhibitor of protein synthesis, suggesting that the adhesion molecules are the products of de novo protein synthesis (Fig. 2). Treatment of endothelial cells with IL-1β stimulates the transcription of chemokines CXCL2, CXCL1, and CXCL12 (Table 1). These molecules are ligands for a family of G-protein-coupled receptors located on the microvilli of leukocytes and hematopoietic stem cells and are responsible for the regulation of cell motility and adhesion. Among them, CXCL12 (stromal-cell-derived factor-1, SDF-1) may play a pivotal role in the regulation of firm adhesion of leukocytes and hematopoietic stem cells. Binding of SDF-1 with its receptor (CXCR4) on the surface of adhering cells triggers rapid conformational changes in integrins and increases affinity of integrins for their ligands [42, 47, 51, 52]. Activation of integrins requires SDF-1 immobilization on the endothelial surface. Although the molecular mechanism of SDF-1-dependent activation of integrins and the role of SDF-1 immobilization on the endothelial surface remain to be elucidated, the physiological consequences are relatively predictable. SDF-1 immobilized on the surface of endothelial cells marks sites of injury and promotes cell adhesion, whereas soluble SDF-1 is inactive and does not trigger cell adhesion at

Table 1 Changes in human umbilical vein endothelial cell (HUVEC) gene expression induced by IL-1β alone or in combination with actinomycin D in comparison with nontreated cells. The top 20 genes upregulated by IL-1β are shown

		Fold change	
Gene symbol	Gene title	Without actinomycin D	With actinomycin D
CXCL2	Chemokine (C–X–C motif) ligand 2	84.8	4.6
CXCL1	Chemokine (C–X–C motif) ligand 1	47.8	1.6
BCL2L14	BCL2-like 14 (apoptosis facilitator)	42.1	2.7
SELE	Selectin E (endothelial adhesion molecule 1)	33.9	−3.3
IL8	Interleukin 8	31.9	1.9
VCAM1	Vascular cell adhesion molecule 1	30.5	−3.7
CXCL6	Chemokine (C–X–C motif) ligand 6	24.1	1.7
POSTN	Periostin, osteoblast specific factor, abundant in the infarct border of human hearts	23.9	1.2
SOD2	Superoxide dismutase 2, mitochondrial	19.4	1.1
ICAM1	Intercellular adhesion molecule 1 (CD54)	22.2	−1.2
CD47	CD47, receptor for the C-terminal cell binding domain of thrombospondin	18.2	1.1
NCAM1	Neural cell adhesion molecule 1	18.1	1.2
IL6	Interleukin 6	15.9	−2.1
NRXN1	Neurexin 1	11.5	1.2
CCL2	Chemokine (C–C motif) ligand 2	8.9	1.3
CD44	CD44, receptor for hyaluronic acid, osteopontin, collagens, and matrix metalloproteinases	8.0	1.0
ITGA6	Integrin, alpha 6	7.3	1.54
COL4A3	Collagen, type IV, alpha 3	5.5	1.2
ITGA8	Integrin, alpha 8	5.3	−1.1
CXCL12	Chemokine (C–X–C motif) ligand 12 (stromal-cell-derived factor 1)	5.1	−1.31

Exemplar DNA microarray analysis of gene expression in HUVECs was performed using an Affymetrix HG-U133-Plus2 array before and after treatment of the cells with IL-1β or the combination of IL-1β and actinomycin D for 4 h

distant sites. In this regard, SDF-1 is similar to von Willebrand factor (vWF), which circulates in the bloodstream as an inactive dimer and activates after binding with the endothelial surface or extracellular matrix [53]. After firm adhesion, leukocytes transmigrate through the endothelial barrier and enter underlying tissue (Fig. 1b).

Phenomenologically, the homing of MSCs follows the three-step paradigm formulated for leukocytes. During adhesion, MSCs display coordinated rolling and adhesion behavior which depends on the expression of P-selectin and VCAM-1/VLA-4 on the surface of endothelial cells [54]. After firm adhesion, MSCs

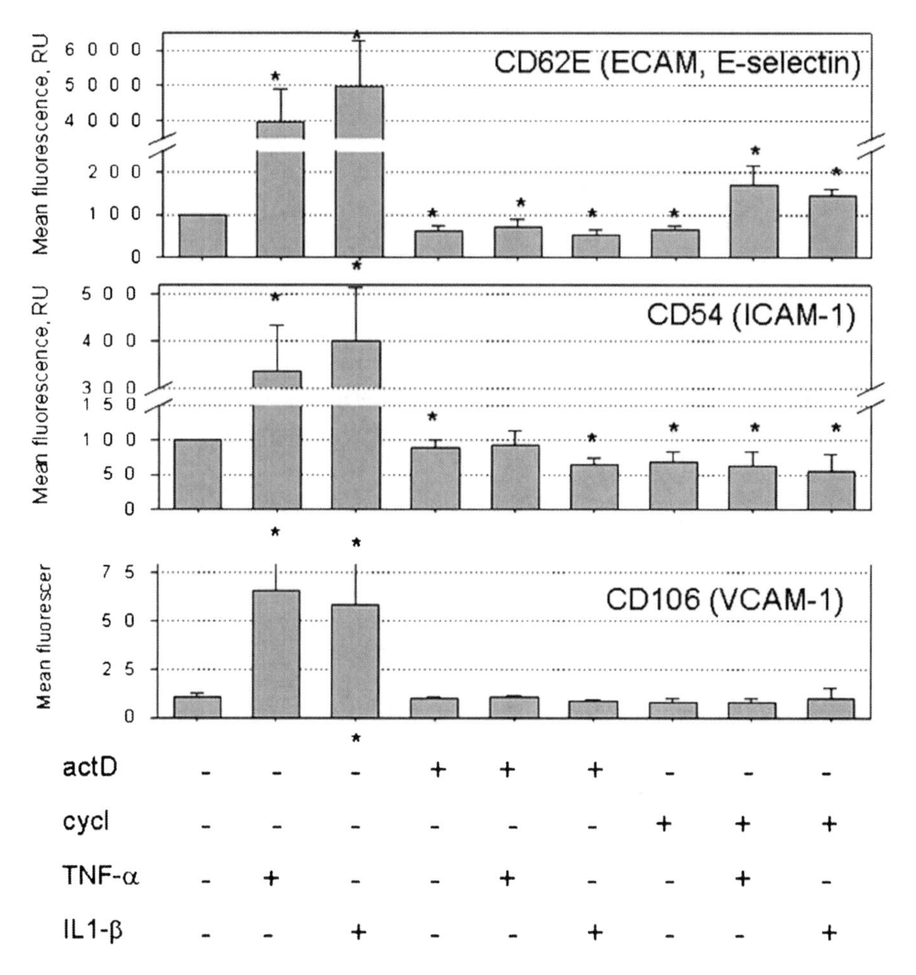

Fig. 2 Cell surface expression of E-selectin, ICAM-1 and vascular cell adhesion molecule 1 in human umbilical vein endothelial cells (HUVECs) treated with TNF-α and IL-1β in the presence and absence of actinomycin D (*actD*) or cycloheximide (*cycl*) (from [50])

transmigrate through the endothelial barrier [55]. However, the application of the theory of leukocyte adhesion to MSCs faces a number of challenges. One of them is heterogeneity in primary isolates of MSCs which contain a relatively high fraction of hematopoietic stem cells [3, 56–59]. Direct comparison of isolation protocols reveals that the percentage of colony-forming cells (self-renewal MSCs) may vary from 65 to 77% of the total cell population [60]. Therefore, primary isolates of MSCs are sometimes referred as bone marrow mononuclear cells. The surface markers of cell-culture-expanded MSCs include CD105, CD73, CD44, CD90, CD71 as well as the adhesion molecules CD29, CD106 (VCAM-1), activated leukocyte cell adhesion molecule (ALCAM or CD166) and CD54 (ICAM-1). At the same time, MSCs do not express hematopoietic markers CD45, CD34, CD14, or

CD11 and are negative for the expression of CD80, CD86, CD40, and platelet endothelial cell adhesion molecule 1 (PECAM-1 or CD31) [5, 61–64]. None of the markers expressed by MSCs are MSC-specific, which creates substantial difficulties with identification of MSCs in tissues and characterization of homing properties of MSCs from primary isolates.

The molecular mechanism that regulates MSC firm adhesion remains unknown and may not correspond to that described for leucocytes. Difficulty in the identification of the molecular mechanism of MSC firm adhesion is related to progressive loss of chemokine receptors by MSCs during subculturing (Fig. 3) [65]. One report indicates that the percentage of CXCR4-positive cells was reduced from 45 to 14% in cell-culture-expanded human MSCs (hMSCs) [65]. Other report suggests that CXCR4-positive cells may disappear from the isolates of MSCs within 24 h [60]. Wynn et al. [66] reported that less than 1% of hMSCs express CXCR4 on the cell surface, with most CXCR4 located in the intracellular compartments of MSCs. Functional expression of CCR1, CCR4, CCR7, CCR10, CXCR5 but not CXCR4 in hMSCs was reported by von Luttichau et al. [67]. Others reported the expression of CXCR4 in at least 25% [64] and possibly as many as 96% of hMSCs [68]. CXCR4-positive hematopoietic stem cells disappear from MSC culture within 2–3 weeks and it is possible to obtain a clean culture of MSCs; however, the adhesion properties of cell-culture-expanded MSCs may not match that of primary MSCs. Since the expression of chemokine receptors may vary substantially depending on the isolation protocol, the cell culture conditions, and the species origin, it is difficult to define a primary regulator of integrin affinity in MSCs, and researchers often suggest custom solutions that are specific to particular isolate of MSCs [69–74].

The precise reason for the loss of chemokine receptors by cell-culture-expanded MSCs remains to be established. The question whether CXCR4 is lost permanently was addressed by Potapova et al. [75]. It was suggested that maintenance of hMSCs as a monolayer places the cells in abnormal conditions where cells are provided with limited cell-to-cell and cell-to-extracellular matrix contacts [39]. The ability of hMSCs to form dense cellular aggregates, spheroids, in the absence of surfaces to which to adhere was used to reestablish the lost contacts [39, 75]. The pattern of surface antigen expression by hMSCs from the spheroids was generally similar to that of hMSCs from a monolayer. Substantial changes, however, were observed in the expression of proteins responsible for cell adhesion and motility. hMSCs from the spheroids expressed the α_2 integrin subunit (CD49b), CXCR4 (CD184), and lost the expression of the α_4 integrin subunit (CD49d) [39, 75]. A similar set of changes in the expression of the α_2 integrin subunit was reported for the spheroid cell culture of ovarian cancer cells [76] and T cells incubated in 3D collagen lattices [77]. Both the α_2 and the α_4 integrin subunits form a complex with the β_1 integrin subunit, however, and display different ligand specificities. The $\alpha_4\beta_1$ integrin interacts with fibronectin and VCAM-1, and participates in hematopoietic stem cell homing. The $\alpha_2\beta_1$ integrin is a receptor for collagen and laminin and is involved in hematopoietic, endothelial, and T-cell migration [78]. The changes in integrin expression may affect homing properties of hMSCs; however, the effect of these changes on the homing properties of hMSCs remains to be investigated [75]. The percentage of CXCR4-positive cells was increased in hMSCs cultured as

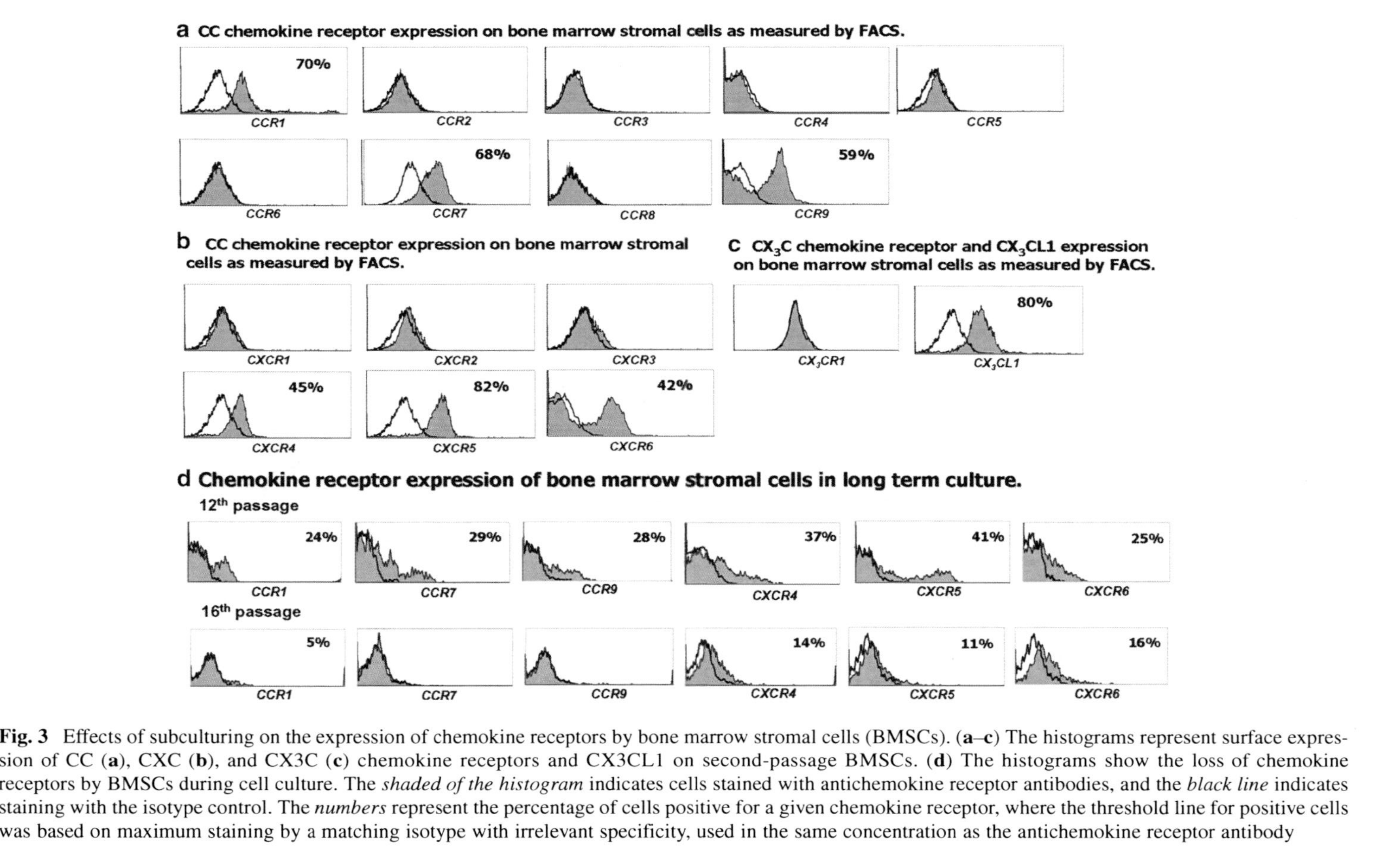

Fig. 3 Effects of subculturing on the expression of chemokine receptors by bone marrow stromal cells (BMSCs). (**a–c**) The histograms represent surface expression of CC (**a**), CXC (**b**), and CX3C (**c**) chemokine receptors and CX3CL1 on second-passage BMSCs. (**d**) The histograms show the loss of chemokine receptors by BMSCs during cell culture. The *shaded of the histogram* indicates cells stained with antichemokine receptor antibodies, and the *black line* indicates staining with the isotype control. The *numbers* represent the percentage of cells positive for a given chemokine receptor, where the threshold line for positive cells was based on maximum staining by a matching isotype with irrelevant specificity, used in the same concentration as the antichemokine receptor antibody

spheroids from 2 to 35% by flow cytometry. Immunocytochemical analysis indicated that the number of CXCR4-positive cells in hMSC spheroids may be substantially higher (Fig. 4a) [75]. CXCR4 expressed by hMSCs from the spheroids was a functional receptor that internalizes after treatment with SDF-1 (Fig. 4b) and regulates hMSC adhesion to hypoxic endothelial cells [75]. Internalized CXCR4 was detected in the perinuclear space and in lamellipodia of SDF-1-treated hMSCs (Fig. 4c). A similar internalization pattern was reported for CXCR4 expressed by leukocytes [79]. Most CXCR4 detected in lamellipodia was positioned along the F-actin cytoskeleton. Some portion of the receptor was colocalized with the tips of growing F-actin filaments (Fig. 4c). The tips of F-actin filaments are the sites of attachment for focal adhesion complexes. Colocalization of CXCR4 with the tips of F-actin filaments is a strong indicator that CXCR4 may regulate motility and cell adhesion of MSCs [65, 80–82].

The expression of the α_2 integrin subunit (CD49b) and CXCR4 (CD184) by hMSCs from the spheroids was nonsustainable, and the cells quickly returned to a monolayer phenotype after dissociation of the spheroids (Fig. 5) [75]. The expression of CXCR4 inversely correlated with the expression of SDF-1 that was expressed by a monolayer of hMSCs and was downregulated in hMSC spheroids (Fig. 6) [75]. This observation suggested an intriguing hypothesis that endogenous SDF-1 is responsible for the desensitization and downregulation of CXCR4 in a monolayer of MSCs [75]. This hypothesis is in good agreement with general properties of G-protein-coupled receptors which may experience desensitization, internalization, proteasomal degradation, and downregulation at the transcriptional level after sustained exposure to an agonist [83]. Overall, the experiments with hMSC spheroids showed that the loss of CXCR4 expression is not permanent and represents an adaptation of MSCs to cell culture conditions [75].

The great degree of flexibility in the expression of chemokine receptors and dynamic regulation of their expression by environmental factors may explain the variations in the results of experiments aimed at establishing the homing capacity of MSCs. Although some research groups observed the homing of MSCs to sites of injury [24, 25, 28–35], others argue that the loss of CXCR4 and progressive decrease in the capacity of cultured MSCs to home to injured tissues indicates that cell-culture-expanded MSCs may be homing-impaired [32, 60, 84]. Overall, the role of CXCR4 in the regulation of MSC homing to injured tissues remains unclear. Analysis of hMSCs circulating in the bloodstream of patients with acute myocardial infarction (MI) showed that only 11% of the cells are CXCR4-positive. The disappearance of circulating hMSCs from the bloodstream after MI presumably reflects the homing of MSCs to the infarcted heart; however, it does not show a correlation with the elevation of SDF-1 levels in the bloodstream [85]. At the same time, data from animal models suggest that the CXCR4–SDF-1 axis may play a role in the homing of MSCs to ischemic brain lesions [86] and bone marrow [66]. Nevertheless, experiments with MSCs transfected with CXCR4 showed that the expression of CXCR4 boosts MSC homing to injured tissues, suggesting that MSCs have all the necessary components for the activation of integrins via CXCR4 [87–89].

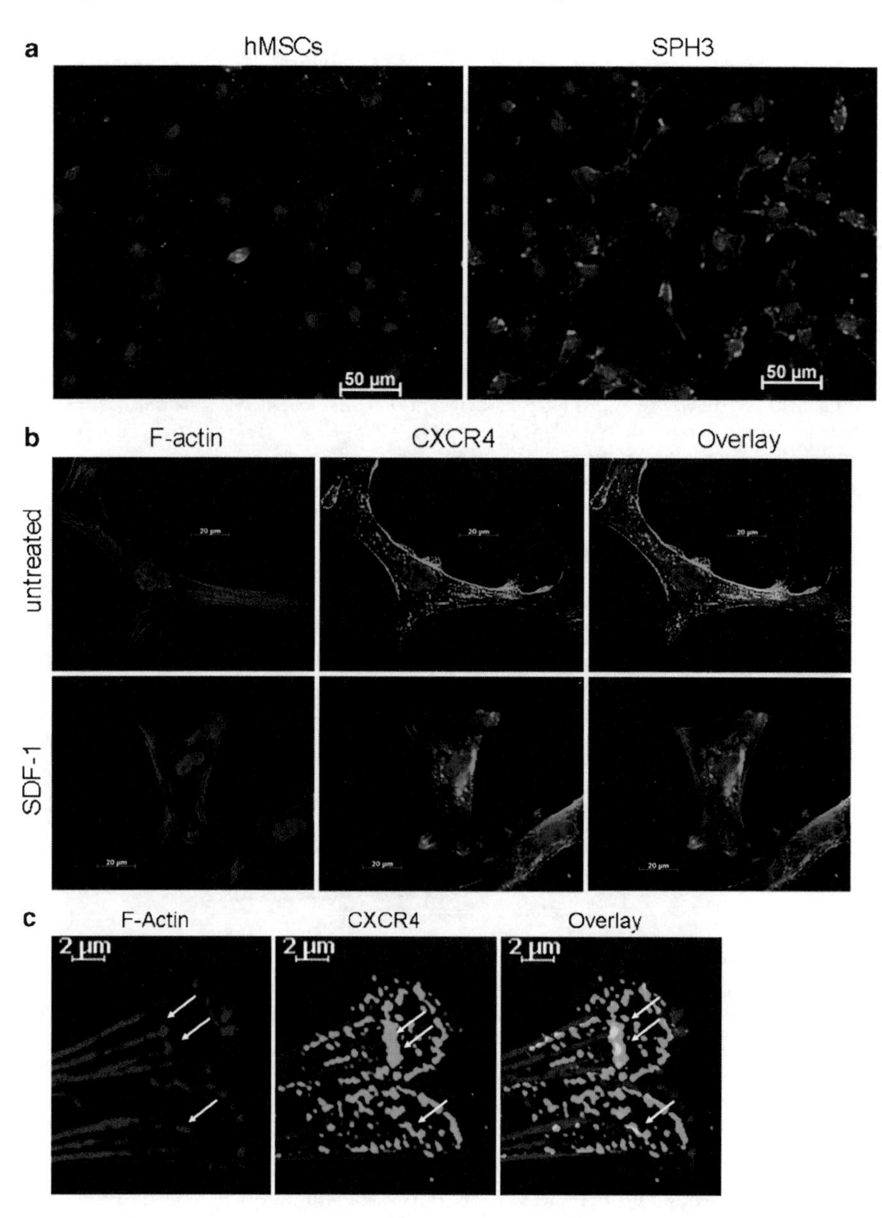

Fig. 4 Intracellular localization of CXCR4. (**a**) CXCR4 staining (*green*) in human mesenchymal stem cells (hMSCs) from a monolayer and hMSCs from a 3-day-old spheroid (*SPH3*). (**b**) CXCR4 (*green*) and F-actin (*red*) staining in SPH3 cells before (untreated) and after treatment with stromal cell-derived factor (SDF)-1α. (**c**) Staining for CXCR4 (*green*) and F-actin (*red*) in lamellipodia of cells from SPH3 treated with SDF-1. Sites of CXCR4 colocalization with focal adhesion complexes are denoted by *white arrows*

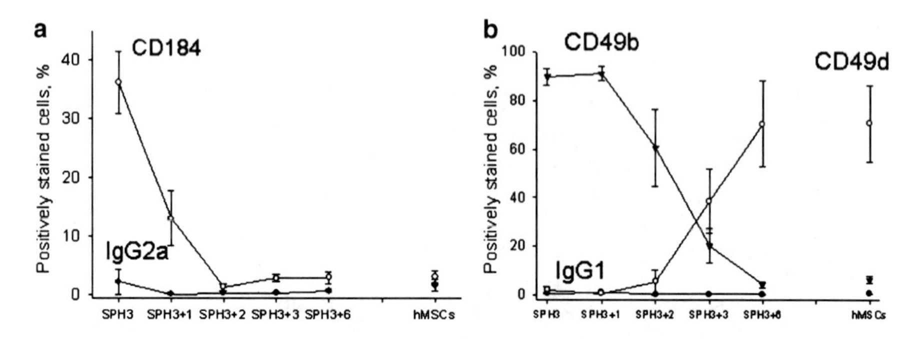

Fig. 5 Changes in CD184, CD49d, and CD49b expression by hMSCs from the spheroids after plating them as a monolayer. Cells were dissociated from 3-day-old hMSC spheroids (*SPH3*) and cultured in a monolayer for 1–6 days (*SPH3+1, SPH3+2, SPH3+3, SPH3+6*). (**a**) Changes in the expression of CD184. (**b**) Changes in the expression of CD49b and CD49d

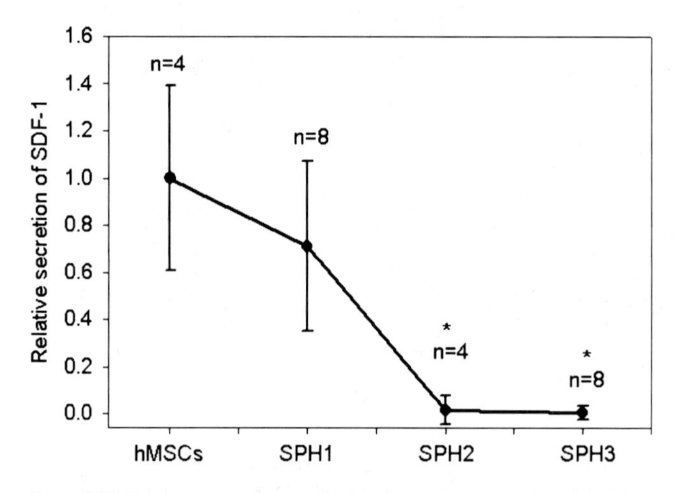

Fig. 6 Secretion of SDF-1 by a monolayer of hMSCs and hMSC spheroids. Concentrations of SDF-1 in media conditioned by a monolayer of hMSCs and 1-day old (*SPH1*), 2-day-old (*SPH2*) and 3-day-old (*SPH3*) hMSC spheroids were measured by ELISA and normalized to the total number of cells. Mean values of the relative secretion ± the standard deviation of SDF-1 by hMSCs ($n=4$), SHP1 ($n=8$), SPH2 ($n=4$), and SPH3 ($n=4$) are shown. Statistically significant changes in comparison with hMSCs are denoted by asterisks (t test, $p < 0.05$)

Therapeutic application of MSCs requires determination of the optimal time when an infusion of MSCs leads to their homing to sites of injury. In the absence of an adequate theory, researchers and clinicians rely on development of heuristic models of stem cell homing. In this process, results from animal models offer a little help. Conducted under controllable conditions, these data may introduce both negative and positive trends which otherwise may not exist in a random population. Most researchers assumed that the susceptibility of tissue to home MSCs is a monotonic function that probably declines over time; therefore, usually a few time points are arbitrarily selected for analysis. However, if the affinity of tissue for MSCs

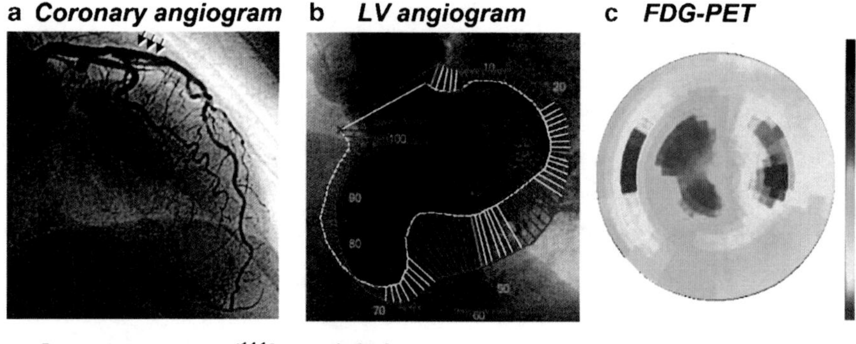

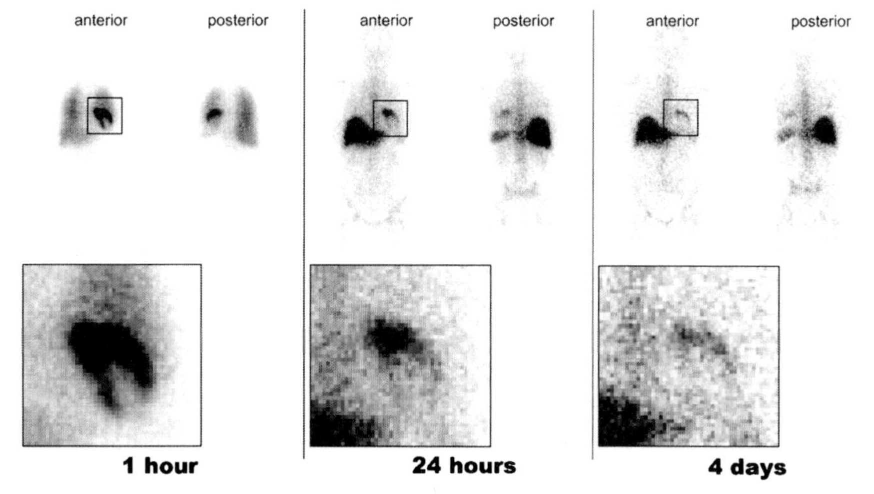

Fig. 7 Homing of bone marrow mononuclear cells into infarcted human heart ^{111}In-labeled mononuclear bone marrow stem cell uptake in a patient after an anterior acute myocardial infarction (cell administration 5 days after acute percutaneous coronary intervention). (**a**) The coronary angiogram before cell infusion. *Arrows* denote the site of cell infusion. (**b**) The left ventricular (*LV*) angiogram. (**c**) ^{18}F-Fluorodeoxyglucose *(FDG)* PET imaging. The *dark areas* indicate low tissue viability. (**d**) ^{111}In distribution at 1 h, 24 h, and 4 days after infusion of indium-111 oxine labeled bone marrow mononuclear cells. Anterior and posterior whole-body scans were acquired. The *inserts* show the heart at a higher magnification (from [90])

changes rapidly over a short period of time, this approach may introduce substantial variations in the observed results. The results of the pilot clinical trial aimed at establishing the optimal timing for stem cell infusion have shown that the homing of bone marrow mononuclear cells to the infarcted heart indeed exhibits a negative trend over a prolonged period of time (Figs. 7 and 8). These results are in good agreement with the data from animal models which suggest that the homing of MSCs requires the presence of injury [91]. At the same time, the data from the clinical trial showed a poor correlation between post-MI time and the homing of the stem cells within the first 10 days after MI (Fig. 8). The data from the clinical trial

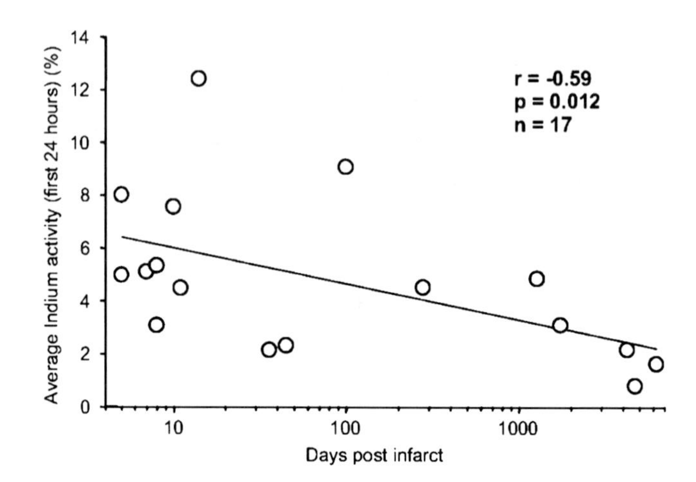

Fig. 8 Correlation between average homing of ^{111}In-labeled mononuclear bone marrow stem cells to the infarcted human heart versus age of infarct (logarithmic scale) (from [90])

suggest that individual differences in the physiological state of endothelium may account for the poor correlation between the homing of stem cells and post-MI time and that the "windows of opportunity" for the infusion of the stem cells may not exist within a time domain in a random population of patients.

If MSC homing varies within a population of patients, then how can the optimal timing for the infusion of stem cells be established? The answer to this question might be obtained from analysis of factors that regulate MSC adhesion. These factors might be found by an experiment or predicted. Considering the plasticity of MSC expression of chemokine receptors, the theoretical prediction of the factors that determine the optimal timing for MSC delivery faces significant challenges. Nevertheless, the viability of this approach was demonstrated on two separate occasions. To establish the optimal time for infusion of MSCs, the spectrum of chemokine receptors on hMSCs was matched with the spectrum of chemokines expressed by infarcted myocardium [70, 92]. It was established that the MSCs express the receptors for monocyte chemoattractant protein (MCP)-1 and MCP-3, and that their homing correlated with the expression of MCP-1 or MCP-3 in the infarcted myocardium. Since, the expression of MCP-1 and MCP-3 in the infarcted heart is transient, the "windows of opportunity" for the infusion of MSCs were short and sustained homing of MSCs to the infarcted heart required the delivery of MCP-1 or MCP-3 genes to the infarcted tissue [70, 92]. The experiments that matched the spectrum of chemokine receptors on MSCs with the spectrum of chemokines expressed in the injured tissue emphasize that the optimal time for infusion of stem cells can be predicted from the analysis of factors in the bloodstream that are involved in the regulation of MSC adhesion. However, development of such semiheuristic models may require the inclusion of many more cytokines in the analysis and clear understanding of the molecular mechanisms of MSC homing to injured tissue.

The spectrum of paracrine and autocrine factors that are responsible for the regulation of MSC homing remains to be established. Identification of tissue conditions that favor MSC homing may provide valuable information both for identification of these factors and for development of a heuristic model for the prediction of MSC homing. The data from the pilot clinical trial provide an important hint in this direction [90]. Figure 11 shows a reliable correlation between the homing of bone marrow mononuclear cells and the viability of cardiac tissue within the first 10 days after MI [90]. The data from the clinical trial correlate with the observation that MSCs tend to localize at the sites of endothelial apoptosis and are not found at the sites of healthy endothelium [93]. Potapova et al. [50] have studied adhesion of hMSCs to apoptotic endothelial cells. Apoptosis in human umbilical vein endothelial cells (HUVECs) was induced by treatment of endothelial cells with the combination of TNF-α or IL-1β with actinomycin D or cycloheximide under conditions when de novo synthesis and delivery of adhesion molecules to the surface of endothelial cells are inhibited (Table 1, Fig. 2). Alternatively, apoptosis of endothelial cells was induced by treatment of cells with okadaic acid, wortmannin, or staurosporine (Fig. 9) [50]. Presumably, owing to the lack of CXCR4 expression, cell-culture-expanded hMSCs only mildly responded to activation of endothelial cells with TNF-α or IL-1β (Fig. 9). The adhesion of hMSCs was potentiated by the combination of actinomycin D or cycloheximide with TNF-α or IL-1β. Okadaic acid, wortmannin, and staurosporine potentiated the hMSC adhesion even further. The adhesion of hMSCs to distressed/apoptotic HUVECs strictly correlated with the decrease in the mitochondrial transmembrane potential of endothelial cells and

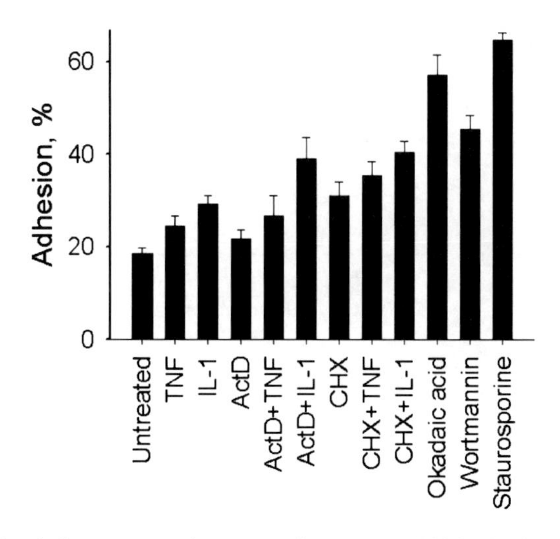

Fig. 9 Effects of proinflammatory and proapoptotic agents on hMSC adhesion to HUVECs. Firm adhesion of hMSCs to HUVECs was assayed after treatment of endothelial cells with 10 ng/ml TNF-α, 10 ng/ml IL-1β, 10 μg/ml actinomycin D (*ActD*), 20 μg/ml cycloheximide (*CHX*), 0.05 μM okadaic acid, 5 μM wortmannin, 0.25 μM staurosporine, or mixtures of TNF-α with actinomycin D, IL-1β with actinomycin D, TNF-α with cycloheximide, or IL-1β with cycloheximide for 4 h

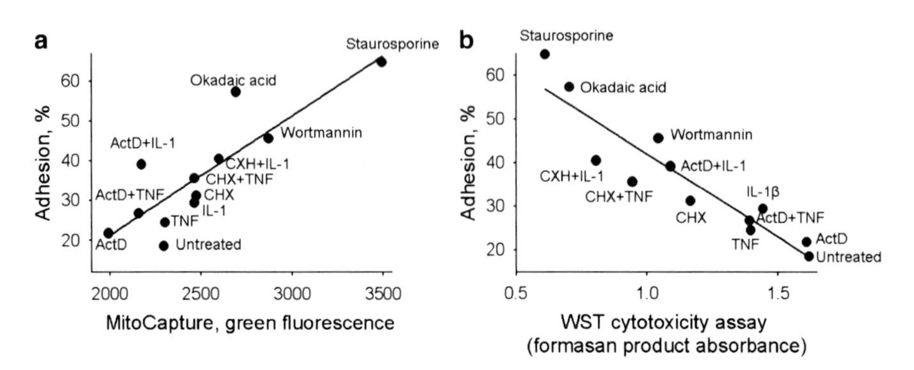

Fig. 10 Correlation between the hMSC adhesion and the inhibition of mitochondrial transmembrane potential and the activity of NADH dehydrogenases in HUVECs treated with inflammatory factors and proapoptotic agents. Adhesion of hMSCs to HUVECs displays a linear correlation with the inhibition of transmembrane mitochondrial potential (**a** $R=0.85$, $p=0.005$) and the inhibition of NADH dehydrogenases (**b** $R=-0.93$, $p=0.00001$) in endothelial cells treated with 10 ng/ml TNF-α, 10 ng/ml IL-1β, 10 μg/ml actinomycin D (*ActD*), 20 μg/ml cycloheximide (*CHX*), or mixtures of TNF-α with actinomycin D, IL-1β with actinomycin D, TNF-α with cycloheximide, or IL-1β with cycloheximide for 4 h

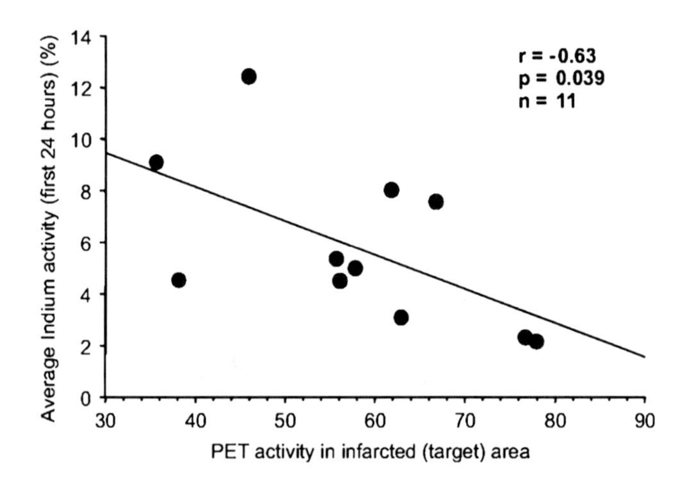

Fig. 11 Correlation between FDG-PET activity and average ^{111}In uptake in the heart tissue of patients with less than 2-week-old infarct. The ^{111}In uptake represents the degree of stem cell homing. FDG-PET activity reflects metabolic activity of infarcted tissue (from [90])

the inhibition of oxidative phosphorylation pathways (Fig. 10) [50]. Thus, despite the difference in the mechanism of action, all these agents have at least one feature in common – they decrease mitochondrial transmembrane potential of endothelial cells, causing the induction of apoptosis via intrinsic apoptotic pathways (Fig. 11). Induction of apoptosis is triggered by the increase in permeability of the mitochondrial membrane, leading to the inhibition of oxidative phosphorylation pathways and the release of cytochrome *c* into the cytosol [94, 95]. The degree of endothelial

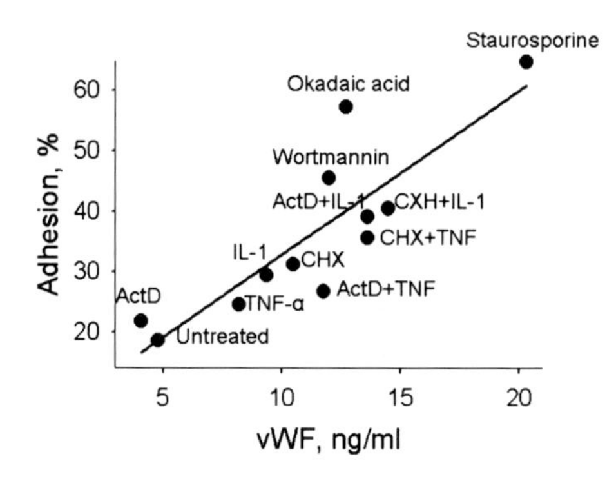

Fig. 12 Correlation between the adhesion of hMSCs and concentrations of von Willebrand factor (*vWF*) in media conditioned by HUVECs. Adhesion of hMSCs to HUVECs displays a linear correlation ($R=0.86$, $p=0.0003$) with the secretion of von Willebrand factor by endothelial cells treated with 10 ng/ml TNF-α, 10 ng/ml IL-1β, 10 μg/ml actinomycin D (*ActD*), 20 μg/ml cycloheximide (*CHX*), or mixtures of TNF-α with actinomycin D, IL-1β with actinomycin D, TNF-α with cycloheximide, or IL-1β with cycloheximide for 4 h

distress could be also assessed by analysis of vWF in media conditioned by endothelial cells. The data in Fig. 12 show that the adhesion of MSCs strongly correlated with the secretion of vWF by distressed/apoptotic endothelial cells [50]. Overall, in vitro data on hMSC adhesion suggest that cell-culture-expanded hMSCs may possess impaired recognition of endothelial cells stimulated with the inflammatory factors; however, they are capable of surveying mitochondrial dysfunctions of endothelial cells and specifically adhere to distressed/apoptotic endothelial cells [50]. The data from the in vitro analysis of MSC adhesion to endothelial cells [50] and the data from the pilot clinical trial [90] suggest that measurements of endothelium distress may serve as predictors of the optimal time for stem cell infusion; however, the applicability of this approach requires further validation (Table 2).

Accumulation of MSCs at sites of injury in response to cell death may represent a natural mechanism of wound healing. A growing body of evidence suggests that reprogramming of tissue microenvironments is one of the major biological functions of MSCs. Positive paracrine effects of MSCs are evident within 24 h after stem cell delivery [40] and include the modulation of the immune response and stimulation of cell survival [96–107]. Long-term paracrine effects of MSCs are manifested in the stimulation of angiogenesis [41, 108–123] and scar remodeling [124–126]. Since the major therapeutic effects of MSCs are attributed to the secretion of paracrine factors, it was not surprising that media conditioned by MSCs displayed therapeutic effects similar to those of MSCs. Delivery of media conditioned by MSCs increases the thickness of the cardiac wall [127], attenuates cardiac fibroblast proliferation, affects collagen deposition [128], facilitates vascular regeneration [113], and appears to be effective in the healing of corneal and cutaneous wounds [129, 130]. The concentration of paracrine factors in media conditioned by MSCs is relatively low. To increase the therapeutic potency of a cocktail of paracrine factors produced by MSCs, it should be concentrated. This is not a trivial biochemical task

Table 2 Changes in antigen expression on the surface of HUVECs treated with proinflammatory and proapoptotic agents

	None	TNF-α	IL-1β	Actinomycin D	Actinomycin D+TNF-α	Actinomycin D+IL-1β	CHX	CHX+TNF-α	CHX+IL-1β
CD31 (PECAM-1)	1.0	0.9±0.2	1.0±1.0	0.95±0.05[a]	0.97±0.2	0.8±0.2[a]	0.7±0.1[a]	0.7±0.1[a]	0.6±0.1[a]
CD34	1.0	1.0±0.2	1.0±0.1	0.86±0.2[a]	1.0±0.3	0.6±0.2[a]	0.74±0.3[a]	0.6±0.2[a]	0.4±0.2[a]
CD44	1.0	0.8±0.2[a]	0.8±0.2[a]	1.0±0.2	0.9±0.1	0.8±0.1[a]	0.8±0.2[a]	0.7±0.1[a]	0.8±0.1[a]
CD51/61 ($\alpha_V\beta_3$ complex)	1.0	0.8±0.1[a]	1.0±0.1	1.00±0.03	0.95±0.03[a]	0.9±0.2	0.81±0.04[a]	0.79±0.02[a]	0.7±0.1[a]
CD54 (ICAM-1)	1.0	3.0±1 [a]	4.0±1[a]	0.9±0.1[a]	0.9±0.2	0.6±0.1[a]	0.7±0.2[a]	0.6±0.2[a]	0.5±0.2[a]
CD62E (E-selectin)	1.0	40±10[a]	50±13[a]	0.6±0.1[a]	0.7±0.2[a]	0.5±0.1[a]	0.66±0.09[a]	2. 0±0.5[a]	1.5±0.2[a]
CD99	1.0	0.85±0.08	0.92±0.06	1.0±0.08	1.02±0.05	0.87±0.02[a]	0.8±0.2	0.86±0.08	0.70±0.03[a]
CD106 (VCAM-1)[b]	11±2	66±40[a]	58±31[a]	10.4±0.6	10.9±0.5	8.8±0.6	8.5±1.5	8.3±1.6	10.0±5.3
CD107a (LAMP-1)[b]	18±2	15±3	17.3±0.9	17±3	18±2	19±4	15±3	17±5	18±6
CD114	1.0	0.8±0.3	1.0±0.2	0.8±0.2	0.8±0.1	0.63±0.08[a]	0.7±0.1[a]	0.61±0.08[a]	0.55±0.06[a]
CD142 (tissue factor)[b]	19±10	48±9[a]	74±12[a]	14±6	16±6	14±7	14±8	16±6	18±8
CD146 (MCAM)	1.0	0.8±0.2	0.9±0.1	0.8±0.2[a]	0.9±0.1	0.75±0.06[a]	0.8±0.1[a]	0.8±0.2	0.9±0.3
CD166 (ALCAM)	1.0	0.83±0.05[a]	0.90±0.06[a]	1.0±0.1	0.98±0.04[a]	0.9±0.1	0.88±0.05[a]	0.9±0.1	0.82±0.07[a]

The data represent an average of at least four independent experiments

PECAM-1 platelet endothelial cell adhesion molecule 1, *ICAM-1* intercellular adhesion molecule 1, *VCAM-1* vascular cell adhesion molecule 1, *LAMP-1* lysosomal-associated membrane protein 1, *MCAM* melanoma cell adhesion molecule, *ALCAM* activated leukocyte cell adhesion molecule, *CHX* cycloheximide

[a]Statistically significant changes in comparison with nontreated HUVECs ($p < 0.05$)

[b]Antigens CD106, CD107a, and CD142 were not expressed on the surface of nontreated HUVECs. For these antigens mean fluorescent values ± the standard deviation are shown. For other antigens, changes relative to nontreated HUVECs are presented

considering the diversity of paracrine factors secreted by MSCs. One obvious solution to this problem was to reduce the ratio between the volume of cell culture medium and the number of cells. Reduction of the medium volume by at least ten times can be achieved by culturing of hMSCs in hanging drops (Table 3) [39]. Culturing of hMSCs in hanging drops increases concentrations of vascular endothelial growth factor (VEGF), angiogenin and procathepsin B in media conditioned by MSCs approximately ten times. Concentrations of bone morphogenetic protein 2, angiogenin, and basic fibroblast growth factor (FGF) were increased more than ten times, suggesting that the production and the secretion of these cytokines was stimulated by the formation of hMSC spheroids. No substantial increases in the concentrations of MCP-1, IL-6, IL-8, and IL-11 were detected in media conditioned by hMSC spheroids [39].

The spectrum of paracrine factors produced by MSCs is very diverse and includes hundreds of proteins and an unknown number of low molecular weight compounds whose structure and therapeutic properties remains to be determined. The secretome of MSCs was studied extensively both at the mRNA and at the protein level [39, 40, 129, 131]. These analyses have established that MSCs emit an array of paracrine factors that includes factors involved in the regulation of angiogenesis, immune response, and inflammation. Paracrine factors secreted by MSCs are characterized by a high degree of redundancy in the spectrum of their action. Modulation of endothelial cells by MSCs represents an example of such redundancy. Delivery of MSCs or media conditioned by them stimulates the perfusion of ischemic tissues and induces the formation of tubular structures by endothelial cells [132, 133]. Media conditioned by MSCs support the migration, extracellular matrix invasion, proliferation, and survival of endothelial cells in vitro [39]. It was established that MSCs produce growth factors that are involved in the regulation of angiogenesis such as VEGF, basic FGF, hepatocyte growth factor, insulin-like growth factor 1, MCP-2, and MCP-3 [14, 110, 113, 133, 134]. Neutralization of VEGF and basic FGF or both inhibits but does not completely eliminate the stimulation of endothelial cell migration [132, 133]. The diversity in the spectrum of paracrine factors secreted by MSCs has led to the hypothesis that MSCs are responsible for the turnover and maintenance of mesenchymal tissues [41, 135]. Such diversity might be caused by the necessity to activate different receptor types expressed on the surface of the various cell types. Overlapping patterns of paracrine factor actions also secure the redundancy required for a robust response.

Production of paracrine factors by MSCs is regulated by environmental conditions. Culturing of hMSCs as the spheroids upregulates the transcription and secretion of factors involved in the regulation of cell survival and angiogenesis [39]. Exposure of MSCs to hypoxia substantially modulates the survival of MSCs and increases the secretion of angiogenic factors [136–141]. The exact effect of the tissue environment and, in particular, the environment at the site of injury on the biosynthesis and secretion of paracrine factors by MSCs remains to be investigated. The diversity and dynamic nature of the cocktail of paracrine factors secreted by MSCs determines MSCs as a novel type of therapeutic agent. In contrast to traditional therapeutics that have defined targets and, subsequently, predetermined

Table 3 Concentrations of cytokines in media conditioned by human mesenchymal stem cells (*hMSCs*) or hMSC spheroids as determined by a cytometric bead array and ELISA

	VEGF (ng/ml)	Basic FGF (pg/ml)	Angiogenin (ng/ml)	Cathepsin B (ng/ml)	BMP-2 (pg/ml)	IL-6 (ng/ml)	IL-8 (ng/ml)	MCP-1 (ng/ml)	IL-11 (ng/ml)
30% FBS	0.3±0.1[a]	24±3	ND	ND	ND	ND	ND	0.16±0.03[a]	(0.20±0.02[a])
hMSC	3.8±0.5 (3.6±0.4)	30±3	0.9±0.3	(4.9±0.7)	(12±3)	7.8±0.8	4.7±0.3 (2.3±0.2)	4.4±0.7	(3±1)
hMSC SPH1	28±1[a] (40±2)	46±10	17±3[a]	(61±2[a])	(99±8[a])	15±2[a]	1.3±0.1[a] (1.3±0.3)	5.0±0.8	(8±3[a])
hMSC SPH2	25±1[a] (43±2[a])	39±22	7.2±0.6[a]	(59±1[a])	(152±17[a])	3.8±0.3[a]	0.6±0.2[a] (0.5±0.2)	3.9±0.4	(14±4[a])
hMSC SPH3	27±8[a] (45±2[a])	171±48[a]	4.2±0.4[a]	(61±2[a])	(221±6[a])	2.7±0.3[a]	6±2 (4±2)	3.9±0.5	(15±3[a])
Hypoxic hMSCs	4.7±0.2 (7±1[a])	19±3	0.78±0.05	(5.9±1)	(9±2)	8.1±0.4	6.6±0.6 (5±1)	3.7±0.5	(2±1)

Data are shown as the mean ± the standard error. Concentrations of factors determined by ELISA are shown in *parentheses*

VEGF vascular endothelial growth factor, *FGF* fibroblast growth factor, *BMP-2* bone morphogenetic protein 2, *MCP-1* monocyte chemoattractant protein 1, *FBS* fetal bovine serum, *ND* not detectable, *SPH1* 1-day-old spheroid, *SPH2* 2-day-old spheroid, *SPH3* 3-day-old spheroid

[a]Statistically significant changes ($p < 0.05$) relative to hMSCs from a monolayer cultured under normoxic conditions

therapeutic effects, MSCs seem not to have a specific tissue or cell type as their target. No MSC-specific receptor has been identified that determines the final outcome of therapeutic interventions based on the use of MSCs or media conditioned by them. Administration of MSCs results in their homing to the injured tissues as well as to healthy tissues, primarily lungs, liver, and spleen [36–38]. Despite the accumulation in the lungs, liver, spleen, and, to a small degree, in other organs, MSCs have little adverse effect on function of healthy tissues and systemic administration of MSCs is relatively safe [142]. Once MSCs have reached their destination, they release an array of paracrine factors with diverse mechanisms of action. Traditional analysis of therapeutic properties of a compound assumes a study of its impact on targeted signal transduction or biochemical pathways with some analysis of cross talk between neighboring signal transduction or biochemical pathways. This approach will fall short in application to the analysis of the therapeutic impact of MSCs. Owing to the diversity of paracrine factors released by MSCs and the dynamic nature of MSC homing, the analysis of affected individual pathways and their interactions is a task of extreme combinatorial complexity. However, studies of MSC interactions with cells that come in contact with them as well as the response of MSCs to their local microenvironment will enhance the thoughtful therapeutic application of MSCs.

References

1. Friedenstein, A. J., Gorskaja, J. F. & Kulagina, N. N. (1976) Fibroblast precursors in normal and irradiated mouse hematopoietic organs, *Exp Hematol. 4*, 267–74.
2. Friedenstein, A. J., Chailakhyan, R. K. & Gerasimov, U. V. (1987) Bone marrow osteogenic stem cells: in vitro cultivation and transplantation in diffusion chambers, *Cell Tissue Kinet. 20*, 263–72.
3. Caplan, A. I. (1991) Mesenchymal stem cells, *J Orthop Res. 9*, 641–50.
4. Prockop, D. J. (1997) Marrow stromal cells as stem cells for nonhematopoietic tissues, *Science. 276*, 71–4.
5. Pittenger, M. F., Mackay, A. M., Beck, S. C., Jaiswal, R. K., Douglas, R., Mosca, J. D., Moorman, M. A., Simonetti, D. W., Craig, S. & Marshak, D. R. (1999) Multilineage potential of adult human mesenchymal stem cells, *Science. 284*, 143–7.
6. Horwitz, E. M., Gordon, P. L., Koo, W. K., Marx, J. C., Neel, M. D., McNall, R. Y., Muul, L. & Hofmann, T. (2002) Isolated allogeneic bone marrow-derived mesenchymal cells engraft and stimulate growth in children with osteogenesis imperfecta: implications for cell therapy of bone, *Proc Natl Acad Sci U S A. 99*, 8932–7.
7. Le Blanc, K., Gotherstrom, C., Ringden, O., Hassan, M., McMahon, R., Horwitz, E., Anneren, G., Axelsson, O., Nunn, J., Ewald, U., Norden-Lindeberg, S., Jansson, M., Dalton, A., Astrom, E. & Westgren, M. (2005) Fetal mesenchymal stem-cell engraftment in bone after in utero transplantation in a patient with severe osteogenesis imperfecta, *Transplantation. 79*, 1607–14.
8. Wakitani, S., Nawata, M., Tensho, K., Okabe, T., Machida, H. & Ohgushi, H. (2007) Repair of articular cartilage defects in the patello-femoral joint with autologous bone marrow mesenchymal cell transplantation: three case reports involving nine defects in five knees, *J Tissue Eng Regen Med. 1*, 74–9.
9. Lazarus, H. M., Koc, O. N., Devine, S. M., Curtin, P., Maziarz, R. T., Holland, H. K., Shpall, E. J., McCarthy, P., Atkinson, K., Cooper, B. W., Gerson, S. L., Laughlin, M. J., Loberiza, F.

R., Jr., Moseley, A. B. & Bacigalupo, A. (2005) Cotransplantation of HLA-identical sibling culture-expanded mesenchymal stem cells and hematopoietic stem cells in hematologic malignancy patients, *Biol Blood Marrow Transplant. 11*, 389–98.
10. Koc, O. N., Gerson, S. L., Cooper, B. W., Dyhouse, S. M., Haynesworth, S. E., Caplan, A. I. & Lazarus, H. M. (2000) Rapid hematopoietic recovery after coinfusion of autologous-blood stem cells and culture-expanded marrow mesenchymal stem cells in advanced breast cancer patients receiving high-dose chemotherapy, *J Clin Oncol. 18*, 307–16.
11. Le Blanc, K. & Ringden, O. (2005) Immunobiology of human mesenchymal stem cells and future use in hematopoietic stem cell transplantation, *Biol Blood Marrow Transplant. 11*, 321–34.
12. Ball, L. M., Bernardo, M. E., Roelofs, H., Lankester, A., Cometa, A., Egeler, R. M., Locatelli, F. & Fibbe, W. E. (2007) Cotransplantation of ex vivo expanded mesenchymal stem cells accelerates lymphocyte recovery and may reduce the risk of graft failure in haploidentical hematopoietic stem-cell transplantation, *Blood. 110*, 2764–7.
13. Le Blanc, K., Samuelsson, H., Gustafsson, B., Remberger, M., Sundberg, B., Arvidson, J., Ljungman, P., Lonnies, H., Nava, S. & Ringden, O. (2007) Transplantation of mesenchymal stem cells to enhance engraftment of hematopoietic stem cells, *Leukemia. 21*, 1733–8.
14. Chen, J., Li, Y., Katakowski, M., Chen, X., Wang, L., Lu, D., Lu, M., Gautam, S. C. & Chopp, M. (2003) Intravenous bone marrow stromal cell therapy reduces apoptosis and promotes endogenous cell proliferation after stroke in female rat, *J Neurosci Res. 73*, 778–86.
15. Chen, J., Li, Y., Wang, L., Zhang, Z., Lu, D., Lu, M. & Chopp, M. (2001) Therapeutic benefit of intravenous administration of bone marrow stromal cells after cerebral ischemia in rats, *Stroke. 32*, 1005–11.
16. Chen, J., Zhang, Z. G., Li, Y., Wang, L., Xu, Y. X., Gautam, S. C., Lu, M., Zhu, Z. & Chopp, M. (2003) Intravenous administration of human bone marrow stromal cells induces angiogenesis in the ischemic boundary zone after stroke in rats, *Circ Res. 92*, 692–9.
17. Mahmood, A., Lu, D., Lu, M. & Chopp, M. (2003) Treatment of traumatic brain injury in adult rats with intravenous administration of human bone marrow stromal cells, *Neurosurgery. 53*, 697–702; discussion 702–3.
18. Mahmood, A., Lu, D., Wang, L., Li, Y., Lu, M. & Chopp, M. (2001) Treatment of traumatic brain injury in female rats with intravenous administration of bone marrow stromal cells, *Neurosurgery. 49*, 1196–203; discussion 1203–4.
19. Seyfried, D., Ding, J., Han, Y., Li, Y., Chen, J. & Chopp, M. (2006) Effects of intravenous administration of human bone marrow stromal cells after intracerebral hemorrhage in rats, *J Neurosurg. 104*, 313–8.
20. Orlic, D., Kajstura, J., Chimenti, S., Bodine, D. M., Leri, A. & Anversa, P. (2003) Bone marrow stem cells regenerate infarcted myocardium, *Pediatr Transplant. 7 Suppl 3*, 86–8.
21. Dai, W., Hale, S. L., Martin, B. J., Kuang, J. Q., Dow, J. S., Wold, L. E. & Kloner, R. A. (2005) Allogeneic mesenchymal stem cell transplantation in postinfarcted rat myocardium: short- and long-term effects, *Circulation. 112*, 214–23.
22. Gnecchi, M., He, H., Liang, O. D., Melo, L. G., Morello, F., Mu, H., Noiseux, N., Zhang, L., Pratt, R. E., Ingwall, J. S. & Dzau, V. J. (2005) Paracrine action accounts for marked protection of ischemic heart by Akt-modified mesenchymal stem cells, *Nat Med. 11*, 367–8.
23. Assmus, B., Honold, J., Schachinger, V., Britten, M. B., Fischer-Rasokat, U., Lehmann, R., Teupe, C., Pistorius, K., Martin, H., Abolmaali, N. D., Tonn, T., Dimmeler, S. & Zeiher, A. M. (2006) Transcoronary transplantation of progenitor cells after myocardial infarction, *N Engl J Med. 355*, 1222–32.
24. Mosca, J. D., Hendricks, J. K., Buyaner, D., Davis-Sproul, J., Chuang, L. C., Majumdar, M. K., Chopra, R., Barry, F., Murphy, M., Thiede, M. A., Junker, U., Rigg, R. J., Forestell, S. P., Bohnlein, E., Storb, R. & Sandmaier, B. M. (2000) Mesenchymal stem cells as vehicles for gene delivery, *Clin Orthop Relat Res*, 379 Suppl, S71–90.
25. Devine, M. J., Mierisch, C. M., Jang, E., Anderson, P. C. & Balian, G. (2002) Transplanted bone marrow cells localize to fracture callus in a mouse model, *J Orthop Res. 20*, 1232–9.

26. Ramirez, M., Lucia, A., Gomez-Gallego, F., Esteve-Lanao, J., Perez-Martinez, A., Foster, C., Andreu, A. L., Martin, M. A., Madero, L., Arenas, J. & Garcia-Castro, J. (2006) Mobilisation of mesenchymal cells into blood in response to skeletal muscle injury, *Br J Sports Med. 40*, 719–22.
27. Mansilla, E., Marln, G. H., Drago, H., Sturla, F., Salas, E., Gardiner, C., Bossi, S., Lamonega, R., Guzmőn, A., Nucez, A., Gil, M. A., Piccinelli, G., Ibar, R. & Soratti, C. (2006) Bloodstream cells phenotypically identical to human mesenchymal bone marrow stem cells circulate in large amounts under the influence of acute large skin damage: new evidence for their use in regenerative medicine, *Transplant Proc. 38*, 967–9.
28. Wang, L., Li, Y., Chen, J., Gautam, S. C., Zhang, Z., Lu, M. & Chopp, M. (2002) Ischemic cerebral tissue and MCP-1 enhance rat bone marrow stromal cell migration in interface culture, *Experimental Hematology. 30*, 831–6.
29. Barbash, I. M., Chouraqui, P., Baron, J., Feinberg, M. S., Etzion, S., Tessone, A., Miller, L., Guetta, E., Zipori, D., Kedes, L. H., Kloner, R. A. & Leor, J. (2003) Systemic delivery of bone marrow-derived mesenchymal stem cells to the infarcted myocardium: feasibility, cell migration, and body distribution, *Circulation. 108*, 863–8.
30. Bittira, B., Shum-Tim, D., Al-Khaldi, A. & Chiu, R. C. J. (2003) Mobilization and homing of bone marrow stromal cells in myocardial infarction, *Eur J Cardiothorac Surg. 24*, 393–8.
31. Chapel, A., Bertho, J. M., Bensidhoum, M., Fouillard, L., Young, R. G., Frick, J., Demarquay, C., Cuvelier, F., Mathieu, E., Trompier, F., Dudoignon, N., Germain, C., Mazurier, C., Aigueperse, J., Borneman, J., Gorin, N. C., Gourmelon, P. & Thierry, D. (2003) Mesenchymal stem cells home to injured tissues when co-infused with hematopoietic cells to treat a radiation-induced multi-organ failure syndrome, *J Gene Med. 5*, 1028–38.
32. Rombouts, W. J. C. & Ploemacher, R. E. (2003) Primary murine MSC show highly efficient homing to the bone marrow but lose homing ability following culture, *Leukemia. 17*, 160–70.
33. Ji, J. F., He, B. P., Dheen, S. T. & Tay, S. S. W. (2004) Interactions of chemokines and chemokine receptors mediate the migration of mesenchymal stem cells to the impaired site in the brain after hypoglossal nerve injury, *Stem Cells. 22*, 415–27.
34. Kraitchman, D. L., Tatsumi, M., Gilson, W. D., Ishimori, T., Kedziorek, D., Walczak, P., Segars, W. P., H. Chen, H., Fritzges, D., Izbudak, I., Young, R. G., Marcelino, M., Pittenger, M. F., Solaiyappan, M., Boston, R. C., Tsui, B. M. W., Wahl, R. L. & Bulte, J. W. M. (2005) Dynamic imaging of allogeneic mesenchymal stem cells trafficking to myocardial infarction, *Circulation. 112*, 1451–61.
35. Mouiseddine, M., Francois, S., Semont, A., Sache, A., Allenet, B., Mathieu, N., Frick, J., Thierry, D. & Chapel, A. (2007) Human mesenchymal stem cells home specifically to radiation-injured tissues in a non-obese diabetes/severe combined immunodeficiency mouse model, *Br J Radiol. 80*, S49–55.
36. Allers, C., Sierralta, W. D., Neubauer, S., Rivera, F., Minguell, J. J. & Conget, P. A. (2004) Dynamic of distribution of human bone marrow-derived mesenchymal stem cells after transplantation into adult unconditioned mice, *Transplantation. 78*, 503–8.
37. Devine, S. M., Cobbs, C., Jennings, M., Bartholomew, A. & Hoffman, R. (2003) Mesenchymal stem cells distribute to a wide range of tissues following systemic infusion into nonhuman primates, *Blood. 101*, 2999–3001.
38. Gao, J., Dennis, J. E., Muzic, R. F., Lundberg, M. & Caplan, A. I. (2001) The dynamic in vivo distribution of bone marrow-derived mesenchymal stem cells after infusion, *Cells Tissues Organs. 169*, 12–20.
39. Potapova, I. A., Gaudette, G. R., Brink, P. R., Robinson, R. B., Rosen, M. R., Cohen, I. S. & Doronin, S. V. (2007) Mesenchymal stem cells support migration, extracellular matrix invasion, proliferation, and survival of endothelial cells in vitro, *Stem Cells. 25*, 1761–8.
40. Gnecchi, M., Zhang, Z., Ni, A. & Dzau, V. J. (2008) Paracrine mechanisms in adult stem cell signaling and therapy, *Circ Res. 103*, 1204–19.
41. Caplan, A. I. & Dennis, J. E. (2006) Mesenchymal stem cells as trophic mediators, *J Cell Biochem. 98*, 1076–84.

42. Springer, T. A. (1994) Traffic signals for lymphocyte recirculation and leukocyte emigration: the multistep paradigm, *Cell. 76*, 301–14.
43. Barreiro, O. & Sanchez-Madrid, F. (2009) Molecular basis of leukocyte-endothelium interactions during the inflammatory response, *Rev Esp Cardiol. 62*, 552–62.
44. Eriksson, E. E., Xie, X., Werr, J., Thoren, P. & Lindbom, L. (2001) Importance of primary capture and L-selectin-dependent secondary capture in leukocyte accumulation in inflammation and atherosclerosis in vivo, *J Exp Med. 194*, 205–18.
45. Hidalgo, A., Peired, A. J., Wild, M. K., Vestweber, D. & Frenette, P. S. (2007) Complete identification of E-selectin ligands on neutrophils reveals distinct functions of PSGL-1, ESL-1, and CD44, *Immunity. 26*, 477–89.
46. Berlin, C., Berg, E. L., Briskin, M. J., Andrew, D. P., Kilshaw, P. J., Holzmann, B., Weissman, I. L., Hamann, A. & Butcher, E. C. (1993) Alpha 4 beta 7 integrin mediates lymphocyte binding to the mucosal vascular addressin MAdCAM-1, *Cell. 74*, 185–95.
47. Alon, R., Kassner, P. D., Carr, M. W., Finger, E. B., Hemler, M. E. & Springer, T. A. (1995) The integrin VLA-4 supports tethering and rolling in flow on VCAM-1, *J Cell Biol. 128*, 1243–53.
48. Berlin, C., Bargatze, R. F., Campbell, J. J., von Andrian, U. H., Szabo, M. C., Hasslen, S. R., Nelson, R. D., Berg, E. L., Erlandsen, S. L. & Butcher, E. C. (1995) Alpha 4 integrins mediate lymphocyte attachment and rolling under physiologic flow, *Cell. 80*, 413–22.
49. Rao, R. M., Yang, L., Garcia-Cardena, G. & Luscinskas, F. W. (2007) Endothelial-dependent mechanisms of leukocyte recruitment to the vascular wall, *Circ Res. 101*, 234–47.
50. Potapova, I. A., Cohen, I. S. & Doronin, S. V. (2009) Apoptotic endothelial cells demonstrate increased adhesiveness for human mesenchymal stem cells, *J Cell Physiol. 219*, 23–30.
51. Campbell, J. J., Hedrick, J., Zlotnik, A., Siani, M. A., Thompson, D. A. & Butcher, E. C. (1998) Chemokines and the arrest of lymphocytes rolling under flow conditions, *Science. 279*, 381–4.
52. Shamri, R., Grabovsky, V., Gauguet, J. M., Feigelson, S., Manevich, E., Kolanus, W., Robinson, M. K., Staunton, D. E., von Andrian, U. H. & Alon, R. (2005) Lymphocyte arrest requires instantaneous induction of an extended LFA-1 conformation mediated by endothelium-bound chemokines, *Nat Immunol. 6*, 497–506.
53. Reininger, A. J. (2008) Function of von Willebrand factor in haemostasis and thrombosis, *Haemophilia. 14 Suppl 5*, 11–26.
54. Ruster, B., Gottig, S., Ludwig, R. J., Bistrian, R., Muller, S., Seifried, E., Gille, J. & Henschler, R. (2006) Mesenchymal stem cells display coordinated rolling and adhesion behavior on endothelial cells, *Blood. 108*, 3938–44.
55. Schmidt, A., Ladage, D., Steingen, C., Brixius, K., Schinkuthe, T., Klinz, F.-J., Schwinger, R. H. G., Mehlhorn, U. & Bloch, W. (2006) Mesenchymal stem cells transmigrate over the endothelial barrier, *Eur J Cell Biol. 85*, 1179–88.
56. Baddoo, M., Hill, K., Wilkinson, R., Gaupp, D., Hughes, C., Kopen, G. C. & Phinney, D. G. (2003) Characterization of mesenchymal stem cells isolated from murine bone marrow by negative selection, *J Cell Biochem. 89*, 1235–49.
57. Cheng, S. L., Yang, J. W., Rifas, L., Zhang, S. F. & Avioli, L. V. (1994) Differentiation of human bone marrow osteogenic stromal cells in vitro: induction of the osteoblast phenotype by dexamethasone, *Endocrinology. 134*, 277–86.
58. Clark, B. R. & Keating, A. (1995) Biology of bone marrow stroma, *Ann N Y Acad Sci. 770*, 70–8.
59. Keating, A., Horsfall, W., Hawley, R. G. & Toneguzzo, F. (1990) Effect of different promoters on expression of genes introduced into hematopoietic and marrow stromal cells by electroporation, *Exp Hematol. 18*, 99–102.
60. Seeger, F. H., Tonn, T., Krzossok, N., Zeiher, A. M. & Dimmeler, S. (2007) Cell isolation procedures matter: a comparison of different isolation protocols of bone marrow mononuclear cells used for cell therapy in patients with acute myocardial infarction, *Eur Heart J. 28*, 766–72.
61. Galmiche, M. C., Koteliansky, V. E., Briere, J., Herve, P. & Charbord, P. (1993) Stromal cells from human long-term marrow cultures are mesenchymal cells that differentiate following a vascular smooth muscle differentiation pathway, *Blood. 82*, 66–76.

62. Haynesworth, S. E., Baber, M. A. & Caplan, A. I. (1992) Cell surface antigens on human marrow-derived mesenchymal cells are detected by monoclonal antibodies, *Bone. 13*, 69–80.
63. Le Blanc, K., Tammik, C., Rosendahl, K., Zetterberg, E. & Ringden, O. (2003) HLA expression and immunologic properties of differentiated and undifferentiated mesenchymal stem cells, *Exp Hematol. 31*, 890–6.
64. Sordi, V., Malosio, M. L., Marchesi, F., Mercalli, A., Melzi, R., Giordano, T., Belmonte, N., Ferrari, G., Leone, B. E., Bertuzzi, F., Zerbini, G., Allavena, P., Bonifacio, E. & Piemonti, L. (2005) Bone marrow mesenchymal stem cells express a restricted set of functionally active chemokine receptors capable of promoting migration to pancreatic islets, *Blood. 106*, 419–27.
65. Honczarenko, M., Le, Y., Swierkowski, M., Ghiran, I., Glodek, A. M. & Silberstein, L. E. (2006) Human bone marrow stromal cells express a distinct set of biologically functional chemokine receptors, *Stem Cells. 24*, 1030–41.
66. Wynn, R. F., Hart, C. A., Corradi-Perini, C., O'Neill, L., Evans, C. A., Wraith, J. E., Fairbairn, L. J. & Bellantuono, I. (2004) A small proportion of mesenchymal stem cells strongly expresses functionally active CXCR4 receptor capable of promoting migration to bone marrow, *Blood. 104*, 2643–5.
67. von Luttichau, I., Notohamiprodjo, M., Wechselberger, A., Peters, C., Henger, A., Seliger, C., Djafarzadeh, R., Huss, R. & Nelson, P. J. (2005) Human adult CD34- progenitor cells functionally express the chemokine receptors CCR1, CCR4, CCR7, CXCR5, and CCR10 but not CXCR4, *Stem Cells Dev. 14*, 329–36.
68. Chamberlain, G., Wright, K., Rot, A., Ashton, B. & Middleton, J. (2008) Murine mesenchymal stem cells exhibit a restricted repertoire of functional chemokine receptors: comparison with human, *PLoS One. 3*, e2934.
69. Zhang, F., Tsai, S., Kato, K., Yamanouchi, D., Wang, C., Rafii, S., Liu, B. & Kent, K. C. (2009) Transforming growth factor-beta promotes recruitment of bone marrow cells and bone marrow-derived mesenchymal stem cells through stimulation of MCP-1 production in vascular smooth muscle cells, *J Biol Chem. 284*, 17564–74.
70. Schenk, S., Mal, N., Finan, A., Zhang, M., Kiedrowski, M., Popovic, Z., McCarthy, P. M. & Penn, M. S. (2007) Monocyte chemotactic protein-3 is a myocardial mesenchymal stem cell homing factor, *Stem Cells. 25*, 245–51.
71. Brooke, G., Tong, H., Levesque, J. P. & Atkinson, K. (2008) Molecular trafficking mechanisms of multipotent mesenchymal stem cells derived from human bone marrow and placenta, *Stem Cells Dev. 17*, 929–40.
72. Hwang, J. H., Shim, S. S., Seok, O. S., Lee, H. Y., Woo, S. K., Kim, B. H., Song, H. R., Lee, J. K. & Park, Y. K. (2009) Comparison of cytokine expression in mesenchymal stem cells from human placenta, cord blood, and bone marrow, *J Korean Med Sci. 24*, 547–54.
73. Ringe, J., Strassburg, S., Neumann, K., Endres, M., Notter, M., Burmester, G. R., Kaps, C. & Sittinger, M. (2007) Towards in situ tissue repair: human mesenchymal stem cells express chemokine receptors CXCR1, CXCR2 and CCR2, and migrate upon stimulation with CXCL8 but not CCL2, *J Cell Biochem. 101*, 135–46.
74. Dwyer, R. M., Potter-Beirne, S. M., Harrington, K. A., Lowery, A. J., Hennessy, E., Murphy, J. M., Barry, F. P., O'Brien, T. & Kerin, M. J. (2007) Monocyte chemotactic protein-1 secreted by primary breast tumors stimulates migration of mesenchymal stem cells, *Clin Cancer Res. 13*, 5020–7.
75. Potapova, I. A., Brink, P. R., Cohen, I. S. & Doronin, S. V. (2008) Culturing of human mesenchymal stem cells as 3D-aggregates induces functional expression of CXCR4 that regulates adhesion to endothelial cells, *J. Biol. Chem. 283*, 13100–7.
76. Shield, K., Riley, C., Quinn, M. A., Rice, G. E., Ackland, M. L. & Ahmed, N. (2007) Alpha2beta1 integrin affects metastatic potential of ovarian carcinoma spheroids by supporting disaggregation and proteolysis, *J Carcinog. 6*, 11.
77. Friedl, P., Noble, P. B. & Zanker, K. S. (1995) T lymphocyte locomotion in a three-dimensional collagen matrix. Expression and function of cell adhesion molecules, *Journal of immunology. 154*, 4973–85.

78. Sixt, M., Bauer, M., Lammermann, T. & Fassler, R. (2006) Beta1 integrins: zip codes and signaling relay for blood cells, *Curr Opin Cell Biol. 18*, 482–90.
79. van Buul, J. D., Voermans, C., van Gelderen, J., Anthony, E. C., van der Schoot, C. E. & Hordijk, P. L. (2003) Leukocyte-endothelium interaction promotes SDF-1-dependent polarization of CXCR4, *J Biol Chem. 278*, 30302–10.
80. Glodek, A. M., Le, Y., Dykxhoorn, D. M., Park, S. Y., Mostoslavsky, G., Mulligan, R., Lieberman, J., Beggs, H. E., Honczarenko, M. & Silberstein, L. E. (2007) Focal adhesion kinase is required for CXCL12-induced chemotactic and pro-adhesive responses in hematopoietic precursor cells, *Leukemia. 21*, 1723–32.
81. Honczarenko, M., Le, Y., Glodek, A. M., Majka, M., Campbell, J. J., Ratajczak, M. Z. & Silberstein, L. E. (2002) CCR5-binding chemokines modulate CXCL12 (SDF-1)-induced responses of progenitor B cells in human bone marrow through heterologous desensitization of the CXCR4 chemokine receptor, *Blood. 100*, 2321–9.
82. Le, Y., Honczarenko, M., Glodek, A. M., Ho, D. K. & Silberstein, L. E. (2005) CXC chemokine ligand 12-induced focal adhesion kinase activation and segregation into membrane domains is modulated by regulator of G protein signaling 1 in pro-B cells, *Journal of immunology. 174*, 2582–90.
83. Neel, N. F., Schutyser, E., Sai, J., Fan, G. H. & Richmond, A. (2005) Chemokine receptor internalization and intracellular trafficking, *Cytokine Growth Factor Rev. 16*, 637–58.
84. Seeger, F. H., Zeiher, A. M. & Dimmeler, S. (2007) Cell-enhancement strategies for the treatment of ischemic heart disease, *Nat Clin Pract Cardiovasc Med. 4 Suppl 1*, S110–3.
85. Wang, Y., Johnsen, H. E., Mortensen, S., Bindslev, L., Ripa, R. S., Haack-Sorensen, M., Jorgensen, E., Fang, W. & Kastrup, J. (2006) Changes in circulating mesenchymal stem cells, stem cell homing factor, and vascular growth factors in patients with acute ST elevation myocardial infarction treated with primary percutaneous coronary intervention, *Heart. 92*, 768–74.
86. Wang, Y., Deng, Y. & Zhou, G. Q. (2008) SDF-1alpha/CXCR4-mediated migration of systemically transplanted bone marrow stromal cells towards ischemic brain lesion in a rat model, *Brain Res. 1195*, 104–12.
87. Bhakta, S., Hong, P. & Koc, O. (2006) The surface adhesion molecule CXCR4 stimulates mesenchymal stem cell migration to stromal cell-derived factor-1 in vitro but does not decrease apoptosis under serum deprivation, *Cardiovasc Revasc Med. 7*, 19–24.
88. Cheng, Z., Ou, L., Zhou, X., Li, F., Jia, X., Zhang, Y., Liu, X., Li, Y., Ward, C. A., Melo, L. G. & Kong, D. (2008) Targeted migration of mesenchymal stem cells modified with CXCR4 gene to infarcted myocardium improves cardiac performance, *Mol Ther. 16*, 571–9.
89. Zhang, D., Fan, G. C., Zhou, X., Zhao, T., Pasha, Z., Xu, M., Zhu, Y., Ashraf, M. & Wang, Y. (2008) Over-expression of CXCR4 on mesenchymal stem cells augments myoangiogenesis in the infarcted myocardium, *J Mol Cell Cardiol. 44*, 281–92.
90. Schachinger, V., Aicher, A., Dobert, N., Rover, R., Diener, J., Fichtlscherer, S., Assmus, B., Seeger, F. H., Menzel, C., Brenner, W., Dimmeler, S. & Zeiher, A. M. (2008) Pilot trial on determinants of progenitor cell recruitment to the infarcted human myocardium, *Circulation. 118*, 1425–32.
91. Abbott, J. D., Huang, Y., Liu, D., Hickey, R., Krause, D. S. & Giordano, F. J. (2004) Stromal cell-derived factor-1alpha plays a critical role in stem cell recruitment to the heart after myocardial infarction but is not sufficient to induce homing in the absence of injury, *Circulation. 110*, 3300–5.
92. Belema-Bedada, F., Uchida, S., Martire, A., Kostin, S. & Braun, T. (2008) Efficient homing of multipotent adult mesenchymal stem cells depends on FROUNT-mediated clustering of CCR2, *Cell Stem Cell. 2*, 566–75.
93. Abedin, M., Tintut, Y. & Demer, L. L. (2004) Mesenchymal stem cells and the artery wall, *Circ Res. 95*, 671–6.
94. Riedl, S. J. & Salvesen, G. S. (2007) The apoptosome: signalling platform of cell death, *Nat Rev Mol Cell Biol. 8*, 405–13.
95. Orrenius, S., Gogvadze, V. & Zhivotovsky, B. (2007) Mitochondrial oxidative stress: implications for cell death, *Annu Rev Pharmacol Toxicol. 47*, 143–83.

96. Semedo, P., Palasio, C. G., Oliveira, C. D., Feitoza, C. Q., Goncalves, G. M., Cenedeze, M. A., Wang, P. M., Teixeira, V. P., Reis, M. A., Pacheco-Silva, A. & Camara, N. O. (2009) Early modulation of inflammation by mesenchymal stem cell after acute kidney injury, *Int Immunopharmacol. 9*, 677–82.
97. Caplan, A. I. (2009) Why are MSCs therapeutic? New data: new insight, *J Pathol. 217*, 318–24.
98. Bian, L., Guo, Z. K., Wang, H. X., Wang, J. S., Wang, H., Li, Q. F., Yang, Y. F., Xiao, F. J., Wu, C. T. & Wang, L. S. (2009) In vitro and in vivo immunosuppressive characteristics of hepatocyte growth factor-modified murine mesenchymal stem cells, *In Vivo. 23*, 21–7.
99. Bexell, D., Gunnarsson, S., Tormin, A., Darabi, A., Gisselsson, D., Roybon, L., Scheding, S. & Bengzon, J. (2009) Bone marrow multipotent mesenchymal stroma cells act as pericyte-like migratory vehicles in experimental gliomas, *Mol Ther. 17*, 183–90.
100. Asari, S., Itakura, S., Ferreri, K., Liu, C. P., Kuroda, Y., Kandeel, F. & Mullen, Y. (2009) Mesenchymal stem cells suppress B-cell terminal differentiation, *Exp Hematol. 37*, 604–15.
101. Jones, B. J. & McTaggart, S. J. (2008) Immunosuppression by mesenchymal stromal cells: from culture to clinic, *Exp Hematol. 36*, 733–41.
102. Le Blanc, K. & Ringden, O. (2007) Immunomodulation by mesenchymal stem cells and clinical experience, *J Intern Med. 262*, 509–25.
103. Gur-Wahnon, D., Borovsky, Z., Beyth, S., Liebergall, M. & Rachmilewitz, J. (2007) Contact-dependent induction of regulatory antigen-presenting cells by human mesenchymal stem cells is mediated via STAT3 signaling, *Exp Hematol. 35*, 426–33.
104. Gerdoni, E., Gallo, B., Casazza, S., Musio, S., Bonanni, I., Pedemonte, E., Mantegazza, R., Frassoni, F., Mancardi, G., Pedotti, R. & Uccelli, A. (2007) Mesenchymal stem cells effectively modulate pathogenic immune response in experimental autoimmune encephalomyelitis, *Ann Neurol. 61*, 219–27.
105. Fibbe, W. E., Nauta, A. J. & Roelofs, H. (2007) Modulation of immune responses by mesenchymal stem cells, *Ann N Y Acad Sci. 1106*, 272–8.
106. Beggs, K. J., Lyubimov, A., Borneman, J. N., Bartholomew, A., Moseley, A., Dodds, R., Archambault, M. P., Smith, A. K. & McIntosh, K. R. (2006) Immunologic consequences of multiple, high-dose administration of allogeneic mesenchymal stem cells to baboons, *Cell Transplant. 15*, 711–21.
107. Beyth, S., Borovsky, Z., Mevorach, D., Liebergall, M., Gazit, Z., Aslan, H., Galun, E. & Rachmilewitz, J. (2005) Human mesenchymal stem cells alter antigen-presenting cell maturation and induce T-cell unresponsiveness, *Blood. 105*, 2214–9.
108. Fukunaga, A., Uchida, K., Hara, K., Kuroshima, Y. & Kawase, T. (1999) Differentiation and angiogenesis of central nervous system stem cells implanted with mesenchyme into ischemic rat brain, *Cell Transplant. 8*, 435–41.
109. Furumatsu, T., Shen, Z. N., Kawai, A., Nishida, K., Manabe, H., Oohashi, T., Inoue, H. & Ninomiya, Y. (2003) Vascular endothelial growth factor principally acts as the main angiogenic factor in the early stage of human osteoblastogenesis, *J Biochem. 133*, 633–9.
110. Nagaya, N., Fujii, T., Iwase, T., Ohgushi, H., Itoh, T., Uematsu, M., Yamagishi, M., Mori, H., Kangawa, K. & Kitamura, S. (2004) Intravenous administration of mesenchymal stem cells improves cardiac function in rats with acute myocardial infarction through angiogenesis and myogenesis, *Am J Physiol Heart Circ Physiol. 287*, H2670–6.
111. Tang, Y. L., Zhao, Q., Zhang, Y. C., Cheng, L., Liu, M., Shi, J., Yang, Y. Z., Pan, C., Ge, J. & Phillips, M. I. (2004) Autologous mesenchymal stem cell transplantation induce VEGF and neovascularization in ischemic myocardium, *Regul Pept. 117*, 3–10.
112. Mayer, H., Bertram, H., Lindenmaier, W., Korff, T., Weber, H. & Weich, H. (2005) Vascular endothelial growth factor (VEGF-A) expression in human mesenchymal stem cells: autocrine and paracrine role on osteoblastic and endothelial differentiation, *J Cell Biochem. 95*, 827–39.
113. Tang, Y. L., Zhao, Q., Qin, X., Shen, L., Cheng, L., Ge, J. & Phillips, M. I. (2005) Paracrine action enhances the effects of autologous mesenchymal stem cell transplantation on vascular regeneration in rat model of myocardial infarction, *Ann Thorac Surg. 80*, 229–36; discussion 236–7.

114. Ghajar, C. M., Blevins, K. S., Hughes, C. C., George, S. C. & Putnam, A. J. (2006) Mesenchymal stem cells enhance angiogenesis in mechanically viable prevascularized tissues via early matrix metalloproteinase upregulation, *Tissue Eng. 12*, 2875–88.
115. Han, S. K., Chun, K. W., Gye, M. S. & Kim, W. K. (2006) The effect of human bone marrow stromal cells and dermal fibroblasts on angiogenesis, *Plast Reconstr Surg. 117*, 829–35.
116. Shyu, K. G., Wang, B. W., Hung, H. F., Chang, C. C. & Shih, D. T. (2006) Mesenchymal stem cells are superior to angiogenic growth factor genes for improving myocardial performance in the mouse model of acute myocardial infarction, *J Biomed Sci. 13*, 47–58.
117. Tang, J., Xie, Q., Pan, G., Wang, J. & Wang, M. (2006) Mesenchymal stem cells participate in angiogenesis and improve heart function in rat model of myocardial ischemia with reperfusion, *Eur J Cardiothorac Surg. 30*, 353–61.
118. Kasper, G., Dankert, N., Tuischer, J., Hoeft, M., Gaber, T., Glaeser, J. D., Zander, D., Tschirschmann, M., Thompson, M., Matziolis, G. & Duda, G. N. (2007) Mesenchymal stem cells regulate angiogenesis according to their mechanical environment, *Stem Cells. 25*, 903–10.
119. Kim, Y., Kim, H., Cho, H., Bae, Y., Suh, K. & Jung, J. (2007) Direct comparison of human mesenchymal stem cells derived from adipose tissues and bone marrow in mediating neovascularization in response to vascular ischemia, *Cell Physiol Biochem. 20*, 867–76.
120. Wu, Y., Chen, L., Scott, P. G. & Tredget, E. E. (2007) Mesenchymal stem cells enhance wound healing through differentiation and angiogenesis, *Stem Cells. 25*, 2648–59.
121. Askarov, M. B. & Onischenko, N. A. (2008) Multipotent mesenchymal stromal cells of autologous bone marrow stimulate neoangiogenesis, restore microcirculation, and promote healing of indolent ulcers of the stomach, *Bull Exp Biol Med. 146*, 512–6.
122. Egana, J. T., Fierro, F. A., Kruger, S., Bornhauser, M., Huss, R., Lavandero, S. & Machens, H. G. (2009) Use of human mesenchymal cells to improve vascularization in a mouse model for scaffold-based dermal regeneration, *Tissue Eng Part A. 15*, 1191–200.
123. Huang, N. F., Lam, A., Fang, Q., Sievers, R. E., Li, S. & Lee, R. J. (2009) Bone marrow-derived mesenchymal stem cells in fibrin augment angiogenesis in the chronically infarcted myocardium, *Regen Med. 4*, 527–38.
124. Bellini, A. & Mattoli, S. (2007) The role of the fibrocyte, a bone marrow-derived mesenchymal progenitor, in reactive and reparative fibroses, *Lab Invest. 87*, 858–70.
125. Dixon, J. A., Gorman, R. C., Stroud, R. E., Bouges, S., Hirotsugu, H., Gorman, J. H., 3rd, Martens, T. P., Itescu, S., Schuster, M. D., Plappert, T., St John-Sutton, M. G. & Spinale, F. G. (2009) Mesenchymal cell transplantation and myocardial remodeling after myocardial infarction, *Circulation. 120*, S220–9.
126. Molina, E. J., Palma, J., Gupta, D., Torres, D., Gaughan, J. P., Houser, S. & Macha, M. (2009) Reverse remodeling is associated with changes in extracellular matrix proteases and tissue inhibitors after mesenchymal stem cell (MSC) treatment of pressure overload hypertrophy, *J Tissue Eng Regen Med. 3*, 85–91.
127. Dai, W., Hale, S. L. & Kloner, R. A. (2007) Role of a paracrine action of mesenchymal stem cells in the improvement of left ventricular function after coronary artery occlusion in rats, *Regen Med. 2*, 63–8.
128. Ohnishi, S., Sumiyoshi, H., Kitamura, S. & Nagaya, N. (2007) Mesenchymal stem cells attenuate cardiac fibroblast proliferation and collagen synthesis through paracrine actions, *FEBS Lett. 581*, 3961–6.
129. Chen, L., Tredget, E. E., Wu, P. Y. & Wu, Y. (2008) Paracrine factors of mesenchymal stem cells recruit macrophages and endothelial lineage cells and enhance wound healing, *PLoS One. 3*, e1886.
130. Oh, J. Y., Kim, M. K., Shin, M. S., Lee, H. J., Ko, J. H., Wee, W. R. & Lee, J. H. (2008) The anti-inflammatory and anti-angiogenic role of mesenchymal stem cells in corneal wound healing following chemical injury, *Stem Cells. 26*, 1047–55.
131. Estrada, R., Li, N., Sarojini, H., An, J., Lee, M. J. & Wang, E. (2009) Secretome from mesenchymal stem cells induces angiogenesis via Cyr61, *J Cell Physiol. 219*, 563–71.
132. Gruber, R., Kandler, B., Holzmann, P., Vogele-Kadletz, M., Losert, U., Fischer, M. B. & Watzek, G. (2005) Bone marrow stromal cells can provide a local environment that favors migration and formation of tubular structures of endothelial cells, *Tissue Eng. 11*, 896–903.

133. Kinnaird, T., Stabile, E., Burnett, M. S., Lee, C. W., Barr, S., Fuchs, S. & Epstein, S. E. (2004) Marrow-derived stromal cells express genes encoding a broad spectrum of arteriogenic cytokines and promote in vitro and in vivo arteriogenesis through paracrine mechanisms, *Circ Res. 94*, 678–85.
134. Kamihata, H., Matsubara, H., Nishiue, T., Fujiyama, S., Tsutsumi, Y., Ozono, R., Masaki, H., Mori, Y., Iba, O., Tateishi, E., Kosaki, A., Shintani, S., Murohara, T., Imaizumi, T. & Iwasaka, T. (2001) Implantation of bone marrow mononuclear cells into ischemic myocardium enhances collateral perfusion and regional function via side supply of angioblasts, angiogenic ligands, and cytokines, *Circulation. 104*, 1046–52.
135. Caplan, A. I. (2005) Review: mesenchymal stem cells: cell-based reconstructive therapy in orthopedics, *Tissue Eng. 11*, 1198–211.
136. Tang, Y. L., Zhu, W., Cheng, M., Chen, L., Zhang, J., Sun, T., Kishore, R., Phillips, M. I., Losordo, D. W. & Qin, G. (2009) Hypoxic preconditioning enhances the benefit of cardiac progenitor cell therapy for treatment of myocardial infarction by inducing CXCR4 expression, *Circ Res. 104*, 1209–16.
137. Ma, T., Grayson, W. L., Frohlich, M. & Vunjak-Novakovic, G. (2009) Hypoxia and stem cell-based engineering of mesenchymal tissues, *Biotechnol Prog. 25*, 32–42.
138. Wang, J. A., Chen, T. L., Jiang, J., Shi, H., Gui, C., Luo, R. H., Xie, X. J., Xiang, M. X. & Zhang, X. (2008) Hypoxic preconditioning attenuates hypoxia/reoxygenation-induced apoptosis in mesenchymal stem cells, *Acta Pharmacol Sin. 29*, 74–82.
139. Rosova, I., Dao, M., Capoccia, B., Link, D. & Nolta, J. A. (2008) Hypoxic preconditioning results in increased motility and improved therapeutic potential of human mesenchymal stem cells, *Stem Cells. 26*, 2173–82.
140. Mylotte, L. A., Duffy, A. M., Murphy, M., O'Brien, T., Samali, A., Barry, F. & Szegezdi, E. (2008) Metabolic flexibility permits mesenchymal stem cell survival in an ischemic environment, *Stem Cells. 26*, 1325–36.
141. Hu, X., Yu, S. P., Fraser, J. L., Lu, Z., Ogle, M. E., Wang, J. A. & Wei, L. (2008) Transplantation of hypoxia-preconditioned mesenchymal stem cells improves infarcted heart function via enhanced survival of implanted cells and angiogenesis, *J Thorac Cardiovasc Surg. 135*, 799–808.
142. Hare, J. M., Traverse, J. H., Henry, T. D., Dib, N., Strumpf, R. K., Schulman, S. P., Gerstenblith, G., DeMaria, A. N., Denktas, A. E., Gammon, R. S., Hermiller, J. B., Jr., Reisman, M. A., Schaer, G. L. & Sherman, W. (2009) A randomized, double-blind, placebo-controlled, dose-escalation study of intravenous adult human mesenchymal stem cells (prochymal) after acute myocardial infarction, *J Am Coll Cardiol. 54*, 2277–86.

Bone Marrow Cell Therapy After Myocardial Infarction: What have we Learned from the Clinical Trials and Where Are We Going?

Kai C. Wollert

Abstract Modern reperfusion strategies and advances in pharmacological management have resulted in an increasing proportion of patients surviving after an acute myocardial infarction (AMI). Cell transplantation has been developed as a potential treatment for these patients. Multiple clinical trials have been conducted with various cell types to restore function in AMI patients. Although some clinical trials have provided a signal that cell therapy improves cardiac function after AMI, parallel experimental investigations of the mechanisms involved have shattered the concept of adult stem cell plasticity and have highlighted instead the importance of paracrine effects. The data that have created this scientific roller-coaster ride, which has generated excitement and confusion, are reviewed in this chapter.

Keywords Myocardial infarction • Bone marrow stem cells • Clinical trials • Bone marrow cells • Heart failure • Cell therapy

1 Introduction

Modern reperfusion strategies and advances in pharmacological management have resulted in an increasing proportion of patients surviving after an acute myocardial infarction (AMI). Unfortunately, this decrease in early mortality has resulted in an increased incidence of chronic heart failure (HF) in the survivors [1]. Because none of our current therapies address the central problem in ischemic HF, i.e. the massive loss of cardiomyocytes, vascular cells, and interstitial cells, these patients continue to experience frequent hospitalizations and premature death [2].

Cell transplantation was conceptualized more than 10 years ago as a means to augment myocyte numbers and to improve cardiac function after AMI [3, 4].

K.C. Wollert (✉)
Hans-Borst Center for Heart and Stem Cell Research, Department of Cardiology and Angiology, Hannover Medical School, Hannover, Germany
e-mail: wollert.kai@mh-hannover.de

I.S. Cohen and G.R. Gaudette (eds.), *Regenerating the Heart*, Stem Cell Biology and Regenerative Medicine, DOI 10.1007/978-1-61779-021-8_8,
© Springer Science+Business Media, LLC 2011

Regeneration of the infarcted heart is a daunting task, however, considering (1) that the myocyte deficiency in infarction-induced human HF may be on the order of one billion cardiac myocytes [5], (2) that not only cardiomyocytes but also supporting (e.g. vascular) cells have to be supplied, and (3) that the environmental cues that are required to guide transplanted cells into multicellular 3D heart structures may be lacking in infarcted and scarred myocardium [6]. In many of the early experimental studies, fetal, neonatal, and adult cardiomyocytes were transplanted and were shown to form stable grafts in injured hearts [7]. Because of their limited availability, differentiated cardiomyocytes are not a realistic cell source for large-scale clinical applications, however.

Because of their capacities for self-renewal, infinite ex vivo proliferation, and differentiation into mature specialized cells, stem cells have now emerged as the prime cell source for regenerative therapies. Pluripotent stem cells can differentiate into cells derived from all three germ layers, a typical example being embryonic stem cells that can be isolated from the inner cell mass of blastocysts and that can give rise to all cardiac cell types. Adult stem cells, by contrast, are multipotent and restricted in their differentiation potential to cell lineages of the organ in which they are located [e.g. hematopoietic stem cells (HSCs) giving rise to mature hematopoietic cells or mesenchymal stem cells (MSCs) giving rise to osteoblasts, chondrocytes, and adipocytes]. Progenitor cells are even more restricted in their differentiation potential and have a limited capacity to self-renew [e.g. endothelial progenitor cells (EPCs) or skeletal myoblasts]. Experiments conducted at the beginning of this millennium appeared to challenge the traditional concept that adult stem and progenitor cells are lineage-restricted and suggested instead that these cells can transdifferentiate into cell types outside their original lineage (e.g. HSCs differentiating into neurons or cardiomyocytes) [8, 9]. This new concept of adult stem cell "plasticity," combined with a large body of animal data indicating that adult stem and progenitor cell transplantation can improve the contractile function of the infarcted heart, provided the scientific rationale to treat cardiac patients with adult stem and progenitor cells [10].

Although culture-expanded skeletal myoblasts were the first cell type to enter the clinic, most subsequent trials have used unfractionated bone marrow cells (BMCs) as an easily accessible source of HSCs, MSCs, and EPCs. The field has made rapid progress. Although some clinical trials have provided a signal that BMC therapy improves cardiac function after AMI, parallel experimental investigations of the mechanisms involved have shattered the concept of adult stem cell plasticity and have highlighted instead the importance of paracrine effects. This scientific roller-coaster ride has generated excitement and confusion. Confusion is caused also by the lack of a universally accepted nomenclature for stem and progenitor cells, by the occasional imprecise use of terminology (e.g. "stem cell" therapy instead of BMC therapy), by the large number of different cell types that are undergoing clinical testing, and by a lack of standardization in cell isolation protocols that would facilitate a comparison of clinical trial results from different institutions. Here, we use the cell nomenclature that has been used by the respective investigators. The reader is encouraged, however, to go back to the original publications and learn more about the specific cell isolations protocols.

2 What Have We Learned from the First Clinical Trials?

We will start our discussion with a brief review of the largest randomized-controlled clinical trials that have explored the prospects of BMC therapy after myocardial infarction. With the exception of the TOPCARE-CHD trial, which included patients with ischemic HF, all other randomized trials have been performed in patients with AMI (Table 1).

In the BOOST trial, 60 patients were randomized to intracoronary nucleated BMC (nBMC) transfer, on average 4.8 days after acute percutaneous coronary intervention (PCI), or to a control group in which neither bone aspiration nor placebo injection was performed. Cardiac function was assessed in a blinded fashion using magnetic resonance imaging (MRI) at serial time points before cell transfer, and after 6, 18, and 61 months. As compared with the control group, the nBMC group showed an enhanced left ventricular (LV) contractility in the infarct border zone, and a significant improvement of global LV ejection fraction (LVEF) by 6.0 percentage points after 6 months [11]. In the overall study cohort, the differences in LVEF between the nBMC and control groups were diminished and no longer statistically significant after 18 months (2.8 percentage points) and 61 months (0.8 percentage points) [12, 13]. In a post hoc analysis, patients with greater infarcts and an infarct transmurality greater than the median appeared to benefit from nBMC transfer, with a sustained improvement of LVEF also at the later time points [12, 13]. After 6 and 18 months, the control group had developed echocardiographic signs of mild diastolic dysfunction, which were significantly attenuated in the nBMC group [14].

The Leuven AMI trial randomized 67 patients to intracoronary mononucleated BMC (mnBMC) or placebo infusion within 24 h after acute PCI [15]. MRI assessment of LVEF at 3–4 days and 4 months after PCI did not demonstrate a significant impact of mnBMC therapy on LVEF recovery, the primary end point of the trial. Notably, however, the reduction of infarct volume after 4 months, as measured by serial contrast-enhanced MRI, was greater in mnBMC-treated patients than in controls [15]. Moreover, a significant improvement in regional contractility was observed by MRI in the mnBMC group, with the greatest infarct transmurality at the baseline [15]. Echocardiographic strain rate imaging confirmed that mnBMC infusions improved the recuperation of myocardial function in the infarct region, suggesting that quantitative assessment of regional systolic function may be more sensitive than measuring global LVEF for the evaluation of cell therapy after myocardial infarction [16].

In the REPAIR-AMI trial, 204 patients were randomized to an intracoronary infusion of mnBMCs or a placebo, on average 4.4 days after acute PCI. LV function was assessed by contrast angiography. mnBMC infusion promoted an increase in LVEF of 2.5 percentage points after 4 months as compared with the control group [17]. In a subgroup of 54 patients who underwent serial MRI investigations, the treatment effect of mnBMC infusion on LVEF amounted to 2.8 percentage points ($P=0.26$) at 12 months [18], similar to what was observed in the BOOST trial after 18 months [12].

Table 1 Randomized bone marrow cell therapy trials after myocardial infarction

Study	Patients	Cell type	Dose (mL)	Timing of cell delivery	Primary end point
Acute myocardial infarction					
BOOST	60	nBMC	128	Day 6 ± 1	LVEF ↑
REPAIR-AMI	187	mnBMC	50	Days 3–6	LVEF ↑
Leuven-AMI	66	mnBMC	130	Day 1	LVEF ↔
ASTAMI	97	mnBMC	50	Day 6 ± 1	LVEF ↔
FINCELL	77	mnBMC	80	Day 3	LVEF ↑
REGENT	117	mnBMC (unselected vs. selected)	50–70 (unselected) 100–120 (selected)	Days 3–12	LVEF ↑ with both cell types
HEBE	189	mnBMC vs. mnPBC		Days 3–8	Regional contractility ↔
Ischemic heart failure					
TOPCARE-CHD	58	mnBMC vs. CPC	50	Month 81 ± 72	LVEF ↑ (mnBMC) LVEF ↔ (CPC)

Only patients with complete imaging studies are considered here. Dose refers to the average amount of bone marrow that was harvested. Cells were delivered intracoronarily in all studies. See the text for details and references

nBMC nucleated bone marrow cells, *mnBMC* mononucleated bone marrow cells, *mnPBC* mononucleated peripheral blood cells, *CPC* circulating-blood-derived progenitor cells, *LVEF* left ventricular ejection fraction, ↑ improved, ↔ no significant change

In the ASTAMI trial, 100 patients with anterior myocardial infarction were randomized to intracoronary infusion of mnBMCs or a control group in which neither bone aspiration nor placebo injection was performed. Cells were infused, on average, 6 days after acute PCI. After 6 and 12 months, no significant effects of mnBMC therapy on LVEF, LV volumes, or infarct size were observed by single-photon-emission computed tomography, echocardiography, and MRI [19, 20]. Fewer mnBMCs were infused in the ASTAMI trial as compared with the REPAIR-AMI trial (median, 68×10^6 vs. 198×10^6 cells) [17, 19]. It has also been proposed that the cell isolation protocol used in the ASTAMI trial may have recovered a mnBMC population with impaired functionality, as assessed by in vitro migratory and colony-forming capacities, as well as in vivo capacity to promote blood flow recovery in a mouse model of hind-limb ischemia [21].

The HEBE trial randomized 200 patients with AMI to an intracoronary infusion of mnBMCs or mononucleated peripheral blood cells (mnPBCs) or to a control group that did not undergo bone marrow aspiration or placebo injection [22]. The final results of the HEBE trial showed that intracoronary infusion of mnBMCs or mnPBCs did not improve global or regional LV systolic function at 4 months as assessed by MRI [23]. The reason(s) for these negative findings are not clear at the moment.

In the FINCELL trial, 80 patients with myocardial infarction treated with thrombolytic therapy followed by PCI were randomized to intracoronary mnBMC or placebo infusions. Cells were infused immediately after PCI, which was performed 2–6 days after thrombolysis. As shown by LV contrast angiography before and 6 months after cell transfer, mnBMC infusions improved LVEF recovery by 5.0 percentage points as compared with the control group. Paired echocardiographic investigations yielded similar results [24].

In the REGENT trial, 200 patients with anterior myocardial infarction were randomized to an intracoronary infusion of unselected mnBMCs or $CXCR4^+$ $CD34^+$ mnBMCs, on average 7 days after acute PCI, or to a control group in which neither bone aspiration nor placebo injection was performed [25]. Paired MRI images to assess LVEF at the baseline and after 6 months were available in 117 patients. Significant improvements in LVEF from the baseline to follow-up were noted within the unselected and selected mnBMC groups (3.0 percentage points, each), but not in the control group (no change in LVEF). However, changes in LVEF between the groups, the primary end point of the trial, were not significantly different between the mnBMC and control groups. This trial is limited by imbalances in baseline LVEF and incomplete follow-up. Nevertheless, the REGENT trial indicates that a specific BMC population expressing progenitor cell surface markers may carry much of the treatment effect observed after mnBMC transfer in post-AMI patients.

In the TOPCARE-CHD trial, 75 patients with ischemic HF were randomized to receive no cell infusion or intracoronary infusions of mnBMC or circulating-blood-derived progenitor cells into the patent coronary artery supplying the most dyskinetic LV area. To obtain circulating-blood-derived progenitor cells, mononucleated cells were isolated from the peripheral blood by Ficoll density gradient centrifugation and cultured ex vivo in medium containing vascular endothelial growth factor, a statin, and autologous serum. After 3 months, the absolute change in LVEF, as assessed by contrast angiography, was significantly greater among patients receiving mnBMCs (+2.9 percentage points) than among those receiving circulating-blood-derived progenitor cells (–0.4 percentage points) or no cell infusion (–1.2 percentage points) [26].

So far, no safety concerns related to intracoronary BMC infusions have emerged. An increased risk of in-stent restenosis has been observed in a small nonrandomized study after intracoronary infusion of $CD133^+$ mnBMCs [27]. In the placebo-controlled FINCELL trial, no increased risk of in-stent restenosis was observed by intravascular ultrasound after 6 months [24]. In two meta-analyses, the risks of target vessel restenosis or repeat revascularization were not increased in cell-treated patients [28, 29]. Moreover, none of the clinical trials reported an increased incidence of symptomatic arrhythmias after intracoronary BMC transfer. An electrophysiological study performed in the BOOST trial [11] and a careful assessment of microvolt T-wave alternans and signal-averaged ECG measures in the FINCELL trial [24] provide further assurance in this respect.

3 What Can We Expect from Ongoing Clinical Trials?

Considering the heterogeneity in cell isolation protocols, trial design, and the methods used for outcome measurement, and given the fact that autologous cell preparations represent a medical product whose complexity far exceeds that of any drug that is currently prescribed to patients with coronary heart disease, it may not be surprising that mixed results have emerged from the first clinical trials in this emerging field. In the setting of AMI, the fairly large database provides encouraging evidence that intracoronary BMC transfer after AMI has the potential to enhance the recovery of LV systolic function beyond what can be achieved by current interventional and medical therapies. This conclusion is supported by two meta-analyses of randomized and non-randomized controlled clinical trials of intracoronary BMC transfer in patients with AMI [28, 29]. In both of these analyses, cell transfer was associated with a significant improvement in global LVEF by 3.0 percentage points that was accompanied by a significant reduction in LV end-systolic volume and a nonsignificant reduction in LV end-diastolic volume, suggesting that cell transfer exerts a beneficial effect on LV systolic function and remodeling [28, 29]. It has been pointed out that the magnitude of these effects is quite comparable to what is achieved by established therapies after AMI, including acute PCI and ACE inhibitor or β-blocker therapy [30]. The effects of BMC transfer on LV function and remodeling beyond an observation period of about 4–6 months remain poorly characterized at this time, emphasizing the need to obtain long-term follow-up data in clinical trials [13, 31].

Because of the favorable safety profile and the promising efficacy data, several clinical trials are under way to further explore the prospects of cell therapy after myocardial infarction. As discussed below, important issues are addressed in these trials in an attempt to maximize patient benefit (Table 2). Considering that previous clinical trials used apparently similar cell isolation protocols yet observed largely different outcomes [17, 19, 23, 24], it will be critical to establish assays to assess cell functionality and the quality of the cell product. This, however, requires a better understanding of which cellular functions determine clinical benefit.

3.1 Patient Selection

For safety reasons, initial studies in patients with AMI have included mostly individuals with moderately depressed baseline LVEF. These patients, however, have a favorable prognosis and may not be in need of cell therapy [32]. Moreover, it appears that patients with more extensive infarct damage benefit more from intracoronary BMC transfer in terms of LVEF improvement; these patients have been identified by more severely depressed baseline LVEF or stroke volumes [17, 25, 26], or a greater transmural extent of the infarct [12, 13, 15]. Many of the ongoing trials focus on these higher-risk patients. Conversely, the presence of microvascular obstruction in the reperfused infarct territory, as identified by late-enhancement MRI, may identify a patient population that does not respond to intracoronary BMC therapy (Stefan Janssens, personal communication) [15].

3.2 Procedural Details

Procedural details such as the timing of cell transfer, cell dose, cell type and cell isolation protocols, and the mode of cell delivery need to be tailored to the specific disease setting, resulting in hundreds of possible permutations, thus highlighting the complexity of optimizing cell therapy protocols.

The success of BMC therapy after AMI may critically depend on the timing of cell transfer and cell dose. As indicated by a subgroup analysis of the REPAIR-AMI trial and one meta-analysis, intracoronary BMC transfer in the first days after AMI may be associated with less improvement in LVEF as compared with later cell delivery [17, 29]. The same meta-analysis also suggested that the improvement in LVEF may correlate with the administered BMC dose [29]. Results from a small randomized-controlled clinical trial comparing the effects of low-dose and high-dose mnBMC transfer after AMI support the hypothesis that there is a dose–response relationship. This hypothesis is currently being addressed in the BOOST 2 trial.

Table 2 Ongoing bone marrow cell therapy trials after myocardial infarction

Study identifier	Principal investigator(s)	Acronym	Patients	Cells	Primary end point	Route of cell delivery
Non-ST-elevation acute coronary syndrome						
ClinicalTrials NCT00711542	Andreas M Zeiher	REPAIR-ACS	100	Bone-marrow-derived progenitor cells	Coronary flow reserve	Intracoronary
Acute myocardial infarction						
ControlledTrials ISRCTN17457407	Helmut Drexler Kai C Wollert	BOOST 2	200	Bone marrow cells Low vs. high cell number Nonirradiated vs. irradiated cells	LVEF	Intracoronary
ClinicalTrials NCT00355186	Roberto Corti	SWISS-AMI	150	Bone-marrow-derived stem cells	LVEF	Intracoronary Days 5–7 vs. days 21–28
ClinicalTrials NCT00684021	Carl Pepine et al.	TIME	120	Bone marrow mononuclear cells	LVEF	Intracoronary Day 3 vs. day 7 after myocardial infarction
ClinicalTrials NCT00684060	Carl Pepine et al.	Late TIME	87	Bone marrow mononuclear cells	LVEF	Intracoronary 2–3 weeks after myocardial infarction
ClinicalTrials NCT00501917	Hyo-Soo Kim	MAGIC Cell-5	116	Peripheral blood stem cells mobilized with G-CSF vs. G-CSF and darbepoetin	LVEF	Intracoronary
ClinicalTrials NCT00877903	Anthony DeMaria et al.		220	Allogeneic mesenchymal stem cells	LVESV	Intravenous
ClinicalTrials NCT00677222	Marc Penn Warren Sherman		28	Allogeneic mesenchymal stem cells	Safety	Perivascular
ClinicalTrials NCT00442806	Patrick Serruys	APOLLO	48	Adipose-tissue-derived stem cells	Safety	

Ischemic heart failure						
ClinicalTrials NCT00824005	Carl Pepine et al.	FOCUS	87	Bone marrow mononuclear cells	MVO_2, LVESV Ischemic area	Transendocardial
ClinicalTrials NCT00747708	Anthony Mathur	REGENERATE-IHD	165	G-CSF-stimulated bone-marrow-derived stem/progenitor cells	LVEF	Transendocardial vs. intracoronary
ClinicalTrials NCT00326989	Andreas M Zeiher	Cellwave	100	Bone marrow mononuclear cells	LVEF	Extracorporal shock wave, then intracoronary cell therapy
ClinicalTrials NCT00285454	Eric W Alton		60	Bone marrow mononuclear cells	Safety, perfusion Systolic function	Retrograde coronary venous delivery
ClinicalTrials NCT00462774	Boris Nasseri Christof Stamm	Cardio133	60	$CD133^+$ bone marrow cells	LVEF	Transepicardial during CABG
ClinicalTrials NCT00810238	Jozef Bartunek André Terzic	C-Cure	240	Bone-marrow-derived cardiopoietic cells	LVEF	Transendocardial
ClinicalTrials NCT00768066	Joshua M Hare et al.	TAC-HFT	60	Bone marrow cells vs. mesenchymal stem cells	Safety	Transendocardial
ClinicalTrials NCT00644410	Jens Kastrup		60	Mesenchymal stem cells	LVEF	Transendocardial
ClinicalTrials NCT00587990	Joshua M Hare et al.	PROMETHEUS	45	Mesenchymal stem cells	Safety	Transepicardial during CABG
ClinicalTrials NCT00721045	Nabil Dib et al.		60	Allogeneic mesenchymal precursor cells	Safety	Transendocardial
ClinicalTrials NCT00474461	Roberto Bolli		40	Cardiac stem cells harvested from right atrial appendage	Safety	Intracoronary

Unless otherwise stated, autologous cell sources are used

G-CSF granulocyte colony-stimulating factor, *LVEF* left ventricular ejection fraction, *LVESV* left ventricular end-systolic volume, MVO_2 maximal oxygen consumption, *CABG* coronary artery bypass grafting

Head-to-head comparisons of cell delivery strategies (REGENERATE-IHD trial) and cell types (TAC-HF trial), exploration of alternative cell delivery methods (e.g. transcoronary venous infusion or transcoronary arterial injection into the perivascular space), and improvements in intramyocardial injection needle design (e.g. needles which limit immediate washout and promote cell dispersion) may help to optimize existing cell therapy protocols (Table 2) [33–36]. Progress can also be expected from the use of more comprehensive imaging techniques that help to characterize the target tissue (especially in patients with advanced and chronic disease) and to facilitate delivery of cells to tissue sites on the basis of their physiological characteristics and anatomic location [36, 37]. As recently shown in a swine model of myocardial infarction of different ages, fusion of 3D magnetic resonance images with 2D fluoroscopic images can be used to precisely target transendocardial cell injections to the remote, peri-infarct, and infarct locations. This technique can be applied without the need for a combined X-ray/MRI suite and may be combined with electroanatomic mapping, if desired [37].

State-of-the art imaging techniques and end point evaluation by external core laboratories are required to unequivocally demonstrate moderate functional effects of cell therapy. LV dimensions and systolic function, for example, should be evaluated by MRI rather than echocardiography or angiography.

3.3 Clinical End Points

Previous trials have not been powered to assess the impact of cell therapy on mortality and other clinical end points. In the REPAIR-AMI trial, the cumulative end point of death, recurrent myocardial infarction, or necessity for revascularization was significantly reduced in the mnBMC group as compared with the placebo group after 12 months. Likewise, the combined end point of death, AMI, and HF hospitalizations was significantly reduced after mnBMC transfer [38]. In a recently completed study in AMI patients, intracoronary mnBMC transfer was associated with a significant reduction of all-cause mortality after 5 years [39]. It should be noted that patients who refused cell treatment served as controls in this nonrandomized study [39]. Trends in favor of BMC therapy with regard to the end points of death, risk of recurrent AMI, and HF hospitalizations have also emerged from the meta-analyses [28, 29]. Ultimately, outcome trials will have to be conducted.

3.4 Strategies to Address the Current Limitations of Cardiac Cell Therapy

Current cell therapy strategies are limited by low rates of cell engraftment after intracoronary delivery [36, 40, 41], and by poor cell survival after intramyocardial injections [5, 42]. Moreover, advanced age, cardiovascular risk factors (in particular diabetes), and HF appear to have a negative impact on the number and functional activity of bone-marrow-derived and blood-derived progenitor cells [43–46]. EPCs and mnBMCs isolated from patients with diabetes or HF, for example, display reduced activities in promoting reendothelialization of denuded arteries and blood flow recovery after ischemia when transplanted into nude mice [46–50]. The exact functional deficits that account for these reduced in vivo activities remain poorly characterized. However, markers of reduced functionality, such as reduced in vitro migratory or colony-forming capacities of mnBMCs, have been associated with reduced clinical benefit in cell therapy trials [51, 52]. Accordingly, cell enhancement strategies may be required to realize the full therapeutic potential of cell therapy [42, 53]. Some cell enhancement strategies are already being explored in clinical trials (Table 2). On the basis of experimental studies showing that tissue preconditioning with low-energy shock waves enhances EPC recruitment into ischemic tissues [54], the Cellwave trial combines extracorporal shock wave treatment with intracoronary mnBMC transfer in patients with ischemic HF. In the C-Cure trial, a combination of growth factors is used to stimulate the expression of cardiomyocyte genes in MSCs prior to transplantation in patients with ischemic HF.

4 Future Directions

Although bone-marrow-derived and blood-derived cells appear to have a favorable impact on systolic function and remodeling of the infarcted heart, it is increasingly being recognized that these cells cannot replenish lost cardiomyocytes and vascular cells to a meaningful extent. It has been proposed that human bone marrow contains multipotent stem cell population(s) with the potential to differentiate into cells that express vascular and cardiomyocyte markers; isolation of these rare stem cells, however, requires serial culture steps and clonal expansion [55, 56]. Genetic fate mapping studies indicate that freshly isolated unfractionated BMCs fail to transdifferentiate into cardiomyocytes when transplanted into infarcted mouse hearts [57]. The benefits of MSC transplantation in rodent infarct models are independent of the differentiation of these cells into cardiomyocytes [58–61]. Moreover, it has become clear that only a small subpopulation of culture-expanded EPCs behave as true progenitor cells that can differentiate into mature endothelial cells in situ [62]. A large body of evidence suggests that the beneficial effects mediated by these cell types in cardiac injury models are related, instead, to a secretion of soluble factors acting in a paracrine manner [63]. In line with this conclusion, human peripheral-blood-derived EPCs and nBMCs have been found to secrete large arrays of bioactive molecules that are distinct from the secretory profiles of other cell types such as blood leukocytes or fibroblasts [64, 65]. Conceptually, paracrine factors may exert their actions via several mechanisms, including myocardial protection, neovascularization, modulation of inflammatory and fibrogenic processes, cardiac metabolism, and cardiac contractility, enhancement of cardiomyocyte proliferation, and activation of resident stem and progenitor cells. The relative importance of these proposed paracrine actions will depend on the age of the infarct (e.g. direct cytoprotective effects may be more important early after reperfusion). Cytoprotective and proangiogenic effects have been studied most extensively in experimental models [63]. Cytoprotective effects may salvage cardiomyocytes at risk and lead to a reduction in infarct size [66–68]. Increased angiogenesis has been postulated to improve infarct healing, energy metabolism, and contractility in the infarct border zone [55, 69–71]. Notably, cell transplantation may induce secondary humoral effects in the infarcted heart, which are sustained by the host tissue after the transplanted cells have been eliminated [68].

In the clinical setting, data from the REPAIR-AMI trial indicate that intracoronary mnBMC transfer leads to an improvement in regional microvascular function and tissue perfusion [72]. The BOOST investigators have shown that conditioned nBMC supernatants promote proangiogenic effects in cultured human coronary artery endothelial cells and protect cultured cardiomyocytes from ischemia/reperfusion-induced apoptosis [65]. With use of ProteinChip and GeneChip array analyses, nBMC were shown to secrete more than 100 soluble factors, some of them with known proangiogenic and cytoprotective activities [65]. Although these data indicate that intracoronary infusion of BMCs delivers a cocktail of cytokines and growth factors to the infarcted heart, experimental studies suggest that individual soluble factors, when applied at sufficient dosages, may carry much of the therapeutic effects. In this regard, the Wnt signaling modulator secreted frizzled-related protein 2 (Sfrp-2) has been shown to play a key role in mediating the cytoprotective effects of Akt-transduced rat MSCs [73], and interleukin-10 has been found to contribute significantly to the antiremodeling effects of mouse mnBMCs [74].

The human genome encodes more than 1,400 secreted proteins, many with as yet unknown biological functions [75]. A comprehensive functional analysis of the BMC, MSC, or EPC secretomes might therefore lead to the identification of paracrine factors with therapeutic potential after AMI, e.g. factors with cytoprotective and/or proangiogenic activities. As it has recently been proposed that individual soluble factors can stimulate cardiomyocyte proliferation or reactivate tissue-resident (epicardial) progenitor cells in the adult heart [76, 77], secretome analyses may also lead to the identification of soluble factors promoting tissue regeneration. These efforts may eventually enable therapeutic approaches based on the application of specific paracrine factors (Fig. 1). The most obvious challenge for protein therapy is the necessity to maintain therapeutic concentrations for the necessary length of time [63]. New strategies are emerging to address this problem and to allow sustained therapeutic delivery of recombinant proteins [78–80].

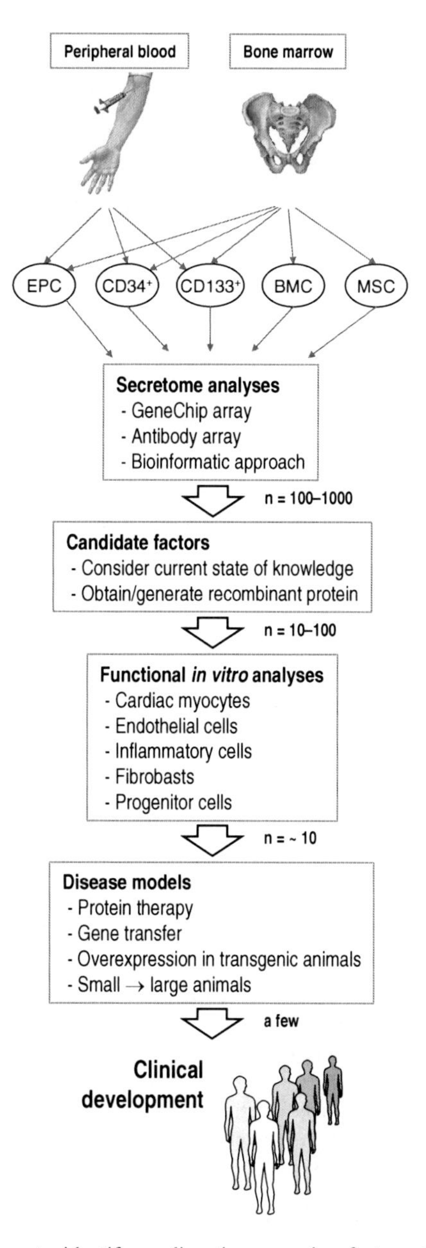

Fig. 1 Secretome analyses to identify cardioactive paracrine factors. Genome-wide screening coupled with bioinformatic approaches and functional analyses is used to identify paracrine factors with therapeutic potential in patients with cardiovascular disease. *EPC* endothelial progenitor cells, *BMC* bone marrow cells, *MSC* mesenchymal stem cells, *n* the numbers of factors entering the next step of exploration

References

1. Velagaleti RS, Pencina MJ, Murabito JM, Wang TJ, Parikh NI, D'Agostino RB, Levy D, Kannel WB, Vasan RS. Long-term trends in the incidence of heart failure after myocardial infarction. Circulation. 2008;118:2057–2062.
2. McMurray JJ, Pfeffer MA. Heart failure. Lancet. 2005;365:1877–1889.
3. Soonpaa MH, Koh GY, Klug MG, Field LJ. Formation of nascent intercalated disks between grafted fetal cardiomyocytes and host myocardium. Science. 1994;264:98–101.
4. Taylor DA, Atkins BZ, Hungspreugs P, Jones TR, Reedy MC, Hutcheson KA, Glower DD, Kraus WE. Regenerating functional myocardium: improved performance after skeletal myoblast transplantation. Nat Med. 1998;4:929–933.
5. Robey TE, Saiget MK, Reinecke H, Murry CE. Systems approaches to preventing transplanted cell death in cardiac repair. J Mol Cell Cardiol. 2008;45:567–581.
6. Chien KR, Domian IJ, Parker KK. Cardiogenesis and the complex biology of regenerative cardiovascular medicine. Science. 2008;322:1494–1497.
7. Dowell JD, Rubart M, Pasumarthi KB, Soonpaa MH, Field LJ. Myocyte and myogenic stem cell transplantation in the heart. Cardiovasc Res. 2003;58:336–350.
8. Blau HM, Brazelton TR, Weimann JM. The evolving concept of a stem cell: entity or function? Cell. 2001;105:829–841.
9. Wagers AJ, Weissman IL. Plasticity of adult stem cells. Cell. 2004;116:639–648.
10. Wollert KC, Drexler H. Clinical applications of stem cells for the heart. Circ Res. 2005;96: 151–163.
11. Wollert KC, Meyer GP, Lotz J, Ringes-Lichtenberg S, Lippolt P, Breidenbach C, Fichtner S, Korte T, Hornig B, Messinger D, Arseniev L, Hertenstein B, Ganser A, Drexler H. Intracoronary autologous bone-marrow cell transfer after myocardial infarction: the BOOST randomised controlled clinical trial. Lancet. 2004;364:141–148.
12. Meyer GP, Wollert KC, Lotz J, Steffens J, Lippolt P, Fichtner S, Hecker H, Schaefer A, Arseniev L, Hertenstein B, Ganser A, Drexler H. Intracoronary bone marrow cell transfer after myocardial infarction: eighteen months' follow-up data from the randomized, controlled BOOST (BOne marrOw transfer to enhance ST-elevation infarct regeneration) trial. Circulation. 2006;113:1287–1294.
13. Meyer GP, Wollert KC, Lotz J, Pirr J, Rager U, Lippolt P, Hahn A, Fichtner S, Schaefer A, Arseniev L, Ganser A, Drexler H. Intracoronary bone marrow cell transfer after myocardial infarction: 5-year follow-up from the randomized-controlled BOOST trial. Eur Heart J. 2009;30(24):2978–2984 [Epub ahead of print]
14. Schaefer A, Meyer GP, Fuchs M, Klein G, Kaplan M, Wollert KC, Drexler H. Impact of intracoronary bone marrow cell transfer on diastolic function in patients after acute myocardial infarction: results from the BOOST trial. Eur Heart J. 2006;27:929–935.
15. Janssens S, Dubois C, Bogaert J, Theunissen K, Deroose C, Desmet W, Kalantzi M, Herbots L, Sinnaeve P, Dens J, Maertens J, Rademakers F, Dymarkowski S, Gheysens O, Van Cleemput J, Bormans G, Nuyts J, Belmans A, Mortelmans L, Boogaerts M, Van de Werf F. Autologous bone marrow-derived stem-cell transfer in patients with ST-segment elevation myocardial infarction: double-blind, randomised controlled trial. Lancet. 2006;367:113–121.
16. Herbots L, D'Hooge J, Eroglu E, Thijs D, Ganame J, Claus P, Dubois C, Theunissen K, Bogaert J, Dens J, Kalantzi M, Dymarkowski S, Bijnens B, Belmans A, Boogaerts M, Sutherland G, Van de Werf F, Rademakers F, Janssens S. Improved regional function after autologous bone marrow-derived stem cell transfer in patients with acute myocardial infarction: a randomized, double-blind strain rate imaging study. Eur Heart J. 2009;30:662–670.
17. Schachinger V, Erbs S, Elsasser A, Haberbosch W, Hambrecht R, Holschermann H, Yu J, Corti R, Mathey DG, Hamm CW, Suselbeck T, Assmus B, Tonn T, Dimmeler S, Zeiher AM. Intracoronary bone marrow-derived progenitor cells in acute myocardial infarction. N Engl J Med. 2006;355:1210–1221.
18. Dill T, Schachinger V, Rolf A, Mollmann S, Thiele H, Tillmanns H, Assmus B, Dimmeler S, Zeiher AM, Hamm C. Intracoronary administration of bone marrow-derived progenitor cells

improves left ventricular function in patients at risk for adverse remodeling after acute ST-segment elevation myocardial infarction: results of the Reinfusion of Enriched Progenitor cells And Infarct Remodeling in Acute Myocardial Infarction study (REPAIR-AMI) cardiac magnetic resonance imaging substudy. Am Heart J. 2009;157:541–547.

19. Lunde K, Solheim S, Aakhus S, Arnesen H, Abdelnoor M, Egeland T, Endresen K, Ilebekk A, Mangschau A, Fjeld JG, Smith HJ, Taraldsrud E, Grogaard HK, Bjornerheim R, Brekke M, Muller C, Hopp E, Ragnarsson A, Brinchmann JE, Forfang K. Intracoronary injection of mononuclear bone marrow cells in acute myocardial infarction. N Engl J Med. 2006;355: 1199–1209.
20. Lunde K, Solheim S, Forfang K, Arnesen H, Brinch L, Bjornerheim R, Ragnarsson A, Egeland T, Endresen K, Ilebekk A, Mangschau A, Aakhus S. Anterior myocardial infarction with acute percutaneous coronary intervention and intracoronary injection of autologous mononuclear bone marrow cells: safety, clinical outcome, and serial changes in left ventricular function during 12-months' follow-up. J Am Coll Cardiol. 2008;51:674–676.
21. Seeger FH, Tonn T, Krzossok N, Zeiher AM, Dimmeler S. Cell isolation procedures matter: a comparison of different isolation protocols of bone marrow mononuclear cells used for cell therapy in patients with acute myocardial infarction. Eur Heart J. 2007;28:766–772.
22. van der Laan A, Hirsch A, Nijveldt R, van der Vleuten PA, van der Giessen WJ, Doevendans PA, Waltenberger J, Ten Berg JM, Aengevaeren WR, Zwaginga JJ, Biemond BJ, van Rossum AC, Tijssen JG, Zijlstra F, Piek JJ. Bone marrow cell therapy after acute myocardial infarction: the HEBE trial in perspective, first results. Neth Heart J. 2008;16:436–439.
23. Hirsch A, Nijveldt R, van der Vleuten PA, Tijssen JG, van der Giessen WJ, Tio RA, Waltenberger J, ten Berg JM, Doevendans PA, Aengevaeren WR, Zwaginga JJ, Biemond BJ, van Rossum AC, Piek JJ, Zijlstra F, and on behalf of the HEBE investigators. Intracoronary infusion of mononuclear cells from bone marrow or peripheral blood compared with standard therapy in patients after acute myocardial infarction treated by primary percutaneous coronary intervention: results of the randomized controlled HEBE trial. Eur Heart J. 2010; Epub ahead of print.
24. Huikuri HV, Kervinen K, Niemela M, Ylitalo K, Saily M, Koistinen P, Savolainen ER, Ukkonen H, Pietila M, Airaksinen JK, Knuuti J, Makikallio TH. Effects of intracoronary injection of mononuclear bone marrow cells on left ventricular function, arrhythmia risk profile, and restenosis after thrombolytic therapy of acute myocardial infarction. Eur Heart J. 2008;29:2723–2732.
25. Tendera M, Wojakowski W, Ruzyllo W, Chojnowska L, Kepka C, Tracz W, Musialek P, Piwowarska W, Nessler J, Buszman P, Grajek S, Breborowicz P, Majka M, Ratajczak MZ. Intracoronary infusion of bone marrow-derived selected CD34+ CXCR4+ cells and non-selected mononuclear cells in patients with acute STEMI and reduced left ventricular ejection fraction: results of randomized, multicentre Myocardial Regeneration by Intracoronary Infusion of Selected Population of Stem Cells in Acute Myocardial Infarction (REGENT) Trial. Eur Heart J. 2009;30:1313–1321.
26. Assmus B, Honold J, Schachinger V, Britten MB, Fischer-Rasokat U, Lehmann R, Teupe C, Pistorius K, Martin H, Abolmaali ND, Tonn T, Dimmeler S, Zeiher AM. Transcoronary transplantation of progenitor cells after myocardial infarction. N Engl J Med. 2006;355:1222–1232.
27. Mansour S, Vanderheyden M, De Bruyne B, Vandekerckhove B, Delrue L, Van Haute I, Heyndrickx G, Carlier S, Rodriguez-Granillo G, Wijns W, Bartunek J. Intracoronary delivery of hematopoietic bone marrow stem cells and luminal loss of the infarct-related artery in patients with recent myocardial infarction. J Am Coll Cardiol. 2006;47:1727–1730.
28. Lipinski MJ, Biondi-Zoccai GG, Abbate A, Khianey R, Sheiban I, Bartunek J, Vanderheyden M, Kim HS, Kang HJ, Strauer BE, Vetrovec GW. Impact of intracoronary cell therapy on left ventricular function in the setting of acute myocardial infarction: a collaborative systematic review and meta-analysis of controlled clinical trials. J Am Coll Cardiol. 2007;50:1761–1767.
29. Martin-Rendon E, Brunskill SJ, Hyde CJ, Stanworth SJ, Mathur A, Watt SM. Autologous bone marrow stem cells to treat acute myocardial infarction: a systematic review. Eur Heart J. 2008;29:1807–1818.

30. Reffelmann T, Konemann S, Kloner RA. Promise of blood- and bone marrow-derived stem cell transplantation for functional cardiac repair: putting it in perspective with existing therapy. J Am Coll Cardiol. 2009;53:305–308.
31. Cao F, Sun D, Li C, Narsinh K, Zhao L, Li X, Feng X, Zhang J, Duan Y, Wang J, Liu D, Wang H. Long-term myocardial functional improvement after autologous bone marrow mononuclear cells transplantation in patients with ST-segment elevation myocardial infarction: 4 years follow-up. Eur Heart J. 2009;30:1986–1994.
32. Moller JE, Hillis GS, Oh JK, Reeder GS, Gersh BJ, Pellikka PA. Wall motion score index and ejection fraction for risk stratification after acute myocardial infarction. Am Heart J. 2006;151:419–425.
33. Yokoyama S, Fukuda N, Li Y, Hagikura K, Takayama T, Kunimoto S, Honye J, Saito S, Wada M, Satomi A, Kato M, Mugishima H, Kusumi Y, Mitsumata M, Murohara T. A strategy of retrograde injection of bone marrow mononuclear cells into the myocardium for the treatment of ischemic heart disease. J Mol Cell Cardiol. 2006;40:24–34.
34. Silva SA, Sousa AL, Haddad AF, Azevedo JC, Soares VE, Peixoto CM, Soares AJ, Issa AF, Felipe LR, Branco RV, Addad JA, Moreira RC, Tuche FA, Mesquita CT, Drumond CC, Junior AO, Rochitte CE, Luz JH, Rabischoffisky A, Nogueira FB, Vieira RB, Junior HS, Borojevic R, Dohmann HF. Autologous bone-marrow mononuclear cell transplantation after acute myocardial infarction: comparison of two delivery techniques. Cell Transplant. 2009;18:343–352.
35. Perin EC, Lopez J. Methods of stem cell delivery in cardiac diseases. Nat Clin Pract Cardiovasc Med. 2006;3 Suppl 1:S110–S113.
36. Bartunek J, Sherman W, Vanderheyden M, Fernandez-Aviles F, Wijns W, Terzic A. Delivery of biologics in cardiovascular regenerative medicine. Clin Pharmacol Ther. 2009;85:548–552.
37. de Silva R, Gutierrez LF, Raval AN, McVeigh ER, Ozturk C, Lederman RJ. X-ray fused with magnetic resonance imaging (XFM) to target endomyocardial injections: validation in a swine model of myocardial infarction. Circulation. 2006;114:2342–2350.
38. Schachinger V, Erbs S, Elsasser A, Haberbosch W, Hambrecht R, Holschermann H, Yu J, Corti R, Mathey DG, Hamm CW, Suselbeck T, Werner N, Haase J, Neuzner J, Germing A, Mark B, Assmus B, Tonn T, Dimmeler S, Zeiher AM. Improved clinical outcome after intracoronary administration of bone-marrow-derived progenitor cells in acute myocardial infarction: final 1-year results of the REPAIR-AMI trial. Eur Heart J. 2006;27:2775–2783.
39. Yousef M, Schannwell CM, Kostering M, Zeus T, Brehm M, Strauer BE. The BALANCE study: clinical benefit and long-term outcome after intracoronary autologous bone marrow cell transplantation in patients with acute myocardial infarction. J Am Coll Cardiol. 2009;53:2262–2269.
40. Hofmann M, Wollert KC, Meyer GP, Menke A, Arseniev L, Hertenstein B, Ganser A, Knapp WH, Drexler H. Monitoring of bone marrow cell homing into the infarcted human myocardium. Circulation. 2005;111:2198–2202.
41. Schachinger V, Aicher A, Dobert N, Rover R, Diener J, Fichtlscherer S, Assmus B, Seeger FH, Menzel C, Brenner W, Dimmeler S, Zeiher AM. Pilot trial on determinants of progenitor cell recruitment to the infarcted human myocardium. Circulation. 2008;118:1425–1432.
42. Haider H, Ashraf M. Strategies to promote donor cell survival: combining preconditioning approach with stem cell transplantation. J Mol Cell Cardiol. 2008;45:554–566.
43. Kissel CK, Lehmann R, Assmus B, Aicher A, Honold J, Fischer-Rasokat U, Heeschen C, Spyridopoulos I, Dimmeler S, Zeiher AM. Selective functional exhaustion of hematopoietic progenitor cells in the bone marrow of patients with postinfarction heart failure. J Am Coll Cardiol. 2007;49:2341–2349.
44. Spyridopoulos I, Erben Y, Brummendorf TH, Haendeler J, Dietz K, Seeger F, Kissel CK, Martin H, Hoffmann J, Assmus B, Zeiher AM, Dimmeler S. Telomere gap between granulocytes and lymphocytes is a determinant for hematopoietic progenitor cell impairment in patients with previous myocardial infarction. Arterioscler Thromb Vasc Biol. 2008;28:968–974.
45. Werner N, Kosiol S, Schiegl T, Ahlers P, Walenta K, Link A, Bohm M, Nickenig G. Circulating endothelial progenitor cells and cardiovascular outcomes. N Engl J Med. 2005;353:999–1007.

46. Dimmeler S, Leri A. Aging and disease as modifiers of efficacy of cell therapy. Circ Res. 2008;102:1319–1330.
47. Aicher A, Heeschen C, Mildner-Rihm C, Urbich C, Ihling C, Technau-Ihling K, Zeiher AM, Dimmeler S. Essential role of endothelial nitric oxide synthase for mobilization of stem and progenitor cells. Nat Med. 2003;9:1370–1376.
48. Landmesser U, Engberding N, Bahlmann FH, Schaefer A, Wiencke A, Heineke A, Spiekermann S, Hilfiker-Kleiner D, Templin C, Kotlarz D, Mueller M, Fuchs M, Hornig B, Haller H, Drexler H. Statin-induced improvement of endothelial progenitor cell mobilization, myocardial neovascularization, left ventricular function, and survival after experimental myocardial infarction requires endothelial nitric oxide synthase. Circulation. 2004;110:1933–1939.
49. Sasaki K, Heeschen C, Aicher A, Ziebart T, Honold J, Urbich C, Rossig L, Koehl U, Koyanagi M, Mohamed A, Brandes RP, Martin H, Zeiher AM, Dimmeler S. Ex vivo pretreatment of bone marrow mononuclear cells with endothelial NO synthase enhancer AVE9488 enhances their functional activity for cell therapy. Proc Natl Acad Sci U S A. 2006;103:14537–14541.
50. Sorrentino SA, Bahlmann FH, Besler C, Muller M, Schulz S, Kirchhoff N, Doerries C, Horvath T, Limbourg A, Limbourg F, Fliser D, Haller H, Drexler H, Landmesser U. Oxidant stress impairs in vivo reendothelialization capacity of endothelial progenitor cells from patients with type 2 diabetes mellitus: restoration by the peroxisome proliferator-activated receptor-gamma agonist rosiglitazone. Circulation. 2007;116:163–173.
51. Britten MB, Abolmaali ND, Assmus B, Lehmann R, Honold J, Schmitt J, Vogl TJ, Martin H, Schachinger V, Dimmeler S, Zeiher AM. Infarct remodeling after intracoronary progenitor cell treatment in patients with acute myocardial infarction (TOPCARE-AMI): mechanistic insights from serial contrast-enhanced magnetic resonance imaging. Circulation. 2003;108:2212–2218.
52. Assmus B, Fischer-Rasokat U, Honold J, Seeger FH, Fichtlscherer S, Tonn T, Seifried E, Schachinger V, Dimmeler S, Zeiher AM. Transcoronary transplantation of functionally competent BMCs is associated with a decrease in natriuretic peptide serum levels and improved survival of patients with chronic postinfarction heart failure: results of the TOPCARE-CHD Registry. Circ Res. 2007;100:1234–1241.
53. Chavakis E, Urbich C, Dimmeler S. Homing and engraftment of progenitor cells: a prerequisite for cell therapy. J Mol Cell Cardiol. 2008;45:514–522.
54. Aicher A, Heeschen C, Sasaki K, Urbich C, Zeiher AM, Dimmeler S. Low-energy shock wave for enhancing recruitment of endothelial progenitor cells: a new modality to increase efficacy of cell therapy in chronic hind limb ischemia. Circulation. 2006;114:2823–2830.
55. Yoon YS, Wecker A, Heyd L, Park JS, Tkebuchava T, Kusano K, Hanley A, Scadova H, Qin G, Cha DH, Johnson KL, Aikawa R, Asahara T, Losordo DW. Clonally expanded novel multipotent stem cells from human bone marrow regenerate myocardium after myocardial infarction. J Clin Invest. 2005;115:326–338.
56. Aranguren XL, McCue JD, Hendrickx B, Zhu XH, Du F, Chen E, Pelacho B, Penuelas I, Abizanda G, Uriz M, Frommer SA, Ross JJ, Schroeder BA, Seaborn MS, Adney JR, Hagenbrock J, Harris NH, Zhang Y, Zhang X, Nelson-Holte MH, Jiang Y, Billiau AD, Chen W, Prosper F, Verfaillie CM, Luttun A. Multipotent adult progenitor cells sustain function of ischemic limbs in mice. J Clin Invest. 2008;118:505–514.
57. Nygren JM, Jovinge S, Breitbach M, Sawen P, Roll W, Hescheler J, Taneera J, Fleischmann BK, Jacobsen SE. Bone marrow-derived hematopoietic cells generate cardiomyocytes at a low frequency through cell fusion, but not transdifferentiation. Nat Med. 2004;10:494–501.
58. Dai W, Hale SL, Martin BJ, Kuang JQ, Dow JS, Wold LE, Kloner RA. Allogeneic mesenchymal stem cell transplantation in postinfarcted rat myocardium: short- and long-term effects. Circulation. 2005;112:214–223.
59. Wollert KC, Drexler H. Mesenchymal stem cells for myocardial infarction: promises and pitfalls. Circulation. 2005;112:151–153.
60. Noiseux N, Gnecchi M, Lopez-Ilasaca M, Zhang L, Solomon SD, Deb A, Dzau VJ, Pratt RE. Mesenchymal stem cells overexpressing Akt dramatically repair infarcted myocardium and

improve cardiac function despite infrequent cellular fusion or differentiation. Mol Ther. 2006;14:840–850.
61. Field LJ. Unraveling the mechanistic basis of mesenchymal stem cell activity in the heart. Mol Ther. 2006;14:755–756.
62. Prater DN, Case J, Ingram DA, Yoder MC. Working hypothesis to redefine endothelial progenitor cells. Leukemia. 2007;21:1141–1149.
63. Gnecchi M, Zhang Z, Ni A, Dzau VJ. Paracrine mechanisms in adult stem cell signaling and therapy. Circ Res. 2008;103:1204–1219.
64. Urbich C, Aicher A, Heeschen C, Dernbach E, Hofmann WK, Zeiher AM, Dimmeler S. Soluble factors released by endothelial progenitor cells promote migration of endothelial cells and cardiac resident progenitor cells. J Mol Cell Cardiol. 2005;39:733–742.
65. Korf-Klingebiel M, Kempf T, Sauer T, Brinkmann E, Fischer P, Meyer GP, Ganser A, Drexler H, Wollert KC. Bone marrow cells are a rich source of growth factors and cytokines: implications for cell therapy trials after myocardial infarction. Eur Heart J. 2008;29:2851–2858.
66. Gnecchi M, He H, Liang OD, Melo LG, Morello F, Mu H, Noiseux N, Zhang L, Pratt RE, Ingwall JS, Dzau VJ. Paracrine action accounts for marked protection of ischemic heart by Akt-modified mesenchymal stem cells. Nat Med. 2005;11:367–368.
67. Uemura R, Xu M, Ahmad N, Ashraf M. Bone marrow stem cells prevent left ventricular remodeling of ischemic heart through paracrine signaling. Circ Res. 2006;98:1414–1421.
68. Cho HJ, Lee N, Lee JY, Choi YJ, Ii M, Wecker A, Jeong JO, Curry C, Qin G, Yoon YS. Role of host tissues for sustained humoral effects after endothelial progenitor cell transplantation into the ischemic heart. J Exp Med. 2007;204:3257–3269.
69. Kamihata H, Matsubara H, Nishiue T, Fujiyama S, Tsutsumi Y, Ozono R, Masaki H, Mori Y, Iba O, Tateishi E, Kosaki A, Shintani S, Murohara T, Imaizumi T, Iwasaka T. Implantation of bone marrow mononuclear cells into ischemic myocardium enhances collateral perfusion and regional function via side supply of angioblasts, angiogenic ligands, and cytokines. Circulation. 2001;104:1046–1052.
70. Kawamoto A, Tkebuchava T, Yamaguchi J, Nishimura H, Yoon YS, Milliken C, Uchida S, Masuo O, Iwaguro H, Ma H, Hanley A, Silver M, Kearney M, Losordo DW, Isner JM, Asahara T. Intramyocardial transplantation of autologous endothelial progenitor cells for therapeutic neovascularization of myocardial ischemia. Circulation. 2003;107:461–468.
71. Zeng L, Hu Q, Wang X, Mansoor A, Lee J, Feygin J, Zhang G, Suntharalingam P, Boozer S, Mhashilkar A, Panetta CJ, Swingen C, Deans R, From AH, Bache RJ, Verfaillie CM, Zhang J. Bioenergetic and functional consequences of bone marrow-derived multipotent progenitor cell transplantation in hearts with postinfarction left ventricular remodeling. Circulation. 2007;115:1866–1875.
72. Erbs S, Linke A, Schachinger V, Assmus B, Thiele H, Diederich KW, Hoffmann C, Dimmeler S, Tonn T, Hambrecht R, Zeiher AM, Schuler G. Restoration of microvascular function in the infarct-related artery by intracoronary transplantation of bone marrow progenitor cells in patients with acute myocardial infarction: the Doppler Substudy of the Reinfusion of Enriched Progenitor Cells and Infarct Remodeling in Acute Myocardial Infarction (REPAIR-AMI) trial. Circulation. 2007;116:366–374.
73. Mirotsou M, Zhang Z, Deb A, Zhang L, Gnecchi M, Noiseux N, Mu H, Pachori A, Dzau V. Secreted frizzled related protein 2 (Sfrp2) is the key Akt-mesenchymal stem cell-released paracrine factor mediating myocardial survival and repair. Proc Natl Acad Sci U S A. 2007;104:1643–1648.
74. Burchfield JS, Iwasaki M, Koyanagi M, Urbich C, Rosenthal N, Zeiher AM, Dimmeler S. Interleukin-10 from transplanted bone marrow mononuclear cells contributes to cardiac protection after myocardial infarction. Circ Res. 2008;103:203–211.
75. Gilchrist A, Au CE, Hiding J, Bell AW, Fernandez-Rodriguez J, Lesimple S, Nagaya H, Roy L, Gosline SJ, Hallett M, Paiement J, Kearney RE, Nilsson T, Bergeron JJ. Quantitative proteomics analysis of the secretory pathway. Cell. 2006;127:1265–1281.
76. Bersell K, Arab S, Haring B, Kuhn B. Neuregulin1/ErbB4 signaling induces cardiomyocyte proliferation and repair of heart injury. Cell. 2009;138:257–270.

77. Smart N, Risebro CA, Melville AA, Moses K, Schwartz RJ, Chien KR, Riley PR. Thymosin beta4 induces adult epicardial progenitor mobilization and neovascularization. Nature. 2007;445:177–182.
78. Malik DK, Baboota S, Ahuja A, Hasan S, Ali J. Recent advances in protein and peptide drug delivery systems. Curr Drug Deliv. 2007;4:141–151.
79. Zhang G, Nakamura Y, Wang X, Hu Q, Suggs LJ, Zhang J. Controlled release of stromal cell-derived factor-1 alpha in situ increases c-kit+ cell homing to the infarcted heart. Tissue Eng. 2007;13:2063–2071.
80. Segers VF, Tokunou T, Higgins LJ, MacGillivray C, Gannon J, Lee RT. Local delivery of protease-resistant stromal cell derived factor-1 for stem cell recruitment after myocardial infarction. Circulation. 2007;116:1683–1692.

Evidence for the Existence of Resident Cardiac Stem Cells

Isotta Chimenti, Roberto Gaetani, Lucio Barile, Elvira Forte, Vittoria Ionta, Francesco Angelini, Elisa Messina, and Alessandro Giacomello

Abstract The heart has traditionally been considered a terminally differentiated organ. In the past 10 years, though, this paradigm has been challenged and proved questionable, starting from the evidence of cycling myocytes in the adult heart, both in physiological and in pathological conditions. In addition, the discovery and isolation of cells from the adult heart with progenitor-like and stem-like features has started a new field of research. These topics are reviewed in this chapter.

Keywords Cardiac myocytes • Cell cycle • Cardiac stem cells • Myocardial infarction • Mytosis • Bone marrow stem cells

1 Introduction

The mammalian heart is a very peculiar organ: it is able to generate an impressive mechanical strength on the basis of the synchronous contraction of billions of cardiomyocytes, and this approximately 35 million times in each year of our life. This ability is based on its tight and coordinated structure and on electromechanical coupling. It is not surprising then that if parenchymal cells integrated in this structure were continuously worn out and in need of turnover, the overall macroscopic function would be impaired and discontinuous. During fetal heart development in mammals, proliferation and differentiation happen simultaneously. Fetal cardiomyocytes are able to go through mitosis by continuous disassembling and reassembling of their contractile apparatus. Gradually there is cell cycle withdrawal within the end of gestation, and this corresponds to the maturation of the cardiovascular system, resulting in higher blood pressure. After birth, the increase in heart mass (about 30–50-fold from birth to adulthood [1]) is due to hypertrophy rather

A. Giacomello (✉)
Department of Molecular Medicine and Pathology,
Cenci-Bolognetti Foundation, University "Sapienza" of Rome, Italy
e-mail: alessandro.giacomello@uniroma1.it

I.S. Cohen and G.R. Gaudette (eds.), *Regenerating the Heart*, Stem Cell Biology and Regenerative Medicine, DOI 10.1007/978-1-61779-021-8_9,
© Springer Science+Business Media, LLC 2011

than cell division [2], whereas DNA replication followed by karyokinesis in adult cardiomyocytes is a known phenomenon associated with division of nuclei [3, 4] or with the presence of polyploid nuclei [2]. Therefore, the heart has traditionally been considered a terminally differentiated organ.

In the past 10 years, though, this paradigm has been challenged and proved questionable, starting from the evidence of cycling myocytes in the adult heart, both in physiological and in pathological conditions [5–8]. Thereafter, many different groups successfully isolated and characterized resident progenitor cells from the adult heart of different species, from mouse to rat to human. These cells are able to proliferate *ex vivo*, express early-to-late multilineage markers, and differentiate either spontaneously or in co-culture with cardiomyocytes [9]. Finally, two recent elegant studies, based either on genetic fate mapping in mice [10] or on ^{14}C dating of cardiomyocytes in humans [11], were able to detect, track, and quantify their low turnover during adulthood.

How much of these observed phenomena, both *in vitro* and *in vivo*, is due to the direct contribution of a resident stem cell pool, extracardiac sources (not addressed in this chapter), or alternatively dedifferentiation of mature cardiomyocytes is still debated. All these options raise a paradox though: How are all the discoveries about cardiac stem cells and myocyte turnover and cycling compatible with the fact that heart failure is the leading cause of mortality and morbidity in the western world? During adulthood, normal aging and disease reduce the number and the functionality of cardiomyocytes (and resident progenitors?) owing to multiple mechanisms, such as senescence, apoptosis, and susceptibility to oxidative stress [12]. The resistance to damage of an organ depends on its ability to compensate functional loss and/or to renew the tissue. It is obvious that the heart is unable to repair itself after major injuries. The healing happens mainly through hypertrophy and fibrosis, but both these phenomena cannot overcome the underlying parenchymal loss. Many disease-related issues are probably responsible for this inability: the lack of a functional vascular system, the hostile inflammatory environment, the block of diffusion, and/or the shortage of humoral factors and the lack of an easily available and efficient stem cell compartment.

These are actually the features observed in the mammalian heart. In fact, regeneration of tissues and organs is a widely diffuse repair mechanism in vertebrates, for example, in amphibians and reptiles. It often involves dedifferentiation of parenchymal mature cells to form the so-called blastema, with subsequent proliferation and differentiation via a complete morphogenetic process. Regeneration seems to be an ancestral feature: the more complex the physiologic processes of an organism, the less convenient it would be to spend time to reconstruct the complexity of an organ, which might have taken years to develop [13]. Moreover, evolutionary drive favored the most rapid healing response through inflammation and fibrosis. Indeed, the complex and sustained hemodynamic system evolved in mammalians differs from the low-pressure circulation and incomplete blood oxygenation typical of small vertebrates and embryonic/fetal (mammalian) hearts. In this context bleeding could seriously compromise survival. Therefore, there is no regeneration for mammals, but hemostasis and scarring.

In this scenario a growing body of studies are trying to focus on cardiac cell renewal in the absence of, or following, a pathologic insult, and the possible existence of true or putative cardiac stem/progenitor cells and their potential role. These issues will be addressed in this chapter.

2 Evidence of Postnatal Cardiac Turnover and Renewal

Despite cardiomyocytes proliferating extensively during fetal life, it is generally accepted that they become permanently quiescent cells soon after their differentiation. Many evidences have always suggested the absence of any turnover in the adult cardiomyocyte pool, including the decrease in the number of cardiomyocytes with normal aging, the inability to routinely detect mitosis, and the very limited endogenous capacity to maintain homeostasis after major pathological events [12]. Despite this, the phenomenon of multinucleation through karyokinesis (or nuclear mitosis) during myocyte maturation has been well known for years, giving rise in adulthood to approximately one-quarter of binucleated cardiomyocytes in the human heart. The fact that karyokinesis is not always accompanied by cytokinesis may be explained by the presence of additional cell cycle checkpoints at G2/M phase which must be circumvented for progress in the cell division process [14]. It is also very likely that the dense sarcomeric structure of the mature cell acts as a strong physical impediment to cytokinesis [14]. However, the suggestion that sarcomeric organization is an unavoidable cell cycle checkpoint is not the case for neonatal or newt cardiomyocytes.

Moreover, there are many other known molecular checkpoints coresponsible for cardiomyocyte cell cycle exit, mainly at the G0/G1 transition phase, where most of them are arrested. Many *in vitro* studies have suggested multiple mechanisms are possibly involved.

In mammalian organisms the cell cycle progression through G1 phase and the initiation of the DNA synthesis phase is controlled by cooperative interaction of cyclins and cyclin-dependent kinases (CDKs) [15]. The positive regulators cyclin D1 and cyclin D2 are expressed in cardiomyocytes only upon serum treatment [16]. Whereas cyclin D2 alone seems unable to induce mitosis in these cells [16] (one possible cause is that the CDK inhibitors p21 and p27 are highly accumulated in cardiomyocytes), the nuclear expression of cyclin D1 by viral transduction is sufficient to induce cell cycle progression leading to cell division of neonatal cardiomyocytes [17]. These observations correlate to the report of high abundance of the constitutively active form of the G1/S blocker Rb in adult cardiomyocytes. Cycling myocytes have also been induced by overexpressing the multifunctional DNA tumor virus oncoprotein (SV40 large T-antigen oncoprotein), which is able to disrupt Rb function. In those cells the pathways of two T-antigen binding proteins (namely, p53 and p193) are circumvented and all cycling checkpoints are bypassed [18–20].

Starting from 1998, the first reports were published describing the observation of mitotic figures compatible with cytokinesis in human adult cardiomyocytes. Technical advances in immunocytochemistry and confocal laser fluorescence microscopy had an important role in this reexamination of the terminally differentiated state of the adult heart. Kajstura et al. [6] in 1998 first demonstrated the presence of nuclei undergoing mitosis in myocytes of hearts obtained from patients waiting for cardiac transplantation, affected by chronic ischemic heart disease and dilated cardiomyopathy. Sections of myocardium were labeled with propidium iodide and alpha-sarcomeric actin antibody, and a myocyte mitotic index (the ratio of the number of nuclei undergoing mitosis to the number not undergoing mitosis) of 0.015% was measured in explanted hearts by confocal analysis [6]. That value was confirmed and amplified in a study by Beltrami et al. [7] in which the reentry of myocytes into the cell cycle resulted in mitotic indexes of 0.08 and 0.03% in regions adjacent to the infarcts and distant from the infarcts, respectively. These data were obtained in extensive myocardial infarcted tissue that may cause more myocytes to reenter the cycle than during chronic heart failure. The mitotic myocytes were evidenced using antibody against Ki-67, which is a nuclear antigen expressed in all phases of the cell cycle, except G0.

Many data were thereafter collected, analyzed, and compared, with multiple approaches. Sometimes inconsistent results were obtained, but since the physiologic turnover seems to be so low, it is not surprising that different techniques might have insufficient sensitivity. In this regard, the study of telomerase provided further insights into the natural regenerative potential of cardiac tissue. For example, Borges and Liew [21], in a study performed in rats, analyzed telomerase activity in several tissues at fetal and adult stages. As expected, it was not detectable in adult tissues including the heart, except for the adult liver, which preserves a substantial activity (more than 60% of that of fetal liver) [21]. By 5 days after birth, telomerase activity in the heart was only 20% of the activity in 10-day-gestation fetal tissue. The activity was undetectable in 20-day-old hearts and remained below detection limits until 4 months of age [21]. Despite this, measurement of the length distribution of telomeres in myocytes with aging has provided information concerning the capacity of adult myocytes to undergo numerous mitotic divisions before reaching replicative senescence [22]. Thus, those observations alone cannot exclude the possibility of some mitotic rounds.

The most recent and so far most striking study on cardiomyocyte turnover quantification is that by Bergman et al. [11], who applied an elegant approach based on DNA radiocarbon dating. The rationale is that, at any given time in the postnuclear world, the concentration of ^{14}C in the human body reflects that in the atmosphere. Since DNA is stable in G0 stage and since the concentration of ^{14}C in the atmosphere exponentially increased in the decade after 1950, the analysis of the concentration of ^{14}C in the DNA of cardiomyocytes of individuals born before those years allows detection of *de novo* karyokinesis in human adult heart and retrodating of the age of the nuclei. In all cases studied, in people born up to 22 years before the onset of nuclear bomb tests ^{14}C concentrations were higher compared with the levels at the time of their birth. Similarly, in all individuals born near or after the time of the nuclear bomb tests, the ^{14}C concentrations in cardiomyocyte DNA corresponded to the atmospheric concentrations several years after their birth [11]. Thus, at least some of the DNA of myocardial cells is synthesized much later postnatally, indicating that cells in the human heart do renew during adulthood. Mathematical modeling of the kinetics of DNA synthesis and ^{14}C integration suggests that the measured concentration in cardiomyocyte DNA could not be a result of polyploidization during adulthood. This model also predicts that cardiomyocytes are renewed at a rate of about 1% per year at the age of 25 and 0.45% at the age of 75. At the age of 50, 55% of the cardiomyocytes' nuclei are from the time around birth and 45% were synthesized later. Overall, the age of cardiomyocytes is on average 6 years younger than the individual [11]. Although it leaves open the question about the correspondence between karyokinesis and cytokinesis, this study provided new reliable and accurate insights into the debated subject of cardiomyocyte turnover. It was not able though to answer the question of the origin of the cycling and renewed myocytes.

Therefore, besides terminally differentiated adult myocytes, there might be a small subpopulation of cycling myocytes produced by the differentiation of cardiac progenitor cells. The balance between dedifferentiation and *de novo* differentiation of new myocytes is of great interest and is currently under investigation.

3 Evidence of the Presence of Stem-like Cells in the Adult Heart

After the first evidence against the classic paradigm of the heart as a postmitotic organ had been obtained, starting from 2002 a new field of research was born with the discovery and isolation of cells from the adult heart with progenitor-like and stem-like features. Over the last few years, several groups have identified different populations of cells in the postnatal and adult heart with expression profiles and/or biological functions typical of immature committed cells [23–27].

These cells are positive for various embryonic stem/progenitor cell markers [c-kit, stem cell antigen-1 (Sca-1), islet-1 (Isl-1), stage-specific embryonic antigen-1 (SSEA-1), and side population properties], propagate *in vitro*, and develop features of heart cells after differentiation *in vitro* or *in vivo* [23, 26, 28]. The cardiac stem and progenitor cell types characterized so far exhibit significant differences in immunophenotypic, developmental, and biological properties. The heart probably contains various types of stem and progenitor cells. This is in agreement with the emerging consensus that more than one stem cell may be present in a particular tissue [29].

The isolation of these cells can be performed on the basis of the expression of surface markers or on the basis of functional properties. The first criterion includes c-kit+, sca-1+ and isl-1+ cells, whereas the second mostly concerns the isolation of cells with high cardiac regenerative potential from heart biopsy samples. In this regard, the term "cardiac regenerative cells" will be used from now on to address cells isolated with the latter approach, which is based on a functional criterion.

3.1 *Surface-Marker-Based Identification and Isolation of Resident Cardiac Regenerative Cells*

In 2003, Beltrami et al. [23] first reported the isolation of stem cells from the heart expressing c-kit, the tyrosine kinase receptor for the stem cell factor. This population of cells isolated by sorting is very heterogeneous, with rare cells expressing cardiac markers (Nkx2.5, GATA4, Mef2), whereas skeletal muscle, hematopoietic, and neural markers were undetectable. Cardiac c-kit+ cells are clonogenic, self-renewing, and multipotent, showing signs of biochemical differentiation into cardiomyocytes, smooth muscle, or endothelial cells when cultured in differentiating conditions, although these conditions did not result in a mature phenotype. When injected in the infarcted heart of syngeneic rats, these cells or their clonal progeny were able to reconstitute well-differentiated myocardium, formed by functional new vessels and myocytes with an immature phenotype, encompassing 70% of the ventricle [23, 30]. c-kit+ cells were isolated also from human biopsies by enzymatic dissociation of the tissue and immunomagnetic sorting, or alternatively using the primary explant technique, thus potentially acquiring, in the last case, some of the features typical of the "cardiac regenerative cells," as defined above. c-kit+ cells from the human heart are self-renewing, clonogenic, and multipotent *in vitro* and *in vivo* [31].

The initial enthusiasm aroused by these studies was partially reduced by the identification of c-kit+ cells in tissues of other solid organs. These cells supposedly leave the bone marrow in small numbers to scavenge pathogenic molecules as part of the mechanisms to activate local innate immune responses [32]. Furthermore, by means of in situ detection, c-kit+ cells from biopsy samples of human heart have recently been reported to coexpress markers of mast cells and to lack cardiac markers, such as Nkx2.5 and isl-1 [33]. A possible change in cellular phenotype during the culture time course needs to be taken into consideration to explain these conflicting results [34].

Using transgenic mice expressing enhanced green fluorescent protein (eGFP) under the c-kit promoter, Tallini et al. [35] observed c-kit cells at different stages of differentiation in the embryonic heart, peaking at postnatal day 2; thereafter the number of eGFP+ cells declined and they were rarely detected in adult hearts. C-kit-eGFP+ cells isolated from postnatal days 0–5 were able to differentiate into endothelial cells, smooth muscle cells, and beating myocytes. Cryoablation in the adult heart resulted in increased expression of c-kit–eGFP, peaking after 7 days. c-kit expression occurred in endothelial and smooth muscle cells in the revascularizing infarct area and in terminally differentiated cardiomyocytes in the border zone surrounding the infarct. Thus, the authors suggested that *in vivo* c-kit expression is associated with neovascularization, but not with *de novo* myogenesis. Other groups [36] have recently addressed a similar issue by a different experimental setting that will be discussed later in this chapter. However, further studies are required to assess the actual contribution of cardiac progenitor cells, cardiomyocyte dedifferentiation, or cell cycle reentry in cardiomyocyte replacement after infarction [37].

One year later, Schneider's group [26] reported the presence of a population of cells expressing sca-1 in the non-myocyte fraction of the heart. These cells colocalized with vasculature and expressed CD31, but had a phenotype distinct from that of endothelial and hematopoietic stem cells: they were negative for CD45, CD34, flk1, c-kit, vascular endothelial cadherin, and von Willebrand factor. The exposure to 5-azacytidine induced cardiac differentiation in culture. The capability to adopt a cardiac muscle fate in embryogenesis was substantiated by blastocyst injection. When injected intravenously after ischemia–reperfusion, they homed and functionally differentiated in the host myocardium. The differentiation was shown to rely both on fusion-independent and on fusion-associated mechanisms.

Matsuura et al. [38] isolated similar sca-1+ cells from hearts of 12-week-old mice. When treated with oxytocin, these cells expressed genes of cardiac transcription factors and contractile proteins, and showed sarcomeric structure and spontaneous beating. They may be able to differentiate into different cell types when exposed to different environments; however, multipotency of sca-1+ cells has not been proven on the progeny of single cells. Recently, they demonstrated that the transplantation of sheets of clonally expanded sca-1+ cells ameliorates cardiac function after myocardial infarction in mice. Clonal sca-1+ cells efficiently differentiated into cardiomyocytes and secreted cytokines, including soluble vascular cell adhesion molecule 1, which induced migration of endothelial cells and cardiac progenitor cells, and prevented cardiomyocyte death from oxidative stress through activation of Akt, ERK, and p38 mitogen-activated protein kinase [39].

The group of Doevendans [40] identified cardiomyocyte progenitor cells (CMPCs) in fetal and adult human hearts using an antibody against mouse sca-1. Human CMPCs are localized within the atria, atrioventricular region, and epicardial layer, and they can be induced to differentiate *in vitro* into cardiomyocytes and form spontaneously beating aggregates, after stimulation with 5-azacytidine. Recently this group investigated the effect of intramyocardial injection of human CMPCs or predifferentiated CMPC-derived cardiomyocytes (CM-CMPCs) into immunodeficient infarcted mice. The results were higher ejection fraction and reduced left ventricular remodeling up to 3 months after myocardial infarction, when compared with controls. Both CMPCs and CM-CMPCs were able to generate new cardiac tissue consisting of human cardiomyocytes and blood vessels. This excludes the need for *in vitro* predifferentiation, making CMPCs a promising source for autologous cell therapy [41].

Laugwitz et al. [24] reported that early after birth the mammalian heart harbors a rare subset of cells positive for the LIM homeodomain transcription factor isl-1, which disappears soon after the neonatal period. These cells are mostly found in the outflow tract, atria, and right ventricle, suggesting that they could be remnants of the embryonic secondary heart field. Postnatal isl1+ murine cells can be expanded *in vitro* on mesenchymal feeder layers, and they undergo terminal differentiation when co-cultured with neonatal cardiac cardiomyocytes, acquiring electromechanical properties similar to those of adult cardiomyocytes, thereby fulfilling the criteria for endogenous cardioblasts. Moreover their bona fide identity as true cardiac progenitors comes from extensive and detailed genetic fate mapping studies *in vitro* and *in vivo*, providing strong evidence of their cardiac-specific multipotency [42, 43]. However, these cells were isolated only from very young animals, including human neonatal specimens, and the number of isl-1+ cells dramatically decreases over the first weeks of life. Furthermore, it remains to be addressed if these cells are multipotent and able to engraft and regenerate the myocardium.

Despite the importance of all the studies described above on cardiac progenitor cells, whose isolation is mostly based on expression profiles, when translating to preclinical and clinical applications, there are still many unresolved questions, particularly those related to the methods employed: for example, the relationship between the different progenitors described, the variables in defining the differentiation pathway, and the marker expression. All these cells, in fact, display distinct and overlapping traits based on their multilineage diversification, on cell surface markers, on proliferative capacity, and on localization within the heart. Knowledge regarding the relationships of these populations or the molecular networks that regulate their proliferation, self-renewal, or lineage differentiation levels is still incomplete. These studies tell us what is possible in a tissue culture dish, but not necessarily what actually happens *in vivo* during development and disease. In other words, it is difficult to claim that one's own isolated cell is "the" only cardiac stem cell.

In view of this, a functional approach to the isolation and evaluation of "cardiac regenerative cells" may provide important information on the real potential of this low-fertility and very low renewing organ.

3.2 Detection and Isolation of Resident Cardiac Regenerative Cells Based on Functional Properties

As in several other tissues, the heart contains a side population of progenitors cells, isolated for their ability to efflux DNA-binding dyes (such as Hoechst 33342), due to the expression of the ATP-binding cassette transporter Abcg2. Abcg2+ cells were identified during embryogenesis and they persist as a small pool in multiple organs [44]. Cardiac side population cells express stem cell markers, such as sca-1 and c-kit, and are CD34+, CD31−, CD45−. They are capable of proliferation and differentiation into mesodermal derivatives: they express α-actinin after co-culture with cardiac cells in cardiomyocyte-specific media, and they are able to proliferate and form hematopoietic colonies when plated into methylcellulose medium. The greatest cardiomyogenic potential is restricted to cells negative for CD31 and positive for sca-1 [27]. Their ability to differentiate *in vivo* has not been extensively evaluated yet.

A second possible approach for functional selection was developed in 2004 by Messina et al. [25], who debugged a method for the isolation and expansion of progenitor cells from murine heart samples and human cardiac surgical or biopsy samples. They exploited the functional feature of putative immature, regenerative cells to spontaneously migrate and shed from partially digested small pieces of tissue. The small phase-bright round cells moving over and within the fibroblast outgrowth surrounding the explant culture can be collected and further selected by plating them on poly(D-lysine)-coated wells. Here these cells self-organize in three-dimensional structures called cardiospheres, consisting of proliferating c-kit+ cells primarily in their core and of differentiating cells expressing cardiac cell (troponin I, myosin heavy chain) and vascular cell markers on their periphery. Moreover, both cardiospheres and cardiosphere-derived cells (CDCs) expanded as monolayer express connexin-43 in discrete gap-junctional patterns, suggesting their potential to electrically couple to myocytes. This heterogeneous population of cells is capable of terminal differentiation: in fact, they can beat either in co-culture with neonatal rat myocytes, or spontaneously when derived from embryonic murine hearts, suggesting that these cells could perform effective electrical coupling with the cardiac syncytium.

By lineage tracing experiments to explore the origin of these promising therapeutic cells, Davis et al. [45] have recently demonstrated that CDCs are clonogenic, and that cloned CDCs exhibit spontaneous multilineage potential (multipotency). Furthermore, a portion of cardiac progenitor cells arose from myocyte dedifferentiation, most originated from the proliferation of endogenous progenitors, and none derived from contaminating cardiomyocytes located within the explant outgrowth.

Cardiospheres can be expanded as three-dimensional structures by mechanical or enzymatic partial dissociation. To increase the cellular yield, CDCs can be more extensively expanded as monolayers [46]: this latter method is so far the most used method of culture for these cells, despite some reduction in cardiac differentiation, in preclinical *in vivo* experiments and in the ongoing phase I clinical trial CADUCEUS (see http://www.clinicaltrials.gov for details). In fact, compared with growth in monolayers, when embedded in a three-dimensional environment, cells are expected to develop more complex tissue-like intercellular structures (e.g., cell–cell contacts, extracellular matrix), that can orchestrate many cellular functions in a niche-like fashion, including proliferation, differentiation, and paracrine activity (see below). *In vivo* injection of cardiospheres or CDCs into the peri-infarct zone of SCID mice led to cell engraftment, proliferation, and multilineage differentiation, overall resulting in bands of regenerating myocardium in the scar area, with consequent functional improvement at 3 weeks [46].

In contrast to single-marker selection criteria, the main advantage of this isolation method is the exploitation of the intrinsic functional property of the cells to spread out of the tissue and grow in three-dimensional structures, representing a more physiological culture condition compared with monolayers, where natural spatial cellular connections cannot be promoted.

The single-marker and the function-based selection approaches have been combined by several groups and used to obtain regenerating cells from primary cultures of human cardiac or muscle biopsies, subsequently sorted for specific antigens (c-kit, sca-1, etc.) [31, 41]. Interestingly, when CDCs were sorted for c-kit and injected in a mouse model of acute myocardial infarction, the functional improvement was significantly lower than that of the whole population [47], suggesting that cooperative and synergistic effects among multiple cell types in the CDC pool play an important role in the overall beneficial outcome.

The sca-1+ CMPCs obtained by Doevendans et al. (previously discussed in this chapter) are an example of this combined procedure, as are cardiosphere-derived,

Lin−, c-kit+ progenitor cells [48] (which have been shown to be able to activate the CXCR4/sdf axis by hypoxia preconditioning) and the so-called mesoangioblast.

These latter cells have been proposed as a further potential source for cardiac regeneration associated with the vasculature. Mesoangioblasts were isolated by Galvez et al. [49] from juvenile mouse heart, using a technical procedure similar to that previously described by Messina et al. [25]. A unique feature of these cells is the coexpression of early endothelial (CD31, CD34) and pericyte markers (NG2 and alkaline phosphatase); this argues for a vascular origin or at least a close association with the vessel wall. They express cardiac transcription factors Nkx2.5 and GATA4, and upon removal of growth factors they can differentiate spontaneously into beating cardiomyocytes that assemble mature sarcomeres and express the typical pattern of cardiac ion channels. Cells similarly isolated from the atria do not spontaneously differentiate. When injected into the ventricles after coronary ligation, cardiac mesoangioblasts colonize the ventricular wall and differentiate into cardiomyocytes. However, their migratory ability is limited: they can migrate inside the necrotic area just for a few millimeters. The rapid cardiac differentiation may limit the proliferation, migration, and paracrine activity of these cells *in vivo*. Furthermore, the mesoangioblasts tend to disappear in older mouse heart, thus suggesting for these cells a potential role in the treatment of juvenile cardiac diseases without the presence of massive necrosis or loss of cardiac tissue.

4 What Cells Contribute to Cardiac Turnover?

There are at least four potential resident sources that could account for the turnover and limited repair of the heart [50]: (1) self-renewing differentiated cardiomyocytes, (2) blood-derived, heart-homed cells, (3) cardiac progenitor cells, and (4) epicardially derived cells (see Fig. 1):

1. As discussed before in this chapter, a very low percentage of adult cardiomyocytes might be able to reenter the cell cycle and divide, exploiting the ancient regenerative program typical of lower invertebrates. Cardiomyocyte proliferation has been induced by transfecting cardiomyocytes with cell cycle activators (e.g., SV40 large T antigen [51], cyclin A2 [52], and cyclin D2 [53]), by locally administering insulin-like growth factor 1, fibroblast growth factor 1, or periostin, and recently by systemic administration of neuregulin 1 (NRG1) [54]. In particular, Bersell et al. [54], by an elegant series of *in vitro* and *in vivo* fate mapping studies, demonstrated the proliferative capability of differentiated mononucleated cardiomyocytes through a receptor-activated mechanism (ErbB4), involving NRG1 as the exogenous activator, therefore modifying the concept of proliferating cardiomyocytes as "facultative" stem-like cells. In fact the authors excluded the possibility that NRG1 could induce undifferentiated cardiac progenitors to proliferate and then to differentiate into mature cardiomyocytes. By *in vitro/in vivo* fate mapping experiments, dynamic *in vitro* genetic labeling, and by excluding the migration of circulating c-kit cells, they provided four lines of evidence that NRG1 and its receptor ErbB4 control the

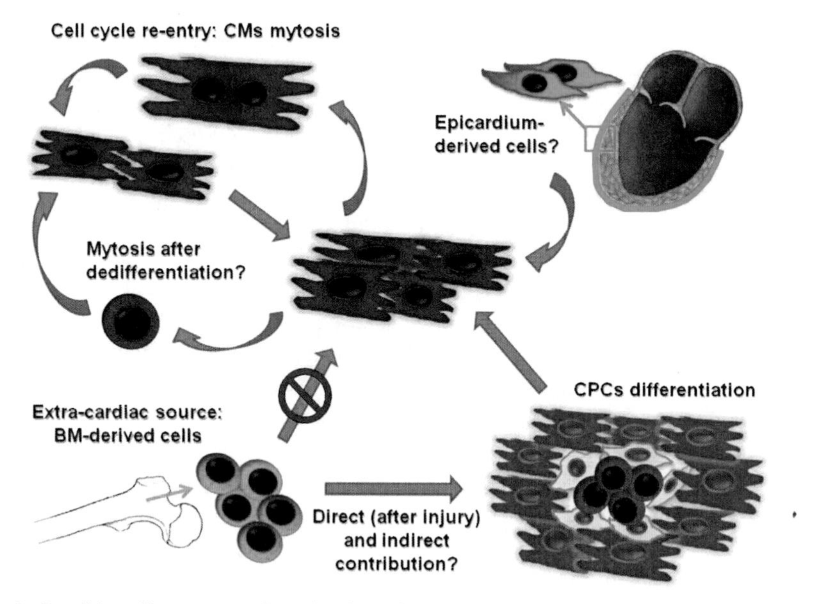

Fig. 1 Possible cell sources and mechanisms for cardiomyocyte turnover and regeneration in physiological and pathological conditions. *CM* cardiomyocyte, *BM* bone marrow, *CPC* cardiac progenitor cell

proliferation of a portion of mononucleated cardiomyocytes, whose contractile apparatus does not prohibit karyokinesis or cytokinesis by means of permanent sarcomeric dedifferentiation. On the other hand, in addition to the postnatal control of cardiomyocyte proliferation, the NRG1/ErbB4 signaling is directly involved in the regulation of cardiac differentiation of embryonic stem cells [55] and during development [56], suggesting a pleiotropic effect of the loop that could be environment- and age-dependent.

2. Blood-derived regenerating cells have been extensively indicated, mostly in the first years of the "cardiac stem cell" age, as a potential source for cardiac regeneration. Circulating stem/progenitor cells from bone marrow have the ability to differentiate into myocyte-like cells *in vitro* and to repopulate the heart *in vivo*. However, it remains unclear whether these cells are actually able to generate functional myocardium since clinical studies only resulted in short-term functional improvement, likely due to a paracrine effect. Recently, the potential contribution of the bone marrow to a resident cell pool in the heart with cardiogenic and vasculogenic properties, in particular in the form of cardiac progenitors isolated from heart biopsy and cultured as cardiospheres, has been suggested [36], giving rise to several basic and translational scientific implications.
3. Recently, another paper has unequivocally demonstrated that there is cellular turnover in the adult heart, in particular after acute infarction, and has also given an indirect assessment of the contribution from a putative undifferentiated pool [10]. In this study the authors used a double transgenic mouse in which, upon tamoxifen administration, only cardiomyocytes were induced to express the green fluorescent protein (GFP). By evaluating the percentage of green cells over the

total number of cardiomyocytes in infarcted or healthy mice, the authors established that the percentage of GFP+ cardiomyocytes significantly decreased after myocardial infarction, implying the formation of new GFP– cardiomyocytes. The authors could not detect any newly formed myocytes during normal aging though, but considering the very low turnover rate and the actual baseline efficiency of the reporter recombination (not higher than 80%), one could speculate that this system was not sensitive enough. This genetic fate-labeling approach represented an important and potent methodological step forward in this difficult field, and provided a precious contribution to the understanding of heart renewal dynamics.

4. Another potential source for cardiac regeneration is the epicardium. During embryogenesis, the epicardium is the source of multipotent mesenchymal cells, which give rise to most of the coronary smooth muscle, a subset of the coronary endothelium, and cardiac fibroblasts. Two recent studies, using similar Cre-lineage tracing approaches, have shown that epicardial progenitor cells expressing Tbx18 [57] and Wt1 [58] are also able to differentiate into functional cardiomyocytes during normal heart development, and they are possibly derived from a Nkx2.5+, isl-1+ common precursor with the rest of the heart [59]. These results suggest a potential role of the epicardium in adult heart regeneration, not only as a source of cardioprotective signals, but also as a source of multipotent progenitor cells, as is the case for zebrafish. In 2007, Limana et al. [60] showed that keeping the pericardial sac intact during surgical infarction prevents cardiac function impairment and left ventricular remodeling, and promotes foci of regeneration. They also observed that epicardial mesothelial cells transduced with a lentivirus expressing GFP could be detected in the left ventricular wall of an infarcted mouse after 1–3 weeks, and that they had acquired a cardiac phenotype. Finally, they identified two different populations of cells in the epicardium expressing c-kit and CD34. Both subsets were negative for the hematopoietic marker CD45, expressed early and late cardiac transcription factors (Nkx2.5, GATA4), and were able to acquire an endothelial phenotype *in vitro*. Following myocardial infarction, epicardial c-kit+ cells proliferated and were identified after 1 week in the subepicardium, coexpressing cardiac (GATA4), endothelial, and smooth muscle cell markers.

The relationship among all these progenitor cells and their origin are not clear: they may be remnants from fetal life, or they could come from extracardiac sources, such as the bone marrow.

Recently Ott et al. [28] identified a population of uncommitted precursor cells (UPCs) that could be upstream of the other cardiac progenitor cells previously identified. The UPCs are characterized by the expression of SSEA-1, which is expressed in early heart development. These adult SSEA-1+ cells, unlike the neonatal ones, do not express early cardiac markers such as Nkx2.5 and GATA4, suggesting that only uncommitted progenitors persist in the adult heart. Adult UPCs can be expanded *in vitro* on a cardiac-derived mesenchymal feeder layer, and they undergo a temporal maturation from an uncommitted SSEA+/Oct4+ state to a mesodermally directed Flk1 state, and from Flk1+/sca-1+ to the cardiac-committed Nkx2.5+/GATA4+ or isl-1+ state. Ultimately they can differentiate into cardiomyocytes, endothelial cells, and smooth muscle cells. Beating colonies could be obtained in differentiating media or co-culture with neonatal cardiomyocytes. After ischemic injury, UPCs contribute to the formation of vascular and cardiomyocyte-like cells.

5 What Is the Origin of Cardiac Progenitor Cells?

The origin of cardiac progenitors is still a debated issue, yet to be clarified. Theoretically these cells could derive from a pool of stem cells of embryonic, resident, or extracardiac origin. However, the acquisition of the functional attitude of the cardiac progenitors by other cardiac (not embryonic) cells (such as those derived from the pericardium) and the possibility of a dedifferentiation process have also been suggested and partially investigated.

In the first hypothesis, cardiac stem cells could represent remnants of embryonic cardiogenic populations, such as the side population [44] or postnatal isl1+ cardioblasts.

Alternatively, evidence of a plausible extracardiac origin initially derived from the analysis of female-donor hearts transplanted into male recipients (sex-mismatch transplant), which showed the presence of male cardiomyocytes and vascular cells in the female organ [61]. The cells analyzed, probably of extracardiac origin, may represent a circulating pool of endogenous stem cells recruited in case of cardiac damage. This second hypothesis is supported by evidence of chimeric hearts in patients undergoing allogeneic bone marrow transplantation, confirming that cells from the bone marrow are able to home into the myocardium [62].

Finally, the possibility that resident cardiac progenitors might derive from dedifferentiation of mature cardiomyocytes, as in urodele amphibians, represents another interesting area of investigation. In fact, these animals have a dedifferentiation mechanism at the damaged site (e.g., a limb) that leads to the formation of a blastema, made up of dedifferentiated progenitor cells, that can proliferate and redifferentiate to generate all the different tissues needed for recovering the original structure [63]. In mammals these regenerative processes are limited by irreversible terminal differentiation, and neither transdifferentiation nor dedifferentiation phenomena are routinely detectable as naturally occurring. However, recent *in vitro* studies suggest that differentiated C2C12 myotubes can be induced to dedifferentiate into mesenchymal progenitor cells by ectopic expression of msx1 [64], or by treatment with reversine [65], opening new exciting areas of investigation to unravel the puzzle of cardiac renewal dynamics.

6 Conclusions

In conclusion, the mammalian heart could be considered as a slow-turnover organ as it seems to be endowed with several cardiac regenerative pathways, including: the ability of some mature myocytes to undergo karyokinesis/cytokinesis by activation through a receptor-dependent exogenous (or endogenous) factor; the presence of resident cells with evident features of progenitor/stem-like cells; and the potential homing of bone-marrow-derived elements able to adopt, at least in part, repairing properties (Fig. 1). Why, in the case of chronic or acute injuries, are these pathways not capable of achieving cardiac tissue regeneration? Maybe the number of regenerative cells is not enough to allow the correct healing of a large ischemic area. The rapid formation of fibrotic tissue, which is the major healing mechanism in the adult injured mammalian heart, could represent an impediment for the migration of cardiac stem cells from the vital cardiac tissue to the damaged area. Alternatively, the inflammatory environment and the lack of oxygen may limit the survival, proliferation, and differentiation of the endogenous regenerative pool [13], in terms of both cycling myocytes and activated progenitor cells.

The intrinsic activation and efficiency of these individual mechanisms could be dependent on the physiological or pathological environment, and will certainly be the targets of several studies.

Concerning the regeneration properties of cardiac progenitor cells, these are very limited after major damage, probably as a negative counterpart of the specific evolutionary advantage achieved by mammals, as previously highlighted. On the other hand, the contribution of bone-marrow-derived c-kit+ cells to the pool of biopsy-derived "regenerative cells" may be prevalent only after ischemic damage, whereas induction of mitosis in mature myocytes could be part of a normal physiological tissue renewal.

In this scenario, slowing down the fibrotic response, promoting cardiomyocyte reentry into the cell cycle, stimulating homing, proliferation, and differentiation of stem cells, and improving their isolation and expansion protocols are the main targets to pursue, in order to shift from healing to regeneration of the myocardium.

References

1. Gruenwald, P. and Minh, H.N., *Evaluation of body and organ weights in perinatal pathology. II. Weight of body and placenta of surviving and of autopsied infants.* Am J Obstet Gynecol, 1961. **82**: p. 312–9.
2. Adler, C.P. and Friedburg, H., *Myocardial DNA content, ploidy level and cell number in geriatric hearts: post-mortem examinations of human myocardium in old age.* J Mol Cell Cardiol, 1986. **18**(1): p. 39–53.
3. Claycomb, W.C. and Bradshaw, H.D., Jr., *Acquisition of multiple nuclei and the activity of DNA polymerase alpha and reinitiation of DNA replication in terminally differentiated adult cardiac muscle cells in culture.* Dev Biol, 1983. **99**(2): p. 331–7.
4. Nag, A.C. and Cheng, M., *DNA synthesis of adult mammalian cardiac muscle cells in long-term culture.* Tissue Cell, 1986. **18**(4): p. 491–7.
5. Anversa, P. and Kajstura, J., *Ventricular myocytes are not terminally differentiated in the adult mammalian heart.* Circ Res, 1998. **83**(1): p. 1–14.
6. Kajstura, J., et al., *Myocyte proliferation in end-stage cardiac failure in humans.* Proc Natl Acad Sci U S A, 1998. **95**(15): p. 8801–5.
7. Beltrami, A.P., et al., *Evidence that human cardiac myocytes divide after myocardial infarction.* N Engl J Med, 2001. **344**(23): p. 1750–7.

8. Nadal-Ginard, B., et al., *Myocyte death, growth, and regeneration in cardiac hypertrophy and failure.* Circ Res, 2003. **92**(2): p. 139–50.
9. Gaetani, R., et al., *New perspectives to repair a broken heart.* Cardiovasc Hematol Agents Med Chem, 2009. **7**(2): p. 91–107.
10. Hsieh, P.C., et al., *Evidence from a genetic fate-mapping study that stem cells refresh adult mammalian cardiomyocytes after injury.* Nat Med, 2007. **13**(8): p. 970–4.
11. Bergmann, O., et al., *Evidence for cardiomyocyte renewal in humans.* Science, 2009. **324**(5923): p. 98–102.
12. Buja, L.M. and Vela, D., *Cardiomyocyte death and renewal in the normal and diseased heart.* Cardiovasc Pathol, 2008. **17**(6): p. 349–74.
13. Ausoni, S. and Sartore, S., *From fish to amphibians to mammals: in search of novel strategies to optimize cardiac regeneration.* J Cell Biol, 2009. **184**(3): p. 357–64.
14. Pasumarthi, K.B. and Field, L.J., *Cardiomyocyte cell cycle regulation.* Circ Res, 2002. **90**(10): p. 1044–54.
15. Sherr, C.J. and Roberts, J.M., *CDK inhibitors: positive and negative regulators of G1-phase progression.* Genes Dev, 1999. **13**(12): p. 1501–12.
16. Tamamori-Adachi, M., et al., *Differential regulation of cyclin D1 and D2 in protecting against cardiomyocyte proliferation.* Cell Cycle, 2008. **7**(23): p. 3768–774.
17. Tamamori-Adachi, M., et al., *Critical role of cyclin D1 nuclear import in cardiomyocyte proliferation.* Circ Res, 2003. **92**(1): p. e12–9.
18. Steinhelper, M.E., et al., *Proliferation in vivo and in culture of differentiated adult atrial cardiomyocytes from transgenic mice.* Am J Physiol, 1990. **259**(6 Pt 2): p. H1826–34.
19. Daud, A.I., et al., *Identification of SV40 large T-antigen-associated proteins in cardiomyocytes from transgenic mice.* Am J Physiol, 1993. **264**(5 Pt 2): p. H1693–700.
20. Tsai, S.C., et al., *Simian virus 40 large T antigen binds a novel Bcl-2 homology domain 3-containing proapoptosis protein in the cytoplasm.* J Biol Chem, 2000. **275**(5): p. 3239–46.
21. Borges, A. and Liew, C.C., *Telomerase activity during cardiac development.* J Mol Cell Cardiol, 1997. **29**(10): p. 2717–24.
22. Kajstura, J., et al., *Telomere shortening is an in vivo marker of myocyte replication and aging.* Am J Pathol, 2000. **156**(3): p. 813–9.
23. Beltrami, A.P., et al., *Adult cardiac stem cells are multipotent and support myocardial regeneration.* Cell, 2003. **114**(6): p. 763–76.
24. Laugwitz, K.L., et al., *Postnatal isl1+ cardioblasts enter fully differentiated cardiomyocyte lineages.* Nature, 2005. **433**(7026): p. 647–53.
25. Messina, E., et al., *Isolation and expansion of adult cardiac stem cells from human and murine heart.* Circ Res, 2004. **95**(9): p. 911–21.
26. Oh, H., et al., *Cardiac progenitor cells from adult myocardium: homing, differentiation, and fusion after infarction.* Proc Natl Acad Sci U S A, 2003. **100**(21): p. 12313–8.
27. Pfister, O., et al., *CD31- but Not CD31+ cardiac side population cells exhibit functional cardiomyogenic differentiation.* Circ Res, 2005. **97**(1): p. 52–61.
28. Ott, H.C., *The adult human heart as a source for stem cells: repair strategies with embryonic-like progenitor cells.* Nat Clin Pract Cardiovasc Med, 2006. **4**(Suppl 1): p. S27–39.
29. Cai, J., Weiss, M.L., and Rao, M.S., *In search of "stemness".* Exp Hematol, 2004. **32**(7): p. 585–98.
30. Rota, M., et al., *Local activation or implantation of cardiac progenitor cells rescues scarred infarcted myocardium improving cardiac function.* Circ Res, 2008. **103**(1): p. 107–16.
31. Bearzi, C., et al., *Human cardiac stem cells.* Proc Natl Acad Sci U S A, 2007. **104**(35): p. 14068–73.
32. Massberg, S., et al., *Immunosurveillance by hematopoietic progenitor cells trafficking through blood, lymph, and peripheral tissues.* Cell, 2007. **131**(5): p. 994–1008.
33. Pouly, J., et al., *Cardiac stem cells in the real world.* J Thorac Cardiovasc Surg, 2008. **135**(3): p. 673–8.
34. Koninckx, R., et al., *Cardiac stem cells in the real world.* J Thorac Cardiovasc Surg, 2008. **136**(3): p. 797–8; author reply 798.

35. Tallini, Y.N., et al., *c-Kit expression identifies cardiovascular precursors in the neonatal heart.* Proc Natl Acad Sci U S A, 2009. **106**(6): p. 1808–13.
36. Barile, L., et al., *Bone marrow-derived cells can acquire cardiac stem cells properties in damaged heart.* J Cell Mol Med, 2009. In press
37. Chimenti, I., et al., *c-Kit cardiac progenitor cells: what is their potential?* Proc Natl Acad Sci U S A, 2009. **106**(28): p. E78; author reply E79.
38. Matsuura, K., et al., *Adult cardiac Sca-1-positive cells differentiate into beating cardiomyocytes.* J Biol Chem, 2004. **279**(12): p. 11384–91.
39. Matsuura, K., et al., *Transplantation of cardiac progenitor cells ameliorates cardiac dysfunction after myocardial infarction in mice.* J Clin Invest, 2009. **119**(8): p. 2204–17.
40. van Vliet, P., et al., *Progenitor cells isolated from the human heart: a potential cell source for regenerative therapy.* Neth Heart J, 2008. **16**(5): p. 163–9.
41. Smits, A.M., et al., *Human cardiomyocyte progenitor cell transplantation preserves long-term function of the infarcted mouse myocardium.* Cardiovasc Res, 2009. **83**(3): p. 527–35.
42. Chien, K.R., Domian, I.J., and Parker, K.K., *Cardiogenesis and the complex biology of regenerative cardiovascular medicine.* Science, 2008. **322**(5907): p. 1494–7.
43. Wu, S.M., Chien, K.R., and Mummery, C., *Origins and fates of cardiovascular progenitor cells.* Cell, 2008. **132**(4): p. 537–43.
44. Martin, C.M., et al., *Persistent expression of the ATP-binding cassette transporter, Abcg2, identifies cardiac SP cells in the developing and adult heart.* Dev Biol, 2004. **265**(1): p. 262–75.
45. Davis D.R., Zhang, Y., Smith, R.R., Cheng, K., Terrovitis, J., Malliaras, K., Li, T., White, A., Makkar, R., and Marban, E., *Validation of the cardiosphere method to culture cardiac progenitor cells from myocardial tissue.* PLoS One, 2009. **4**(9): p. e7195.
46. Smith, R.R., et al., *Regenerative potential of cardiosphere-derived cells expanded from percutaneous endomyocardial biopsy specimens.* Circulation, 2007. **115**(7): p. 896–908.
47. Smith, R.R., Chimenti, I., and Marban, E., *Unselected human cardiosphere-derived cells are functionally superior to c-Kit- or CD90-purified cardiosphere-derived cells.* in *AHA scientific session 2008, Orlando, FL, USA*. Circulation, 2008. **118**:p. S_420.
48. Tang, Y.L., et al., *Hypoxic preconditioning enhances the benefit of cardiac progenitor cell therapy for treatment of myocardial infarction by inducing CXCR4 expression.* Circ Res, 2009. **104**(10): p. 1209–16.
49. Galvez, B.G., et al., *Cardiac mesoangioblasts are committed, self-renewable progenitors, associated with small vessels of juvenile mouse ventricle.* Cell Death Differ, 2008. **15**(9): p. 1417–28.
50. Parmacek, M.S. and Epstein, J.A., *Cardiomyocyte renewal.* N Engl J Med, 2009. **361**(1): p. 86–8.
51. Katz, E.B., et al., *Cardiomyocyte proliferation in mice expressing alpha-cardiac myosin heavy chain-SV40 T-antigen transgenes.* Am J Physiol, 1992. **262**(6 Pt 2): p. H1867–76.
52. Chaudhry, H.W., et al., *Cyclin A2 mediates cardiomyocyte mitosis in the postmitotic myocardium.* J Biol Chem, 2004. **279**(34): p. 35858–66.
53. Pasumarthi, K.B., et al., *Targeted expression of cyclin D2 results in cardiomyocyte DNA synthesis and infarct regression in transgenic mice.* Circ Res, 2005. **96**(1): p. 110–8.
54. Bersell, K., et al., *Neuregulin1/ErbB4 signaling induces cardiomyocyte proliferation and repair of heart injury.* Cell, 2009. **138**(2): p. 257–70.
55. Wang, Z., et al., *Neuregulin-1 enhances differentiation of cardiomyocytes from embryonic stem cells.* Med Biol Eng Comput, 2009. **47**(1): p. 41–8.
56. Gassmann, M., et al., *Aberrant neural and cardiac development in mice lacking the ErbB4 neuregulin receptor.* Nature, 1995. **378**(6555): p. 390–4.
57. Cai, C.L., et al., *A myocardial lineage derives from Tbx18 epicardial cells.* Nature, 2008. **454**(7200): p. 104–8.
58. Zhou, B., et al., *Epicardial progenitors contribute to the cardiomyocyte lineage in the developing heart.* Nature, 2008. **454**(7200): p. 109–13.
59. Zhou, B., et al., *Nkx2-5- and Isl1-expressing cardiac progenitors contribute to proepicardium.* Biochem Biophys Res Commun, 2008. **375**(3): p. 450–3.
60. Limana, F., et al., *Identification of myocardial and vascular precursor cells in human and mouse epicardium.* Circ Res, 2007. **101**(12): p. 1255–65.

61. Quaini, F., et al., *Chimerism of the transplanted heart.* N Engl J Med, 2002. **346**(1): p. 5–15.
62. Thiele, J., et al., *Mixed chimerism of cardiomyocytes and vessels after allogeneic bone marrow and stem-cell transplantation in comparison with cardiac allografts.* Transplantation, 2004. **77**(12): p. 1902–5.
63. Brockes, J.P., *Amphibian limb regeneration: rebuilding a complex structure.* Science, 1997. **276**(5309): p. 81–7.
64. Odelberg, S.J., Kollhoff, A., and Keating, M.T., *Dedifferentiation of mammalian myotubes induced by msx1.* Cell, 2000. **103**(7): p. 1099–109.
65. Chen, S., et al., *Dedifferentiation of lineage-committed cells by a small molecule.* J Am Chem Soc, 2004. **126**(2): p. 410–1.

Multiple Sources for Cardiac Stem Cells and Their Cardiogenic Potential

Antonio Paolo Beltrami, Daniela Cesselli, and Carlo Alberto Beltrami

Abstract The belief that the heart is a terminally differentiated organ was a very well established notion among the scientific community until the early 2000s, although several authors tried to challenge this dogma over the years. Nonetheless, myocyte turnover was only accepted after the demonstration of the intense proliferation that occurs, in human hearts, acutely after myocardial infarction. The first clues indicating that myocytes could originate from unsuspected cell sources, characterized by migratory and differentiation capabilities, were provided by studying the chimerism of transplanted hearts. Following these studies, several classes of cardiac resident primitive cells endowed with cardiomyogenic potential were discovered. Specifically, murine cells expressing c-Kit, Sca1, Abcg2, Isl1, Tbx18, or Wt1 demonstrated their ability to differentiate into cardiac myocytes. Regarding human hearts, cardiospheres, c-Kit^{+} cells, multipotent adult stem cells, and possibly epicardial cells can differentiate into cardiac myocytes. However, cardiac stem cell (CSC) biology is at its beginning and critical questions such as the origin of CSCs and the relationships existing between different stem/progenitor cell classes still need to be answered.

Keywords Cardiac stem cells • Myocytes • Differentiation • Myocardial infarction • Stem cells

Abbreviations

ABC ATP-binding cassette
BrdU 5-bromo-2′-deoxyuridine
Bry brachyury
CDC cardiosphere-derived cell

C.A. Beltrami (✉)
Centro Interdipartimentale di Medicina Rigenerativa (CIME),
Università degli Studi di Udine, Udine, Italy
e-mail: beltrami@uniud.it

I.S. Cohen and G.R. Gaudette (eds.), *Regenerating the Heart*, Stem Cell Biology and Regenerative Medicine, DOI 10.1007/978-1-61779-021-8_10,
© Springer Science+Business Media, LLC 2011

CSC	cardiac stem cell
E	embryonic day
EGFP	enhanced green fluorescence protein
EMT	epithelial–mesenchymal transition
HGF	hepatocyte growth factor
MASC	multipotent adult stem cell
MDR1	multi-drug-resistance 1
MI	myocardial infarction
NRG1	neuregulin 1
SCF	stem cell factor
SP	side population
VEGF	vascular endothelial growth factor

1 A Paradigm Shift in Cardiac Cell Turnover

The belief that the heart is a terminally differentiated organ was a very well established notion among the scientific community until the early 2000s [1], although several authors tried to challenge this dogma over the years [2]. Nonetheless, myocyte turnover was only accepted after the demonstration of the intense proliferation that occurs, in human hearts, acutely after myocardial infarction (MI) [3]. Sophisticated genetic fate-mapping studies [4] and radiocarbon dating of myocyte DNA compellingly confirmed this finding [5]. Recently, Anversa's group [6], by examining the percentage of myocytes, endothelial cells, and fibroblasts labeled by iododeoxyuridine in postmortem samples obtained from cancer patients who received the thymidine analog for therapeutic purposes, established that the lifespan of new myocytes, fibroblasts, and endothelial cells was approximately 4.5, 5, and 8 years, respectively. Moreover, utilizing two independent mathematical models (the hierarchical and population dynamics models), the same group [7] calculated that in the female heart, myocyte turnover occurs at a rate of 10%, 14%, and 40% per year at 20, 60, and 100 years of age, respectively. Corresponding values in the male heart were 7%, 12%, and 32% per year, documenting that cardiomyogenesis involves a large and progressively increasing number of parenchymal cells with aging. Once accepted that cardiac myocytes could be generated postnatally, the next fundamental question that was addressed by the scientific community, over the last 5–10 years, regarded the origin of these renewing myocytes. Two alternative, but not mutually exclusive, possibilities were put forward: either a class of nonterminally differentiated myocytes could persist in human hearts [1], or newly formed myocytes could arise from stem/progenitor cells [8]. Studying the chimerism of sex-mismatched cardiac transplants, scientists had the first clues indicating that myocytes could originate from unsuspected cell sources, characterized by migratory and differentiation capabilities [8, 9]. These early studies demonstrated that both poorly differentiated cells, expressing stem cell markers, and fully mature cells of recipient origin could be identified within transplanted hearts [8, 9]. This finding was regarded as the first indirect indication of the possible existence of cardiac resident stem cells.

2 Cardiac Cell Diversity: May One Cell Maintain It All?

If considered as a whole organ, the heart can be viewed mainly as a pump. However, to correctly perform its functions, it requires a perfectly integrated set of many different cell types that include endothelial cells, smooth muscle cells, fibroblasts, and myocytes. Focusing on the latter, they show profound differences in their biological properties, depending on whether they belong to the atria, ventricles, or pacemaker and conduction systems. Multipotent cardiac stem/progenitor cells could account for this diversity, having the potential to generate multiple cell types. However, knowledge of cardiac stem cell (CSC) biology has been accumulated so rapidly in the last few years that it has generated some degree of confusion [10]. Specifically, several cardiac resident cell types with stem cell properties have been described, although the precise relationship between all these putative CSCs is unknown [10]. In this regard, a central question is whether a "master" CSC exists, located at the top of a putative hierarchy, which is able to maintain this diversity through a series of committed progenitor cells [10, 11]. In spite of its importance, this issue is just starting to be addressed by experimental studies that employ either prospective isolation and cloning of specific cell subsets or lineage-tracing techniques [10]. However, data accumulated over the last decade employing embryonic stem cells demonstrate that pluripotent cells can generate different cardiomyocyte and vascular cell types [12–15]. Nonetheless, the dissection of the putative cardiovascular stem/progenitor cell hierarchy is still incomplete. Finally, recent studies that focused on cardiac development demonstrated that the heart forms from two separate progenitor cell populations (i.e. the primary and secondary heart fields) that seem to diverge from a common progenitor at gastrulation [16, 17]. Whether this putative "master" CSC is present in adults is still an open issue.

In this chapter we will review present knowledge of candidate cardiac stem/progenitor cells, but we will also discuss, for each cell type, evidence of cardiomyocyte differentiation.

3 How Many Types of Stem Cells Have Been Identified in the Heart?

As previously mentioned, once the notion that cardiac myocytes could be formed postnatally had been accepted, several investigators tried to identify the origin of these newly formed cells. Since a distinguishing cardiac resident stem/progenitor cell marker was not available at that time, antigens or enzymatic activities expressed by other stem cell types were initially utilized to prospectively isolate primitive cells from mammalian hearts [10]. Specifically, Hoechst effluxing, multidrug-resistance 1 (MDR1) positive, c-Kit^{+}, and Sca1^{+} cells were isolated from adult mammalian hearts that displayed stem cell features, both in vitro and in vivo [18]. Alternatively, candidate stem cells were cultured under stringent conditions able to discourage the expansion of differentiated cell contaminants. This way, cardiosphere-forming cells and multipotent adult stem cells (MASCs) were identified in human hearts [19, 20]. A possible criticism to the above-mentioned approaches is that long-term culture of cells dispersed from intact tissue sources may alter their nature, thus resulting in tissue culture artifacts [10]. This observation has been generalized and very elegantly summarized by Kirkland [21], who considers that, in stem cell biology, a form of uncertainty principle is in play by which it is nearly impossible to simultaneously measure the capability of a cell to proliferate and differentiate. In an attempt to circumvent this problem, researchers used lineage-tracing analysis as an alternative method. This way, $Isl1^+$, $Flk1^+$, $Wt1^+$, and $Tbx18^+$ multipotent cells were identified [22–24]. We will describe in detail these cell types, discussing their cardiogenic potential.

3.1 Cardiac Side Population

The first reports on the prospective isolation of putative CSCs concerned investigations of the existence of a cardiac "side population" (SP).

This latter cell class was first identified in the bone marrow by Goodell et al. [25] while experimenting on staining with the vital dye Hoechst 33342. When analyzed by flow cytometry, vitally stained bone marrow cells show a complex staining profile. Specifically, when dye fluorescence was analyzed simultaneously at two wavelengths, several distinct populations could be identified. Among these, SP was characterized by low levels of fluorescence, and was greatly enriched in viable, highly primitive $Sca1^+Lin^-$ hematopoietic stem cells. This typical SP staining profile could be masked by verapamil, a calcium channel blocker known to inhibit ATP-binding cassette (ABC) transporters, pumps able to extrude toxic substances and dyes from primitive stem cells. Therefore, MDR1 and other members of this family of proteins, such as ABCG2, were considered to be responsible for this stem cell property [25–27].

Great interest arose from an exciting paper from 2001 by Goodell's group [28], showing that in mice whose hematopoietic system was reconstituted by genetically labeled bone marrow SP cells, donor cells could migrate into infarcted myocardium and differentiate into both cardiac myocytes and endothelial cells.

Hierlihy et al. [29], however, were among the first researchers who investigated whether cardiac resident SP cells could be identified in adult hearts. Cardiac SP accounted for approximately 1% of freshly dissociated cardiac cells, could be induced to differentiate into cells expressing cardiomyocyte markers, when cocultured with neonatal myocytes, and formed hematopoietic colonies in suspension, when cultured in a semisolid medium. Furthermore, the frequency of cardiac SP cells was reduced in MEF2C dominant negative mice, displaying a hypoplastic myocardium. In addition, cardiac SP cells obtained from mutant mice formed a significantly higher number of hematopoietic colonies. Interestingly, according to these authors, cardiac SP cells did not express stem cell markers, such as Sca1, c-Kit, CD34, or Flk-2 [29]. Although groundbreaking, in this early report, the authors neither performed clonal analysis nor evaluated directly the in vivo differentiation potential of cardiac SP.

A few years later, Martin et al. [30] studied the expression of Abcg2 (utilized as a marker of the SP) during murine embryogenesis and in adulthood. During embryonic life, Abcg2 was identified in the yolk sac, the early forming blood islands, and the primitive erythrocyte precursor population. In addition, Abcg2 was recognized in the developing liver, whereas a focal expression was observed in most developing organs, including heart, at mid-gestational ages. However, this protein was robustly expressed throughout the myocardium at embryonic day (E) 8.5. Finally, in adulthood, Abcg2 expression persisted in a small population of cardiac mononuclear cells. These cells, which would account for approximately 2% of the dissociated cells, were distinct from endothelial cells and showed the abilities both to differentiate in vitro into α-actinin positive cells and to form colonies in suspension, when grown in semisolid media. The surface immunophenotype of cardiac SP represented an important discrepancy with previous reports, being described as $Sca1^{high}$, $c\text{-}Kit^{low}$, $CD34^{low}$, and $CD45^{low}$ [30]. In addition, Martin et al. did not perform clonal analysis, nor did they study the in vivo differentiation potential of Abcg2 cells.

Pfister et al. [31] were the first to compare the characteristics of cardiac SP cells with those of bone marrow SP cells. The cell surface immunophenotype of bone marrow SP cells differed from that of cardiac SP cells in the expression of CD45, CD34, and CD44. Although c-Kit was expressed at low levels by cardiac SP cells, this could be attributed to enzymatic cleavage during the digestion process. As a way to identify a cardiomyogenic subpopulation within the cardiac SP, the authors decided to employ Sca1 and CD31. $CD31^-Sca1^+$ cells, which constituted approximately 10% of the cardiac SP (i.e. 500–1,000 cells per adult – 8–12-week-old – mouse heart), were located in the sharp upper part of the cardiac SP cytofluorimetric profile. When grown in vitro, cardiac SP cells adhered and spread to the culture dish, expressed GATA4, MEF2C, α-actinin, and troponin I, and, if cocultured with adult cardiomyocytes, presented connexin 43 at the intercalated discs, showed organized sarcomeres, and displayed both spontaneous contractions and intracellular calcium transients synchronized with neighboring cells. It is noteworthy that differentiation occurred in the absence of cell fusion. In contrast, $CD31^+Sca1^+$ cardiac SP cells did not display any of these properties. The major weakness of this work is the lack of both clonal analysis and in vivo evidence of cardiac SP differentiation. However, this is the most convincing work showing that a rare subfraction of cardiac SP cells has the ability to differentiate into morphologically and, most importantly, functionally competent cardiomyocytes [31].

Two central questions in the SP research field concern the relative contribution of MDR1 and Abcg2 to the SP phenotype and the biological significance of these proteins in the maintenance of stem cell homeostasis [27]. Experiments performed on *Mdr1a/1b*$^{-/-}$ mice demonstrated that the expression of this pump was not essential for the bone marrow SP phenotype, whereas *Bcrp1(Abcg2)*$^{-/-}$ mice showed a reduction in the number of SP cells both in bone marrow and in skeletal muscle [26, 32]. Conversely, forced expression of MDR1 in bone marrow cells increased the SP to 3.6% of the total population, whereas similar experiments utilizing ABCG2 overexpression increased the SP to 62.5% of the total population [26, 33]. Finally, mice lacking both Mdr1a/b and Bcrp1 had normal numbers of peripheral blood cells, bone marrow colony-forming cells, and showed a normal hematopoietic development [34]. However, the hematopoietic cells of these animals were more sensitive to mitoxantrone [34]. In conclusion, in the hematopoietic system, these transporters seem not to be essential in conferring stem cell activity, but may act more in providing environmental protection.

Similarly, Pfister et al. [35] tried to dissect the roles played by Mdr1 and Abcg2 in cardiac SP. As opposed to the bone marrow, in *Bcrp1(Abcg2)*$^{-/-}$ adult mice, cardiac SP cells were clearly detectable, although significantly reduced in number. In contrast, *Mdr1a/1b*$^{-/-}$ hearts exhibited a severe depletion of cardiac SP cells. Importantly, both reverse transcriptase PCR analysis, and fluorescence-activated cell sorting analysis on early postnatal *Mdr1a/1b*$^{-/-}$ and *Bcrp1(Abcg2)*$^{-/-}$ mouse hearts demonstrated that cardiac SP phenotype was mediated by *Abcg2* and *Mdr1a/b* in an age-dependent fashion. Utilizing gain-of-function and loss-of function approaches, these authors also demonstrated that the overexpression of *Abcg2* increased the proliferative capacity of cardiac SP cells, whereas its reduction impaired their expandability in vitro. Additionally, *Abcg2* protected cardiac SP cells from apoptosis and necrosis, especially under conditions of increased oxidative stress. Last, *Abcg2* overexpression prevented cardiac SP cells from undergoing differentiation, consistent with what has been demonstrated both in hematopoietic and in retinal stem cells [26, 36].

A final issue that was addressed by Oyama et al. [37] was the in vivo differentiation potential of cardiac SP cells. When isolated from neonatal hearts and injected into uninjured mice, these cells showed their ability to home to and functionally integrate into several adult tissues, such as liver and skeletal muscle. However, cardiac SP cells were significantly recruited to adult hearts following injury. In this case, they showed a particular tropism for the border zone and started to express cardiac myocyte, smooth muscle cell, and endothelial cell markers. Although in this work a clonal analysis was lacking, the investigators showed evidence of in vivo expression of differentiation markers, suggesting cardiac SP multipotency.

In conclusion, several lines of evidence show that cardiac SP cells have the potential to differentiate into cardiac myocytes both in vitro and, even though less convincingly, in vivo. Nonetheless, the lack of studies conducted at a clonal level cannot exclude that the claimed cardiac SP multipotency may be due to the existence of different cardiac resident progenitor cell populations characterized by drug efflux capacity.

3.2 c-Kit^{+} Cells

After having demonstrated that cardiomyocytes can divide following an acute myocardial injury [3], Anversa's group investigated whether these cells could be replenished by cardiac resident stem cells.

For this purpose, putative cardiac resident stem cells were prospectively isolated using c-Kit as a marker [38]. c-Kit, which is the receptor for stem cell factor (SCF), was chosen since it is expressed in a wide range of primitive cells, including hematopoietic stem cells [39], primordial germ cells [40], embryonic stem cells [41], and melanoblasts [42]. In addition, several different cell types, including neural-crest-derived cells (which seem to rely on the SCF/c-Kit axis for their migration) colonize the heart during development [43, 44]. Consistently, c-Kit$^+$ cells, which do not express hematopoietic markers (Lin$^-$), could be identified in murine myocardium at a frequency of about one cell per 10^4 myocytes [38]. Once isolated from adult rat hearts, these cells displayed the major stem cell characteristics, such as clonogenicity, in vitro self-renewal, multipotency, and the ability to repair infarcted hearts. Specifically, clonogenic c-Kit$^+$ CSCs were able to differentiate into endothelial cells, smooth muscle cells, fibroblasts, and myocytes in vitro. When injected into infarcted rat hearts, c-Kit$^+$ cells contributed to myocardial regeneration in the absence of cell fusion. c-Kit$^+$ cells generated capillaries, arterioles, and striated muscle cells in vivo, thus determining the reappearance of contractions in the infarcted area. Additionally, newly formed myocytes, isolated from regenerated tissue, showed mechanical properties similar to those of fully mature spared myocytes.

Although the work described above demonstrated for the first time that cells characterized by the major stem cell properties could be identified in adult hearts, it did not prove the functional integration of c-Kit$^+$ cells in the myocardium nor their role in tissue homeostasis. Anversa's group [45] investigated this issue a few years later. Specifically, they searched for stem cell niches localized in specific areas of the myocardium, demonstrating that these specialized structures contained CSCs and lineage-committed cells, which were connected to supporting cells, represented by myocytes and fibroblasts. CSC-supporting cell interaction was mediated by cadherins and connexins, whereas undifferentiated cells interacted with the α_2-chain of laminin and fibronectin through α_4 integrin. CSC niches were bigger and more frequent in atrial and apical myocardium, but they could be identified in the base midregion as well. Additionally, 5-bromo-2′-deoxyuridine (BrdU) pulse-chase experiments demonstrated that, although BrdU-positive CSCs declined over time, label-retaining CSCs were present throughout the entire myocardium, being more abundant in the atria and apex. Additionally, the parallel increase in frequency of BrdU-dim myocytes over time supported the notion that these could have been originated from BrdU-bright primitive cells during the chase period.

These findings were corroborated by very elegant work by Orkin's group [46] dealing with cardiac development, which focused on Nkx2.5-expressing cells. This transcription factor was easily detected from E8.5 and was usually not present in cells expressing the transcription factor Isl1 (see below). The authors argued that such a low level of coexpression could be due to the fact that Nkx2.5$^+$ cells were mostly derivatives of the primary heart field (i.e. the population of cardiac progenitors which mainly contributes to the left ventricle and atria) or because expression of Isl1 could have declined prior to the onset of expression of Nkx2.5 in the second heart field (i.e. those progenitors that mainly contribute to the right ventricle and outflow tract) or both. Additionally, Nkx2.5$^+$ cells could be differentiated into smooth muscle cells, functionally competent Purkinje cells, atrioventricular node cells, and atrial and ventricular myocytes in vitro. Clonal analysis was performed utilizing an embryonic stem cell in vitro differentiation system. The authors scoured more than 30 different surface proteins for candidate molecules that could enrich

Nkx2.5$^+$ cells of progenitors, and observed that c-Kit-expressing cells appeared less differentiated and possessed 40-fold higher capacity for proliferation with respect to c-Kit$^-$ cells. This was not the case for Sca1-expressing cells. In addition, c-Kit$^+$ cells were shown to be clonogenic and multipotent, being able to differentiate into smooth muscle cells and beating myocytes at the single-cell level. Last, c-Kit$^+$Nkx2.5$^+$ cells were documented in the developing heart field in vivo. c-Kit$^+$ cells accounted for 13 ± 6% of E8.5 Nkx2.5$^+$ cells, but this expression decreased with myocardial maturation. Interestingly, c-Kit$^+$Nkx2.5$^+$ cells were localized in the outer layer of compact myocardium, suggesting that the niche for these progenitors is adjacent to the epicardium.

The above-mentioned study, together with more recent ones [47, 48], suggested that mammalian heart development involves mesoderm specification to a multipotent precursor stage. To identify this putative master CSC and to establish its differentiation potential, Kotlikoff's group developed a bacterial artificial chromosome transgenic mouse in which the enhanced green fluorescent protein (EGFP) was placed under the control of the c-Kit locus. Following this strategy, they observed the presence of c-Kit-EGFP$^+$ cells in the atrial and ventricular walls at 14.5 days after coitus. The total number of these cells increased as the heart expanded in size, peaked shortly after birth (postnatal days 2–3), and decreased in adulthood. In postnatal hearts, c-Kit-EGFP$^+$ cells were localized in the atrioventricular region, in the atrial and ventricular walls, and in the epicardial border. Some of these cells costained with α-actinin and Flk1, whereas a few days after dissociation more than 50% of c-Kit-EGFP$^+$ cells expressed nestin. Interestingly, no coexpression of Isl1 was observed. Quantitatively, c-Kit-EGFP$^+$ epicardial cells represented 0.101 ± 0.004% of all heart cells, whereas nonepicardial c-Kit-EGFP$^+$ cells constituted 0.46 ± 0.14% of all cells. Regarding cell differentiation, postnatal c-Kit-EGFP$^+$ cells could generate endothelial, smooth muscle, and cardiac myocytes at a clonal level. To investigate the involvement of adult c-Kit-EGFP$^+$ cells in tissue repair, a cryoinjury model was studied, showing that cardiac damage resulted in an increased c-Kit-EGFP expression, which peaked 7 days after cryolesion and which occurred in endothelial cells, smooth muscle cells, and mature myocytes of the border zone. Using a similar model, Sussman's group [48] obtained similar results regarding cardiac development, but demonstrated that, after acute myocardial injury, c-Kit$^+$ cells were involved not only in vascular regeneration, but also in cardiomyogenesis.

An unexpected function for c-Kit was recently described by Husain's group [49] utilizing the *W/W^v* c-Kit-deficient mouse model. The authors observed that, as opposed to the wild type, *W/W^v* mice did not adapt to suprarenal aortic constriction by developing myocyte hypertrophy, but predominantly adapted by developing myocyte hyperplasia. In addition, gene expression and immunofluorescence studies performed on cardiomyocytes obtained from *W/W^v* mice suggested that a higher fraction of these cells were able to reenter the cell cycle when the cells were exposed to pressure overload, thus resulting in an improvement of left ventricular function and survival. Interestingly, in *W/W^v* mice, pressure overload caused a more than tenfold increase in the number of c-Kit$^+$ interstitial cells and a profound reduction in the number of c-Kit$^+$/GATA4$^+$ progenitor cells. Altogether, these findings suggest that, similarly to what has been described for smooth muscle cells [50], melanoblasts [51], and embryonic stem cells [52], c-Kit plays a relevant role in stem cell differentiation.

Finally, CSCs were also identified in human hearts. c-Kit$^+$ cells, isolated from small samples of myocardium, displayed the fundamental properties of stem cells in vitro (i.e. self-renewal, clonogenicity, and multipotency) together with the ability to form arterioles, capillaries, and cardiac myocytes when locally injected in the infarcted myocardium of immunocompromised rodents [53]. Importantly, c-Kit$^+$ cells differentiated into functionally competent beating cardiomyocytes, without cell fusion. More recently, human c-Kit$^+$ cells were further characterized, demonstrating that they comprise two sub-classes. A first KDR$^+$c-Kit$^+$one, mainly responsible for vascular differentiation and a second KDR-c-Kit$^+$one, more prone to differentiate into cardiomyocytes [54].

Altogether, these results indicate that the heart is seeded with c-Kit$^+$ cells during development and that these cells may persist until adulthood. Additionally, c-Kit$^+$ cells isolated from adult hearts are clonogenic, self-renewing, multipotent, and contribute to myocardial regeneration after injury.

3.3 *Sca1$^+$ Cells*

Sca1, a member of the Ly-6 family of glycophosphatidylinositol-anchored membrane proteins thought to be involved in cellular adhesion and signaling, has been extensively used for enrichment of murine hematopoietic stem cells and, more recently, to identify tissue-resident and cancer stem cells [55]. In spite of this, its function has been started to be decrypted only recently [55].

A cardiac resident progenitor cell population expressing this stem cell marker was identified by Schneider's group [56] almost simultaneously with the identification of c-Kit$^+$ CSCs. These authors isolated a population of Sca1$^+$ cells, which constituted 14–17% of the nonmyocyte adult cardiac cell population, and was negative for blood cell lineage markers, c-Kit, Flt-1, Flk1, and vascular endothelial cadherin, whereas it was in large part positive for CD31 and CD38. Furthermore, although cardiac SP was highly enriched in Sca1$^+$ cells, selection for Sca1 was enriched nearly 100-fold for cardiac SP cells (although only 3% of Sca1$^+$ cells were cardiac SP cells). Sca1$^+$ cells possessed telomerase activity, expressed the cardiac transcription factors GATA4, MEF2-C, TEF-1, and were negative for Nkx2.5, α-myosin heavy chain, β-myosin heavy chain, atrial and ventricular myosin light chain-2, cardiac and skeletal α-actin, and CRP3. Following a treatment with the demethylating agent 5-azacytidine, Sca1$^+$ cells started to express Nkx2.5, α-myosin heavy chain, β-myosin heavy chain, α-sarcomeric actin, and troponin I. Finally, when injected in a mouse ischemia/reperfusion injury model, Sca1$^+$ cells showed their ability to differentiate mainly (more than 60% of injected cells) into α-sarcomeric actin positive cells in vivo. However, utilizing the Cre/Lox donor/recipient pair to test the issue of cell fusion, the authors showed that half of the labeled myocytes were actually formed by this mechanism. Although this work was groundbreaking, the authors did not evaluate the differentiation potential of adult progenitor cells into cardiac cell types other than myocytes, nor did they perform clonal analysis.

To address clonogenicity and multipotency in an unbiased fashion, Tateishi et al. used a single-cell clonogenic isolation technique to separate a proliferative cell population from freshly dissociated cardiac cells [57]. For this purpose, clones were first expanded in vitro and then characterized. Freshly isolated cardiac cells displayed a low (0.03%) cloning efficiency. Expanded clones were strongly positive for Sca1, expressed mesenchymal stem cell markers (CD90, CD105, CD29, CD44, CD106, CD73, and CD13), did not express hematopoietic and endothelial markers (CD45, CD34, and CD31), and contained rare c-Kit$^+$ cells. Additionally, most of the clones expressed Abcg2, Bmi1, telomerase reverse transcriptase, nestin, as well as some transcription factors associated with pluripotency (i.e. Nanog and SOX2). Moreover, the authors investigated the role of Sca1 in cell proliferation, survival, and differentiation in a mouse model defective for this protein, documenting that Sca1 deficiency impaired clonogenicity, slowed growth kinetics, reduced telomerase activity, and increased H_2O_2-induced apoptosis. However, Sca1 did not impair the ability of clonogenic cells to differentiate into endothelial cells, smooth muscle cells, and cardiomyocytes in vitro. Finally, the transplantation of clonogenic cells in an acute MI animal model demonstrated that Sca1-deficient clones failed to prevent cardiac remodeling owing to their reduced ability to engraft the ischemic area and, possibly, to secrete paracrine factors (i.e. hepatocyte growth factor (HGF) and vascular endothelial growth factor (VEGF)).

In conclusion, murine hearts can harbor a population of clonogenic, multipotent mesenchymal stem cells that express high levels of Sca1. Interestingly, others have demonstrated that a large fraction of uncultured c-Kit$^+$ cells coexpress Sca1 [45, 58]. Whether the first antigen is downregulated upon clonal expansion remains to be determined. A potential limitation of these studies is that, although a number of potential orthologs for this protein have been identified in several animals, the nonmurine Ly-6 proteins bind distinct ligands and appear to have different cellular functions [59].

3.4 *Isl1$^+$ Cells*

As mentioned already, recent work in developmental biology has defined that two fields of cardiac progenitors (i.e. the primary and secondary or anterior heart fields) contribute to cardiac development. Although evidence that the outflow tract of the heart is not present at the linear heart tube stage comes from experiments that go back many years [60], only recently have investigators succeeded in defining from where it originates. Specifically, Buckingham's group [61] identified a transgenic mouse in which β-galactosidase activity was observed in embryonic right ventricle and outflow tract of the heart and in contiguous splanchnic and pharyngeal mesoderm.

Studying these animals, the researchers observed that the *nlacz* transgene had integrated upstream of the *Fgf10* gene and its expression was driven by *Fgf10* gene regulatory sequences. Additionally, cell tracking studies demonstrated that these cardiac structures originated from cells migrating from the pharyngeal arch region into the growing heart tube between E8.25 and E10.5 [61].

Moving from these observations, Evans' group [62] investigated whether both the right ventricle and the outflow tract originated from a common developmental field. In support of this theory, they observed that mice homozygously null for the transcription factor *Isl1* exhibited growth retardation, died at approximately E10.5, and lacked both an outflow tract and a right ventricle. Regarding the pattern of expression of this factor between E7.25 and E8.75, *Isl1*-expressing cells were found to be contiguous to, but medial and dorsal to, differentiating atrial myosin light chain 2^+ myocytes. Additionally, lineage-tracing analysis demonstrated that *Isl1*-expressing cells would contribute to 97% of cells within the outflow tract, 92% of cells within right ventricle, 65% of cells within left atrium, 70% of cells within right atrium and less than 20% of cells within left ventricle. Functionally, Isl1 seemed to be required for proliferation and survival in both pharyngeal endoderm and splanchnic mesoderm, and for migration of cardiac progenitors into the heart. *Isl1* downstream targets, such as bone morphogenetic proteins 2, 4, 6, and 7 and fibroblast growth factors 4, 8, and 10, possibly mediated these effects. Although these very intriguing lineage-tracing experiments demonstrated that a progenitor population could give rise to different cell types in the embryo, they did not address if these cells were able to generate fully functional cells at a clonal level.

Two works from Chien's group [22, 63] addressed this issue. In the first [63], the authors demonstrated the presence of Isl1-expressing cells in fetal and newborn animals (mouse, rat, and human), showing that their organ distribution matched the defined contribution of $Isl1^+$ embryonic precursors. Additionally, lineage-tracing analysis demonstrated that *Isl1* expressing cells generated a substantial proportion of myocytes in vivo. Finally, the authors developed a coculture system able to expand *Isl1*$^+$ cells on a cardiac mesenchymal cell fraction. These culture-expanded cells did not efflux Hoechst 33342 nor did they express c-Kit. Importantly, $Isl1^+$ cardioblasts were able to differentiate into mechanically and electrically functional cells. In a second work, this group examined the differentiation potential of *Isl1*$^+$ cells at a clonal level [22]. For many years, cardiomyocytes, smooth muscle cells, and endothelial cells in the heart were considered to originate from nonoverlapping embryonic precursors derived from distinct origins [22, 64]. However, a possible alternative, already described for *Nkx2.5*$^+$ cells [46], is that a primitive multipotent cardiovascular stem cell may exist and give rise to a hierarchy of downstream intermediates that eventually differentiate into fully mature heart cells. In support of this possibility, lineage-tracing analysis demonstrated that *Isl1* expressing cells could differentiate into endothelial cells and smooth muscle cells, in addition to cardiomyocytes. Clonal analysis, utilizing embryonic-stem-cell-derived cells cultured on cardiac mesenchymal cell feeder layers, revealed the existence of a hierarchical organization, where $Isl1^+Flk1^+Nkx2.5^+$ triple positive cells differentiated into all three lineages, whereas $Nkx2.5^+$ cells differentiated mainly into cardiac myocytes and smooth muscle cells, and Isl1⁺Flk1⁺cells generated endothelial cells and smooth muscle cells. Moreover, these findings could be replicated, at a clonal level, utilizing E8.0 and E8.5 embryos as a source of *Isl1* cells.

Regarding human cardiogenesis, Chien's group [65] identified *Isl1*-expressing cells in fetal hearts at 11–18 weeks of gestation. Additionally, these authors demonstrated, by multiple immunofluorescence labeling, double-positive cells coexpressing *Isl1* together with markers of cardiomyocyte, smooth muscle cell, and endothelial cell differentiation, suggesting that primitive *Isl1* cells could contribute to second heart field derivatives through a series of multiple lineage-restricted progenitors. These results were corroborated utilizing the human embryonic stem cell differentiation system.

In conclusion, *Isl1*$^+$ cells behave as multipotent cardiovascular progenitors of the secondary heart field, contributing to endothelial cells, smooth muscle cells, and cardiomyocytes.

3.5 Flk1$^+$ or KDR$^+$ Cells

To identify the putative master CSC, able to generate both primary and secondary heart field derivatives, researchers studied early developmental events involved in cardiogenesis. Utilizing a retrospective clonal analysis, Buckingham's group [17] demonstrated that both fields segregated early from a common precursor. For these reasons, research focused on early embryonic events associated with the induction and specification of mesoderm, studying brachyury (Bry) expression. Bry is a transcription factor expressed in all nascent mesoderm, which becomes downregulated when cells undergo patterning and specification into derivative tissues [66]. Targeting green fluorescent protein to the Bry locus and utilizing an embryonic stem cell differentiation system, Keller's group [67] demonstrated that hematopoietic and cardiac development proceeded from distinct Bry$^+$ mesoderm subpopulations that developed, within 72 h (day 3.25), in sequential waves from premesoderm cells. According to these authors, Flk1 (VEGF receptor 2) discriminated between hemangioblast precursors (Flk1$^+$) and cardiomyocyte precursors (Flk1$^-$). Additionally, both sets of mesodermal precursors would develop from Bry$^-$Flk1$^-$ epiblast-like cells. However, this finding seems only to be contradictory with the demonstration that Flk1 marks a broad spectrum of mesodermal cell types, including cardiomyocytes [68]. To reconcile these two findings, Keller's group [15] followed the maturation of day 3.25 Bry$^+$Flk1$^-$ cells, analyzing their developmental progression in more detail. To do so, day 3.25 Bry$^+$Flk1$^-$ cells were reaggregated for 24 h, to enable the continued differentiation of Bry-expressing cells, and analyzed thereafter. Upon differentiation, these cells upregulated Flk1. Cell sorting and differentiation experiments demonstrated that cells expressing Flk1 at day 4.25 contained virtually every cardiomyocyte precursor, whereas clonal experiments demonstrated that day 4.25 Flk1$^+$ cells could differentiate into myocytes, smooth muscle cells, and endothelial cells. Additionally, the authors were able to replicate these experiments utilizing freshly isolated cells obtained from the head-fold-stage embryo.

Last, Keller's group [69] generalized this differentiation paradigm, utilizing human embryonic stem cells. In fact, differentiating human embryoid bodies generated a $KDR^{low}/c\text{-}KIT^{-}$ population able to generate cardiomyocytes, endothelial cells, and vascular smooth muscle cells in vitro and, after transplantation, in vivo.

Altogether, these results indicate that a master CSC exists in development. This cell is the progenitor of both heart fields, but its persistence in adulthood is still an open question.

3.6 *Epicardial Cells*

Great interest has arisen from the hypothesis that epicardium could be considered as a pool of cardiomyocyte progenitors [10], complementing the earlier contribution of the primary and the secondary heart fields [70], and suggesting a possible novel cell source for cardiac regeneration.

In the adult heart, the epicardium is a simple squamous epithelium, similar to other mesothelia (e.g. pleural epithelium and peritoneum), and its major function is to provide a smooth surface on which the heart slides during contraction [71]. In addition, this epithelium serves as a seal in the production and sequestration of pericardial fluid into the pericardial space [71]. However, during the past 2 decades, many scientists have focused their attention on the role played by the epicardium in heart development, structure, and function [10, 72]. During the looping stage of heart development, epicardial cells migrate from the proepicardium, an outgrowth of the septum transversum, to the surface of the developing heart tube to form the epicardium and pericardium [10, 71, 72]. Once the heart is covered by the epicardial mesothelial sheet, a subset of these cells undergo epithelial–mesenchymal transition (EMT) and migrate into the subadjacent trabecular myocardium, where they contribute to cardiac fibroblasts, coronary vascular support cells, and adventitial fibroblasts, therefore playing a crucial role in coronary development, valve development, Purkinje fiber differentiation, and the establishment and maintenance of myocardial architecture [72]. Although no consensus has been established regarding the origin of coronary endothelial cells [72, 73], the generation of vasculogenic cells from a mesothelium is a unique mechanism of blood vessel formation in vertebrate development. Thus, scientists started to investigate whether this embryonic property persisted in postnatal hearts. In line with this hypothesis, Bader's group [71] demonstrated that a rat epicardial mesothelial cell line retained many characteristics of the intact epithelium, including the ability to form a polarized epithelium, and expressed many epicardial genes. When subjected to specific growth factors, these cells retain the ability to produce mesenchyme and to differentiate into smooth muscle cells. Similarly, de Vries's group [74] demonstrated that adult human-derived epicardial cells recapitulated, at least in part, the differentiation potential of their embryonic counterparts, being able to spontaneously undergo EMT early during ex vivo culture and, after infection with an adenoviral vector encoding the transcription factor myocardin or after treatment with transforming growth factor β_1 or bone morphogenetic protein 2, to differentiate into smooth muscle cells. Mouse epicardial explants treated with thymosin-β_4 showed an extensive outgrowth of cells that differentiated into both endothelial and smooth muscle cells [75]. These papers, although mainly based on in vitro studies, suggested the possible persistence, within the pericardium, of a pool of primitive cells. Nonetheless, it was van den Hoff's group [76] who provided the first evidence that proepicardium-derived cells could differentiate into cardiomyocytes, showing that cells of the chicken proepicardium were able to spontaneously differentiate into cardiac muscle cells in vitro. However, we had to wait for two innovative papers published in 2008 in *Nature* to achieve the proof of the existence of epicardial progenitor cells able to contribute to the cardiomyocyte lineage [23, 24].

In the first publication, Zhou et al. [24] identified a novel cardiogenic precursor that expressed Wt1, a transcription factor usually restricted to the proepicardium and epicardium during normal heart development. Lineage-tracing experiments demonstrated that most Wt1-derived cells adopted a smooth muscle cell fate, whereas a minority of $Wt1^+$ cells differentiated into endothelial cells. In fetal hearts, 7–10% of ventricular cardiomyocytes and 18% of atrial cardiomyocytes were derived from $Wt1^+$ cells. Importantly, these Wt1-derived myocytes possessed both structural and functional properties of fully mature cardiomyocytes. The authors, taking advantage of several independent methods to control the temporal and spatial window during which Wt1 cells were labeled, inferred the different roles played by these novel cardiac precursors at different developmental stages. Specifically, when tracked from E10.5 to E11.5, $Wt1^+$ cells were able to differentiate into cardiomyocytes, endothelial cells, and smooth muscle cells. This in vivo ability was further confirmed, showing that E11.5 $Wt1^+$ cells differentiate into cardiomyocytes in vitro. Moreover, when the epicardium of E11.5 explanted hearts was selectively labeled by a brief incubation in culture medium containing 5-chloromethylfluorescein diacetate, and the heart was fixed after 2 days of explant culture, labeled cells were found within the myocardium, and a subset, again, expressed cardiomyocyte markers. The authors, by complex fate-mapping studies, determined that $Wt1^+$ proepicardium and epicardial cells were derived from progenitors that expressed Nkx2.5 and Isl1, suggesting that they shared a developmental origin with multipotent $Nkx2.5^+$ and $Isl1^+$ progenitors that funneled the development of the secondary heart field.

Cai et al. [23], on the other hand, reported the existence, in the proepicardium, of a cardiac progenitor cell, which expresses the transcription factor Tbx18, and is able to contribute not only to cardiac fibroblasts and coronary smooth muscle cells, but also to myocytes in the ventricular septum and in the atrial and ventricular walls. LacZ expression in Tbx18:nlacZ knock-in mice showed that this transcription factor was not expressed, within the myocardium, up to E11.5. However, utilizing a lineage-tracing strategy, the authors demonstrated the presence of myocardial Tbx18-derived cells by E9.75 in the region of forming ventricular septum and in scattered regions within both ventricular walls and atria [23]. These cells exhibited a cardiomyocyte fate, since they expressed cardiac troponin T, cardiac troponin I, Gata4, and Nkx2.5. The ability of epicardial cells to migrate in vivo into the myocardium and to differentiate into myocytes was further confirmed by performing an experiment similar to the one described above [24]. Specifically E11.5 explanted hearts were selectively labeled by a brief incubation in culture medium containing carboxyfluorescein succinimidyl ester, and the heart was fixed after 18–24 h of explant culture. Again, labeled cells were found within the myocardium, and expressed cardiac troponin T. Whereas the first cells to infiltrate the heart from the proepicardium gave rise to myocyte lineages only, the Tbx18^{+} epicardial cells that generated vascular support cells and fibroblasts entered the heart at approximately E12.5. Although in the adult heart Tbx18 expression is restricted to the epicardium, some smooth muscle cells of the coronary vasculature, and cardiac fibroblasts, lineage-tracing analysis demonstrated that smooth muscle cells of coronary vessels, 30% of cardiac fibroblasts, and numerous ventricular, septal, and atrial myocytes originated from the Tbx18^{+} precursors. These Tbx18-derived myocytes were morphologically and functionally indistinguishable from their nonlabeled counterparts. Importantly, Tbx18^{+} precursor cells seemed not to produce endothelial cells; these latter cells could originate from an independent proepicardial population, expressing Flk1. Notably, proepicardial cells retained the ability to differentiate, at a clonal level, into cardiac myocytes and smooth muscle cells, whereas postnatal and adult epicardium lacked this potential.

More recently, Kispert's group [77] criticized the genetic lineage study adopted by Evan's group to demonstrate the epicardial origin of some cardiomyocytes in vivo. Specifically, they showed that Tbx18 would be expressed in cardiomyocytes of the interventricular septum and the ventricular wall from E10.5 onwards. The data were confirmed when they evaluated the Cre recombination pattern (i.e. lacZ expression) in Tbx18$^{Cre/+}$;R26^{lacZ} embryos. For this reason, the presence of lacZ-labeled myocytes within the myocardium could not be unequivocally attributed to the migration of proepicardial Tbx18-expressing cells.

Alongside these sophisticated lineage-tracing studies aimed at identifying and defining the contribution of proepicardial/epicardial precursors to cardiomyocyte lineage in vivo, other investigators focused their attention on the identification of myocardial and vascular precursor cells in postnatal human and mouse myocardium. Capogrossi's group [78] showed that both human and mouse epicardial/subepicardial compartments contained two independent populations of cells expressing either c-Kit or CD34, in the absence of the common leukocyte antigen CD45. Some of these cells already expressed the cardiac-specific transcription factors GATA4 and Nkx2.5 and in vitro were able to acquire some endothelial features, such as the ability to actively take up acetylated-LDL. Finally, the authors utilized a murine model of coronary artery ligation to determine the role played by epicardial cells after MI. Focusing on proliferation and differentiation of epicardial c-Kit^{+} cells at different time points after MI, they verified that, after 3 days, the absolute number of epicardial c-Kit^{+} and c-Kit^{+} GATA4^{+} cells was significantly increased, with respect to sham-operated mice. One week after MI, small blood vessels were found in the subepicardial space, some of which included c-Kit^{+} cells expressing either von Willebrand factor or smooth muscle actin, suggesting the ability of epicardial c-Kit$^+$ cells to proliferate and differentiate into endothelial and smooth muscle cells after an acute MI [78]. Castaldo et al. [79], instead, focused their attention on the spatial and temporal distribution of c-Kit$^+$ cells both in normal and in pathological human hearts and observed that c-Kit$^+$ cells mainly accumulated in the atria, with respect to ventricles, and in the epicardium and subepicardium rather than in the myocardium. However, pathological hearts were characterized by a significant increase in the absolute number of c-Kit$^+$ positive cells, especially those located in the subepicardial region [79]. When the authors analyzed the expression of both α_6 integrin on c-Kit$^+$ cells and laminin-1 and laminin-2 in the heart, they verified that α_6 integrin was expressed in most of the c-Kit$^+$ cells in the subepicardium of pathological hearts. Laminin-1 was found predominantly in the subepicardium of adult hearts, especially normal left atrium and pathological left ventricle, whereas laminin-2 was more abundant in pathological hearts. Further in vitro studies demonstrated that both laminin isoforms reduced apoptosis and increased proliferation and migration of human c-Kit$^+$ cells. Moreover, signaling mediated by α_6 integrin was involved in the migration and protection from apoptosis. The presence of c-Kit$^+$ cells in the subepicardial region and the expression of laminin-1 and integrin α_6 could suggest the possible activation, in pathological hearts, of regenerative processes involving an EMT. More recently, the same group showed that adult human hearts affected by ischemic cardiomyopathy lacked epicardial cells at their surface, whereas cells with both epithelial and mesenchymal markers accumulated in the subepicardial region [80]. The authors concluded that these findings may correspond to the activation of EMT in chronic pathological conditions requiring cardiac cell regeneration, eventually leading to epicardial cell pool exhaustion [80].

In conclusion, the crucial role played by epicardium-derived cells in heart development has been clearly demonstrated, and accumulated evidence suggests a significant role of these cells, not only in the development of coronary vessels and heart architecture, but also in the formation of ventricular and atrial myocytes. However, evidence of the persistence, in postnatal hearts, of epicardium-derived cells characterized by self-renewal, clonogenicity, multipotency, and in vivo regenerative potential is still lacking. Moreover, the possibility that, whether present, this cell pool could undergo cell exhaustion in failing hearts will raise new questions on the possible clinical relevance of this putative novel cell source.

3.7 In Vitro Selection of Adult Cardiac Stem Cells

As stated earlier, an approach utilized to select for putative cardiac resident stem cells was to expand primitive cells by applying culture techniques able to discourage the expansion of differentiated cells.

Giacomello's group [19] identified a complex protocol, based on the combined use of a low-serum medium (supplemented with epidermal growth factor, basic fibroblast growth factor, cardiotrophin-1, and thrombin) and polylysine-coated dishes. Utilizing this method, the authors were able to obtain the in vitro expansion of loosely adherent or free-floating multicellular clusters (similar to those described starting from c-Kit$^+$ rat cells [38]), from prenatal and postnatal mouse hearts and adult human hearts. These clusters were named "cardiospheres." Murine cardiospheres began to beat spontaneously soon after their generation, whereas human cardiospheres did so only when cocultured with rat cardiomyocytes. Importantly, cardiospheres could be generated clonally and subcloned, demonstrating clonogenicity and self-renewal in vitro. Regarding the occurrence of primitive cells within cardiospheres, the authors observed that c-Kit$^+$ cells were present from the beginning and were retained even in expanded cardiospheres. Additionally, cardiospheres contained cells undergoing spontaneous cardiac differentiation. In vivo injection of cultured cardiospheres demonstrated the ability of these cells to engraft and differentiate into cardiac myocytes and vascular cells. However, when implanted in a murine model of acute MI, cardiospheres improved cardiac remodelling and function only partially, possibly because the way cells were administered was suboptimal. To circumvent this problem, Marbán's group modified the original method [81]. Specifically, once cardiospheres were obtained (in a way similar to that in [19]), the researchers expanded cardiosphere-derived cells (CDCs) onto fibronectin-coated dishes in a medium containing 20% fetal bovine serum. CDCs exhibited antigenic and cytochemical similarities to cardiospheres as well as some differences. After two passages in vitro, most of the CDCs were CD105$^+$; additionally a large number of cells expressed CD90 and c-Kit. CDCs also contained differentiating cells that expressed α-sarcomeric actin, connexin 43, and coexpressed CD34 and CD31. Importantly, CDCs cocultured with neonatal rat cardiomyocytes showed electrophysiological properties typical of differentiated cells, in the absence of cell fusion. Finally, delivery of CDCs into infarcted hearts resulted in myocardial regeneration and in a significant improvement of cardiac function.

Last, stimulated by the possibility that widely multipotent cells, with mesenchymal features, could reside in multiple organs [82–84], we expanded in culture and compared cells obtained by applying an almost identical method to several adult human tissues [20]. For this purpose, we utilized a culture medium containing 2% serum, epidermal growth factor, and platelet-derived growth factor to generate a population of MASCs from adult human hearts, livers, and bone marrow. MASCs possess every characteristic that a stem cell should display in vitro: clonogenicity, self-renewal, and multipotency, expression of several pluripotent state-specific transcription factors (i.e. OCT4, Nanog, and Sox2), high levels of telomerase activity, and a highly similar gene expression profile, irrespective of the tissue of origin. Immunophenotypically, MASCs have a mesenchymal immunophenotype, expressing high levels of CD13, CD29, CD90, and CD105, and being negative to CD45, CD34, or CD38. Importantly, MASCs can differentiate in vitro into functionally competent derivatives of the three germ layers, including beating cardiomyocytes.

Although both these cell populations are clinically relevant, one of the major biological limitations of these studies is that the nature of both cardiosphere- and MASC-initiating cells is still unknown. Regarding MASCs, gene expression studies indicate that they may be very primitive, epiblast-like cells [20, 85]. However, when implanted in vivo into immunosuppressed animals, MASCs did not generate teratomas excluding their identity with bona fide pluripotent embryonic stem cells [86].

4 How Are Cardiac Stem/Progenitor Cells Related to One Another?

The picture emerging from the evidence described herein is the existence of a hierarchical organization of cardiac stem/progenitor cells during development. $Bry^{-}Flk1^{-}OCT4^{+}Nanog^{+}Rex1^{-}Fgf5^{+}$ epiblast-like cells (an intermediate developmental phase between embryonic stem cells and mesoderm) would be located at the top of this putative hierarchy [67]. These cells generate a $Bry^{+}Flk1^{-}$ primitive streak mesodermic cell, which is able to differentiate both into day 3.25 $Bry^{+}Flk1^{+}$ hemangioblasts and, at a later stage, into day 4.25 $Bry^{+}Flk1^{+}$ cardiac progenitors. These latter cells are the founders of both the primary heart field and the secondary heart field. Finally, *Nkx2.5* identifies muscle cell progenitors, whereas c-Kit defines the most clonogenic among these.

In regard to adult CSCs, the picture is much more complex. The possibility exists that some cardiovascular (or more undifferentiated) progenitors may be retained throughout adulthood and that these are responsible for cardiac turnover. The embryological origin of these cells is still debated, and possibly is a mixture of primary-heart-field-, secondary-heart-field-, pericardial-, and neural-crest-derived cells. The persistence of cells bearing relevant stem cell markers (e.g. c-Kit and Sca1) and the presence of highly immature, epiblast-like cells (expressing Oct4 and Nanog) may be in favor of this hypothesis.

Finally, a possibility that has been put forward by a series of stem cell biologists is that the source of cells for continuous turnover is one or more pluripotent stem cell populations that can circulate around the body and turn into multiple cell types, depending on the local environment [87]. Although fascinating, this hypothesis is still not proven.

5 Back to the Past: Are Terminally Differentiated Cardiomyocytes Able to Proliferate?

As in every respectable thriller, a sensational development cannot be lacking. Just as the scientific community has accepted the end of the dogma of the heart as a terminally differentiated organ, and scientists are discussing the origin of CSCs (sometimes fighting over the real entity of cardiac turnover), a postrevolutionary paper by Bersell et al. [88] claiming that differentiated myocytes can be induced to proliferate and regenerate was published. Specifically, rat ventricular myocyte proliferation in vitro would be controlled by a molecular mechanism involving the growth factor neuregulin 1 (NRG1) and its tyrosine kinase receptor, ErbB4. NRG1 induced only mononucleated cardiomyocytes to divide. Consistently, ErbB4 regulated cardiomyocyte proliferation in vivo. The injection of NRG1 in infarcted adult mice induced cardiomyocyte cell-cycle activity and promoted myocardial regeneration, leading to improved function. Notably, the collected evidence excluded a contribution of undifferentiated progenitor cells to NRG1-induced cardiomyocyte proliferation. Despite its apparent groundbreaking news, the paper is not in conflict with what was previously shown. The presence of dividing myocytes was one of the first direct proofs that the heart is not a terminally differentiated organ [1, 3]. At that time, the authors speculated that dividing myocytes could originate from two, not mutually exclusive, cell sources. Either they could derive from a pool of myocytes able to reenter the cell cycle or they could differentiate from more primitive cells. Although accumulated evidence has clearly demonstrated the role played by this latter source in cardiac regeneration, Bersell et al.'s paper established for the first time the existence of a small fraction of mononucleated myocytes that can reenter the cell cycle upon the activation of NRG1/ErbB4 signaling. This pathway is not effective in activating undifferentiated CSCs to proliferate and differentiate into myocytes. In accordance, other pathways have been shown to be involved in CSC migration, proliferation, differentiation, and survival, (i.e. HGF-cMet, SCF-c-Kit, stromal-cell-derived factor 1/CXCR4, high-mobility-group protein B1, insulin-like growth factor 1–insulin-like growth factor 1 receptor) [89].

In conclusion, this paper is not denying the existence of CSCs, but is rather indicating the presence of another molecular strategy, based on NRG1/ErbB4 signaling, that could provide myocardial regeneration. The possible contribution of different cell sources to tissue regeneration has already been demonstrated in other tissues such as liver [90] and skin [91], where the specific contribution of each cell source is determined by the extent of the damage.

References

1. Anversa P, Kajstura J (1998) Ventricular myocytes are not terminally differentiated in the adult mammalian heart. Circ Res, 83(1), 1–14.
2. Anversa P, Leri A, Kajstura J et al. (2002) Myocyte growth and cardiac repair. J Mol Cell Cardiol, 34(2), 91–105.
3. Beltrami AP, Urbanek K, Kajstura J et al. (2001) Evidence that human cardiac myocytes divide after myocardial infarction. N Engl J Med, 344(23), 1750–1757.
4. Hsieh PC, Segers VF, Davis ME et al. (2007) Evidence from a genetic fate-mapping study that stem cells refresh adult mammalian cardiomyocytes after injury. Nat Med, 13(8), 970–974.
5. Bergmann O, Bhardwaj RD, Bernard S et al. (2009) Evidence for cardiomyocyte renewal in humans. Science, 324(5923), 98–102.
6. Kajstura J, Urbanek K, Perl S et al. (2010) Cardiomyogenesis in the adult human heart. Circ Res 107, 305–315.
7. Kajstura J, Gurusamy N, Ogorek B et al. (2010) Myocyte turnover in the aging human heart. Circ Res 107, 1374–1386.

8. Quaini F, Urbanek K, Beltrami AP et al. (2002) Chimerism of the transplanted heart. N Engl J Med, 346(1), 5–15.
9. Muller P, Pfeiffer P, Koglin J et al. (2002) Cardiomyocytes of noncardiac origin in myocardial biopsies of human transplanted hearts. Circulation, 106(1), 31–35.
10. Martin-Puig S, Wang Z, Chien KR (2008) Lives of a heart cell: tracing the origins of cardiac progenitors. Cell Stem Cell, 2(4), 320–331.
11. Chien KR, Domian IJ, Parker KK (2008) Cardiogenesis and the complex biology of regenerative cardiovascular medicine. Science, 322(5907), 1494–1497.
12. Kolossov E, Lu Z, Drobinskaya I et al. (2005) Identification and characterization of embryonic stem cell-derived pacemaker and atrial cardiomyocytes. FASEB J, 19(6), 577–579.
13. Kehat I, Kenyagin-Karsenti D, Snir M et al. (2001) Human embryonic stem cells can differentiate into myocytes with structural and functional properties of cardiomyocytes. J Clin Invest, 108(3), 407–414.
14. Sachinidis A, Fleischmann BK, Kolossov E et al. (2003) Cardiac specific differentiation of mouse embryonic stem cells. Cardiovasc Res, 58(2), 278–291.
15. Kattman SJ, Huber TL, Keller GM (2006) Multipotent flk-1+ cardiovascular progenitor cells give rise to the cardiomyocyte, endothelial, and vascular smooth muscle lineages. Dev Cell, 11(5), 723–732.
16. Garry DJ, Olson EN (2006) A common progenitor at the heart of development. Cell, 127(6), 1101–1104.
17. Meilhac SM, Esner M, Kelly RG et al. (2004) The clonal origin of myocardial cells in different regions of the embryonic mouse heart. Dev Cell, 6(5), 685–698.
18. Leri A, Kajstura J, Anversa P (2005) Cardiac stem cells and mechanisms of myocardial regeneration. Physiol Rev, 85(4), 1373–1416.
19. Messina E, De Angelis L, Frati G et al. (2004) Isolation and expansion of adult cardiac stem cells from human and murine heart. Circ Res, 95, 911–921.
20. Beltrami AP, Cesselli D, Bergamin N et al. (2007) Multipotent cells can be generated in vitro from several adult human organs (heart, liver and bone marrow). Blood, 110(9), 3438–3446.
21. Kirkland MA (2004) A phase space model of hemopoiesis and the concept of stem cell renewal. Exp Hematol, 32(6), 511–519.
22. Moretti A, Caron L, Nakano A et al. (2006) Multipotent embryonic isl1+ progenitor cells lead to cardiac, smooth muscle, and endothelial cell diversification. Cell, 127(6), 1151–1165.
23. Cai CL, Martin JC, Sun Y et al. (2008) A myocardial lineage derives from Tbx18 epicardial cells. Nature, 454(7200), 104–108.
24. Zhou B, Ma Q, Rajagopal S et al. (2008) Epicardial progenitors contribute to the cardiomyocyte lineage in the developing heart. Nature, 454(7200), 109–113.
25. Goodell MA, Brose K, Paradis G et al. (1996) Isolation and functional properties of murine hematopoietic stem cells that are replicating in vivo. J Exp Med, 183(4), 1797–1806.
26. Zhou S, Schuetz JD, Bunting KD et al. (2001) The ABC transporter Bcrp1/ABCG2 is expressed in a wide variety of stem cells and is a molecular determinant of the side-population phenotype. Nat Med, 7(9), 1028–1034.
27. Challen GA, Little MH (2006) A side order of stem cells: the SP phenotype. Stem Cells, 24(1), 3–12.
28. Jackson KA, Majka SM, Wang H et al. (2001) Regeneration of ischemic cardiac muscle and vascular endothelium by adult stem cells. J Clin Invest, 107(11), 1395–1402.
29. Hierlihy AM, Seale P, Lobe CG et al. (2002) The post-natal heart contains a myocardial stem cell population. FEBS Lett, 530(1–3), 239–243.
30. Martin CM, Meeson AP, Robertson SM et al. (2004) Persistent expression of the ATP-binding cassette transporter, Abcg2, identifies cardiac SP cells in the developing and adult heart. Dev Biol, 265(1), 262–275.
31. Pfister O, Mouquet F, Jain M et al. (2005) CD31- but not CD31+ cardiac side population cells exhibit functional cardiomyogenic differentiation. Circ Res, 97(1), 52–61.
32. Zhou S, Morris JJ, Barnes Y et al. (2002) Bcrp1 gene expression is required for normal numbers of side population stem cells in mice, and confers relative protection to mitoxantrone in hematopoietic cells in vivo. Proc Natl Acad Sci U S A, 99(19), 12339–12344.

33. Bunting KD, Zhou S, Lu T et al. (2000) Enforced P-glycoprotein pump function in murine bone marrow cells results in expansion of side population stem cells in vitro and repopulating cells in vivo. Blood, 96(3), 902–909.
34. Zhou S, Zong Y, Lu T et al. (2003) Hematopoietic cells from mice that are deficient in both Bcrp1/Abcg2 and Mdr1a/1b develop normally but are sensitized to mitoxantrone. Biotechniques, 35(6), 1248–1252.
35. Pfister O, Oikonomopoulos A, Sereti KI et al. (2008) Role of the ATP-binding cassette transporter Abcg2 in the phenotype and function of cardiac side population cells. Circ Res, 103(8), 825–835.
36. Bhattacharya S, Das A, Mallya K et al. (2007) Maintenance of retinal stem cells by Abcg2 is regulated by notch signaling. J Cell Sci, 120(Pt 15), 2652–2662.
37. Oyama T, Nagai T, Wada H et al. (2007) Cardiac side population cells have a potential to migrate and differentiate into cardiomyocytes in vitro and in vivo. J Cell Biol, 176(3), 329–341.
38. Beltrami AP, Barlucchi L, Torella D et al. (2003) Adult cardiac stem cells are multipotent and support myocardial regeneration. Cell, 114(6), 763–776.
39. Bryder D, Rossi DJ, Weissman IL (2006) Hematopoietic stem cells: the paradigmatic tissue-specific stem cell. Am J Pathol, 169(2), 338–346.
40. Kerr CL, Hill CM, Blumenthal PD et al. (2008) Expression of pluripotent stem cell markers in the human fetal testis. Stem Cells, 26(2), 412–421.
41. Zwaka TP, Thomson JA (2005) A germ cell origin of embryonic stem cells? Development, 132(2), 227–233.
42. Motohashi T, Yamanaka K, Chiba K et al. (2009) Unexpected multipotency of melanoblasts isolated from murine skin. Stem Cells, 27(4), 888–897.
43. Kirby ML (1989) Plasticity and predetermination of mesencephalic and trunk neural crest transplanted into the region of the cardiac neural crest. Dev Biol, 134(2), 402–412.
44. Brito FC, Kos L (2008) Timeline and distribution of melanocyte precursors in the mouse heart. Pigment Cell Melanoma Res, 21(4), 464–470.
45. Urbanek K, Cesselli D, Rota M et al. (2006) Stem cell niches in the adult mouse heart. Proc Natl Acad Sci U S A, 103(24), 9226–9231.
46. Wu SM, Fujiwara Y, Cibulsky SM et al. (2006) Developmental origin of a bipotential myocardial and smooth muscle cell precursor in the mammalian heart. Cell, 127(6), 1137–1150.
47. Tallini YN, Greene KS, Craven M et al. (2009) c-Kit expression identifies cardiovascular precursors in the neonatal heart. Proc Natl Acad Sci U S A, 106(6), 1808–1813.
48. Fransioli J, Bailey B, Gude NA et al. (2008) Evolution of the c-kit-positive cell response to pathological challenge in the myocardium. Stem Cells, 26(5), 1315–1324.
49. Li M, Naqvi N, Yahiro E et al. (2008) c-Kit is required for cardiomyocyte terminal differentiation. Circ Res, 102(6), 677–685.
50. Davis BN, Hilyard AC, Nguyen PH et al. (2009) Induction of microRNA-221 by platelet-derived growth factor signaling is critical for modulation of vascular smooth muscle phenotype. J Biol Chem, 284(6), 3728–3738.
51. Hirobe T, Osawa M, Nishikawa S (2003) Steel factor controls the proliferation and differentiation of neonatal mouse epidermal melanocytes in culture. Pigment Cell Res, 16(6), 644–655.
52. Bashamboo A, Taylor AH, Samuel K et al. (2006) The survival of differentiating embryonic stem cells is dependent on the SCF-KIT pathway. J Cell Sci, 119(Pt 15), 3039–3046.
53. Bearzi C, Rota M, Hosoda T et al. (2007) Human cardiac stem cells. Proc Natl Acad Sci U S A, 104(35), 14068–14073.
54. Bearzi C, Leri A, Lo Monaco F et al. (2009) Identification of a coronary vascular progenitor cell in the human heart. Proc Natl Acad Sci USA 106, 15885–15890.
55. Holmes C, Stanford WL (2007) Concise review: stem cell antigen-1: expression, function, and enigma. Stem Cells, 25(6), 1339–1347.
56. Oh BH, Bradfute SB, Gallardo TD et al. (2003) Cardiac progenitor cells from adult myocardium: homing, differentiation, and fusion after infarction. Proc Natl Acad Sci U S A, 100, 12313–12318.

57. Tateishi K, Ashihara E, Takehara N et al. (2007) Clonally amplified cardiac stem cells are regulated by Sca-1 signaling for efficient cardiovascular regeneration. J Cell Sci 120, 1791–1800.
58. Cesselli D, Jakoniuk I, Beltrami AP et al. (2002) Cardiac stem cells are endowed in niches of the mouse heart and possess the ability to divide and differentiate in the various cardiac lineages [Abstract]. Circulation, 106 (Suppl II), 206.
59. Bamezai A (2004) Mouse Ly-6 proteins and their extended family: markers of cell differentiation and regulators of cell signaling. Arch Immunol Ther Exp (Warsz), 52(4), 255–266.
60. de la Cruz MV, Sanchez Gomez C, Arteaga MM et al. (1977) Experimental study of the development of the truncus and the conus in the chick embryo. J Anat, 123(Pt 3), 661–686.
61. Kelly RG, Brown NA, Buckingham ME (2001) The arterial pole of the mouse heart forms from Fgf10-expressing cells in pharyngeal mesoderm. Dev Cell, 1(3), 435–440.
62. Cai CL, Liang X, Shi Y et al. (2003) Isl1 identifies a cardiac progenitor population that proliferates prior to differentiation and contributes a majority of cells to the heart. Dev Cell, 5(6), 877–889.
63. Laugwitz KL, Moretti A, Lam J et al. (2005) Postnatal isl1+ cardioblasts enter fully differentiated cardiomyocyte lineages. Nature, 433(7026), 647–653.
64. Mikawa T, Gourdie RG (1996) Pericardial mesoderm generates a population of coronary smooth muscle cells migrating into the heart along with ingrowth of the epicardial organ. Dev Biol, 174(2), 221–232.
65. Bu L, Jiang X, Martin-Puig S et al. (2009) Human ISL1 heart progenitors generate diverse multipotent cardiovascular cell lineages. Nature, 460(7251), 113–117.
66. Kispert A, Herrmann BG (1994) Immunohistochemical analysis of the Brachyury protein in wild-type and mutant mouse embryos. Dev Biol, 161(1), 179–193.
67. Kouskoff V, Lacaud G, Schwantz S et al. (2005) Sequential development of hematopoietic and cardiac mesoderm during embryonic stem cell differentiation. Proc Natl Acad Sci U S A, 102(37), 13170–13175.
68. Ema M, Takahashi S, Rossant J (2006) Deletion of the selection cassette, but not cis-acting elements, in targeted Flk1-lacZ allele reveals Flk1 expression in multipotent mesodermal progenitors. Blood, 107(1), 111–117.
69. Yang L, Soonpaa MH, Adler ED et al. (2008) Human cardiovascular progenitor cells develop from a KDR+ embryonic-stem-cell-derived population. Nature, 453(7194), 524–528.
70. Buckingham M, Meilhac S, Zaffran S (2005) Building the mammalian heart from two sources of myocardial cells. Nat Rev Genet, 6(11), 826–835.
71. Wada AM, Smith TK, Osler ME et al. (2003) Epicardial/mesothelial cell line retains vasculogenic potential of embryonic epicardium. Circ Res, 92(5), 525–531.
72. Lie-Venema H, van den Akker NM, Bax NA et al. (2007) Origin, fate, and function of epicardium-derived cells (EPDCs) in normal and abnormal cardiac development. Sci World J, 7, 1777–1798.
73. Winter EM, Gittenberger-de Groot AC (2007) Epicardium-derived cells in cardiogenesis and cardiac regeneration. Cell Mol Life Sci, 64(6), 692–703.
74. van Tuyn J, Atsma DE, Winter EM et al. (2007) Epicardial cells of human adults can undergo an epithelial-to-mesenchymal transition and obtain characteristics of smooth muscle cells in vitro. Stem Cells, 25(2), 271–278.
75. Smart N, Risebro CA, Melville AA et al. (2007) Thymosin beta4 induces adult epicardial progenitor mobilization and neovascularization. Nature, 445(7124), 177–182.
76. Kruithof BP, van Wijk B, Somi S et al. (2006) BMP and FGF regulate the differentiation of multipotential pericardial mesoderm into the myocardial or epicardial lineage. Dev Biol, 295(2), 507–522.
77. Christoffels VM, Grieskamp T, Norden J et al. (2009) Tbx18 and the fate of epicardial progenitors. Nature, 458(7240), E8–9; discussion E9–10.
78. Limana F, Zacheo A, Mocini D et al. (2007) Identification of myocardial and vascular precursor cells in human and mouse epicardium. Circ Res, 101(12), 1255–1265.

79. Castaldo C, Di Meglio F, Nurzynska D et al. (2008) CD117-positive cells in adult human heart are localized in the subepicardium, and their activation is associated with laminin-1 and alpha6 integrin expression. Stem Cells, 26(7), 1723–1731.
80. Di Meglio F, Castaldo C, Nurzynska D et al. (2009) Epicardial cells are missing from the surface of hearts with ischemic cardiomyopathy: a useful clue about the self-renewal potential of the adult human heart? Int J Cardiol.
81. Smith RR, Barile L, Cho HC et al. (2007) Regenerative potential of cardiosphere-derived cells expanded from percutaneous endomyocardial biopsy specimens. Circulation 115, 896–908.
82. Jiang Y, Jahagirdar BN, Reinhardt RL et al. (2002) Pluripotency of mesenchymal stem cells derived from adult marrow. Nature, 418(6893), 41–49.
83. Jiang Y, Vaessen B, Lenvik T et al. (2002) Multipotent progenitor cells can be isolated from postnatal murine bone marrow, muscle, and brain. Exp Hematol, 30(8), 896–904.
84. Krause DS, Theise ND, Collector MI et al. (2001) Multi-organ, multi-lineage engraftment by a single bone marrow-derived stem cell. Cell, 105(3), 369–377.
85. Cesselli D, Beltrami AP, Rigo S et al. (2009) Multipotent progenitor cells are present in human peripheral blood. Circ Res, 104(10), 1225–1234.
86. Beltrami AP, Cesselli D, Beltrami CA (2009) Pluripotency rush! Molecular cues for pluripotency, genetic reprogramming of adult stem cells, and widely multipotent adult cells. Pharmacol Ther 124, 23–30.
87. Slack JM (2008) Origin of stem cells in organogenesis. Science, 322(5907), 1498–1501.
88. Bersell K, Arab S, Haring B et al. (2009) Neuregulin1/ErbB4 signaling induces cardiomyocyte proliferation and repair of heart injury. Cell, 138(2), 257–270.
89. Dimmeler S, Leri A (2008) Aging and disease as modifiers of efficacy of cell therapy. Circ Res, 102(11), 1319–1330.
90. Fausto N, Campbell JS (2003) The role of hepatocytes and oval cells in liver regeneration and repopulation. Mech Dev, 120(1), 117–130.
91. Ito M, Liu Y, Yang Z et al. (2005) Stem cells in the hair follicle bulge contribute to wound repair but not to homeostasis of the epidermis. Nat Med, 11(12), 1351–1354.

Skeletal Muscle Stem Cells in the Spotlight: The Satellite Cell

Zipora Yablonka-Reuveni and Kenneth Day

Abstract The formation of striated muscle during embryogenesis involves morphogenetic programs that orchestrate the development of the functional units of skeletal muscle – the myofibers. These multinucleated syncytia share with cardiomyocytes some structural and contractile characteristics that commonly define skeletal and cardiac muscles as sarcomeric muscles. However, distinctive to adult skeletal muscle is its repair capacity that is contributed by satellite cells. These cells are specified during development for response to muscle trauma in the adult. Satellite cells are myogenic progenitors that reside on the surface of skeletal muscle myofibers beneath the basal lamina and provide a supply of myoblasts that readily contribute to muscle regeneration. Activated satellite cells give rise to myoblast progeny that fuse into new myofibers or with damaged myofibers. This differentiation process of satellite cell progeny integrates temporal gene expression with cell cycle control. Furthermore, satellite cells also may self-renew, which qualifies their identity as muscle stem cells. Similarities in the structure of the contractile units of skeletal and cardiac muscle have led cardiologists to consider novel approaches toward the use of skeletal muscle stem cells for improving cardiac regeneration. In this chapter, we discuss the basic biology of satellite cells and introduce in brief the hopes and obstacles for repairing muscle deterioration with donor satellite cells.

Keywords Satellite cells • Myogenesis • Pax7 • MRFs • Myofibers • Intrafusal fibers • Aging • Muscular dystrophy • Skeletal muscle repair • Myocardium repair

---

Z. Yablonka-Reuveni (✉)
Department of Biological Structure, University of Washington
School of Medicine, Seattle, WA, USA
e-mail: reuveni@u.washington.edu

I.S. Cohen and G.R. Gaudette (eds.), *Regenerating the Heart*, Stem Cell Biology and Regenerative Medicine, DOI 10.1007/978-1-61779-021-8_11,
© Springer Science+Business Media, LLC 2011

1 Introduction

1.1 Relevance of Skeletal Muscle Satellite Cells to Cardiac Repair

Similarities in the structure of the contractile units of skeletal and cardiac muscle have led cardiologists to consider novel approaches toward the use of skeletal muscle myogenic progenitors and myoblasts for repair of failing myocardium [1, 2]. However, the actual use of such cells to restore cardiac tissue has not progressed as fast as it had been hoped [3, 4]. Recent studies have again raised hopes that the use of skeletal muscle myogenic cells might indeed be feasible for cardiac repair [5–11].

1.2 About This Chapter

In this chapter we provide an overview of the basic biology of skeletal muscle satellite cells, myogenic progenitors reside on the myofiber surface and contributing to muscle homeostasis and repair [12–14]. Other cell types isolated from skeletal muscle, such as mesoangioblasts, pericytes, and myoendothelial cells also seem to have myogenic potency, but it remains unknown if these additional cell types participate normally in muscle maintenance and repair [15–17]. This chapter focuses exclusively on the satellite cell, a bona fide myogenic progenitor defined initially by electron microcopy on the basis of its localization underneath the myofiber basal lamina [18, 19]. More recent studies have identified molecular markers of satellite cells, especially the expression of Pax7, allowing the distinction of satellite cells by standard immunofluorescent microscopy [20] (Fig. 1).

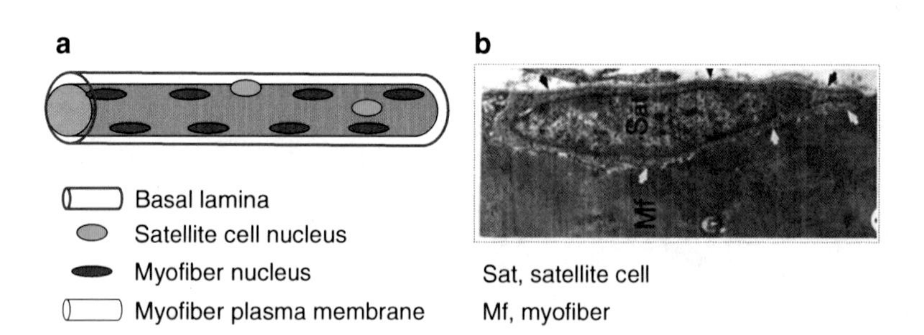

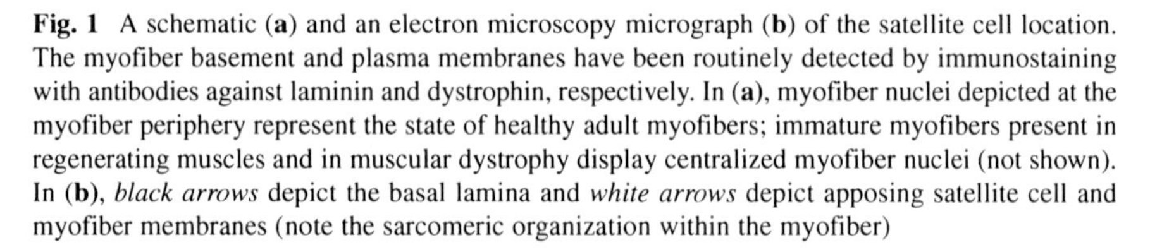

Fig. 1 A schematic (**a**) and an electron microscopy micrograph (**b**) of the satellite cell location. The myofiber basement and plasma membranes have been routinely detected by immunostaining with antibodies against laminin and dystrophin, respectively. In (**a**), myofiber nuclei depicted at the myofiber periphery represent the state of healthy adult myofibers; immature myofibers present in regenerating muscles and in muscular dystrophy display centralized myofiber nuclei (not shown). In (**b**), *black arrows* depict the basal lamina and *white arrows* depict apposing satellite cell and myofiber membranes (note the sarcomeric organization within the myofiber)

Herein, we review classic and contemporary studies in satellite cell biology. We first discuss the niche and function of these cells within the intact adult muscle, followed by details on approaches for satellite cell detection, isolation, and characterization using cell culture. We then focus on the molecular signature of satellite cells and their progeny, emphasizing the topics of satellite cell specification and the dynamics of satellite cell progeny as they transit through proliferation, differentiation, and self-renewal. We conclude with a brief description of the status of cell-based therapy within the context of satellite cell delivery to host skeletal muscle. This chapter is not an exhaustive review; we cover topics mainly in accordance with our current research interests. There is an immense amount of literature on areas not discussed here, including satellite cell regulation by extracellular factors, mitogens, autocrine/paracrine factors, mechanical loading, and more; the reader should refer to other, original reviews on such subjects (e.g., [12, 21–27]).

2 Location and Function of Satellite Cells

Satellite cells are myogenic progenitors that are required for the growth, repair, and maintenance of adult skeletal muscle [12, 28]. The anatomically distinct position of the satellite cell on the surface of the myofiber beneath the basal lamina provides immediacy and sensitivity to a postmitotic tissue such as skeletal muscle that is critically dependent on mechanical, structural, and functional integrity [12, 21]. In the juvenile growth phase, when muscles enlarge, the satellite cells are proliferative and add nuclei to growing myofibers [29–32]. In adult muscle, satellite cells are typically quiescent, until their activation is invoked by muscle injury [32, 33]. Subtle injuries may lead to minimal proliferation of activated satellite cells, whereas major trauma can recruit greater numbers of satellite cells and promote a prolonged episode of proliferation prior to differentiation. Activation of these myogenic precursors is controlled by proximal signals from the muscle niche, microvasculature, and inflammation [24, 26, 34–38]. Systemic factors may also regulate satellite cell activation [39, 40]. Following their activation, satellite cells may contribute to repair of damaged myofibers and also may generate new myofibers following rounds of cell division and fusion of myoblast progeny [41, 42]. The behavior of satellite cells is under stringent regulatory control to orchestrate balance and decisions among their maintenance in quiescence, entry into proliferation and continuity of the cell cycle, and terminal differentiation [23, 43]. Furthermore, apart from their ability to fortify myofibers and contribute to muscle regeneration, satellite cells have the capacity to replenish a reserve pool and self-renew, which also classifies these progenitors as adult stem cells [41].

During early growth, muscle satellite cells may represent about 30% of the nuclei, whereas in the healthy adult, satellite cells represent just approximately 2–7% of nuclei within skeletal muscle [13, 30]. The numbers of satellite cells per myofiber or per cross-sectional area may vary immensely between different muscles. For example, the fast twitch extensor digitorum longus (EDL) contains fewer satellite cells compared with the slow twitch soleus [13, 44, 45]. Additionally, myofiber ends may display a higher concentration of satellite cells than the bulk of the myofiber [46]. There are also reports on an age-associated decline in satellite cell numbers and the extent (if any) of such a decline may vary between muscles [44, 47–49]. There is also evidence that the performance of satellite cells declines in the aging environment, but their activity can be rejuvenated upon exposure of old muscle to a juvenile environment by cross-transplantation or by parabiosis of young and old mice [35, 40, 50]. The decline in satellite cell performance in old age might be a contributory factor to age-associated muscle deterioration, also known as sarcopenia [51]. However, additional studies suggest that the initial performance of skeletal muscle progenitors is delayed but not necessarily impaired in old age, and that factors beyond merely satellite cell activity may also play a role in reducing muscle repair in old age [38, 52]. Muscle-wasting disorders are also thought to lead to exhaustion of satellite cells owing to the continuous need for supply of myogenic cells to repair myofibers [53–55]. Overall, satellite cells are vital to skeletal muscle homeostasis and regeneration throughout the lifespan.

3 Isolation and Culture Approaches for Satellite Cell Analysis

Satellite cell behavior has often been investigated by transplantation of isolated populations of satellite cells into host muscles of experimental animals, with assessment of donor cell contribution to myofiber formation shown by expression of donor-derived genes [41, 47, 56–58]. However, most of our understanding of satellite cell biology has arisen from studies using primary cultures of satellite cells. Studies with myogenic cell lines further permitted extensive biochemical and molecular analyses, although these models do not always fully adhere to the biology of satellite cells. The closing paragraph of this section details some of the most commonly used myogenic cell lines.

Satellite cell populations are routinely isolated through enzymatic digestion of skeletal muscles [59]. Depending on the enzymatic procedure that is used and the purpose for cell isolation, enrichment for satellite cells beyond the basic isolation protocol is often unnecessary. For example, see our studies, with Pronase digestion of skeletal muscle [30, 43, 44, 59]. Alternatively, satellite cells can be enriched from whole muscle cell suspensions by various approaches that remove myofibril debris present in the initial cell suspension and reduce the presence of fibroblastic cells. Such approaches have included (1) initial short-term plating on uncoated tissue culture dishes that results in removal of fast adhering cells followed by culturing of the remaining nonadhering cells (i.e., differential plating) [60–62], (2) fractionation on Percoll density gradients [63–66], and (3) cell sorting by forward and side scatter [67].

In studies where further enrichment of satellite cells is warranted, cells can be isolated by fluorescence-activated cell sorting (FACS) using antibodies that react with satellite cell surface antigens [58]. First, cells are released from the muscle tissue using collagenase–dispase, which is an enzyme mixture that preserves cell surface antigens (whereas Pronase or trypsin digestion methods may not necessarily preserve such cell surface antigens). Studies from various laboratories (performed mainly with mouse tissue) have established that satellite cells can be isolated on the basis of being negative for CD45, CD31, and Sca1, and positive for CD34 and α_7 integrin [57, 58]. Additional cell surface antigens, including CXCR4, β_1 integrin, and syndecan-4 have also been utilized for isolation of myogenic progenitors from adult muscle [56, 68]. A range of fluorescence-based reporter systems in genetically manipulated mouse strains have also permitted reliable isolation and study of satellite cells. For example, we demonstrated that satellite cells could be isolated from different muscle groups of transgenic nestin-green fluorescent protein (GFP) mice where strong GFP expression in satellite cells is driven by nestin regulatory elements [48, 69]. Pax3- and Pax7-driven reporter expression has also been utilized to sort mouse satellite cells by FACS [57, 70]. As discussed in greater detail later in this chapter, Pax7 is commonly expressed by satellite cells across different muscle groups, but Pax3 expression is limited to select satellite cells. Mutant mice with a GFP reporter gene targeted into the myogenic factor 5 (Myf5) locus also permitted isolation of myogenic cells from such mice by FACS, but GFP expression is below the detection level in many of the satellite cells in adult mice and reduces the efficiency of satellite cell isolation [36, 71, 72]. Cre-Lox mouse models can also be used to isolate satellite cells on the basis of permanent fluorescent reporter expression in cells derived from myogenic progenitor cells expressing Cre recombinase driven by genes relevant to the myogenic lineage such as Pax3, Myf5, and MyoD [73–75]. When working with such Cre-Lox mouse models to sort satellite cells, one should be careful to ensure that the reporter is not expressed in additional cell types during embryogenesis. For example, see the report on Myf5-Cre expression in nonmyogenic sites [76].

The ability to isolate and culture single live myofibers (especially from rodent models) has also provided the advantage of following individual resident satellite cell activity in the presence of the parent myofiber [77–81]. In adherent myofiber culture models, satellite cells typically emanate from their position on the myofiber and give rise to proliferating myoblasts that differentiate and generate myotubes [44, 45]. Depending on culture and media conditions, rapid activation, proliferation, and differentiation of satellite cells can take place within the adherent parent myofiber unit [25, 26, 63, 81, 82]. Alternatively, culture of myofibers in suspension permits the unique ability to observe the behavior of satellite cells as they proceed through multiple rounds of cell division in cell clusters while remaining in contact with the myofiber [83]. Isolated single myofibers also allow tracing and recording of satellite cell numbers per myofiber on the basis of specific marker expression, with endogenous Pax7 expression becoming a common and direct approach to monitor satellite cells by immunofluorescence [44, 45, 48, 84]. Individual progeny of satellite cells can also be analyzed in clonal studies. We reported on such clonal analyses of satellite cells upon isolation of cells from whole muscles and also from triturated individual myofibers [44, 48, 66].

Many studies with myogenic cell lines have provided important insight into satellite cell biology with the caveat that not all findings with these models are completely applicable. The C2C12 mouse cell line is the most commonly used model for myogenesis and permits biochemical and molecular analyses when larger numbers of cells are needed. The C2C12 line (available from American Tissue Culture Collection, ATCC), was developed by subcloning of the parental C2 cell line that originated from adult mouse muscle [85–88]. The MM14 mouse cell line has also been used some studies [89, 90]. The rat myogenic cell lines L6 and L8 (available from ATCC) have also been used in some studies [91, 92]. In such cell models, myoblasts remain proliferative for a longer time when placed in a serum-rich environment and rapidly differentiate when placed in a serum-poor environment. However, it is important to recognize that the immediate progeny of satellite cells from adult skeletal muscle typically cannot be blocked from entering differentiation, regardless of medium composition or cell density [44, 48, 66]. Furthermore, long-term passages of cell lines typically have led to major variations in regulatory feedback loops that may not be reflective of cell signaling pathways that operated within the ancestor cells used to generate the cell line. Thus, the behavior of the long-term passaged myoblasts, which constitute the myogenic cell lines, may reflect only some of the features of satellite cell progeny. We previously provided a detailed summary of each of the aforementioned myogenic cell lines with regard to expression patterns of myogenic regulatory factors (MRFs), growth factor receptor expression, and response to growth factors [23]. We emphasized some of the main deviations of these cell lines from freshly isolated satellite cells placed into primary culture [23], and the reader can refer to this reference for further details.

4 Satellite Cell Specification and Pax3/Pax7 Transcription Factors

Satellite cells in body and limb muscles are thought to originate from the same compartments of the somite from where early muscle progenitors arise for development of trunk and limb muscles [75, 93]. These studies, relying on contemporary approaches of lineage tracing, have affirmed the original report on the somitic origin of satellite cells based on quail-chick chimeras [94]. Early studies demonstrated that progeny of myogenic cells isolated from adult muscles displayed various morphological and biochemical features that were distinct from those of cells derived from myogenic progenitors present in the fetal phase of embryogenesis. These investigations further identified the emergence of "adult-type" myogenic progenitors in the late fetal phase, corresponding with the development of mature basal lamina around myofibers [19, 95–99]. In the absence of a direct means to trace the satellite cells themselves, the latter studies established the concept that myogenic progenitors in the adult could be distinct from those present in fetal muscle on the basis of the features of cells cultured from adult muscle.

Satellite cells in body and limb muscles are specified from the embryonic and fetal muscle progenitor populations that express the paired box transcription factors Pax3 and Pax7 in trunk and limb muscle [75, 100–102]. Pax3 and Pax7 are paralogs and members of a larger family of Pax transcription factors involved in cell type and organ determination during embryogenesis of multicellular animals. The Pax proteins are characterized by their DNA-binding domains that include a paired domain and a homeodomain [103, 104]. Both Pax3 and Pax7 are expressed during muscle development, and both possess distinct and overlapping roles [105–109]. Some muscles, such as craniofacial muscles, may develop completely with the absence of Pax3 gene activity [110–113]. The lack of Pax3 expression in head muscle is reflective of the differences in precursor cells (i.e., head mesenchyme) that contribute to development of these muscles [111, 114].

Pax3 cannot compensate for the essential role of Pax7 in postnatal muscle, although embryonic muscle development progresses normally in the absence of Pax7 [115–118]. During embryogenesis, Pax3 plays a vital role in the transcriptional upregulation of c-met/ hepatocyte growth factor receptor that is required for the migration of muscle precursors to the limb bud [110]. Early studies indicated that *Pax7* is generally required for specification of satellite cells and the survival of mice in which the *Pax7* gene was inactivated [118]. However, more detailed studies of *Pax7* gene inactivation in mice showed that Pax7 may control survival of satellite cells beyond the neonatal period as there was a rapid decline in the number of detectable satellite cells postnatally, with some exhibiting markers of apoptosis when Pax7 was absent [116, 117]. A recent report demonstrated that Pax3 and Pax7 are only required up to the juvenile stage for myogenic performance and suggested that these Pax genes are necessary for the initial entry of specified satellite cells into quiescence, but not for satellite cell function and maintenance in adult muscle [119]. This study reaffirms the findings of the early studies and indicates that satellite cells (or adult-type myogenic progenitors) may represent a distinct population of myogenic progenitors that replaces the fetal myogenic population [19].

5 Pax3/Pax7 Expression by Satellite Cells in Different Muscle Groups

Satellite cells in postnatal and adult muscle express Pax7, providing a molecular marker for their identification by immunostaining and by stable Pax7-driven reporter expression [30, 41, 44, 118, 119]. Pax3-driven reporter expression and endogenous Pax3 expression is only detected in satellite cells of some adult muscles, whereas all satellite cells express Pax7 [57, 69, 117, 120]. We demonstrated that preparations of freshly isolated satellite cells from diaphragm muscle express relatively high levels of Pax3 messenger RNA (mRNA) compared with satellite cells from hind-limb muscles that show very little expression; Pax7 expression levels are similar, regardless of the muscle origin of isolated satellite cells [69]. We found that this distinction between satellite cells from limb and diaphragm muscles is maintained in old mice; this difference is also retained in the proliferating progeny as shown by real time reverse transcription PCR (RT-PCR) (unpublished results). An independent study also reported on elevated Pax3 expression in cultures of satellite cells from diaphragm compared with limb muscles [70].

The role that Pax3 and Pax7 play in skeletal muscle development and in satellite cells remains unclear. Reports on the possible roles of Pax3 and Pax7 during myogenesis suggest involvement of Pax7 in prevention of cell cycle progression while also suppressing differentiation, thereby playing a possible role in satellite cell self-renewal [121, 122]. A different study suggested that Pax7 may be involved in maintaining proliferation and preventing precocious differentiation, but does not promote quiescence [123]. Studies on integrated expression of Pax3 and Pax7 have suggested that these factors function to maintain expression of myogenic regulatory factors (MRFs) and promote myoblast expansion, but are also required for myogenic differentiation to proceed [124]. The regulation of *Pax3*/*Pax7* gene expression and the functional targets of these genes may vary not only between muscles, but also during satellite cell specification, myoblast proliferation, and satellite cell maintenance in quiescence within adult muscle. Some genome-wide approaches to map Pax7 binding sites using chromatin immunoprecipitation with Pax7 antibodies have shown that Pax7 may potentiate the expression of genes important in the skeletal muscle lineage such as the myogenic transcription factor Myf5 by inducing chromatin modifications [125]. However, Pax7 may rather suppress the entry of the satellite cell into myogenesis by upregulating *Id* gene family members that negatively regulate the myogenic basic helix–loop–helix transcription factors [126]. Pax3 and Pax7 may also work coordinately with other factors as the Pax family members in general have been shown to possess weak transactivation domains on their own [106, 127]. Chromatin immunoprecipitation for Pax7 also yielded some target genes whereby their expression is regulated by alternative splicing of Pax7 transcripts [128]. Alternatively spliced Pax3 and Pax7 mRNAs have been detected in various tissues and cells [129–133]. Overexpression studies have suggested both inducing and inhibitory roles of Pax3 isoforms identified in myogenic cultures [129]. Further studies are required to determine the unique functions of these alternatively spliced Pax3 and Pax7 isoforms during myogenesis.

6 Satellite Cells and the Role of the Myogenic Regulatory Factors

Activation, proliferation, and differentiation of satellite cells correlates with distinct temporal expression profiles of Pax7 and the myogenic regulatory factors (MRFs) [14, 20]. The muscle-specific MRF family of basic loop–helix transcription factors includes myogenic determination factor 1 (MyoD), Myf5, myogenin, and myogenic regulatory factor 4 (MRF4), which are also all involved in determination of myogenic fate during early embryogenesis [134–138].

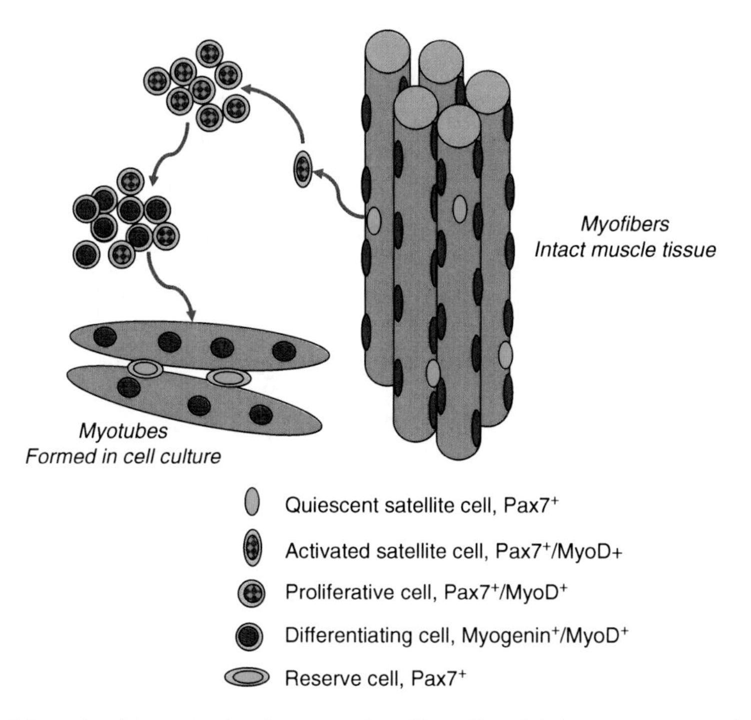

Fig. 2 Schematic of the molecular signatures of satellite cells and their progeny upon activation, proliferation, differentiation, and self-renewal (based on primary cell culture studies)

Quiescent satellite cells commonly express the paired-homeobox transcription factor Pax7 in contrast to their proliferating progeny that coexpress Pax7 and MyoD. Myogenin induction along with a decline in Pax7 mark myoblasts that have entered into the differentiation phase. Emergence of cells that express Pax7, but not MyoD, in cultures derived from satellite cells demarcate a self-renewing population, also referred to as reserve cells [30, 43, 44, 48, 63, 81, 83]. This dynamic molecular signature of satellite cells and their progeny is depicted in Fig. 2.

Quiescent satellite cells were also shown to express the *Myf5* gene on the basis of β-galactosidase detection in mutant mice in which one of the Myf5 alleles was modified to direct lacZ expression in $Myf5^{nlacZ/+}$ mice [14, 48, 49, 84]. Endogenous Myf5 mRNA expression has also been demonstrated in quiescent satellite cells [48, 63, 69]. However, the detection of Myf5 protein has not been reported in quiescent satellite cells, although proliferating progeny of satellite cells do express Myf5 protein together with Pax7 [43, 139]. Thus, it is possible that Myf5 protein is expressed only at a low level in quiescent satellite cells and therefore cannot be detected by standard immunostaining approaches.

We found that Myf5 expression declines when myoblasts enter differentiation, whereas MyoD expression persists well into the differentiation stage, suggesting that these two MRFs have different roles during myogenesis of satellite cells [43]. Muscles of MyoD-null mice contain functional satellite cells that can contribute to normal (albeit slightly delayed) muscle regeneration [139, 140]. However, it was initially reported that MyoD is essential for muscle regeneration [141]. Indeed, the delayed differentiation observed in the absence of MyoD can potentially impact the dynamics of muscle repair in vivo [139]. Recent studies from our laboratory have also demonstrated that progeny of individual satellite cells from MyoD-null mice differentiate poorly when the progenitors are cultured at clonal density (unpublished data). Overall, it is possible that absence of MyoD may impact muscle regeneration only in certain kinds of injury approaches and this may lead to the different conclusions of the published in vivo regeneration studies with MyoD-null mice [140, 141].

Early studies that developed Myf5-null mice demonstrated neonatal lethality as the Myf5 mutations impaired the function of an additional gene [142]. However, studies with recently developed Myf5-null mice, which survive to adulthood and are fertile, suggest that Myf5 is not essential for satellite cell function in adult muscle [72, 143]. Myogenin expression is critical for muscle formation during embryogenesis. However, conditional inactivation of the myogenin gene in the adult muscle does not interfere with myogenesis, and further raises questions about the actual role of myogenin in adult life [144, 145]. The role of the fourth MRF, namely, MRF4, remains unknown in myogenesis of satellite cells, but it is the only MRF that is expressed at a high level in adult muscle, whereas MyoD, Myf5, and myogenin expression levels are relatively lower [146]. MRF4 involvement in down-regulating myogenin expression has been reported [147]. In two different studies on MRF4 mRNA expression in primary myogenic cultures, this factor was detected before, after, or concurrently with myogenin expression [148, 149].

6.1 MyoD and Satellite Cell Activation

Pax7 is coordinately expressed with the myogenic regulatory factor MyoD upon satellite cell activation and entry into proliferation (see Fig. 2) [30, 43, 44, 83]. MyoD serves as the master transcription factor required to upregulate genes associated with myogenic differentiation [150, 151] and most studies on the role of MyoD during proliferation addressed mechanisms that suppress MyoD function to avoid premature differentiation [152]. However, the reason for the presence of MyoD in proliferating progeny of satellite cells (while spending cellular resources on suppressing its function) remains unknown. An earlier study identified Id3 and NP1 genes as targets of MyoD in proliferating myoblasts and suggests an active role for this factor prior to differentiation [153]. A recent study has shed new light on this subject by defining a mechanism whereby MyoD allows satellite cells to enter into the first round of DNA replication after transitioning out of quiescence [154]. The latter study demonstrated in C2C12 and primary mouse myoblasts that MyoD can occupy an E-box within the promoter of Cdc6 (which is an essential regulator of DNA replication) and that this association, along with E2F3a, is required for transcriptional activity of the Cdc6 gene. After being stimulated for growth, C2C12 myoblasts or quiescent satellite cells (in association with parent myofibers) express MyoD and Cdc6, but MyoD is detected at least 2–3 h earlier than Cdc6.

Our laboratory demonstrated that MyoD expression represents one of the earliest events of satellite cell activation [81]. MyoD protein synthesis was detected before the first round of cell division by using parallel immunodetection of 5-bromo-2-deoxyuridine (BrdU) incorporation and cell cycle markers [20]. Therefore, it is apparent that MyoD protein expression may play an important role in the activation of satellite cells and entry into the cell cycle. Rapid upregulation of MyoD expression within several hours after muscle injury as demonstrated by in situ hybridization [155] further provides in vivo support for the hypothesis that satellite cells in adult muscle are already committed to entering the MyoD-expressing state. MyoD transcripts are detectable in freshly isolated satellite cells by RT-PCR [63, 156]. Our unpublished real-time RT-PCR analysis demonstrated that such MyoD mRNA levels in freshly isolated satellite cells were only about 20% of those in proliferating progeny of satellite cells. Furthermore, with use of a MyoD-Cre mouse model, it has been established that satellite cells in adult muscles derived from cells that had already expressed MyoD during development [73]. Hence, on the basis of such historical expression, satellite cells might easily upregulate MyoD expression to maintain the balance between the quiescent state and occasional activation required for supporting adult muscle. MyoD expression might be suppressed in quiescent satellite cells by cues from the satellite cell niche. Once there is a modification of this niche, even a subtle one, MyoD expression may no longer be suppressed. Various studies have suggested that fibroblast growth factors, hepatocyte growth factor, and myostatin could all be involved in satellite cell activation [25–27, 44, 63], in addition to other factors [24, 77, 157, 158].

Notably, MyoD expression levels also seem to be influenced by cell culture conditions. In a serum-rich, mitogen-depleted medium, satellite cells undergo only one or two rounds of proliferation, after which their progeny enter differentiation and express myogenin, but not MyoD [25, 26, 81]. In contrast, the progeny of satellite cells remain proliferative for a long time (more than 1 week) and express MyoD even upon fusion into myotubes when grown in our rich growth medium [44, 159]. Although expression of MyoD is observed during proliferation, its main role has been well characterized during differentiation, where it may induce myogenin expression, along with a concomitant decline in Pax7 and Myf5 expression, cell cycle withdrawal, and subsequent fusion of myoblasts into multinucleated myotubes [43, 151]. Satellite cells from MyoD-null mice do not display obvious impairment in myogenesis, but the transition into the differentiation phase is delayed both in culture and in vivo [139, 140]. Thus, it is thought that the role of MyoD as a regulator of myogenic differentiation may be at least partially compensated by Myf5 [160, 161]. Mouse models in which MyoD and Myf5 gene function can be eliminated conditionally in adult muscle are needed for further studies on the exact roles of these MRFs in adult myogenesis.

6.2 MyoD and Myf5 Detection During the Cell Cycle

MyoD protein was detected in proliferating and differentiated progeny of satellite cells, but Myf5 protein was only detected in proliferating cells and not upon differentiation and fusion into myotubes [43, 87, 139, 162]. The temporal presence of Myf5 in Pax7^{+} myoblasts during S phase in early culture stages and its absence in differentiated cells reinforces the notion that Myf5 has a unique role apart from MyoD that is linked to cell cycle kinetics [43, 162]. Previous reports showed that Myf5 and MyoD levels oscillated at different phases of the cell cycle in mouse myoblasts, with the Myf5 levels being highest in quiescent, nondifferentiated cells (i.e., G0) [163, 164]. Other studies demonstrated that phosphorylated Myf5 was degraded between mid S phase and mitosis, but MyoD levels were maintained through mitosis and degraded at late G1 phase [162, 165, 166]. Furthermore, by blocking of Myf5 degradation, the passage of cells through cell division was perturbed [167, 168]. Thus, decreased levels of Myf5 during S phase may be regulated by proteolysis for potential entry into differentiation. Myf5 levels may also be highest before S phase to prevent continuation of myoblasts through the cell cycle and to direct myoblasts toward reservation and quiescence [43]. A minor Pax7^{+}/MyoD^{-} population at the late culture stage may also represent the cells entering reservation in which MyoD levels have decreased in late G1 phase [30, 83, 166, 169]. However, it remains unclear if the Pax7^{+}/MyoD^{-} cells also express Myf5. Notably, Myf5 was detected in a proposed reserve pool in primary cultures, although Pax7 was not examined [170]. The recent finding that Pax7 directly regulates Myf5 expression is also in accordance with our detection of Myf5 protein strictly within Pax7^{+} cells [43, 125].

Many studies have investigated the regulation of myoblast proliferation, yet it is not entirely clear what regulates the continuity of cell proliferation before the decision to enter differentiation or self-renewal is made. The maintenance of Pax7 may be required for cell cycle continuity in individual myoblasts as Pax7 was detected in the cytoplasm of individual mouse myoblasts even during mitosis, in contrast to some of the MRFs [171]. Pax7-deficient satellite cells do not appear to appropriately progress in the cell cycle, and may alternatively initiate apoptosis [117]. However, as mentioned above, a recent study showed that regeneration in adult muscle can progress in the absence of Pax7; this observation suggests that Pax7 may be important for cell proliferation only in the early postnatal stage during the robust growth phase of skeletal muscle myofibers [119].

Pulse labeling of cells with BrdU permits detection of cells that undergo DNA replication during the pulse labeling period as this thymidine analog incorporates into the replicating DNA. Our detection of a minor BrdU^{+}/Pax7^{+}/myogenin^{+} cell population during early culture stages that decreased with BrdU pulse duration suggests that myogenin is incompatible with S phase, and the elimination of Pax7 outside S phase is just as critical to complete differentiation as is the induction of myogenin [43]. One report showed that sustained Pax7 expression with simultaneous upregulation of myogenin in Rb-deficient myoblasts prevented their terminal differentiation [172]. Therefore, cell cycle withdrawal and terminal differentiation in myogenin$^+$ cells may not be complete until the cell cycle inhibitor p21 is present at a sufficient level [173]. Furthermore, there may be negative reciprocal inhibition between Pax7 and myogenin that may be the mechanism of decision between renewal and differentiation [122].

7 Differentiation of Satellite Cell Progeny

7.1 Myogenin Expression, Cell Cycle Withdrawal, and Fusion into Multinucleated Myotubes

The onset of myogenin expression marks entry of satellite cell progeny into myogenic differentiation [81, 87, 173]. Differentiation is also associated with a decline in Pax7 and Myf5 expression, a withdrawal from the cell cycle, and subsequent fusion of myoblasts into multinucleated myotubes [30, 43, 44, 83, 87, 174, 175]. Under normal culture conditions, only a small number of cells entering the onset of differentiation (i.e., myogenin$^+$) also show expression of differentiation-specific structural proteins such as sarcomeric myosin that is a characteristic of striated muscle [44, 159]. Once cells are fused into myotubes, there is an increase in the expression of differentiation-linked structural and regulatory genes [63]. If fusion is arrested, but terminal differentiation is permitted to progress (e.g., by manipulating calcium levels in the culture medium), there is an increase in the number of differentiated myoblasts expressing differentiation-linked proteins even without fusion into myotubes [99]. In some instances, cells expressing differentiation-linked markers were reported to be able to reenter proliferation [176]. However, in general, mononucleated myoblast expression of differentiation-linked genes such as sarcomeric myosin is considered to be terminally withdrawn from the cell cycle under normal conditions.

Myoblasts are able to proliferate at the onset of differentiation when cells transition into the myogenin-expressing state, whereas terminal differentiation requires permanent withdrawal from the cell cycle [164, 173]. The evaluation of BrdU$^{\pm}$/Pax7$^{\pm}$/MRF$^{\pm}$ populations with regard to the length of the BrdU incubation period revealed a population of myoblasts actively entering differentiation at the late culture stage and suggests a transitional state (Pax7$^+$/MyoD$^+$ to Pax7$^-$/MyoD$^+$) during differentiation in which myogenin upregulation in Pax7$^-$/MyoD$^+$ cells occurs [43]. MyoD may set the pace of differentiation by upregulating myogenin expression as Pax7 levels decrease. The delayed appearance of myogenin$^+$ cells in myogenic cultures from MyoD$^{-/-}$ mice promotes this hypothesis [139]. MyoD may govern the induction of the cell cycle regulators involved in cell cycle withdrawal and differentiation [177, 178]. Therefore, MyoD function must be kept inactive until differentiation-inducing signals are present to prevent premature cell cycle arrest [179–183]. Cyclin-dependent kinases, cyclin-dependent kinase inhibitors, cyclin D3, and retinoblastoma protein control terminal myogenic differentiation by regulating cell cycle withdrawal and expression of genes encoding for structural proteins [172, 184]. Differentiating progeny of satellite cells also exhibit enhanced expression of myocyte enhancer factor 2A in correlation with the onset of myogenin expression [63, 87, 159, 185]. Members of the myocyte enhancer factor 2 transcription factor family act in concert with MRFs to direct the late stages of myogenesis by inducing the expression of muscle-specific structural proteins [186, 187].

7.2 *Expression of Muscle-Specific Contractile Protein: Intrafusal Versus Extrafusual Myofibers*

In this section, we briefly introduce the subject of skeletal muscle structural genes within the context of satellite cell biology. Depending on the species from which satellite cells were isolated and the culture conditions, both slow and fast isoforms of differentiation-linked gene families of sarcomeric proteins might be detected in cultures derived from satellite cells upon differentiation and fusion into myotubes [63, 188–191]. However, the differentiation programs lead to full expression of differentiation-linked genes more effectively in vivo because they are influenced by innervation [12, 192–194]. The temporal expression of desmin, an intermediate filament protein that is expressed at high level in myofibers in all species analyzed, has also been investigated extensively in cell culture, where it was expressed by myotubes similar to sarcomeric myosin. However, desmin can also be expressed by proliferating myoblasts and this distinction seems to be species-specific. Mouse and rat myoblasts express desmin, whereas bovine myoblasts do not [139, 195]. In the chicken, some infrequent proliferating myoblasts may express desmin, whereas progeny of fetal myoblasts do not express desmin in culture [196].

During early stages of skeletal muscle embryogenesis, developing muscles express isoforms of sarcomeric proteins, including alpha slow myosin heavy chain, which is the predominant form of cardiac myosin throughout life. However, as skeletal muscle development progresses, cardiac gene expression is diminished [194]. Nevertheless, there is a strong expression of cardiac myosin (i.e., alpha slow) in intrafusal myofibers [197]. Intrafusal myofibers are within special structures in skeletal muscles known as spindles, which function as proprioceptive receptors [197, 198]. Intrafusal myofibers make up a small subpopulation of all muscle myofibers; the bulk of skeletal muscle myofibers that are typically studied by myogenesis investigators are termed "extrafusal myofibers." Even in mature animals, the intrafusal myofibers are distinct; their diameter remains small from an early postnatal age, and they retain expression of developmental myosins as well as expression of LacZ in myofiber nuclei of Myf5$^{nLacZ/+}$ knock-in mice [120, 197–200]. Intrafusal myofibers also contain Pax7^{+} satellite cells and display retention of Pax3 expression at higher levels than satellite cells on extrafusal myofibers [120, 198]. However, it is unknown if the satellite cells associated with the two types of myofibers are functionally unique and/or are derived from distinct myogenic populations. It may be of potential benefit to investigators of both skeletal and cardiac regenerative medicine to better understand the myogenic program of intrafusal versus extrafusal myofibers.

8 Satellite Cell Self-Renewal/Reservation

Homeostasis of skeletal myofibers requires nuclear replacement and myofiber turnover throughout life [201], and therefore without renewal, the satellite cell pool would be depleted. Our laboratory has some evidence that the capacity of individual satellite cells to self-renew may be impaired as animals age [48]. In this section, we will discuss how progeny of satellite cells that are "reserved" for future muscle maintenance are detected, and potential mechanisms that drive the self-renewal process.

The unique patterns of Pax7 and MRF detection within myoblasts reflects the stages of proliferation and terminal differentiation during myogenesis, but much less is understood about how new satellite cells are generated or reserved from the pool of proliferating myoblasts. Following transplantation of donor satellite cells to host muscles, satellite cells are detected in vivo with donor reporter markers, indicating that at least some of the satellite cells underwent self-renewal [41]. Such in vivo studies cannot provide accurate feedback about the proliferative dynamics that precedes the formation of the renewed satellite cells. However, on the basis of engraftment of single satellite cells, it is sensible to suggest that the donor cells first go through an amplification phase before some of the cells enter the satellite cell niche [58]. Our laboratory and others have demonstrated the presence of Pax7$^+$/MyoD$^-$ cells in myogenic cultures developed from satellite cells and suggests that these cells reflect the "reserve compartment" [20, 30, 44, 83]. BrdU-labeling studies suggested that such Pax7$^+$/MyoD$^-$ cells are derived from Pax7$^+$/MyoD$^+$ myoblasts that have downregulated MyoD expression and do not continue in the cell cycle or transit into differentiation, thus demonstrating the self-renewing population [69, 83]. There is some evidence that satellite cells may self-renew through asymmetric cell divisions that are controlled by the Notch signaling pathway [74, 202]. Cosegregation of the Notch inhibitor Numb with template DNA strands during asymmetric divisions in vivo has also been reported as a way to identify self-renewal of satellite cells [171, 203].

In view of the close association between cell cycle withdrawal and myogenic differentiation, it is anticipated that the mechanisms of satellite cell self-renewal may involve signals that are required to repress the myogenic program toward terminal differentiation while driving select myoblasts to leave the cell cycle and self-renew. The Notch pathway has been shown to play a role in regulating myogenic differentiation during embryonic development and activation of Notch in adult muscle prevents satellite cell differentiation [202, 204]. Although it is unclear how the Notch pathway directly controls the self-renewal of satellite cells, inhibition of

Notch activity demonstrated fewer $Pax7^+/MyoD^-$ reserve cell progeny of satellite cells [74]. Following activation of satellite cells, pathways such as Wnt and Notch that converge on degradation or stabilization of β-catenin may control decisions toward proliferation and differentiation [205]. Stabilization of β-catenin was shown to direct satellite cell progeny toward the self-renewal pathway [169]. A recent report demonstrated that presenilin-1, a catalytic subunit of γ-secretase that is crucial for the cleavage and nuclear passage of the Notch intracellular domain, may also regulate the development of reserve cells both dependent on and independent of protease activity [206]. Notably, other factors that interact with Notch signaling, such as the transmembrane protein Megf10 that is expressed in satellite cells, may also play a role in decisions to proliferate, differentiate, or self-renew [207]. The close association of satellite cells with muscle capillaries also may provide satellite cells with direct signals from pathways traditionally shown to be involved in angiogenesis and vascular integrity [36]. Interestingly, the angiopoeitin-1/Tie2 receptor signaling pathway may directly control development of reserve cells and entry into quiescence [208].

Our laboratory has found that transgenic expression of GFP driven by regulatory elements of the rat nestin gene (nestin-GFP) in mice may also be used to follow and record satellite cells and their self-renewing, $Pax7^+/MyoD^-$ reserve progeny that develop in vitro [48, 69]. Satellite cells express nestin-GFP, but this expression is diminished in proliferating and differentiating myoblasts and myotubes. However, cultures with dense myotube networks eventually develop mononuclear, nestin-GFP^+ cells that no longer proliferate (on the basis of BrdU incorporation studies). Examination of satellite cells in clonal cultures showed that nestin-GFP^+ reserve progeny develop only in such dense clones, which accounts for only some of the clones [48, 69]. We also demonstrated an age-linked decline in such clones containing nestin-GFP^+ reserve cells [48]. This age-linked decline in clones containing reserve cells is concomitant with a decline in satellite cell numbers on myofibers isolated from old mice [48]. Therefore, impairment in satellite cell renewal in old age may lead to the observed decline in satellite cell numbers during aging, which also results in a significant number of myofibers that lack satellite cells as shown in myofibers from EDL muscles [48]. It is possible that such myofibers that lack satellite cells are more prone to age-associated myofiber atrophy, but this issue has not been studied further. We are currently using this cell culture model of reserve cell development to investigate what factors may regulate satellite cell renewal. In addition to secreted factors, contact with the myotubes might provide a signal for self-renewal as also suggested in studies with isolated myofibers maintained in suspension where the satellite cells remain on the myofiber surface [74, 209].

Apart from the evidence that the self-renewing population is reserved from the pool of proliferating myoblasts is the demonstration that satellite cells may include a minor subpopulation of stem cells that can self-renew while also providing committed satellite cells within a single proliferative round [74, 209]. According to this asymmetric cell division model, most of the satellite cells are "committed" and only provide committed progeny. The minor stem cell population did not display Myf5 gene activity on the basis of Myf5-Cre-driven Rosa26-yellow fluorescent protein (YFP) expression, whereas the committed cells expressed YFP [74]. However, recent studies with $Myf5^{nLacZ/+}$ mice have suggested that essentially all satellite cells do express the Myf5 reporter, but the cells with weaker expression may only be detected with increased sensitivity of β-galactosidase detection [48]. Studies reporting some of the satellite cells in isolated myofibers from EDL muscle that did not express the Myf5-driven β-galactosidase reporter could potentially have failed to detect the weakly expressing satellite cells [49, 69, 84]. A recent study with the MyoD-Cre/Rosa26-YFP mouse model has suggested that all satellite cells are derived from cells that "historically" expressed MyoD, and therefore are all committed to myogenesis [73]. Further studies based on different strains of reporter mice are required to resolve questions about satellite cell functional heterogeneity and self-renewal.

9 Donor Satellite Cell Engraftment for Therapy of Muscle-Wasting Disorders: Hopes and Hypes

Satellite cell behavior has often been investigated by transplantation of cells into host muscles of experimental animals and assessment of donor cell contribution to myofiber formation shown by expression of donor-derived genes [41, 47, 56–58]. The long-term goal of such studies is the improvement of muscle quality and performance in cases of severe muscle-wasting disorders such as Duchenne muscular dystrophy [210–213].

Donor satellite cells contribute myofiber nuclei while also establishing additional satellite cells through self-renewal in experimental mouse models, enabling further understanding of satellite cell biology in the in vivo environment [47, 58]. However, issues of limited migration of satellite cells and their progeny, donor/host histocompatability aspects, death of donor cells, and the need to deliver cells to many host muscles at once represent some of the major challenges that have not yet been resolved for cell-based therapies of skeletal-muscle-wasting disorders [16, 214–218]. Additionally, some reports have alluded to the failure of satellite cells injected into the circulation to reach target muscles for whole body treatment, although detailed evidence is not available [15, 219]. Myoblasts expanded from satellite cells in culture to multiply the number of donor cells available also cannot be delivered systemically to target muscles and lose regenerative capacity compared with freshly isolated satellite cells upon direct intramuscular delivery [57, 220–222]. It also appears that there is a need to injure and irradiate muscle prior to donor cell delivery to enhance engraftment, as shown in experimental mouse models [41, 58, 223]. Therefore, cell-based therapy for skeletal muscle remains a challenge despite the availability and our understanding of satellite cells. Although repair of specific muscles might soon be possible, genetic disorders that affect whole body muscles such as muscular dystrophies may await effective means to deliver genes by combining cell-based therapies with viral transduction approaches [212]. However, compared to whole body approach that is required for the repair of many muscles in the case of muscular dystrophy, rejuvenation of myocardium requires only repair of a single organ and therefore, there is much anticipation that this will eventually be possible [224]. Lessons learned to date from satellite cell engraftment to host skeletal muscles could potentially accelerate the use of such cells for repairing the myocardium.

Acknowledgements Current research in the Z.Y.-R. laboratory is supported by grants from the National Institutes of Health (AG021566, AG013798, AG035377, AR057794) and the Muscular Dystrophy Association (award 135908). K.D. was supported by the Genetic Approaches to Aging Training Program (T32 AG000057) during his postdoctoral research in this laboratory. Z.Y.-R. is additionally grateful to the American Heart Association and the USDA Cooperative State Research, Education, and Extension Service for past support that facilitated some of the research described in this chapter.

References

1. Horackova M, Arora R, Chen R, Armour JA, Cattini PA, Livingston R, Byczko Z (2004) Cell transplantation for treatment of acute myocardial infarction: unique capacity for repair by skeletal muscle satellite cells. Am J Physiol Heart Circ Physiol 287:H1599–1608
2. Murry CE, Wiseman RW, Schwartz SM, Hauschka SD (1996) Skeletal myoblast transplantation for repair of myocardial necrosis. J Clin Invest 98:2512–2523
3. Reinecke H, Poppa V, Murry CE (2002) Skeletal muscle stem cells do not transdifferentiate into cardiomyocytes after cardiac grafting. J Mol Cell Cardiol 34:241–249
4. Suzuki K, Murtuza B, Beauchamp JR, Smolenski RT, Varela-Carver A, Fukushima S, Coppen SR, Partridge TA, Yacoub MH (2004) Dynamics and mediators of acute graft attrition after myoblast transplantation to the heart. FASEB J 18:1153–1155
5. Aharinejad S, Abraham D, Paulus P, Zins K, Hofmann M, Michlits W, Gyongyosi M, Macfelda K, Lucas T, Trescher K, Grimm M, Stanley ER (2008) Colony-stimulating factor-1 transfection of myoblasts improves the repair of failing myocardium following autologous myoblast transplantation. Cardiovasc Res 79:395–404
6. Menasche P (2008) Towards the second generation of skeletal myoblasts? Cardiovasc Res 79:355–356
7. Menasche P (2008) Skeletal myoblasts and cardiac repair. J Mol Cell Cardiol 45:545–553
8. Okada M, Payne TR, Zheng B, Oshima H, Momoi N, Tobita K, Keller BB, Phillippi JA, Peault B, Huard J (2008) Myogenic endothelial cells purified from human skeletal muscle improve cardiac function after transplantation into infarcted myocardium. J Am Coll Cardiol 52:1869–1880
9. Formigli L, Zecchi-Orlandini S, Meacci E, Bani D (2010) Skeletal myoblasts for heart regeneration and repair: state of the art and perspectives on the mechanisms for functional cardiac benefits. Curr Pharm Des 16:915–928
10. Larose E, Proulx G, Voisine P, Rodes-Cabau J, De Larochelliere R, Rossignol G, Bertrand OF, Tremblay JP (2010) Percutaneous versus surgical delivery of autologous myoblasts after chronic myocardial infarction: an in vivo cardiovascular magnetic resonance study. Catheter Cardiovasc Interv 75:120–127
11. Schabort EJ, Myburgh KH, Wiehe JM, Torzewski J, Niesler CU (2009) Potential myogenic stem cell populations: sources, plasticity, and application for cardiac repair. Stem Cells Dev 18:813–830
12. Ciciliot S, Schiaffino S (2010) Regeneration of mammalian skeletal muscle. Basic mechanisms and clinical implications. Curr Pharm Des 16:906–914

13. Hawke TJ, Garry DJ (2001) Myogenic satellite cells: physiology to molecular biology. J Appl Physiol 91:534–551
14. Zammit PS, Partridge TA, Yablonka-Reuveni Z (2006) The skeletal muscle satellite cell: the stem cell that came in from the cold. J Histochem Cytochem 54:1177–1191
15. Dellavalle A, Sampaolesi M, Tonlorenzi R, Tagliafico E, Sacchetti B, Perani L, Innocenzi A, Galvez BG, Messina G, Morosetti R, Li S, Belicchi M, Peretti G, Chamberlain JS, Wright WE, Torrente Y, Ferrari S, Bianco P, Cossu G (2007) Pericytes of human skeletal muscle are myogenic precursors distinct from satellite cells. Nat Cell Biol 9:255–267
16. Tedesco FS, Dellavalle A, Diaz-Manera J, Messina G, Cossu G (2010) Repairing skeletal muscle: regenerative potential of skeletal muscle stem cells. J Clin Invest 120:11–19
17. Zheng B, Cao B, Crisan M, Sun B, Li G, Logar A, Yap S, Pollett JB, Drowley L, Cassino T, Gharaibeh B, Deasy BM, Huard J, Peault B (2007) Prospective identification of myogenic endothelial cells in human skeletal muscle. Nat Biotechnol 25:1025–1034
18. Mauro A (1961) Satellite cell of skeletal muscle fibers. J Biophys Biochem Cytol 9:493–495
19. Yablonka-Reuveni Z (1995) Development and postnatal regulation of adult myoblasts. Microsc Res Tech 30:366–380
20. Yablonka-Reuveni Z, Day K, Vine A, Shefer G (2008) Defining the transcriptional signature of skeletal muscle stem cells. J Anim Sci 86:E207–E216
21. Anderson JE (2006) The satellite cell as a companion in skeletal muscle plasticity: currency, conveyance, clue, connector and colander. J Exp Biol 209:2276–2292
22. Luo D, Renault VM, Rando TA (2005) The regulation of Notch signaling in muscle stem cell activation and postnatal myogenesis. Semin Cell Dev Biol 16:612–622
23. Shefer G, Yablonka-Reuveni Z (2008) Ins and outs of satellite cell myogenesis: the role of the ruling growth factors. In Schiaffino S, Partridge T, eds. Skeletal Muscle Repair and Regeneration (Advances in Muscle Res, vol 3). Springer, Dordrecht, 107–143
24. Wozniak AC, Kong J, Bock E, Pilipowicz O, Anderson JE (2005) Signaling satellite-cell activation in skeletal muscle: markers, models, stretch, and potential alternate pathways. Muscle Nerve 31:283–300
25. Yablonka-Reuveni Z (1997) Proliferative dynamics and the role of FGF2 during myogenesis of rat satellite cells on isolated fibers. Basic Appl Myol 7:176–189
26. Yablonka-Reuveni Z, Seger R, Rivera AJ (1999) Fibroblast growth factor promotes recruitment of skeletal muscle satellite cells in young and old rats. J Histochem Cytochem 47:23–42
27. Yamada M, Tatsumi R, Yamanouchi K, Hosoyama T, Shiratsuchi SI, Sato A, Mizunoya W, Ikeuchi Y, Furuse M, Allen RE (2010) High concentrations of HGF inhibit skeletal muscle satellite cell proliferation in vitro by inducing expression of myostatin. Am J Physiol Cell Physiol 298:C465–C476
28. Zammit PS (2008) All muscle satellite cells are equal, but are some more equal than others? J Cell Sci 121:2975–2982
29. Campion DR (1984) The muscle satellite cell: a review. Int Rev Cytol 87:225–251
30. Halevy O, Piestun Y, Allouh MZ, Rosser BW, Rinkevich Y, Reshef R, Rozenboim I, Wleklinski-Lee M, Yablonka-Reuveni Z (2004) Pattern of Pax7 expression during myogenesis in the posthatch chicken establishes a model for satellite cell differentiation and renewal. Dev Dyn 231:489–502
31. Moss FP, Leblond CP (1971) Satellite cells as the source of nuclei in muscles of growing rats. Anat Rec 170:421–435
32. Schultz E (1996) Satellite cell proliferative compartments in growing skeletal muscles. Dev Biol 175:84–94
33. Snow MH (1978) An autoradiographic study of satellite cell differentiation into regenerating myotubes following transplantation of muscles in young rats. Cell Tissue Res 186:535–540
34. Bischoff R (1989) Analysis of muscle regeneration using single myofibers in culture. Med Sci Sports Exerc 21:S164–S172
35. Carlson ME, Hsu M, Conboy IM (2008) Imbalance between pSmad3 and Notch induces CDK inhibitors in old muscle stem cells. Nature 454:528–532

36. Christov C, Chretien F, Abou-Khalil R, Bassez G, Vallet G, Authier FJ, Bassaglia Y, Shinin V, Tajbakhsh S, Chazaud B, Gherardi RK (2007) Muscle satellite cells and endothelial cells: close neighbors and privileged partners. Mol Biol Cell 18:1397–1409
37. Gopinath SD, Rando TA (2008) Stem cell review series: aging of the skeletal muscle stem cell niche. Aging Cell 7:590–598
38. Shavlakadze T, McGeachie J, Grounds MD (2010) Delayed but excellent myogenic stem cell response of regenerating geriatric skeletal muscles in mice. Biogerontology 11:363–376
39. Carlson ME, Conboy MJ, Hsu M, Barchas L, Jeong J, Agrawal A, Mikels AJ, Agrawal S, Schaffer DV, Conboy IM (2009) Relative roles of TGF-beta1 and Wnt in the systemic regulation and aging of satellite cell responses. Aging Cell 8:676–689
40. Conboy IM, Conboy MJ, Wagers AJ, Girma ER, Weissman IL, Rando TA (2005) Rejuvenation of aged progenitor cells by exposure to a young systemic environment. Nature 433:760–764
41. Collins CA, Olsen I, Zammit PS, Heslop L, Petrie A, Partridge TA, Morgan JE (2005) Stem cell function, self-renewal, and behavioral heterogeneity of cells from the adult muscle satellite cell niche. Cell 122:289–301
42. Grounds MD, Yablonka-Reuveni Z (1993) Molecular and cell biology of skeletal muscle regeneration. Mol Cell Biol Hum Dis Ser 3:210–256
43. Day K, Paterson B, Yablonka-Reuveni Z (2009) A distinct profile of myogenic regulatory factor detection within Pax7+ cells at S phase supports a unique role of Myf5 during posthatch chicken myogenesis. Dev Dyn 238:1001–1009
44. Shefer G, Van de Mark DP, Richardson JB, Yablonka-Reuveni Z (2006) Satellite-cell pool size does matter: defining the myogenic potency of aging skeletal muscle. Dev Biol 294:50–66
45. Zammit PS, Heslop L, Hudon V, Rosenblatt JD, Tajbakhsh S, Buckingham ME, Beauchamp JR, Partridge TA (2002) Kinetics of myoblast proliferation show that resident satellite cells are competent to fully regenerate skeletal muscle fibers. Exp Cell Res 281:39–49
46. Allouh MZ, Yablonka-Reuveni Z, Rosser BW (2008) Pax7 reveals a greater frequency and concentration of satellite cells at the ends of growing skeletal muscle fibers. J Histochem Cytochem 56:77–87
47. Collins CA, Zammit PS, Ruiz AP, Morgan JE, Partridge TA (2007) A population of myogenic stem cells that survives skeletal muscle aging. Stem Cells 25:885–894
48. Day K, Shefer G, Shearer A, Yablonka-Reuveni Z (2010) The depletion of skeletal muscle satellite cells with age is concomitant with reduced capacity of single progenitors to produce reserve progeny. Dev Biol 340:330–343
49. Ono Y, Boldrin L, Knopp P, Morgan JE, Zammit PS (2010) Muscle satellite cells are a functionally heterogeneous population in both somite-derived and branchiomeric muscles. Dev Biol 337:29–41
50. Carlson BM, Faulkner JA (1989) Muscle transplantation between young and old rats: age of host determines recovery. Am J Physiol 256:C1262–C1266
51. Thompson LV (2009) Age-related muscle dysfunction. Exp Gerontol 44:106–111
52. Grounds MD (1998) Age-associated changes in the response of skeletal muscle cells to exercise and regeneration. Ann N Y Acad Sci 854:78–91
53. Aguennouz M, Vita GL, Messina S, Cama A, Lanzano N, Ciranni A, Rodolico C, Di Giorgio RM, Vita G (2010) Telomere shortening is associated to TRF1 and PARP1 overexpression in Duchenne muscular dystrophy. Neurobiol Aging
54. Blau HM, Webster C, Pavlath GK (1983) Defective myoblasts identified in Duchenne muscular dystrophy. Proc Natl Acad Sci USA 80:4856–4860
55. Webster C, Blau HM (1990) Accelerated age-related decline in replicative life-span of Duchenne muscular dystrophy myoblasts: implications for cell and gene therapy. Somat Cell Mol Genet 16:557–565
56. Cerletti M, Jurga S, Witczak CA, Hirshman MF, Shadrach JL, Goodyear LJ, Wagers AJ (2008) Highly efficient, functional engraftment of skeletal muscle stem cells in dystrophic muscles. Cell 134:37–47
57. Montarras D, Morgan J, Collins C, Relaix F, Zaffran S, Cumano A, Partridge T, Buckingham M (2005) Direct isolation of satellite cells for skeletal muscle regeneration. Science 309:2064–2067

58. Sacco A, Doyonnas R, Kraft P, Vitorovic S, Blau HM (2008) Self-renewal and expansion of single transplanted muscle stem cells. Nature 456:502–506
59. Yablonka-Reuveni Z (2004) Isolation and culture of myogenic stem cells. In: Lanza R, Blau H, Melton D, Moore M, Thomas ED, Verfaillie C, Weissman IL, West M, eds. Handbook of Stem Cells–Vol 2: Adult and Fetal Stem Cells. San Diego: Elsevier, 571–580
60. Qu-Petersen Z, Deasy B, Jankowski R, Ikezawa M, Cummins J, Pruchnic R, Mytinger J, Cao B, Gates C, Wernig A, Huard J (2002) Identification of a novel population of muscle stem cells in mice: potential for muscle regeneration. J Cell Biol 157:851–864
61. Rando TA, Blau HM (1994) Primary mouse myoblast purification, characterization, and transplantation for cell-mediated gene therapy. J Cell Biol 125:1275–1287
62. Richler C, Yaffe D (1970) The in vitro cultivation and differentiation capacities of myogenic cell lines. Dev Biol 23:1–22
63. Kastner S, Elias MC, Rivera AJ, Yablonka-Reuveni Z (2000) Gene expression patterns of the fibroblast growth factors and their receptors during myogenesis of rat satellite cells. J Histochem Cytochem 48:1079–1096
64. Morgan JE (1988) Myogenicity in vitro and in vivo of mouse muscle cells separated on discontinuous Percoll gradients. J Neurol Sci 85:197–207
65. Yablonka-Reuveni Z, Nameroff M (1987) Skeletal muscle cell populations. Separation and partial characterization of fibroblast-like cells from embryonic tissue using density centrifugation. Histochemistry 87:27–38
66. Yablonka-Reuveni Z, Quinn LS, Nameroff M (1987) Isolation and clonal analysis of satellite cells from chicken pectoralis muscle. Dev Biol 119:252–259
67. Yablonka-Reuveni Z (1988) Discrimination of myogenic and nonmyogenic cells from embryonic skeletal muscle by 90 degrees light scattering. Cytometry 9:121–125
68. Tanaka KK, Hall JK, Troy AA, Cornelison DD, Majka SM, Olwin BB (2009) Syndecan-4-expressing muscle progenitor cells in the SP engraft as satellite cells during muscle regeneration. Cell Stem Cell 4:217–225
69. Day K, Shefer G, Richardson JB, Enikolopov G, Yablonka-Reuveni Z (2007) Nestin-GFP reporter expression defines the quiescent state of skeletal muscle satellite cells. Dev Biol 304:246–259
70. Bosnakovski D, Xu Z, Li W, Thet S, Cleaver O, Perlingeiro RC, Kyba M (2008) Prospective isolation of skeletal muscle stem cells with a Pax7 reporter. Stem Cells 26:3194–3204
71. Biressi S, Tagliafico E, Lamorte G, Monteverde S, Tenedini E, Roncaglia E, Ferrari S, Ferrari S, Cusella-De Angelis MG, Tajbakhsh S, Cossu G (2007) Intrinsic phenotypic diversity of embryonic and fetal myoblasts is revealed by genome-wide gene expression analysis on purified cells. Dev Biol 304:633–651
72. Gayraud-Morel B, Chretien F, Flamant P, Gomes D, Zammit PS, Tajbakhsh S (2007) A role for the myogenic determination gene Myf5 in adult regenerative myogenesis. Dev Biol 312:13–28
73. Kanisicak O, Mendez JJ, Yamamoto S, Yamamoto M, Goldhamer DJ (2009) Progenitors of skeletal muscle satellite cells express the muscle determination gene, MyoD. Dev Biol 332:131–141
74. Kuang S, Kuroda K, Le Grand F, Rudnicki MA (2007) Asymmetric self-renewal and commitment of satellite stem cells in muscle. Cell 129:999–1010
75. Schienda J, Engleka KA, Jun S, Hansen MS, Epstein JA, Tabin CJ, Kunkel LM, Kardon G (2006) Somitic origin of limb muscle satellite and side population cells. Proc Natl Acad Sci USA 103:945–950
76. Gensch N, Borchardt T, Schneider A, Riethmacher D, Braun T (2008) Different autonomous myogenic cell populations revealed by ablation of Myf5-expressing cells during mouse embryogenesis. Development 135:1597–1604
77. Bischoff R (1986) Proliferation of muscle satellite cells on intact myofibers in culture. Dev Biol 115:129–139
78. Bischoff R (1975) Regeneration of single skeletal muscle fibers in vitro. Anat Rec 182: 215–235

79. Konigsberg UR, Lipton BH, Konigsberg IR (1975) The regenerative response of single mature muscle fibers isolated in vitro. Dev Biol 45:260–275
80. Rosenblatt JD, Lunt AI, Parry DJ, Partridge TA (1995) Culturing satellite cells from living single muscle fiber explants. In Vitro Cell Dev Biol Anim 31:773–779
81. Yablonka-Reuveni Z, Rivera AJ (1994) Temporal expression of regulatory and structural muscle proteins during myogenesis of satellite cells on isolated adult rat fibers. Dev Biol 164:588–603
82. Shefer G, Yablonka-Reuveni Z (2005) Isolation and culture of skeletal muscle myofibers as a means to analyze satellite cells. Methods Mol Biol 290:281–304
83. Zammit PS, Golding JP, Nagata Y, Hudon V, Partridge TA, Beauchamp JR (2004) Muscle satellite cells adopt divergent fates: a mechanism for self-renewal? J Cell Biol 166:347–357
84. Beauchamp JR, Heslop L, Yu DS, Tajbakhsh S, Kelly RG, Wernig A, Buckingham ME, Partridge TA, Zammit PS (2000) Expression of CD34 and Myf5 defines the majority of quiescent adult skeletal muscle satellite cells. J Cell Biol 151:1221–1234
85. Kwiatkowski BA, Kirillova I, Richard RE, Israeli D, Yablonka-Reuveni Z (2008) FGFR4 and its novel splice form in myogenic cells: Interplay of glycosylation and tyrosine phosphorylation. J Cell Physiol 215:803–817
86. Yablonka-Reuveni Z, Balestreri TM, Bowen-Pope DF (1990) Regulation of proliferation and differentiation of myoblasts derived from adult mouse skeletal muscle by specific isoforms of PDGF. J Cell Biol 111:1623–1629
87. Yablonka-Reuveni Z, Rivera AJ (1997) Influence of PDGF-BB on proliferation and transition through the MyoD-myogenin-MEF2A expression program during myogenesis in mouse C2 myoblasts. Growth Factors 15:1–27
88. Yaffe D, Saxel O (1977) Serial passaging and differentiation of myogenic cells isolated from dystrophic mouse muscle. Nature 270:725–727
89. Clegg CH, Linkhart TA, Olwin BB, Hauschka SD (1987) Growth factor control of skeletal muscle differentiation: commitment to terminal differentiation occurs in G1 phase and is repressed by fibroblast growth factor. J Cell Biol 105:949–956
90. Fedorov YV, Rosenthal RS, Olwin BB (2001) Oncogenic Ras-induced proliferation requires autocrine fibroblast growth factor 2 signaling in skeletal muscle cells. J Cell Biol 152:1301–1305
91. Graves DC, Yablonka-Reuveni Z (2000) Vascular smooth muscle cells spontaneously adopt a skeletal muscle phenotype: a unique Myf5(–)/MyoD(+) myogenic program. J Histochem Cytochem 48:1173–1193
92. Yaffe D (1969) Cellular aspects of muscle differentiation in vitro. Curr Top Dev Biol 4:37–77
93. Buckingham M, Montarras D (2008) Skeletal muscle stem cells. Curr Opin Genet Dev 18:330–336
94. Armand O, Boutineau AM, Mauger A, Pautou MP, Kieny M (1983) Origin of satellite cells in avian skeletal muscles. Arch Anat Microsc Morphol Exp 72:163–181
95. Cossu G, Eusebi F, Grassi F, Wanke E (1987) Acetylcholine receptor channels are present in undifferentiated satellite cells but not in embryonic myoblasts in culture. Dev Biol 123:43–50
96. Cossu G, Molinaro M (1987) Cell heterogeneity in the myogenic lineage. Curr Top Dev Biol 23:185–208
97. Feldman JL, Stockdale FE (1992) Temporal appearance of satellite cells during myogenesis. Dev Biol 153:217–226
98. Hartley RS, Bandman E, Yablonka-Reuveni Z (1992) Skeletal muscle satellite cells appear during late chicken embryogenesis. Dev Biol 153:206–216
99. Hartley RS, Bandman E, Yablonka-Reuveni Z (1991) Myoblasts from fetal and adult skeletal muscle regulate myosin expression differently. Dev Biol 148:249–260
100. Gros J, Manceau M, Thome V, Marcelle C (2005) A common somitic origin for embryonic muscle progenitors and satellite cells. Nature 435:954–958
101. Relaix F, Marcelle C (2009) Muscle stem cells. Curr Opin Cell Biol 21:748–753
102. Relaix F, Rocancourt D, Mansouri A, Buckingham M (2005) A Pax3/Pax7-dependent population of skeletal muscle progenitor cells. Nature 435:948–953

103. Barber TD, Barber MC, Cloutier TE, Friedman TB (1999) PAX3 gene structure, alternative splicing and evolution. Gene 237:311–319
104. Vorobyov E, Horst J (2006) Getting the proto-Pax by the tail. J Mol Evol 63:153–164
105. Buckingham M, Bajard L, Daubas P, Esner M, Lagha M, Relaix F, Rocancourt D (2006) Myogenic progenitor cells in the mouse embryo are marked by the expression of Pax3/7 genes that regulate their survival and myogenic potential. Anat Embryol (Berl) 211 Suppl 1:51–56
106. Buckingham M, Relaix F (2007) The role of Pax genes in the development of tissues and organs: Pax3 and Pax7 regulate muscle progenitor cell functions. Annu Rev Cell Dev Biol 23:645–673
107. Galli LM, Willert K, Nusse R, Yablonka-Reuveni Z, Nohno T, Denetclaw W, Burrus LW (2004) A proliferative role for Wnt-3a in chick somites. Dev Biol 269:489–504
108. Lamey TM, Koenders A, Ziman M (2004) Pax genes in myogenesis: alternate transcripts add complexity. Histol Histopathol 19:1289–1300
109. Relaix F, Rocancourt D, Mansouri A, Buckingham M (2004) Divergent functions of murine Pax3 and Pax7 in limb muscle development. Genes Dev 18:1088–1105
110. Epstein JA, Shapiro DN, Cheng J, Lam PY, Maas RL (1996) Pax3 modulates expression of the c-Met receptor during limb muscle development. Proc Natl Acad Sci USA 93:4213–4218
111. Harel I, Nathan E, Tirosh-Finkel L, Zigdon H, Guimaraes-Camboa N, Evans SM, Tzahor E (2009) Distinct origins and genetic programs of head muscle satellite cells. Dev Cell 16:822–832
112. Sambasivan R, Gayraud-Morel B, Dumas G, Cimper C, Paisant S, Kelly RG, Tajbakhsh S (2009) Distinct regulatory cascades govern extraocular and pharyngeal arch muscle progenitor cell fates. Dev Cell 16:810–821
113. Tajbakhsh S, Rocancourt D, Cossu G, Buckingham M (1997) Redefining the genetic hierarchies controlling skeletal myogenesis: Pax-3 and Myf-5 act upstream of MyoD. Cell 89:127–138
114. Noden DM, Francis-West P (2006) The differentiation and morphogenesis of craniofacial muscles. Dev Dyn 235:1194–1218
115. Kuang S, Charge SB, Seale P, Huh M, Rudnicki MA (2006) Distinct roles for Pax7 and Pax3 in adult regenerative myogenesis. J Cell Biol 172:103–113
116. Oustanina S, Hause G, Braun T (2004) Pax7 directs postnatal renewal and propagation of myogenic satellite cells but not their specification. EMBO J 23:3430–3439
117. Relaix F, Montarras D, Zaffran S, Gayraud-Morel B, Rocancourt D, Tajbakhsh S, Mansouri A, Cumano A, Buckingham M (2006) Pax3 and Pax7 have distinct and overlapping functions in adult muscle progenitor cells. J Cell Biol 172:91–102
118. Seale P, Sabourin LA, Girgis-Gabardo A, Mansouri A, Gruss P, Rudnicki MA (2000) Pax7 is required for the specification of myogenic satellite cells. Cell 102:777–786
119. Lepper C, Conway SJ, Fan CM (2009) Adult satellite cells and embryonic muscle progenitors have distinct genetic requirements. Nature 460:627–631
120. Kirkpatrick LJ, Yablonka-Reuveni Z, Rosser BW (2010) Retention of Pax3 expression in satellite cells of muscle spindles. J Histochem Cytochem 58:317–327
121. Olguin HC, Olwin BB (2004) Pax-7 up-regulation inhibits myogenesis and cell cycle progression in satellite cells: a potential mechanism for self-renewal. Dev Biol 275:375–388
122. Olguin HC, Yang Z, Tapscott SJ, Olwin BB (2007) Reciprocal inhibition between Pax7 and muscle regulatory factors modulates myogenic cell fate determination. J Cell Biol 177:769–779
123. Zammit PS, Relaix F, Nagata Y, Ruiz AP, Collins CA, Partridge TA, Beauchamp JR (2006) Pax7 and myogenic progression in skeletal muscle satellite cells. J Cell Sci 119:1824–1832
124. Collins CA, Gnocchi VF, White RB, Boldrin L, Perez-Ruiz A, Relaix F, Morgan JE, Zammit PS (2009) Integrated functions of Pax3 and Pax7 in the regulation of proliferation, cell size and myogenic differentiation. PLoS One 4:e4475.
125. McKinnell IW, Ishibashi J, Le Grand F, Punch VG, Addicks GC, Greenblatt JF, Dilworth FJ, Rudnicki MA (2008) Pax7 activates myogenic genes by recruitment of a histone methyltransferase complex. Nat Cell Biol 10:77–84
126. Kumar D, Shadrach JL, Wagers AJ, Lassar AB (2009) Id3 is a direct transcriptional target of Pax7 in quiescent satellite cells. Mol Biol Cell 20:3170–3177

127. Czerny T, Busslinger M (1995) DNA-binding and transactivation properties of Pax-6: three amino acids in the paired domain are responsible for the different sequence recognition of Pax-6 and BSAP (Pax-5). Mol Cell Biol 15:2858–2871
128. White RB, Ziman MR (2008) Genome-wide discovery of Pax7 target genes during development. Physiol Genomics 33:41–49
129. Pritchard C, Grosveld G, Hollenbach AD (2003) Alternative splicing of Pax3 produces a transcriptionally inactive protein. Gene 305:61–69
130. Vorobyov E, Horst J (2004) Expression of two protein isoforms of PAX7 is controlled by competing cleavage-polyadenylation and splicing. Gene 342:107–112
131. Ziman MR, Fletcher S, Kay PH (1997) Alternate Pax7 transcripts are expressed specifically in skeletal muscle, brain and other organs of adult mice. Int J Biochem Cell Biol 29:1029–1036
132. Ziman MR, Kay PH (1998) Differential expression of four alternate Pax7 paired box transcripts is influenced by organ- and strain-specific factors in adult mice. Gene 217:77–81
133. Ziman MR, Pelham JT, Mastaglia FL, Kay PH (2000) Characterization of the alternate allelic forms of human PAX7. Mamm Genome 11:332–337
134. Kassar-Duchossoy L, Gayraud-Morel B, Gomes D, Rocancourt D, Buckingham M, Shinin V, Tajbakhsh S (2004) Mrf4 determines skeletal muscle identity in Myf5:Myod double-mutant mice. Nature 431:466–471
135. Ludolph DC, Konieczny SF (1995) Transcription factor families: muscling in on the myogenic program. FASEB J 9:1595–1604
136. Rawls A, Morris JH, Rudnicki M, Braun T, Arnold HH, Klein WH, Olson EN (1995) Myogenin's functions do not overlap with those of MyoD or Myf-5 during mouse embryogenesis. Dev Biol 172:37–50
137. Rawls A, Valdez MR, Zhang W, Richardson J, Klein WH, Olson EN (1998) Overlapping functions of the myogenic bHLH genes MRF4 and MyoD revealed in double mutant mice. Development 125:2349–2358
138. Valdez MR, Richardson JA, Klein WH, Olson EN (2000) Failure of Myf5 to support myogenic differentiation without myogenin, MyoD, and MRF4. Dev Biol 219:287–298
139. Yablonka-Reuveni Z, Rudnicki MA, Rivera AJ, Primig M, Anderson JE, Natanson P (1999) The transition from proliferation to differentiation is delayed in satellite cells from mice lacking MyoD. Dev Biol 210:440–455
140. White JD, Scaffidi A, Davies M, McGeachie J, Rudnicki MA, Grounds MD (2000) Myotube formation is delayed but not prevented in MyoD-deficient skeletal muscle: studies in regenerating whole muscle grafts of adult mice. J Histochem Cytochem 48:1531–1544
141. Megeney LA, Kablar B, Garrett K, Anderson JE, Rudnicki MA (1996) MyoD is required for myogenic stem cell function in adult skeletal muscle. Genes Dev 10:1173–1183
142. Olson EN, Arnold HH, Rigby PW, Wold BJ (1996) Know your neighbors: three phenotypes in null mutants of the myogenic bHLH gene MRF4. Cell 85:1–4
143. Ustanina S, Carvajal J, Rigby P, Braun T (2007) The myogenic factor Myf5 supports efficient skeletal muscle regeneration by enabling transient myoblast amplification. Stem Cells 25:2006–2016
144. Knapp JR, Davie JK, Myer A, Meadows E, Olson EN, Klein WH (2006) Loss of myogenin in postnatal life leads to normal skeletal muscle but reduced body size. Development 133:601–610
145. Meadows E, Cho JH, Flynn JM, Klein WH (2008) Myogenin regulates a distinct genetic program in adult muscle stem cells. Dev Biol 322:406–414
146. Hinterberger TJ, Sassoon DA, Rhodes SJ, Konieczny SF (1991) Expression of the muscle regulatory factor MRF4 during somite and skeletal myofiber development. Dev Biol 147:144–156
147. Zhang W, Behringer RR, Olson EN (1995) Inactivation of the myogenic bHLH gene MRF4 results in up-regulation of myogenin and rib anomalies. Genes Dev 9:1388–1399
148. Smith CK, II, Janney MJ, Allen RE (1994) Temporal expression of myogenic regulatory genes during activation, proliferation, and differentiation of rat skeletal muscle satellite cells. J Cell Physiol 159:379–385
149. Smith TH, Block NE, Rhodes SJ, Konieczny SF, Miller JB (1993) A unique pattern of expression of the four muscle regulatory factor proteins distinguishes somitic from embryonic, fetal and newborn mouse myogenic cells. Development 117:1125–1133

150. Berkes CA, Tapscott SJ (2005) MyoD and the transcriptional control of myogenesis. Semin Cell Dev Biol 16:585–595
151. Tapscott SJ (2005) The circuitry of a master switch: Myod and the regulation of skeletal muscle gene transcription. Development 132:2685–2695
152. Gillespie MA, Le Grand F, Scime A, Kuang S, von Maltzahn J, Seale V, Cuenda A, Ranish JA, Rudnicki MA (2009) p38-{gamma}-dependent gene silencing restricts entry into the myogenic differentiation program. J Cell Biol 187:991–1005
153. Wyzykowski JC, Winata TI, Mitin N, Taparowsky EJ, Konieczny SF (2002) Identification of novel MyoD gene targets in proliferating myogenic stem cells. Mol Cell Biol 22:6199–6208
154. Zhang K, Sha J, Harter ML (2010) Activation of Cdc6 by MyoD is associated with the expansion of quiescent myogenic satellite cells. J Cell Biol 188:39–48
155. Grounds MD, Garrett KL, Lai MC, Wright WE, Beilharz MW (1992) Identification of skeletal muscle precursor cells in vivo by use of MyoD1 and myogenin probes. Cell Tissue Res 267:99–104
156. Gnocchi VF, White RB, Ono Y, Ellis JA, Zammit PS (2009) Further characterisation of the molecular signature of quiescent and activated mouse muscle satellite cells. PLoS One 4:e5205
157. Bischoff R (1990) Control of satellite cell proliferation. Adv Exp Med Biol 280:147–157; discussion 157–148
158. Bischoff R (1986) A satellite cell mitogen from crushed adult muscle. Dev Biol 115:140–147
159. Yablonka-Reuveni Z, Paterson BM (2001) MyoD and myogenin expression patterns in cultures of fetal and adult chicken myoblasts. J Histochem Cytochem 49:455–462
160. Rudnicki MA, Braun T, Hinuma S, Jaenisch R (1992) Inactivation of MyoD in mice leads to up-regulation of the myogenic HLH gene Myf-5 and results in apparently normal muscle development. Cell 71:383–390
161. Rudnicki MA, Schnegelsberg PN, Stead RH, Braun T, Arnold HH, Jaenisch R (1993) MyoD or Myf-5 is required for the formation of skeletal muscle. Cell 75:1351–1359
162. Lindon C, Montarras D, Pinset C (1998) Cell cycle-regulated expression of the muscle determination factor Myf5 in proliferating myoblasts. J Cell Biol 140:111–118
163. Kitzmann M, Carnac G, Vandromme M, Primig M, Lamb NJ, Fernandez A (1998) The muscle regulatory factors MyoD and myf-5 undergo distinct cell cycle-specific expression in muscle cells. J Cell Biol 142:1447–1459
164. Kitzmann M, Fernandez A (2001) Crosstalk between cell cycle regulators and the myogenic factor MyoD in skeletal myoblasts. Cell Mol Life Sci 58:571–579
165. Batonnet-Pichon S, Tintignac LJ, Castro A, Sirri V, Leibovitch MP, Lorca T, Leibovitch SA (2006) MyoD undergoes a distinct G2/M-specific regulation in muscle cells. Exp Cell Res 312:3999–4010
166. Tintignac LA, Leibovitch MP, Kitzmann M, Fernandez A, Ducommun B, Meijer L, Leibovitch SA (2000) Cyclin E-cdk2 phosphorylation promotes late G1-phase degradation of MyoD in muscle cells. Exp Cell Res 259:300–307
167. Doucet C, Gutierrez GJ, Lindon C, Lorca T, Lledo G, Pinset C, Coux O (2005) Multiple phosphorylation events control mitotic degradation of the muscle transcription factor Myf5. BMC Biochem 6:27
168. Lindon C, Albagli O, Domeyne P, Montarras D, Pinset C (2000) Constitutive instability of muscle regulatory factor Myf5 is distinct from its mitosis-specific disappearance, which requires a D-box-like motif overlapping the basic domain. Mol Cell Biol 20:8923–8932
169. Perez-Ruiz A, Ono Y, Gnocchi VF, Zammit PS (2008) beta-Catenin promotes self-renewal of skeletal-muscle satellite cells. J Cell Sci 121:1373–1382
170. Friday BB, Pavlath GK (2001) A calcineurin- and NFAT-dependent pathway regulates Myf5 gene expression in skeletal muscle reserve cells. J Cell Sci 114:303–310
171. Shinin V, Gayraud-Morel B, Gomes D, Tajbakhsh S (2006) Asymmetric division and cosegregation of template DNA strands in adult muscle satellite cells. Nat Cell Biol 8:677–687
172. Huh MS, Parker MH, Scime A, Parks R, Rudnicki MA (2004) Rb is required for progression through myogenic differentiation but not maintenance of terminal differentiation. J Cell Biol 166:865–876

173. Andres V, Walsh K (1996) Myogenin expression, cell cycle withdrawal, and phenotypic differentiation are temporally separable events that precede cell fusion upon myogenesis. J Cell Biol 132:657–666
174. Horsley V, Pavlath GK (2004) Forming a multinucleated cell: molecules that regulate myoblast fusion. Cells Tissues Organs 176:67–78
175. Jansen KM, Pavlath GK (2008) Molecular control of mammalian myoblast fusion. Methods Mol Biol 475:115–133
176. Devlin BH, Konigsberg IR (1983) Reentry into the cell cycle of differentiated skeletal myocytes. Dev Biol 95:175–192
177. Cenciarelli C, De Santa F, Puri PL, Mattei E, Ricci L, Bucci F, Felsani A, Caruso M (1999) Critical role played by cyclin D3 in the MyoD-mediated arrest of cell cycle during myoblast differentiation. Mol Cell Biol 19:5203–5217
178. Halevy O, Novitch BG, Spicer DB, Skapek SX, Rhee J, Hannon GJ, Beach D, Lassar AB (1995) Correlation of terminal cell cycle arrest of skeletal muscle with induction of p21 by MyoD. Science 267:1018–1021
179. Kitzmann M, Vandromme M, Schaeffer V, Carnac G, Labbe JC, Lamb N, Fernandez A (1999) cdk1- and cdk2-mediated phosphorylation of MyoD Ser200 in growing C2 myoblasts: role in modulating MyoD half-life and myogenic activity. Mol Cell Biol 19:3167–3176
180. Novitch BG, Spicer DB, Kim PS, Cheung WL, Lassar AB (1999) pRb is required for MEF2-dependent gene expression as well as cell-cycle arrest during skeletal muscle differentiation. Curr Biol 9:449–459
181. Perry RL, Parker MH, Rudnicki MA (2001) Activated MEK1 binds the nuclear MyoD transcriptional complex to repress transactivation. Mol Cell 8:291–301
182. Puri PL, Iezzi S, Stiegler P, Chen TT, Schiltz RL, Muscat GE, Giordano A, Kedes L, Wang JY, Sartorelli V (2001) Class I histone deacetylases sequentially interact with MyoD and pRb during skeletal myogenesis. Mol Cell 8:885–897
183. Song A, Wang Q, Goebl MG, Harrington MA (1998) Phosphorylation of nuclear MyoD is required for its rapid degradation. Mol Cell Biol 18:4994–4999
184. De Falco G, Comes F, Simone C (2006) pRb: master of differentiation. Coupling irreversible cell cycle withdrawal with induction of muscle-specific transcription. Oncogene 25:5244–5249
185. Yablonka-Reuveni Z, Anderson JE (2006) Satellite cells from dystrophic (mdx) mice display accelerated differentiation in primary cultures and in isolated myofibers. Dev Dyn 235:203–212
186. Black BL, Olson EN (1998) Transcriptional control of muscle development by myocyte enhancer factor-2 (MEF2) proteins. Annu Rev Cell Dev Biol 14:167–196
187. Molkentin JD, Black BL, Martin JF, Olson EN (1995) Cooperative activation of muscle gene expression by MEF2 and myogenic bHLH proteins. Cell 83:1125–1136
188. Bonavaud S, Agbulut O, Nizard R, D'Honneur G, Mouly V, Butler-Browne G (2001) A discrepancy resolved: human satellite cells are not preprogrammed to fast and slow lineages. Neuromuscul Disord 11:747–752
189. Dusterhoft S, Pette D (1993) Satellite cells from slow rat muscle express slow myosin under appropriate culture conditions. Differentiation 53:25–33
190. Dusterhoft S, Yablonka-Reuveni Z, Pette D (1990) Characterization of myosin isoforms in satellite cell cultures from adult rat diaphragm, soleus and tibialis anterior muscles. Differentiation 45:185–191
191. Rosenblatt JD, Parry DJ, Partridge TA (1996) Phenotype of adult mouse muscle myoblasts reflects their fiber type of origin. Differentiation 60:39–45
192. DiMario JX, Stockdale FE (1997) Both myoblast lineage and innervation determine fiber type and are required for expression of the slow myosin heavy chain 2 gene. Dev Biol 188:167–180
193. Esser K, Gunning P, Hardeman E (1993) Nerve-dependent and -independent patterns of mRNA expression in regenerating skeletal muscle. Dev Biol 159:173–183
194. Gunning P, Hardeman E (1991) Multiple mechanisms regulate muscle fiber diversity. FASEB J 5:3064–3070

195. Allen RE, Rankin LL, Greene EA, Boxhorn LK, Johnson SE, Taylor RG, Pierce PR (1991) Desmin is present in proliferating rat muscle satellite cells but not in bovine muscle satellite cells. J Cell Physiol 149:525–535
196. Yablonka-Reuveni Z, Nameroff M (1990) Temporal differences in desmin expression between myoblasts from embryonic and adult chicken skeletal muscle. Differentiation 45:21–28
197. Walro JM, Kucera J (1999) Why adult mammalian intrafusal and extrafusal fibers contain different myosin heavy-chain isoforms. Trends Neurosci 22:180–184
198. Kirkpatrick LJ, Allouh MZ, Nightingale CN, Devon HG, Yablonka-Reuveni Z, Rosser BW (2008) Pax7 shows higher satellite cell frequencies and concentrations within intrafusal fibers of muscle spindles. J Histochem Cytochem 56:831–840
199. Kozeka K, Ontell M (1981) The three-dimensional cytoarchitecture of developing murine muscle spindles. Dev Biol 87:133–147
200. Zammit PS, Carvajal JJ, Golding JP, Morgan JE, Summerbell D, Zolnerciks J, Partridge TA, Rigby PW, Beauchamp JR (2004) Myf5 expression in satellite cells and spindles in adult muscle is controlled by separate genetic elements. Dev Biol 273:454–465
201. Schmalbruch H, Lewis DM (2000) Dynamics of nuclei of muscle fibers and connective tissue cells in normal and denervated rat muscles. Muscle Nerve 23:617–626
202. Conboy IM, Rando TA (2002) The regulation of Notch signaling controls satellite cell activation and cell fate determination in postnatal myogenesis. Dev Cell 3:397–409
203. Shinin V, Gayraud-Morel B, Tajbakhsh S (2009) Template DNA-strand co-segregation and asymmetric cell division in skeletal muscle stem cells. Methods Mol Biol 482:295–317
204. Vasyutina E, Lenhard DC, Birchmeier C (2007) Notch function in myogenesis. Cell Cycle 6:1451–1454
205. Brack AS, Conboy MJ, Roy S, Lee M, Kuo CJ, Keller C, Rando TA (2007) Increased Wnt signaling during aging alters muscle stem cell fate and increases fibrosis. Science 317:807–810
206. Ono Y, Gnocchi VF, Zammit PS, Nagatomi R (2009) Presenilin-1 acts via Id1 to regulate the function of muscle satellite cells in a gamma-secretase-independent manner. J Cell Sci 122:4427–4438
207. Holterman CE, Le Grand F, Kuang S, Seale P, Rudnicki MA (2007) Megf10 regulates the progression of the satellite cell myogenic program. J Cell Biol 179:911–922
208. Abou-Khalil R, Le Grand F, Pallafacchina G, Valable S, Authier FJ, Rudnicki MA, Gherardi RK, Germain S, Chretien F, Sotiropoulos A, Lafuste P, Montarras D, Chazaud B (2009) Autocrine and paracrine angiopoietin 1/Tie-2 signaling promotes muscle satellite cell self-renewal. Cell Stem Cell 5:298–309
209. Kuang S, Gillespie MA, Rudnicki MA (2008) Niche regulation of muscle satellite cell self-renewal and differentiation. Cell Stem Cell 2:22–31.
210. Gussoni E, Blau HM, Kunkel LM (1997) The fate of individual myoblasts after transplantation into muscles of DMD patients. Nat Med 3:970–977
211. Miller RG, Sharma KR, Pavlath GK, Gussoni E, Mynhier M, Lanctot AM, Greco CM, Steinman L, Blau HM (1997) Myoblast implantation in Duchenne muscular dystrophy: the San Francisco study. Muscle Nerve 20:469–478
212. Muir LA, Chamberlain JS (2009) Emerging strategies for cell and gene therapy of the muscular dystrophies. Expert Rev Mol Med 11:e18.
213. Tremblay JP, Skuk D (2008) Another new "super muscle stem cell" leaves unaddressed the real problems of cell therapy for duchenne muscular dystrophy. Mol Ther 16:1907–1909
214. Lafreniere JF, Caron MC, Skuk D, Goulet M, Cheikh AR, Tremblay JP (2009) Growth factor coinjection improves the migration potential of monkey myogenic precursors without affecting cell transplantation success. Cell Transplant 18:719–730
215. Mouly V, Aamiri A, Perie S, Mamchaoui K, Barani A, Bigot A, Bouazza B, Francois V, Furling D, Jacquemin V, Negroni E, Riederer I, Vignaud A, St Guily JL, Butler-Browne GS (2005) Myoblast transfer therapy: is there any light at the end of the tunnel? Acta Myol 24:128–133
216. Partridge TA (2004) Stem cell therapies for neuromuscular diseases. Acta Neurol Belg 104:141–147

217. Richard PL, Gosselin C, Laliberte T, Paradis M, Goulet M, Tremblay JP, Skuk D (2010) A first semi-manual device for clinical intramuscular repetitive cell injections. Cell Transplant 19:67–78.
218. Tremblay JP, Skuk D, Palmieri B, Rothstein DM (2009) A case for immunosuppression for myoblast transplantation in duchenne muscular dystrophy. Mol Ther 17:1122–1124
219. Sampaolesi M, Torrente Y, Innocenzi A, Tonlorenzi R, D'Antona G, Pellegrino MA, Barresi R, Bresolin N, De Angelis MG, Campbell KP, Bottinelli R, Cossu G (2003) Cell therapy of alpha-sarcoglycan null dystrophic mice through intra-arterial delivery of mesoangioblasts. Science 301:487–492
220. Beauchamp JR, Morgan JE, Pagel CN, Partridge TA (1999) Dynamics of myoblast transplantation reveal a discrete minority of precursors with stem cell-like properties as the myogenic source. J Cell Biol 144:1113–1122
221. Cossu G, Sampaolesi M (2007) New therapies for Duchenne muscular dystrophy: challenges, prospects and clinical trials. Trends Mol Med 13:520–526
222. Morgan JE, Pagel CN, Sherratt T, Partridge TA (1993) Long-term persistence and migration of myogenic cells injected into pre-irradiated muscles of mdx mice. J Neurol Sci 115:191–200
223. Gross JG, Bou-Gharios G, Morgan JE (1999) Potentiation of myoblast transplantation by host muscle irradiation is dependent on the rate of radiation delivery. Cell Tissue Res 298:371–375
224. Chien KR, Domian IJ, Parker KK (2008) Cardiogenesis and the complex biology of regenerative cardiovascular medicine. Science 322:1494–1497

Regenerating Mechanical Function In Vivo with Skeletal Myoblasts

Todd K. Rosengart and Muath Bishawi B.S.

Abstract Cellular cardiomyoplasty describes the administration of extraneous, relatively undifferentiated cells which may integrate structurally and/or functionally into infarcted myocardium for the purpose of improving ventricular function or stabilizing/reversing adverse ventricular remodeling. The first cell type to enter clinical trials was skeletal myoblast. Preclinical studies demonstrated improved ventricular function in small-animal and large-animal models. The results from clinical trials have been mixed, with many studies documenting improved ventricular function. Preclinical and clinical studies are reviewed in this chapter, along with guidance for the next steps and future directions.

Keywords Cellular cardiomyoplasty • Skeletal myoblasts • Clinical trials • Heart failure • Myocardial infarction • Ejection fraction

1 Introduction

Cellular cardiomyoplasty (CCM) describes the administration of extraneous, relatively undifferentiated (stem) cells which may integrate structurally and/or functionally into infarcted myocardium for the purpose of improving (systolic) ventricular function or stabilizing/reversing adverse ventricular remodeling (diastolic function) [1–4]. In theory, the implanted cells, typically harvested autogenously and expanded in culture, "repopulate" the area of myocardial scar, presumably with viable myocytes. This technique has been shown in a number of animal models to improve myocardial function in the setting of ventricular infarction, and as a consequence advanced to clinical trials [1–13].

T.K. Rosengart (✉)
Department of Surgery, Stony Brook University Medical Center,
Stony Brook, NY, USA
e-mail: todd.rosengart@stonybrook.edu

I.S. Cohen and G.R. Gaudette (eds.), *Regenerating the Heart*, Stem Cell Biology and Regenerative Medicine, DOI 10.1007/978-1-61779-021-8_12,
© Springer Science+Business Media, LLC 2011

It has been considered that CCM may provide for the regeneration of cardiac function in the setting of myocardial infarction by (1) "repopulating" scarred myocardium with contractile myocytes that are electrically and/or functionally integrated into the host myocardium, (2) providing a biologic "scaffolding" to improve ventricular wall stresses and, potentially through such a mechanism, diminish further remodeling of the thinned, injured ventricle, or (3) given evidence of the relatively sparse deposition of the cell implants as a realistic substrate for such mechanical support, acting merely as secretory vehicles to stimulate angiogenesis and the perfusion of recoverable peri-infarct border zones [14–20].

The paradox of CCM, however, as illustrated by data generated in studies of skeletal myoblasts as a candidate cell implant type, is that functional (mechanical) benefits have repeatedly been reported, although the mechanisms of action underlying the putative efficacy of this therapy remains poorly understood. More specifically, it is intriguing that skeletal myoblasts have recurringly demonstrated functional benefits in animal myocardial infarction models, despite evidence that these reserve skeletal muscle cells do not appear to possess the two capacities that have conventionally been considered to be essential to the successful application of CCM, namely, differentiation into a contractile cardiomyocyte and electromechanical connectivity with host myocytes.

In regard to the "appropriateness" of the use of skeletal myoblasts as a candidate cell type for CCM, whereas it has been claimed that this cell type is incapable of differentiation toward a "cardiomyocyte-like" phenotype, it is important to note that skeletal myoblasts in at least some studies have demonstrated postmyocardial implant to (1) express cardiac proteins, such as troponin, α-actin, actinin, and myosin light and heavy chain proteins, (2) develop desmosomes, fascia adherens, gap junctions, and a sarcomeric structure, and (3) align with host cardiomyocyte [14–19]. Conversely, although it has been postulated that skeletal myoblasts may, alternatively, create a "biologic bandage" that prevents/reverses remodeling by virtue of the biologic "mass" of the cell implants, most studies have demonstrated the extremely inefficient (less than 10%) engraftment of cells into areas of myocardial scar, and the subsequent presence of only scattered "islands" of such cells that would be unlikely to provide the "scaffolding" suggested above [14–21]. Finally, whereas more pluripotent stem cells may be capable of secreting angiogenic growth factors and other antiapoptotic cytokines, no such capacity is thought to exist for skeletal myoblast [16–20].

The available evidence thus makes it difficult to clearly ascertain the mechanisms underlying the functional (mechanical) effects of myoblasts once they are implanted or the mechanisms underlying such activities. Nevertheless, as presented in this chapter, it appears clear that CCM mediated via skeletal myoblast implantation clearly does confer such benefits to the failing heart. Furthermore, it would appear that the efficacy of skeletal myoblast implantation is not significantly different from the efficacy after implantation of other cell types presumed to be more competent to differentiate into a cardiomyocyte phenotype, rendering even greater confusion as to the function and mechanisms of action of these and other cell types following myocardial implantation [16–21]. Given such contradictions, it is conceivable that there may exist alternative mechanisms of action underlying these observations. This conclusion would render moot many of the current controversies in this field and "condemnations" of the skeletal myoblast as a possible CCM candidate cell implant. Such a finding would potentially be very advantageous to the field, because skeletal myoblasts possess a number of characteristics that are extremely conductive to their clinical use. Namely, these cells carry none of the ethical, legal, and donor issues associated with other stem cell types derived from embryonic or other like origins: they are easily procurable, they can easily be expanded in vitro from autogenous sources (via skeletal muscle biopsy), and they may be less susceptible to ischemic insult (and thus possess greater "survivability") after implantation into the hostile milieu of the ischemic myocardium compared with other more delicate candidate cell types.

A careful review of data derived from animal and clinical studies involving the use of skeletal myoblasts therefore provides an opportunity to better understand the therapeutic physiology of this potential new treatment for heart failure. Ideally, insights gained from such an investigation, as is undertaken in this chapter, might provide better guidance for future planned or proposed clinical cell implant trials, with skeletal myoblast or even with other cell lines, specifically as related to an idealized cell type and method of delivery.

2 Animal Studies

Data suggesting the efficacy of skeletal myoblast administration as a candidate cell type for CCM have been derived from a number of studies in the animal model, and represent some of the earliest data suggesting the efficacy of a cell delivery into the infarcted myocardium for the purpose of improving cardiac function [1–12]. One of the earliest reports of such benefit was by Murry et al. [3] in a rat coronary ligation/infarct model, which provided evidence of improvement in architectural integrity and *systolic* work. This study also suggested that this newly generated "muscle" expressed β-myosin heavy chain, a component of (cardiac) slow twitch responses. Furthermore, Murry et al. reported that skeletal myoblasts could contract when stimulated electrically, and therefore potentially could connect electrically with the host myocardium. As noted below, evidence of such electromechanical competency has been a source of great confusion and criticism with the use of skeletal myoblasts.

Taylor et al. [4, 5] popularized and provided confirmation of Murry et al.'s work, noting in a rat cryoinjury model that myoblast implantation improved myocardial performance and generated "islands of …elongated, striated cells that retained characteristics of both skeletal and cardiac cells." Aside from demonstrating (predominantly diastolic) functional improvements in myoblast-treated versus untreated animals, this study provided the first evidence suggesting the engraftment of myoblasts into (elongated) implants aligned with the host myocardium in such a way as to support mechanical function. Importantly, such improvements seemed to be conferred without benefit of any apparent differentiation of myoblasts into a cardiomyocyte phenotype. Criticism of this work, parenthetically, immediately focused on the use of cryoinjury as an artifactual source of myocardial infarction, and most subsequent work involved coronary occlusion/ligation schema in small-animal or (subsequently) large-animal models.

Confirmation of this work was provided by the studies of Hutcheson et al. [6], who in 2000 published results in a rabbit infarction model. The mechanism underlying the benefits conferred by myoblast implantation appeared to be further elucidated in this study comparing the implantation of skeletal myoblasts versus dermal fibroblasts. Specifically, although diastolic performance was reported to be improved with both skeletal myoblast and dermal fibroblast groups, only the administration of skeletal myoblasts appeared to improve systolic performance. These investigators therefore concluded that although "contractile and noncontractile cells can improve regional material properties or structural integrity of terminally injured heart," only skeletal myoblasts "improved systolic performance in the damaged region, supporting the role of myogenic cells in augmenting contraction."

Further support for the functional (systolic) benefits conferred by skeletal myoblasts was provided by the work of Jain et al. [7] in a rat (1-h) coronary occlusion model. In this work, a 92% rate of myoblast engraftment was observed following direct myoblast administration into the area of scar when cell administration was performed 1 week after coronary occlusion (an interval selected to allow scar "stabilization" and abatement of the inflammation that is thought to potentially be injurious to implanted cells). Whereas rats that underwent coronary occlusion without cell injection had a progressive decline in maximum exercise tolerance over time, deterioration in post-myocardial infarction exercise capacity was prevented in those animals that received cell injections ($p<0.05$). Importantly, left ventricular (LV) *systolic* pressures were also improved in the cell therapy group upon maximal exercise testing. Likewise, this report also provided evidence of a significant decrease in LV size, consistent with improved ventricular compliance and (*diastolic*) hemodynamics.

Ghostine et al. [8] were able to provide more detailed hemodynamic assessments of the effects of myoblast implantation into areas of infarction in a larger animal (ovine) model taken to extended (4- and 12-month) postimplant time points. These investigators observed a reduction in the increase of postinfarction LV end-diastolic volume (LVEDV) in animals receiving myoblasts versus control animals (LVEDV 72 ± 8 ml vs 105 ± 13 ml, $p<0.05$), with a decrease in the deterioration of ejection fraction also noted in cell-treated versus control animals ($48\pm5\%$ vs. $33\pm3\%$, respectively; $p=0.006$). This report also specifically provided evidence of improved global and regional contractile (i.e., systolic) function in the transplanted group, as demonstrated by improved wall motion score compared with the controls (5.4 ± 1.2 vs. 13 ± 2.2, $p<0.01$) and improved systolic myocardial velocity gradient. Consistent with these findings, collagen density was significantly decreased in the transplanted versus control animals ($30\pm2\%$ vs. $73\pm3\%$, $p<0.0001$), suggesting that the implant technique had resulted in a reversal of the long-term sequelae of congestive heart failure. Finally, these investigators noted the presence of myotubes and myoblasts in implanted territories, as well as the coexpression of fast and *slow* isoforms of the myosin heavy chain, the latter a potential marker of cardiomyocyte differentiation.

Although evidence of enhancement of only diastolic performance has been generated in both porcine and ovine models [9, 10], the above-noted series of animal studies, taken together, suggest that skeletal myoblast implantation may improve myocardial contractile (systolic) mechanisms, rather than only inducing stabilization of ventricular geometry and diastolic mechanisms – i.e., "the bandage effect."

Further refinement of these observations was provided by Tambara et al.[11], who performed (neonatal) skeletal myoblast transplantation into the area of myocardial infarction in a rat coronary ligation model. In their study, LV diastolic dimension was decreased, fractional area change was increased, and myocardial infarct size was also decreased in the groups that received cell transplantation compared with controls. Importantly, these investigators related observations of a "dose response" in their study, with improved outcomes noted as a function of the number of cells administered. As compared with animals receiving lesser amounts of administered cells, animals in which the greatest number of cells (5×10^7 cells) were administered into infarct regions demonstrated nearly normal wall thickness and volume (40 mm^3), as determined by histology studies wherein donor cells were tracked via positive staining for antibodies against fast skeletal myosin. These data thereby provide evidence of an important "dose response" that may help explain some of the contradictions of other studies (i.e., ineffectual systolic responses may have been observed in some studies if inadequate cell administration was performed). It is unclear from this work, however, whether the use of neonatal cells, which were harvested and injected without in vitro expansion, is relevant to other studies using cultured cells derived from adult sources.

Significant to the issue of dose response and the adequacy of implanted cell number, Gulbins et al. [22] using a vital fluorescent dye (PKH-26) for labeling implanted cells have also provided evidence that implanted myoblasts retained the capacity for cell division. If cell division does occur after, implant, as has been suggested by others (see Table 1; Maurel et al.), then the amount of an initial cell dose may be of less consequence than the ultimate ability of these implants to replicate and repopulate areas of myocardial scar.

Finally, the concept that effective delivery of an adequate number of cells to the infarct may impact outcome is also supported by the work of Zhong et al.[23], who, using 4′,6-diamidino-2-phenylindone-labeled canine skeletal myoblasts in a canine infarct model, suggested better ordering/alignment of injected cells that were delivered through coronary artery perfusion, versus by direct injection. Furthermore, these investigators reported evidence of cell differentiation into what appeared to be striated muscle fibers connected to intercalated discs, consistent with the concept that such cells are capable of some degree of differentiation sufficient to support contractile function (and even electrical connectivity).

Along these lines, other investigators have also provided evidence of the postimplant differentiation of myoblasts toward a "cardiomyocyte-like" phenotype. In a study by Robinson et al. [24] that used a transendocardial delivery technique, implanted

Table 1 Selected animal model studies of mechanical function after skeletal myoblast administration

Authors	Year	Animal model	Delivery/intervention follow-up	Findings
Murry et al. [3]	1996	Rat/cryoinjury	3 months	Engraftment of partially differentiated cells
Taylor et al. [4]	1998/1999	Rat/cryoinjury	Up to 6 weeks	Engraftment of partially differentiated cells Improvement in diastolic performance
Jain et al. [7]	2001			
Ghostine et al. [8]	2002	Ovine/coronary embolization	Direct, 1 year	Improved increases in end-diastolic volume and decreases in EF Improved wall motion score; histologic evidence of engraftment with slow isoforms; decreased collagen density
McConnell et al. [9]	2005	Ovine/coronary embolization		Improved diastolic function
Tambara [11]	2003			
Leobon et al. [12]	2003	Rat	GFP-labeled cells	Implanted cells electrically uncoupled from host myocardium
Dersimonian [14]	2001			
Scorsin et al. [15]	2000			
Chedrawy et al. [17]	2002			
Suzuki et al. [20]	2001			
Gulbins et al. [22]	2002	Rat	Labeled implants	Implanted cells retain proliferative capacity
Zhong et al. [23]				
Robinson et al. [24]	1996			
Miyagawa et al. [37]	2005			
Al Attar et al. [38]	2003	Rat	1 year	Improved systolic and diastolic function up to 1 year; decreased number of implanted cells decreased over time
Scorsin et al. [15]	2000			
Reinecke et al. [16]	1999/2002/2003			

Toh et al. [40]	2004	Rat	Direct injection	Cells do not express cardiac-specific genes after transplantation
Maurel et al. [39]	2005	Rat/LV hypertrophy	Heat shock/ cryopreservation	Significant early cell loss; postimplant replication; therapies failed to improve engraftment
Winitsky et al. [41]	2005	Mouse	Tail vein	Evidence of postimplant differentiation to "striated heart muscle"
He [42]	2005	Canine/embolization	Direct 10 weeks	Increased dP/dt (105 vs. 97); increased EF (46 vs. 40%) No evidence of engraftment
Ye [43]	2005	Rat/coronary microembolization	Direct and percutaneous	LVEF increased by 5–7%, reversal of ventricular remodeling observed, increased dP/dt(max) of about 10%, possible decrease in EDP, and improved ESPVR and EDPVR
Brasselet et al. [44]	2005	Ovine/ coil	Transcoronary sinus, 2 months	Improved EF (50 vs. 39%) Evidence of implantation (3 of 7 animals)
Ott et al. [45]	2005	Sheep/anterior MI	Catheter	Improved LVEF (from 35 to 50%), clusters of myoblasts identified by histology in 3 animals, significantly smaller increase in end-systolic LV volume compared with controls, overall decrease in regional score index (better kinetics)
Gavira et al. [46]	2006	Pigs	Direct injection vs. percutaneous	Improved EF Increased vasculogenesis and decreased fibrosis
Payne et al. [47]	2007	Pig/LV MI	Percutaneous	Trend toward improved viability, EF, and cardiac inject. No arrhythmias noted. No skeletal muscles found on histology
Retuerto et al. [34]	2007	Mice	Direct injection	Decreased LV enlargement compared with controls in SM and SM+ VEGF25 (overexpression). Better LV contractility and systolic function compared with controls
Farahmand et al. [48]	2008	Rat	Direct injection	Improved EF and infarct perfusion with the use of SM and pretreatment of engraftment site with AdVEGF21

(continued)

Table 1 (continued)

Authors	Year	Animal model	Delivery/intervention follow-up	Findings
Pedrotty et al. [49]	2008	Rat	Intramyocardial injections	Small improvements in function, poor graft survival, and integration of cells to the cardiomyocytes, VPC observed
Zhu et al. [50]	2009	Rat	In vitro	Connexin 43 and N-cadherin expression between skeletal myoblasts and cardiomyocytes pairs was 22% and 14%, respectively
Khan et al. [51]	2008	Rabbit	Intracoronary	LVEF improved, higher number of neoangiogenesis, reduced cardiac remodeling , improved LF end-diastolic diameter
Hoashi et al. [52]	2009	Rat/pulmonary banding	Myoblast sheets 4 weeks	Skeletal myoblast sheet transplantation improved diastolic function Increased capillary density and decreased fibrosis
Patila et al. [10]	2009	Porcine/ameroid	Direct injection 4 weeks	Improved peak filling, ejection rate, and duration of diastole No difference in EF or local thickening Increase in microvessel density.

All data are expressed compared with saline or other appropriate controls
LV left ventricular, *MI* myocardial infarct, *GFP* green fluorescent protein, *EF* ejection fraction, *LVEF* left ventricular ejection fraction, *EDP* end-diastolic pressure, *ESPVR* end-systolic pressure–volume relationship, *EDPVR* end-diastolic pressure–volume relationship, *SM* smooth muscle, *VEGF25* diluted vascular endothelial growth factor 1 in 3 dilution (25%), *AdVEGF121* adenovirus encoding the 121–amino acid isoform of vascular endothelial growth factor, *VPC* ventricular premature contractions

cells were demonstrated to express fast-twitch skeletal muscle sarcoendoplasmic reticulum calcium ATPase 1 and were observed to be beginning to form myofilaments. Four months after implant, cell implants were also observed to be expressing slow-twitch/cardiac protein phospholamban. Most significantly, evidence also exists for the expression by implanted skeletal-muscle-derived cells of connexin 43, a key component for the electrical connectivity of implants with the host myocardium and synergized contractile function.

In contrast, Leobon et al. [12] used intracellular recordings coupled to video and fluorescence microscopy to track excitable and contractile properties of skeletal myoblasts labeled with green fluorescent protein. These investigators reported that the myoblasts differentiate into "peculiar hyperexcitable myotubes with a contractile activity fully *independent* of neighboring cardiomyocytes," refuting the possibility of electrical connectivity to the host myocardium with the use of skeletal myoblasts.

Taken together, these studies suggest that myoblast administration is capable of enhancing ventricular function following infarction, and that such cells confer such benefits to a greater degree than contractile-incompetent cells such as fibroblasts. Whereas the absence of evidence of observed cell contractility or improvement in systolic function in some studies supports a "scaffolding" mechanism of action for the cell implants, other studies have demonstrated improved systolic function, potentially as a function of the survival and subsequent proliferation of cell implants, but albeit without to date the observance of implanted cell contractility.

Whereas animal studies have not yet provided clear evidence of myoblast differentiation into a cardiomyocyte-like cell type that would presumably be advantageous to providing such myocardial functionality, some evidence does exist for some degree of such differentiation. Such data blur the boundary between the conceptual legitimacy of labeling myoblasts as "nondifferentiating" as compared with other candidate stem cells. Conversely, observations of functional benefits without evidence of differentiation suggest that such differentiation might not be necessary to the efficacy of CCM. This premise is supported by the observation that no functional advantages have yet been demonstrated with the use of stem cell versus skeletal myoblasts for CCM, and the reality that no study has yet demonstrated the contractility of any cell implants. It is therefore also conceivable that the efficacy of CCM is provided by mechanisms distinct from contractile function or myocardial integration – an unknown variable that myoblasts might also possess.

It must finally be considered that the efficacy demonstrated in animal studies led to the (some claim too early) launch of clinical trials some 10 years ago, despite the fact that these studies have yielded surprisingly few good data as to parameters and mechanisms underlying these outcomes. Such important parameters, including the optimum number of cells for implant, the survivability of these cells, and their capacity for differentiation and electrical integration, are potentially vital "pieces of the puzzle" which, as discussed next, may have significantly impaired the potential for success and appropriate analysis of the results of these ultimately disappointing endeavors.

3 Clinical Data

The era of clinical testing of stem cell therapies (using skeletal myoblasts) was marked in January of 2001, when *Lancet* published a letter by Menasche et al. reporting the first successful ("adjunct-to-CABG [coronary artery bypass graft]") injection of autologous skeletal myoblasts, which had been previously harvested from the same patient and expanded in vitro [13]. At 5 month follow-up, postimplantation echocardiography and PET data suggested improved regional viability and contractility in the implant zone in this first, 72-year-old patient with refractory heart failure and inferior wall myocardial infarct.

A second report by Menasche et al. of nine patients receiving adjunct-to-CABG myoblast administration demonstrated improved ejection fractions in these patients, and that 60% of injected (infarcted) segments displayed new-onset systolic thickening following cell grafting at 1 month follow-up, suggesting (systolic) functional benefit [25].

The Menasche group later provided less compelling (essentially negative) data in their publication of the results of their entire phase I examination of skeletal myoblast administration, structured as a randomized, placebo-controlled, adjunct-to-CABG trial ("MAGIC" trial) [26]. This study included three treatment arms encompassing a total of 97 patients: low dose (400-million-cell injection), high dose (800-million-cell injection), and controls. Although those patients given an 800-million-cell injection displayed decreased LV volumes compared with controls, suggesting the possibility of reverse ventricular remodeling, ejection fraction and regional function assessed by echocardiography *did not* improve in treated versus control groups, highlighting the importance of conducting appropriate placebo-controlled trials to analyze newly proposed treatment modalities.

In analyzing this work, one must first consider that as an adjunct-to-CABG study, it is impossible to sort out the effects of new perfusion conferred by CABG from the potential benefits of cell administration, a constraint shared by all such adjunctive treatment trials. Further, although the authors first suggested that the mechanism of benefit conferred by their intervention was the active contraction of implanted cells, as opposed to passive motion derived from the enhanced contractility of adjacent (viable) myocardium, evidence supporting such a conclusion was ultimately not established. Moreover, fueling criticisms that the electrical integration of implanted cells with host myocardium is critical for clinical safety because of the risk of aberrant reentrant circuits, this study also reported sustained, monomorphic ventricular tachycardia in a number of the patients, although it is unclear whether this complication was induced by the cell transplant or the diseased myocardial substrate in this severely compromised cohort of patients.

Consistent with the limitations ultimately reported for the MAGIC trial, an autopsy analysis of one patient from this trial disappointingly demonstrated only six relatively small subepicardial islets of transplanted cells, suggesting that insufficient transplanted myocardial mass had been generated to be clinically relevant, and again raising questions regarding the putative mechanisms underlying the efficacy of this strategy [27]. On the other hand, myotubes embedded within areas of fibrosis that were aligned in parallel to the adjacent cardiomyocytes were noted in these harvested specimens. Furthermore, troponin T and CD56 were demonstrated in these cells, and 35% of the cells were noted to have been expressing the fast myosin heavy chain isoform, 32% the slow isoform, with the rest of the cells expressing both isoforms.

The expression of these markers rarely seen in skeletal muscle could be ascribed to cardiac-milieu-driven differentiation of the transplanted myoblasts. On the other hand, the transplanted cells did not demonstrate expression of connexin 43, desmosomes, or pan-cadherin, important cardiac-specific antigens relevant to cell–cell interactions. Thus, these findings suggest the possibility of myoblast differentiation toward a cardiomyocyte phenotype, but highlight the challenges of creating a plausible mechanism for transplanting a sufficient mass of "cardiac-like" cells to provide a clinically relevant functional effect. Further, even if these cells to were to undergo appropriate differentiation, their isolation amid a "sea of scar" makes it difficult to explain the efficacy of such a limited yield of implanted cells in a relatively large area of essentially inert myocardial scar.

Multiple additional studies with progressively more instructive trial design both added to and detracted from a clear picture of a clinical effect. A second such clinical trial was reported in 2004 by Chachques et al. [28]. In this trial, 20 patients underwent autologous (vastus lateralis) myoblast harvest and successful expansion of the cells in vitro prior to implantation of these cells as part also of an "adjunct-to-CABG" procedure. Each patient received approximately six injections into and around infarcted areas of myocardium, and was followed for 14 ± 5 months without mortality. LV ejection fraction improved from $28 \pm 3\%$ to $52 \pm 5\%$ ($p = 0.03$) and regional wall motion score index improved from 3.1 to 1.4 ($p = 0.04$) in the cell-treated segments. Myocardial viability tests showed areas of regeneration. Patients moved from mean New York Heart Association (NYHA) class 2.5 to class 1.2. Importantly, no cardiac arrhythmias or other cardiac adverse events were seen in this cohort, similar to other subsequent observations.

The shortcoming of this early trial, however, was that it did not include a control group, and that it again could not separate the effects of cell implantation from the effects of CABG. Likewise, Dib et al. presented the results of an adjunct-to-CABG trial involving 30 patients that demonstrated improved ejection fraction, new areas of glucose uptake within infarct scar by PET, no increase in the expected incidence of arrhythmias, and myoblast engraftment in harvested left ventricular assist device (LVAD) specimens in a six-patient subgroup analysis [29], but these results were again confounded by the potential effects of CABG and other procedures.

In 2003, however, Smits et al. [30] reported 6-month results of the first percutaneous, standalone delivery of autologous skeletal myoblasts – addressing the issue of interference from adjunctive CABG in these first trials. In the Smits et al. study, five patients with heart failure due to an anterior wall infarction received autologous skeletal myoblasts that had previously been cultured from quadriceps muscle biopsies. With use of a NOGA-guided catheter system (Biosense-Webster, Waterloo, Belgium), nearly 200 million cells were injected transendocardially into the infarcted area.

At 3-month follow-up, LV ejection fraction increased in these patients from $36 \pm 11\%$ to $41 \pm 9\%$ ($p = 0.009$), and further increased to $45 \pm 8\%$ at 6 months. Wall analysis using MRI also demonstrated a significant increase in wall thickening in the target areas, with lesser wall thickening noted in remote areas ($p = 0.008$).

Despite the encouraging findings in this report and the elimination of the confounding effects of adjunctive treatments, this study remained limited by its lack of control patients – raising the question as to whether these outcomes were the result of the natural history of the disease, rather than the effect of the intervention, a shortcoming of a number of subsequent inadequately controlled early-phase trials (Table 2). For example, in 2005 Siminiak et al. [31] presented the results of the POZNAN trial of catheter-based deliver of myoblasts to infarcted myocardium in nine patients. At 6-month follow-up, NYHA class improved in all patients and ejection fraction increased by 3–8% in six patients, but no control patients were included in this study. The group of Serruys presented similar evidence of improved regional and global LV systolic function during dobutamine infusion at 1-year follow-up in a ten-patient (uncontrolled) transendocardial myoblast trial [32].

Finally, however, the first prospective, case-controlled, standalone trial of myoblast administration was reported by Ince et al. [33], involving six patients receiving transendocardial cell delivery versus six untreated, matched controls. Ejection fraction in this trial increased from $25 \pm 7\%$ to $32 \pm 10\%$ at 12 months in the treated group, whereas there was a decline from $25 \pm 5\%$ to $21 \pm 4\%$ in controls, consistent with improvements in walking distance and NYHA class in the treated group. The trial still, however, remained limited by the absence of a placebo-control group.

Taken together, these studies suggest the potential of mechanical benefit following myoblast implantation, although the adjunctive performance of CABG or the lack of an appropriate control in these studies limits their validity. Definitive proof of benefit in the clinical arena still awaits the performance of a standalone, placebo-controlled trial.

Nevertheless, these clinical data also suggest that mechanical benefit could be derived from myoblasts that also undergo some degree of differentiation toward a cardiomyocyte phenotype. These clinical data offer little to better elucidate the mechanisms underlying such outcomes, however, or to provide a clear direction in terms of the optimal route and the number of dells needed for delivery, the role of electrical connectivity to the host, and the degree of cell survival and differentiation in this process. These questions lead to the opportunity (and need) for important additional laboratory efforts.

4 Next Steps

In light of the previous considerations, we hypothesized that a lack of vascular support in the relatively ischemic environment of a myocardial infarction may be an important challenge to the survival and function of cell implants. We were accordingly able to demonstrate that pretreatment of myocardial scar with angiogenic

Table 2 Clinical trials using autologous skeletal myoblasts

Authors	Year	Trial design	Cells administered ($\times 10^6$)	Delivery	*N*	Follow-up	Findings
Ince et al. [33]	2004	Standalone matched controls	200	Endocardial/catheter	6	12 months	Improved LVEF (from 24 to 32%) Improved walking distance Improved NYHA class
Siminiak et al. [31]	2005	Standalone uncontrolled	100	Endocardial/catheter	10	6 months	Improved NYHA class Improved EF (in 6 out of 9 patients; 3–8%)
Dib et al. [29]	2005	Adjunct to CABG Uncontrolled	Escalating; 100–300	Direct	30	4 years	Improved EF (from 28 to 36%) New uptake by PET Histologic engraftment
Gavira [53]	2006	Matched control Adjunct to CABG	250	Direct	12	12 months	Improved EF (from 36 to 55%) Improved WMSI (from 3.0 to 1.4) Decreased LV volumes
Biagini et al. [32]	2006	Standalone uncontrolled	200	Endocardial/Catheter	10	12 months	Improved NYHA class Improved EF (from 40 to 46%) Improved (decreased) ESV Improved systolic velocity
Menasche et al. [26]	2008	Randomized, placebo-controlled, double-blind Adjunct CABG	400, 800	Direct	97	6 months	No improvement regional or global LV function Decreased (improved) LV volumes (high dose vs. control)

CABG coronary artery bypass graft, *NYHA* New York Heart Association, *WMSI* Wall Motion Score Index, *ESV* end-systolic volume growth factors (vascular endothelial growth factor delivered via an adenoviral vector) prior, but not coincident, to cell implantation significantly enhanced scar vascularization and perfusion, and improved the survival and functional efficacy of implanted cells, as demonstrated by histology, echocardiography, and animal exercise testing [34]. Such data suggest that options exist for improving the survivability of cell implants.

Others have likewise suggested a variety of techniques and strategies for improving implanted myoblast survival and function. For example, tissue-engineered cell sheets have been used by Miyagawa et al., among others, in an effort to improve survival, decrease inflammation but yet maintain graft–host interactions as a means of enhancing implanted myoblast mechanical integration and function in the host myocardium.

Menasche et al. have also reported work with latissimus dorsi derived skeletal muscle cells that undergo fiber-type switching when conditioned for dynamic cardiomyoplasty. When using repeated electrical stimulation, the group noted a conversion from fast fiber-easy fatigability to a slow fiber phenotype. This is a potentially important finding in that the slow fiber phenotype is associated with a better mechanical function profile for a cardiac workload.

In considering the need for the development of gap junctions to allow the mechanical integration of skeletal myoblasts into the host myocardium, Suzuki et al. [35] overexpressed connexin 43 in rat skeletal myoblasts by transfection and clonal selection. Western blotting demonstrated a 17-fold increase in connexin 43 expression in these cells. These investigators were then able to demonstrate enhancement in intercellular dye transfer in these myoblasts overexpressing connexins 43, suggesting that myoblasts could be transfected in vitro with connexins 43, or other transgenes, to improve their functionality in the myocardial environment. Related work by Abraham et al. [36] in a coculture assay noted that genetic modification of myoblasts to express connexin 43 decreased the arrhythmogenicity of these cells, suggesting another pathway for enhancing the function of the cells as a CCM implant candidate.

5 Conclusion and Future Direction

Whereas many investigators in the field have largely discounted the use of skeletal myoblasts as a candidate implant type for CCM, the available data reviewed herein suggest that myoblasts may ultimately prove to be a useful implant cell type. Should such prove to be the case, myoblasts would possess a number of advantages in terms of clinical utility as noted herein, although they may be theoretically disadvantaged by their putative lack of capacity to differentiate (and thereby integrate) into host myocardium. In light of these considerations, and given the lack of definitive (but potentially premature) clinical trials, it must be concluded that the use of skeletal myoblasts may be a viable clinical strategy for CCM, and additional laboratory investigations, as outlined herein, are indicated as both reasonable and appropriate.

References

1. Marelli D, Ma F, Chiu RC. Satellite cell implantation for neo-myocardial regeneration. Transplant Proc. 1992; 24: 2995–2999.
2. Yoon PD, Kao RL, Magovern GJ. Myocardial regeneration. Transplanting satellite cells into damaged myocardium. Tex Heart Inst J. 1995; 22: 119–125.
3. Murry CE, Wiseman RW, Schwartz SM, Hauschka SD. Skeletal myoblast transplantation for repair of myocardial necrosis. J Clin Invest. 1996; 98: 2512–2523.
4. Taylor, DA, Atkins BZ, Hungspreugs P, Jones TR, Reedy MC, Hutcheson KA, Glower DD, Kraus, WE. Regenerating functional myocardium: improved performance after skeletal myoblast transplantation. Nat Med. 1998; 4: 929–933.
5. Atkins BZ, Lewis CW, Kraus WE, Hutcheson KA, Glower DD, Taylor DA. Intracardiac transplantation of skeletal myoblasts yields two populations of striated cells in situ. Ann Thorac Surg. 1999; 67: 124–129.
6. Hutcheson KA, Atkins BZ, Hueman MT, Hopkins MB, Glower DD, Taylor DA. Comparison of benefits on myocardial performance of cellular cardiomyoplasty with skeletal myoblasts and fibroblasts. Cell Transplant. 2000; 9: 359–368.
7. Jain M, DerSimonian H, Brenner DA, Ngoy S, Teller P, Edge ASB, Zawadzka A, Wetzel K, Sawyer DB, Colucci WS, Apstein CS, Liao R. Cell therapy attenuates deleterious ventricular remodeling and improves cardiac performance after myocardial infarction. Circulation. 2001; 103: 1920–1927.
8. Ghostine S, Carrion C, Souza LCG, Richard P, Bruneval P, Vilquin J-T, Pouzet B, Schwarts K, Menasche P, Hagege AA. Long-term efficacy of myoblast transplantation on regional structure and function after myocardial infarction. Circulation. 2002; 100 [Suppl I]: I-131–I-136.
9. McConnell PI, del Rio CL, Jacoby DB, Pavlicova M, Kwiatkowski P, Zawadzka A, Dinsmore JH, Astra A, Wisel S, Michler RE. Correlation of autologous skeletal myoblast survival with changes in left ventricular remodeling in dilated ischemic heart failure. J Thorac Cardiovasc Surg. 2005; 130: 1001–1009.
10. Patila T, Ikonen T, Patila T, Ikonen T, Kankuri E, et al. Improved diastolic function after myoblast transplantation in a model of ischemia-infarction. Scand Cardiovasc J. 2009; 43: 100–109.
11. Tambara K, Sakakibara Y, Sakaguchi G, Lu F, Premaratne GU, Lin X, Nishimura K, Komeda M. Transplanted skeletal myoblasts can fully replace the infracted myocardium when they survive in the host in large numbers. Circulation. 2003; 108 [Suppl II]: II-259–II-263.
12. Leobon B, Garcin I, Menasche P, Vilquin J-T, Audinat E, Charpak S. Myoblasts transplanted into rat infarcted myocardium are functionally isolated from their host. Proc Natl Acad Sci USA. 2003; 100: 7808–7811.
13. Menasche P, Hagege AA, Scorsin M, Pouzet B, Desnos M, Duboc D, Schwartz K, Vilquin JT, Marolleau JP. Myoblast transplantation for heart failure. Lancet. 2001; 357: 279–280.
14. DerSimonian H, Brenner DA, Ngoy S, Teller P, Edge ASB, Zawadzka A, Wetzel K, Sawyer DB, Colucci WS, Apstein CS, Liao R. Cell therapy attenuates deleterious ventricular remodeling and improves cardiac performance after myocardial infarction. Circulation. 2001; 103: 1920–1927.
15. Scorsin M, Hagege A, Vilquin JT, Fiszman M, Marotte F, Samuel JL, Rappaport L, Schwartz K, Menasche P. Comparison of the effects of fetal cardiomyocyte and skeletal myoblast transplantation on postinfarction left ventricular function. J Thorac Cardiovasc Surg. 2000; 119: 1169–1175.
16. Reinecke H, Zhang M, Bartosek, T, Murry CE. Survival, integration, and differentiation of cardiomyocyte grafts: a study in normal and injured rat hearts. Circulation. 1999; 100: 193–202.
17. Chedrawy EG, Wang JS, Nguyen DM, Shum-Tim D, Chiu RC. Incorporation and integration of implanted myogenic and stem cells into native myocardial fibers: anatomic basis for functional improvements. J Thorac Cardiovasc Surg. 2002; 124: 584–590.
18. Pagani FD, DerSimonian H, Zawadzka A, et al. Autologous skeletal myoblasts transplanted to ischemia-damaged myocardium in humans: histological analysis of cell survival and differentiation. J Am Coll Cardiol. 2003; 41: 879–888.

19. Minami E, Reinecke H, Murry CE. Skeletal muscle meets cardiac muscle. J Am Coll Cardiol. 2003; 41: 1084–1086.
20. Suzuki K, Murtuza B, Smolenski RT, Sammut IA, Suzuki N, Kaneda Y, Yacoub MH. Cell transplantation for the treatment of acute myocardial infarction using vascular endothelial growth factor-expressing skeletal myoblasts. Circulation. 2001; 104 [Suppl I]: I-207–I-212.
21. Reinecke H, Murry CE. Taking the toll after cardiomyocyte grafting: a reminder of the importance of quantitative biology. J Mol Cell Cardiol. 2002; 34: 251–253.
22. Gulbins H, Schrepfer S, Uhlig A, et al. Myoblasts survive intracardiac transfer and divide further after transplantation. Heart Surg Forum. 2002; 5: 340–344.
23. Zhong H, Zhu H, Wei H, et al. Influence of skeletal muscle satellite cells implanted into infarcted myocardium on remnant myocyte volumes. Chin Med J (Engl). 2003; 116: 1088–1091.
24. Robinson SW, Cho PW, Levitsky HI, et al. Arterial delivery of genetically labeled skeletal myoblasts to the murine heart: long-term survival and phenotypic modification of implanted myoblasts. Cell Transplant. 1996; 5: 77–91.
25. Menasche P, Hagege AA, Vilquin JT, Denos M, Abergel E, Pouzet B, Bel A, Sarateanu S, Scorsin M, Schwartz K, Bruneval P, Benbunan M, Marolleau JP, Duboc D. Autologous skeletal myoblast transplantation for severe postinfarction left ventricular dysfunction. J Am Coll Cardiol. 2003; 41: 1078–1083.
26. Menasche P, Alfieri O, Janssens S, et al. The myoblast autologous grafting in ischemic cardiomyopathy (MAGIC) trial: first randomized placebo-controlled study of myoblast transplantation. Circulation. 2008; 117: 1189–1200.
27. Hagege AA, Carrion C, Menasche P, Vilquin JT, Duboc D, Marolleau JP, Desnos M, Bruneval P. Viability and differentiation of autologous skeletal myoblast grafts in ischaemic cardiomyopathy. Lancet. 2003; 361: 491–492.
28. Chachques JC, Jesus Herreros J, Jorge Trainini J, Alberto Juffe A, Esther Rendal E, Felipe Prosper F, Genovese J. Autologous human serum for cell culture avoids the implantation of cardioverter-defibrillators in cellular cardiomyoplasty. Int J Cardiol. 2004; 95: S29–S33.
29. Dib N, Michler RE, Pagani FD, et al. Safety and feasibility of autologous myoblast transplantation in patients with ischemic cardiomyopathy: four year follow-up. Circulation. 2005; 112: 1748–1755.
30. Smits PC, van Geuns RJ, Poldermans D, Bountioukos M, Onderwater EE, Lee CH, Maat AP, Serruys PW. Catheter-based intramyocardial injection of autologous skeletal myoblasts as a primary treatment of ischemic heart failure: clinical experience with six-month follow-up. J Am Coll Cardiol. 2003; 42: 2063–2069.
31. Siminiak T, Fiszer D, Jerzykowska O, et al. Percutaneous trans-coronary-venous transplantation of autologous skeletal myoblasts in the treatment of post-infarction myocardial contractility impairment: the POZNAN trials. Eur Heart J. 2005; 26: 1188–1195.
32. Biagini E, Valgimigli M, Smits PC, et al. Stress and tissue Doppler echocardiographic evidence of effectiveness of myoblast transplantation in patients with ischaemic heart failure. Eur J Heart Fail. 2006; 8: 641–648.
33. Ince H, Petzsch M, Rehders TC, Chatterjee T, Nienaber CA. Transcatheter transplantation of autologous skeletal myoblasts in postinfarction patients with severe left ventricular dysfunction. J Endovasc Ther. 2004; 11: 695–704.
34. Retuerto MA, Beckmann JT, Carbray J, et al. Angiogenic pretreatment to enhance myocardial function after cellular cardiomyoplasty with skeletal myoblasts. J Thorac Cardiovasc Surg. 2007; 133: 478–484.
35. Suzuki K, Brand NJ, Allen S, et al. Overexpression of connexin 43 in skeletal myoblasts: relevance to cell transplantation to the heart. J Thorac Cardiovasc Surg. 2001; 122: 759–766.
36. Abraham MR, Henrikson CA, Tung L, Chang MG, Aon M, Xue T, Li RA, Rourke BO, Marban E. Antiarrhythmic engineering of skeletal myoblasts for cardiac transplantation. Circ Res. 2005; 97: 159–167.
37. Miyagawa S, Sawa Y, Sakakida S, et al. Tissue cardiomyoplasty using bioengineered contractile cardiomyocyte sheets to repair damaged myocardium: their integration with recipient myocardium. Transplantation. 2005; 80: 1586–1595.

38. Al Attar N, Carrion C, Ghostine S, Garcin I, Vilquin JT, Hagege AA, Menasche P. Long-term (1 year) functional and histological results of autologous skeletal muscle cells transplantation in rat. Cardiovasc Res. 2003; 58: 142–148.
39. Maurel A, Azarnoush K, Sabbah L et. al. Can cold or heat shock improve skeletal myoblasts engraftment in infarcted myocardium? Transplantation. 2005; 80: 660–665.
40. Toh R, Kawashima S, Kawai M, Sakoda T, Ueyama T, Satomi-Kobayashi S, Hirayama S, Yokoyama M. Transplantation of cardiotrophin-1-expressing myoblasts to the left ventricular wall alleviates the transition from compensatory hypertrophy to congestive heart failure in dahl salt-sensitive hypertensive rats. J Am Coll Cardiol. 2004; 43: 2337–2347.
41. Winitsky SO, Gopal TV, Hassanzadeh S, Takahashi H, Gryder D, Rogawski MA, Takeda K, Yu ZX, Xu YH, Epstein ND. Adult murine skeletal muscle contains cells that can differentiate into beating cardiomyocytes in vitro. PLoS Biol. 2005 Apr; 3: e87.
42. He KL, Yi GH, Sherman W, Zhou H, Zhang GP, Gu A, Kao R, Haimes HB, Harvey J, Roos E, White D, Taylor DA, Wang J, Burkhoff D. Autologous skeletal myoblast transplantation improved hemodynamics and left ventricular function in chronic heart failure dogs. J Heart Lung Transplant. 2005; 24: 1940–1949.
43. Ye L, Haider HKh, Jiang S, Ling LH, Ge R, Law PK, Sim EK. Reversal of myocardial injury using genetically modulated human skeletal myoblasts in a rodent cryoinjured heart model. Eur J Heart Fail. 2005; 7: 945–952.
44. Brasselet C, Morichetti MC, Messas E, Carrion C, Bissery A, Bruneval P, Vilquin JT, Lafont A, Hagège AA, Menasché P, Desnos M. Skeletal myoblast transplantation through a catheter-based coronary sinus approach: an effective means of improving function of infarcted myocardium. Eur Heart J. 2005; 26: 1551–1556.
45. Ott HC, Kroess R, Bonaros N, Marksteiner R, Margreiter E, Schachner T, Laufer G, Hering S. Intramyocardial microdepot injection increases the efficacy of skeletal myoblast transplantation. Eur J Cardiothorac Surg. 2005; 27: 1017–1021.
46. Gavira JJ, Perez-Ilzarbe M, Abizanda G, García-Rodríguez A, Orbe J, Páramo JA, Belzunce M, Rábago G, Barba J, Herreros J, Panizo A, de Jalón JA, Martínez-Caro D, Prósper F. A comparison between percutaneous and surgical transplantation of autologous skeletal myoblasts in a swine model of chronic myocardial infarction. Cardiovasc Res. 2006; 71: 744–753.
47. Payne TR, Oshima H, Okada M, Momoi N, Tobita K, Keller BB, Peng H, Huard J. A relationship between vascular endothelial growth factor, angiogenesis, and cardiac repair after muscle stem cell transplantation into ischemic hearts. J Am Coll Cardiol. 2007; 50: 1677–1684.
48. Farahmand P, Lai TY, Weisel RD, Fazel S, Yau T, Menasche P, Li RK. Skeletal myoblasts preserve remote matrix architecture and global function when implanted early or late after coronary ligation into infarcted or remote myocardium. Circulation. 2008; 118(14 Suppl): S130–S137.
49. Pedrotty DM, Klinger RY, Badie N, Hinds S, Kardashian A, Bursac N. Structural coupling of cardiomyocytes and noncardiomyocytes: quantitative comparisons using a novel micropatterned cell pair assay. Am J Physiol Heart Circ Physiol. 2008; 295: H390–H400.
50. Zhu H, Song X, Jin LJ, Jin P, Guan R, Liu X, Li XQ. Comparison of intra-coronary cell transplantation after myocardial infarction: Autologous skeletal myoblasts versus bone marrow mesenchymal stem cells. J Int Med Res. 2009; 37: 298–307.
51. Khan M, Kutala VK, Wisel S, Chacko SM, Kuppusamy ML, Kwiatkowski P, Kuppusamy P. Measurement of oxygenation at the site of stem cell therapy in a murine model of myocardial infarction. Adv Exp Med Biol. 2008; 614: 45–52.
52. Hoashi T, Matsumiya G, Miyagawa S, Ichikawa H, Ueno T, Ono M, Saito A, Shimizu T, Okano T, Kawaguchi N, Matsuura N, Sawa Y. Skeletal myoblast sheet transplantation improves the diastolic function of a pressure-overloaded right heart. J Thorac Cardiovasc Surg. 2009; 138: 460–467.
53. Gavira JJ, Herreros J, Perez A, Garcia-Velloso MJ, Barba J, Martin-Herrero F, Cañizo C, Martin-Arnau A, Martí-Climent JM, Hernández M, López-Holgado N, González-Santos JM, Martín-Luengo C, Alegria E, Prósper F. Autologous skeletal myoblast transplantation in patients with nonacute myocardial infarction: 1-year follow-up. J Thorac Cardiovasc Surg. 2006; 131: 799–804.

Methods for Inducing Pluripotency

Raymond L. Page, Christopher Malcuit, and Tanja Dominko

Abstract Induced pluripotent stem (iPS) cells are embryonic stem-like cells produced by forcing expression of a minimal number of key factors in differentiated somatic cells. In many ways, they are indistinguishable from embryonic stem cells in that they can differentiate into any cell type in the body. This development has led to worldwide excitement over the possibility to develop cell-based therapies for a variety of degenerative diseases using cells derived from and thus genetically matched to the patient they are aimed to treat. This chapter reviews the scientific foundation that has led to the ability to create iPS cells and the current methods used to make them, as well as the studies that have been conducted to help decipher the molecular pathways involved. The evolution of steps developed in recent years to improve both the efficiency and the safety of the process for clinical and in vitro diagnostic use is also presented.

1 Introduction

The pluripotent state in cell biology is defined by the ability of the cell to differentiate into any type of cell in the body. This state contrasts with the earlier state of totipotency, in which the cell can develop into the whole organism, including the extraembryonic, placental tissues, and the later state of multipotency, where the cells become lineage-restricted. In preimplantation embryonic development, pluripotent cells reside in the inner cell mass (ICM) of the blastocysts in mammals, which corresponds to about the fourth day of development in mice and the fifth to sixth days in humans. This represents the stage of development in which if the cells of the ICM are harvested and cultured using special conditions, embryonic stem cells (ESCs), or pluripotent cells, can be derived. Theoretically, these cells can be

R.L. Page (✉)
Department of Biomedical Engineering, Department of Biology and Biotechnology, Bioengineering Institute, Worcester Polytechnic Institute, Worcester, MA 01609, USA
and
Cellthera, Inc., Southbridge, MA 01550, USA
e-mail: rpage@wpi.edu

I.S. Cohen and G.R. Gaudette (eds.), *Regenerating the Heart*, Stem Cell Biology and Regenerative Medicine, DOI 10.1007/978-1-61779-021-8_13,
© Springer Science+Business Media, LLC 2011

maintained in this state indefinitely as long as the culture conditions are not altered substantially. For potential therapeutic applications, human pluripotent ESCs have remained at the center of experimental attention since their original derivation owing to the theoretical ability to direct the differentiation of these cells into any cell type in the body. Thus, it is possible to think about using pluripotent ESCs to derive cells or tissues to replace those lost to degenerative disease or trauma. This notion has represented a great hope for the potential cure of a host of intractable diseases or accidents which result in irreparable damage to functional tissue.

The downside to widespread usage of pluripotent ESCs for cell therapy is that they must be derived from a viable developing human embryo. Hence, therapeutic cells made in this way would in most cases present a genetic mismatch between the cells used for tissue replacement and the person into which they are transplanted; requiring lifelong immune-suppression therapy. To sidestep this issue, the notion of therapeutic cloning was introduced, where the embryo used to derive the pluripotent cells is made by somatic cell nuclear transfer (SCNT; or cloning) using a donor cell nucleus from the patient. The donor oocyte for this procedure is enucleated (having its genomic DNA removed), thus producing patient-specific ESCs. This procedure has been demonstrated successfully in a variety of laboratory and domestic animals. However, attempts in nonhuman and human primates have not been successful. In 2006, a landmark work in the field changed what was known about cell potentiality and our ability to control it with the invention of induced pluripotency using ordinary somatic cells derived from skin tissue. In an extremely elegant set of observations leading to a series of rigorous experiments, Takahashi and Yamanaka demonstrated that as few as four genes could be used to reset the epigenetic program of an ordinary skin fibroblast to that of a pluripotent cell with differentiation potential similar to that of ESCs [1]. This work has led to a flurry of effort from scientists all over the world to understand and improve this process to not only make it suitable for therapeutic applications, but also to enrich our understanding of disease onset and progression, and to develop better targeted therapeutic agents. It is now possible to take a small sample of cells from a skin biopsy and through the process of introducing a core set of factors into them produce a self-renewing and virtually unlimited supply of cells capable of differentiating into any cell of the body. Because cells made in this manner have essentially all of the features of ESCs, they are referred to as induced pluripotent stem cells (iPS cells).

Although this discovery represents a game changer for the potential of customized medicine, it is not without its potential drawbacks when contemplating clinical application. The original and still most efficient method for producing iPS cells requires the use of a retrovirus to insert genes encoding the reprogramming factors into the target cells. Thus, the last 3 years has seen a rush of activity in the development of a method that either removes the reprogramming transgenes once they have done their job, avoiding the process of transgene integration, or identifies small-molecule drugs that mimic the activity of the reprogramming factors. Although the science of these methods is enlightening and exciting, clinical applications using iPS cells bear the same concerns as those using ESCs in the risk of potential transfer of undifferentiated cells that might lead to tumor formation or spontaneous differentiation of undesired cell types. In fact, the methods used to create iPS cells are arguably more invasive than those producing ESCs; thus there are concerns over the potential difficulty to detect mutagenicity owing to nonnatural use of exogenous agents to achieve the epigenetic reprogramming response. This chapter will summarize the current state of the art of methods used to produce these iPS cells as well as some other technologies that may find utility in making this process more efficient or clinically applicable.

2 Reprogramming at Fertilization and Molecular Analysis in Preimplantation Embryos

Epigenetic changes that lead to a cell's identity being changed to that of a less differentiated cell are generally referred to as cellular reprogramming. In normal mammalian development, reprogramming of the sperm and egg nuclei occurs on a massive scale, and virtually all of the factors responsible reside in the oocyte. Reprogramming events persist through the early cleavage stages and there are differences between species as to the timing of these events. Reprogramming the entire genome to the totipotent state takes approximately 30 h in mice (two-cell stage [2]), 72 h in primates (four- to eight-cell stage [3]), and about 96 h in cattle (eight- to 16-cell stage [4]), although some genes associated with pluripotency, such as Nanog, are actively demethylated within 30 min following fertilization [5]. At these early stages, if the blastomeres are separated and transferred individually, they can lead to development of normal offspring, thus demonstrating totipotency at this stage [6–8].

This rapid process, known as the maternal to zygotic transition (MZT), is largely independent of RNA transcription [9], but DNA replication is required [10–12]. At the molecular level, these mitotically heritable changes in chromatin structure are mediated by DNA methylation, chromatin assembly proteins, histone proteins, and their methylation or acetylation. These proteins are produced largely from maternal transcripts and include histone acetyltransferases (CBP, P300), histone deacetylases (HDAC2, HDAC6), the SWI/SNF-related transcriptional regulators (SMARCA2, ARID1A, BR140), regulators of chromatin accessibility High Mobility Group proteins (HMGs, ACF1, SMARCA5), and modifiers of transcription (YY1, YAF2, RY-1 CREB, YAP65) [10, 13–15]. The decrease in maternal RNA during early cleavage (reprogramming) stages is followed by "activation" of the zygotic genome [13, 14]. Interestingly, these MZT events including zygotic genome activation also coincide with loss of totipotency and "differentiation" of blastomeres into cells destined to constitute either the trophectoderm (fetal placental tissues) or the ICM (fetus). The transcription factors Oct4, Sox2, and Nanog are among the genes activated following the MZT [16, 17]. These transcription factors are essential for post-MZT development [18] and dysregulation of their expression leads to abnormal differentiation of cells to pluripotent cells of the ICM or trophectoderm [19]. As development proceeds beyond the blastocyst stage, cells of the ICM become more specialized and form the three primordial germ layers: ectoderm, endoderm, and mesoderm. Coincidently, the most efficient stage of development from which to derive ESCs is the blastocyst stage. This stage represents a relatively small window in time where cells transition from a totipotent stage to cells of specialized tissues. Nearly all of the human ESC lines have been produced from the isolated ICM of surplus in vitro fertilized embryos from fertility clinics, first described by Thomson et al. [20]. The same morphological appearance, molecular expression profile, and in vitro and in vivo differentiation characteristics used in the original studies remain the gold standard by which all putatively pluripotent stem cells are evaluated. Pluripotent ESCs are able to be propagated in this state in vitro, in effect, by artificially creating an environment that maintains the expression profile of genes transcribed in the ICM cells at the blastocyst stage. Consequently, many studies have focused on evaluating the gene expression patterns at these developmental stages, aiming to identify master regulators (or markers) of pluripotency. In particular, appropriate levels of Oct4 [21] and Nanog [17] expression seem to be at the core of regulating formation of the ICM.

Interrogation of complementary DNA libraries created from different preimplantation stage oocytes and embryos showed that OCT4 and LIN28 are expressed at all stages, whereas NANOG, SOX2, KLF4, and c-myc are activated after the MZT [19, 20]. For OCT4, both messenger RNA and protein have been detected in mouse and human oocytes and preimplantation embryos [23–26]. In humans, analysis of OCT4 is complicated by splice isoforms that result in messenger RNA's coding for nuclear transcription factor OCT4A and an apparently cytoplasmic variant OCT4B [27], whose function is unknown. Analysis using antibodies that distinguish OCT4A and OCT4B isoforms [25] showed that OCT4B is present in the oocyte cytoplasm and the preimplantation development stages, whereas OCT4A was detected in nuclei of post MZT morula and blastocyst stage embryos [25]. The expression pattern of these genes also differs according to species. In mice, Oct4 is quite strikingly restricted to the ICM and upon further development is restricted to the germ cells [28, 29], but in humans and other large mammals, Oct4 is expressed in both the ICM and the trophectoderm compartments of blastocyst stage embryos [25, 29]. So for animals other than mice, OCT4 alone is not a sufficient marker to identify pluripotent cells. However, a promoter occupancy analysis of pluripotent human ESCs showed that OCT, SOX2, and NANOG individually or in combination bind to promoters of a large number of downstream target genes and collectively establish a master regulatory circuit that maintains the pluripotent state [30].

3 Lessons About Nuclear Reprogramming from Somatic Cell Nuclear Transfer

Until now we have focused on defining the characteristics of either pluripotent cells as they occur naturally in development (the ICM) or as they are held in this state by defined culture conditions, but still the cells reached such a point naturally by fertilization. In 1958, John Gurdon demonstrated that in *Xenopus laevis*, up to a point in development (tadpole stage), a somatic cell transplanted into the cytoplasm of an egg could lead to development of the resulting organism to the adult stage [31]. This experiment not only demonstrated that the oocyte cytoplasm could reprogram nuclei from sperm and eggs, but also demonstrated that the oocyte cytoplasm could reprogram nuclei from later developmental stages. It was not until the 1980s that the micromanipulation and microscopy tools required to perform similar experiments in mammals would be available. The first successes in mammalian nuclear transfer that produced viable offspring were only done with donor nuclei from preimplantation stage blastomeres (reviewed in [32]), and it was widely thought that the mammalian genome was so differentiated that the epigenetic changes were irreversible. Then, in the 1990s Keith Campbell and Ian Wilmut embarked on a successive set of experiments where they systematically studied the loss of totipotency as cells derived from sheep embryos were cultured (differentiated) further in vitro and transplanted into enucleated oocytes. The need to synchronize the cell cycle phase of the oocyte with that of the donor nucleus led first to serum depriving the cell cultures from accumulating the cells in G1/G0 phase [33]. For both differentiated late-stage cultures from embryos and cells derived from an adult sheep, viable offspring were produced [34]. These hallmark studies showed that the genomic potential of mammalian cells was not fixed and the oocyte contained all of the factors required to reverse cell fate. The ability to reprogram differentiated adult somatic cells to produce viable mammalian offspring was rapidly recognized as a potential alternative to isolating ESCs from fertilized embryos, whereby cells could be derived from the individual in which the ESCs were desired. This opened the door for virtually any autologous cell therapy using ESCs as the source producing the therapeutic cell population and the term "therapeutic cloning" was introduced [35]. This approach was surrounded by ethical controversy since a human embryo would, in effect, be produced by somatic cell cloning technology and then destroyed to harvest the stem cells [36]. Nonetheless, several groups attempted the procedures using human cells and donor oocytes, but never succeeded. Initial work done in nonhuman primates using similar procedures to those developed for cattle showed a variety of chromosomal segregation problems in the first and subsequent early cell divisions, which seemed to suggest that cloning of primates was nearly impossible [37, 38]. However, recently, pluripotent stem cells have been derived from nonhuman primate SCNT embryos [39, 40], but ESCs from SCNT embryos in humans remain elusive. Therapeutic cloning models developed in mice indeed showed that the therapeutically useful cells could be produced from ESCs derived from SCNT embryos [41–45], which made the case that somatic cells reprogrammed entirely in vitro, albeit using an egg, could provide a functional autologous cell source for therapy.

The low developmental efficiency of SCNT-derived embryos has been largely attributed to incomplete reprogramming of the somatic cell genome in the early cleavage stages leading to aberrant gene expression in the early embryo. This can lead to abnormal differentiation events following MZT where dysregulated partitioning between blastomeres to the ICM and trophectoderm occurs.

More specifically, DNA methylation and histone modifications appear to be critical for stage-appropriate transcription of the embryonic genome [46, 47]. One of the genes used to model these epigenetic events is Oct4, where its promoter is hypermethylated in somatic cells, unmethylated in the early embryo, and aberrantly methylated in SCNT embryos [46–49]. Inappropriate expression (either too high or too low) of Oct4 leads to differentiation of all cells of the developing blastocyst to the extraembryonic trophectoderm lineage; thus, pluripotent cells of the ICM are not formed [18]. In addition to transcription factor promoter demethylation, histone acetylation plays a role in somatic cell reprogramming in SCNT. Inhibition of histone deacetylase activity using chemical inhibitors such as trichostatin A has been shown to increase reprogramming efficiency in SCNT embryos [50, 51]. The resultant greater degree of acetylated histone proteins produced by trichostatin A treatment presumably enables increased access of transcription factors to their target genes by modifying chromatin structure, which can also lead to demethylation of transcription factor promoters as has been shown for the embryonic transcription factor Sox2 [51]. Studies comparing the transcriptional profile expression of ESCs derived from fertilized versus SCNT embryos suggest that, when successful, complete genomic reprogramming can be achieved using SCNT [40, 52, 53]. Owing to supply issues surrounding obtaining human oocytes with which to reprogram human somatic cells, animal oocytes have been employed as a potential source of "reprogramming cytoplasm." Many of the mechanisms by which donor cell chromatin is remodeled to a transcriptionally permissive state are conserved among species. *Xenopus. laevis* frog and bovine oocytes are readily available sources owing to the large size and number of frog eggs and the availability of in vitro matured bovine oocytes from ovaries obtained from slaughterhouses. Indeed, frog egg extracts have been shown to activate OCT4 expression in mammalian nuclei, even though this species lacks an OCT4 ortholog [54], and bovine oocytes have been shown to support the development of interspecies SCNT embryos beyond the critical MZT developmental stage [55]. Activation of embryonic genes and at least partial reprogramming has been demonstrated in embryos produced by SCNT between primate (including human) human donor cells and bovine oocytes [56, 57]. Because of these results, interspecies SCNT has recently been approved as a method to be employed in an attempt to derive therapeutically useful human pluripotent cells in the UK [58]. Undoubtedly, continued research aimed at understanding the problems with SCNT in humans will likely lead to techniques being developed to circumvent the developmental problems of the embryos, and ultimately leading to the derivation of pluripotent ESCs. Notwithstanding the technical feasibility of reprogramming by SCNT in humans or human animal hybrid embryos, the process will likely remain incredibly inefficient, leaving standing controversies over where very large number of human oocytes will come from to treat all of the patients that could benefit from autologous cell therapy, or whether we should be creating hybrid embryos between humans and other species.

4 Cytoplasmic Extracts from Pluripotent Cells and Reprogramming Somatic Cells

For decades we have known that cell fusion between different cell types (cybrids) can result in the cytoplasm of one cell type influencing the genomic plasticity of the other. Cell-extract-mediated transdifferentiation has been employed to reprogram fibroblasts to T-cell function [59] and cardiomyocytes have been used to induce formation of cardiac muscle from endothelial cells [60]. Cell extracts have also been employed to facilitate differentiation of ESCs [61]. For dedifferentiation, cytoplasmic extracts derived from *Xenopus* oocytes [62–64], embryonal carcinoma cells [65–67], embryonic germ cells [68], and ESCs [66, 67–71] have been studied extensively as potential reprogramming vehicles for differentiated somatic cells.

Nuclear structural remodeling activity is mediated by nucleosomal ATPase imitation switch (ISWI) present in frog egg extracts [62]. Additional chromatin remodeling occurs by nucleoplasmin-mediated chromatin decondensation through centromeric DNA decondensation, removal of histone H1 protein, loss of histone H3K9 trimethylation, and activation of the oocyte-specific genes c-mos, Msy-2, and H1foo [72]. Upon injection of human lymphocytes into frog eggs, OCT4 activation and inactivation of the somatic cell-specific THY1 gene shows that reprogramming activity maintains function across species [54]. Successful reprogramming requires active DNA demethylation and occurs without replication, as shown by OCT4 activation [49]. However, there is still no clear evidence of a mechanism of such DNA demethylation, with the exception of one report. In 2008, Rai et al. [73] demonstrated for the first time that zebra fish embryos utilize a deaminase to convert 5-methylcytosine to thymine. The resulting mismatch (G:T) is then corrected by Mbd4, a mismatch-specific thymine glycosylase, and this reaction is stabilized/enhanced by the presence of Gadd45, a stress-response DNA repair enzyme. Whether a similar mechanism functions in mammals through orthologous proteins remains to be investigated.

It is also possible that reactivation of embryonic-specific genes during reprogramming may result from activity of methyl-binding domain (MBD) proteins and/or other ATP-dependent chromatin remodeling factors. In permeabilized human leukocytes exposed to frog egg extract, OCT4 and the germ- and ESC-specific alkaline phosphatases were activated and detected after 1 week of culture [63]. In this case, Brahma-related group 1 (BRG1) activity, which has been implicated in chromatin remodeling activity through the MZT in primate development [13], was required. Differences in remodeling activity between cytoplasmic extract from frog oocytes and eggs suggest that some of the reprogramming processes may depend on activities acquired during oocyte maturation [64]. Extracts from both oocytes and eggs were able to remove somatic nuclear lamin A/C, whereas polymerase I and polymerase II transcriptional activity was only detected in oocyte extracts, indicating that egg extracts silence transcriptional activity.

Cytoplasmic extracts from undifferentiated mammalian cell lines have also been shown to alter the gene expression profile of somatic cells. To at least a certain degree, dedifferentiation of 293T cells and NIH3T3 fibroblasts following permeabilization and incubation with extracts derived from embryonic carcinoma (EC) cells and ESCs has been shown [66]. Incubation of somatic cells with these extracts resulted in formation of cell colonies with morphological characteristics similar to those of ESCs with high nuclear to cytoplasm ratio and cell shape similar to that of ESCs. Additionally, several genes related to pluripotency were upregulated and somatic-cell-associated genes were downregulated, whereby this phenotype was maintained for up to 4 weeks of culture following extract treatment. The extract-treated cells could be induced to differentiate into mesoderm and ectoderm lineages, suggesting at least a transdifferentiation potential in EC cell and ESC cytoplasmic extracts.

In mouse ESC–thymocyte hybrids, global hyperacetylation and hypermethylation of histones H3 and H4 occurs, which is consistent with the ESC and not the thymocyte phenotype [70]. Furthermore, extract treatment resulted in demethylation at the promoter of the ESC-related gene Oct4 and its concomitant activation. Transcriptionally active Oct4 was further characterized by hyperacetylation of histone H3 and hypomethylation of histone H3-K9 [70]. Interestingly, histone H3 K4me2 and histone H3 K4me3 are associated with Isw1p ATPase activity in regulation of transcriptionally active chromatin [74], and histone H3 K4 methylation has been implicated as a potential epigenetic target for recruitment of Brm-associated SWI/SNF chromatin remodeling factors [75]. Additional analyses of the chromatin structure at the regulatory regions of OCT4 revealed that posttranslational modifications of histone H3 play a role in regulating active transcription of OCT4 [65]. Transient exposure of human 293T epithelial cells to EC cell extract resulted in acetylation of histone H3 at K9 and demethylation at K9me2, K9me3, and K27me3. The absence of changes in methylation at histone H3 K4me2 and K4me3 suggests a potential threshold in somatic cells between a transcriptionally permissive and a transcriptionally active state for OCT4. In mouse somatic cells, Oct4 has been described as transcriptionally silent [70], whereas low levels of expression have been reported in some human somatic cells [67, 76].

In summary, the cytoplasmic extracts of less differentiated cells can be used to induce epigenetic modifications and transcriptional activation in differentiated somatic cells. This has been demonstrated through remodeling of chromatin structure [67, 70, 77], upregulation and downregulation of gene expression [54, 71], X-chromosome reactivation [78, 79], and acquisition of the pluripotent cell phenotype and functionality, which includes teratoma formation with contributions to all three germ layers [69, 80]. From a therapeutic standpoint, the use of transformed cells such as EC cells or nonhuman cells such as mouse ESCs posses risks that are desired to be avoided. Furthermore, in the case of permeabilized somatic cells incubated with cell extracts, incomplete reprogramming may result from interference with the cytoplasm from the somatic cells itself being mixed with factors from the "reprogramming" extract. This potential for "molecular confusion" makes it very difficult to completely isolate and control all of the activities desired for complete reprogramming to pluripotency.

5 Transcription Factor Control of Cell Fate: Induced Pluripotent Stem Cells

The ability to directly reprogram somatic cells to a pluripotent stem-cell-like state is an attractive alternative to reprogramming using either SCNT or undifferentiated cell extract incubation. It is well known that maintenance of the ESC phenotype depends on appropriate expression of the transcription factors OCT4, SOX2, and NANOG; and these factors are often referred to as representing the master regulators of "stemness" [81, 82]. Together, OCT4, SOX2, and NANOG cooperate to form what is known as the master regulatory circuitry for the maintenance of pluripotency [30]. Thus, it can be supposed that faithful reactivation and maintenance of this master regulatory circuit in somatic cells might lead to acquisition of pluripotency in somatic cells.

Guided by the ESC extract reprogramming work [66, 80], Takahashi and Yamanaka hypothesized that the genes which play the most prominent roles in maintenance of ESC identity may be able to induce pluripotency if forcibly expressed. On the basis of whole transcriptome analysis in ESCs, 24 candidate genes expressed specifically by ESCs were selected and each one was inserted into mouse embryonic fibroblasts which contained a double knock-in construct in the Fbx15 locus coding for a β-galactosidase/neomycin fusion gene (Fbx15$^{\beta geo/\beta geo}$) using a highly efficient retroviral transduction system [1]. Fbx15 is expressed only in embryos and ESCs, but is not necessary for maintaining pluripotency. Since only mouse cells or ESCs express Fbx15 [83], addition of G418 to the culture medium will kill any nonexpressing cells. This enabled positive selection of cells that had acquired an ESC-like gene expression pattern. Cells transduced with any of the single candidate genes failed to confer G418 resistance, whereas cells transduced with all 24 genes acquired an ESC-like phenotype, although at a low frequency (0.01%). Systematic removal of candidate genes, one at a time, and evaluating the formation of ESC-like colonies enabled the determination of each factor's role in inducing the phenotype. In some cases, the elimination of the candidate led to complete failure to produce ESC-like colonies, suggesting that this factor was indispensible. This selective reduction of redundant genes led to identification of "four-factor" programming and the term "induced pluripotent stem cells" (iPS cells) with factors c-myc, Oct4, Sox2, and Klf4. The resultant cells displayed all of the features of ESCs, including both embryoid-body-mediated differentiation in vitro and formation of teratomas in vivo, with resultant cells representing ectodermal, endodermal, and mesodermal lineages. Similar experiments using fibroblasts isolated from tail tips of adult FBX15$^{\beta geo/\beta geo}$ mice showed that adult cells could also be reprogrammed to pluripotency using the four factors. Gene expression profiling studies showed that stem cell genes could only be expressed in cell populations which included all four factors. Increased histone H3 acetylation and decreased methylation of histone H3 at K9me2 were associated with Oct4 activation, although its promoter remained partially methylated [1]. This partial demethylation of the Oct4 promoter was similar to that found with reprogramming using EC cell extract [67]. Interestingly, although the iPS cells were similar in phenotype and differentiation potential to ESCs, global gene expression analysis showed that there were distinctly different transcriptional profiles. However, not all of the iPS cell lines created could produce teratomas and/or chimeric animals, indicating that these cells were not truly pluripotent. However, the same laboratory next employed the same four-factor approach, but this time used donor cells in which the Nanog locus contained the selection marker. These iPS cells displayed a more ESC-like transcriptome and were indeed able to produce germline chimeric mice following blastocyst injection, which is the ultimate proof of pluripotency [84]. A similar four-factor approach was rapidly applied to the human system where c-MYC and KLF4 were replaced by NANOG and LIN28, and iPS cells from both embryonic and adult fibroblasts were produced [85, 86]. Although the dedifferentiation efficiency of this technique is not quite as rapid as that of SCNT in that it takes up to 30 days of continuous forced expression of the preprogramming factors for the somatic cells to acquire the ESC phenotype, these experiments marked a paradigm shift in stem cell biology toward "transcription-factor-mediated manipulation of cell fate."

With four-factor reprogramming, endogenous stem-cell-related genes are activated gradually at different times after transduction. Using mouse embryonic fibroblasts modified with green fluorescent protein (GFP)–internal ribosome entry site (IRES)–puromycin at the Nanog locus and transduction with Oct4, Sox2, c-Myc, and Klf4, Maherali et al. [87] showed that selection as early as 3 days after transduction failed to produce ESC-like colonies, whereas selection applied after 7 days yielded colonies. This study suggested that the endogenous Fbx15 and Nanog genes were activated at different times. It was also shown that selection delayed for up to 30 days after induction yielded only homogeneous colonies with ESC morphology and SSEA1 and CD9 expression, whereas early (7-day) selection gave rise to different types of colonies within the population. With the use of antibiotic-inducible transgene constructs where the expression of individual reprogramming factors could be manipulated, it was further shown that the process of endogenous stem cell gene activation following iPS cell induction procedures is gradual and occurs over several weeks [88]. In Oct4-GFP transgenic mouse fibroblasts transduced with doxycycline-inducible Oct4, Sox2, Klf4, and c-Myc transgenes yielded weakly GFP positive cells (indicating endogenous Oct4 activation) at 9 days after transduction. After this time, iPS cell colonies could be produced and maintained without doxycycline [89]. This timeframe was also consistent with the reactivation of stem-cell-associated genes Oct4, Sox2, and telomerase, along with activation of the silent X chromosome. Downregulation of the differentiated cell marker Thy1 was detectable as early as day 3, whereas upregulation of the stem cell marker SSEA1 occurred around day 5. Continued culture with doxycycline produced cell populations with gradually more SSEA1-positive cells and fewer Thy1-positive cells until day 12 and after day 12, when doxycycline was removed, colonies remained Thy1-negative and mostly SSEA1-positive. This timeframe seems to mark an iPS cell commitment point at which the reprogramming events are complete and the cells are capable of maintaining the pluripotent phenotype in the absence of exogenous gene expression. Retrovirus inactivation of embryonic cells

[90], which has been implicated as a mechanism to prevent insertional mutagenesis in the germline, is also observed in established iPS cell cultures [87, 91, 92] and interestingly occurs concomitantly with the activation of late-expressing stem cell genes in iPS cells [88]. At about the same time after the iPS cell induction protocol as the endogenous stem cell genes are activated, cell morphology changes are detected where small ESC-like colonies begin to form. However, it takes about 6 weeks of culture using conditions similar to those used for ESCs before the iPS cell colonies take on a morphological appearance that is indistinguishable from that of ESCs. Once the conversion to iPS cells is complete and the cultures are sustained as ESC-like colonies, expression of the reprogramming transgenes is often downregulated, suggesting that the endogenous ESC maintenance regulatory circuitry has indeed been reactivated. Taken together, these studies help to define a temporal pattern of morphological and molecular events necessary for induction of pluripotency in mouse cells using integrating viral vectors. Consequently, this knowledge makes it possible to identify cells heading toward the iPS cell phenotype without relying on expression of transgenic reporters and can lead to procedures for producing genetically unmodified human iPS cells for clinical applications [93–96].

In the early studies reported above, induction of iPS cells seemed to rely on integration of the retrovirus into the genome perhaps to achieve the sustained expression levels necessary for conversion to pluripotency. It goes without saying that the more transgene copies there are, the more chance there is for deleterious disruption of the genome. In attempting to reduce the number of genomic integrations, which range between one and four per transgene in iPS cells derived from mouse liver and stomach cells and between one and nine per transgene in mouse embryonic fibroblasts [97], combinations of delivery vectors were employed. In Nanog-IRES-puromycin primary hepatocytes, nonintegrating adenovirus could be used to deliver either Sox2 or Klf4, whereas integrating retrovirus was used to deliver Oct4 and c-Myc [91]. Indeed the number of integration sites was reduced since none of the adenoviral sequences were detected in the iPS cells by quantitative PCR or Southern blotting. In the same report, use of retrovirus was avoided altogether by using transfection of a single polycistronic vector with the 2A self-cleaving peptide [98] separating Oct4, Klf4, and Sox2, followed by transfection of a separate plasmid to deliver c-Myc. The resultant iPS cells were capable of producing adult chimeric mice without evidence of plasmid integration, although the efficiency was reduced about tenfold from 0.001 to 0.0001%.

Since cMyc is a well characterized oncogene, and its reactivation would surely provide concern for unwanted tumor formation, much effort has been devoted to developing methods to remove it from the iPS cell protocols. When c-Myc was omitted and only three factors (Oct4, Sox2, and Klf4) were used in mouse and human fibroblasts, ESC-like colonies took about 30 days to first appear; however, the resultant iPS cells morphologically and functionally more closely resembled those of ESCs [99]. Although omission of c-Myc reduced the efficiency, the result seemed to produce iPS cells of higher quality. With c-Myc^{-} mouse iPS cells, chimeric mice did not develop tumors and the human c-Myc^{-} iPS cells displayed all the hallmarks of human ESCs, including SSEA3, SSEA4, TRA-1-60, and TRA-1-81 expression and differentiation into cells of all three germ layers. It is noted that the human iPS cells produced using the four-factor approach described by Yu et al. [100] did not contain c-MYC, establishing very early that its presence was not required to generate iPS cells, at least when starting from ESC-derived mesenchymal cells or fetal-tissue-derived IMR90 fibroblasts. This work showed that substituting LIN28 and NANOG for c-MYC and KLF4 did not compromise iPS cell production efficiency, suggesting that neither KLF4 nor c-MYC is required for reprogramming human cells.

To contemplate applying this technology to the clinical setting, several groups have established methods for either removing the reprogramming transgenes after their role has been fulfilled or introducing them using systems that avoid genomic integration. Methods used to accomplish this objective include nonintegrating adenoviruses [89], polycistronic plasmids [91, 101–104], and episomal vectors [95]. In all cases, the iPS cell induction efficiency using these approaches was lower than that using genomic integrating systems. Perhaps the "cleanest" iPS strategies employed so far are those using piggyBAC retrotransposons [105, 106] and protein-only systems where the reprogramming factors are delivered as recombinant proteins [107, 108]. The piggyBAC system allows for complete removal of the transgenes upon transposase induction with no detectable genomic footprint left behind. However, although many of the iPS cell lines created using this technique fail to produce chimeric offspring, the cells display very convincing pluripotent characteristics in vitro. This leads to the speculation that perhaps there are undetectable DNA mutations that arise during the reprogramming process, potentially during a specific phase in this process characterized by an extremely high rate of cell proliferation. This may lead to inaccurate DNA replication, and thus mutations that cannot be detected by traditional karyotype analysis.

Fueled by the desire to omit transgenes altogether from reprogramming protocols, much effort has been devoted to screening small-molecule libraries for their ability to mimic the activity of some of the reprogramming factors. The most obvious chemical methods initially chosen were those that inhibit specific chromatin modifications that silence transcription by inhibiting histone methyltransferase [109], histone deacetylase [108, 110], and DNA methyltransferase [93]. Indeed human iPS cells using only OCT4 and SOX2 as reprogramming transgenes coupled with the histone deacetylase inhibitor valproic acid have been produced [111]. Inhibition of chromatin-modifying enzyme activity using chemical inhibitors has led to improved efficiency and/or reduced the number of reprogramming factors required, but has not yet resulted in complete elimination of transgenes. These studies suggest that genome-wide chromatin structure modifications alone are not sufficient to activate endogenous stem cell genes. However, perhaps the relaxed chromatin structure facilitates access of the exogenous transcription factors to their targets (reviewed in [112]). Ichida et al. [113] employed a combined approach consisting of three factors (Oct4, Klf4, and c-Myc) plus the histone deacetylase inhibitor valproic acid in the Oct4-GFP transgenic mouse embryonic fibroblast cell line to screen for chemical compounds that could replace Sox2. This strategy proved fruitful in that they discovered a compound that when added to the reprogramming protocol generated iPS cells. Further investigation of the mechanism revealed that the molecule inhibited transforming growth factor β (TGF-β) signaling and complemented the induction process by inducing endogenous Nanog expression. This work further emphasized the importance of Nanog as a central regulator of the pluripotent state [19]. However, reprogramming mouse [114] and human [115] neural stem cells to pluripotency with OCT4 alone has been demonstrated, suggesting a vital role for OCT4.

Other approaches to eliminating transgenes have focused on producing and delivering reprogramming factors as recombinant proteins. With use of the short basic peptide segment at amino acids 48–60 of the transactivator of transcription (TAT) protein from HIV [116, 117] and a His-tag to facilitate purification fused to OCT4 and SOX2, it was shown that recombinant forms of these transcription factors could bind DNA and penetrate the cell membrane [118]. In mouse ESCs, the transducible recombinant Oct4 and Sox2 proteins could functionally replace endogenous Oct4 and Sox2 knocked down by RNA interference [118]. Similar versions of this approach were recently employed to reprogram human fibroblasts to the iPS cell state using solely recombinant forms of OCT4, SOX2, KLF4, and c-MYC [107, 108]. The reprogramming cocktail containing the factors had to be repeatedly applied to the cells and the emergence of ESC-like colonies took about 2 months, with a severe amount of cell death in the process. Although the efficiency was low, these latest "vector-free" approaches for the derivation of iPS cells omit the requirement for transgenes, and thus genetic modifications such as insertional mutagenesis and unregulated transgene reactivation (reviewed in [119]).

With the various iPS cell induction procedures that have been developed, the efficiency remains low, and even with SCNT the efficiency of producing live offspring is lower that what would be expected if the starting cell population were uniform. This raises questions about the epigenetic uniformity of the starting population of cell cultures. Array-based transcriptome analyses are conducted on whole cell populations and the results are essentially an average for all of the cells in the population, making it impossible to determine if there are small populations of cells with a fibroblast culture, per se, that are inherently more stem-like than the bulk of the culture. In fact, subpopulations of human fibroblasts express stem cell markers such as SSEA3 [96]. Interestingly, a purified SSEA3-positive subpopulation of human fibroblasts led to an eightfold increase in efficiency of deriving iPS cells compared with the unsorted population. At the transcription level in adult human fibroblasts, basal levels of OCT4, SOX2, and NANOG have been reported which under certain culture conditions can lead to translation of these transcripts into protein [76]. In this study, several factors were reported to affect the expression of stem cell-related genes and proteins, but reduced oxygen tension in the culture environment and fibroblast growth factor 2 (FGF2) signaling were the predominant players. However, at the protein level, only subpopulations of the cells expressed nuclear localized transcription factors OCT4, SOX2, and NANOG. Although the cells cultured under these conditions survived much longer in vitro before reaching senescence, the morphological changes associated with iPS cells were never observed, and sustained nuclear detection of the stem-cell-associated transcription factors was not observed. These results suggest that there might be subpopulations of cells within whole primary cell cultures that are more amenable to genomic reprogramming than others and perhaps these subpopulations represent tissue resident stem cells, although this speculation remains to be proven. It is of note that "four-factor" or modified reduced factor reprogramming by forced expression alone is not sufficient to maintain pluripotency. In all of the cases studied, at some stage during the forced reprogramming process, the environment of the cells had to be changed to one that supports ESC self renewal. Continued culture using a medium that supports adherent cells such as Dulbecco's modified Eagle's medium with serum does not support complete iPS cell induction. However, transfer to an ESC medium too early in the process leads to cell death. Culture media designed to support ESC self-renewal do not support survival and proliferation of primary anchorage-dependent cells such as fibroblasts. From the perspective of maintaining self-renewal of already pluripotent cells, the specific factors required have been worked out in detail. It is worth mentioning that these regulatory networks do indeed differ between mouse and human systems. In human ESCs it has been shown that FGF2 and TGF-β/activin signaling synergize to inhibit bone morphogenetic protein (BMP) signaling and that both TGF-β- and BMP-responsive Smads bind to the NANOG proximal promoter to help maintain self-renewal [120]. In mouse ESCs, BMP4 enables self-renewal, but only in combination with leukocyte inhibitory factor leukemia inhibitor factor (LIF) [121], which has no effect on suppressing differentiation in human ESCs [122].

Mouse ESCs can be maintained in vitro without the use of serum or mitotically inactivated feeders by addition of LIF and BMP to the medium, whereas human feeder-free culture is dependent on high levels of FGF2 along with Matrigel®. Given these species differences, those interested in iPS cells for model development in species other than mice or humans may have to perform further developmental work to identify factors that can maintain the pluripotent state. Interestingly, procedures for ESC line derivation developed for mouse and primate systems have not yet been successfully applied to agriculture animal species such as rabbits, sheep, cattle, and pigs. There may be delicate differences between species in cellular responses to signaling pathways involved during different phases of the reprogramming process and in the maintenance of pluripotency that will need to be addressed.

Somatic cell types that have been successfully reprogrammed into iPS cells to date include keratinocytes [102], liver and stomach cells [97], lymphocytes [123], pancreatic β cells [123], adipose stem cells [125], CD34+ cells [124], and adult neural stem cells [127, 128]. Augmenting iPS cell induction protocols with selection of reprogramming permissive cells [96] and alteration of in vitro culture conditions [76, 129, 130] promises to increase the efficiency of iPS cell methods. The ultimate view to the future of iPS cell induction technology would be to have a detailed understanding of the temporal nature in which the signaling pathways can be manipulated in the various intermediate cell types created along the path to pluripotency. We can then conceive of a scenario where pluripotent cells can be induced by careful manipulation of the culture conditions alone, without the need for forced transgene expression or foreign chemicals.

6 Conclusions

Since the first reports of iPS cells from human and mouse somatic cells, many groups have further simplified, refined, and optimized the technique. iPS cells have been generated from a variety of different somatic cell types and from patients with degenerative disease, offering the potential to study disease progression in vitro and develop therapies either by using these cells as models for drug screening and development or by using the cells themselves as therapeutic agents. From the inherent differences in the regulation of biological function in stem cells between mouse and human systems, along with the ability to reliably create iPS cells from human somatic cells, one might suggest there is no longer a need for further iPS cell technology development in mice. However, with the recent developments showing discrepancies between iPS cells and ESCs, the chimeric animal assay for pluripotency will remain paramount to any available assays with human cells with which to evaluate the potentiality and viability in vivo of iPS cells produced by different methods. Additionally, examining the developmental potential of tissues and whole organs in vivo, starting with iPS cells has enormous potential for drug development and to identify and treat genetic diseases. Although the published iPS cell lines in general exhibit all of the properties of ESCs, there are emerging reports of some potential differences from ESCs upon redifferentiation [131]. However, as the methods become more defined and perhaps transgenes with the potential for spontaneous reactivation are eliminated completely, the field of regenerative medicine will be forever changed.

The application of iPS cell technology to clinically relevant problems has already been reported. Transplantation of iPS cells differentiated into neurons and glia has been shown to yield functional improvements in rat models of Parkinson's disease [132]. Motor neurons differentiated from iPS cells derived from patients with amyotrophic lateral sclerosis [133] and cardiac myocytes differentiated from iPS cells [134–136] are among the most advanced clinical targets. Development of patient-specific cells for therapeutic applications for traumatic tissue loss and degenerative diseases may someday be a clinical reality with the emergence of iPS cell technology. In less than 3 years from the first report of "four-factor reprogramming," transgene integration has been either reduced or eliminated altogether by substituting recombinant proteins. The efficiency of the procedures will likely improve in a short time. The remaining clinical concerns over the use of cells differentiated from a completely undifferentiated cell population, as with ESCs, will remain for iPS cells. The generation of a pluripotent cell population for therapeutic purposes carries with it the inherent pitfalls of target cell purity in the context of tumorigenicity. Ensuring a cell therapy product free of contaminating stem cells is key to resolving this issue. In consideration of the exorbitant manufacturing costs required to purify target cells under cGMPs, we predict that the concept of transcription factor reprogramming will be employed on a less grandiose scale than for pluripotent cells, and emphasis will be placed on generating therapeutically useful intermediate cell populations with lineage- or tissue-type-restricted differentiation potential.

References

1. Takahashi K, Yamanaka S (2006) Induction of pluripotent stem cells from mouse embryonic and adult fibroblast cultures by defined factors. Cell 126: 663–676
2. Schultz RM (1993) Regulation of zygotic gene activation in the mouse. Bioessays 15: 531–538
3. Telford NA, Watson AJ, et al. (1990) Transition from maternal to embryonic control in early mammalian development: a comparison of several species. Mol Reprod Dev 26: 90–100
4. Frei RE, Schultz GA, et al. (1989) Qualitative and quantitative changes in protein synthesis occur at the 8–16-cell stage of embryogenesis in the cow. J Reprod Fertil 86: 637–641
5. Farthing CR, Ficz G, et al. (2008) Global mapping of DNA methylation in mouse promoters reveals epigenetic reprogramming of pluripotency genes. PloS Genetics 4:e1000116
6. Tagawa M, Matoba S, et al. (2008) Production of monozygotic twin calves using the blastomere separation technique and well of the well culture system. Theriogenology 69: 574–582
7. Johnson WH, Loskutoff NM, et al. (1995) Production of four identical calves by the separation of blastomeres from an in vitro derived four-cell embryo. Vet Rec 137: 15–16
8. Chan AW, Dominko T, et al. (2000) Clonal propagation of primate offspring by embryo splitting. Science 287: 317–319
9. Bilodeau-Goeseels S, Schultz GA (1997) Changes in the relative abundance of various housekeeping gene transcripts in in vitro-produced early bovine embryos. Mol Reprod Dev 47: 413–420
10. Stein P, Worrad DM, et al. (1997) Stage-dependent redistributions of acetylated histones in nuclei of the early preimplantation mouse embryo. Mol Reprod Dev 47: 421–429
11. Memili E, First NL (1999) Control of gene expression at the onset of bovine embryonic development. Biol Reprod 61: 1198–1207
12. Wangh LJ, DeGrace D, et al. (1995) Efficient reactivation of *Xenopus* erythrocyte nuclei in *Xenopus* egg extracts. J Cell Sci 108 (Pt 6): 2187–2196
13. Zheng P, Patel B, et al. (2004) Expression of genes encoding chromatin regulatory factors in developing rhesus monkey oocytes and preimplantation stage embryos: possible roles in genome activation. Biol Reprod 70: 1419–1427
14. Vigneault C, McGraw S, et al. (2004) Transcription factor expression patterns in bovine in vitro-derived embryos prior to maternal-zygotic transition. Biol Reprod 70: 1701–1709
15. Misirlioglu M, Page GP, et al. (2006) Dynamics of global transcriptome in bovine matured oocytes and preimplantation embryos. Proc Natl Acad Sci USA 103: 18905–18910
16. Pesce M, Scholer HR (2001) Oct-4: gatekeeper in the beginnings of mammalian development. Stem Cells 19: 271–278
17. Chambers I, Colby D, et al. (2003) Functional expression cloning of nanog, a pluripotency sustaining factor in embryonic stem cells. Cell 113: 643–655
18. Nichols J, Zevnik B, et al. (1998) Formation of pluripotent stem cells in the mammalian embryo depends on the POU transcription factor Oct4. Cell 95: 379–391
19. Silva J, Nichols J, et al. (2009) Nanog is the gateway to the pluripotent ground state. Cell 138: 722–737
20. Thomson JA, Itskovitz-Eldor J, et al. (1998) Embryonic stem cell lines derived from human blastocysts. Science 282: 1145–1147
21. Niwa H, Miyazaki J, et al. (2000) Quantitative expression of Oct-3/4 defines differentiation, dedifferentiation or self-renewal of ES cells. Nat Genet 24: 372–376
22. de Vries WN, Evsikov AV, et al. (2008) Reprogramming and differentiation in mammals: motifs and mechanisms. Cold Spring Harb Symp Quant Biol 73: 33–38
23. Scholer HR, Balling R, et al. (1989) Octamer binding proteins confer transcriptional activity in early mouse embryogenesis. EMBO J 8: 2551–2557
24. Rosner MH, Vigano MA, et al. (1990) A POU-domain transcription factor in early stem cells and germ cells of the mammalian embryo. Nature 345: 686–692

25. Cauffman G, Liebaers I, et al. (2006) Pou5f1 isoforms show different expression patterns in human embryonic stem cells and preimplantation embryos. Stem Cells 24: 2685–2691
26. Cauffman G, Van de Velde H, et al. (2005) Oct-4 mRNA and protein expression during human preimplantation development. Mol Hum Reprod 11: 173–181
27. Takeda J, Seino S, et al. (1992) Human Oct3 gene family: cDNA sequences, alternative splicing, gene organization, chromosomal location, and expression at low levels in adult tissues. Nucleic Acids Res 20: 4613–4620
28. Pesce M, Wang X, et al. (1998) Differential expression of the Oct-4 transcription factor during mouse germ cell differentiation. Mech Dev 71: 89–98
29. Kirchhof N, Carnwath JW, et al. (2000) Expression pattern of Oct-4 in preimplantation embryos of different species. Biol Reprod 63: 1698–1705
30. Boyer LA, Lee TI, et al. (2005) Core transcriptional regulatory circuitry in human embryonic stem cells. Cell 122: 947–956
31. Gurdon JB, Elsdale TR, et al. (1958) Sexually mature individuals of *Xenopus laevis* from the transplantation of single somatic nuclei. Nature 182: 64–65
32. First NL, Prather RS (1991) Genomic potential in mammals. Differentiation 48: 1–8
33. Campbell KH, Loi P, et al. (1996) Cell cycle co-ordination in embryo cloning by nuclear transfer. Rev Reprod 1: 40–46
34. Wilmut I, Schnieke AE, et al. (1997) Viable offspring derived from fetal and adult mammalian cells. Nature 385: 810–813
35. Lanza RP, Cibelli JB, et al. (1999) Human therapeutic cloning. Nat Med 5: 975–977
36. Burley J (1999) The ethics of therapeutic and reproductive human cloning. Semin Cell Dev Biol 10: 287–294
37. Simerly C, Dominko T, et al. (2003) Molecular correlates of primate nuclear transfer failures. Science 300: 297
38. Simerly C, Navara C, et al. (2004) Embryogenesis and blastocyst development after somatic cell nuclear transfer in nonhuman primates: overcoming defects caused by meiotic spindle extraction. Dev Biol 276: 237–252
39. Byrne JA, Pedersen DA, et al. (2007) Producing primate embryonic stem cells by somatic cell nuclear transfer. Nature 450: 497–502
40. Sparman M, Dighe V, et al. (2009) Epigenetic reprogramming by somatic cell nuclear transfer in primates. Stem Cells 27: 1255–1264
41. Jiang W, Bai Z, et al. (2008) Differentiation of mouse nuclear transfer embryonic stem cells into functional pancreatic beta cells. Diabetologia 51: 1671–1679
42. Mombaerts P (2003) Therapeutic cloning in the mouse. Proc Natl Acad Sci USA 100 Suppl 1: 11924–11925
43. Munsie MJ, Michalska AE, et al. (2000) Isolation of pluripotent embryonic stem cells from reprogrammed adult mouse somatic cell nuclei. Curr Biol 10: 989–992
44. Sviridova-Chailakhyan TA, Chailakhyan LM (2005) Mouse embryo reconstruction as an adequate model for developing the principles of therapeutic cloning. Dokl Biol Sci 404: 399–401
45. Rideout WM, 3rd, Hochedlinger K, et al. (2002) Correction of a genetic defect by nuclear transplantation and combined cell and gene therapy. Cell 109: 17–27
46. Dean W, Santos F, et al. (2001) Conservation of methylation reprogramming in mammalian development: aberrant reprogramming in cloned embryos. Proc Natl Acad Sci USA 98: 13734–13738
47. Santos F, Zakhartchenko V, et al. (2003) Epigenetic marking correlates with developmental potential in cloned bovine preimplantation embryos. Curr Biol 13: 1116–1121
48. Yamazaki Y, Fujita TC, et al. (2006) Gradual DNA demethylation of the Oct4 promoter in cloned mouse embryos. Mol Reprod Dev 73: 180–188
49. Simonsson S, Gurdon J (2004) DNA demethylation is necessary for the epigenetic reprogramming of somatic cell nuclei. Nat Cell Biol 6: 984–990
50. Kishigami S, Mizutani E, et al. (2006) Significant improvement of mouse cloning technique by treatment with trichostatin A after somatic nuclear transfer. Biochem Biophys Res Commun 340: 183–189

51. Li X, Kato Y, et al. (2008) The effects of trichostatin A on mRNA expression of chromatin structure-, DNA methylation-, and development-related genes in cloned mouse blastocysts. Cloning Stem Cells 10: 133–142
52. Brambrink T, Hochedlinger K, et al. (2006) ES cells derived from cloned and fertilized blastocysts are transcriptionally and functionally indistinguishable. Proc Natl Acad Sci USA 103: 933–938
53. Wakayama S, Mizutani E, et al. (2005) Mice cloned by nuclear transfer from somatic and ntES cells derived from the same individuals. J Reprod Dev 51: 765–772
54. Byrne JA, Simonsson S, et al. (2003) Nuclei of adult mammalian somatic cells are directly reprogrammed to Oct-4 stem cell gene expression by amphibian oocytes. Curr Biol 13: 1206–1213
55. Dominko T, Mitalipova M, et al. (1999) Bovine oocyte cytoplasm supports development of embryos produced by nuclear transfer of somatic cell nuclei from various mammalian species. Biol Reprod 60: 1496–1502
56. Li F, Cao H, et al. (2008) Activation of human embryonic gene expression in cytoplasmic hybrid embryos constructed between bovine oocytes and human fibroblasts. Cloning Stem Cells 10: 297–305
57. Wang K, Beyhan Z, et al. (2009) Bovine ooplasm partially remodels primate somatic nuclei following somatic cell nuclear transfer. Cloning Stem Cells 11: 187–202
58. Vogel G (2008) Bioethics. U.K. approves new embryo law. Science 322: 663
59. Hakelien AM, Landsverk HB, et al. (2002) Reprogramming fibroblasts to express t-cell functions using cell extracts. Nat Biotechnol 20: 460–466
60. Condorelli G, Borello U, et al. (2001) Cardiomyocytes induce endothelial cells to transdifferentiate into cardiac muscle: implications for myocardium regeneration. Proc Natl Acad Sci USA 98: 10733–10738
61. Qin M, Tai G, et al. (2005) Cell extract-derived differentiation of embryonic stem cells. Stem Cells 23: 712–718
62. Kikyo N, Wade PA, et al. (2000) Active remodeling of somatic nuclei in egg cytoplasm by the nucleosomal ATPase ISWI. Science 289: 2360–2362
63. Hansis C, Barreto G, et al. (2004) Nuclear reprogramming of human somatic cells by *Xenopus* egg extract requires brg1. Curr Biol 14: 1475–1480
64. Alberio R, Johnson AD, et al. (2005) Differential nuclear remodeling of mammalian somatic cells by *Xenopus laevis* oocyte and egg cytoplasm. Exp Cell Res 307: 131–141
65. Shimazaki T, Okazawa H, et al. (1993) Hybrid cell extinction and re-expression of Oct-3 function correlates with differentiation potential. EMBO J 12: 4489–4498
66. Taranger CK, Noer A, et al. (2005) Induction of dedifferentiation, genomewide transcriptional programming, and epigenetic reprogramming by extracts of carcinoma and embryonic stem cells. Mol Biol Cell 16: 5719–5735
67. Freberg CT, Dahl JA, et al. (2007) Epigenetic reprogramming of Oct4 and nanog regulatory regions by embryonal carcinoma cell extract. Mol Biol Cell 18: 1543–1553
68. Tada M, Tada T, et al. (1997) Embryonic germ cells induce epigenetic reprogramming of somatic nucleus in hybrid cells. EMBO J 16: 6510–6520
69. Tada M, Takahama Y, et al. (2001) Nuclear reprogramming of somatic cells by in vitro hybridization with ES cells. Curr Biol 11: 1553–1558
70. Kimura H, Tada M, et al. (2004) Histone code modifications on pluripotential nuclei of reprogrammed somatic cells. Mol Cell Biol 24: 5710–5720
71. Bru T, Clarke C, et al. (2008) Rapid induction of pluripotency genes after exposure of human somatic cells to mouse es cell extracts. Exp Cell Res 314: 2634–2642
72. Tamada H, Van Thuan N, et al. (2006) Chromatin decondensation and nuclear reprogramming by nucleoplasmin. Mol Cell Biol 26: 1259–1271
73. Rai K, Huggins IJ, et al. (2008) DNA demethylation in zebrafish involves the coupling of a deaminase, a glycosylase, and gadd45. Cell 135: 1201–1212
74. Santos-Rosa H, Schneider R, et al. (2003) Methylation of histone h3 k4 mediates association of the ISW1p ATPase with chromatin. Mol Cell 12: 1325–1332

75. Kingston RE, Narlikar GJ (1999) ATP-dependent remodeling and acetylation as regulators of chromatin fluidity. Genes Dev 13: 2339–2352
76. Page RL, Ambady S, et al. (2009) Induction of stem cell gene expression in adult human fibroblasts without transgenes. Cloning Stem Cells 11: 417–426
77. Kikyo N, Wolffe AP (2000) Reprogramming nuclei: insights from cloning, nuclear transfer and heterokaryons. J Cell Sci 113 (Pt 1): 11–20
78. Kimura H, Tada M, et al. (2002) Chromatin reprogramming of male somatic cell-derived Xist and Tsix in ES hybrid cells. Cytogenet Genome Res 99: 106–114
79. Tada T, Tada M (2001) Toti-/pluripotential stem cells and epigenetic modifications. Cell Struct Funct 26: 149–160
80. Cowan CA, Atienza J, et al. (2005) Nuclear reprogramming of somatic cells after fusion with human embryonic stem cells. Science 309: 1369–1373
81. Ivanova NB, Dimos JT, et al. (2002) A stem cell molecular signature. Science 298: 601–604
82. Ramalho-Santos M, Yoon S, et al. (2002) "Stemness": transcriptional profiling of embryonic and adult stem cells. Science 298: 597–600
83. Tokuzawa Y, Kaiho E, et al. (2003) Fbx15 is a novel target of Oct3/4 but is dispensable for embryonic stem cell self-renewal and mouse development. Mol Cell Biol 23: 2699–2708
84. Okita K, Ichisaka T, et al. (2007) Generation of germline-competent induced pluripotent stem cells. Nature 448: 313–317
85. Takahashi K, Tanabe K, et al. (2007) Induction of pluripotent stem cells from adult human fibroblasts by defined factors. Cell 131: 861–872
86. Takahashi K, Okita K, et al. (2007) Induction of pluripotent stem cells from fibroblast cultures. Nat Protoc 2: 3081–3089
87. Maherali N, Sridharan R, et al. (2007) Directly reprogrammed fibroblasts show global epigenetic remodeling and widespread tissue contribution. Cell Stem Cell 1: 55–70
88. Stadtfeld M, Maherali N, et al. (2008) Defining molecular cornerstones during fibroblast to iPS cell reprogramming in mouse. Cell Stem Cell 2: 230–240
89. Stadtfeld M, Nagaya M, et al. (2008) Induced pluripotent stem cells generated without viral integration. Science 322: 945–949
90. Wolf D, Goff SP (2007) Trim28 mediates primer binding site-targeted silencing of murine leukemia virus in embryonic cells. Cell 131: 46–57
91. Okita K, Nakagawa M, et al. (2008) Generation of mouse induced pluripotent stem cells without viral vectors. Science 322: 949–953
92. Wernig M, Meissner A, et al. (2007) In vitro reprogramming of fibroblasts into a pluripotent ES-cell-like state. Nature 448: 318–324
93. Meissner A, Wernig M, et al. (2007) Direct reprogramming of genetically unmodified fibroblasts into pluripotent stem cells. Nat Biotechnol 25: 1177–1181
94. Lowry WE, Richter L, et al. (2008) Generation of human induced pluripotent stem cells from dermal fibroblasts. Proc Natl Acad Sci USA 105: 2883–2888
95. Yu J, Hu K, et al. (2009) Human induced pluripotent stem cells free of vector and transgene sequences. Science 324: 797–801
96. Byrne JA, Nguyen HN, et al. (2009) Enhanced generation of induced pluripotent stem cells from a subpopulation of human fibroblasts. PLoS One 4: e7118
97. Aoi T, Yae K, et al. (2008) Generation of pluripotent stem cells from adult mouse liver and stomach cells. Science 321: 699–702
98. Hasegawa K, Cowan AB, et al. (2007) Efficient multicistronic expression of a transgene in human embryonic stem cells. Stem Cells 25: 1707–1712
99. Nakagawa M, Koyanagi M, et al. (2008) Generation of induced pluripotent stem cells without Myc from mouse and human fibroblasts. Nat Biotechnol 26: 101–106
100. Yu J, Vodyanik MA, et al. (2007) Induced pluripotent stem cell lines derived from human somatic cells. Science 318: 1917–1920
101. Gonzales C, Pedrazzini T (2009) Progenitor cell therapy for heart disease. Exp Cell Res 315: 3077–3085

102. Carey BW, Markoulaki S, et al. (2009) Reprogramming of murine and human somatic cells using a single polycistronic vector. Proc Natl Acad Sci U S A 106: 157–162
103. Sommer CA, Stadtfeld M, et al. (2009) Induced pluripotent stem cell generation using a single lentiviral stem cell cassette. Stem Cells 27: 543–549
104. Chang CW, Lai YS, et al. (2009) Polycistronic lentiviral vector for "hit and run" reprogramming of adult skin fibroblasts to induced pluripotent stem cells. Stem Cells 27: 1042–1049
105. Kaji K, Norrby K, et al. (2009) Virus-free induction of pluripotency and subsequent excision of reprogramming factors. Nature 458: 771–775
106. Woltjen K, Michael IP, et al. (2009) Piggybac transposition reprograms fibroblasts to induced pluripotent stem cells. Nature 458: 766–770
107. Kim D, Kim CH, et al. (2009) Generation of human induced pluripotent stem cells by direct delivery of reprogramming proteins. Cell Stem Cell 4: 472–476
108. Zhou H, Wu S, et al. (2009) Generation of induced pluripotent stem cells using recombinant proteins. Cell Stem Cell 4: 381–384
109. Shi Y, Do JT, et al. (2008) A combined chemical and genetic approach for the generation of induced pluripotent stem cells. Cell Stem Cell 2: 525–528
110. Huangfu D, Maehr R, et al. (2008) Induction of pluripotent stem cells by defined factors is greatly improved by small-molecule compounds. Nat Biotechnol 26: 795–797
111. Huangfu D, Osafune K, et al. (2008) Induction of pluripotent stem cells from primary human fibroblasts with only Oct4 and Sox2. Nat Biotechnol 26(11): 1269–1275
112. Feng B, Ng JH, et al. (2009) Molecules that promote or enhance reprogramming of somatic cells to induced pluripotent stem cells. Cell Stem Cell 4: 301–312
113. Ichida JK, Blanchard J, et al. (2009) A small-molecule inhibitor of TGF-beta signaling replaces Sox2 in reprogramming by inducing nanog. Cell Stem Cell 5(5): 491–503
114. Kim JB, Sebastiano V, et al. (2009) Oct4-induced pluripotency in adult neural stem cells. Cell 136: 411–419
115. Kim JB, Greber B, et al. (2009) Direct reprogramming of human neural stem cells by Oct4. Nature 461: 649–653
116. Frankel AD, Bredt DS, et al. (1988) Tat protein from human immunodeficiency virus forms a metal-linked dimer. Science 240: 70–73
117. Frankel AD, Pabo CO (1988) Cellular uptake of the tat protein from human immunodeficiency virus. Cell 55: 1189–1193
118. Bosnali M, Edenhofer F (2008) Generation of transducible versions of transcription factors Oct4 and Sox2. Biol Chem 389: 851–861
119. O'Malley J, Woltjen K, et al. (2009) New strategies to generate induced pluripotent stem cells. Curr Opin Biotechnol 20(5): 516–521
120. Xu RH, Sampsell-Barron TL, et al. (2008) Nanog is a direct target of tgfbeta/activin-mediated smad signaling in human ESCs. Cell Stem Cell 3: 196–206
121. Ying QL, Nichols J, et al. (2003) BMP induction of Id proteins suppresses differentiation and sustains embryonic stem cell self-renewal in collaboration with STAT3. Cell 115: 281–292
122. Humphrey RK, Beattie GM, et al. (2004) Maintenance of pluripotency in human embryonic stem cells is STAT3 independent. Stem Cells 22: 522–530
123. Hanna J, Carey BW, et al. (2008) Reprogramming of somatic cell identity. Cold Spring Harb Symp Quant Biol 73: 147–155
124. Stadtfeld M, Brennand K, et al. (2008) Reprogramming of pancreatic beta cells into induced pluripotent stem cells. Curr Biol 18: 890–894
125. Sun N, Panetta NJ, et al. (2009) Feeder-free derivation of induced pluripotent stem cells from adult human adipose stem cells. Proc Natl Acad Sci USA 106: 15720–15725
126. Ye Z, Zhan H, et al. (2009) Human induced pluripotent stem cells from blood cells of healthy donors and patients with acquired blood disorders. Blood 114(27): 5473–5480
127. Kim JB, Zaehres H, et al. (2008) Pluripotent stem cells induced from adult neural stem cells by reprogramming with two factors. Nature 454: 646–650
128. Kim JB, Zaehres H, et al. (2009) Generation of induced pluripotent stem cells from neural stem cells. Nat Protoc 4: 1464–1470

129. Yoshida Y, Takahashi K, et al. (2009) Hypoxia enhances the generation of induced pluripotent stem cells. Cell Stem Cell 5: 237–241
130. Silvan U, Diez-Torre A, et al. (2009) Hypoxia and pluripotency in embryonic and embryonal carcinoma stem cell biology. Differentiation 78: 159–168
131. Feng Q, Lu SJ, et al. Hemangioblastic derivatives from human induced pluripotent stem cells exhibit limited expansion and early senescence. Stem Cells 28(4): 704–712
132. Wernig M, Zhao JP, et al. (2008) Neurons derived from reprogrammed fibroblasts functionally integrate into the fetal brain and improve symptoms of rats with Parkinson's disease. Proc Natl Acad Sci USA 105: 5856–5861
133. Dimos JT, Rodolfa KT, et al. (2008) Induced pluripotent stem cells generated from patients with ALS can be differentiated into motor neurons. Science 321: 1218–1221
134. Mauritz C, Schwanke K, et al. (2008) Generation of functional murine cardiac myocytes from induced pluripotent stem cells. Circulation 118: 507–517
135. Narazaki G, Uosaki H, et al. (2008) Directed and systematic differentiation of cardiovascular cells from mouse induced pluripotent stem cells. Circulation 118: 498–506
136. Gai H, Leung EL, et al. (2009) Generation and characterization of functional cardiomyocytes using induced pluripotent stem cells derived from human fibroblasts. Cell Biol Int 33(11): 1184–1193

Inducible Pluripotent Stem Cells for Cardiac Regeneration

Naama Zeevi-Levin and Joseph Itskovitz-Eldor

Abstract The adult human heart has limited intrinsic regenerative capacity, and so the myocardium lost after a myocardial infarction is typically replaced by noncontractile scar tissue, often initiating congestive heart failure. Current treatment options are limited in preventing ventricular remodeling because of their inability to repair or replace damaged myocardium. Novel therapeutic modalities for improving cardiac function, such as cell therapy, are imperative. With the first derivation of human embryonic stem cells from preimplantation embryos more than a decade ago it seemed that a new potential source of cells for the treatment of many degenerative diseases had emerged, including cardiomyocytes for cardiac repair. Recently, another breakthrough took place. By the overexpression of four retrovirally transduced transcription factors in mouse dermal fibroblasts, grown in conditions favoring embryonic stem cells expansion, inducible pluripotent stem (iPS) cell could be generated. The promise of applying this technology to human cells was rapidly realized with the successful generation of human iPS cells from human fibroblasts. These cells have the potential to regenerate functional myocardium. In addition, they may be a useful model for studying genetic diseases. The development and potential of these cells for treating and studying cardiac diseases is reviewed in this chapter.

Keywords Inducible pluripotent stem cells • Embryonic stem cells • Cell therapy • Differentiation • Cardiomyocytes

1 Introduction

The adult human heart has limited intrinsic regenerative capacity, and so the myocardium lost after a myocardial infarction is typically replaced by noncontractile scar tissue, often initiating congestive heart failure (HF). According to the American

J. Itskovitz-Eldor (✉)
Chairman, Department of Ob-Gyn, Rambam Health Care Campus Head, Stem Cell Center, Technion – Israel Institute of Technology, PoB 9602 Haifa 31096, Israel
e-mail: itskovitz@rambam.health.gov.il

I.S. Cohen and G.R. Gaudette (eds.), *Regenerating the Heart*, Stem Cell Biology and Regenerative Medicine, DOI 10.1007/978-1-61779-021-8_14,
© Springer Science+Business Media, LLC 2011

Heart Association, an estimated 550,000 new cases of HF occur each year [1]. More than five million Americans have HF and it is the leading cause of hospitalization. From many roots, HF leads to common pathways of significant morbidity and mortality (one of every 2.8 deaths in the USA in 2004; [2]). Cardiovascular risk factors and diseases are on the rise not only in developed countries but also in developing countries, leading to this growing epidemic becoming a major national and global public health threat.

The current pharmacotherapy for congestive HF improves clinical outcomes. Other treatment options, including various interventional and surgical therapeutic methods, are limited in preventing ventricular remodeling because of their inability to repair or replace damaged myocardium. The only standard therapy for HF that addresses the fundamental problem of cardiomyocyte loss is cardiac transplantation. Given the high morbidity and mortality rates associated with congestive HF, shortage of donor hearts for transplantation, complications resulting from immunosuppression, and long-term failure of transplanted organs, novel therapeutic modalities for improving cardiac function and preventing HF are imperative [3]. Hence, regeneration or repair of infarcted or ischemic myocardium may be accomplished by the use of cell therapy – the transplantation of healthy, functional, and propagating cells to restore the viability or function of deficient tissues [4, 5, 71]).

A number of cell types have been considered as candidates for cell transplantation strategies, including skeletal myoblasts [6–8], bone-marrow-derived hematopoietic stem cells [9], mesenchymal stem cells [10–13], intrinsic cardiac stem cells [14–17], and embryonic stem cells (ESCs) [18–22]. Of these, skeletal myoblasts and bone-marrow-derived cells have undergone the most extensive testing in humans, but phase 1 and phase 2 clinical trials have yielded mixed results [23–29]. However, with the first derivation of human ESCs (hESCs) from preimplantation embryos more than a decade ago [30], it seemed that a new potential source of cells for the treatment of many degenerative diseases had emerged, including cardiomyocytes for cardiac repair. These pluripotent cells can be isolated and maintained by well-established protocols, with scalable options; they have unquestioned cardiomyogenic potential [31–33]; and most importantly, a number of recent reports have shown that hESC-derived cardiomyocytes (hESC-CMs) survive after transplantation into infarcted rodent hearts, form stable cardiac implants, and result in preserved contractile function [18, 19, 22].

Recently, another breakthrough took place when Takahashi and Yamanaka [34] discovered that by the overexpression of four retrovirally transduced transcription factors (Oct4, Sox2, c-Myc, and Klf4) in mouse dermal fibroblasts, grown in conditions favoring ESC expansion, they were able to generate inducible pluripotent stem (iPS) cells. Oct4 and Sox2 are part of the core transcriptional network required for pluripotency. c-Myc is a protooncogene required for cell cycle progression. Klf4 is a cell cycle regulator that may control self-renewal of ESCs and block apoptotic pathways induced by c-Myc. The promise of applying this technology to human cells was rapidly realized with the successful generation of human iPS cells from human fibroblasts using either the same combination of transcription factors [35, 36] or a slightly different cocktail of lentivirally transduced genes (Oct4, Sox2,

Nanog, and Lin28) [38]. This approach has also been recently applied to generate disease-specific iPS cell lines from patients with a variety of diseases [37, 39]. Although there are still critical issues such as viral integration and combination of reprogramming genes, the progress in this emerging field will be valuable not only for the application of stem cells in patient-matched lines for future therapy that would avoid the need for immunosuppression, but also for the usage of lines bearing disease traits as excellent models to study mechanisms of disease and for drug and toxicology screening. We discuss here the current knowledge regarding cardiomyocytes derived from human pluripotent stem cells and their potential in therapy and disease models.

2 Pluripotency of the Cells

Like mouse ESCs, mouse iPS cells generate teratomas (tumors comprising ectoderm, mesoderm, and endoderm derivatives) when injected into adult mice, and additionally their pluripotency has been proven by generation of chimeric animals following blastocyst injection with subsequent germline transmission [40, 41]. Obviously, ethical constraints preclude human embryo experiments, and therefore formation of teratomas in immunocompromised mice has provided evidence for the pluripotency of human iPS cells on the basis of the existence of derivatives of the three embryonic germ layers. In addition, in vitro differentiation studies are critically important for demonstrating the pluripotency of human iPS cells. Indeed, the first studies in which induction of iPS cells was achieved, identified derivatives of the three primary germ layers in embryoid body (EB) differentiation, in addition to specifically demonstrating the ability of the cells to differentiate into neural progenitors and cardiomyocytes [35, 36, 38].

3 Cardiac Differentiation of iPS Cells

Cardiac differentiation of ESCs is mainly achieved by three approaches: (1) spontaneous differentiation through EBs grown in suspension, during which contracting areas containing cardiomyocytes start to appear within a few days [31]; (2) coculture with endoderm-like cells or cells secreting endoderm-like signals [33]; and (3) direct addition of growth factors, hormones, or small molecules thought to be involved in heart development to ESCs grown as EBs or in monolayers [19, 21, 42]. Since the generation of the iPS cells, an increasing number of studies have been dedicated to exploring the ability of these cells to generate cardiomyocytes. For this purpose, the approaches developed for the differentiation of ESCs are utilized for iPS cells. In studies with mouse iPS cells, Mauritz et al. induced differentiation by forming EBs in the presence of serum [43] whereas Narazaki et al. differentiated the cells in monolayers using serum and selected by fluorescence-activated cell sorting for mesodermal cells expressing the vascular endothelial growth factor receptor Flk1. They then obtained cardiomyocytes by co-culture with OP9 cells [44]. Schenke-Layland et al. compared mouse ESCs and iPS cells for their ability to differentiate into cardiovascular cells using EB differentiation or isolated Flk1$^+$ from cells grown on collagen IV [45]. Recently, Pfannkuche et al. performed cardiac differentiation using a mass culture system for EB production starting with a single-cell suspension and followed by plating the EBs on gelatin [46].

In regard to the efficiency of cardiac differentiation of mouse iPS cells, Mauritz et al. reported delayed iPS-cell-derived cardiomyocyte (iPSC-CM) differentiation compared with mouse ESCs, by measurement of both the percentage of spontaneously contracting EBs and expression of cardiomyocyte-specific messenger RNA (mRNA) for troponin (threefold to fivefold lower in iPS cell EBs) [43]. Yet, it is important to note that the line of ESCs used in that work showed an unusually efficient rate of differentiation into cardiomyocytes (100% of EBs beating by day 8), lines. Along these lines, Narazaki et al. found that cardiac differentiation of iPS cells was within the range of the different lines of ESCs they studied in terms of both beating activity and cardiac gene expression [44]. Similarly, Schenke-Layland et al. found that induction of cardiomyocytes occurred with comparable efficiency in ESCs and iPS cells [45], whereas the differentiation protocol of Pfannkuche et al. needed some slight modifications for beating areas to be formed at an efficient rate. Having that done, the frequency of beating areas in iPS cell EBs was comparable to that of ESCs [46].

Takashaki et al. were the first to report cardiac differentiation of human iPS cells. They used a directed differentiation protocol established for hESCs which utilizes activin A and bone morphogenetic protein 4 [35]. A detailed evaluation of the cardiac differentiation potential of recently described human iPS cell lines in comparison with well-studied hESC lines, was provided by Zhang et al., exploiting EB formation in the presence of serum [47]. The ability of human iPSC-CMs from beating EBs to serve as an in vitro drug screening model was also recently evaluated [48, 49].

As was reported for mouse iPS cells, Zhang et al. demonstrated that there was no observable difference in the time course for the development of contraction for human-iPS- and hESC-derived EBs, although the efficiency of forming contracting EBs varied for both iPS cell (4.2% and about 10%) and ESC (22% and about 10%) lines [47].

The studies performed so far for cardiac differentiation from both mouse and human iPS cell lines are summarized in Table 1. These studies give the impression that the cardiac differentiation protocols employed for ESCs are applicable for the differentiation of iPS cells, although slight adjustments may be required. Additionally, given that cardiac differentiation efficiency and yield vary significantly between ESC lines, it seems reasonable to conclude that cardiogenesis from iPS cells is efficient to an extent comparable to that for ESCs. It would be interesting to further systematically evaluate the variability of differentiation of various iPS cell lines and to understand if it is affected by the source of the reprogrammed cells (e.g., fetal vs. adult; healthy vs. diseased).

Table 1 Summary of published studies on cardiomyocytes derived from inducible pluripotent stem (iPS) cells

Reference	Species	Reprogrammed cells/line name	Reprogramming factors	Differentiation method	Contracting areas (%)	Cardiac characterization
Mauritz et al. [43]	Mouse	MEFs: O9	Oct3/4, Sox2, c-Myc, Klf4	EBs (hanging drops)	55 ± 4.2	RT-PCR Immunostaining Intracellular Ca measurements MEA measurements
Narazaki et al. [44]	Mouse	MEFs: 20D17 38C2 38D2	Oct3/4, Sox2, c-Myc, Klf4	Flk1+ cells co-cultured with OP9 cells	16.1 ± 1.6 17.4 ± 3.1 18.4 ± 2.6	RT-PCR Immunostaining Patch clamp recordings
Schenke-Layland et al. [45]	Mouse	MEFs: 2D4	Oct3/4, Sox2, c-Myc, Klf4	EBs (dynamic suspension) or Flk1^{+} from cells grown on collagen IV		RT-PCR Immunostaining Intracellular Ca measurements
Pfannkuche et al. [46]	Mouse	MEFs: O9 N10	Oct3/4, Sox2, c-Myc, Klf4	EBs (mass culture system)	43 ± 15 40 ± 11	RT-PCR Immunostaining MEA measurements Patch clamp recordings Force measurements
Nelson et al. [59]	Mouse	MEFs	Oct3/4, Sox2, c-Myc, Klf4			RT-PCR in vivo
Takashaki et al. [35]	Human	Adult HDF	Oct3/4, Sox2, c-Myc, Klf4	Activin A + BMP4		RT-PCR
Zhang et al. [47]	Human	Fetal fibroblasts: IMR90 Clone 1 IMR90 Clone 4 Newborn foreskin fibroblasts: Forskin Clone 1 Forskin Clone 2	Oct4, Sox2, Nanog, Lin28	EBs	 ~4.5 ~10 4.2 ~1	RT-PCR Immunostaining Patch clamp recordings

(continued)

Table 1 (continued)

Reference	Species	Reprogrammed cells/line name	Reprogramming factors	Differentiation method	Contracting areas (%)	Cardiac characterization
Tanaka et al. [48]	Human	Adult HDF: 201B7	Oct3/4, Sox2, c-Myc, Klf4	EBs		RT-PCR Immunostaining MEA measurements
Yokoo et al. [49]	Human	Adult HDF: 201B7	Oct3/4, Sox2, c-Myc, Klf4	EBs		RT-PCR Immunostaining MEA measurements Beating rate and contractility

MEFs mouse embryonic fibroblasts, *HDF* human dermal fibroblasts, *EBs* embryoid bodies, *BMP4* bone morphogenetic protein 4, *RT-PCR* reverse transcription polymerase chain reaction, *MEA* multielectrode array

4 Properties of iPSC-CMs

To improve the prospects of using cardiac cells for transplantation as well as for other therapeutic applications, it is widely realized that the functional properties as well as the hormonal and pharmacological responsiveness of iPSC-CMs should be thoroughly investigated.

4.1 Transcriptional Profile

To characterize the differentiation pathway of undifferentiated murine iPS cells making the transition to functional cardiomyocytes, reverse transcription (RT) PCR analyses were performed. The differentiating iPS cells were found to express the mesodermal markers brachyury and mesoderm posterior factor 1 (Mesp1), followed by increased expression of the early cardiac-specific transcription factors including cardiac mesodermal marker genes friend of GATA2 (FOG-2), GATA-binding protein 4 (GATA4), NK2 transcription factor related locus 5 (Nkx2.5), myocyte enhancer factor 2C (Mef2c), T-box 5 (Tbx5), and T-box 20 (Tbx20). Differentiated mouse iPS cells expressed the cardiomyocyte markers myosin light chain 2 atrial (MLC2a) and myosin light chain 2 ventricular (MLC2v) isoforms, α-myosin heavy chain (α-MHC), and atrial natriuretic factor (ANF) [43–45]. Additionally, the mRNAs of slow skeletal troponin I and cardiac troponin T (cTnT) were strongly expressed in beating cardiac EBs [46]. These expression patterns were compared and found to be in agreement with those in differentiating mouse ESCs.

Similarly, RT-PCR analyses of human iPSC-CMs demonstrated the expression of the full range of cardiac genes: the transcription factors Nkx2.5, GATA4, Mef2c and the myofilament protein genes cTnT, α-MHC, α-actinin (ACTN2), MLC2a, MLC2v, and ANF. This robust expression was comparable to that observed in hESC-CMs and in human heart tissue [47–49].

4.2 Structural Properties

Immunofluorescence analyses further characterized the iPSC-CMs for their structural and functional properties. Mouse iPSC-CMs exhibit regular sarcomeric organization with well-organized cross-striation and express myocyte markers, including α-sarcomeric actinin, titin, cTnT, and MLC2v. Immunostaining of single beating clusters demonstrated the presence of the gap junction protein connexin 43 in iPSC-CMs that expressed cTnT [43–46].

The expression of myofilament proteins and the sarcomeric organization was evaluated in human iPSC-CMs as well. Immunofluorescence analysis revealed a clear striated pattern for α-actinin and MLC2a labeling, comparable to that in hESC-CMs. Overlap of α-actinin and MLC2a labeling demonstrated an alternating pattern in the sarcomeres in agreement with the known localization of MLC2a to the A-band of the sarcomere, which lies between the Z-lines highlighted by the α-actinin labeling. Immunolabeling for cTnT, which is a highly cardiac specific myofilament protein, yielded comparable sarcomeric labeling of human iPSC-CMs and hESC-CMs. Likewise, immunolabeling with an antibody to the ventricular specific protein MLC2v detected presumed ventricular cardiomyocytes derived from iPS cells and ESCs [47]. Other studies also reported sarcomeric structure of human iPSC-CMs on the basis of the expression of α-actinin, MHC, and tropomyosin [48] in addition to the expression of the cardiac markers cardiac troponin I, MLC2v, and ANP [49]. Thus, immunolabeling of multiple myofilament proteins indicates that a well-organized sarcomeric structure can similarly develop in human iPSC-CMs and hESC-CMs.

4.3 Electrophysiological Properties

Previous electrophysiology studies of both mouse-ESC-CMs and hESC-CMs demonstrated that the derived cardiomyocytes exhibit diverse electrophysiologic signatures that include cells with distinct nodal/pacemaker-, atrial-, and ventricular-like action potential properties [33, 50, 51]. This classification was based on the shape and properties of the action potential, such as resting potential, upstroke velocity, amplitude and duration.

In regard to mouse iPSC-CMs, Narazaki et al. identified that some of the isolated cardiomyocytes showed typical pacemaker-like potential and could depolarized spontaneously [44]. Pfannkuche et al., who did a detailed electrophysiological characterization of mouse iPSC-CMs, reported that in the iPS cell line they studied the cells mainly differentiated toward ventricular-like cells. They also reported lack of ANP expression that could have reflected the low proportion of atrial-like cells in the total EB cell mass or delayed appearance of matured atrial cells in the course of ESC and iPS cell differentiation [46]. More studies are needed to gain a broader understanding of the mouse iPSC-CM phenotype. This issue in human iPSC-CMs was addressed comprehensively by Zhang et al., who performed sharp microelectrode recordings from spontaneously contracting EBs from two iPS cell clones (of fetal and newborn origin). Three major types of action potential were observed: ventricular, atrial, and nodal. Cells with ventricular-like action potentials were the most frequently encountered, and typically displayed a more negative maximum diastolic potential, a rapid action potential upstroke, and a distinct plateau phase. Atrial-like cells were distinguished from ventricular-like cells by the absence of a distinct plateau during repolarization and typically also exhibited spontaneous activity that was higher in frequency than that observed for ventricular cells. Finally, nodal-like cells were distinguished by maximum diastolic potentials that were less negative than those of ventricular- and atrial-like cells, smaller amplitude action potentials, a slower action potential upstroke, and a pronounced phase-4 depolarization preceding the action potential upstroke. Comparison of recordings from cardiomyocytes within the same EB revealed that a given action potential phenotype was predominant in each EB, as has previously been shown for hESC-derived EBs [47].

4.4 Ion Channel Expression

Voltage-clamp studies indicate that several cardiac-specific currents are present in hESC-CMs, including fast sodium current, L-type calcium current, pacemaker currents, as well as transient outward and inward rectifier potassium currents [33, 52, 53]. Immunostaining of mouse iPSC-CMs showed the expression of the hyperpolarization-activated cyclic-nucleotide-gated channel 4 (HCN4) and the T-type calcium channel Cav3.2, which were expressed in mouse sinoatrial node and were important for automaticity of ESC-CMs, in cTnT positive iPSC-CMs. The atrial and ventricular ion channel, Kir2.1, was also observed in these cells [44]. Application of different ion channel blockers (lidocaine, CsCl, SEA0400, and verapamil plus nifedipine) to plated iPS cell EBs during multielectrode array measurements of extracellular field potentials and intracellular sharp electrode recordings of action potentials revealed the presence of Na^+ channel current (I_{Na}), funny current (I_f), Na^+/Ca^{2+} exchanger current (I_{NCX}) and L-type Ca^{2+} channel current (I_{CaL}), respectively, and suggested their involvement in cardiac pacemaking. Since the blockage of L-type Ca^{2+} channels led to a full halt of beating, it was suggested it is of major importance for generation and uphold of spontaneous electrical activity of iPSC-CMs [46].

Tanaka et al. investigated the levels of transcripts for cardiac ion channel genes by RT-PCR in human iPSC-CMs. The gene for the sodium channel α-subunit SCN5A, which determines cardiac excitability and conduction velocity, was expressed in differentiated human iPS cells. The cells also expressed the genes for Cav1.2, which is an L-type Ca^{2+} channel α-subunit that contributes to cardiac contraction, and KCNH2 (hERG), which is a rapidly activating delayed rectifier K^+ channel (I_{Kr}) that contributes to action potential repolarization. The genes for these ion channels were not expressed in the undifferentiated iPS cells. Further, the functionality of the channels was examined using ion-channel-specific inhibitors during field potential recordings of beating EBs using a multielectrode array system. The dose-dependent changes to the field potential waveform indicated the expression of functional sodium channel (blocked with quinidine), calcium channel (blocked with verapamil), and the potassium channel I_{Kr} (blocked with E-4031) [48]. Similarly, Yokoo et al. showed by RT-PCR that these cells expressed the I_f channel (HCN4), the L-type calcium channel (Cav1.2), the sodium channel (SCN5A), the inward rectifier (Kir2.1), the transient outward channel (Kv4.3), and the delayed rectifier I_{Kr} (hERG). These authors explored the effects of different drugs, including ion channels blockers on beating frequency and contractility of the iPSC-CMs. The effects of sodium channel blockers (procaineamide, mexiletine, and flecainide), calcium channel blocker (verapamil), and potassium channel blocker (amiodarone) were broadly similar and occurred within the same range of drug concentrations as on hESC-CMs, as well as with clinical empirical results [49]. Collectively, these results suggest that the main cardiac depolarizing and repolarizing ion channels are functionally expressed in human iPSC-CMs.

4.5 Excitation–Contraction Coupling

Cardiomyocytes derived from ESCs are known to exhibit calcium transients. Nevertheless, considerable controversy exists regarding the mechanisms of excitation–contraction coupling in hESC-CMs. So far, only few data have been obtained regarding excitation–contraction coupling in iPSC-CMs. Mauritz et al. assessed spontaneous intracellular Ca^{2+} fluctuations in mouse iPSC-CMs. The cells displayed spontaneous rhythmic intracellular Ca^{2+} fluctuations with amplitudes of Ca^{2+} transients comparable to those of cardiomyocytes derived from mouse ESCs. Ca^{2+} concentration increased homogeneously throughout the cell, pointing to a finely regulated Ca^{2+} release from intracellular Ca^{2+} stores, most likely the sarcoplasmic reticulum. Simultaneous Ca^{2+} release within clusters of iPSC-CMs indicated functional coupling of the cells. Moreover, calcium release after caffeine treatment suggested stores of calcium in functional sarcoplasmic reticulum [43]. Schenke-Layland et al. also reported onset of intracellular Ca^{2+} transients that pointed to an intact cellular function with electrically triggered Ca^{2+} release [45]. As already mentioned, Pfannkuche et al. described the existence of the I_{NCX} in these cells [46]. Besides these data, the mRNA expression of ryanodine receptor 2 in mouse iPSC-CMs [46] and phosphlamban in human iPSC-CMs [47], both associated with calcium handling in cardiomyocytes, has been reported. Additional studies are needed to obtain comprehensive information about the excitation–contraction coupling of iPSC-CMs, particularly for human iPSC-CMs, an issue that may affect their integration in the regenerating heart.

4.6 β-Adrenergic and Muscarinic Signaling Pathways

Because one of the most critical determinants of normal cardiac electrophysiologic function is the intact response to hormones and transmitters of the central nervous system, and since ESCs respond to both adrenergic and cholinergic agents [31, 33], this issue was addressed by several studies which characterized iPSC-CMs. Studies with multielectrode arrays demonstrated functionality and the presence of the β-adrenergic and muscarinic signaling cascade in mouse iPSC-CMs, as the beating rates of the cells increased with isoproterenol and decreased with carbachol in a manner similar to that reported in ESCs [43, 46]. Similar responsiveness to β-adrenergic stimulation with isoproterenol manifested by an increase in spontaneous rate and a decrease in action potential duration was demonstrated in human iPSC-CMs [47, 48]. This was in addition to an increase in contractility of the cells in response to isoproteranol [49].

5 iPS Cells for Cardiac Cell Therapy

5.1 Cardiac Regeneration

The derivation of hESCs and their cardiomyocyte derivatives offers a number of advantages for myocardial cell replacement and tissue engineering strategies. hESCs can be isolated and maintained by well-established protocols, and are scalable. hESCs have unquestioned cardiomyogenic potential and can potentially provide ex vivo an unlimited number of human cardiomyocytes. hESCs can also be differentiated into noncardiac cell types present in the myocardium (e.g., endothelial cells), offering the possibility to repopulate all myocardial tissue elements, not just cardiomyocytes. Because of their inherent cardiac phenotype, hESC-CMs are more likely to achieve a functional connection with the host myocardium. The feasibility of using ESCs for myocardial regeneration has already been demonstrated in animal models. It was initially evaluated with the mouse ESC model, but several recent studies demonstrated that transplantation of hESC-CMs into uninjured hearts of immunocompromised mice, rats, and pigs results in the formation of stable grafts of human myocardium, with beneficial effects on left ventricular remodeling and function. Most studies used hESC derivatives containing a significant proportion of cardiomyocytes but also other cell types [18, 19, 22], and some have used hESCs pretreated with growth factor to be predisposed to form cardiac progenitors [21].

The therapeutic value of reprogramming remains largely unknown. In neurodegenerative diseases such as amyotrophic lateral sclerosis, Parkinson's disease, and spinal muscular atrophy [39, 54, 55], as well as in a variety of other genetic diseases [37], patient-specific iPS cells have already been established. To date, only three noncardiac disease models have been treated with iPS-cell-derived strategies – sickle cell anemia [56], Parkinson's disease [57], and hemophilia A [58]; all have been limited to lineages prespecified in vitro.

As described herein, the cardiac differentiation potential and the cardiomyogenic phenotype of both murine and human iPS cells were recently intensively studied, and although some findings still require further investigation, it seems that iPS cells hold all the aforementioned advantages as cell candidates for cardiac repair. This most recently led Nelson et al. [59] to expand the therapeutic indications of iPS cells by providing the first evidence for repair in the context of heart disease. In this study, mouse embryonic fibroblasts were transduced with human stemness factors Oct3/4, Sox2, Klf4, and c-Myc to generate valid iPS clones. The cardiac in vitro potential was demonstrated by the expression of precardiac mesoderm and cardiac transcription factors (Mesp1, Tbx5, Mef2c, GATA4, and myocardin) during EB differentiation. Furthermore, in utero, iPS-cell-derived chimera demonstrated de novo organogenesis and patterning of cardiogenic tissue within the developing embryo. The labeled iPS cell progeny demonstrated a robust contribution to the heart field, including cardiac inflow and outflow tracts as well as left and right ventricles of the embryonic heart parenchyma.

Table 2 Cardiac parameters in the model of acute myocardial infarction and cell transplantation

	Basis values		Fibroblasts	iPS cells
Parameter	Before infraction	1d after infraction	After 4 weeks	After 4 weeks
Ejection fraction (%)	82±3	38±3	37±4	50±5
Fractional shortening (%)	–	20±1	Lack of recovery	31±3
Systolic wall thickness (mm)	–	–	0.88±0.06	1.31±0.11
LVDd (mm)	3.2±0.1	–	4.9±0.1	4.2±0.2
QT interval (ms)	28.9±1.4	–	55.9±1.3	40.8±1.5

LVDd left ventricular end–diastolic dimension. Data was collected from [59]

The efficacy of iPS cells to treat acute myocardial infarction was evaluated in adult mice with left coronary artery ligation, in the context of immunocompetent allogeneic transplantation. The intramyocardial delivery of iPS cells (200,000 cells) yielded progeny that properly engrafted without disrupting the cytoarchitecture in the immunocompetent recipients. Monitored by echocardiography, irreversible occlusion of the epicardial coronary blood flow consistently impaired anterior wall motion, depressed global cardiac function, and halved the ejection fraction. Whereas blinded transplantation with parental fibroblasts demonstrated persistent functional decline, iPS cell intervention improved cardiac contractility, reflected by improved ejection fraction and fractional shortening (see Table 2 for details). Moreover, the regional septal wall thickness in systole was significantly rebuilt with iPS cells but not with fibroblasts. Impaired cardiac contractility in the injured anterior wall resulted in akinetic regions with paradoxical motion in systole indicative of aneurysms in fibroblast-treated hearts, in contrast to coordinated concentric contractions in response to iPS cell treatment visualized by long-axis and short-axis echocardiography imaging. In addition to the restoration of myocardial functional performance, iPS cell therapy halted the progression of the pathological remodeling in the infarcted hearts. iPS-cell-based intervention attenuated global left ventricular end-diastolic dimensions. Furthermore, echocardiography demonstrated regional structural deficits with deleterious wall thinning and chamber dilation in fibroblast-treated hearts, rescued by iPS intervention. The prolongation of the QT interval, which results from pathological structural remodeling and increases the risk of arrhythmia, was abrogated with iPS cell treatment. These real-time surrogates for tissue remodeling were confirmed on autopsy on inspection of a gross specimen that demonstrated reduced heart size, lack of aneurysmal formation, and absence of severe wall thinning in iPS-cell-treated hearts compared with fibroblast-treated hearts. Collectively, the favorable remodeling at global, regional, and electric levels demonstrates the overall benefit of iPS cell therapy in the setting of myocardial infarction. Thus, in contrast to parental nonreparative fibroblasts, iPS cell treatment restored postischemic contractile performance, ventricular wall thickness, and electric stability. It was demonstrated that the iPS cells contributed to tissue reconstruction with synchronized cardiovasculogenesis, composed predominantly of cardiac lineage with accompanying smooth muscle and endothelium [59].

This study was the first to show a proof-of-principle for the usage of iPS cells in the treatment of heart disease. It is important to note that iPS cells, rather than iPSC-CMs, were used in this study and achieved cardiac function improvement. The mechanisms underlying this improvement must be studied and the question of cell choice for transplantation must be further considered. More transplantation experiments with iPS cells and particularly with human iPSC-CMs will be needed to evaluate the practicability of these cells, whose limitations will be discussed in the following, for this crucial application.

5.2 Genetic Heart Diseases

Besides the potential role of iPSC-CMs for the replacement of lost cardiomyocytes through injury, iPS cells from patients with a genetic heart disease will help to develop new therapies based on genetic engineering. The potential for using iPS cells for the cure of genetic disorders was recently demonstrated in a mouse model of sickle cell anemia [56]. The cause of this hematopoietic disorder is a mutation in the hemoglobin gene leading to symptoms such as severe anemia due to the abnormal sicklelike morphology of the erythrocytes, splenic infarcts, etc. iPS cells were generated from humanized sickle cell mouse fibroblasts. The human sickle hemoglobin allele was corrected by homologous recombination. The modified iPS cells were differentiated in vitro into hematopoietic cells and transplanted into irradiated sickle cell mice to correct the disease. Thus, this work serves as a proof-of-principle for a therapy based on genetically engineered derivatives of iPS cells.

Many of the cardiac gene disorders are found in proteins that are important for force generation (cardiomyopathy) and electrical stimulation (channelopathy). One of the most common forms of cardiomyopathy, disease of the heart muscle, is hypertrophic cardiomyopathy, which is the most common inherited cardiovascular disorder, affecting one in 500 individuals in the general population. Patients with hypertrophic cardiomyopathy have an increased risk of arrhythmias, myocardial infarction, and sudden cardiac death [60]. The genetic cause underlying the disease is often a mutation in one of 11 different sarcomeric genes (e.g., actin, β-myosin heavy chain). Malfunction of the proteins results in sarcomere and myofibrillar disarray, changes in calcium sensitivity, and impaired contractile performance. Genetic alterations leading to disturbed functions of cardiac ion channels are referred to as cardiac channelopathy (reviewed in [61]). Cardiac channelopathies can be either congenital (e.g., mutations in encoding genes) or acquired (e.g., induced by electrolytes, drugs, genetic predisposition). The congenital form can be caused by mutations in several different genes. Disturbances of the depolarization–repolarization events by ion channel mutations are more accentuated by prolongation of the action potential, which is characterized by a long QT interval on an electrocardiogram. Patients with a long QT interval (long QT syndrome, LQTS) have an increased risk of arrhythmias, which may lead to sudden cardiac death. Most of the known mutations in ion channels are associated with LQTS. In particular, genes encoding for potassium, sodium, and calcium channels are most commonly affected. Replacement of the mutated gene by the functional sequence using homologous recombination will be the first step for therapy in such diseases. After genetic engineering of iPS cells in vitro, the repaired differentiated cardiomyocytes could be transplanted into the heart of the recipient. Gene therapies have been the subject of much clinical interest; however, random virus-mediated insertion of constructs rather than targeted integration has presented risks of oncogenic transformation, in addition to patients' strong immune response due to exposure to the viral vectors which prevent efficient gene delivery. However, iPS cells can be engineered by site-specific targeting ex vivo and tested prior to their transplantation, thus providing enormous progress in this therapeutic field.

6 iPS Cells for Human Disease Models

For genetic diseases, iPS cells derived from patients provide a new opportunity to analyze the pathways that lead to disease pathogenesis based on a particular genetic trait at the cellular level.

Over the years, many different animal models related to cardiac diseases, in the mouse in particular, have been generated and widely used as a genetic in vivo model. Although these models are important for our understanding of signaling pathways involved in the onset and progression of cardiac diseases, it is questionable how predictive these models are. Models were generated for both cardiomyopathies, such as the hypertrophic disease caused by mutations in the NFAT/calcineurin pathway [62], as well as channelopathies. For example, several mouse models now exist with gain-of-function and loss-of-function mutations of the SCN5A gene, which encodes a subunit of a cardiac sodium ion channel and the malfunction of which leads to LQTS (reviewed in [63]). Several mice models also exist for mutation in the KCNQ1 gene, encoding a subunit of the voltage-gated potassium channel, also associated with LQTS [64]. However, the models do not always recapitulate the phenotype seen in patients, a discrepancy that might be related to species differences or to the nature of the mutation causing the loss of function (point mutation vs. complete deletion). In addition, because the heart rates of humans and mice differ, the rapidly beating rodent heart may override the effect of arrhythmias which in a human heart could have severe consequences. The genetic background of mice can also influence the phenotype resulting from the mutations. Moreover, humans are, by definition, of mixed genetic background, resulting in phenotypical variations of genetic diseases, and indeed a single SCN5A mutation resulted in multiple rhythm disturbances within the same family [65]. Therefore, data from studies with transgenic mice on a single genetic background should be complemented by human models.

Pluripotent stem cells, both genetically modified hESCs and human iPS cells derived from patients, are candidates on which to base human models. Cardiac progenitors from adult human heart biopsies obtained during bypass surgery are alternatives, but with reduced availability. The recently improved protocols for genetic modification in hESCs will facilitate the generation of clinically relevant cardiac disease models. Yet, the advantage of human iPS cells derived from the patient suffering from a genetic disease lies in the fact that these stem cells do not need to undergo genetic modification. Moreover, the genetic disorder does not have to be known to generate an in vitro model and perform functional assays. In general, it still needs to be demonstrated that derivatives of iPS cells can indeed serve as in vitro models of the disease. Recently it has been shown that a human iPS cell model of spinal muscular atrophy can recapitulate at least some aspects of this genetic disorder [54]. For the future, it will be interesting to study mechanisms of disease in iPS cell lines from different patients carrying the same mutation to discover the effect of genetic background in humans. The diseased cardiomyocytes, perfectly matching the patient's genetics, could improve our understanding of genotype–phenotype correlations.

Furthermore, iPS cells represent promising targets for drug screening, particularly if the production of derivative cardiomyocytes can be scaled up. Disease-specific human cardiomyocytes may facilitate high-throughput screening of molecules for drug development. This is important, since it is known that developing effective drug therapies for arrhythmic diseases is hampered by the fact that the same drug has different effects depending on the patient. Patient-specific cells could thus become a useful tool for selecting the best drug to suit the individual patient, especially for lethal arrhythmic diseases such as LQTS where this task is often very difficult [49]. iPSC-CMs can also serve for screening of drugs for their cardiac toxicity, known to hamper the clinical use of many developed drugs. The lethal ventricular arrhythmia torsades de pointes, for instance, is the most common reason for the withdrawal or restricted use of many cardiovascular and noncardiovascular drugs [48]. Two studies have already been published demonstrating that current cardioactive substances affect the functional properties of human iPSC-CMs in much the same way as they affect hESC-CMs. Thus, human iPS cells could become attractive human models for cardiac drug toxicity testing and as drug screening and discovery platforms, even at the individual level [48, 49].

7 iPS Cells: Benefits and Limitations

7.1 Advantages

The reprogramming of adult cells to ESC-like pluripotent states is one of the most exciting advances in stem cell biology in the last decade and provides far-reaching possibilities. It enables the derivation of pluripotent cells from specific individuals, including patients with diseases of interest, without the requirement for donor oocytes. The potential for derivation of patient-matched cells for therapy will overcome one of the major concerns with hESC derivatives, which is immune rejection. This is in addition to the potential of creating pluripotent stem cells bearing disease traits. Since the derivation of these pluripotent cells does not require the utilization of human embryos, the legal and ethical issues still associated with hESC derivation in many countries can also be avoided. Although still derived only at low efficiency, iPS cells have now been generated from a range of different somatic cells, including human keratinocytes from hair and cells from bone marrow [36, 37, 66], thus increasing their accessibility. However multiple issues have to be addressed before this technology can be used for patient therapy.

7.2 *Usage of Viruses*

One of the first concerns in the utilization of iPS cells is the usages of viruses in the derivation of the first-generation iPS cells. Although the viruses are a potent tool to deliver genes into a target cell, they have the disadvantage of random integration into the host genome. This can lead to alterations in the expression of endogenous genes and increase the risk of malignant transformations. Recent progress indicates that c-Myc is not essential for iPS cell generation [67] and that iPS cells can be selected on the basis of morphology rather than the use of virally encoded antibiotic resistance genes [68]. This eliminates the number of transgenes needed, and the absence of c-Myc markedly reduced tumor incidence in iPS-cell-derived mice [67]. Depending on the expression levels of endogenous transcription factors, some cell types require even fewer reprogramming factors. Adult mouse neuronal stem cells, for example, already express Sox and c-Myc and Klf4 and only need additional Oct3/4 to be reprogrammed [69]; thus, the choice of the cells to be reprogrammed should also be considered. The most desirable scenario would be to achieve high-efficiency reprogramming using only transient expression of the relevant factors and without using any oncogenes. Work is under way in multiple laboratories to identify alternative reprogramming techniques such as cell-permeant protein reagents, nonintegrating viruses, and the use of specific small molecules. In the meantime, it is essential that investigators using iPS cells screen for expression of the virally encoded reprogramming factors.

7.3 *Teratoma Formation*

Incomplete downregulation of transgenes during differentiation may lead to iPS cell derivatives with an unstable or tumorigenic phenotype. It is therefore important to assess how fully pluripotent cells differentiate because residual undifferentiated cells can form teratomas. This was addressed in the study of Nelson et al. [59]. On permanent occlusion of epicardial coronary vasculature and microsurgical transfer into the ischemic myocardium, mouse iPS cells remained within injected hearts and produced gradual tumor outgrowth between 2 and 4 weeks. Autopsy in immunodeficient recipients verified consistent teratoma formation with extension beyond the myocardial wall and tumor infiltration within the postinjured myocardium. In contrast to tumorigenesis that compromised the safety within immunodeficient environments, subcutaneous transplantation of iPS cells into immunocompetent hosts demonstrated a persistent absence of tumor growth in all animals even at 8 weeks of follow-up. Furthermore, intramyocardial transplantation of iPS cells into the heart produced stable engraftment without detectable tumor formation. Differentiated iPS cells within ischemic immunocompetent hearts were detectable within 2 weeks after transplantation, without metastatic dissemination after 4 weeks of engraftment. Immunostaining of hearts at 4 weeks demonstrated rare iPS cell progeny positive for SSEA-1 expression within the postischemic myocardium. The researchers concluded that the immunocompetent adult host provided a permissive environment for differentiation, offering the opportunity to test the therapeutic potential of iPS cell clones [59]. These results are very encouraging but will have to be further evaluated by additional studies. Directing differentiation toward the desired cell type will maximize the yield of differentiated cells and may reduce the chances for tumor formation. Thus, as for hESCs, efforts should be made on the optimization of guided differentiation protocols with fully defined components that will provide highly purified preparations of derived cardiomyocytes, free of undifferentiated and undesirable noncardiac derivatives.

7.4 *Choice of Somatic Cells*

The ability to differentiate iPS cells efficiently into a particular cell type may be influenced by the choice of the somatic cell type used for reprogramming. Supporting this notion is the case in which the gene expression pattern of a somatic cell was maintained after reprogramming the nucleus to an embryonic-like state [70]. As a consequence, not only the efficiency of reprogramming of an individual somatic cell type and the accessibility as donated tissue but also the tissue origin of the cell used for reprogramming may require consideration for each particular application [73].

7.5 *Other Challenges Toward Regenerative Medicine*

Other challenges to the clinical application of iPSC-CMs are not unique to these cell preparations and are common to those raised for the clinical use of hESC-CMs (reviewed by Zhu et al. [72]). These include the need for host–graft electromechanical integration owing to the potential for arrhythmogenesis, issues concerning survival of transplanted cells, cell delivery, and tissue engineering.

7.6 *In Vitro Modeling*

Potential caveats to the use of mutated iPS cells or genetically modified hESCs as in vitro models for disease may be that many genetic diseases are only manifested in fully differentiated cells in later life (e.g., Parkinson's disease and amyotrophic lateral sclerosis) and stem cell derivatives are generally immature. hESC-CMs, for example, have sarcomeric organization resembling that of primary human fetal cardiomyocytes. The aspects of maturation of iPSC-CMs would have to be studied, and might also be affected by the source of the somatic cell.

8 Future Perspectives

A great advantage of the iPS cell technology will be the possibility to generate isogenic cardiomyocytes which are genetically equivalent to the cells of the recipient receiving the transplant (in contrast to allogenic transplants from an unrelated donor). Isogenic transplants do not require suppression of the immune system. By contrast, allogenic transplants are rejected because their human leukocyte antigens (HLA) trigger the response of the recipient's immune system. It is important to note that individualized therapy of this nature is unlikely to be cost-effective and that the time taken to derive and characterize iPS cells from one patient (about 4 months) is too long to treat anything but chronic ailments. In addition, like hESC lines, different iPS cell lines vary in their ability to differentiate into a certain cell type [47]. Thus, it may even be necessary to derive several iPS cell lines from the same patient and to test these for their differentiation potential. Applications in which a step of repairing a genetic defect is required will be even more complicated, time-consuming, and expensive. To provide suitable cardiomyocytes for a patient in reasonable time, a bank of hESC lines which cover the HLA types of most individuals of a population and which are prescreened for their ability to differentiate could be a better option. This could be complemented by banks of iPS cells for rare genotypes for which no hESC lines are available, in the way banked umbilical cord blood from rare genetic backgrounds presently supplements bone marrow donations in the treatment of hematopoietic diseases [73].

References

1. Schocken DD, Benjamin EJ, Fonarow GC et al (2008) Prevention of heart failure: a scientific statement from the American Heart Association Councils on Epidemiology and Prevention, Clinical Cardiology, Cardiovascular Nursing, and High Blood Pressure Research; Quality of Care and Outcomes Research Interdisciplinary Working Group; and Functional Genomics and Translational Biology Interdisciplinary Working Group. Circulation 117(19):2544–2565.
2. Rosamond W, Flegal K, Friday G et al (2007) Heart disease and stroke statistics: 2007 update: a report from the American Heart Association Statistics Committee and Stroke Statistics Subcommittee. Circulation 115:e69–e171.
3. Hunt SA (1998) Current status of cardiac transplantation. JAMA 280:1692.
4. Fukuda K, Yuasa S (2006) Stem cells as a source of regenerative cardiomyocytes. Circ Res 98:1002.
5. van Laake LW, Hassink R, Doevendans PA et al (2006) Heart repair and stem cells. J Physiol 577:467.

6. Murry CE, Wiseman RW, Schwartz SM et al (1996) Skeletal myoblast transplantation for repair of myocardial necrosis. J Clin Invest 98:2512–2523.
7. Taylor DA, Atkins BZ, Hungspreugs P et al (1998) Regenerating functional myocardium: improved performance after skeletal myoblast transplantation. Nat Med 4:929–933.
8. Menasche P, Hagege AA, Vilquin JT et al (2003) Autologous skeletal myoblast transplantation for severe postinfarction left ventricular dysfunction. J Am Coll Cardiol 41:1078–1083.
9. Orlic D, Kajstura J, Chimenti S et al (2001) Bone marrow cells regenerate infarcted myocardium. Nature 410:701–705.
10. Min JY, Sullivan MF, Yang Y et al (2002) Significant improvement of heart function by cotransplantation of human mesenchymal stem cells and fetal cardiomyocytes in postinfarcted pigs. Ann Thorac Surg 74:1568–1575.
11. Shake JG, Gruber PJ, Baumgartner WA et al (2002) Mesenchymal stem cell implantation in a swine myocardial infarct model: engraftment and functional effects. Ann Thorac Surg 73:1919–1925.
12. Toma C, Pittenger MF, Cahill KS et al (2002) Human mesenchymal stem cells differentiate to a cardiomyocyte phenotype in the adult murine heart. Circulation 105:93–98.
13. Mangi AA, Noiseux N, Kong D et al (2003) Mesenchymal stem cells modified with Akt prevent remodeling and restore performance of infarcted hearts. Nat Med 9:1195–1201.
14. Beltrami AP, Barlucchi L, Torella D et al (2003) Adult cardiac stem cells are multipotent and support myocardial regeneration. Cell 114:763–776.
15. Oh H, Bradfute SB, Gallardo TD et al (2003) Cardiac progenitor cells from adult myocardium: homing, differentiation, and fusion after infarction. Proc Natl Acad Sci USA 100:12313–12318.
16. Laugwitz KL, Morretti A, Lam J et al (2005) Postnatal isl1+ cardioblasts enter fully differentiated cardiomyocyte lineages. Nature 433:647–653.
17. Smith RR, Barile L, Cho HC et al (2007) Regenerative potential of cardiosphere-derived cells expanded from percutaneous endomyocardial biopsy specimens. Circulation 115:896–908.
18. Caspi O, Huber I, Kehat I et al (2007) Transplantation of human embryonic stem cell-derived cardiomyocytes improves myocardial performance in infarcted rat hearts. J Am Coll Cardiol 50:1884–1893.
19. Laflamme MA, Chen KY, Naumova AV et al (2007) Cardiomyocytes derived from human embryonic stem cells in pro-survival factors enhance function of infarcted rat hearts. Nat Biotechnol 25:1015–1024.
20. Leor J, Gerecht S, Cohen S et al (2007) Human embryonic stem cell transplantation to repair the infarcted myocardium. Heart 93:1278–1284.
21. Tomescot A, Leschik J, Bellamy V et al (2007) Differentiation in vivo of cardiac committed human embryonic stem cells in postmyocardial infarcted rats. Stem Cells 25:2200–2205.
22. van Laake LW, Passier R, Doevendans PA et al (2007) Human embryonic stem cell-derived cardiomyocytes survive and mature in the mouse heart and transiently improve function after myocardial infarction. Stem Cell Res 1:9–24.
23. Janssens S, Dubois C, Bogaert J et al (2006) Autologous bone marrow-derived stem-cell transfer in patients with ST-segment elevation myocardial infarction: double-blind, randomised controlled trial. Lancet 367:113–121.
24. Lunde K, Solheim S, Aakhus S et al (2006) Intracoronary injection of mononuclear bone marrow cells in acute myocardial infarction. N Engl J Med 355:1199–1209.
25. Meyer GP, Wollert KC, Lotz J et al (2006) Intracoronary bone marrow cell transfer after myocardial infarction: eighteen months' follow-up data from the randomized, controlled BOOST (BOne marrOw transfer to enhance ST-elevation infarct regeneration) trial. Circulation 113:1287–1294.
26. Schachinger V, Erbs S, Elsasser A et al (2006) Intracoronary bone marrow-derived progenitor cells in acute myocardial infarction. N Engl J Med 355:1210–1221.
27. Abdel-Latif A, Bolli R, Tleyjeh IM et al (2007) Adult bone marrow-derived cells for cardiac repair: a systematic review and meta-analysis. Arch Intern Med 167:989–997.
28. Menasche P (2007) Skeletal myoblasts as a therapeutic agent. Prog Cardiovasc Dis 50:7–17.
29. Schuleri KH, Boyle AJ, Hare JM (2007) Mesenchymal stem cells for cardiac regenerative therapy. Handb Exp Pharmacol (180):195–218.

30. Thomson JA, Itskovitz-Eldor J, Shapiro SS et al (1998) Embryonic stem cell lines derived from human blastocysts. Science 282(5391):1145–1147.
31. Kehat I, Kenyagin-Karsenti D, Snir M et al (2001) Human embryonic stem cells can differentiate into myocytes with structural and functional properties of cardiomyocytes. J Clin Invest 108:407–414.
32. Xu C, Police S, Rao N et al (2002) Characterization and enrichment of cardiomyocytes derived from human embryonic stem cells. Circ Res 91:501–508.
33. Mummery C, Ward-van Oostwaard D, Doevendans P et al (2003) Differentiation of human embryonic stem cells to cardiomyocytes: role of coculture with visceral endoderm-like cells. Circulation 107:2733–2740.
34. Takahashi K, Yamanaka S (2006) Induction of pluripotent stem cells from mouse embryonic and adult fibroblast cultures by defined factors. Cell 126:663–676.
35. Takahashi K, Tanabe K, Ohnuki M et al (2007) Induction of pluripotent stem cells from adult human fibroblasts by defined factors. Cell 131:861–872.
36. Park IH, Zhao R, West JA et al (2008) Reprogramming of human somatic cells to pluripotency with defined factors. Nature 451:141–146.
37. Park IH, Arora N, Huo H et al (2008) Disease-specific induced pluripotent stem cells. Cell 134:877–886.
38. Yu J, Vodyanik MA, Smuga-Otto K et al (2007) Induced pluripotent stem cell lines derived from human somatic cells. Science 318:1917–1920.
39. Dimos JT, Rodolfa KT, Niakan KK et al (2008) Induced pluripotent stem cells generated from patients with ALS can be differentiated into motor neurons. Science 321:1218–1221.
40. Okita K, Ichisaka T, Yamanaka S (2007) Generation of germline-competent induced pluripotent stem cells. Nature 448:313–317.
41. Wernig M, Meissner A, Foreman R et al (2007) In vitro reprogramming of fibroblasts into a pluripotent ES-cell-like state. Nature 448:318–324.
42. Yang L, Soonpaa MH, Adler ED et al (2008) Human cardiovascular progenitor cells develop from a KDR+ embryonic-stem-cell-derived population. Nature 453(7194):524–528.
43. Mauritz C, Schwanke K, Reppel M et al (2008) Generation of functional murine cardiac myocytes from induced pluripotent stem cells. Circulation 118:507–517.
44. Narazaki G, Uosaki H, Teranishi M et al (2008) Directed and systematic differentiation of cardiovascular cells from mouse induced pluripotent stem cells. Circulation 118:498–506.
45. Schenke-Layland K, Rhodes KE, Angelis E et al (2008) Reprogrammed mouse fibroblasts differentiate into cells of the cardiovascular and hematopoietic lineages. Stem Cells 26:1537–1346.
46. Pfannkuche K, Liang H, Hannes T et al (2009) Cardiac myocytes derived from murine reprogrammed fibroblasts: intact hormonal regulation, cardiac ion channel expression and development of contractility. Cell Physiol Biochem 24(1–2):73–86.
47. Zhang J, Wilson GF, Soerens AG et al (2009) Functional cardiomyocytes derived from human induced pluripotent stem cells. Circ Res 104(4):e30–e41.
48. Tanaka T, Tohyama S, Murata M et al (2009) In vitro pharmacologic testing using human induced pluripotent stem cell-derived cardiomyocytes. Biochem Biophys Res Commun 385(4):497–502.
49. Yokoo N, Baba S, Kaichi S et al (2009) The effects of cardioactive drugs on cardiomyocytes derived from human induced pluripotent stem cells. Biochem Biophys Res Commun 387(3):482–488.
50. Maltsev VA, Rohwedel J, Hescheler J et al (1993) Embryonic stem cells differentiate in vitro into cardiomyocytes representing sinusnodal, atrial and ventricular cell types. Mech Dev 44:41–50.
51. He JQ, Ma Y, Lee Y et al (2003) Human embryonic stem cells develop into multiple types of cardiac myocytes: action potential characterization. Circ Res 93:32–39.
52. Satin J, Kehat I, Caspi O et al (2004) Mechanism of spontaneous excitability in human embryonic stem cell derived cardiomyocytes. J Physiol 559:479–496.
53. Sartiani L, Bettiol E, Stillitano F et al (2007) Developmental changes in cardiomyocytes differentiated from human embryonic stem cells: a molecular and electrophysiological approach. Stem Cells 25:1136–1144.

54. Ebert AD, Yu J, Rose Jr FF et al (2009) Induced pluripotent stem cells from a spinal muscular atrophy patient. Nature 457:277–280.
55. Soldner F, Hockemeyer D, Beard C et al (2009) Parkinson's disease patient-derived induced pluripotent stem cells free of viral reprogramming factors. Cell 136:964–977.
56. Hanna J, Wernig M, Markoulaki S et al (2007) Treatment of sickle cell anemia mouse model with iPS cells generated from autologous skin. Science 318:1920 –1923.
57. Wernig M, Zhao JP, Pruszak J et al (2008) Neurons derived from reprogrammed fibroblasts functionally integrate into the fetal brain and improve symptoms of rats with Parkinson's disease. Proc Natl Acad Sci U S A 105:5856–5861.
58. Xu D, Alipio Z, Fink LM et al (2009) Phenotypic correction of murine hemophilia A using an iPS cell-based therapy. Proc Natl Acad Sci U S A 106:808–813.
59. Nelson TJ, Martinez-Fernandez A, Yamada S et al (2009) Repair of acute myocardial infarction by human stemness factors induced pluripotent stem cells. Circulation 120(5):408–416.
60. Soor GS, Luk A, Ahn E, Abraham JR et al (2009) Hypertrophic cardiomyopathy: current understanding and treatment objectives. J Clin Pathol 62:226–235.
61. Marban E (2002) Cardiac channelopathies. Nature 415:213–218.
62. Bourajjaj M, Armand AS, da Costa Martins PA et al (2008) NFATc2 is a necessary mediator of calcineurin-dependent cardiac hypertrophy and heart failure. J Biol Chem 283(32):22295–22303.
63. Charpentier F, Bourge A, Merot J (2008) Mouse models of SCN5A-related cardiac arrhythmias. Prog Biophys Mol Biol 98(2–3):230–237.
64. Casimiro MC, Knollmann BC, Yamoah EN et al (2004) Targeted point mutagenesis of mouse Kcnq1: phenotypic analysis of mice with point mutations that cause Romano-Ward syndrome in humans. Genomics 84:555–564.
65. Remme CA, Wilde AA, Bezzina CR (2008) Cardiac sodium channel overlap syndromes: different faces of SCN5A mutations. Trends Cardiovasc Med 18:78–87.
66. Aasen T, Raya A, Barrero MJ et al (2008) Efficient and rapid generation of induced pluripotent stem cells from human keratinocytes. Nat Biotechnol 26:1276–1284.
67. Nakagawa M, Koyanagi M, Tanabe K et al (2008) Generation of induced pluripotent stem cells without Myc from mouse and human fibroblasts. Nat Biotechnol 26:101–106.
68. Meissner A, Wernig M, Jaenisch R (2007) Direct reprogramming of genetically unmodified fibroblasts into pluripotent stem cells. Nat Biotechnol 25:1177–1181.
69. Kim JB, Sebastiano V, Wu G et al (2009) Oct4-induced pluripotency in adult neural stem cells. Cell 136:411–419.
70. Ng RK, Gurdon JB (2008) Epigenetic memory of an active gene state depends on histone H3.3 incorporation into chromatin in the absence of transcription. Nat Cell Biol 10:102–109.
71. Nir SG, David R, Zaruba M et al (2003) Human embryonic stem cells for cardiovascular repair. Cardiovasc Res 58:313.
72. Zhu WZ, Hauch KD, Xu C et al (2009) Human embryonic stem cells and cardiac repair. Transplant Rev 23(1):53–68.
73. Freund C, Mummery CL (2009) Prospects for pluripotent stem cell-derived cardiomyocytes in cardiac cell therapy and as disease models. J Cell Biochem 107(4):592–599.

Induced Pluripotent Cells for Myocardial Infarction Repair

Timothy J. Nelson and Andre Terzic

Abstract Managing the rapidly expanding scope of chronic degenerative heart disease is a major clinical challenge. Owing to progressive cellular destruction and loss of functional tissues, chronic degenerative diseases are responsible for many of the disabilities suffered throughout lifespan. This creates an ever-growing need for new therapies capable of repairing the underlying pathophysiologic changes and restoring native cellular architecture. The emergence of regenerative medicine expands the therapeutic armamentarium, establishing new paradigms to address disease management unmet by traditional strategies. Stem-cell-based regenerative medicine drives the evolution of medical sciences from palliation, which mitigates symptoms, to curative therapy aimed at treating the root cause of degenerative disease and thus promoting long-term wellness.

With the optimization of acute hospital care, more individuals are surviving life-threatening myocardial infarction and deferring the burden of ischemic heart disease as they accumulate incremental insults leading to progressive heart failure. The incidence of long-term complications of inadequate cardiac performance is increasingly recognized as an unmet need within health care systems. An attractive alternative to address this expanding challenge has been documented with stem cell therapeutics. Specialized stem cell populations demonstrate a unique aptitude to differentiate into cardiac progenitors, and form new tissue. Cell-based strategies that promote, augment, and reestablish repair are at the core of translating the science of stem cell biology into the practice of cardiovascular regenerative medicine. Here, the latest evolution of stem-cell-based therapy, with emphasis on a bioengineered pluripotent stem cell platform, is outlined from discovery science to potential clinical use in the setting of ischemic heart disease.

A. Terzic (✉)
Marriott Heart Disease Research Program, Division of Cardiovascular Diseases, Department of Medicine, Molecular Pharmacology and Experimental Therapeutics and Medical Genetics, Mayo Clinic, Rochester, MN 55905, USA
e-mail: terzic.andre@mayo.edu

I.S. Cohen and G.R. Gaudette (eds.), *Regenerating the Heart*, Stem Cell Biology and Regenerative Medicine, DOI 10.1007/978-1-61779-021-8_15,
© Springer Science+Business Media, LLC 2011

Keywords Inducible pluripotent stem cells • Cell therapy • Myocardial infarction • Nuclear reprogramming • Differentiation • Myocardial ischemia

1 Tools and Goals of Cardiovascular Regenerative Medicine

1.1 Source of Progenitor Cells

Stem cell sources are present throughout lifespan [90–94]. Naturally derived stem cells include embryonic stem cells (ESCs), perinatal stem cells (e.g. umbilical-cord-derived stem cells), and adult stem cells [1]. Collectively, these sources provide the foundation for cardiovascular development and self-repair from in utero stages throughout adulthood [2–4, 86–89, 95]. However, the supply of cardiovascular stem cells is limited in many disease conditions such as ischemic heart disease that results in severe tissue damage and loss of functioning tissue. Augmentation of regenerative capacity in diseased hearts with exogenous stem cells, such as adult bone marrow derived stem cells, has been advanced to early clinical stage testing.

ESCs are present only within the inner cell mass of preimplantation blastocyst during normal development [5–7]. These pluripotent stem cells are capable of giving rise to all tissues, including the complete spectrum of mesoderm, endoderm, and ectoderm derivatives that are the building blocks of the whole adult body [8]. These quintessential stem cell types have been revered as a promising platform to produce a large quantity of unlimited transplantable cell types, despite the logistical limitations of requiring nonself, embryonic tissue sources [9–15].

Perinatal stem cells derived from umbilical cord blood (UCB) offer another available platform for regenerative tissue [16]. Transplantation of UCB has been clinically successful for hematopoietic stem cell applications, resulting in a high degree of engraftment, favorable immunotolerance, and limited evidence for graft-versus-host disease compared with adult bone marrow stem cell transplantation [17]. Beyond hematology, UCB-derived stem cells are capable of in vitro expansion, long-term maintenance, and differentiation into limited cells of all three embryonic germinal layers [18]. These dynamic stem cells, despite lacking universal differentiation potential, create an opportunity to generate multilineage stem cells from a readily available primitive tissue source.

Adult stem cells derived from bone marrow, circulation, or tissue-specific progenitor cells have been developed for clinical applications for decades [19]. Adult stem cells have been a practical source for regenerative applications owing to their accessibility, autologous status, and favorable proliferative potential. Although there is increased understanding of the regenerative potential, current experience primarily relies on bone-marrow-derived stem cells for hematological conditions, similarly to UCB, given the ability to recapitulate the entire hematopoietic system [20]. Mesenchymal stem cells, a specialized subpopulation of adult stem cells, are well known for their nonhematopoietic differentiation potential and have provided a cornerstone of contemporary regenerative medicine applications [21–25]. In fact, both autologous and allogeneic mesenchymal stem cells have been tested in recent clinical trials, including for treatment of multiple diseases [25]. Presently, the use of autologous mesenchymal stem cell therapy in the management of myocardial infarction has been advanced to early-stage clinical trials. Derivation and characterization of mesenchymal stem cells, and their application to specific disease conditions is an emerging strategy for enhanced therapeutic outcome.

Beyond natural sources, the newest platform is bioengineered stem cells that offer the ability to provide an unlimited supply of progenitor cells at any time point for virtually all cell type and tissue of the adult body starting from ordinary self-derived tissues (Fig. 1). By exploiting epigenetics, biotechnology platforms aim to reverse cell fate of common cell types that are readily available to achieve nuclear reprogramming and conversion to the embryonic ground state [26, 27]. Such platforms bypass the need for embryo extraction to generate true pluripotent stem cell phenotypes from autologous sources [28–30]. In the mouse, such an approach has yielded induced pluripotent stem (iPS) cells sufficient for complete de novo embryogenesis and in humans has given rise to all three germ layers that are required for comprehensive tissue differentiation. These self-derived iPS cell lines should largely eliminate the concern of immune rejection, but require a new

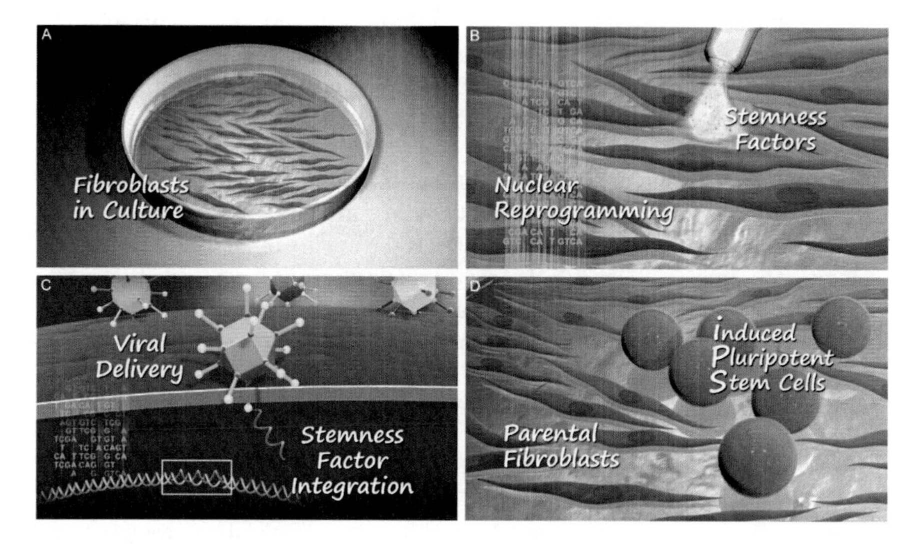

Fig. 1 Bioengineering induced pluripotent stem (iPS) cells. (**a**) Diverse somatic cell types can provide the parental source for nuclear reprogramming, including fibroblasts. (**b**) Various techniques for nuclear reprogramming have been established, including transduction with infectious particles carrying stemness factors. (**c**) Delivery of stemness factors to parental somatic cells leads to stable integration into the host genome. (**d**) Nuclear reprogramming induces metamorphosis of ordinary cells to yield bioengineered pluripotent stem cells

level of protection from dysregulated growth owing to the autologous pluripotent status. Moreover, iPS-cell-based technology will facilitate the production of cell line biobanks that closely reflect the genetic diversity of a population, enabling the discovery, development, and validation of therapies tailored for each individual. Alleviating technical limitations, including mutation through viral integration, incorporation of oncogenic genes, and lack of reliable differentiation protocols, is required prior to applications in clinical transplantation medicine [31]. The next generation of bioengineered stem cells will likely include specialized properties to improve stress tolerance, streamline differentiation capacity, and increase engraftment/survival to improve the regenerative potential for ischemic heart disease.

2 Restore Structure and Function Within Diseased Hearts

Stem cells are the active ingredient of a regenerative regimen that ultimately responds to injured tissue and contributes to effective wound healing [32–34]. Stem cells maintain an autonomous self-renewal potential to expand within appropriate environments, and retain the ability to differentiate into replacement tissues upon exposure to tissue-specific signals [35]. By healing postischemic cardiac injury, stem cells restore structure and function of damaged heart tissue. Thus, cardiovascular regeneration offers a sustained therapeutic advantage for ischemic heart disease [36]. However, the outcome of cell-based therapies depends on the aptitude of the stem cell population to secure maximal, tissue-specific differentiation, and the production of a nurturing niche environment that enables therapeutic repair.

Cardiac structure and function is the product of multidimensional variables that depends on three principal components: heart muscle, vascular supply, and electrical coordination. Ischemic injury affects the vitality of all three components and creates a complex deficiency that requires multilineage repair to restore native structure and function. Therefore, having stem cells or tissue-specific progenitor cells that are sufficient for multilineage differentiation is a prerequisite for de novo regeneration and comprehensive repair [37]. Beyond the tissue-specific lineages that must work in synchrony, paracrine mediators are actively involved with maintaining cardiac tissue homeostasis and promoting endogenous healing of native cardiac tissue [38]. Together these integrated mechanisms require stem cell recruitment, engraftment, expansion, differentiation, and physiological regulation within the postischemic host tissue [39]. Therefore, the goal of structural and functional restoration is dependent on the tissue-specific capacity of stem cells, the timing of delivery, and the responsiveness of the progenitor cells to the host environment [40]. Ultimately, coordinated tissue repair by replacement, regeneration, and/or rejuvenation of damaged cardiac tissues enables improvements in cardiac structure and function.

3 Bioengineering Autologous Pluripotent Stem Cells

3.1 *Traditional Stem-Cell-Based Therapeutics: Bench to Bedside*

Stem cells and their natural or engineered products, collectively known as "biologics," provide the functional components of a regenerative therapeutic regimen [34]. Stem cells have a unique ability to differentiate into specialized cell types and form new tissue [34, 41]. The type of stem cell used must have the capacity to differentiate into target tissues in vivo, engraft within the disease organ, and survive the host environment. Cell-based therapy includes autologous and allogeneic strategies.

"Autologous" refers to transplantation of tissues, cells, or proteins from one part of a patient's body to another, whereas "allogeneic" refers to transplantations between genetically distinct individuals, albeit belonging to the same species. Because autologous stem cells are derived from self, these cells avoid immune intolerance. Applications for autologous stem cells are typically limited to chronic or subacute conditions, given the time required to recycle stem cells from patients through the stages of mobilization, collection, expansion, and preparation for transplantation back to the same patient. The advantages of autologous strategies is the avoidance of immune intolerance, tissue rejection, and graft-versus-host disease as the transplanted tissue is recognized as "self." Alternatively, allogeneic stem cells are derived from a donor who is different from the recipient. Stem cells from multiple donors can be obtained independently, and stored as a tissue biorepository. In addition to the ability to stockpile on-demand therapeutics, allogeneic tissue offers unique advantages when a patient has a genetically based disease that hinders the therapeutic potential of his/her autologous stem cells. However, graft-versus-host disease becomes a major limitation for this type of transplantation in which engrafted stem cells produce progeny that attack the host recognized as foreign or nonself tissues. These distinct strategies are important to consider in the design of clinical applications of stem cell technology.

Cell-based medicinal products are commonly heterogeneous with respect to cell origin, cell type, and complexity of the final product. Clinical development of new applications mandates pharmacodynamic and pharmacokinetic documentation of all cell-based products, along with clinical efficacy and safety studies [42, 43]. Suitable markers of biological activity are needed to adequately identify primary pharmacodynamic properties, whereas function tests are required to demonstrate that tissue function is restored, even if the mechanism of action is only partially understood. In addition, establishing the optimal amounts of cell-based medicinal products needed to achieve the desired effects is a critical component of proving overall efficacy within the therapeutic window. Pharmacokinetic considerations include parameters of cell biodistribution, cell viability, and cell proliferation after administration of single or multiple doses. Clinical studies are designed to provide an adequate demonstration of feasibility and efficacy in the target patient population, to demonstrate an appropriate dose schedule for optimal therapeutic effect, and to evaluate the duration of therapeutic effect for risk–benefit assessment. Clinical safety databases are required to annotate adverse events, including procedural risk, immune response, infection, malignant transformation, and long-term safety [42, 43].

Cell-based medicinal products often involve cell samples of limited amount, mostly to be used in a patient-specific manner. This raises specific issues pertaining to quality-control testing designed for each product under examination. The manufacture of cell-based medicinal products is designed and validated to ensure product consistency and traceability [42]. Control and management of manufacturing and quality-control testing are carried out according to good manufacturing practice requirements [44]. Screening for purity, screening for potency, screening for infectious contamination, and screening for karyotype stability have become necessary elements, i.e. release criteria, in compliance with standard operating practices for production and banking of cells used as autologous or "off-the-shelf" allogeneic therapy. Accordingly, the US Food and Drug Administration and the European Medicines Agency impose regulatory guidelines for risk assessment, quality of manufacturing, preclinical and clinical development, and postmarketing surveillance pharmacovigilance of stem cell biologics for translation from bench to bedside to populations [40, 41].

3.2 Nuclear Reprogramming Self-Derived Stem Cells: Emerging Platform

Nuclear reprogramming through ectopic gene expression offers a revolutionary framework to derive truly pluripotent stem cells from somatic biopsy, independently of any embryo source [45, 46]. The emerging technology demonstrates that transient expression of stemness-related genes is sufficient to reset parental cell fate, unlocking the potential of unlimited patient-specific regenerative therapies [47–51]. Nuclear reprogramming technology enables access to a genuine, autologous, pluripotent cell population that is available for design of therapeutic applications [1]. Recognition that the genetic makeup of "terminally differentiated" tissue is fully reversible to the pluripotent ground state paved the way toward this embryo-independent bioengineering strategy that could be, in principle, applied to tissue-specific personalized regenerative solutions.

Transcription factors sets, Oct4, Sox2, c-Myc, and Klf4 [45] or alternatively Oct4, Sox2, Nanog, and Lin28 [48], were originally demonstrated to reprogram somatic cells to pluripotent stem cells that exhibit the essential characteristics of ESCs, including maintenance of an unlimited developmental potential to give rise to all cell types and tissues of the adult body (Fig. 1). This technology has advanced by matching the minimal requirements of ectopic transgenes for the target tissue, and has reduced the number of transgenes to a simple Oct3/4 factor in selected cell types [52, 53]. The process of nuclear reprogramming requires controlled expression of stemness factors in the proper stoichiometry for a transient time period [54]. The balanced exposure of ectopic factors is sufficient to induce a cascade of mechanisms that lead to telomere elongation [55], histone modifications [56], secondary gene expression profiles [57], and cellular metamorphosis that collectively reestablish a self-stabilizing phenotype of stemness [39]. This process occurs typically within weeks of coerced equilibrium of the *trans*-acting factors that can be delivered to the nucleus by plasmids, viruses, or bioengineered proteins. Thereby, ectopic transgene expression initiates a sequence of stochastic events that eventually transform a small fraction of cells (less than 0.5%) to acquire this reversible pluripotent state characterized by a stable epigenetic environment, essentially indistinguishable from blastocyst-derived natural ESCs. Because the retroviral-based vector systems originally used have built-in sequences that silence the process of transcription upon pluripotent induction, persistent exposure to ectopic gene expression is temporally restricted, indicating that successful self-maintenance of the pluripotent state is possible without long-term transgene expression. This concept of autonomous self-renewal within the pluripotent ground state was more recently definitively confirmed with iPS cells that had been produced with traceless approaches that use "cut-and-paste" features of transposon/transposase interactions to remove all transgenic sequences [58, 59]. Collectively, these pioneering studies pave the way for stem cell platforms that enable production, for the first time, of autologous pluripotent stem cells.

Two models have been used to explain the reprogramming process: the elite model, in which a small number of partially preprogrammed progenitor cells are able to respond to additional stemness factors, and the stochastic model, in which virtually any ordinary cell type can be fully reprogrammed in the presence of the proper combination of conditions depending on both the nature and the environment of the target cell [60]. Although both are plausible models that have support from experimental data, the stochastic model received additional credibility when evidence was recently presented that bona fide mature tissue such as B lymphocytes are capable of dedifferentiating into stable pluripotent stem cells [61]. This supports the model that cell fate is indeed fully reversible, even from mature tissue sources, upon exposure to the proper intracellular and extracellular stimuli.

4 Cardiogenic Potential of iPS Cells

4.1 Ground-State Pluripotency Gives Way to Tissue-Specific Differentiation

Discovery of induced pluripotency was achieved through exposure of discrete sets of qualified pluripotent genes that reset the phenotypic fate of an ordinary parental source and promoted reacquisition of embryonic traits, such as germline transmission, previously believed to be unique to ESCs [46]. The resulting iPS cells have met multiple levels of pluripotent stringency with anachronic cellular features that recapitulated primitive morphological marks of stem cells, expressed biomarkers consistent with authentic pluripotency, produced teratoma in vivo, contributed to chimeric offspring, and gave rise to germline transmission or completely iPS cell derived embryos using tetraploid aggregation [62, 63]. To date, therapeutic benefit of iPS-cell-based technology has been tested in four disease models, namely, sickle cell anemia [64], Parkinson's disease [65], hemophilia A [66], and ischemic heart disease [39]. As differentiation protocols are refined to produce "on-demand" tissue-specific progeny, additional preclinical disease models will be screened to address the full regenerative value of iPS cell technology.

4.2 Cardiac Tissue Generated from iPS Cell Sources

In regard to cardiogenesis, ESCs have a spontaneous propensity for cardiac differentiation that fulfills an early requirement for heart formation during embryonic development. Compared with the gold standard of ESCs, iPS cells have demonstrated a similar capacity for in vitro cardiac differentiation. Using methodology established for ESC-derived cardiogenesis, iPS cells differentiating in embryoid bodies or aggregates of tissue starting with 400–500 cells systematically produce mesoderm lineages and precardiac cytotypes according to gene expression profiles [67]. Within appropriate timeframes, mouse and human tissue give rise to early cardiomyocytes with spontaneous beating activity (Fig. 2). This tissue expresses contractile proteins, such as troponin and actinin. Furthermore, the cardiac-like tissue is regulated according to excitable inputs through gap junctions, and calcium from extracellular sources and intracellular depots. As the cardiac tissue matures in vitro,

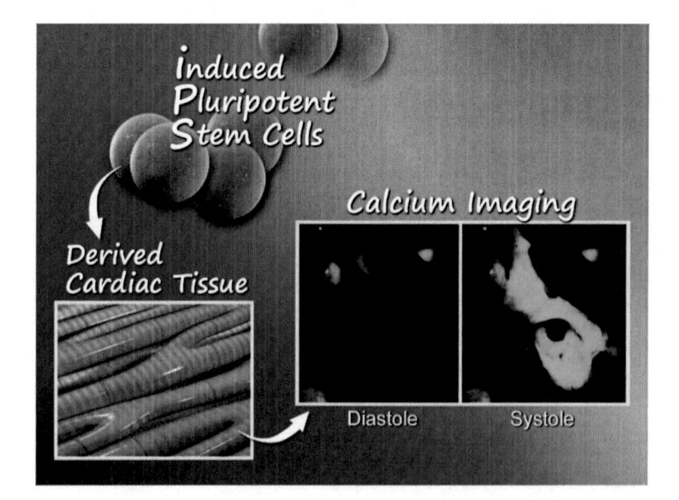

Fig. 2 Functional cardiac tissue derived from iPS cells. iPS cells have demonstrated cardiogenic capacity and provide a source for new heart muscle tissue harboring calcium-dependent contractile sarcomeres and functional excitation–contraction machinery

specialized heart muscle cells become evident with assembly of characteristic ion channel sets responsible for physiological regulation of cardiac contractions and electrical conductance within ventricular, atrial, and pacemaker cell types [68–71].

5 IPS-Cell-Derived Repair Following Myocardial Infarction

5.1 Environment of Postischemic Myocardium

Acute myocardial infarction is the sudden loss of oxygenated blood flow to the heart muscle leading to cell death. This event rapidly leads to a cascade of signaling pathways with cytokines and growth factors that aim to isolate the damaged area, prevent bleeding or tissue breakdown, and recruit circulatory cell types to clean and repair the myocardial wound. The beginning of irreversible tissue damage is produced after 20 min without oxygenated blood flow; between 60 and 180 min there is a progressive loss of tissue that spreads from the endocardial (inside) to the epicardial (outside) surface. Within days following myocardial infarction, ensuing inflammatory changes are characterized by the accumulation of mast cells, macrophages, and neutrophils, the subsequent degradation of the extracellular matrix, and formation of granulation tissue. This inflammatory process also stimulates neoangiogenesis, important for the maintenance and survival of existing cardiomyocytes. Following myocardial infarction, the cytokines that are released from the ischemic tissue into the circulation are used for the recruitment of regenerative cell types [72]. Principal cytokines and growth factors such as stromal-derived factor-1, granulocyte colony-stimulating factor, erythropoietin, vascular endothelial growth factor, hepatic growth factor, and stem cell factor (c-kit ligand) have been associated with the ongoing process of innate regeneration following myocardial infarction.

6 Cardiovascular Tissue Regeneration Within Postischemic Hearts

The therapeutic value of iPS cells requires optimization of tissue-specific regeneration in response to injury in regard to remuscularization and vascularization leading to restoration of damaged cardiac structure and function (Fig. 3). Recently, postischemic cardiac performance was compared in randomized cohorts into which parental fibroblasts versus iPS cells had been transplanted [39]. As quantified by echocardiography, occlusion of anterior epicardial coronary blood flow permanently impaired regional wall motion and cardiac function, despite transplantation with parental fibroblasts. In contrast, iPS cell intervention in the acute stages of myocardial infarction improved cardiac contractility by 4 weeks after transplantation. Functional benefit in response to iPS cell therapy was verified by the improvement

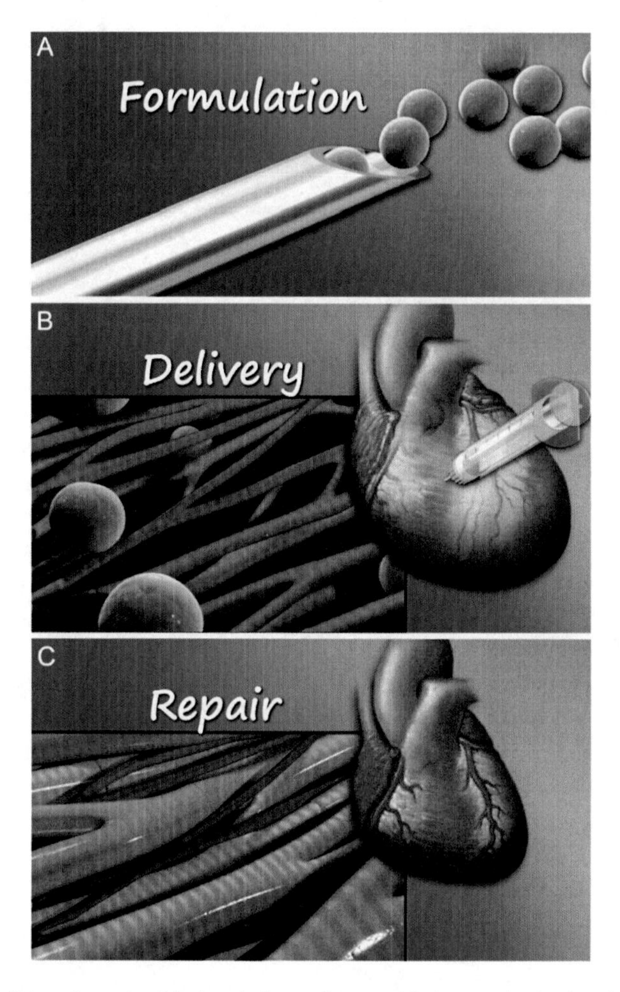

Fig. 3 iPS-cell-based repair of ischemic heart disease. (**a**) Appropriately dosed and fulfilling quality-control criteria, iPS cells are formulated for transplantation. (**b**) Direct intramyocardial delivery leads to iPS engraftment within diseased heart. (**c**) The potential for therapeutic benefit and repair of iPS has so far been documented in models of myocardial infarction

in fractional shortening and regional septal wall thickness during contraction that demonstrate coordinated concentric contractions visualized by long-axis and short-axis 2D imaging. Beyond functional deterioration, maladaptive remodeling with detrimental structural changes prognosticates poor outcome following ischemic injury to the heart. iPS-cell-based intervention can also attenuate global left ventricular diastolic diameter, predictive of decompensated heart disease. A consequence of pathologic structural remodeling is evident by prolongation of the QT interval, which increases risk of life-threatening arrhythmias [39]. Successful iPS cell treatment prevents structural remodeling and avoids the deleterious effects on electrical conductivity.

These real-time surrogates for tissue remodeling are confirmed by gross inspection of specimens that can be physically measured and quantified. Autopsy allows histological analysis to determine the extent of scar tissue formation within the postischemic region of the anterior circulation distal to the coronary ligation [39]. In contrast to parental fibroblasts, iPS cell treatment halts structural deterioration, with decreased fibrotic scarring and induction of remuscularization along with evidence for angiogenesis according to vascular endothelial markers. Surgical dissection is required to further verify the absence of tumor infiltration or dysregulated cell expansion following iPS cell transplantation in the myocardium itself, as well as in organs with high metastatic risk such as the liver, lung, and spleen. Collectively, iPS-cell-derived regeneration of the ischemic heart has been demonstrated at multiple levels of stringency that include cellular, tissue, structural, functional, and metabolic levels [39].

7 Clinical Applicability of iPS Cell Technology for Heart Disease

Clinical translation of iPS cell technology faces similar challenges that have in part been addressed by natural stem cell applications, including ESCs that were approved early in 2009 by the Food and Drug Administration in the United States for trials involving patients with incurable spinal cord injuries. The first universal obstacle for clinical translation of pluripotent stem cell technology is unregulated tumor formation [73]. Even a limited contamination of undifferentiated cells can, in theory, result in the formation of dysregulated tumors. Therefore, a critical milestone is to secure differentiation of iPS cells into the required cell type, purifying them away from residual undifferentiated cells prior to transplantation [27, 73]. This becomes a unique challenge for iPS cell technology when the immune system is no longer involved in the elimination process of dysregulated foreign tissue, active with ESC applications. The second issue that is unique to iPS cells is the accuracy of complete reprogramming of ordinary cells into pluripotent progeny. Inadequate conversion according to nuclear reprogramming strategies could result in impaired differentiation of iPS cells into target tissues required for specific applications [27]. Third, the issue of persistent transgene expression in iPS cell progeny requires careful consideration. Traditionally, iPS cells have been produced by transduction of ordinary cells with retroviruses or lentiviruses carrying ectopic transgenes to efficiently transfer stable expression into the host nucleus. This creates the risk of not only continuous expression of transgenes that are known to promote dysregulated tumor growth, but also involves permanent genomic modifications that raise concern for insertion mutagenesis of endogenous loci.

In regard to safety of iPS cell production, evolution of the reprogramming technology has rapidly advanced to design safer yet efficient processes to convert ordinary cells into bona fide stem cells. Retroviral and lentiviral approaches provided the initial method that launched the field, and established the technological basis of nuclear reprogramming with rapid confirmation across integrating vector systems [45–53]. Owing to retroviral-based vector systems having inherent sequences that become inactive in the pluripotent state, continuous expression of transgenes is temporally restricted despite persistent maintenance of the pluripotent state. Thereby, systems were designed for transient production of stemness-related genes without integration into the genome. The first proof-of-principle was achieved by nonintegrating viral vector systems, such as adenovirus [74], and was confirmed by repeated exposure to extrachromosomal plasmid-based transgenes [75]. Importantly, these reports demonstrated that expression of stemness-related factors was required for only a limited timeframe until progeny developed autonomous self-renewal, establishing nuclear reprogramming as a bioengineered process that resets a sustainable pluripotent cell fate independent of permanent genomic modifications. The latest innovation that advances efficient iPS-cell-based technology toward clinical applications involves nonviral approaches capable of iPS cell production [58, 59]. These newest approaches are dependent on short sequences of mobile genetic elements that can be used to integrate transgenes into host cell genomes and provide a genetic tag to remove flanked genomic DNA sequences. The piggyBac system couples enzymatic cleavage with sequence-specific recognition using a transposon/transposase interaction to ensure high-efficiency removal of flanked DNA without a residual footprint. Importantly, this technology achieves a traceless transgenic approach in which nonnative genomic sequences that are transiently required for nuclear reprogramming can be removed upon induction of pluripotency. Specifically, using the piggyBac transposition system with randomly integrated stemness-related transgenes, studies have demonstrated that disposal of ectopic genes could be efficiently regulated upon induction of self-maintaining pluripotency according to expression of the transposase enzyme without infringement on genomic stability [58]. This state-of-the-art system is uniquely qualified to allow safe integration and removal of ectopic transgenes, improving the efficiency of iPS production. Alternatively, the security of unmodified genomic intervention can be achieved with nonintegrating episomal vectors [76]. Collectively, these recent strategies are accelerating the discovery science of regenerative medicine, bringing it a step closer to clinical applicability by allowing genetically unmodified progenitor cells to acquire the capacity of pluripotency and produce unlimited, autologous replacement tissues.

Cardiac tissue specificity from stem cells has been investigated for more than a decade and as of yet no single gene or cluster of genes has been identified to secure cardiac differentiation. However, recent studies have significantly enriched the cardiac propensity with exogenous growth factors [77], cell sorting of cardiac progenitors [78–81], or genetically engineering precardiac pathways, all to encourage and promote cardiogenesis from primitive stem cell pools [82]. Collectively, these technologies offer the rational basis to design strategies to ensure cardiogenesis and avoidance of undifferentiated subpopulations prior to transplantation. The crucial balance between cardiac specification and progenitor cell proliferation [83] will be essential to develop a robust manufacturing process that can be scaled up and applied to clinical-grade production of a cardiac stem-cell-based product.

To translate iPS cell technology into clinical reality for heart disease, additional milestones will need to be considered. First, the target patient population will need to be identified on the basis of disease severity and lack of alternative options to justify inclusion into a first-in-man study. Many patients are too severely deconditioned or have significant comorbidities to allow consideration for heart transplant, thus limiting treatment strategies to palliative medicines and procedures. This category of patients needs to be considered a priority in terms of experimental cell-based interventions. An advantage with iPS cell technology is that no toxic immunosuppression is required and sick individuals, who may be elderly, have poor natural stem cell pools. Thus, iPS-cell-based products should be considered in these patients with no other options to decrease symptoms and decrease the need for hospitalization and expensive yet invasive palliative management strategies such as destination left ventricular assist devices. Next, a good manufacturing practice (GMP) production process and facility will need to be developed and implemented to ensure not only clinical-grade production of patient-derived iPS cells, but also standardized differentiation for targeted applications. Finally, regulatory agencies will require evidence of proper engraftment, survival, and safety of transplanted iPS-cell-derived progeny. This will require proof-of-principle studies using clinical-grade cell product in disease model systems.

8 Clinical Prospective

Regenerative medicine, built on emerging discoveries in stem cell biology [84], has begun to define the scope of future clinical practice. Regenerative medicine and stem cell biology cross all disciplines of medicine, and provide a universal paradigm of curative goals based on scientific discovery and clinical translation. The challenges to realize the full potential of stem cell biology remain substantial even for the most optimistic physicians/scientists, and require integration of multidisciplinary expertise to form a dedicated regenerative pharmacology community of practice [85]. Building on the foundation of transplant medicine, regenerative medicine will continue to expand and implement technologies to treat new diseases at earlier stages with safer and more effective outcomes, not possible with current therapies. Progress in this field will proceed with significant interest and support from the public, biotechnology and pharmaceutical industry, governmental agencies, and academic institutions, but the pace at which discovery, development, validation, and regulation will impact translation to clinical practice needs to be safely expedited for the sake of the patients. Individualized treatment algorithms for regenerative medicine will require quantification of the inherent reparative potential to determine patients who would benefit from stem cell therapy to target safe and effective therapies at the earliest stage of disease in the new era of individualized regenerative medicine.

Triggered nuclear reprogramming through ectopic transgene expression of stemness factors offers a revolutionary strategy for embryo-independent derivation of autologous pluripotent stem cells from an ordinary adult source. In this way, iPS cells have attained functions previously demonstrated only by natural ESCs to independently produce all tissue types and develop the complete organism within an embryonic environment. The reprogrammed iPS cell progeny have established the therapeutic value for cardiac tissue regeneration in a setting of ischemic heart disease. Specifically, transplantation of iPS cells in the acutely ischemic myocardium yielded structural and functional repair to secure performance recovery as qualified clones contributed to in vivo tissue reconstruction with "on-demand" cardiovasculogenesis. Therefore, converting self-derived fibroblasts, the main contributors to postischemic scar, into reparative progenitors can now be considered as a goal of regenerative medicine to individualize treatment algorithms for multilineage cardiovascular repair.

References

1. Nelson TJ, Behfar A, Yamada S, Martinez-Fernandez A, Terzic A. Stem cell platforms for regenerative medicine. Clin Transl Sci 2009;2:222–227.
2. Klimanskaya I, Rosenthal N, Lanza R. Derive and conquer: sourcing and differentiating stem cells for therapeutic applications. Nat Rev Drug Discov 2008;7:131–142.
3. Morrison SJ, Spradling AC. Stem cells and niches: mechanisms that promote stem cell maintenance throughout life. Cell 2008;132:598–611.
4. Surani MA, McLaren A. Stem cells: a new route to rejuvenation. Nature 2006;443:284–285.
5. Rossant J. Stem cells and early lineage development. Cell 2008;132:527–531.
6. Thomson JA, Itskovitz-Eldor J, Shapiro SS, Waknitz MA, Swiergiel JJ, Marshall VS, Jones JM. Embryonic stem cell lines derived from human blastocysts. Science 1998;282:1145–1147.
7. Solter D. From teratocarcinomas to embryonic stem cells and beyond: a history of embryonic stem cell research. Nat Rev Genet 2006;7:319–327.
8. Silva J, Smith A. Capturing pluripotency. Cell 2008;132:532–536.
9. Behfar A, Perez-Terzic C, Faustino RS, Arrell DK, Hodgson DM, Yamada S, Puceat M, Niederländer N, Alekseev AE, Zingman LV, Terzic A. Cardiopoietic programming of embryonic stem cells for tumor-free heart repair. J Exp Med 2007;204:405–420.
10. Murry CE, Keller G. Differentiation of embryonic stem cells to clinically relevant populations: lessons from embryonic development. Cell 2008;132:661–680.
11. Koch CA, Geraldes P, Platt JL. Immunosuppression by embryonic stem cells. Stem Cells 2008;26:89–98.
12. Goldman S. Stem and progenitor cell-based therapy of the human central nervous system. Nat Biotechnol 2005;23:862–871.
13. Kroon E, Martinson LA, Kadoya K, Bang AG, Kelly OG, Eliazer S, Young H, Richardson M, Smart NG, Cunningham J, Agulnick AD, D'Amour KA, Carpenter MK, Baetge EE. Pancreatic endoderm derived from human embryonic stem cells generates glucose-responsive insulin-secreting cells in vivo. Nat Biotechnol 2008;4:443–452.
14. Laflamme MA, Murry CE, Regenerating the heart. Nat Biotechnol 2005;23:845–856.
15. Fraidenraich D, Benezra R. Embryonic stem cells prevent developmental cardiac defects in mice. Nat Clin Pract Cardiovasc Med 2006;3 Suppl 1:S14–17.
16. Van de Ven C, Collins D, Bradley MB, Morris E, Cairo MS. The potential of umbilical cord blood multipotent stem cells for nonhematopoietic tissue and cell regeneration. Exp Hematol 2007;35:1753–1765.

17. McGuckin CP, Forraz N. Potential for access to embryonic-like cells from human umbilical cord blood. Cell Prolif 2008;41 Suppl 1:31–40.
18. De Coppi P, Bartsch G Jr, Siddiqui MM, Xu T, Santos CC, Perin L, Mostoslavsky G, Serre AC, Snyder EY, Yoo JJ, Furth ME, Soker S, Atala A. Isolation of amniotic stem cell lines with potential for therapy. Nat Biotechnol 2007;25:100–106.
19. Wagers AJ, Weissman IL. Plasticity of adult stem cells. Cell 2004;116:639–648.
20. Orkin SH, Zon LI. Hematopoiesis: an evolving paradigm for stem cell biology. Cell 2008;132:631–644.
21. Chamberlain G, Fox J, Ashton B, Middleton J. Mesenchymal stem cells: their phenotype, differentiation capacity, immunological features, and potential for homing. Stem Cells 2007;25:2739–2749.
22. Phinney DG, Prockop DJ. Mesenchymal stem/multipotent stromal cells: the state of transdifferentiation and modes of tissue repair. Stem Cells 2007;25:2896–2902.
23. Caplan AL. Adult mesenchymal stem cells for tissue engineering versus regenerative medicine. J Cell Physiol. 2007;213:341–347.
24. Martinez C, Hofmann TJ, Marino R, Dominici M, Horwitz EM. Human bone marrow mesenchymal stromal cells express the neural ganglioside GD2: a novel surface marker for the identification of MSCs. Blood 2007;109:4245–4248.
25. Le Blanc K, Ringdén O. Immunomodulation by mesenchymal stem cells and clinical experience. J Intern Med 2007;262:509–525.
26. Jaenisch R, Young R. Stem cells, the molecular circuitry of pluripotency and nuclear reprogramming. Cell 2008;132:567–582.
27. Yamanaka, S. A fresh look at iPS cells. Cell 2009;137:13–17.
28. Yang X, Smith SL, Tian XC, Lewin HA, Renard JP, Wakayama T. Nuclear reprogramming of cloned embryos and its implications for therapeutic cloning. Nat Genet 2007;39:295–302.
29. Nakagawa M, Koyanagi M, Tanabe K, Takahashi K, Ichisaka T, Aoi T, Okita K, Mochiduki Y, Takizawa N, Yamanaka S. Generation of induced pluripotent stem cells without Myc from mouse and human fibroblasts. Nat Biotechnol 2008;26:101–106.
30. Park IH, Zhao R, West JA, Yabuuchi A, Huo H, Ince TA, Lerou PH, Lensch MW, Daley GQ. Reprogramming of human somatic cells to pluripotency with defined factors. Nature 2008;451:141–146.
31. Nelson TJ, Terzic A. Induced pluripotent stem cells: reprogrammed without a trace. Regen Med 2009;4:333–355.
32. Leri A, Kajstura J, Anversa P, Frishman WH. Myocardial regeneration and stem cell repair. Curr Probl Cardiol 2008;33:91–153.
33. Dimmeler S, Zeiher AM, Schneider MD. Unchain my heart: the scientific foundations of cardiac repair. J Clin Invest 2005;115:572–583.
34. Nelson TJ, Behfar A, Terzic A. Stem cells: biologics for regeneration. Clin Pharmacol Ther 2008;84:620–623.
35. Chien KR, Domian IJ, Parker KK. Cardiogenesis and the complex biology of regenerative cardiovascular medicine. Science 2008;322:1494–1497.
36. Menasche P. Cell-based therapy for heart disease: a clinically oriented perspective. Mol Ther 2009;17:758–766.
37. Reinecke H, Minami E, Zhu WZ, Laflamme MA. Cardiogenic differentiation and transdifferentiation of progenitor cells. Circ Res 2008;103:1058–1071.
38. Gnecchi M, Zhang Z, Ni A, Dzau VJ. Paracrine mechanisms in adult stem cell signaling and therapy. Circ Res 2008;103:1204–1219.
39. Nelson TJ, Martinez-Fernandez A, Yamada S, Perez-Terzic C, Ikeda Y, Terzic A. Repair of acute myocardial infarction with human stemness factors induced pluirpotent stem cells. Circulation 2009;120:408–416.
40. Bartunek J, Sherman W, Vanderheyden M, Fernandez-Aviles F, Wijns W, Terzic A. Delivery of biologics in cardiovascular regenerative medicine. Clin Pharmacol Ther 2009;85:548–552.
41. Waldman SA, Christensen NB, Moore JE, Terzic A. Clinical pharmacology: the science of therapeutics. Clin Pharmacol Ther 2007;81:3–6.

42. European Medicines Agency. Guideline on human cell-based medicinal products. EMEA 2007;EMEA/CHMP/410869/2006:1–24.
43. Halme DG, Kessler DA. FDA regulation of stem-cell-based therapies. N Engl J Med 2006;355:1730–1735.
44. Dietz AB, Padley DJ, Gastineau DA. Infrastructure development for human cell therapy translation. Clin Pharmacol Ther 2007;82:320–324.
45. Takahashi K, Yamanaka, S. Induction of pluripotent stem cells from mouse embryonic and adult fibroblast cultures by defined factors. Cell 2006;126:663–676.
46. Okita K, Ichisaka T, Yamanaka S. Generation of germline-competent induced pluripotent stem cells. Nature 2007;448:313–317.
47. Takahashi K, Tanabe K, Ohnuki M, Narita M, Ichisaka T, Tomoda K, Yamanaka S. Induction of pluripotent stem cells from adult human fibroblasts by defined factors. Cell 2007;131:861–872.
48. Yu J, Vodyanik MA, Smuga-Otto K et al. Induced pluripotent stem cell lines derived from human somatic cells. Science 2007;318:1917–1920.
49. Yamanaka S. Strategies and new developments in the generation of patient-specific pluripotent stem cells. Cell Stem Cell 2007;1:39–49.
50. Park IH, Lerou PH, Zhao R, Huo H, Daley GQ. Generation of human-induced pluripotent stem cells. Nat Protoc 2008;3:1180–1186.
51. Park IH, Arora N, Huo H, Maherali N, Ahfeldt T, Shimamura A, Lensch MW, Cowan C, Hochedlinger K, Daley GQ. Disease-specific induced pluripotent stem cells. Cell 2008;134:877–886.
52. Kim JB, Zaehres H, Wu G, Gentile L, Ko K, Sebastiano V, Araúzo-Bravo MJ, Ruau D, Han DW, Zenke M, Schöler HR. Pluripotent stem cells induced from adult neural stem cells by reprogramming with two factors. Nature 2008;454:646–650.
53. Kim JB, Sebastiano V, Wu G, Araúzo-Bravo MJ, Sasse P, Gentile L, Ko K, Ruau D, Ehrich M, van den Boom D, Meyer J, Hübner K, Bernemann C, Ortmeier C, Zenke M, Fleischmann BK, Zaehres H, Schöler HR. Oct4-induced pluripotency in adult neural stem cells. Cell 2009;136:411–419.
54. Papapetrou EP, Tomishima MJ, Chambers SM, Mica Y, Reed E, Menon J, Tabar V, Mo Q, Studer L, Sadelain M. Stoichiometric and temporal requirements of Oct4, Sox2, Klf4, and c-Myc expression for efficient human iPSC induction and differentiation. Proc Natl Acad Sci U S A 2009;106:12759–12764 Jun 23 [Epub ahead of print].
55. Marion RM, Strati K, Li H, Tejera A, Schoeftner S, Ortega S, Serrano M, Blasco MA. Telomeres acquire embryonic stem cell characteristics in induced pluripotent stem cells. Cell Stem Cell 2009;4:141–154.
56. Deng J, Shoemaker R, Xie B, Gore A, LeProust EM, Antosiewicz-Bourget J, Egli D, Maherali N, Park IH, Yu J, Daley GQ, Eggan K, Hochedlinger K, Thomson J, Wang W, Gao Y, Zhang K. Targeted bisulfite sequencing reveals changes in DNA methylation associated with nuclear reprogramming. Nat Biotechnol 2009;27:353–360.
57. Mikkelsen TS, Hanna J, Zhang X, Ku M, Wernig M, Schorderet P, Bernstein BE, Jaenisch R, Lander ES, Meissner A. Dissecting direct reprogramming through integrative genomic analysis. Nature 2008;454:49–55.
58. Woltjen K, Michael IP, Mohseni P et al. piggyBac transposition reprograms fibroblasts to induced pluripotent stem cells. Nature 2009;458:766–770.
59. Kaji K, Norrby K, Paca A et al. Virus-free induction of pluripotency and subsequent excision of reprogramming factors. Nature 2009;458:771–775.
60. Yamanaka S. Elite and stochastic models for induced pluripotent stem cell generation. Nature 2009;460:49–52.
61. Hanna J, Markoulaki S, Schorderet P et al. Direct reprogramming of terminally differentiated mature B lymphocytes to pluripotency. Cell 2008;133:250–264.
62. Maherali N, Hochedlinger K. Guidelines and techniques for the generation of induced pluripotent stem cells. Cell Stem Cell 2008;3:595–605.
63. Smith KP, Luong MX, Stein GS. Pluripotency: toward a gold standard for human ES and iPS cells. J Cell Physiol 2009;220:21–29.

64. Hanna J, Wernig M, Markoulaki S, Sun CW, Meissner A, Cassady JP, Beard C, Brambrink T, Wu LC, Townes TM, Jaenisch R. Treatment of sickle cell anemia mouse model with iPS cells generated from autologous skin. Science 2007;318:1920–1923.
65. Wernig M, Zhao JP, Pruszak J, Hedlund E, Fu D, Soldner F, Broccoli V, Constantine-Paton M, Isacson O, Jaenisch R. Neurons derived from reprogrammed fibroblasts functionally integrate into the fetal brain and improve symptoms of rats with Parkinson's disease. Proc Natl Acad Sci USA 2008;105:5856–5861.
66. Xu D, Alipio Z, Fink LM, Adcock DM, Yang J, Ward DC, Ma Y. Phenotypic correction of murine hemophilia A using an iPS cell-based therapy. Proc Natl Acad Sci USA 2009;106:808–813.
67. Schenke-Layland K, Rhodes KE, Angelis E, Butylkova Y, Heydarkhan-Hagvall S, Gekas C, Zhang R, Goldhaber JI, Mikkola HK, Plath K, MacLellan WR. Reprogrammed mouse fibroblasts differentiate into cells of the cardiovascular and hematopoietic lineages. Stem Cells 2008;26:1537–1546.
68. Narazaki G, Uosaki H, Teranishi M, Okita K, Kim B, Matsuoka S, Yamanaka S, Yamashita JK. Directed and systematic differentiation of cardiovascular cells from mouse induced pluripotent stem cells. Circulation 2008;118:498–506.
69. Mauritz C, Schwanke K, Reppel M, Neef S, Katsirntaki K, Maier LS, Nguemo F, Menke S, Haustein M, Hescheler J, Hasenfuss G, Martin U. Generation of functional murine cardiac myocytes from induced pluripotent stem cells. Circulation 2008;118:507–517.
70. Zhang J, Wilson GF, Soerens AG, Koonce CH, Yu J, Palecek SP, Thomson JA, Kamp TJ. Functional cardiomyocytes derived from human induced pluripotent stem cells. Circ Res 2009;104:e30–e41.
71. Yokoo N, Baba S, Kaichi S, Niwa A, Mima T, Doi H, Yamanaka S, Nakahata T, Heike T. The effects of cardioactive drugs on cardiomyocytes derived from human induced pluripotent stem cells. Biochem Biophys Res Commun 2009;387:482–488.
72. Srinivas G, Anversa P, Frishman WH. Cytokines and myocardial regeneration: a novel treatment option for acute myocardial infarction. Cardiol Rev 2009;17:1–9.
73. Li JY, Christophersen NS, Hall V, Soulet D, Brundin P. Critical issues of clinical human embryonic stem cell therapy for brain repair. Trends Neurosci 2008;31:146–153.
74. Stadtfeld M, Nagaya M, Utikal J, Weir G, Hochedlinger K. Induced pluripotent stem cells generated without viral integration. Science 2008;322:945–949.
75. Okita K, Nakagawa M, Hyenjong H, Ichisaka T, Yamanaka S. Generation of mouse induced pluripotent stem cells without viral vectors. Science 2008;322:949–953.
76. Yu J, Hu K, Smuga-Otto K, Tian S, Stewart R, Slukvin II, Thomson JA. Human induced pluripotent stem cells free of vector and transgene sequences. Science 2009;324:797–801.
77. Behfar A, Faustino RS, Arrell DK, Dzeja PP, Perez-Terzic C, Terzic A. Guided stem cell cardiopoiesis: discovery and translation. J Mol Cell Cardiol 2008;45:523–529.
78. Nelson TJ, Faustino RS, Chiriac A, Crespo-Diaz R, Behfar A, Terzic A. CXCR4$^+$/FLK-1$^+$ biomarkers select a cardiopoietic lineage from embryonic stem cells. Stem Cells 2008;26:1464–1473.
79. Moretti A, Caron L, Nakano A, Lam JT, Bernshausen A, Chen Y, Qyang Y, Bu L, Sasaki M, Martin-Puig S, Sun Y, Evans SM, Laugwitz KL, Chien KR. Multipotent embryonic isl1+ progenitor cells lead to cardiac, smooth muscle, and endothelial cell diversification. Cell 2006;127:1151–1165.
80. Kattman SJ, Huber TL, Keller GM. Multipotent flk-1+ cardiovascular progenitor cells give rise to the cardiomyocyte, endothelial, and vascular smooth muscle lineages. Dev Cell 2006;11:723–732.
81. Yang L, Soonpaa MH, Adler ED, Roepke TK, Kattman SJ, Kennedy M, Henckaerts E, Bonham K, Abbott GW, Linden RM, Field LJ, Keller GM. Human cardiovascular progenitor cells develop from a KDR+ embryonic-stem-cell-derived population. Nature 2008; 453:524–852.
82. Takeuchi JK, Bruneau BG. Directed transdifferentiation of mouse mesoderm to heart tissue by defined factors. Nature 2009;459:708–711.

83. Martinez-Fernandez A, Nelson TJ, Yamada S, Reyes S, Alekseev AE, Perez-Terzic C, Ikeda Y, Terzic A. iPS Programmed without c-MYC yield proficient cardiogenesis for functional heart chimerism. Circ Res. 2009;105;648–656.
84. Nelson TJ, Behfar A, Terzic A. Strategies for therapeutic repair: the "R3" regenerative medicine paradigm. Clin Transl Sci 2008;1:168–171.
85. Nelson TJ, Behfar A, Terzic A. Regenerative medicine and stem cell therapeutics. *In*: *Pharmacology and Therapeutics – Principles to Practice*. Edited by SA Waldman and A Terzic, Saunders Elsevier, 2009.
86. Daley GQ, Scadden DT. Prospects for stem cell-based therapy. Cell 2008;132:544–548.
87. Rosenthal N. Prometheus's vulture and the stem-cell promise. N Engl J Med 2003;349:267–274.
88. Segers V, Lee RT. Stem-cell therapy for cardiac disease. Nature 2008;451:937–942.
89. Torella D, Ellison GM, Méndez-Ferrer S, Ibanez B, Nadal-Ginard B. Resident human cardiac stem cells: role in cardiac cellular homeostasis and potential for myocardial regeneration. Nat Clin Pract Cardiovasc Med 2006;3 Suppl 1:S8–13.
90. Quaini F, Urbanek K, Beltrami AP, Finato N, Beltrami CA, Nadal-Ginard B, Kajstura J, Leri A, Anversa P. Chimerism of the transplanted heart. N Engl J Med 2002;346:5–15.
91. Kajstura J, Hosoda T, Bearzi C, Rota M, Maestroni S, Urbanek K, Leri A, Anversa P. The human heart: a self-renewing organ. Clin Transl Sci 2008;1:80–86.
92. Deb A, Wang S, Skelding KA, Miller D, Simper D, Caplice NM. Bone marrow-derived cardiomyocytes are present in adult human heart: a study of gender-mismatched bone marrow transplantation patients. Circulation 2003;107:1247–1249.
93. Kubo H, Jaleel N, Kumarapeli A, Berretta RM, Bratinov G, Shan X, Wang H, Houser SR, Margulies KB. Increased cardiac myocyte progenitors in failing human hearts. Circulation 2008;118:649–657.
94. Rupp S, Koyanagi M, Iwasaki M, Bauer J, von Gerlach S, Schranz D, Zeiher AM, Dimmeler S. Characterization of long-term endogenous cardiac repair in children after heart transplantation. Eur Heart J 2008;29:1867–1872.
95. Urbanek K, Torella D, Sheikh F, De Angelis A, Nurzynska D, Silvestri F, Beltrami CA, Bussani R, Beltrami AP, Quaini F, Bolli R, Leri A, Kajstura J, Anversa P. Myocardial regeneration by activation of multipotent cardiac stem cells in ischemic heart failure. Proc Natl Acad Sci U S A 2005;102:8692–8697.

Part II
Stem Cells for Regeneration of Electrical Function

Substrates of Cardiac Reentrant Arrhythmias: The Possible Role of Tissue Regeneration and Replacement

André G. Kléber

Abstract This chapter describes the basic mechanisms leading to circulating excitation with reentry in the heart, such as the formation of unidirectional conduction block and slowing of electrical propagation velocity. Special emphasis is put on the role of heterogeneity in ion channel expression, heterogeneity in connexin expression, the role of the architecture of the cellular network, and the interaction between myocytes and nonmyocytes. Heterogeneity in ion channel expression has been shown to affect the dynamics of reentrant spiral waves, especially the frequency of rotation. Heterogeneity in connexin expression can lead to local propagation block, a main factor in initiation of reentrant arrhythmias. Interaction between cells of the connective tissue and cardiomyocytes can affect propagation velocity, cause local conduction delays, and initiate premature beats. The role of these factors in regenerative therapy and the arrhythmogenic propensity of interventions aimed at tissue repair are discussed.

Keywords Arrhythmias • Gap junctions • Reentrant circuits • Tissue discontinuity • Myofibroblasts

1 Introduction

Cardiac arrhythmias have fascinated clinicians and scientists since the beginning of the twentieth century. They consist of the rapid transition of the orderly, rhythmic heart beat into disordered electrical excitation of the heart at the level of the ventricles and the atria. These conditions can be associated with acute impairment of health, in the extreme case of ventricular fibrillation, with sudden cardiac death. It is assumed that 300,000–500,000 individuals die each year from such a condition in the USA.

A.G. Kléber (✉)
Department of Physiology, University of Bern, Bühlplatz 5, 3012 Bern, Switzerland
e-mail: kleber@pyl.unibe.ch

I.S. Cohen and G.R. Gaudette (eds.), *Regenerating the Heart*, Stem Cell Biology and Regenerative Medicine, DOI 10.1007/978-1-61779-021-8_16,
© Springer Science+Business Media, LLC 2011

The normal heart is excited rhythmically from its biological clock, the sinoatrial pacemaker. Subsequently, excitation spreads in a strictly predefined pattern, through the atria, the atrioventricular junction, the His/Purkinje system, and eventually through the working myocardium of the ventricles. In each excited cell, the mechanism of electromechanical coupling acts as a transducer to induce and regulate contraction via the preceding electrical excitation. At the biophysical level, the heart can be considered as a so-called reaction–diffusion system, where the local reaction, i.e., the action potential, diffuses or propagates through an excitable network with distinct "diffusion properties," determined by parameters such as the size and shapes of cardiac cells, cell-to-cell coupling, and the architecture determined by the network of cells and interspersed connective tissue. With this basic knowledge as a starting point, the mechanisms of cardiac arrhythmias were classified early on into disturbances of impulse initiation and impulse propagation [1].

From the above conceptual discussion it becomes evident that arrhythmogenesis concerns changes of a complex biological system, which are caused and/or modulated by a multitude of molecular changes contributing to the phenotype. This complexity may make it particularly difficult to target single molecular players in a new therapeutic approach unless the molecular targets act upstream to influence the general deregulatory process leading to maladaptation and disease.

Replacement or regenerative cell therapy as a potential new therapeutic approach aims at the improvement of function of remodeled cell pools or creation of new cell pools either by stimulation of growth of intrinsic cells or by transplantation of new cells. All these concepts, although quite straightforward in their basic strategy, must involve regeneration of tissue in a way that ensures physiological electrical excitation and excitation–contraction coupling. In this chapter, the changes that underlie the formation of the substrate of reentrant arrhythmias will be described, with emphasis put on the role of molecular and/or cellular heterogeneity. Elimination of molecular heterogeneity may represent one of the goals of regenerative therapy. Alternatively, molecular heterogeneity, representing an arrhythmogenic substrate, may occur as a complication of regenerative cell therapy. In both cases, knowledge of the role of heterogeneity in expression of the molecules involved in electrical excitation and propagation, and in tissue architecture is important.

2 The Classical Theory of Reentrant Excitation

Mines [2] formulated the basic biophysical rules leading from normal cardiac excitation spread to circulating excitation and reentry early in the twentieth century. These rules are in essence still valid, although it became evident later from the work of Allessie et al. [3] that the laws governing circulating excitation around an anatomical obstacle, as formulated by Mines, also hold for tissue regions where all cells remain fully excitable. In such tissue, circulating excitation rotates around a zone of functional propagation block. A further improvement in knowledge was made by the observation that circulating excitation waves in the heart assume the shape of spirals by virtue of being limited by a propagation velocity that corresponds to the velocity of a planar wave. The curvature of the wave front and the electrical interaction of the cells in the center of these spirals determine rotor movement and frequency [4–6].

2.1 The Principle of Circulating Excitation

The simplest scheme describing the basic mechanisms of circulating excitation was established by Mines [2] and is shown in Fig. 1. Two main principles underlying reentrant excitation can be derived from this scheme, which has maintained its usefulness throughout the many decades since its definition: (1) propagation needs to be unidirectional (or "asymmetric") to induce a circle of propagation; (2) the wavelength of excitation, defined by the product of the refractory period and the propagation velocity needs to be shorter than the circuit length. The principles underlying asymmetrical or unidirectional block are multifold and will be treated in a separate section. The reasons for the wavelength becoming shorter than the path of excitation follow from the above definition: the wavelength can decrease as a consequence of shortening of the refractory period, a slowing of the propagation velocity, or both. Importantly, changes in the process of propagation play two different roles in the above concept. First, development of asymmetrical or unidirectional excitation spread, which may have structural or functional reasons, is an important

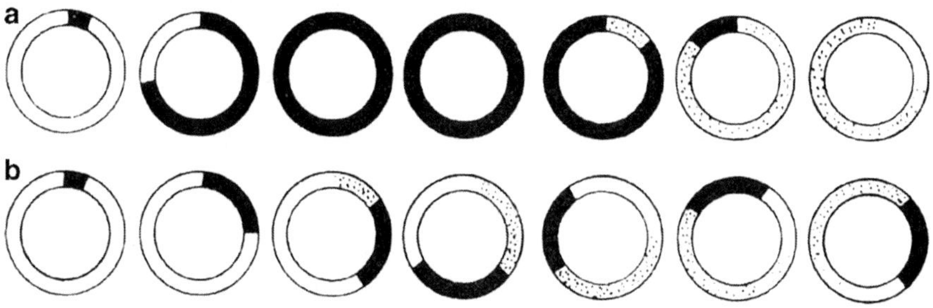

Fig. 1 Concept of circulating excitation with reentry. (**a**) Excited tissue is indicated in *black*, nonexcited tissue is indicated in *white*. A ring of tissue is excited on the top. Owing to unidirectional block (see the text), excitation propagates toward the right part of the ring but fails to propagate to the left part. Subsequently, circulating excitation travels along the ring until it becomes extinct at the site of the original stimulation. Excitability gradually recovers, going through a phase of relative refractoriness, as indicated by the *dotted areas*. (**b**) A ring of tissue is excited on the top. Owing to unidirectional block (see the text), excitation propagates toward the right part of the ring. Excited tissue is indicated in *black*, nonexcited tissue is indicated in *white*. Subsequently, circulating excitation travels along the ring. At the site of the original stimulation the propagation wave can reenter and continue to circle along the ring for an indefinite number of cycles. The difference between (**a**) and (**b**) can be discerned by comparing the length of the excited (*black*) segment that corresponds to the wavelength of excitation. The long wavelength in (**a**) prevents reentry, because the wavelength is longer than the circumference of the ring. (After [2])

prerequisite for initiation of reentrant circuits. Second, slowing of the average propagation velocity is a "scaling factor" of reentrant circuits. For instance, at a velocity of 50 cm/s and with a refractory period of 0.2 s, the wavelength of excitation amounts to 10 cm. A change in the velocity to 1 cm/s will produce a wavelength of 2 mm.

2.2 *Mechanisms of Unidirectional Block*

The network formed by the cardiac myocytes in the atria and the ventricles contains a vast number of structural discontinuities that affect the propagation of the electrical impulse. In the atria such discontinuities are caused by branching and converting muscular trabecula [7]. In the ventricles, layers of anisotropic muscular tissue are wrapped around the cavities, with a continuous change of the angle of the axes of anisotropy. Repetitive thin muscular bridges connect these layers and ensure transmural propagation [8]. Both types of tissue geometries cause propagation to become discontinuous. In addition to such sites occurring in the normal heart, discontinuities are formed in the aging myocardium by the increase in fibrosis [9, 10], which follows the anisotropic architecture of the tissue in its texture. Fibrosis is an important part of the phenotype in cardiac hypertrophy and failure in atrial remodeling [11, 12], and in scars of myocardial infarctions [13].

At sites of structural discontinuities within a network of excitable cells, a mismatch occurs between the tissue upstream that delivers excitatory current and the number of cells downstream that form a load and require input current for excitation and propagation of the electrical impulse. Although many types and many degrees of current-to-load mismatch exist, there is a unifying biophysical principle causing propagation to become asymmetrical at such a site [14]. In the experiment producing the image presented in Fig. 2a, current-to-load mismatch was produced by patterning a 2D culture of rat ventricular myocytes in such a way that the excitatory current produced by the small "street" merging with the large bulk of cells was too small to excite the cells in the bulk. As a consequence, propagation block developed if the excitation wave traveled from the street in the direction of the bulk. Inversely, propagation was even faster than normal if the bulk of cells was excited first and the impulse propagated toward the geometrical transition, where it entered the small street. This example, published together with many others in experimental or theoretical studies (see [1] for references), illustrates the fact that propagation can be unidirectional (or "asymmetrical") merely on the basis of a discontinuous architecture of an excitable network. Whereas the principle illustrated in Fig. 2 seems rather simple and straightforward, it is in reality the source of significant complexities, if one takes into account the fact that propagation is further dependent on ion currents, cell-to-cell coupling by gap junctions, and that ion currents are frequency-dependent. Thus, at a site of moderate current-to-load mismatch where the impulse is propagated bidirectionally at a given frequency, the same impulse might be blocked if the frequency of excitation is increased, owing to the frequency-dependent

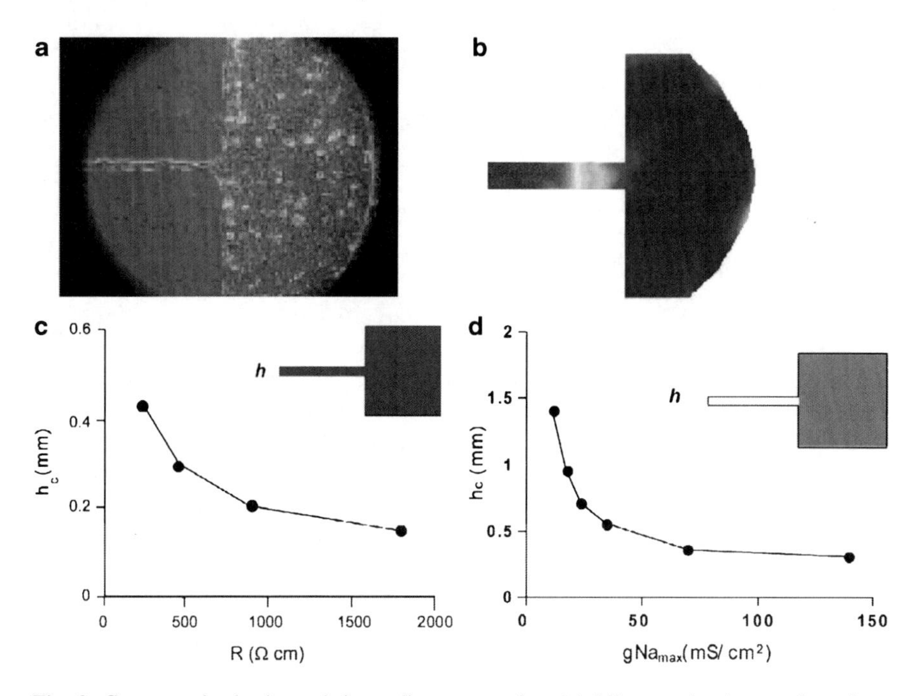

Fig. 2 Current-to-load mismatch in cardiac propagation. (**a**) Microscopic picture of a culture (neonatal rat ventricular myocytes) representing a geometrical discontinuity with a small cell strand emerging abruptly into a large tissue area. (**b**) Propagation in a narrow strand from left to right. Excited tissue is illustrated by *red*. The very narrow strand width ($h = 50$ μm) produces current-to-load mismatch with conduction block at the transition, as illustrated by the fact that the cells in the large bulk area remain in the resting (*blue*) state. Retrograde propagation from the bulk into the strand was normal in this experiment (not shown). (**c**) The dependence of h_c on the coupling resistance between the cells. The parameter h_c is defined as the critical strand width at which block occurs. If the cells are partially uncoupled, as indicated by an increase of R, h_c decreases because of a diminution in downstream load. (**d**) The dependence of h_c on the maximal electrical sodium conductance of the cells. If the sodium inward current decreases, for instance, following inhibition or a decrease in expression of Na^+ channels, h_c increases because of a diminution in excitatory current furnished by the upstream elements in the small strand. (After [14, 16])

partial inactivation of the Na^+ inward current (Fig. 2d) [15] Similarly, the effect of current-to-load mismatch causing unidirectional propagation is sensitive to the local degree of cell-to-cell coupling [14, 16]. In the situation shown in Fig. 2c, increasing the coupling resistance between cells prevents the loss of excitatory currents (or "electrotonic" currents) from the small street into the bulk and can reestablish by directional conduction at a site that shows unidirectional conduction in normal coupling conditions [16]. Cell-to-cell coupling therefore becomes an element that can compensate for a discontinuous geometry of a cellular network. In other words, successful bidirectional propagation requires a match between the discontinuous geometry and the degree of cell-to-cell coupling. This principle was recognized a long time ago by Joyner [17] and has been used to argue that the low degree of cell-to-cell coupling in the sinoatrial node can contribute to successful excitation of the atrial "crista terminalis" by the relatively small pacemaker region [18]. In summary, this discussion illustrates the complexity of the process of propagation and the interaction of all elements contributing to normal propagation and the formation of unidirectional block, respectively.

2.3 Propagation Inhomogeneities in Tissue with Continuous Properties

Whereas heart tissue contains structural and molecular discontinuities, which can promote disturbances of impulse propagation, reentrant circuits can be established in reaction–diffusion systems that are continuous (or "homogeneous") with respect to all their parameters (cellular architecture, cell-to-cell coupling, distribution, and density of ion channels). In such a situation, a spiral can be initiated by an external disturbance, which produces a transient, short-lived functional inhomogeneity in the local electrical behavior. Figure 3 illustrates a result from an early theoretical study [19]. In this simulation a sheet of excitable tissue was first stimulated simultaneously at several locations forming a row at the upper edge. As a consequence, propagation moved from top to bottom, illustrating the continuous behavior of a planar wave. In a second theoretical experiment (Fig. 3, top right), the tissue was stimulated at two sites: First, excitable elements were stimulated in the left middle part of the 2D sheet. Subsequently, the upper row of elements was excited in the same way as in the top-left of Fig. 3, and the wave traveling from the top collided with the wave that was produced with the first, eccentric stimulus. With appropriate timing of the two stimuli, the wave from the top could propagate around the first wave and enter the site of the first stimulus from the back, since the first stimulus was confined to the left part of the tissue. At the location where it entered this site, enough time had elapsed for recovery from refractoriness. In such a way, interaction between two propagating waves could set up a maintained reentrant wave in a system in which heterogeneity was brought about solely by a single stimulus, setting the condition for a transient development of local refractoriness. The top-right panel in Fig. 3 shows that the rotating wave assumes the form of a spiral and that the center of the spiral is moving in the excitable 2D sheet (see [5] for a review of the dynamics of spiral waves).

In summary, rotating spiral waves can emerge in heart issue as a consequence of multiple underlying causes that may be related to structural changes (fibrosis), changes in cell-to-cell coupling, and changes in ion currents. Importantly, a reentrant arrhythmia is the consequence of interaction of a multitude of important contributors to the arrhythmogenic substrate. Moreover, a single transient disturbance, for instance, caused by a premature beat, can initiate reentrant circuits even in the absence of major structural and ionic remodeling. This suggests that it does not seem adequate to speak of the changes in one variable as an arrhythmogenic mechanism, without knowing the exact state of the other potential contributors.

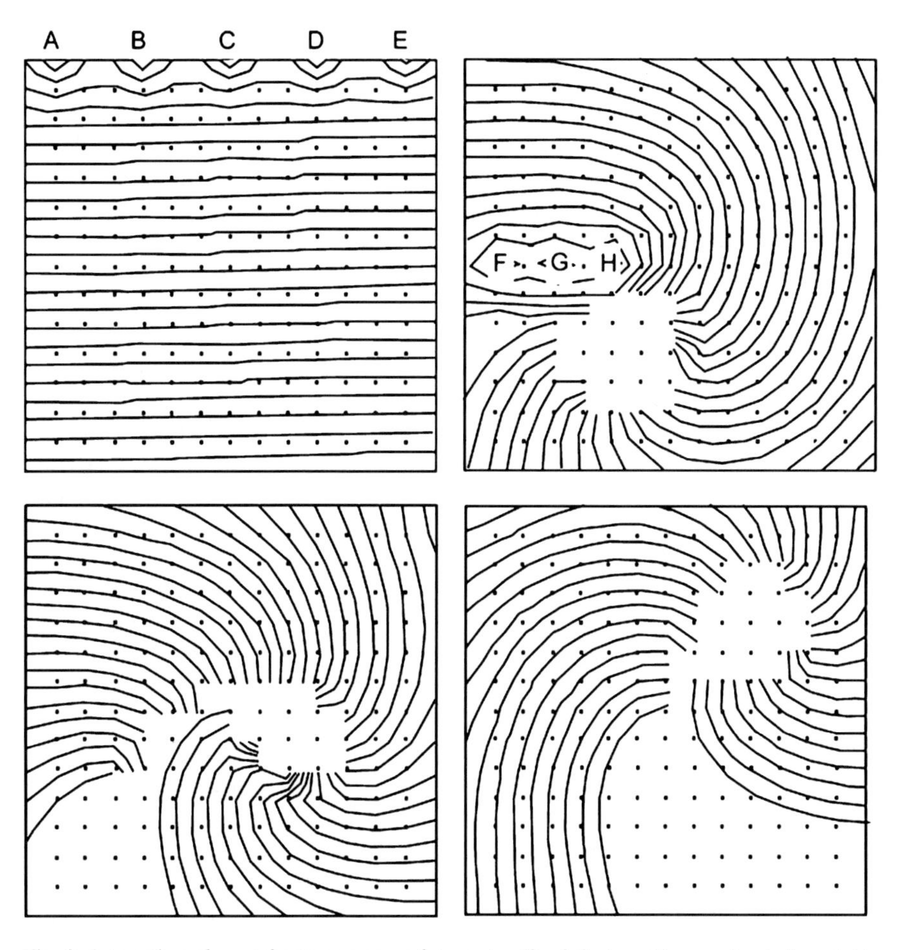

Fig. 3 Interaction of wave fronts as a cause for reentry. *Top left*: A continuous sheet of excitable elements is stimulated simultaneously at sites *A–E*. This produces a planar excitation wave moving from top to bottom. *Top right*: A single excitation is produced by stimulation of sites *F* and *G*. The planar wave traveling from top to bottom is colliding with the premature stimulus, and consequently, circulates around sites *F* and *G*. Recovery from excitation of the premature stimulus enables the circulating wave to reenter from the bottom. The *bottom panels* illustrate a snapshot of the circulating spiral wave that changes its core continuously. (After [19])

2.4 Mechanisms of Propagation Slowing

As already mentioned, propagation velocity is a scaling factor for reentrant circuits. To establish a relationship between the parameters determining the propagation velocity and the velocity itself, the concept of a safety factor (SF) [20] has proven to be useful: during the process of propagation, the depolarizing ion channels (Na^+ channels, L-type Ca^{2+} channels) provide positive electrical charge flowing from the extracellular to the intracellular space. The effects of these ion movements are

(1) to change the electrical charge stored in the membrane capacitance, i.e., to produce the action potential, and (2) to induce a downstream axial current, which will excite the cells downstream to threshold and propagate the electrical impulse. The SF is given by the ratio of (1) the electrical charge produced by the ion channels during cellular excitation and (2) the electrical charge required for excitation of the cells downstream. As long as a cell produces more charge than it needs for its excitation (SF > 1), propagation is safe. Figure 4 illustrates that the effect of changes in depolarizing ion current and cell-to-cell coupling on the SF and propagation are fundamentally different.

In a chain of electrically identical cells, a decrease in the maximal electrical conductance of the Na^+ channels, $\bar{g}Na$, produces a continuous decrease in SF and velocity until propagation is blocked at a $\bar{g}Na$ of approximately 10%. Figure 4 illustrates that inhibition of the Na^+ channels produces a "threshold-like" phenomenon: a very small change in $\bar{g}Na$ produces an instantaneous transition from a relatively high velocity value of approximately 20 cm/s to full block. This predicts that reentrant circuits caused by block of depolarizing ion currents alone must be relatively large (see Fig. 1). Very slow velocities in the range of a few centimeters per second are not observed with inhibition of Na^+ current alone. Moreover, the block caused by inhibition of $\bar{g}Na$ is sensitive to heart rate, because a small change in rate is predicted to change $\bar{g}Na$ (due to increased inactivation of Na^+ channels). The frequencies where $\bar{g}Na$ is affected are relatively high in normal tissue. However, they can decrease to normal ranges if the myocardium is depolarized. Ventricular tachycardia and ventricular fibrillation occurring very early after coronary occlusion represent a typical example of arrhythmias predominantly caused by an ischemia-related inactivation of Na^+ current [21, 22].

The effect of cell-to-cell coupling on propagation velocity is shown in Fig. 4 (bottom) [20]. Two differences with respect to the effect of depolarizing channel inhibition are prominent. First, cell-to-cell uncoupling produces very slow propagation on the order 1 cm/s until propagation block develops at a coupling resistance that has increased more than 100-fold above normal. Second, the SF for propagation shows a biphasic course; it markedly increases with cell-to-cell uncoupling and decreases (to produce propagation block) only above an increase of coupling resistance of more than 100-fold. This effect of cell-to-cell coupling to slow but also to stabilize propagation has been confirmed experimentally [23, 24]. The explanation for the seemingly unexpected effect of cell-to-cell uncoupling to stabilize propagation can be explained as follows. In a model chain of cells uncoupling (increase of intercellular resistance) will introduce two major changes with *opposite* effects. First, the decrease in resistance will decrease the amount of axial current that excites the cell downstream. This decrease will delay the depolarization of the downstream elements and accordingly will be responsible for propagation slowing, and, in the extreme case, for propagation block. Second, the ubiquitous increase of coupling resistance will decrease the overall downstream load. Consequently, the axial current flowing from the front of the excitatory wave into the downstream load will be distributed over a shorter downstream segment and

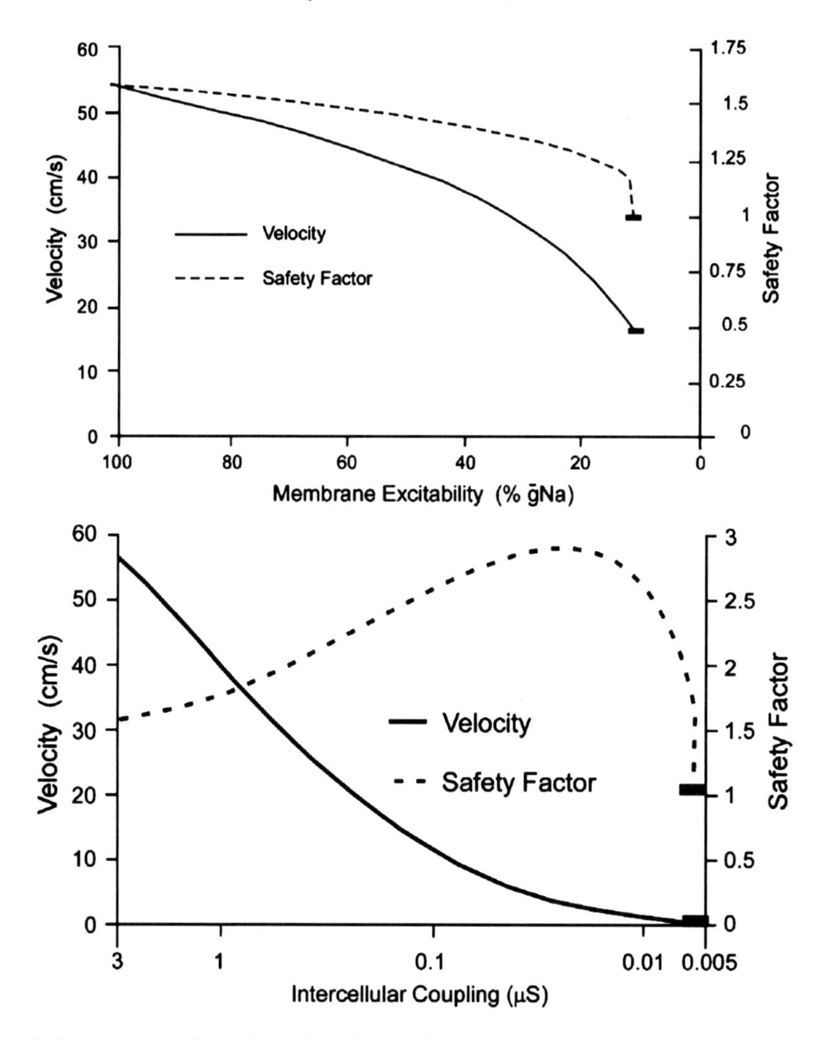

Fig. 4 Propagation safety. The safety factor of propagation and propagation velocity. *Top*: Dependence of propagation velocity and safety factor on membrane excitability, expressed as percentage of Na^+-channel conductance, $\bar{g}Na$. *Bottom*: Dependence of propagation velocity and safety factor on intercellular coupling, expressed as electrical conductance. (After [20])

more current will become available for depolarization of the cells immediately before the wave front. This protection of axial current from being lost in a large region downstream and its concentration to the immediate downstream region is responsible for the increasing propagation safety and the stabilization of propagation. The biphasic course of the SF in Fig. 4 (bottom) therefore reflects the fact that the second mechanism (decrease of downstream sink) dominates in a first phase and that the first mechanism (decrease of upstream source) limits propagation only at a very high level of intercellular resistance. Importantly, the second mechanism is only effective if the downstream load is decreased by the repetitive presence of increased intercellular resistances. If there is only a single boundary between partially coupled and normally coupled tissue there is no decrease in downstream load (no decrease of the "sink") and considerably less cell-to-cell uncoupling is needed to produce propagation block [14, 25]. This finding is likely to be relevant for the discussion of the effect of heterogeneous cell-to-cell coupling and propagation, as discussed below.

3 Integration of Cellular Compartments into Host Tissue and Arrhythmogenesis

The goals of cell and tissue therapy are multifold. Cell therapy may be aimed at the replacement of specific tissue, for instance, pacemaker tissue, or at an increase of the cell mass contributing eventually to improvement of electrical and mechanical function. Attempts to achieve this latter goal may consist in the introduction of cells into the myocardium that have matured in vitro, of cells that will divide and differentiate to reach the desired phenotype in interaction with the host, and/or of cells that will create a signaling environment favorable to growth and/or redifferentiation of intrinsic cells to the desired phenotype. In all these interventions, the desired end point is improvement of electrical and mechanical function without an increased risk for arrhythmogenesis.

Many of the studies testing experimentally the differentiation of implanted cells into a host heart have assessed the microscopic appearance and the presence of molecules of the contractile apparatus, cell junctions, and/or other indicators of the stage of differentiation. Some studies have tested functional integration directly. For instance, it has been shown that sheets of neonatal rat cardiomyocytes couple electrically to adult host rat hearts without producing major conduction delays at the transition of the transplant to the host [26]. Single embryonic rat cardiomyocytes expressing green fluorescent protein (GFP) have been shown to morphologically integrate and to operate in full coordination with the neighboring non-GFP-expressing host cells [27].

An indication of the effect of cells that express a spectrum of ion channel density or connexins different from host tissue (following implantation and/or implantation-induced signaling) can be taken from experimental and theoretical studies that investigated the effect of heterogeneity in channel expression and/or cell-to-cell coupling on the initiation and maintenance of cardiac arrhythmias. There is extensive literature on the relation between ion channel heterogeneity and *the initiation of impulses* (e.g., by early after depolarizations) [28]. It seems clear that relatively subtle changes in the expression spectrum and function of ion channels may initiate premature impulse and/or tachyarrhythmias. In accordance with the first part of this chapter, the next section focuses on the modulation of *reentrant arrhythmias* by molecular and/or structural heterogeneities.

3.1 The Role of Heterogeneity in Channel Expression on Reentrant Circuits

Assessment and prediction of the spatiotemporal changes of electrical activation during tachyarrhythmias uses the tool of biophysics to characterize the nonlinear behavior of excitation waves. The combination of these concepts with molecular biology to assess the effect of heterogeneous expression of ion channels came from the observation that local excitation during ventricular fibrillation in a guinea pig heart occurred "domainwise" in the right and left ventricles, with the highest frequency being present in the anterior wall of the left ventricle [29]. The region of the highest frequency corresponded to the site where a single full rotor was maintained. Centripetal excitation spread from this rotor showed fractionation of the waves in the periphery into multiple rotating wavelets that changed position from beat to beat, never producing full reentry (Fig. 5).

The observation that only one or a few sites were able to excite the tissue at a fast rate and the remainder of the myocardium produced fractionated so-called fibrillatory conduction led to the hypothesis that at least some forms of ventricular fibrillation require a "mother rotor" as a central initiating principle [6]. This finding was attributed to heterogeneous expression of the K^+ channel IK1 (Kir 1.4), with the site of high expression density corresponding to the high-frequency domain and the location of the mother rotor. The explanation of the modulation of spiral waves by a change in density of IK1 was demonstrated in a theoretical study and an experimental study involving transfection of the myocardium with IK1-containing vectors [30].

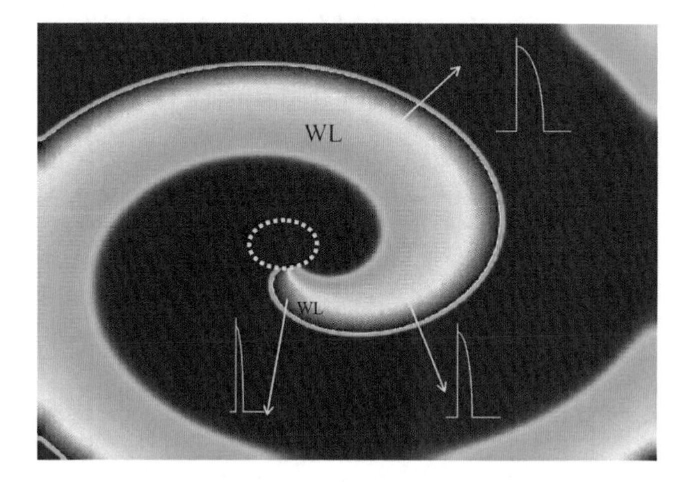

Fig. 5 Effect of IK1 expression on spiral waves. Simulation and illustration of the core of a spiral wave: *Colors* indicate the phase of the excitation status (*blue* resting state, *red-orange* plateau phase, *green* repolarization). Three action potentials are shown, each located at a different distance from the core. The action potential closest to the core is of shortest duration. This is because the electrotonic influence of the resting (*blue*) tissue is largest at the core. This interaction has a marked influence on the frequency of rotation. *WL*. (After [31])

Figure 5 illustrates the relationship between the shape of the spiral in the center and expression of IK1. In the center of the spiral wave, at the so-called phase singularity, the activated region (the "tip of the spiral") is exposed to the electrotonic interaction of the larger resting region. Because of this spatial mismatch, the electrotonic effect of a repolarizing channel in the resting region will be particularly large, larger than during propagation of a planar wave. In the case of IK1, the first effect of increasing the expression of IK1 channels will be to hyperpolarize the cells in the resting state and consequently to remove inactivation from the Na^+ channels. The second effect will be to shorten the action potential duration (acceleration of late repolarization) in the center of the spiral, i.e., to shorten the wavelength of excitation [31]. The first effect will speed up the movement of the spiral in the core; the second effect will stabilize the rotor at a faster speed. If the excitation waves travel from the center of the rotor to regions with lower expression of IK1, the high frequency of local excitation can no longer be maintained and the waves fractionate into a pattern of fibrillatory propagation. Importantly, blockade of IK1 by Ba^{2+} interrupted ventricular fibrillation in these experiments. In a more recent study, Munoz et al. [32] have shown a significant effect of overexpression of IKs on spiral waves. Interventions that modulate expression of other ion channels and either modify the flow of depolarizing or repolarizing ion current are predicted to affect the location and the frequency of rotors as well. Regenerative therapies, which are aimed at increasing the functional cell mass, need to take the potential effect of heterogeneities in ion channel expression into account.

3.2 *The Role of Heterogeneity in Cell-to-Cell Coupling*

Two murine models involving genetic ablation of connexin 43 (Cx43) in ventricular myocardium suggest that heterogeneity in connexins expression may represent an important factor in arrhythmogenesis. In a first model, conditional genetic ablation of Cx43 decreased continuously during birth and amounted to 60% of the control level after 25 days and to 20% of the control level after 45 days (so-called old Cx43 conditional knockout, O-CKO, mice) [33]. The Kaplan–Meier survival curves demonstrated an almost normal survival for 45 days and a rapid increase of sudden death thereafter, with 80% cumulative death rate after approximately 80 days, which occurs 35 days after the beginning of the arrhythmic episodes. After 45 days, propagation velocity had decreased to about 50% of normal. Interestingly, the marked decrease in expression of ventricular Cx43, which is the main ventricular connexin, was associated with a marked increase in heterogeneity of Cx43 reflected by an increase in the coefficients of variance of the average Cx43 signal measured within a relatively small tissue section (region of interest 700×500 μm). In a second model, the same group of authors created a chimeric model of heterogeneous Cx43 expression (CM-Ex) growing from blastocysts consisting of Cx43-deficient stem cells and Cx43 wild-type cells [34]. In total, 30 viable chimeric mice (23 mice with marked mosaicism) developed and showed no signs of illness. Within a period of

5 months, only one of the 23 mice died suddenly. Both analysis of contractility and analysis of propagation behavior showed marked anomalies. In contrast to the model involving O-CKO, contractility was markedly decreased and propagation at the ventricular surface showed a marked irregularity in electrical propagation. Comparison of the O-CKO model with the CM-Ex model raises questions which might also be relevant for the situation of implantation of cells into host tissue. Thus, the effect of heterogeneous reduction of connexin expression depends on the type of heterogeneity and is different in a mosaic, patchy ablation pattern from heterogeneous ablation that occurs ubiquitously at a smaller scale (Fig. 6).

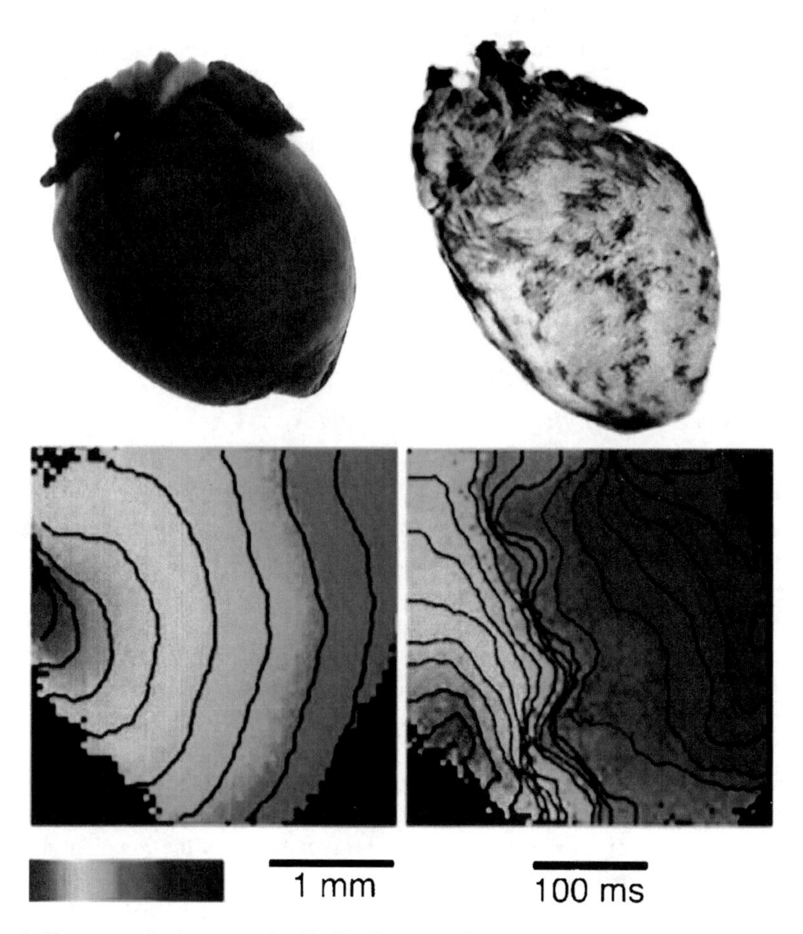

Fig. 6 Heterogeneity in connexin 43 (Cx43) expression and ventricular excitation. *Top left*: "*Blue*" heart with normal Cx43 expression expressing ROSA-26 β-galactosidase transgene. *Top right*: Chimeric heart with *yellowish*-appearing areas devoid of Cx43 expression. *Bottom left*: Epicardial isochrone map from normal (*blue*) mouse heart, showing regular centripetal spread from the stimulus site. *Bottom left*: Centripetal epicardical spread measured on the epicardial surface of a chimeric heart. Note that heterogeneity in Cx43 expression induces "crowding" of isochrones, reflecting slowing of propagation, and irregular isochrones, indicating local heterogeneity in propagation (isochrones are separated by 1 ms). (After [34])

The chimeric, patchy ablation (CM-Ex) appears to be less arrhythmogenic than high-degree ablation at a microscopic level (O-CKO), despite the fact that the first model produces a highly irregular pattern of electrical excitation. Another contrast between the two models relates to the observation that the high degree of Cx43 ablation in O-CKO mice in not associated with a decrease in ventricular contractility, whereas the chimeric ablation (CM-Ex) of Cx43 is. Because of the unexpected observation that a very high degree of ventricular Cx43 ablation (O-CKO mice) can exist in a heart that shows almost normal contractility, it has been postulated that the transfer of the electrical impulse from cell to cell, ensuring excitation and contraction of the totality of cells, may be transmitted by processes other than electrical current flow through gap junction channels (so-called electrical field transmission [35]). Although the existence of electrical field transmission is disputed, its existence would have a very important impact on the concepts used for myocardial regeneration. Presently, many basic questions related to the functional effect of heterogeneous connexin ablation remain unanswered. However, the existing experimental and theoretical data suggest that detailed, quantitative knowledge of cell-to-cell coupling is required to judge the "nonarrhythmogenic safety" in procedures involving regeneration of functional myocardial tissue.

3.3 The Role of Tissue Discontinuity and Interactions Between Myocytes and Nonmyocytes

Electrical cross talk between cells of the working myocardium and nonmyocytes has been known for several decades. It received major attention with the possibility of studying this interaction at high resolution in models of cell cultures (see [36] for references). As already mentioned, it has been shown that embryonic cells can functionally fully integrate into adult myocardium [27]. Also, it has been shown in several models that transfected stem cells can integrate into host myocardium. Moreover, it seems undisputed that intrinsic cardiac tissue can electrically couple to transplanted adult, differentiated tissue in humans [37]. Electrical interaction between recipient and donor hearts was diagnosed in approximately 10% of cardiac transplantations. The fact that these cases were curable by site-specific electrical ablation indicated that the electrical "reconnection" was confined to distinct electrical pathways, whereas the main part of the scar between the recipient and the donor heart functioned as an electrical insulator.

Reconnection to intrinsic myocardium is essential for functional integration of transplanted tissue. However, the presence of connective tissue within the myocyte compartment may also form an arrhythmogenic substrate. Both the potentially insulating effect of cells and the matrix of the connective tissue compartments, and inversely, coupling between the myocyte and nonmyocyte compartment are likely to have an impact on cell implantation and regeneration procedures. In a seminal study, de Bakker et al. [13] studied the details of propagation behavior in excised papillary muscles of patients with myocardial infarction. The anatomical substrate found in these muscles consisted of a marked fibrosis, separating parallel longitudinal tracts of surviving myocardium. Recordings of extracellular electrograms after stimulation indicated a very high degree of fractionation of electrical excitation (so-called longitudinal dissociation). A detailed analysis revealed that propagation followed a "zigzag" pathway changing the tracts of surviving myocardium at distinct sites. The observation that the electrical propagation velocity along the individual tracts was rapid and that the resulting overall propagation time was markedly prolonged led to the conclusion that the prolongation of excitation was mainly due to the *insulating effect of connective tissue* which slowed the excitation process via an increase in the total ("zigzag") length of the excitation path.

In addition to "zigzagging" propagation through surviving tissue in scars of chronic infarction, interaction between cardiomyocytes and specific noncardiomyocytes, so-called myofibroblasts, may furnish the substrate for or initiate arrhythmias. Myofibroblasts are fibroblast-like cells of mesenchymal origin that express α-smooth muscle actin. They are observed locally in a variety of cardiac diseases (hypertrophy-associated fibrosis, infarction, inflammation). Their origin has not been fully elucidated. In the in vivo heart, they may represent transdifferentiated cells of cardiac origin or cells stemming from the hemopoietic system (see [36] for a detailed review and discussion). In most cell cultures, myofibroblasts appear if cells are cultured in media containing serum and if no special measures are taken for elimination. The exact signaling cascades responsible for the myofibroblast phenotype in vitro and in vivo remain to be defined.

The arrhythmogenic potential of myofibroblasts is due to the fact that they can express connexins (Cx43 and connexin 45) and that they couple to cardiomyocytes. In the noncoupled state, myofibroblast are nonexcitable cells (lacking the elements to produce an action potential) exhibiting a resting membrane potential significantly more positive than that of cardiomyocytes. Therefore, coupling cardiomyocytes to myofibroblasts produces (1) a sink for electrotonic current and (2) a voltage gradient between the myofibroblast and the myocyte compartment. The potential arrhythmogenic effect of this interaction is complex and depends on the size and the spatial relationship of the respective compartments that interact. Single noncardiomyocytes interspersed between myocytes in culture produce small local conduction delays reflecting the effect of the coupled nonexcitable cell to drain electrotonic current ("sink effect") [38]. The effect of interspersion of a whole zone of myofibroblast bridging areas of excitable cardiomyocytes follows the laws of passive transmission of electrical impulses [39]. In patterned cultured strands of cardiomyocytes, it was shown that bridges as long as 300 μm can transmit action potentials successfully with a propagation delay increasing with bridge length. Projection of this finding to a situation of a cell transplant signifies that (1) the interface between the host and the transplanted cell population can consist of nonmyocytes, provided that they electrically couple, and (2) the probability of conduction delays, and thus of the formation of an arrhythmogenic substrate, increases with the width of the nonmyocyte interface. Two other interesting observations that may be relevant for regenerative cell therapy were made in cell culture experiments. First, the effect of myofibroblasts to depolarize cardiomyocytes increases with an increase in the contacting myofibroblast cell mass [40]. For a moderate interaction, depolarization of the myocytes by myofibroblasts produces a depolarization of the resting membrane to an extent that propagation velocity is increased. This is a well-known phenomenon termed "supernormal conduction" in classical electrophysiology [41, 42]. For a more pronounced interaction, propagation velocity decreases because further depolarization of the resting membrane produces inactivation of Na^+ channels. A second effect of the low level of resting membrane potential is that depolarizing electrotonic current is flowing into the myocyte compartment. If the spatial arrangement between the two compartments is such that enough electrotonic current is funneled into cardiomyocytes, spontaneous phase IV depolarization may initiate ectopic beats [43]. This situation is equivalent to the production of spontaneous ectopic activity by longitudinal current flow between two myocyte compartments [44].

4 Conclusion

Normal cardiac propagation results from a subtle equilibrium between the cardiac action potential, cell-to-cell coupling, and the architecture of the network composed of excitable cells. Remodeling of the molecular players determining this equilibrium can form the substrate of circus movement and reentry. This chapter described the basic mechanisms underlying the reentry substrate and emphasized the role of heterogeneities in ion channel and connexin expression, as well as interactions between myocytes and nonmyocytes in reentry. Theoretical and experimental results suggest that regeneration or implantation of functional cardiac tissue has to take these effects into account.

References

1. Kleber AG, Rudy Y (2004) Basic mechanisms of cardiac impulse propagation and associated arrhythmias. Physiol Rev, 84(2), 431–88.
2. Mines GR (1913) On dynamic equilibrium in the heart. J Physiol, 46(4–5), 349–83.
3. Allessie MA, Bonke FI, Schopman FJ (1977) Circus movement in rabbit atrial muscle as a mechanism of tachycardia. III. The "leading circle" concept: a new model of circus movement in cardiac tissue without the involvement of an anatomical obstacle. Circ Res, 41(1), 9–18.
4. Fast VG, Kleber AG (1997) Role of wavefront curvature in propagation of cardiac impulse. Cardiovasc Res, 33(2), 258–71.
5. Fenton FH, Cherry EM, Hastings HM, Evans SJ (2002) Multiple mechanisms of spiral wave breakup in a model of cardiac electrical activity. Chaos, 12(3), 852–92.
6. Jalife J, Berenfeld O (2004) Molecular mechanisms and global dynamics of fibrillation: an integrative approach to the underlying basis of vortex-like reentry. J Theor Biol, 230(4), 475–87.
7. Spach MS, Miller WT, III, Dolber PC, Kootsey JM, Sommer JR, Mosher CE, Jr. (1982) The functional role of structural complexities in the propagation of depolarization in the atrium of the dog. Cardiac conduction disturbances due to discontinuities of effective axial resistivity. Circ Res, 50(2), 175–91.

8. LeGrice IJ, Smaill BH, Chai LZ, Edgar SG, Gavin JB, Hunter PJ (1995) Laminar structure of the heart: ventricular myocyte arrangement and connective tissue architecture in the dog. Am J Physiol, 269(2 Pt 2), H571–82.
9. Dolber PC, Spach MS (1987) Thin collagenous septa in cardiac muscle. Anat Rec, 218(1), 45–55.
10. Spach MS, Heidlage JF, Dolber PC, Barr RC (2007) Mechanism of origin of conduction disturbances in aging human atrial bundles: experimental and model study. Heart Rhythm, 4(2), 175–85.
11. Allessie M, Ausma J, Schotten U (2002) Electrical, contractile and structural remodeling during atrial fibrillation. Cardiovasc Res, 54(2), 230–46.
12. Ausma J, van der Velden HM, Lenders MH, et al. (2003) Reverse structural and gap-junctional remodeling after prolonged atrial fibrillation in the goat. Circulation, 107(15), 2051–8.
13. de Bakker JM, van Capelle FJ, Janse MJ, et al. (1993) Slow conduction in the infarcted human heart. 'Zigzag' course of activation. Circulation, 88(3), 915–26.
14. Fast VG, Kleber AG (1995) Block of impulse propagation at an abrupt tissue expansion: evaluation of the critical strand diameter in 2- and 3-dimensional computer models. Cardiovasc Res, 30(3), 449–59.
15. Cabo C, Pertsov AM, Baxter WT, Davidenko JM, Gray RA, Jalife J (1994) Wave-front curvature as a cause of slow conduction and block in isolated cardiac muscle. Circ Res, 75(6), 1014–28.
16. Rohr S, Kucera JP, Fast VG, Kleber AG (1997) Paradoxical improvement of impulse conduction in cardiac tissue by partial cellular uncoupling. Science, 275(5301), 841–4.
17. Joyner RW (1982) Effects of the discrete pattern of electrical coupling on propagation through an electrical syncytium. Circ Res, 50(2), 192–200.
18. Joyner RW, van Capelle FJ (1986) Propagation through electrically coupled cells. How a small SA node drives a large atrium. Biophys J, 50(6), 1157–64.
19. van Capelle FJ, Durrer D (1980) Computer simulation of arrhythmias in a network of coupled excitable elements. Circ Res, 47(3), 454–66.
20. Shaw RM, Rudy Y (1997) Ionic mechanisms of propagation in cardiac tissue. Roles of the sodium and L-type calcium currents during reduced excitability and decreased gap junction coupling. Circ Res, 81(5), 727–41.
21. Janse MJ, Kleber AG (1981) Electrophysiological changes and ventricular arrhythmias in the early phase of regional myocardial ischemia. Circ Res, 49(5), 1069–81.
22. Janse MJ, Wit AL (1989) Electrophysiological mechanisms of ventricular arrhythmias resulting from myocardial ischemia and infarction. Physiol Rev, 69(4), 1049–169.
23. Beauchamp P, Choby C, Desplantez T, et al. (2004) Electrical propagation in synthetic ventricular myocyte strands from germline connexin43 knockout mice. Circ Res, 95(2), 170–8.
24. Rohr S, Kucera JP, Kleber AG (1998) Slow conduction in cardiac tissue, I: effects of a reduction of excitability versus a reduction of electrical coupling on microconduction. Circ Res, 83(8), 781–94.
25. Wang Y, Rudy Y (2000) Action potential propagation in inhomogeneous cardiac tissue: safety factor considerations and ionic mechanism. Am J Physiol Heart Circ Physiol, 278(4), H1019–29.
26. Furuta A, Miyoshi S, Itabashi Y, et al. (2006) Pulsatile cardiac tissue grafts using a novel three-dimensional cell sheet manipulation technique functionally integrates with the host heart, in vivo. Circ Res, 98(5), 705–12.
27. Rubart M, Pasumarthi KB, Nakajima H, Soonpaa MH, Nakajima HO, Field LJ (2003) Physiological coupling of donor and host cardiomyocytes after cellular transplantation. Circ Res, 92(11), 1217–24.
28. Antzelevitch C, Dumaine R (2002) Electrical heterogeneity in the heart: physiological, pharmacological and clinical implications. In: Page E, Fozzard H, Solaro R, eds. *Handbook of Physiology, Section 2: The Cardiovascular System*. New York: Oxford University Press: 654.
29. Samie FH, Berenfeld O, Anumonwo J, et al. (2001) Rectification of the background potassium current: a determinant of rotor dynamics in ventricular fibrillation. Circ Res, 89(12), 1216–23.
30. Noujaim SF, Pandit SV, Berenfeld O, et al. (2007) Up-regulation of the inward rectifier K+ current (I K1) in the mouse heart accelerates and stabilizes rotors. J Physiol, 578(Pt 1), 315–26.

31. Beaumont J, Davidenko N, Davidenko JM, Jalife J (1998) Spiral waves in two-dimensional models of ventricular muscle: formation of a stationary core. Biophys J, 75(1), 1–14.
32. Munoz V, Grzeda KR, Desplantez T, et al. (2007) Adenoviral expression of IKs contributes to wavebreak and fibrillatory conduction in neonatal rat ventricular cardiomyocyte monolayers. Circ Res, 101(5), 475–83.
33. Danik SB, Liu F, Zhang J, et al. (2004) Modulation of cardiac gap junction expression and arrhythmic susceptibility. Circ Res, 95(10), 1035–41.
34. Gutstein DE, Morley GE, Vaidya D, et al. (2001) Heterogeneous expression of gap junction channels in the heart leads to conduction defects and ventricular dysfunction. Circulation, 104(10), 1194–9.
35. Kucera JP, Rohr S, Rudy Y (2002) Localization of sodium channels in intercalated disks modulates cardiac conduction. Circ Res, 91(12), 1176–82.
36. Rohr S (2009) Myofibroblasts in diseased hearts: new players in cardiac arrhythmias? Heart Rhythm, 6(6), 848–56.
37. Lefroy DC, Fang JC, Stevenson LW, Hartley LH, Friedman PL, Stevenson WG (1998) Recipient-to-donor atrioatrial conduction after orthotopic heart transplantation: surface electrocardiographic features and estimated prevalence. Am J Cardiol, 82(4), 444–50.
38. Fast VG, Darrow BJ, Saffitz JE, Kleber AG (1996) Anisotropic activation spread in heart cell monolayers assessed by high-resolution optical mapping. Role of tissue discontinuities. Circ Res, 79(1), 115–27.
39. Gaudesius G, Miragoli M, Thomas SP, Rohr S (2003) Coupling of cardiac electrical activity over extended distances by fibroblasts of cardiac origin. Circ Res, 93(5), 421–8.
40. Miragoli M, Gaudesius G, Rohr S (2006) Electrotonic modulation of cardiac impulse conduction by myofibroblasts. Circ Res, 98(6), 801–10.
41. Moore EN, Spear JF, Fisch C (1993) "Supernormal" conduction and excitability. J Cardiovasc Electrophysiol, 4(3), 320–37.
42. Spear JF, Moore EN (1974) Supernormal excitability and conduction in the His–Purkinje system of the dog. Circ Res, 35(5), 782–92.
43. Miragoli M, Salvarani N, Rohr S (2007) Myofibroblasts induce ectopic activity in cardiac tissue. Circ Res, 101(8), 755–8.
44. Katzung BG, Hondeghem LM, Grant AO (1975) Letter: cardiac ventricular automaticity induced by current of injury. Pflugers Arch, 360(2), 193–7.

Integration of Stem Cells into the Cardiac Syncytium: Formation of Gap Junctions

Peter R. Brink, Ira S. Cohen, and Richard T. Mathias

Abstract The heart is a functional syncytium. This means that each myocyte is electrically connected to other myocytes in its vicinity. Functional electrical coupling requires gap junctions. Gap junctions are composed of connexins. There are over 20 connexins in the human genome. If either electrical or mechanical regeneration of cardiac function is to be achieved via cell therapy (exclusive of paracrine effects), then the delivery cells must couple to the existing myocytes via gap junctions. In this chapter we review the basic physiology of gap junctions and what is known about their expression in cardiac myocytes and in stem cells. Given that multiple connexins are expressed in myocytes, we consider the types of gap junction channels that can be formed and their prevalence in a calculation of independent assortment of equally expressed connexins. Finally, myocytes are not the only cells present in the heart; there is a substantial presence of fibroblasts and endothelial cells. Fibroblasts do not express the same assortment of connexins as myocytes and do not in general form gap junctions with them. When stem cells are considered for cardiac regeneration, their expression of cardiac connexins is usually confirmed. However, it is just as important to confirm the absence of fibroblast connexins which could potentially create a sink for the local circuit currents that generate the cardiac impulse. Given the excitement generated by induced pluripotent stem cells which are derived from fibroblasts, it will be particularly important to demonstrate that the new myocytes generated from these cells express only cardiac-myocyte-specific connexins.

Keywords Gap junction • Connexin • Fibroblast • Cardiac myocyte • Arrhythmia • Embryonic stem cell • Induced pluripotent stem cell • Human mesenchymal stem cell

P.R. Brink (✉)
Department of Physiology and Biophysics, Stony Brook University,
Stony Brook, NY 11794, USA
e-mail: peter.brink@sunysb.edu

I.S. Cohen and G.R. Gaudette (eds.), *Regenerating the Heart*, Stem Cell Biology and Regenerative Medicine, DOI 10.1007/978-1-61779-021-8_17,
© Springer Science+Business Media, LLC 2011

1 Introduction

Cellular cardiomyoplasty is considered a potential therapy for both mechanical and electrical regeneration of cardiac function. In this paradigm for mechanical repair, cardiac myocyte cell number must increase. If cell division is not involved, then the delivered cells must ultimately become cardiac myocytes and "*integrate*" into the cardiac syncytium. If electrical repair is to be induced, then the delivered cells whether differentiated to myocytes or stem cells simply used as a platform to deliver an electrical gene product must also "*integrate*" into the cardiac syncytium. The cardiac myocytes in the heart form a functional electrical syncytium. This means that each myocyte is electrically connected to all other cardiac myocytes. It is this electrical coupling that guarantees mechanical synchrony necessary to develop intrachamber pressure. In this chapter we consider integration of stem cells or their differentiated products into the cardiac syncytium. Electrical integration occurs via the formation of gap junction channels. These channels are the only pathway from the interior of one cell to the interior of another which excludes the extracellular space. These channels are formed of subunits called connexins. Over 20 different connexins are expressed in the human genome and each cell type expresses a particular subset. Not all connexins will form gap junction channels with each other and so it is important that the delivered cells express the subset necessary for functional coupling with their target cells and also do not express those sufficient to couple to other cell populations within the cardiac parenchyma. It is the purpose of this chapter to review current knowledge of connexins in both cardiac myocytes and the various stem cell types currently under investigation for cardiac regeneration.

2 Gap Junctions Formed by Connexins

Gap junction channels in vertebrates are formed from subunit proteins called connexins. Each gap junction channel is composed of two hemichannels, each of which contains six connexins. When two cells are in close apposition, hemichannels from each cell can link together via the extracellular loops of the component connexins to form gap junction channels. This composite channel represents a unique cell-to-cell pathway because it is the only form of direct communication that excludes the extracellular space. For reasons that are not completely understood, but have been attributed to lipid membrane domains including lipid rafts, gap junction channels tend to aggregate and form plaques containing tens to thousands of channels [1].

3 Unitary Conductance and Permeability: Functional Diversity

The single channel conductance and selectivity/permselectivity of gap junction channels is highly dependent on the type of subunit connexin [2–5]. For example connexin 43 (Cx43) single channel conductance measured in Cs^+ or K^+ salt is approximately 90 pS, while that of connexin 40 (Cx40) is 140 pS. Two other connexins found in a number of tissues are connexin 32 (Cx32) and connexin 26 (Cx26), whose unitary conductances are 50 and 150 pS, respectively. The latter two have been classified as β connexins, whereas Cx43 and Cx40 are classified as α connexins. The α/β designation is based on sequence homology [6]. α connexins do not generally combine with β connexins to form gap junction channels [7].

The selectivity or permselectivity properties of gap junction channels also differ. The permeability ratio for Lucifer yellow (LY) relative to K^+ is 1/40 for Cx43, 1/400 for Cx40, and 1/230 for Cx26 [2, 5, 8]. A number of other publications have shown that channels formed from various connexins have different permeabilities to a variety of probes [3, 9].

The transfer rates of endogenous solutes, such as nucleotides and second messengers, have also been assessed for channels made from a number of different connexins. For example the permeability ratio for cyclic AMP (cAMP) relative to K^+ is 1/6 for Cx43, 1/60 for Cx40, and 1/35 for Cx26 [8]. Again, both the LY and the cAMP data illustrate the point that permeability is highly dependent on the connexin type. Other connexin types commonly found either in the vasculature or in the heart that have not been studied as completely as Cx43 or Cx40 are connexin 37 (Cx37), found most abundantly in endothelium [10], and connexin 45 (Cx45), found in select regions of the heart and vascular wall [11]. Unitary conductances range from 400 pS for Cx37 to 27 pS for Cx45. These data, once again, point to the diversity of properties in the multigene family of connexins. The cell-to-cell transfer of oligonucleotides and small interfering RNA (siRNA) is also highly dependent on connexin type.

Early studies of gap junction channels generated a consensus view that they were not permeable to molecules with molecular masses greater than 1.5 kDa [12, 13] or with minor diameters greater than 1.2–1.3 nm [14]. Recent publications showing the passage of rod-shaped oligonucleotides and siRNA with minor diameters of 1.2 nm or less but with major diameters of 3–8 nm and masses up to 4–5 kDa have redefined the limits of gap junction channel permeability [15, 16]. The ability of siRNA to traverse gap junction channels adds yet another dimension to the role gap junction channels play in coordinated tissue functions: besides allowing the movement of monovalent ions, second messengers, and metabolites, they are also able to facilitate gene silencing and potentially influence cell phenotype. Figure 1 illustrates a semi-log plot of the permeability of Cx43, Cx40, and Cx26 for Na^+,

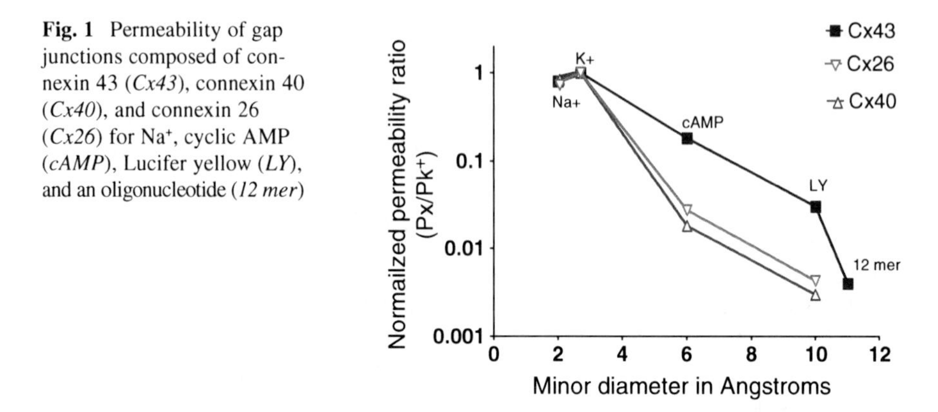

Fig. 1 Permeability of gap junctions composed of connexin 43 (*Cx43*), connexin 40 (*Cx40*), and connexin 26 (*Cx26*) for Na^+, cyclic AMP (*cAMP*), Lucifer yellow (*LY*), and an oligonucleotide (*12 mer*)

cAMP, and LY normalized relative to K^+ and plotted against the minor diameter of the specific solute. For Cx43, its permeability to a morpholino 12 nucleotides long is also plotted (12 mer).

The data shown in Fig. 1 illustrate that connexins possess both size and charge selectivity. For all three, with increased minor diameter of a solute, the permeability is reduced. Charge selectivity is also illustrated when one considers that Cx43 has a unitary conductance of approximately 90 pS and Cx40 and Cx26 have conductances nearly double that but are less permeable to cAMP and LY by an order of magnitude.

4 Connexin Types Within the Heart

The α connexins Cx43, Cx40, and Cx45 are ubiquitously expressed in the human heart [11, 17]. Gap junction channels formed by these connexins conduct monovalent ions efficiently, generating the local circuit currents necessary for action potential propagation. Cx43 is expressed in the ventricles, atria, and Purkinje fibers, whereas Cx40 is expressed in the sinoatrial node, atria, bundle branches, and Purkinje fibers. Cx45 is expressed in the sinoatrial and atrioventricular nodes and the Purkinje fibers [11].

Connexin 30.2 (Cx30.2) is expressed in the atrioventricular node of mice [18] and is under specific regulatory control during development [19]. The human ortholog, connexin 31.9 (Cx31.9), has been identified [20] and White et al. [21] have determined its unitary conductance and gating properties and demonstrated that it is expressed in human heart. The properties thus far demonstrated for Cx30.2/Cx31.9 strongly suggest that it is an essential component of the longitudinal resistance within the atrioventricular node and hence potentially instrumental in determining conduction velocity. However, its exact distribution within the atrioventricular node has yet to be determined. The fact that multiple connexins are expressed in the heart and in specific cell types suggests the possibility of connexin mixing within any one gap junction channel. Would mixing affect function and is there in fact evidence for mixing of cardiac connexins?

5 Coexpression of Connexins in Heteromeric and Heterotypic Channels: Implications for Myocytes and Stem Cells

In general, gap junction channels can be classified as homotypic (all connexin subunits are the same), heterotypic (each hemichannel is made from a different connexin), or heteromeric (each hemichannel is made from two or more connexins). Coexpression of two or more connexins within cells can give rise to heteromeric channels. Cx43 and Cx40 have been shown to form heteromeric channels in vitro [2] and other cardiac connexins, such as Cx30.2, have also been shown to form heteromeric channels with Cx40, Cx43, and Cx45 in vitro [22]. Other noncardiac connexins, including the β connexins Cx32 and Cx26, also have the ability to form heteromeric and heterotypic channels [1, 23].

In general, heteromeric channel formation in vitro has been deduced from macroscopic and microscopic junctional currents as well as from immunostaining, co-immunoprecipitation assays, and western blot analysis. In vivo evidence has relied more heavily on immunostaining, a recent example being the demonstration of heterotypic and possibly heteromeric channels composed of connexin 30 and Cx43 in astrocytes within the CNS.

Stem cells are no exception. Human adult mesenchymal stem cells (MSCs) are known to coexpress Cx43 and Cx40 [24, 25], whereas murine induced pluripotent stem cells (IPSCs) have been shown to express Cx43 [26]. Murine embryonic stem cells (ESCs) are known to coexpress at least three connexins, Cx45, Cx43, and connexin 31 [27]. In human ESCs Cx45, Cx43, and Cx40 expression has been detected using both reverse transcription PCR and immunostaining. Reverse transcription PCR has also revealed the presence of messenger RNA (mRNA) for 18 of the known 20 human connexins [28] in human ESCs.

The expression of connexins in MSCs, ESCs, and IPSCs is essential if they are to be used as a source for a cell-based delivery system in vivo. Likewise, the cells within a target tissue must express compatible connexins. In the case of stem cells and myocytes, both express Cx43 and Cx40, eliminating the compatibility issue, but raising the issue of the role of heteromeric channels and their influence on the efficacy of delivery.

First, for cardiac myocytes a fundamental question is: Do heteromeric channels affect conduction of an impulse within the heart? Does alteration of the composition of heteromeric channels have the potential to be proarrhythmic or conversely antiarrhythmic? There are no experimental data in the literature showing whether coexpressed connexins within a given channel or population of channels between myocytes will significantly affect conduction relative to homotypic channels. Further, in vitro data from nonmyocyte cell types reveal that biophysical properties of heteromeric channels, such as transjunctional voltage dependence and unitary conductance, are not limiting for current passage from cell to cell, or the movement of small solutes or oligonucleotides [2, 15, 29, 30].

Second, for integration of stem cells into a tissue such as the myocardium the issue is whether coexpression of connexins in the delivery cells acts to hamper, optimize, or significantly alter gap-junction-mediated delivery of gene products in the form of either exogenous membrane currents or small molecules.

Although experimental data are not readily available, it is possible to make some predictions about coexpression of connexins in terms of the types and distribution of heteromeric, heterotypic, and homotypic channels found between any two cells coexpressing connexins.

6 A Predictive Model for Heteromeric Gap Junction Hemichannel and Channel Composition

The following is a theoretical analysis where each cell of a pair is coexpressing the same two types of connexins. We assume the two different connexins will form heteromeric, heterotypic, and homotypic channels with the same probability.

6.1 Hemichannels

To calculate the probability of having k CxA subunits out of a total of six subunits per hemichannel, one uses the binomial distribution: where $n=6$ connexins per hemichannel, k is the number of one type of connexin (CxA), p is the probability of that type connexin, $f(k)$ is the probability that k out of six connexins are CxA,

$$f(k) \equiv \binom{n}{k} p^k (1-p)^{n-k}$$

and the combinatorial symbol means

$$\binom{n}{k} = \frac{n!}{k!(n-k)!}.$$

The combinatorial calculates the number of different ways that there can be k CxA subunits out of a total of $n=6$ subunits in a hemichannel.

Example 1. Assume that two connexins are expressed in equal amounts, so $p=0.5$, $1-p=0.5$, and $p^k(1-p)^{n-k}=0.0156=1/64$. There are 64 possible combinations, each with the same probability of 1/64:

$$f(k) \binom{6}{k}$$

$$
\begin{array}{lcccccc}
f(0) & = & 1 & \times & 0.0156 & = & 0.0156 \\
f(1) & = & 6 & \times & 0.0156 & = & 0.0938 \\
f(2) & = & 15 & \times & 0.0156 & = & 0.2344 \\
f(3) & = & 20 & \times & 0.0156 & = & 0.3125 \\
f(4) & = & 15 & \times & 0.0156 & = & 0.2344 \\
f(5) & = & 6 & \times & 0.0156 & = & 0.0938 \\
f(6) & = & \underline{1} & \times & 0.0156 & = & \underline{0.0156} \\
 & & 64 & & & & 1.00
\end{array}
$$

Example 2. Assume CxA is expressed 9/1 over the other connexin, so $p=0.9$ and $1-p=0.1$:

$$f(k)\binom{6}{k} \quad 0.9^k 0.1^{6-k}$$

$$
\begin{array}{lcccccc}
f(0) & = & 1 & \times & 10^{-6} & = & 10^{-6} \\
f(1) & = & 6 & \times & 9\times10^{-6} & = & 5.4\times10^{-5} \\
f(2) & = & 15 & \times & 8.1\times10^{-5} & = & 1.2\times10^{-3} \\
f(3) & = & 20 & \times & 7.3\times10^{-4} & = & 0.015 \\
f(4) & = & 15 & \times & 6.56\times10^{-3} & = & 0.098 \\
f(5) & = & 6 & \times & 0.059 & = & 0.354 \\
f(6) & = & \underline{1} & \times & 0.531 & = & \underline{0.531} \\
 & & 64 & & & & 1.00
\end{array}
$$

The binomial distribution does not distinguish between the different physical arrangements of making a hemichannel. Thus, the above analysis assumes it is the number of CxA subunits that makes a hemichannel distinct and not how they are arranged. There is no compelling experimental evidence for configurational differences, so this may be a valid assumption.

6.2 Whole Channels

Although it seems reasonable to assume the number of CxA subunits is the critical parameter in determining the properties of a hemichannel, the same is not true for whole channels. For example, consider two ways of making a whole channel containing six CxA subunits, as shown in Fig. 2. Most physiologists would expect these two channels to have distinct properties, so the binomial theorem should not

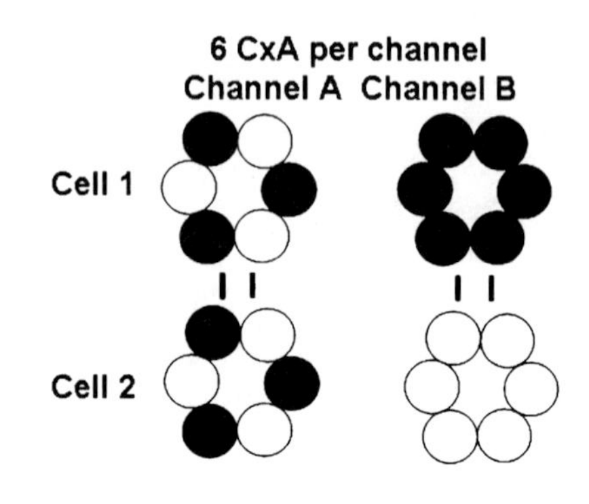

Fig. 2 Two distinct ways of forming a channel that is composed of six subunits of CxA and six of CxB

be applied to whole channels, since it would not distinguish between the two channels shown. An alternative approach is described below.

7 Number of Possible Combinations

Assume one cell can contain N distinct heteromeric hemichannels. If the number of CxA subunits is the criterion for a distinct hemichannel, then there are seven distinct channels containing zero, one, two, three, four, five, and six CxA subunits. If distinctions between physical arrangements of subunits are made, then N would be larger.

$$\text{Cell 1}: \quad 1 \quad 2 \quad 3 \quad 4 \quad 5,\ldots,N$$

$$\text{Cell 2}: \quad 1 \quad 2 \quad 3 \quad 4 \quad 5,\ldots,N$$

Hemichannel 1 in cell 1 can mate with hemichannels 1 to N in cell 2 to form whole channels $1=1$, $1=2,\ldots,$ $1=N$ (=implies mating of hemi channels). This yields N combinations. Similarly, hemichannel 2 in cell 1 can mate with hemichannels 1 to N in cell 2 to from N whole channels. However, with regard to single channel conductance, the appropriate assumption is $2=1$ and $1=2$ are not distinct; in this situation there are $N-1$ distinct combinations for hemichannel 2 in cell 1. By similar reasoning, hemichannel K in cell 1 can form $N-K-1$ distinct combinations with hemichannels in cell 2. Thus, the total possible combinations is given by

$$S_N = \sum_{K=0}^{N-1} N-K = \sum_{K=1}^{N} K = \frac{N(N+1)}{2}.$$

On the other hand, if we assume $1=2$ and $2=1$ are distinct, as might be the case if we measure voltage dependence, the number of possible combinations increases to N^2.

7.1 The Probability of Each Combination

The above analysis says nothing about the probability of a particular combination. The binomial distribution calculates the probability of k subunits out of six in each hemichannel. Thus, if we want the probability of $k=2$ in cell 1 mating with $k=3$ in cell 2, that will be given by the product $f(2)f(3)$. A matrix of probabilities (expressed as a percentage) can be constructed for the case where each connexin is equally expressed and has equal probability of association with each of the other expressed connexin types and is given below for the case where $1=2$ and $2=1$ are distinct.

		k cell 2						
		0	1	2	3	4	5	6
k cell1	0	0.02	0.15	0.37	0.49	0.37	0.15	0.02
	1	0.15	0.88	2.20	2.93	2.20	0.88	0.15
	2	0.37	2.20	5.49	7.33	5.49	2.20	0.37
	3	0.49	2.93	7.33	9.77	7.33	2.93	0.49
	4	0.37	2.20	5.49	7.33	5.49	2.20	0.37
	5	0.15	0.88	2.20	2.93	2.20	0.88	0.15
	6	0.02	0.15	0.37	0.49	0.37	0.15	0.02

Thus, if distinct hemichannels are determined simply by how many subunits of CxA are present, then there are $7^2=49$ distinct whole channels, each with the probability (expressed as a percentage) given in the above matrix.

With this same assumption, but for the case where $1=2$ and $2=1$ are not distinct, the number of distinct whole channels is given by $7(7+1)/2=28$. The probabilities (expressed as a percentage) of these 28 whole channels are given in the matrix below.

		k cell 2						
		0	1	2	3	4	5	6
k cell 1	0	0.02	0.30	0.74	0.98	0.74	0.30	0.04
	1		0.88	4.40	5.86	4.40	1.76	0.30
	2			5.49	14.66	10.98	4.40	0.74
	3				9.77	14.66	5.86	0.98
	4					5.49	4.40	0.74
	5						0.88	0.30
	6							0.02

Note that the probabilities are heavily weighted in the central region of this matrix. The whole channels composed of k, j in cell 1, cell 2 for

1,2	1,3	1,4	
	2,2	2,3	2,4	2,5
		3,3	3,4	3,5
			4,4	4,5

have a total probability of 90.37%, whereas they are just 12 of the 28 possible combinations. Moreover, the channels have a total

$$\begin{matrix} 2,3 & 2,4 \\ 3,3 & 3,4 \end{matrix}$$

probability of 50.07%, whereas they are just four of the 28 possible combinations. Since more than 50% of the channels have this small grouping of subunit compositions, one expects the single channel conductance histogram to have less variance than when two types of connexins make only homotypic channels. This is illustrated by the following example.

7.2 *Single Channel Conductance Histograms*

Assume homotypic channels made of CxA have a conductance of 60 pS whereas homotypic channels made of CxB have a conductance of 120 pS. When CxA and CxB are coexpressed in equal amounts, assume they form heterotypic channels with conductances that are simply proportional to the number of subunits of each connexin, with each subunit of CxA contributing 5 pS and each subunit of CxB contributing 10 pS. For example, a channel containing four CxA and eight CxB subunits would have a conductance of $4 \times 5 + 8 \times 10 = 100$ pS. Obviously, this is highly contrived, but it illustrates the effect of mixing on histograms. The single channel conductance matrix is given below.

		k cell 2						
		0	1	2	3	4	5	6
k cell 1	0	120	115	110	105	100	95	90
	1		110	105	100	95	90	85
	2			100	95	90	85	80
	3				90	85	80	75
	4					80	75	70
	5						70	65
	6							60

Based on the previous matrix of probabilities, the number of observations out of 100 can be assigned to each conductance. This event histogram is shown in Fig. 3. Of course, this analysis does not include variance in experimental measurements of each conductance, but one nevertheless gets the idea.

If CxA and CxB did not form heteromeric channels, there would be just two points, one at 60 pS with an occurrence of 50 and one at 120 pS with an occurrence of 50. If one does the thought experiment of making each point into a Gaussian distribution, then the two homotypic channels would either give a bimodal-like distribution, or if the variance was large enough, give a very broad

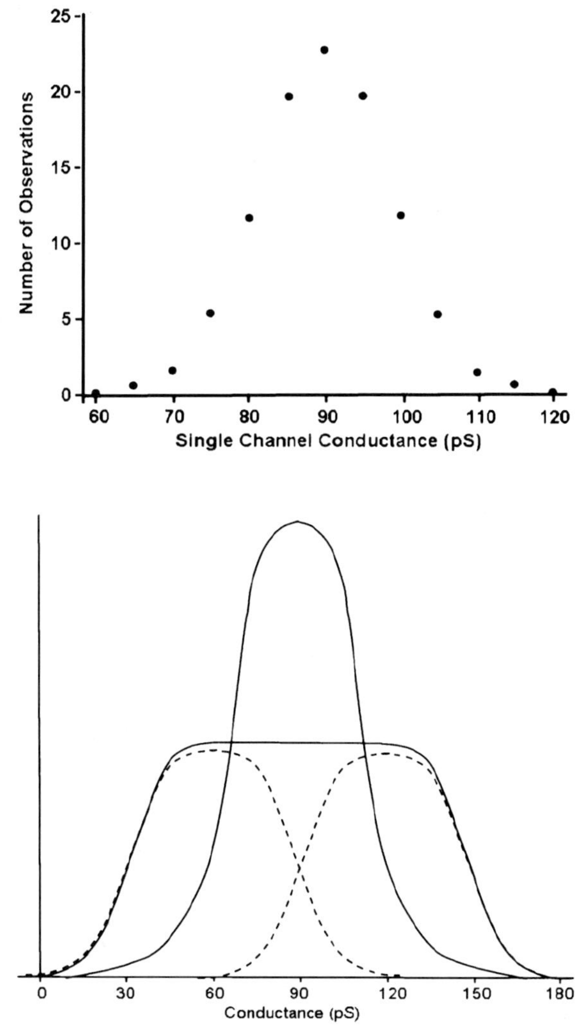

Fig. 3 An idealized single channel conductance histogram for 100 observations of heteromeric channels made from two different types of connexins. In a real experiment, each point would be the mean of a Gaussian distribution of measured conductances. The conductance values are based on assumptions given in the text

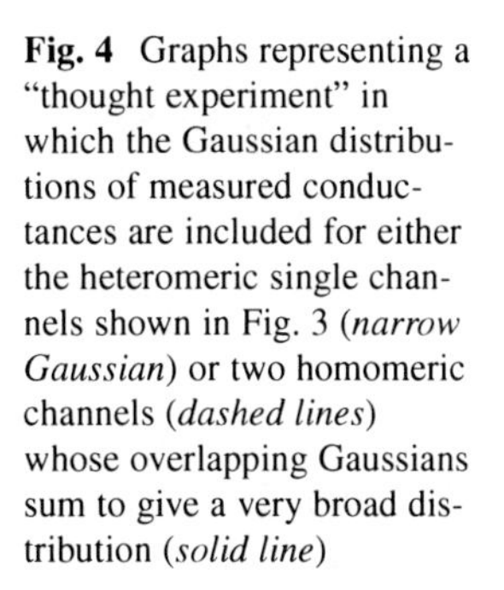

Fig. 4 Graphs representing a "thought experiment" in which the Gaussian distributions of measured conductances are included for either the heteromeric single channels shown in Fig. 3 (*narrow Gaussian*) or two homomeric channels (*dashed lines*) whose overlapping Gaussians sum to give a very broad distribution (*solid line*)

histogram centered at 90 pS, such as shown in the Fig. 4, whereas the heteromeric channels would have a much narrower distribution, again centered at 90 pS as also shown in Fig. 4.

The dashed lines in Fig. 4 are supposed to represent the individual distributions centered at 60 and 120 pS, of single channel conductance for the two homotypic channels. The broad flat distribution is supposed to represent the sum of the individual distributions. This would represent the situation where the two connexins do not form heteromeric channels. The taller, narrower distribution is supposed to represent the sum of all the distributions centered at the points shown in Fig. 4

(i.e. the heteromeric distribution). Functionally, this suggests that heteromeric forms will create an averaged junctional conductance somewhere between the optimum for either homotypic form.

The experimental evidence for heteromeric forms is consistent with the probabilistic model presented here. Further, coexpression results in weak voltage dependence and also appears to have high open probabilities much like homotypic forms [4, 29]. Experimentally derived amplitude histograms of steady-state multichannel activity from coexpressing cell pairs have not yet been obtained, making direct comparison with the model predictions impossible.

8 Gap Junctions and Longitudinal Resistance in the Heart: Their Influence on Action Potential Propagation and Arrhythmias

The classic studies of Barr et al. [31] and Weidmann [32] were the first to establish that currents associated with action potential propagation flow from myocyte to myocyte via gap junctions, and that reductions in gap-junction-mediated communication result in propagation failure. A number of subsequent studies detailed the essential role gap junctions play in the propagation of cardiac action potentials [4, 33–35]. An important factor in the conduction of the cardiac action potential is longitudinal resistance arising from the cytoplasm and gap junctional membranes connecting cellular interiors. In fact, conduction velocity, θ, is inversely proportional to the square root of the longitudinal resistance, R_i ($\theta \propto 1/\sqrt{R_i}$). R_i is composed of both cytoplasmic resistance and junctional membrane resistance such that $R_i = R_{cyto} + R_j$. In a typical ventricular myocyte that is 80 μm in length and 10–15 μm in width, the total longitudinal resistance contributed by the cytoplasm is approximately 1 MΩ or 1,000 nS assuming the resistivity of the cytoplasm to be 100–150 Ωcm, where the junctional resistance at the intercalated disc is approximately 2–5 MΩ (a conductance of approximately 500–200 nS). These values make it clear that junctional resistance at the intercalated disc is the dominant determinant of longitudinal resistance. It has been shown that reducing junctional conductance slows conduction and can be a major determinant in conduction failure [34, 35]. A similar conclusion has been drawn from optical mapping of perfused ventricular wedge preparations and from in vitro studies [36, 37].

Reentrant rhythms are usually associated with abnormalities of conduction and may be further exacerbated by abnormal dispersion of repolarization within the atria or ventricles [38]. A reentrant pathway can be anatomically defined by association with a scar or it can functionally lead to or result in circular reentry [38]. Gap junctions and their component connexins are critical contributors to the conducting properties in the heart in general and in the reentrant circuit in particular. Studies of cardiac failure and reentrant ventricular arrhythmias have shown that reduced expression and abundance of Cx43 results in a reduction in the effective space constant relative to controls [36, 39].

9 Stem Cell Integration into the Myocardium: Consequences for Conduction

9.1 Adult Mesenchymal Stem Cells

Recall that adult MSCs are an autologous or allogenic cell population that forms gap junctions composed of Cx43 and Cx40 [24, 25, 40]. An important question with regard to arrhythmias is: Will the integration of stem cells into the myocardium be proarrhythmic owing to their ability to form gap junctions and act as a sink or possible source of membrane currents? The study by Potapova et al. [40] addressed these concerns by first demonstrating that the delivery of approximately one million adult MSCs to the left ventricular wall of the canine heart did not result in any demonstrable arrhythmia. The stem cells were shown to form gap junctions with ventricular myocytes in vivo and deliver currents generated by HCN2 pacemaker channels expressed in the MSCs. They were able to pace the heart. The control injection of an equivalent number of stem cells without HCN2 did not generate pacing nor did it produce an arrhythmia. These results suggest that genetically engineered MSCs can integrate into the myocardial syncytium without proarrhythmic consequences.

Effective integration of MSCs, able to form gap junctions, is presumably dependent to some degree on the expression of adhesion molecules such as cadherins [41]. In vitro experimental evidence shows that MSCs robustly express cadherins [42]. These data taken in combination with the studies of Potapova et al. [40] are strong indicators of cadherin expression in vivo.

Once exogenous MSCs have been injected into the myocardium, neither rapid cell division nor rapid cell death would be desirable. In vitro evidence for slow proliferation of MSCs [43] demonstrated that over a 44-day interval, five cell cycles occurred. These data imply that cell proliferation would be slow in vivo. If proliferation and apoptosis, in vivo, are in balance, then no net gain or loss in injected cell number would be predicted. In vivo experiments by Potapova et al. [53] have shown an increase in cardiac mechanical function due to a cardiac patch of human MSCs and cardiogenic stem cells in a canine model over a multiweek time interval. These data suggest that these cell types slowly proliferate in vivo, generating a functional phenotype to repair the damaged region of myocardium.

A related issue is accurate determination of the extent of migration of injected MSCs. A number of studies suggest that junctional adhesion molecules are part of a membrane complex that prevents cell migration [44]. To better understand the potential of MSCs, ESCs, or IPSCs to migrate within an organ or organism, the molecular processes involved in cell migration will have to be defined (see Chap. 7). Genetic engineering of one or a number of the junctional adhesion molecules might well render MSCs, ESCs, or IPSCs immobile once they have been injected into a tissue.

In the case of homing where cells are migrating to a specific site, membrane surface receptors, such as CXCR4, are necessary. MSCs express the CXCR4 receptor, which is known to be essential for the sequestration of MSCs to injured tissues [45]. It is possible to alter the expression levels of CxCR4 receptor density on the surface of MSCs in vitro. In fact, Potapova et al. [46] have shown that MSCs downregulate the receptor in culture, but by the use of an embryoid body approach, it can be upregulated. This suggests that MSCs can be fine-tuned for systemic delivery to home to an area of injury (also see Chap. 7).

9.2 Embryonic Stem Cells

The most obvious strength of ESCs is the ability to differentiate into a desired phenotype in vitro prior to being delivered to a target tissue or organ. An important part of the integration of a cell into a tissue is the expression of adhesion molecules and the appropriate connexins. The expression of specific connexins within ESCs is not known. mRNA for 18 of the 20 known connexins has been found to be present, but it is unlikely that ESCs simultaneously express 18 connexins that are all contributing to the functional coupling of an ESC to a neighboring cell. A proteomics approach would be useful to better understand what proteins are being expressed and their relative abundance. In this example, determining the connexin content in the desired cell phenotype would be useful.

Why should we be concerned with the nature of the connexin-based coupling between a delivery cell and a target cell? Delivery of a cell able to generate significant amounts of Cx43, Cx40, and Cx45 or Cx32, for example, might allow the delivered cell to couple not only to a myocyte, but also to other cells present in the interstitial spaces such as endogenous cardiac fibroblasts [47]. This might create a sink for current flow, which would act as an arrhythmogenic source. MSCs expressing Cx43 and Cx40 did not create proarrhythmic behavior in the canine ventricular myocardial wall and have also been shown not to express Cx45 [24, 25, 40]. Unpublished results from our laboratories suggest that they do not express Cx32 or Cx26. This suggests that connexin expression in adult stem cells is not a limiting step nor is it one that creates an arrhythmia, but the result for ESCs is not known.

Other issues associated with the delivery and subsequent integration of ESCs are proliferation and migration. These issues are only now coming to the forefront as serious concerns for cell-based therapies. As already indicated for MSCs, imaging of appropriately tagged delivery cells using tools such as quantum dots or reporter genes represents one approach to address both proliferation and migration [43], but long-term studies have not yet been carried out in vitro or in vivo for ESCs.

9.3 Induced Pluripotent Stem Cells

The ability to induce fibroblasts to become pluripotent cells [48] has great potential for cell-based therapies (see Chaps. 13–15). The strength of IPSCs is their ability to generate the desired phenotype in vitro for introduction into a tissue or organ. In the case of the heart, the generation of cardiac myocytes specific to a region of the heart is a very exciting prospect. Sinoatrial node cells could be generated as a therapy for sick sinus syndrome, or ventricular myocytes could be generated to repair an infarction. However, IPSCs, like MSCs and ESCs, must first be shown to integrate into the myocardium and form gap junction channels with myocytes.

IPSCs derived from fibroblasts have been shown to express Cx43 [26, 49]. Mouse fibroblasts in culture will often express Cx43, but can also express other connexins, such as the β connexins Cx32 and Cx26 [15]. As indicated for ESCs, an IPSC that allows communication between itself and a heart cell and other non-cardiac cells creates the potential for arrhythmias. The problem is compounded if other connexins are expressed, such as β connexins, as it is then possible for IPSCs to couple with surrounding endogenous cells expressing compatible connexins.

Although little is known about IPSCs in terms of connexin expression, there are in vivo data on cardiac fibroblasts that represent one source of IPSCs. Camelliti et al. [47] used immunostaining and dye transfer to study connexin expression and junctional communication. Only three connexins were tested, Cx43, Cx40, and Cx45. The findings showed select expression of Cx40 and Cx45 in cardiac fibroblasts. The expression of other connexins, including the β connexin family, was not determined. LY transfer indicated that the coupling was sparse. Moreover, considering the fingerlike projections and thin cellular processes typical of fibroblasts, the resultant input resistance due to fibroblast coupling is likely to be very high. The functional significance is that the fibroblast network may not create a significant sink for current and thus may not be arrhythmogenic in the normal heart.

Figure 5 illustrates possible coupling of a myocyte with an underlying network of fibroblasts based on the study by Camelliti et al. [47]. In addition, how MSCs, ESCs, and IPSCs might integrate into such tissue is illustrated. The data suggest, as illustrated in the figure, that only heterotypic junctions would occur between the two populations of fibroblasts identified by Camelliti et al. [47]. In vitro studies on heterotypic Cx43–Cx45 and Cx40–Cx45 junctions reveal rectification when the Cx45-expressing cell is hyperpolarized or the adjacent cell is depolarized Valiunas [24, 25, 54], therefore potentially representing a limit to the creation of a current sink with the myocardium during the propagation of an action potential. MSCs, ESCs, and IPSCs also express cardiac connexins. They have the potential to couple with both myocytes and the fibroblast network. The extent of coupling must be considered because of the potential to create proarrhythmic behavior. However, in vivo evidence from Potapova et al. [40] suggests that if coupling is occurring between fibroblasts, injected cells, and myocytes, it is not causing arrhythmias.

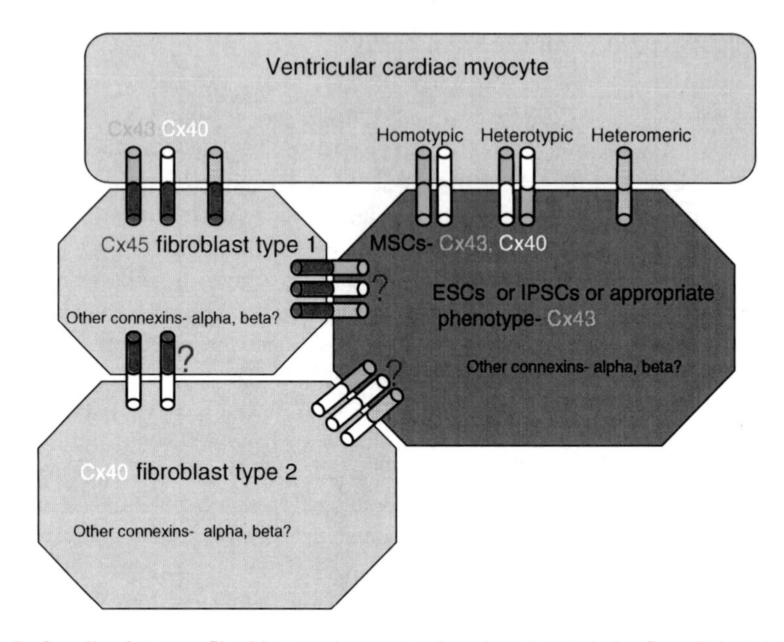

Fig. 5 Coupling between fibroblasts and myocytes based on the study by Camelliti et al. [47]. In addition, the potential for mesenchymal stem cells (*MSCs*), embryonic stem cells (*ESCs*), or induced pluripotent stem cells (*IPSCs*) is illustrated on the basis of the known expression of connexins. The *question marks* illustrate probable connections that have not be verified in vivo or in vitro. *Cx45* connexin 45

Finally, before IPSCs are used as therapeutic agents, it will be necessary to screen for all the connexin proteins. Again like for the ESCs, a proteomics approach is the most reliable first step. As suggested in Fig. 5, depending on the tissue source of the fibroblast, one might find a large diversity of connexin expression where some forms express β connexins and α connexins other than those expressed by cardiac myocytes.

As was the case for MSCs and ESCs, proliferation and migration must also be understood for IPSCs or their derived phenotypes if they are to represent a viable therapeutic approach.

10 Summary

Integration of MSCs, ESCs, and IPSCs into a tissue such as the heart first requires that they are each able to express the appropriate connexins to allow coupling to myocytes. Figure 6 shows the connexins that have thus far been shown to be expressed in the three generic cell types discussed in this review. For all three cell types, it is also important to understand the proliferative capability as well as

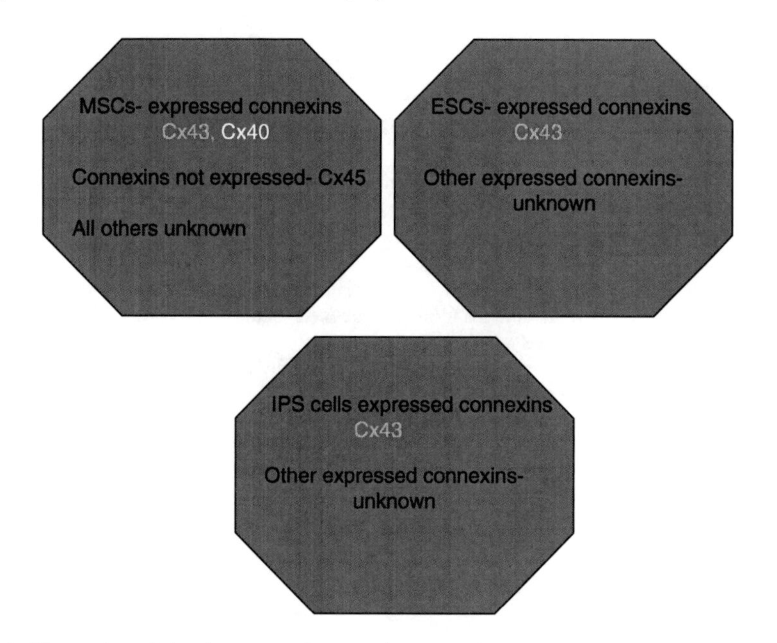

Fig. 6 Illustration of the documented connexin expression or lack therein for MSCs, ESCs, and IPSCs

migratory and homing behavior in vivo. For ESCs and IPSCs, when phenotypes are generated in vitro, it is essential to confirm the expression of desired proteins such as Cx43, but it is of equal importance to verify the lack of expression of proteins that might impair function. For example, the expression of Cx32 in a cardiac phenotype might result in proarrhythmic behavior by allowing significant current flow into surrounding fibroblasts that express Cx32.

11 Significance

Evidence strongly implicates changes in gap-junction-mediated conductance in the myocardium as a major cause of many cardiac arrhythmias [50]. Moreover, many studies have attempted to pharmacologically modulate gap junction channel function without success. Cell therapy approaches provide a unique opportunity and novel approach to address cardiac arrhythmias with the possibility of a cure rather than palliation [51, 52]. For them to be successful, replacement cells must couple with their cell targets by expressing cardiac connexins and must not couple significantly with other cells in the cardiac parenchyma [53, 54]. Whether this is a reliable outcome for each stem cell type in each delivery modality will depend on its connexin expression, which remains to be entirely determined.

References

1. Harris, A., Emerging issues of connexin channels: biophysics fills the gap. Q. Rev. Biophys. 34:325–472, 2001
2. Valiunas, V., E.C. Beyer and P.R. Brink Cardiac gap junction channels show a quantitative difference in selectivity. Circ. Res. 91(2):104–111, 2002.
3. Goldberg, G., Valiunas, V. and Brink, P.R. Selectivity permeability of gap junction channels. Biochem. Biophys. Acta 662: 96–101, 2004.
4. Brink, P.R., Ramanan, S.V. and Christ, G.J. Human connexin43 gap junction channel gating: evidence for mode shifts and/or heterogeneity. Am. J. Physiol. 271: C321–C331, 1996.
5. Ek-Vitorin, J.F. and J.M. Burt. Quantification of gap junction selectivity. Am. J. Physiol. Cell Physiol. 289(6): C1535–C1546, 2005.
6. Kumar, N. Molecular biology of the interactions between connexins. In: Gap junction-mediated intercellular signaling in health and disease. 1999 pages 6–15 Novartis Foundation Symposium 219 Wiley
7. White, T.W. and R. Bruzzone, Multiple connexin proteins in single intercellular channels: connexin compatibility and functional consequences. J. Bioenerg. Biomembr. 28(4): 339–350, 1996.
8. Kanaporis, G, G. Mese, L. Valiuniene, T.W. White, P.R. Brink, V. Valiunas. Gap junction channels exhibit connexin-specific permeability to cyclic nucleotides. J. Gen. Physiol. 131: 293–305, 2008
9. Niessen, H., Harz, H., Bedner, P., Kramer, K., Willecke, K. Selective permeability of different connexin channels to the second messenger IP3. J. Cell Sci. 113: 1365–1372, 2000.
10. Beyer, E.C. Gap junctions. Int. Rev. Cytol. 137C: 1–37, 1993.
11. Van Veen, T.A.B., H.W.M. van Rijen and T. Opthof. Cardiac gap junction channels: modulation of expression and channel properties. Cardiovasc. Res. 51: 217–229, 2001.
12. Simpson, I., B. Rose and W.R. Loewenstein Size limit of molecules permeating the junctional membrane channels. Science 195: 294–296, 1977.
13. Schwarzmann, G., H. Wiegarndt, B. Rose, A. Zimmerman, D. Ben-Haim, W. Loewenstein Diameter of the cell to cell junctional membrane channels as probed with neutral molecules. Science 213: 551–553, 1981.
14. Neijssen, J. Herberts, C., Drijfhout, J.W., Reits, E., Janssen, L. and Neefjes, J. Cross-presentation by intercellular peptide transfer through gap junctions. Nature 434: 84–88, 2005.
15. Valiunas, V., Polosina, Y., Miller, H., Potapova, I., Valiuniene, L., Doronin, S., Mathias, R.T., Robinson, R.B., Rosen, M.R., Cohen, I.S. and Brink, P.R. Connexin-specific cell-to-cell transfer of short interfering RNA by gap junctions. J. Physiol. 568: 459–468, 2005.
16. Wolvetang, E.J., M.F. Pera and K.S. Zuckerman, Gap junction mediated transport of shRNA between human embryonic stem cells. Biochem. Biophys. Res. Commun. 363(3): 610–615, 2007
17. Severs, N.J., E. Dupont, N. Thomas, R. Kaba, S. Rothery, R. Jain, K. Sharpey and C.H. Fry. Alterations in cardiac connexin expression in cardiomyopathies. Adv. Cardiol. 42: 228–242, 2006.
18. Bukauskas, F.F., M.M. Kreuzberg, M. Rackauskas, A. Bukauskiene, M.V.L. Bennett, V.K. Verselis and K. Willecke. Properties of mouse connexin 30.2 and human connexin 31.9 hemichannels: implications for atrioventricular conduction in the heart. Proc. Natl. Acad. Sci. USA. 103(25): 9726–9731, 2006.
19. Munshi, N., J. McAnally, S. Bezprozvannaya, J. Berry, J. Richardson, J. Hill, E. Olson Cx30.2 enhancer analysis identifies Gata4 as a novel regulator of atrioventricular delay. Development 136: 2665–2674, 2009
20. Belluardo, N., T.W. White, M. Srinivas, A. Trovato-Salinaro, H. Ripps, G. Mudo, R. Bruzzone and D.F. Condorelli. Identification and functional expression of HCx31.9. Cell Commun. Adhes. 8(4–6): 173–178, 2001.

21. White, T.W., M. Srinivas, H. Ripps, A. Trovato-Salinaro, D.F. Condorelli and R. Bruzzone. Virtual cloning, functional expression, and gating analysis of human connexin31.9. Am. J. Physiol. Cell Physiol. 283(3): C960–C967, 2002.
22. Gemel, J., X. Lin, R. Collins, R.D. Veenstra, E.C. Beyer. Cx30.2 can form heteromeric gap junction channels with other cardiac connexins. (Abstract). Biochem Biophys Res Commun. 369: 388–394, 2008.
23. Yum, S., J. Zhang, V. Valiunas, G. Kanaporis, P.R. Brink, T. White, S. Scherer. Human connexins 26 and connexin 30 form functional heteromeric and heterotypic channels. Am. J. Physiol. 293: 1032–1048, 2007
24. Valiunas, R., Doronin, S., Valiuniene, L., Potapova, I., Zuckerman, J., Walcott, B., Robinson, R.B., Rosen, M.R., Brink, P.R. and Cohen, I.S. Human mesenchymal stem cells make cardiac connexins and form functional gap junctions. J. Physiol. 555: 617–626, 2004.
25. Valiunas, V., R. Mui, E. McLachan, G. Valdimarsson, P.R. Brink, T. White Biophysical characterization of zebrafish connexin 35 hemichannels. Am. J. Physiol. 287: C1596–C1604, 2004.
26. Mauritz, C., K. Schwanke, M. Reppel, S. Neff, K. Katsirntaki, L. Maier, F. Nguemo, S. Menke, M. Haustein, J. Hescheler, G. Hasenfuss, U. Martin Generation of functional murine cardiac myocytes from induced pluripotent stem cells. Circulation 118: 507–517, 2008
27. Worsdorfer, P., S. Maxeiner, C. Markopoulos, G. Kirfel, V. Wulf, T. Auth, S. Ureschel, J. von Maltzahn, K. Willecke Connexin expression and functional analysis of gap junctional communication in mouse embryonic stem cells 26: 431-439 2008
28. Huettner, J., A. Lu, Y. Qu, Y. Wu, M. Kim, J. McDonald Gap junctions and connexon hemichannels in human embryonic stem cells. Stem Cells 24: 1654–1667, 2006
29. Brink, P.R., K. Cronin, K. Banach, E. Peterson, E. Westphale, K.H. Seul, S.V. Ramanan and E.C. Beyer Evidence of heteromeric gap junction channels formed from rat connexin43 and human connexin37. Am. J. Physiol. 273: C1386–C1396, 1997.
30. Valiunas, V., Gemel, J., Brink, P.R. and Beyer, E.C. Gap junction channels formed by Co-expressed Cx40 and Cx43. Am. J. Physiol. 281: H1675–H1688, 2001.
31. Barr, L., M. Dewey and Berger, W. Propagation of action potentials and the structure of the nexus in cardiac muscle. J. Gen. Physiol. 48: 797–823, 1965.
32. Weidmann, S. Electrical constants of trabecular muscle from mammalian heart. J. Physiol. 210: 1041–1054, 1970.
33. Wilders, R., E.E. Verheijck, R. Kumar, W.N. Goolsby, A.C. van Ginneken, R.W. Joyner and H.J. Jongsma. Model clamp and its application to synchronization of rabbit sinoatrial node cells. Am. J. Physiol. 271: H2168–2182, 1996.
34. Cole, W.C., J.B. Picone and N. Sperelakis. Gap junction uncoupling and discontinuous propagation in the heart. A comparison of experimental data with computer simulations. Biophys. J. 53(5): 809–818, 1988.
35. de Groot, J.R., T. Veenstra, A.O. Verkerk, R. Wilders, J.P. Smits, F.J. Wilms-Schopman, R.F Wiegerinck, J. Bourier, C.N. Belterman, R. Coronel and E.E. Verheijck. Conduction slowing by the gap junctional uncoupler carbenoxolone. Cardiovasc. Res. 60(2): 288–97, 2003.
36. Poelzing, S. and D.S. Rosenbaum. Altered connexin43 expression produces arrhythmia substrate in heart failure. Am. J. Physiol. Heart Circ. Physiol. 287: H1762–H1770, 2004.
37. Beauchamp, P., K. Yamada, A Baertschi, K. Green, E. Kanter, J. Saffitz, A. Kleber. Relative contributions of connexins 40 and 43 to atrial impulse propagation in synthetic strands of neonatal and fetal murine cardiomyocytes. Circ. Res. 99: 1216–1224, 2006
38. Janse, M. and A. Wit Electrophysiological mechanisms of ventricular arrhythmias resulting from myocardial ischemia and infarction. Phys. Rev. 69: 1049–1169, 1989.
39. Gutstein, D.E., G.E. Morley, D. Vaidya, F. Liu, F.L. Chen, H. Stuhlmann, G.I. Fishman Heterogeneous expression of Gap junction channels in the heart leads to conduction defects and ventricular dysfunction. Circulation 104: 1194–1199, 2001
40. Potapova, I., A. Plotnikov, Z. Lu, P. Danilo, Jr., W. Valiunas, J. Qu, S. Doronin, J. Zuckerman, I.N. Shlapakova, J. Gao, Z. Pan, A.J. Herron, R.B. Robinson, P.R. Brink, M.R. Rosen and I.S. Cohen Human mesenchymal stem cells as a gene delivery system to create cardiac pacemakers. Circ. Res. 94: 952–959, 2004.

41. Musil, L., B. Cunningham, G. Edelman, D. Goodenough Differential phosphorylation of the gap junction protein connexin 43 in junctional communication-competent and deficient cell lines J. Cell Biol. 111: 2077–2088, 1990
42. Pedrotty, D., R. Klinger, N. Badie, S. Hinds, A Kardashian, N. Bursac Structural coupling of cardiomyocytes and noncardiomyocytes: quantitative comparisons using a novel micropatterned cell pair assay. Am. J. Physiol. 295: H390–H400, 2008
43. Rosen A.B., D.J. Kelly, A.J. Schuldt , J. Lu, I.A. Potapova, S.V. Doronin, K.J. Robichaud, R.B. Robinson, M.R. Rosen, P.R. Brink, G.R. Gaudette, I.S. Cohen Finding fluorescent needles in the cardiac haystack: tracking human mesenchymal stem cells labeled with quantum dots for quantitative in vivo three-dimensional fluorescence analysis. Stem Cells 8: 2128–2138, 2007
44. Severson, E., C. Parkos Mechanisms of outside-in signaling at the tight junction by junctional adhesion molecule A. Ann. N. Y. Acad. Sci. 1165: 10–18, 2009
45. Chen, Y., L. Xiang, J. Shao, R Pan, Y Wang, X Dong, R, Zhang Recruitment of endogenous bone marrow mesenchymal stem cells towards injured liver. J. Cell Mol. Med. 14: 1494–1508, 2010.
46. Potapova, I., P.R. Brink, I.S. Cohen, S.V. Doronin Culturing of human mesenchymal stem cells as 3D-aggregates induces functional expression of CXCR4 that regulates adhesion to endothelial cells. J. Biol. Chem. 283: 13100–13107, 2008
47. Camelliti, P., C. Green, I. LeGrice, P. Kohl Fibroblast network in rabbit sinoatrial node: structural and functional identification of homogenous and heterogenous cell coupling. Circ. Res. 94: 828–835, 2004
48. Graf, T., T. Enver Forcing cells to change lineage Nature 462: 587–594, 2009
49. Song, H., C. Yoon, S. Kattman, J. Dengler, S. Masse, T. Thavaratnam, M. Gewarges, K. Nanthakumar, M. Rubart, G. Keller, M. Radisic, P. Zandstra. Interrogating functional integration between injected pluripotent stem cell-derived cells and surrogate cardiac tissue. Proc. Natl. Acad. Sci. U S A 107: 3329–3334, 2010
50. Gollob, M.H., D.L. Jones, A.D. Krahn, L. Danis, X. Gong, Q. Shao, X. Liu, J.P. Veinot, A. Tang, A. Stewart, F. Tesson, G. Klein, R. Yee, A. Skanes, G. Guiraudon, L. Ebihara and D. Bai. Somatic mutations in the connexin 40 gene (GJA5) in atrial fibrillation. N. Engl. J. Med. 354: 25, 2006
51. Kelly, D.J., A.B. Rosen, A.J. Schuldt, P.V. Kochupura, S.V. Doronin, I.A. Potapova, E.U. Azeloglu, S.F. Badlyak, P.R. Brink, I.S. Cohen, G.R. Gaudette Increased myocyte content and mechanical function within a tissue-engineered myocardial patch following implantation. Tissue Eng. Part A 15: 2189–2201, 2009
52. Valiunas, V., G. Kanaporis, L. Valiuniene, C Gordon, H. Wang, L. Li, R.B. Robinson, M.R. Rosen, I.S. Cohen, P.R. Brink Coupling an HCN2 expressing cell to a myocyte creates a two cell pacing unit. J. Physiol. 587: 5211–5226, 2009
53. Potapova, I.A., Doronin, S.V., Kelly, D.J., Rosen, A.B., Schuldt, A.J., Lu, Z., Kochupura, P.V., Robinson, R.B., Rosen, MR., Brink, P.R., Gaudette, G.R., Cohen, I.S. Enhanced recovery of mechanical function in the canine heart by seeding an extracellular matrix patch with mesenchymal stem cells committed to a cardiac lineage. Am J Physiol Heart Circ Physiol. 295: H2257–63, 2008
54. Valiunas, V. Biophysical properties of connexin-45 gap junction hemichannels studied in vertebrate cells. J Gen Physiol. 119: 147–64, 2002

Bradyarrhythmia Therapies: The Creation of Biological Pacemakers and Restoring Atrioventricular Node Function

Richard B. Robinson

Abstract In the USA, over 300,000 electronic pacemakers are implanted annually to treat slow heart rates resulting from abnormal sinus or atrioventricular (AV) node function. Although life-saving, limitations and problems with this hardware-based approach have led to interest in using gene and cell delivery methods to develop purely biological based therapies. The goal in creating a biological pacemaker or AV bypass is not to replicate the sinus or AV nodes at a cellular or molecular level, but rather to recreate their functionality. There has been more research on biological pacemakers than on AV bypasses, and the former has explored both implantation of endogenously automatic cells and genetic engineering methods, with the latter involving either direct gene delivery or gene delivery to stem cells that are subsequently implanted into the myocardium. This makes use of all the tools of bioinformatics and molecular biology to engineer custom channels with distinct biophysical properties that may be best suited for a specific patient population or implant site. Although various gene products have been targeted as the basis of a biological pacemaker, most recent efforts have focused on the HCN gene family of pacemaker current. These channels open on hyperpolarization, so they contribute current largely during diastole and have minimal impact on action potential duration. Further, these channels directly bind and respond to the adrenergic second messenger cyclic AMP, so autonomic responsiveness is integral to an HCN-based biological pacemaker. Although proof of principle has been demonstrated by a number of laboratories in several model systems, numerous challenges remain to achieve physiologically appropriate rates with sufficient robustness. Before clinical trials can begin, critical issues of safety and persistence of function must first be addressed in long-term animal studies. Even then, the appropriate patient population will be limited until a biological AV bypass is also developed to ensure AV

R.B. Robinson (✉)
Department of Pharmacology and Center for Molecular Therapeutics, Columbia University College of Physicians and Surgeons, New York, NY 10032, USA
e-mail: rbr1@columbia.edu

I.S. Cohen and G.R. Gaudette (eds.), *Regenerating the Heart*, Stem Cell Biology and Regenerative Medicine, DOI 10.1007/978-1-61779-021-8_18,

© Springer Science+Business Media, LLC 2011

synchrony. However, the possibility of implanting a biologically based bypass tract, perhaps containing a proximal pacemaker, is compelling, and provides the continuing motivation driving these research efforts.

Keywords Electronic pacemaker • Biological pacemaker • HCN • Bradycardia • Gap junctions • Human mesenchymal stem cell • Embryonic stem cell

1 The Need: Limitations of Electronic Pacemakers

Bradyarrhythmias associated with sinoatrial (SA) node or atrioventricular (AV) node disease are a serious and life-threatening condition, resulting in the treatment of over 300,000 people annually in the USA with electronic pacemakers [1]. Although a highly successful palliative, this treatment is not a cure. It is associated with complications related to the need for implanting and maintaining hardware, wires, and batteries. For example, although leadless pacemakers are under development, presently site placement is dictated by the need of ensuring lead stability, which can result in nonoptimal ventricular activation and contraction, and eventually cardiac failure [2]. In addition, electronic pacemakers present issues related to the monitoring and changing of batteries and risks associated with infection, interference, and lead fracture. Recently, the failure rate of one particular cable was reported to be on the order of 5% over 45 months [3].

Lead placement and lead failure are not the only issues. For pediatric patients there are additional problems of mismatches between a child's growth and development and the geometry and size of power packs and electrodes. Further, although progress has been made in developing rate-responsive electronic pacemakers, they still lack sensitivity to neurohumoral agonists and therefore do not fully match the physiological responsiveness of the native heart pacemaker. As a result, numerous laboratories worldwide are exploring the feasibility of developing biological pacemakers and/or biological AV bypasses as alternative therapies that would overcome these limitations [2, 4].

2 The Concept: Advantages of a Biological Pacemaker and AV Bypass

A biological pacemaker is defined, quite simply, as anything that creates an automatic focus somewhere within the heart using entirely biological mechanisms. Thus, it implies the delivery of genes or cells to a localized region of the heart (or conceivably even chemicals that alter the expression of specific endogenous genes) to create such an automatic focus. The location chosen will depend on the nature of the rhythm disturbance in a particular individual. Someone with sinus node dysfunction but a healthy AV node could receive a biological pacemaker implanted in atrial tissue. In contrast, someone with AV block would require placement within the ventricular myocardium or conduction system, distal to the block. Obvious candidates here, at least for first-generation biological pacemakers, would include individuals with AV block combined with atrial fibrillation, since these patients are not candidates for AV synchronous electronic pacemakers.

It is this issue of AV synchronization that naturally leads to the idea of developing a biological AV bypass, which would be a pathway created from biological material (possible along with an inert framework) and capable of conducting electrical activity from the atrium to the ventricle with an appropriate delay. An AV bypass by itself would be sufficient in patients with normal SA node function but AV block, whereas in patients with both SA and AV node disease one could envision an AV bypass that contains a biological pacemaker situated at the proximal end.

The idea behind a biological pacemaker or AV bypass is to recreate the functionality of these structures, but not necessarily by replicating them at a cellular or molecular level. In this context, the longstanding debate concerning the ionic basis of the normal SA node pacemaker [5] is irrelevant except for the insight it provides into what ionic mechanisms may be useful and how they would function (Fig. 1).

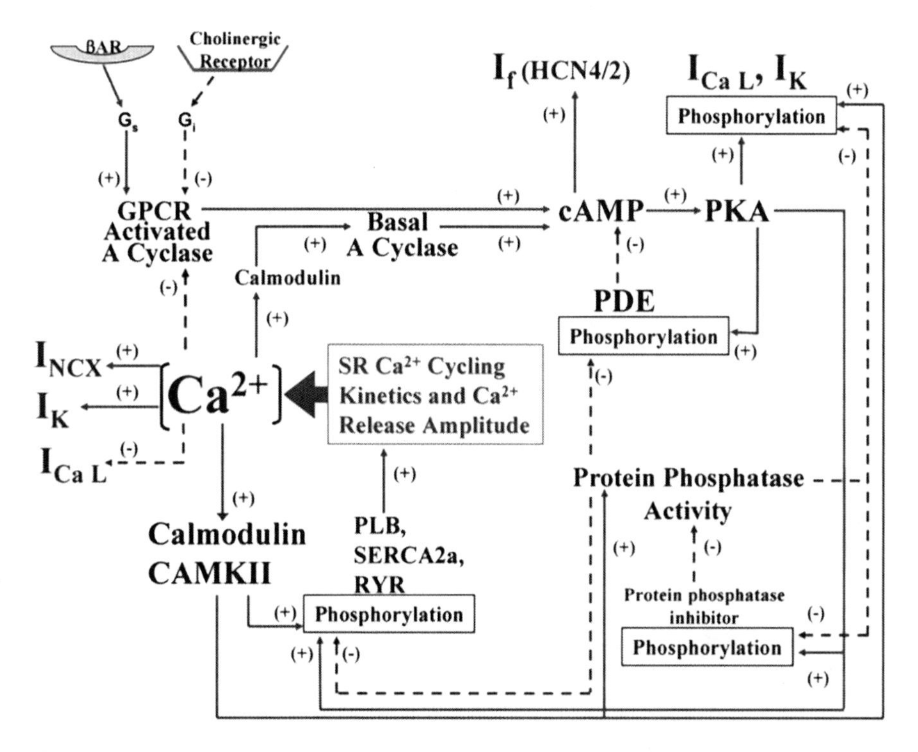

Fig. 1 The contribution to sinoatrial node pacemaking of Ca^{2+}-dependent processes (including sarcoplasmic reticulum Ca^{2+} cycling, Ca^{2+}-activated adenylyl cyclase, L-type Ca^{2+} channels, and Na/Ca exchanger), and other membrane channels including I_f, and the interdependence of the different processes. *βAR* β-adrenergic receptor, *GPCR* G-protein coupled receptor, *cAMP* cyclic AMP, *PDE* phosphodiesterase, *PKA* protein kinase A, *SR* sarcoplasmic reticulum, *PLB* phospholipase B, *SERCA* sarco/endoplasmic reticulum Ca^{2+}-ATPase, *RYR* ryanodine receptor, *CAMKII* calmodulin kinase II (reprinted with permission from [81])

As an example, it does not matter whether the endogenous "pacemaker current" (I_f) is sufficiently large, sufficiently fast, or activates in the correct voltage range to serve as an effective pacemaker, since all those parameters are subject to manipulation when creating a de novo biological pacemaker. Similarly, propagation through the normal AV node is primarily a Ca^{2+}-dependent process, which ensures a sufficient conduction delay while traveling through this small structure. However, an artificial AV bypass may travel along the exterior of the heart and thus encompass a significantly longer path. In that case, a more rapid action potential upstroke, perhaps driven by a Na^+ current, may be required to achieve an equivalent delay.

For these reasons, in designing a biological pacemaker or AV bypass the emphasis is often on the desired behavior or functionality rather than the underlying molecular/cellular mechanisms, and different research groups have explored distinct approaches for achieving the desired objectives. In the case of a biological pacemaker, a primary objective is to incorporate autonomic responsiveness, since this represents a clear potential advantage over existing electronic pacemakers. Given the reliability and effectiveness of electronic pacemakers, to be competitive biological pacemakers must not only provide added value (e.g., autonomic responsiveness) but also need to be virtually 100% reliable and safe. In the short term, this high standard may be moderated by the fact that any initial clinical trial or early usage will certainly include an electronic pacemaker in demand or "backup" mode as a safety feature. As discussed later, such a tandem approach provides not just safety during initial testing, but also the possibility of genuine advantages over a standalone electronic pacemaker. These include extending the time between battery replacement (since the electronic device fires much less frequently) and taking advantage of the memory of the electronic device to monitor the function of the biological pacemaker over time. However, unquestionably, the true commercial appeal of biological pacemakers will only be fully realized when they attain a level of functionality and reliability that allows their use without the extra costs and potential complications of a tandem electronic device.

Similarly, a primary objective of a biological AV bypass is to provide both the appropriate conduction delay and the appropriate use-dependence so as to block excessively high rates from reaching the ventricle. Here also, to realize their full potential, such devices must ultimately demonstrate sufficient reliability so as to obviate the need for the simultaneous implantation of an electronic device solely as a backup.

If biological pacemakers and AV bypasses can achieve these goals, they would represent a disruptive technology that could replace a significant segment of the existing electronic pacemaker market. They would allow maintenance of normal heart rate, with physiologic rate responsiveness and more normal and therefore efficient ventricular contraction, while avoiding the risks and complications associated with the permanent implantation of hardware and leads. The sections that follow focus primarily on biological pacemakers, since this is where much of the research and progress to date has occurred. Existing and potential future efforts related to AV bypasses are addressed in Sect. 5.

3 The Approach: Types of Biological Pacemaker

Research to date has taken either a gene (naked DNA or viral vectors) or cell [embryonic stem cells (ESCs) or adult stem cells or explanted sinus node cells] approach [4]. Despite natural advantages and successful proof-of-principle experiments for both approaches, each is hampered by critical bottlenecks that have kept them from reaching clinical trial. Gene delivery is hindered by the fact that persistent viral vectors carry unacceptable risk, particularly when injected in vivo, where they may travel anywhere in the body. Cell delivery, which is the primary focus of this chapter, is beset by issues such as the possibility of an immune response unless autologous or "immunoprivileged" allogeneic cells are used. These and other technical challenges are discussed in more detail in Sect. 6. For now, we will focus on the relative functionality of the different approaches as demonstrated in short-term proof-of-principle studies.

It is useful to separate the research to date into two broad categories, depending on whether endogenously automatic cells are used or whether some form of genetic engineering is an integral part of the approach. The former, relying on cells that are endogenously automatic, includes work with explanted SA node cells and stem cells driven down a cardiac lineage [4]. This approach has an inherent limitation, in that the outcome is constrained by whatever ion channels are naturally expressed in the selected cell type. Thus, even if problems related to cell source, differentiation, and immunogenicity are resolved, functionality may fall short. The alternative approach employs genetic engineering to introduce genes that can initiate or enhance automaticity [2, 4]. This permits all the tools of bioinformatics and molecular biology to be employed to engineer custom channels with distinct biophysical properties that may be best suited for a specific patient population or implant site. The engineered genes can be delivered directly into myocardial cells as naked DNA or viral vectors, or they can be expressed in another cell type, such as an adult human mesenchymal stem cell (hMSC), that is then implanted into the myocardium. The latter requires that the cell chosen contains or is engineered to express compatible connexin proteins [6] or that methods are employed to cause fusion of the exogenous cell and myocyte [7], in either case ensuring electrical continuity.

4 The Result: Effectiveness of Existing Attempts

4.1 Early Studies with Gene-Based Delivery and Identification of the HCN Gene Family

The earliest attempt to modify cardiac rhythm through gene expression employed the β_2-adrenergic receptor. DNA encoding this receptor was injected into the right atrium of the mouse [8] or pig [9] heart, resulting in a 40–50% increase of basal

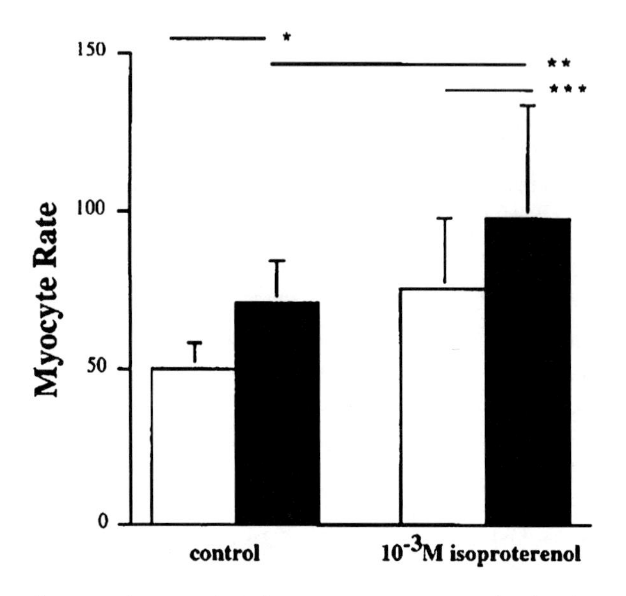

Fig. 2 Adrenergic responsiveness of cardiac myocytes overexpressing β_2-adrenergic receptors. Fetal murine myocytes maintained in culture were exposed to isoproterenol and the effect on rate was determined for LacZ transfected cultures (*white bars*) and β_2-adrenergic receptor transfected cultures (*black bars*). *Asterisks* indicate significant differences. The β_2-adrenergic receptor transfected cultures beat significantly faster than the LacZ transfected cultures, both under basal conditions and after exposure to isoproterenol (reprinted with permission from [8])

heart rate. Although in vivo autonomic responsiveness was not assessed, in vitro overexpression of these receptors in mouse myocytes significantly increased both basal and isoproterenol stimulated spontaneous rate [8]. Figure 2 shows that overexpression increased basal rate by 50%, and a high concentration of a β-adrenergic agonist had a comparable effect on rate (in terms of increased number of beats per minute) in control and overexpressing myocytes.

It is likely that the mechanism underlying the effect of β_2-adrenergic receptor overexpression on rate involved increased basal cyclic AMP (cAMP) concentration, which in turn modified the activity of various ion channels and pumps involved in regulation of spontaneous rate in the sinus node (see Fig. 1). Subsequently, researchers focused on expressing genes that more directly change net current flow and therefore automaticity. The first such effort made use of the fact that many ion channels are composed of multiple (homomeric or heteromeric) subunits and that disrupting the pore sequence of a single subunit can impair or prevent ion flow through the channel. Such mutations are referred to as dominant negative constructs, in that a single "negative" subunit dominates channel function. When an adenovirus containing a dominant negative construct of an inward rectifier channel subunit was injected into the ventricular cavity of guinea pig hearts in vivo, expression was achieved in about 20% of the myocytes, resulting in 80% suppression of inward rectifier current in these myocytes [10]. The subsequent depolarized maximum diastolic potential resulted in phase 4 depolarization and spontaneous action potentials,

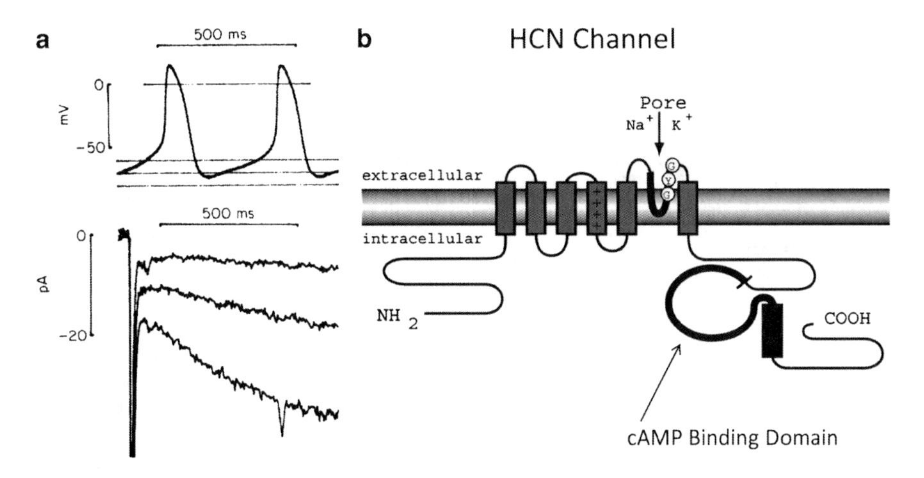

Fig. 3 Characteristics of I_f pacemaker current and its molecular correlate. (**a**) Sinoatrial node action potential and the time-dependent activation of I_f at voltages in the diastolic potential range. (**b**) The HCN subunit illustrating the six transmembrane domains and the cyclic nucleotide binding domain in the cytosolic C-terminal. ((**a**) Reprinted with permission from [82]; (**b**) reprinted with permission from [14])

but also action potential prolongation [10, 11]. A later computer simulation by another group suggested that the primary pacemaking mechanism in this situation is the Na^+/Ca^{2+} exchanger, and indicated that β-adrenergic sensitivity would likely be limited owing to an opposing action on the slow delayed rectifier current [12]. However, the major concern with this approach is the proarrhythmic potential of the observed action potential prolongation, which emphasizes the need to develop strategies that selectively or preferentially impact current flow during diastole.

This was in part the rationale behind our group's decision to develop a biological pacemaker based on the hyperpolarization-activated, cyclic-nucleotide-gated (HCN) gene family. These genes are the molecular correlate of the I_f current, and are highly expressed in the sinus node [4, 13–15]. I_f has several unique characteristics that make it appealing for use in a biological pacemaker (Fig. 3). First, because the channels activate upon hyperpolarization, have a reversal potential around −30 mV (negative to the action potential plateau), and rapidly deactivate upon depolarization, inward current flows primarily during diastole. Therefore, overexpression is unlikely to be associated with potentially arrhythmogenic action potential prolongation. Second, because the channels are strongly modulated by direct cAMP binding, autonomic regulation of rhythm is in effect built into the design. The HCN gene family consists of four mammalian isoforms (HCN1 through HCN4), with regional specificity of cardiac isoform expression [16]. In the sinus node the isoform present at the highest levels is HCN4 [16–18]. However, the native pacemaker current in the SA node activates faster than does HCN4 in heterologous expression systems, suggesting that other factors can affect the biophysical characteristics of the channel. Indeed, it is now known that HCN channel expression and function can be altered by β-subunits [19, 20],

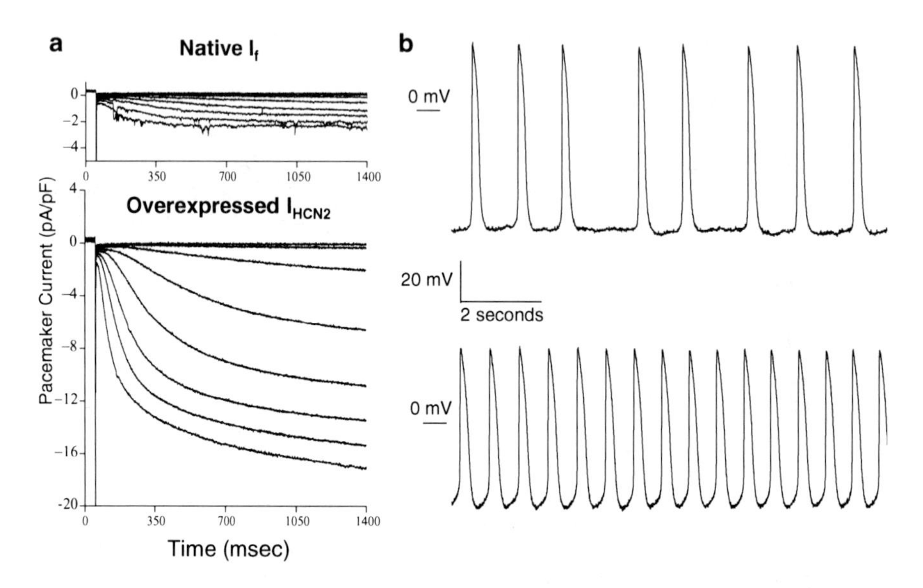

Fig. 4 Native and expressed pacemaker current, and effect on spontaneous rate, in newborn rat ventricular cells in culture. (**a**) Representative pacemaker current in control (*top*) and HCN2-expressing (*bottom*) myocytes. (**b**) Representative spontaneous action potential recordings from control (*top*) and HCN2-expressing (*bottom*) myocytes (portions reprinted with permission from [26])

membrane localization [21], interaction with scaffolding and other cytoskeletal proteins [22–24], and posttranslational processing such as truncation [25].

We chose to initially explore the utility of HCN2 as a biological pacemaker. This isoform is the dominant isoform present in ventricular myocytes [16] and has intermediate kinetics (slower than that of HCN1 but faster than that of HCN4), and a robust cAMP response (comparable to that of HCN4 and markedly greater than that of HCN1). We prepared an adenovirus of HCN2 which we initially characterized in isolated neonatal and adult rat ventricle cells in culture [26]. Since neonatal ventricular cultures are spontaneously active, they are useful in determining the effect of a particular gene on rate (Fig. 4) [27], and since they contain the full array of adrenergic signaling, they also can be used to assess autonomic responsiveness of HCN2-expressing cells. In the canine heart, we next injected the HCN2 adenovirus into either the left atrium [28] or the left bundle branch of the Purkinje system [29], and used transient vagal stimulation to suppress the intrinsic sinus node pacemaker and AV node conduction to unmask the functionality of the exogenous biological pacemaker. When HCN2 was expressed in the left atrium in conjunction with vagal stimulation to induce sinus arrest, a spontaneous left atrial rhythm was observed in all animals. Control (green fluorescent protein expressing) animals never developed a spontaneous left atrial rhythm under these test conditions. In the left bundle branch experiments, right ventricular escape rhythms were observed more often in control than in HCN2-expressing animals. Further, when left ventricular escape rhythms were observed, the mean cycle length was significantly shorter in the HCN2-expressing animals.

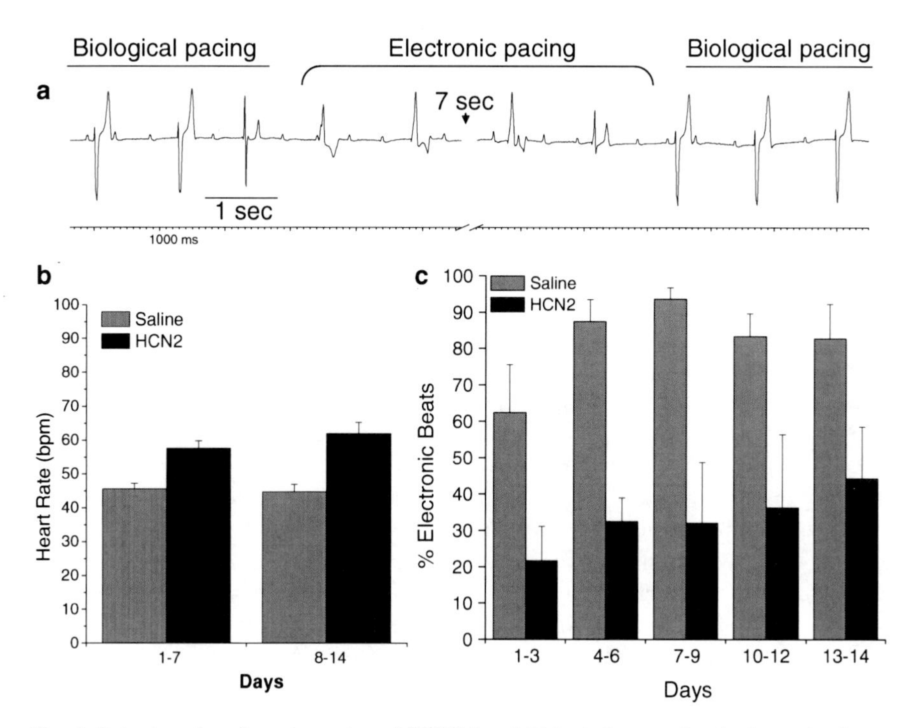

Fig. 5 Behavior of tandem electronic and HCN2-based biological pacemaker in the canine heart. (**a**) Representative tracing of interaction between biological and electronic pacemakers. (**b**) Mean basal heart rate over days 1–7 and 8–14 of groups into which saline or HCN2 adenovirus had been injected. Rates in the HCN2 group were significantly faster than those in the saline group. (**c**) Percentage of electronically paced beats occurring in hearts into which saline had been injected and in which an electronic pacemaker had been implanted or in hearts into which HCN2 had been injected in tandem with an electronic pacemaker. The electronic pacemaker was set at VVI 45 beats/min and throughout the 14-day period, the number of beats initiated electronically was higher in the group into which saline had been injected than in the group into which HCN2 had been injected (reprinted with permission from [30])

Since vagal stimulation was used to suppress the endogenous rhythm and unmask the biological pacemaker function in these experiments, autonomic responsiveness of the biological pacemaker could not be accurately assessed. Therefore, in subsequent experiments we combined left bundle branch injection of HCN2 virus with induction of permanent AV ablation and insertion of a demand electronic pacemaker set at a low escape rate [30]. This approach both allows for more detailed analysis of biological pacemaker function, and provides a template for initial clinical trials, where a backup electronic pacemaker in demand mode will certainly be required. Figure 5a demonstrates how the biological and demand electronic pacemakers operating in tandem smoothly transition back and forth. The efficacy of the biological pacemaker in sustaining a higher heart rate is evident when animals into which the HCN2 virus had been injected are compared with saline control animals (Fig. 5b). Finally, the presence of the tandem electronic pacemaker provides additional information thanks to the existence of its memory function, which allows one to tabulate the percentage of total beats that are generated by the device. Figure 5c illustrates that during the time period of peak adenovirus expression (days 4–12) about 30% of all beats arise from the electronic pacemaker, whereas in the saline control animals this figure is about 90%.

Since one rationale for development of biological pacemakers is to create an autonomically responsive rate, it is necessary to assess the response of implanted biological pacemakers to autonomic modulation or autonomic agonists. This is often done by means of catecholamine infusion [30], but more physiological indices also can be employed. These include measurement of heart rate variability as an index of vagal input, and response to physiological stimuli as a measure of sympathetic input [31]. For example, in dogs with HCN2 virus implanted in the left bundle branch following AV ablation, a 24-h fast followed by presentation of food results in an increased ventricular rate that pace-maps to the injection site.

Although we have focused on adenoviral delivery of HCN2, that is neither the sole option for gene-based delivery of ion channels nor the sole viable HCN isoform that can be employed. For example, Piron et al. [32] delivered a combination of HCN2 and β_2-adrenergic receptors using poloxamine nanospheres rather than virus to achieve expression in the ventricle of mice following AV block. Expression was associated with shorter RR intervals that persisted for at least 45 days (Fig. 6), delayed onset of heart failure, and a more marked rate response to isoproterenol. In addition, several groups have explored the use of the HCN4 subunit as a biological pacemaker in vitro, employing either adenovirus [33] or lentivirus [34] as the delivery system and demonstrating an effect to increase both basal [33, 34] and adrenergically stimulated [34] rate. Tse et al. [35] reported on the adenoviral delivery of a mutated HCN1 subunit to the left atrium of pigs in which the SA node had been ablated. They found an increase in rate and a decrease in the percentage of beats

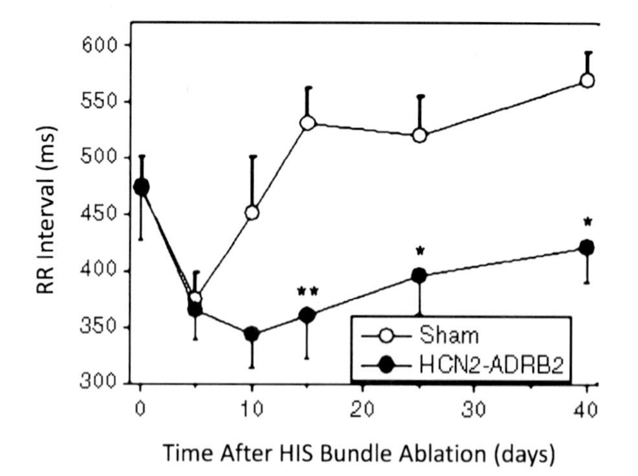

Fig. 6 Long-term ECG follow-up mice treated with HCN2 and β_2-adrenergic receptor. RR interval was monitored for 40 days in treated ($n=9$) and sham ($n=7$) mice after His bundle ablation on day 0. *$P<0.05$ versus the corresponding value in sham mice, **$P<0.01$ versus the corresponding value in sham mice, *ADRB2* β_2-adrenergic receptor (reprinted with permission from [32])

generated by a backup electronic pacemaker. They also reported an increase in rate in response to isoproterenol, but the increase was less than what was observed in control animals. They did not report if the isoproterenol stimulated rate pace-mapped to the injection site. It should be remembered that the HCN1 subunit exhibits relatively little sensitivity to cAMP [13, 14].

Taken together, these results illustrate the potential of an HCN-based biological pacemaker to achieve a physiologically sustainable heart rate and to respond appropriately to autonomic modulation of rate, including responding to emotional stimuli.

4.2 *Use of Endogenously Active Cells*

Several laboratories have transplanted spontaneously active cells into a localized region of the heart to create a biological pacemaker. The cell types evaluated were human ESCs (hESCs) [36, 37] and isolated SA node cells from 14–19 week gestational age aborted human fetuses [38]. It should be noted that yet another potential cell source is induced pluripotent cells, which, like hESCs, can be driven down a cardiac lineage to become spontaneously active [39]. However, since the function and many of the limitations of these cells are similar to those of ESCs, and since to date they have not been used as a biological pacemaker in any studies, they are not further discussed here.

Before using cardiac-like hESCs as the basis of an autonomically responsive biological pacemaker, one should understand the autonomic cascades and ionic currents present in these cells. This is complicated because the impact of long-term in vivo implantation on electrophysiological phenotype is unknown. Further, even in vitro studies have indicated the potential for electrical heterogeneity within this preparation [40], and results can differ by laboratory. For example, one report indicates that the spontaneous rate of these cells in culture is slowed by a specific I_f blocker [37], whereas another reported the failure to observe such I_f blocker dependent slowing [41], leading these authors to suggest a major role of the inward Na^+ current in automaticity. The ionic basis of automaticity in these cells is likely to impact autonomic responsiveness of an hESC-derived biological pacemaker in vivo. In addition, the signaling cascades present in these cells will also impact responsiveness. Reppel et al. [42] found a positive chronotropic response to isoproterenol at physiologically relevant concentrations, as well as a transient cholinergic response. They argued the latter was consistent with an action through the nitric oxide and cyclic GMP pathway, as had been previously reported for developing cardiac cells. The implications of this observation for the in vivo autonomic regulation of an hESC-based biological pacemaker remains to be determined. It should be noted that similar studies have been undertaken with murine ESCs, and these support the involvement of I_f and T-type Ca^{2+} current in ESC-based automaticity [43, 44].

The first exploration of an hESC biological pacemaker in vivo was carried out by Kehat et al. [36], who grew hESCs as embryoid bodies so that they differentiated into spontaneously beating cardiac-like cells. Beating embryoid bodies were injected into the left ventricular wall. The embryoid bodies were injected into the pig heart following induction of complete AV block, resulting in a stable ventricular rhythm originating at the site of injection in six of 13 animals (Fig. 7). The rhythm was catecholamine-responsive, with infusion of 10–20 μg/min of isoproterenol increasing rate from an average of 59 to 94 beats/min. A second group conducted similar experiments in guinea pig [37], using Langendorff perfusion of the isolated hearts for subsequent analysis following cryoablation to disrupt normal AV conduction. Ventricles of sham were quiescent, whereas hearts of animals receiving hESCs exhibited a slow rhythm and slow conduction consistent with epicardial propagation from the injection site.

The other approach involving spontaneously active cells employed isolated SA node myocytes of human aborted fetuses which were implanted into the left ventricular

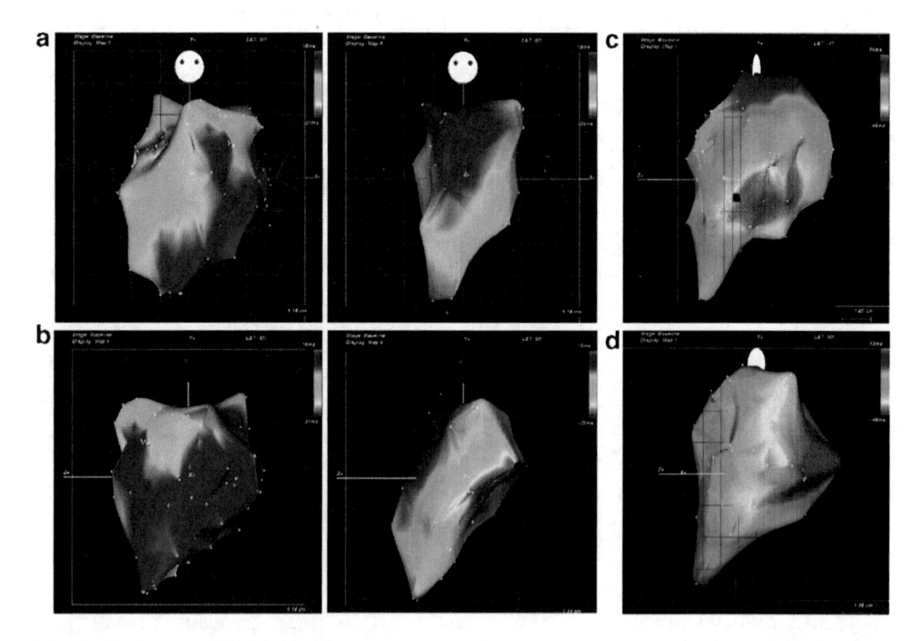

Fig. 7 Electroanatomical mapping of the ectopic rhythm following implantation of human embryonic stem cells in the swine ventricle. (**a**, **b**) Electroanatomical mapping of the junctional (*left*) and new ventricular ectopic (*right*) rhythms, as shown from anteroposterior (**a**) and left lateral (**b**) views. Note that the earliest activation (*red*) during the junctional rhythm originated from the superior septum, with the posterolateral wall activated last (*blue-purple*). In contrast, the earliest activation during the ectopic rhythm was at the posterolateral wall (*red*), with the septum activated last. (**c**, **d**) Reproducibility of electrophysiological findings. The same animal was mapped on two separate occasions and the corresponding electroanatomical maps are shown in a *left* posterior oblique view (reprinted with permission from [36])

wall of the pig heart [38]. Following AV ablation, four of five animals exhibited a stable ventricular rhythm that was significantly faster than in control hearts into which only solution had been injected (86 vs. 30 beats/min). The animals into which cells had been injected, but not the controls, also exhibited an increase in rate in response to isoproterenol infusion (3 μg/min), with rate increasing to 117 beats/min. The electrophysiologic and autonomic characteristics of these human fetal cells are not well defined. However, a study in the murine heart concluded that HCN4 was a major contributor to automaticity and cAMP responsiveness in the embryonic SA node [45].

There are numerous concerns regarding the source and in vivo fate of ESCs (see Sect. 6), but a potential advantage of these cells over direct gene delivery or the use of genetically engineered cells (see the next section) is the fact that genetic modification is not required, and there are therefore no associated complexities and risks. At the same time, the corresponding potential limitation is that the functionality (i.e., the extent of available depolarizing current) is defined and constrained by the cell phenotype. Although a "sinus-node-like" cell may function perfectly as a biological pacemaker within the specialized structure of the sinus node, it is not obvious that such a cell would be adequate to drive pacemaking at appropriate rates when implanted elsewhere in the heart. This is not an insurmountable problem, since these cells can in theory be enhanced through genetic engineering. For example, a recent study compared the function of murine ESCs driven down a cardiac lineage from wild-type and connexin 43 knockout animals [46]. The cells were co-cultured with a spontaneously active myocyte preparation. Although both the wild-type and the connexin 43 knockout ESCs were able to couple to the myocytes and the resulting spontaneous rates were comparable, the connexin 43 knockout co-culture was more likely (40 vs. 13%) to establish pacemaker dominance when the site of impulse initiation was mapped, perhaps because excitation spread from the myocytes into the region of ESCs was hampered. Thus, although these in vitro experiments did not provide a clear advantage of reducing cell coupling for creation of a cell-based biological pacemaker, they do offer a demonstration that genetic engineering can modulate endogenous pacemaker functionality of an ESC-based biological pacemaker. However, if genetic engineering is required, then one may be better off doing so in a cell type that is more readily available and/or carries less risk of neoplasia, such as adult mesenchymal stem cells (MSCs), as discussed in the next section.

4.3 Use of Genetically Engineered Cells

Using an endogenously active cell as a biological pacemaker source is readily understandable. Such cells beat spontaneously, so if they can electrically couple to myocytes when implanted into the heart, then they should be able to drive those myocytes and establish a localized site of impulse initiation. The question is whether they generate sufficient depolarizing current, and transfer it sufficiently efficiently, to drive the heart at a physiologically appropriate rate. In contrast, using genetically engineered passive cells as a biological pacemaker source is perhaps not as readily understandable. Certainly, one could in theory incorporate through genetic engineering all the necessary currents to recreate a spontaneous action potential, and some progress in this direction has been reported [47]. But this is a complex and time-consuming endeavor, and in the reported study the observed spontaneous rate was quite slow. However, such a herculean effort is not necessary. Owing to the space constant characteristics of myocytes and the cells used as the pacemaker source, such as adult hMSCs, a cell pair consisting of an hMSC and a myocyte will function as a single unit if electrically coupled. In other words, it is only necessary to incorporate the channels required to create a diastolic depolarization; the other channels needed to create an action potential already exist in the myocyte.

For this purpose, the HCN gene family is again the ideal candidate because of the unique biophysical characteristics of these channels. An hMSC typically has a resting potential on the order of −30 mV. Therefore, any expressed HCN channels will be closed and effectively nonfunctional since these channels open upon hyperpolarization with a typical threshold of −40 mV or more negative (depending on the isoform and cell environment). However, when electrically coupled to a myocyte, the hMSC will "see" the membrane potential of the myocyte. That is, the cell will hyperpolarize into a potential range where the expressed HCN channels can open. This will result in depolarizing current flowing into the hMSC and, through gap junctions, into the coupled myocyte, driving the membrane potential of the myocyte positive (Fig. 8). If current flow is sufficient, the myocyte will reach threshold and fire an action potential.

What are the requirements for this to occur? First, the hMSCs must be able to electrically couple to myocytes upon implantation into the heart. For that, hMSCs must contain (or be engineered to express) cardiac-compatible connexin subunits. We have demonstrated this to be the case, with both connexin 43 and connexin 40 expressed on the surface of hMSCs (Fig. 9) [6]. Second, one must be able to express the chosen HCN isoform in the hMSCs with reasonable efficiency. We have found that electroporation provides expression of HCN channel genes with approximately 50% efficiency [48]. Others have published results with lentiviral mediated expression of HCN in hMSCs, demonstrating functional biological pacemaking in vitro [49]. Finally, if this cell-based biological pacemaker is to be superior to electronic pacemakers, it should possess autonomic responsiveness, which means essential elements of adrenergic and cholinergic signaling must be present in the cells. Figure 10 demonstrates that expressed HCN2 in hMSCs responds to the β-adrenergic agonist isoproterenol with a positive shift of the activation relation, as would be expected if the drug resulted in an increase in intracellular cAMP level. In contrast, the cholinergic agonist acetylcholine does not, by itself, shift the HCN2 activation relation negative. However, in the presence of isoproterenol it does cause such a shift. Thus, hMSCs possess intact cAMP signaling cascades, but basal cAMP levels are such that acetylcholine, which lowers the cAMP level, only has an effect after its level has been elevated by a stimulatory agonist, a phenomenon referred to as accentuated antagonism.

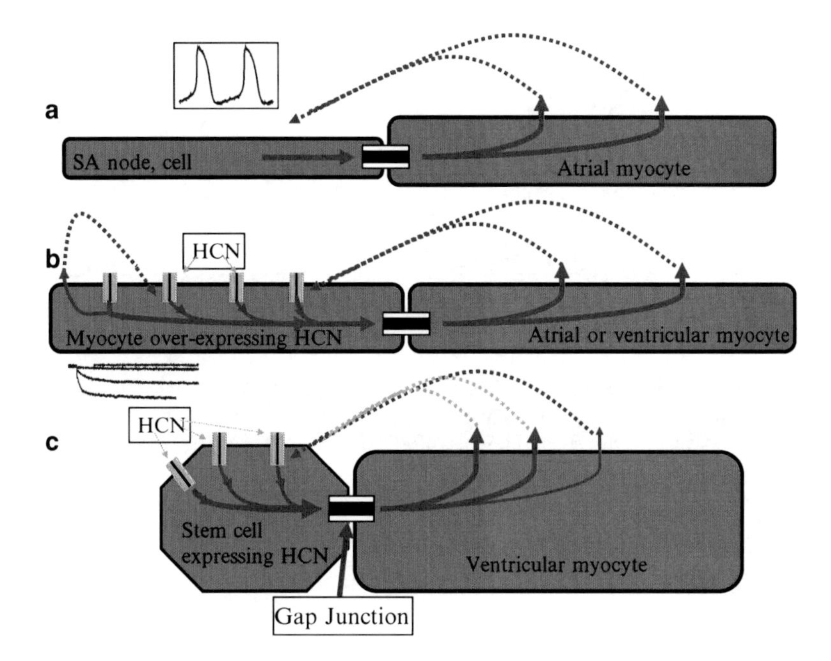

Fig. 8 Initiation of spontaneous rhythms by native pacemaker cells, by genetically engineered myocytes, or by genetically engineered stem cells. (**a**) In a native pacemaker cell or (**b**) in a myocyte engineered to incorporate pacemaker current via HCN2 gene transfer, action potentials (*inset* in (**a**)) are initiated via inward current flowing through transmembrane HCN channels, which open during membrane repolarization following an action potential. Current flows via gap junctions to adjacent myocytes, resulting in their excitation and propagation of excitability. (**c**) HCN2 channels engineered into the membrane of a stem cell can only open, and current can only flow through them (*inset*) when the membrane is hyperpolarized. In this case such hyperpolarization can only be delivered if an adjacent myocyte is tightly coupled to the stem cell via gap junctions. In the presence of such coupling and the opening of the HCN channels to induce local current flow, the adjacent myocyte will be excited and initiate an action potential that then propagates. The depolarization of the action potential will result in the closing of the HCN channels until the next repolarization restores a high negative membrane potential (modified and reprinted with permission from [83])

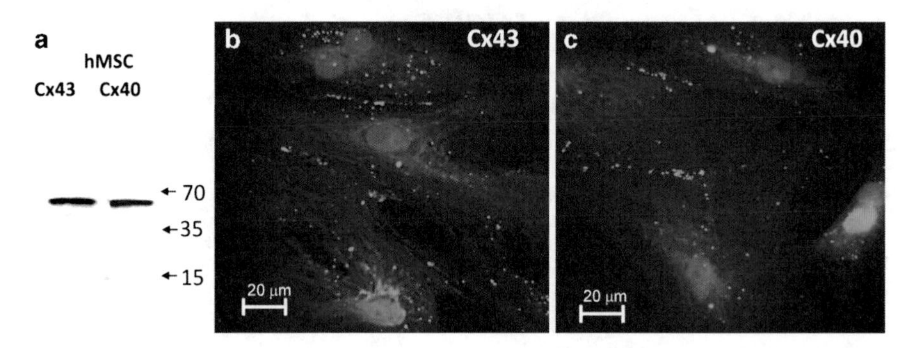

Fig. 9 Identification of connexins in gap junctions of human mesenchymal stem cells (*hMSCs*). (**a**) Immunoblot analysis of connexin 43 (*Cx43*) and connexin 40 (*Cx40*) in hMSCs. (**b**, **c**) Immunostaining of hMSCs for Cx43 and Cx40, respectively. ((**a**) Courtesy of V. Valiunas; (**b**, **c**) reprinted with permission from [6])

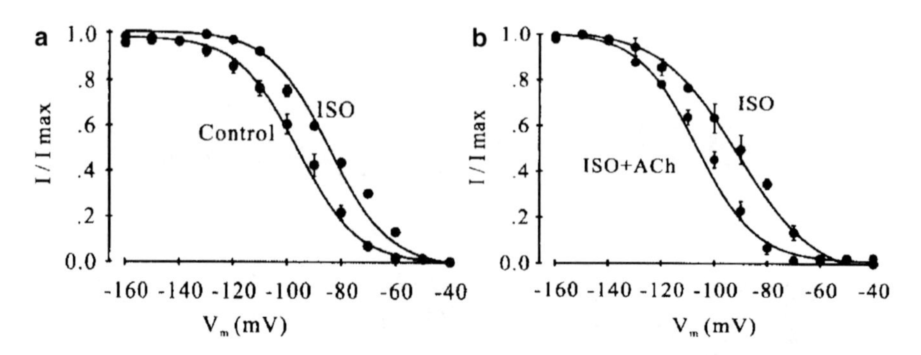

Fig. 10 Adrenergic and cholinergic responsiveness of adult hMSCs transfected with the murine HCN2 gene. (**a**) Activation relation generated from normalized tail current measurements of HCN2 in the absence (*control*) and presence of 10^{-6} M isoproterenol (*ISO*), demonstrating the positive shift of the relation by isoproterenol. (**b**) Modulation of the activation relation of HCN2 by 10^{-6} M acetylcholine (*ACh*) in the presence of isoproterenol, demonstrating the negative shift of the relation by acetylcholine in the continuous presence of isoproterenol. In the absence of isoproterenol, acetylcholine did not shift the activation relation (not shown) (reprinted with permission from [48])

We recently demonstrated that a single hMSC expressing HCN2 and a single ventricular myocyte, when in physical contact, create a spontaneously active cell pair, with the automaticity depending on cell coupling (Fig. 11) [50]. When these HCN2-expressing hMSCs were implanted into the canine heart in tandem with an electronic pacemaker, an impulse was observed that site-mapped to the site of implantation and that could be observed for the entire 6-week duration of the study [51]. If sufficient cells are implanted, then, as with viral gene delivery (see Fig. 5c), the biological pacemaker provides about two-thirds of the heartbeats (Fig. 12).

4.4 Genetic Manipulation of HCN Genes To Enhance Function

In the last 5 years numerous studies have attempted to create a "better" biological pacemaker by using structure–function analysis and mutagenesis to design an HCN channel subunit with more desirable biophysical characteristics. The general philosophy has been to try to develop a subunit that activates at less negative potentials and/or with faster kinetics, with the rationale being that either of these changes should result in more depolarizing current earlier in diastole, and therefore a faster endogenous rate. At the same time, these engineered channels must be expressed at least as well as the native isoforms (i.e., maximal current must be comparable) and maintain a strong response to cAMP (i.e., preserve autonomic responsiveness). There has been significant progress in elucidating the structure–function relation of specific regions of the HCN molecule [13–15], and designing subunits which manifest the

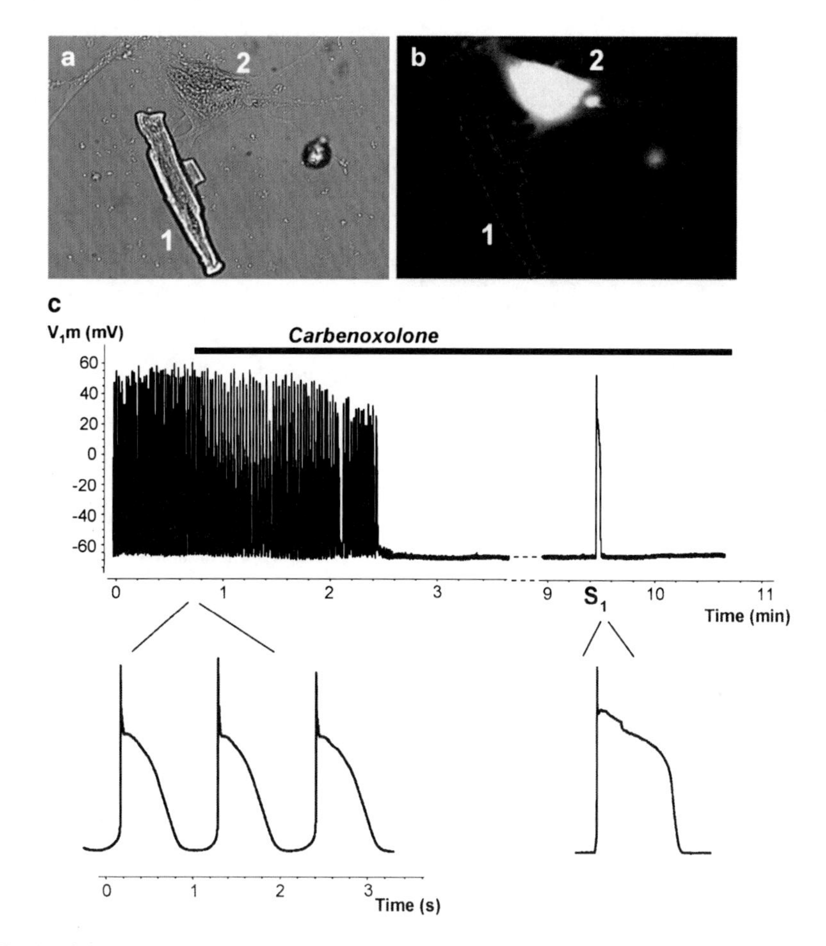

Fig. 11 Spontaneous activity in an hMSC–myocyte cell pair when the hMSC expressed HCN2. (**a**) Transmitted light and (**b**) fluorescent images of a canine ventricle myocyte (cell 1) and an hMSC transfected with HCN2 (cell 2). (**c**) Spontaneous action potentials recorded from the myocyte of this cell pair and the effect of 200 μM carbenoxolone. Carbenoxolone abolished spontaneous activity in a time-dependent manner, but an external stimulus (S_1) evoked a single action potential, demonstrating persistent excitability. The *lower panel* in (**c**) is an expanded timescale of spontaneous activity before carbenoxolone administration and evoked action potential during carbenoxolone administration (reprinted with permission from [50])

desired changes in voltage dependence and kinetics [2, 52]. However, to date these efforts have not resulted in an improved biological pacemaker in vivo.

Studies by us and other groups have drawn on some of the above-mentioned structure–function analyses to test specific constructs as biological pacemakers. In one such study we employed a point mutation previously characterized by Chen et al. [53] to create an HCN2 variation with positively shifted activation [30]. However, the mutated channel was not expressed as well as wild-type HCN2, and so the advantage of the shift in voltage dependence was canceled out by the reduced

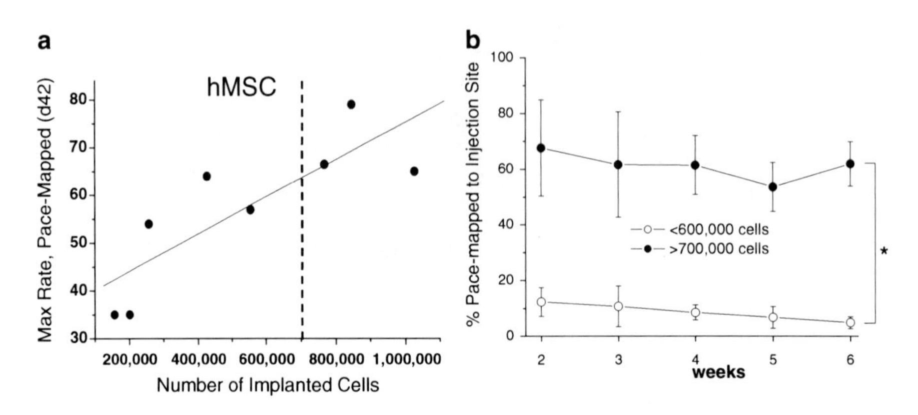

Fig. 12 Function of a biological pacemaker based on hMSCs expressing HCN2. (**a**) Maximum rates recorded on day 42 of pace-mapped beats as a function of the number of cells injected into the canine ventricle. (**b**) Analysis of the percentage of beats pace-mapped to the injection sites for animals that received either more than 700,000 or fewer than 600,000 hMSCs. The hMSCs were implanted into the left ventricle wall and the animals were studied for a total of 42 days (reprinted with permission from [51])

current magnitude. Another group [54] used a deletion strategy to modify HCN1, shifting activation positive to the reversal potential. As might be expected from such a large shift in voltage dependence, highly expressing cells exhibited depolarization and loss of excitability, i.e., the voltage shift exceeded the optimal value. In studies of activation kinetics, rather than voltage dependence, we also found that it was possible to excessively modify a biophysical parameter. We employed a previously constructed chimeric channel [55], using the transmembrane portion of HCN1 (fast gating) and the cytoplasmic portions of HCN2 (strong cAMP response). The resulting channel "overshot" in terms of biological pacemaking, resulting in periods of ventricular tachycardia interspersed with pauses [56]. Although neither approach yielded a better biological pacemaker, these studies and others [57] did demonstrate that the newborn rat ventricle culture preparation served as a reliable in vitro predictor of in vivo function (Fig. 13). This suggests that cell culture and co-culture systems may be useful to develop and test a series of HCN constructs that encompass a range of biophysical values for a specific parameter such as kinetics [58], particularly if combined with computer modeling to identify critical biophysical parameters [27].

5 The Extension: Addressing the AV Node

Creation of an AV bridge to bypass a nonfunctioning AV node requires the ability to create a continuous chain of cells that can both couple to myocytes at either end and propagate an electrical signal from one cardiac region to the other. The ability

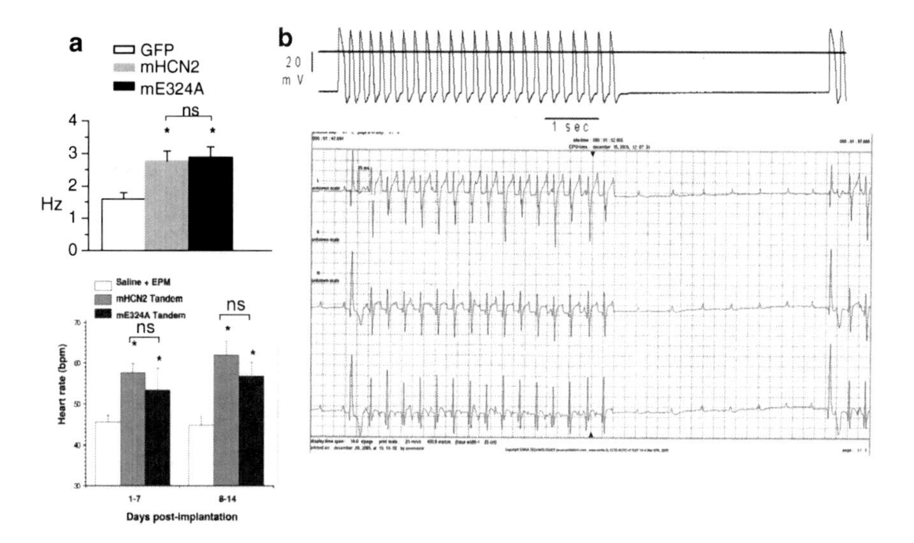

Fig. 13 Comparison of in vitro and in vivo behavior of biological pacemakers based on different HCN constructs. (**a**) Expressed HCN2 and the E324A mutation of HCN2 have equivalent effects on rate in culture (*top*; $n=5$–10) and in intact canine heart (*bottom*; $n=4$–7), suggesting that the benefit of favorable biophysical properties of E324A are limited by the poor expression of this construct. (**b**) An expressed chimeric channel (HCN212) consisting of the transmembrane domain of HCN1 and the N- and C-termini of HCN2 causes bursts of rapid beating and pauses in both culture (*top*) and in the intact canine heart (*bottom*). *EPM* electronic pacemaker, *GFP* green fluorescent protein, *ns* not significant. ((**a**) Reprinted with permission from [30]; (**b**) portion reprinted with permission from [56])

to couple noncardiac cells to myocytes and transmit an electrical signal has been demonstrated by several groups for nonexcitable cells. Gaudesius et al. [59] showed that fibroblasts of cardiac origin express connexins and can preserve impulse propagation in cardiac cultures across regions extending up to 300 μm. Pijnappels et al. [60] demonstrated a similar result when using hMSCs to bridge a similar size gap. Since in both cases the bridging cells were nonexcitable, the distance that could be spanned without loss of propagation was limited. For a viable AV bypass one would need to use or create excitable cells to extend the distance that could be spanned. Kizana et al. [61] used lentivirus to express MyoD and connexin 43 in human dermal fibroblasts, resulting in myogenic conversion. After electrical stimulation, they observed calcium transients, with adjacent cells demonstrating identical thresholds. Thus, they created potentially excitable cells but did not demonstrate that these could be used to synchronize separate regions of cardiac cells. Another group [62, 63] differentiated murine adipose-tissue-derived MSCs into beating myocytes and then injected them into the AV region of mice in complete heart block, achieving sinus rhythm or 2:1 block.

These latter preliminary studies provide proof of principle that excitable cells can be used to improve conduction through a damaged AV node. However, it is unclear that such direct cell injection would provide sustained benefit in the setting of progressive conduction disease. For this reason, a true AV bypass may be the more viable approach. Such a bypass would have to deal with, besides the issue of cell type, the challenge of cell packaging. An AV bypass would certainly require some sort of encapsulation technology (see the next section). In addition, as discussed earlier, an external bypass would entail a longer conduction path than the native AV node and therefore more rapid propagation, most likely requiring a Na-dependent rather than a Ca-dependent action potential.

6 The Challenge: Improving Outcome

HCN subunit mutagenesis to fine-tune channel biophysics, as described earlier, is obviously one way to improve the functional outcome with an implanted biological pacemaker. However, there are other factors that impact outcome besides channel biophysics, as well as other ways to alter the magnitude/time course of current delivery besides introducing mutations into the HCN sequence.

The current generated in early diastole is determined by the number of channels at the cell membrane, the activation kinetics of the channel, and its voltage dependence. Several of these parameters can be influenced, independent of specific subunit sequence, by the expression of cofactors. For example, MiRP1 (KCNE2) acts as a β-subunit for HCN channels, increasing current magnitude and speeding channel kinetics without altering voltage dependence [19, 20, 64]. Coexpression of various Trip8b isoforms either increases or decreases HCN channel expression, but uniformly shifts activation negative; cAMP responsiveness is also impaired [65–67]. HCN1 (but not other isoforms) also interacts with filamin A, resulting in a slowing of kinetics [22]. Thus, coexpression of some of these other proteins with HCN subunits, or selective knockdown of inhibitory subunits if endogenously expressed in the targeted cell type, could result in a more robust pacemaker current.

Studies of HCN as a biological pacemaker have tended to use a ubiquitous promoter such as CMV for transgene expression. There are two concerns with this approach [68, 69]: (1) if expressed in antigenic cells, it can contribute to an immune response; (2) over a period of months gene silencing can occur, possibly via a mechanism involving histone hypoacetylation and/or DNA methylation. Introducing the transgene into hMSCs ex vivo minimizes the risk of the former effect, whereas switching to a cell-specific promoter would minimize the latter. However, using a cell-specific promoter is likely, at least in the short term, to result in a lower expression level and therefore reduced current [70]. In addition, if the hMSC phenotype changes following in vivo implantation, promoter activity could be affected. In summary, the ideal promoter for persistent expression in a cell-based biological pacemaker remains to be determined.

Current magnitude and persistence can also be influenced by the method of transgene expression. In the case of cell-based therapy, a method resulting in 100% expression in the target cells will yield greater current delivery to the surrounding myocardium than a less efficient method. To date, the methods that have been used to achieve gene expression in cell-based biological pacemakers are electroporation and lentivirus. Electroporation achieves expression efficiencies on the order of 50% and has been demonstrated to persist in vivo for at least 6 weeks [48, 51]. In addition, some electroporation devices are reported to result in direct transfer of the DNA into the cell nucleus [71]. Lentiviral expression vectors are effective vehicles for transducing and stably expressing different effector molecules (small interfering RNA, complementary DNA, etc.) in almost any mammalian cell, including cardiac cells both in vitro and in vivo [72–74], and have been used to express HCN subunits in hMSCs [75]. Current methods for producing lentiviral stock typically result in relatively low titer. Consequently, smaller amounts of viral particles can be delivered in vivo compared with adenovirus [72, 76]. Moreover, the process of integration of the transgene into genomic DNA in cardiac myocytes appears less efficient compared with some other cell types, again resulting in low expression [73, 77]. All of this argues against direct lentiviral delivery to the myocardium to create a biological pacemaker. However, none of these limitations apply when used to express HCN ex vivo in hMSCs. Thus, lentivirus may be a suitable alternative to electroporation for high-efficiency and/or persistent expression of HCN subunits in hMSCs.

Regardless of the method of transgene expression, the promoter, and the presence or absence of expression of additional cofactors, ultimately the stem cells must be delivered to a location in the myocardium and persist at that location. This leads to questions of cell fate, which encompasses issues of cell differentiation after implantation (which could affect channel expression or function), cell survival, and cell migration, all of which apply equally to hESCs and hMSCs. The first step in addressing any of these issues is to be able to track the implanted stem cells in vivo. We have reported on a novel approach that incorporates fluorescent quantum dots into hMSCs prior to cardiac implantation [78], and this and other approaches are described in greater detail in chapter "Tracking of Stem Cells In Vivo". When the cells can be reliably identified in vivo, it permits questions of cell survival and differentiation to be pursued and related to the degree of functionality. Perhaps the most critical issue, however, is whether the cells migrate far from the implant site and whether such migration can be controlled or limited. Migration not only would reduce efficacy but also raise safety concerns based on where the migrating cells settle.

An obvious approach to the migration issue would be to encapsulate the stem cells with an inert material prior to implantation. In considering such a method, one must remember that a critical aspect of a biological pacemaker is the formation of gap junctions between the stem cells and myocytes to provide electrical continuity. Thus, an encapsulating mesh would have to provide a pore size too small to allow stem cells to escape but large enough for cellular processes to extend through the mesh and connect to cells beyond the encapsulation. It is also important to remember that, at least in the case of nonexcitable hMSCs, the stem cell resting potential is not sufficiently negative to activate the expressed HCN channels unless the stem cells are close to more negative myocytes. Thus, if an encapsulated sphere of stem cells were implanted into the myocardium, only an outer shell of cells would have sufficiently negative resting potentials to activate the HCN channels and contribute current. For these reasons, a first-generation encapsulation approach may be to grow the stem cells on a membrane that is then implanted on or in the heart in such a way that at least one surface of the cell monolayer is in direct contact with myocytes. Furuta et al. [79] employed such an approach to prepare a cell sheet of biodegradable polymerized fibrin containing neonatal rat ventricular myocytes which was then grafted onto the heat-ablated epicardium of a rat heart in vivo. One to 4 weeks later, there was evidence of functional integration, including bidirectional action potential propagation between the host heart and the grafted cover-sheet. Other groups are pursuing similar research employing different substrates [80]. It remains to be determined if anchoring stem cells within such a sheet will reduce in vivo migration following cardiac implantation. Further, even if such a sheet approach proves successful for biological pacemakers, it will not suffice for an AV bypass. In this case true encapsulation will be required, and the encapsulation must not only allow for electrical continuity along the bridge and at the two ends, but must also provide insulation along the length of the bridge so that the electrical signal is directed at full strength toward the distal cardiac region.

7 Conclusion

Proof of principle for both gene-based and cell-based biological pacemakers is now well established. Since the first reported attempt in 1998 [8], dozens of original articles and reviews have been published on this topic. In the case of cell-based therapies, both excitable cells (hESCs driven down a cardiac lineage) and nonexcitable cells (hMSCs engineered to express HCN channels) have been successfully employed. Despite these results, we are still not yet at the stage where a clinical trial of a biological pacemaker is feasible. Critical issues of safety and persistence of function must first be addressed in long-term animal studies. Once that milestone is reached, biological pacemakers may enter our therapeutic repertoire. Even then, the patient population for which they are appropriate will be limited until a biological AV bypass is also developed. Although many of the technical challenges of an AV bypass are similar to those of a biological pacemaker, and thus development of the latter will accelerate progress on the former, an AV bypass also entails an additional set of engineering challenges that will have to be resolved. However, the possibility of implanting a biologically based bypass tract, perhaps containing a proximal pacemaker, is so compelling that efforts in this area are likely to continue in multiple laboratories.

References

1. Gregoratos G (2005) Indications and recommendations for pacemaker therapy. Am Fam Physician 71:1563–1570
2. Rosen MR, Brink PR, Cohen IS, Robinson RB (2008) Cardiac pacing: from biological to electronic… to biological? Circ Arrhythm Electrophysiol 1:54–61
3. Hauser RG, Hayes DL (2009) Increasing hazard of Sprint Fidelis implantable cardioverter-defibrillator lead failure. Heart Rhythm 6:605–610

4. Robinson RB, Brink PR, Cohen IS, Rosen MR (2006) I(f) and the biological pacemaker. Pharmacol Res 53:407–415
5. Lakatta EG, DiFrancesco D (2009) What keeps us ticking: a funny current, a calcium clock, or both? J Mol Cell Cardiol 47:157–170
6. Valiunas V, Doronin S, Valiuniene L, Potapova I, Zuckerman J, Walcott B, Robinson RB, Rosen MR, Brink PR, Cohen IS (2004) Human mesenchymal stem cells make cardiac connexins and form functional gap junctions. J Physiol (London) 555:617–626
7. Cho HC, Kashiwakura Y, Marban E (2007) Creation of a biological pacemaker by cell fusion. Circ Res 100:1112–1115
8. Edelberg JM, Aird WC, Rosenberg RD (1998) Enhancement of murine cardiac chronotropy by the molecular transfer of the human beta2 adrenergic receptor cDNA. J Clin Invest 101:337–343
9. Edelberg JM, Huang DT, Josephson ME, Rosenberg RD (2001) Molecular enhancement of porcine cardiac chronotropy. Heart 86:559–562
10. Miake J, Marban E, Nuss HB (2002) Gene therapy: biological pacemaker created by gene transfer. Nature 419:132–133
11. Miake J, Marban E, Nuss HB (2003) Functional role of inward rectifier current in heart probed by Kir2.1 overexpression and dominant-negative suppression. J Clin Invest 111:1529–1536
12. Silva J, Rudy Y (2003) Mechanism of pacemaking I_{K1}-downregulated myocytes. Circ Res 92:261–263
13. Biel M, Schneider A, Wahl C (2002) Cardiac HCN channels. Structure, function, and modulation. Trends Cardiovasc Med 12:206–212
14. Robinson RB, Siegelbaum SA (2003) Hyperpolarization-activated cation currents: from molecules to physiological function. Ann Rev Physiol 65:453–480
15. Baruscotti M, Barbuti A, Bucchi A (2010) The cardiac pacemaker current. J Mol Cell Cardiol 48(1):55–64
16. Shi W, Wymore R, Yu H, Wu J, Wymore RT, Pan Z, Robinson RB, Dixon JE, McKinnon D, Cohen IS (1999) Distribution and prevalence of hyperpolarization-activated cation channel (HCN) mRNA expression in cardiac tissues. Circ Res 85:e1–e6
17. Brioschi C, Micheloni S, Tellez JO, Pisoni G, Longhi R, Moroni P, Billeter R, Barbuti A, Dobrzynski H, Boyett MR, DiFrancesco D, Baruscotti M (2009) Distribution of the pacemaker HCN4 channel mRNA and protein in the rabbit sinoatrial node. J Mol Cell Cardiol 47:221–227
18. Liu J, Dobrzynski H, Yanni J, Boyett MR, Lei M (2007) Organisation of the mouse sinoatrial node: structure and expression of HCN channels. Cardiovasc Res 73:729–738
19. Yu H, Wu J, Potapova I, Wymore RT, Holmes B, Zuckerman J, Pan Z, Wang H, Shi W, Robinson RB, El-Maghrabi R, Benjamin W, Dixon J, McKinnon D, Cohen IS, Wymore R (2001) MinK-related protein 1: a β subunit for the HCN ion channel subunit family enhances expression and speeds activation. Circ Res 88:e84–e87
20. Qu J, Kryukova Y, Potapova IA, Doronin SV, Larsen M, Krishnamurthy G, Cohen IS, Robinson RB (2004) MiRP1 modulates HCN2 channel expression and gating in cardiac myocytes. J Biol Chem 279:43497–43502
21. Barbuti A, Gravante B, Riolfo M, Milanesi R, Terragni B, DiFrancesco D (2004) Localization of pacemaker channels in lipid rafts regulates channel kinetics. Circ Res 94:1325–1331
22. Gravante B, Barbuti A, Milanesi R, Zappi I, Viscomi C, DiFrancesco D (2004) Interaction of the pacemaker channel HCN1 with filamin A. J Biol Chem 279:43847–43853
23. Santoro B, Wainger BJ, Siegelbaum SA (2004) Regulation of HCN channel surface expression by a novel C-terminal protein–protein interaction. J Neurosci 24:10750–10762
24. Kimura K, Kitano J, Nakajima Y, Nakanishi S (2004) Hyperpolarization-activated, cyclic nucleotide-gated HCN2 cation channel forms a protein assembly with multiple neuronal scaffold proteins in distinct modes of protein–protein interaction. Genes Cells 9:631–640
25. Ye B, Nerbonne JM (2009) Proteolytic processing of HCN2 and co-assembly with HCN4 in the generation of cardiac pacemaker channels. J Biol Chem 284:25553–25559

26. Qu J, Barbuti A, Protas L, Santoro B, Cohen IS, Robinson RB (2001) HCN2 over-expression in newborn and adult ventricular myocytes: distinct effects on gating and excitability. Circ Res 89:e8–e14
27. Robinson RB (2009) Engineering a biological pacemaker: *in vivo*, *in vitro* and *in silico* models. Drug Discov Today Dis Models 6(3):93–98.
28. Qu J, Plotnikov AN, Danilo PJ, Shlapakova I, Cohen IS, Robinson RB, Rosen MR (2003) Expression and function of a biological pacemaker in canine heart. Circulation 107: 1106–1109
29. Plotnikov AN, Sosunov EA, Qu J, Shlapakova IN, Anyukhovsky EP, Liu L, Janse MJ, Brink PR, Cohen IS, Robinson RB, Danilo PJ, Rosen MR (2004) Biological pacemaker implanted in canine left bundle branch provides ventricular escape rhythms that have physiologically acceptable rates. Circulation 109:506–512
30. Bucchi A, Plotnikov AN, Shlapakova I, Danilo PJ, Kryukova Y, Qu J, Lu Z, Liu H, Pan Z, Potapova I, KenKnight B, Girouard S, Cohen IS, Brink PR, Robinson RB, Rosen MR (2006) Wild-type and mutant HCN channels in a tandem biological-electronic cardiac pacemaker. Circulation 114:992–999
31. Shlapakova IN, Nearing BD, Lau DH, Boink GJJ, Danilo P, Jr., Kryukova Y, Robinson RB, Cohen IS, Rosen MR, Verrier RL (2010) Biological pacemakers in canines exhibit positive chronotropic response to emotional arousal. Heart Rhythm 12:1835–1840
32. Piron J, Quang KL, Briec F, Amirault JC, Leoni AL, Desigaux L, Escande D, Pitard B, Charpentier F (2008) Biological pacemaker engineered by nonviral gene transfer in a mouse model of complete atrioventricular block. Mol Ther 16:1937–1943
33. Er F, Larbig R, Ludwig A, Biel M, Hofmann F, Beuckelmann DJ, Hoppe UC (2003) Dominant-negative suppression of HCN channels markedly reduces the native pacemaker current I(f) and undermines spontaneous beating of neonatal cardiomyocytes. Circulation 107:485–489
34. Boink GJ, Verkerk AO, van Amersfoorth SC, Tasseron SJ, van der Rijt R, Bakker D, Linnenbank AC, van der Meulen J, de Bakker JM, Seppen J, Tan HL (2008) Engineering physiologically controlled pacemaker cells with lentiviral HCN4 gene transfer. J Gene Med 10:487–497
35. Tse HF, Xue T, Lau CP, Siu CW, Wang K, Zhang QY, Tomaselli GF, Akar FG, Li RA (2006) Bioartificial sinus node constructed via in vivo gene transfer of an engineered pacemaker HCN channel reduces the dependence on electronic pacemaker in a sick-sinus syndrome model. Circulation 114:1000–1011
36. Kehat I, Khimovich L, Caspi O, Gepstein A, Shofti R, Arbel G, Huber I, Satin J, Itskovitz-Eldor J, Gepstein L (2004) Electromechanical integration of cardiomyocytes derived from human embryonic stem cells. Nat Biotechnol 22:1282–1289
37. Xue T, Cho HC, Akar FG, Tsang SY, Jones SP, Marban E, Tomaselli GF, Li RA (2005) Functional integration of electrically active cardiac derivatives from genetically engineered human embryonic stem cells with quiescent recipient ventricular cardiomyocytes: insights into the development of cell-based pacemakers. Circulation 111:11–20
38. Lin G, Cai J, Jiang H, Shen H, Jiang X, Yu Q, Song J (2005) Biological pacemaker created by fetal cardiomyocyte transplantation. J Biomed Sci 12:513–519
39. Zwi L, Caspi O, Arbel G, Huber I, Gepstein A, Park IH, Gepstein L (2009) Cardiomyocyte differentiation of human induced pluripotent stem cells. Circulation 120:1513–1523
40. He JQ, Ma Y, Lee Y, Thomson JA, Kamp TJ (2003) Human embryonic stem cells develop into multiple types of cardiac myocytes: action potential characterization. Circ Res 93:32–39
41. Satin J, Kehat I, Caspi O, Huber I, Arbel G, Itzhaki I, Magyar J, Schroder EA, Perlman I, Gepstein L (2004) Mechanism of spontaneous excitability in human embryonic stem cell derived cardiomyocytes. J Physiol (London) 559:479–496
42. Reppel M, Boettinger C, Hescheler J (2004) Beta-adrenergic and muscarinic modulation of human embryonic stem cell-derived cardiomyocytes. Cell Physiol Biochem 14:187–196
43. Yanagi K, Takano M, Narazaki G, Uosaki H, Hoshino T, Ishii T, Misaki T, Yamashita JK (2007) Hyperpolarization-activated cyclic nucleotide-gated channels and T-type calcium channels confer automaticity of embryonic stem cell-derived cardiomyocytes. Stem Cells 25:2712–2719

44. Barbuti A, Crespi A, Capilupo D, Mazzocchi N, Baruscotti M, DiFrancesco D (2009) Molecular composition and functional properties of f-channels in murine embryonic stem cell-derived pacemaker cells. J Mol Cell Cardiol 46:343–351
45. Stieber J, Herrmann S, Feil S, Loster J, Feil R, Biel M, Hofmann F, Ludwig A (2003) The hyperpolarization-activated channel HCN4 is required for the generation of pacemaker action potentials in the embryonic heart. Proc Natl Acad Sci USA 100:15235–15240
46. Fahrenbach JP, Ai X, Banach K (2008) Decreased intercellular coupling improves the function of cardiac pacemakers derived from mouse embryonic stem cells. J Mol Cell Cardiol 45:642–649
47. Marban E, Cho HC (2008) Biological pacemakers as a therapy for cardiac arrhythmias. Curr Opin Cardiol 23:46–54
48. Potapova I, Plotnikov A, Lu Z, Danilo P, Jr., Valiunas V, Qu J, Doronin S, Zuckerman J, Shlapakova IN, Gao J, Pan Z, Herron AJ, Robinson RB, Brink PR, Rosen MR, Cohen IS (2004) Human mesenchymal stem cells as a gene delivery system to create cardiac pacemakers. Circ Res 94:952–959
49. Yang XJ, Zhou YF, Li HX, Han LH, Jiang WP (2008) Mesenchymal stem cells as a gene delivery system to create biological pacemaker cells in vitro. J Int Med Res 36:1049–1055
50. Valiunas V, Kanaporis G, Valiuniene L, Gordon C, Wang HZ, Li L, Robinson RB, Rosen MR, Cohen IS, Brink PR (2009) Coupling an HCN2 expressing cell to a myocyte creates a pacing two cell syncytium. J Physiol (London) 587:5211–5226
51. Plotnikov AN, Shlapakova I, Szabolcs MJ, Danilo P, Jr., Lorell BH, Potapova IA, Lu Z, Rosen AB, Mathias RT, Brink PR, Robinson RB, Cohen IS, Rosen MR (2007) Xenografted adult human mesenchymal stem cells provide a platform for sustained biological pacemaker function in canine heart. Circulation 116:706–713
52. Siu CW, Lieu DK, Li RA (2006) HCN-encoded pacemaker channels: from physiology and biophysics to bioengineering. J Membr Biol 214:115–122
53. Chen J, Mitcheson JS, Tristani-Firouzi M, Lin M, Sanguinetti MC (2001) The S4–S5 linker couples voltage sensing and activation of pacemaker channels. Proc Natl Acad Sci USA 98:11277–11282
54. Lieu DK, Chan YC, Lau CP, Tse HF, Siu CW, Li RA (2008) Overexpression of HCN-encoded pacemaker current silences bioartificial pacemakers. Heart Rhythm 5:1310–1317
55. Wang J, Chen S, Siegelbaum SA (2001) Regulation of hyperpolarization-activated HCN channel gating and cAMP modulation due to interactions of COOH terminus and core transmembrane regions. J Gen Physiol 118:237–250
56. Plotnikov AN, Bucchi A, Shlapakova I, Danilo P, Jr., Brink PR, Robinson RB, Cohen IS, Rosen MR (2008) HCN212-channel biological pacemakers manifesting ventricular tachyarrhythmias are responsive to treatment with I_f blockade. Heart Rhythm 5:282–288
57. Zhao X, Bucchi A, Oren RV, Kryukova Y, Dun W, Clancy CE, Robinson RB (2009) In vitro characterization of HCN channel kinetics and frequency-dependence in myocytes predicts biological pacemaker functionality. J Physiol (London) 587:1513–1525
58. Zhao X, Kryukova Y, Yang X, and Robinson RB (2009) Designing a biological pacemaker by altering HCN2 activation/deactivation kinetics. Heart Rhythm 6:S292 (Abstract)
59. Gaudesius G, Miragoli M, Thomas SP, Rohr S (2003) Coupling of cardiac electrical activity over extended distances by fibroblasts of cardiac origin. Circ Res 93:421–428
60. Pijnappels DA, Schalij MJ, van John T, Ypey DL, de Vries AA, van der Wall EE, van der Laarse A, Atsma DE (2006) Progressive increase in conduction velocity across human mesenchymal stem cells is mediated by enhanced electrical coupling. Cardiovasc Res 72:282–291
61. Kizana E, Ginn SL, Allen DG, Ross DL, Alexander IE (2005) Fibroblasts can be genetically modified to produce excitable cells capable of electrical coupling. Circulation 111:394–398
62. Takahashi T, Narikawa M, Oyama T, Naito AT, Ogura T, Nakaya H, Nakai T, Komuro I (2008) Brown adipose tissue-derived cells differentiate into spontaneous beating cells with both cardiac conduction and pacemaker characteristics. Circ Res 103:e51 (Abstract)
63. Takahashi T, Nagai T, Kanda M, Tokunaga M, Liu M, Naito AT, Ogura T, Lee JK, Kodama I, Komuro I (2009) Brown adipose tissue-derived cells differentiate into cardiac conduction and pacemaker cells in vitro and improve complete AV block in vivo. Circ Res 105:e40 (Abstract)

64. Brandt MC, Endres-Becker J, Zagidullin N, Motloch LJ, Er F, Rottlaender D, Michels G, Herzig S, Hoppe UC (2009) Effects of KCNE2 on HCN isoforms: distinct modulation of membrane expression and single channel properties. Am J Physiol Heart Circ Physiol 297:H355–H363
65. Zolles G, Wenzel D, Bildl W, Schulte U, Hofmann A, Muller CS, Thumfart JO, Vlachos A, Deller T, Pfeifer A, Fleischmann BK, Roeper J, Fakler B, Klocker N (2009) Association with the auxiliary subunit PEX5R/Trip8b controls responsiveness of HCN channels to cAMP and adrenergic stimulation. Neuron 62:814–825
66. Santoro B, Piskorowski RA, Pian P, Hu L, Liu H, Siegelbaum SA (2009) TRIP8b splice variants form a family of auxiliary subunits that regulate gating and trafficking of HCN channels in the brain. Neuron 62:802–813
67. Lewis AS, Schwartz E, Chan CS, Noam Y, Shin M, Wadman WJ, Surmeier DJ, Baram TZ, MacDonald RL, Chetkovich DM (2009) Alternatively spliced isoforms of TRIP8b differentially control h channel trafficking and function. J Neurosci 29:6250–6265
68. Wolff LJ, Wolff JA, Sebestyen MG (2009) Effect of tissue-specific promoters and microRNA recognition elements on stability of transgene expression after hydrodynamic naked plasmid DNA delivery. Hum Gene Ther 20:374–388
69. Choi KH, Basma H, Singh J, Cheng PW (2005) Activation of CMV promoter-controlled glycosyltransferase and beta-galactosidase glycogenes by butyrate, tricostatin A, and 5-aza-2′-deoxycytidine. Glycoconj J 22:63–69
70. Lyon AR, Sato M, Hajjar RJ, Samulski RJ, Harding SE (2008) Gene therapy: targeting the myocardium. Heart 94:89–99
71. Hamm A, Krott N, Breibach I, Blindt R, Bosserhoff AK (2002) Efficient transfection method for primary cells. Tissue Eng 8:235–245
72. Zhao J, Pettigrew GJ, Thomas J, Vandenberg JI, Delriviere L, Bolton EM, Carmichael A, Martin JL, Marber MS, Lever AM (2002) Lentiviral vectors for delivery of genes into neonatal and adult ventricular cardiac myocytes in vitro and in vivo. Basic Res Cardiol 97:348–358
73. Bonci D, Cittadini A, Latronico MV, Borello U, Aycock JK, Drusco A, Innocenzi A, Follenzi A, Lavitrano M, Monti MG, Ross J, Jr., Naldini L, Peschle C, Cossu G, Condorelli G (2003) 'Advanced' generation lentiviruses as efficient vectors for cardiomyocyte gene transduction in vitro and in vivo. Gene Ther 10:630–636
74. Kizana E, Chang CY, Cingolani E, Ramirez-Correa GA, Sekar RB, Abraham MR, Ginn SL, Tung L, Alexander IE, Marban E (2007) Gene transfer of connexin43 mutants attenuates coupling in cardiomyocytes: novel basis for modulation of cardiac conduction by gene therapy. Circ Res 100:1597–1604
75. Zhou YF, Yang XJ, Li HX, Han LH, Jiang WP (2007) Mesenchymal stem cells transfected with HCN2 genes by LentiV can be modified to be cardiac pacemaker cells. Med Hypotheses 69:1093–1097
76. Fleury S, Driscoll R, Simeoni E, Dudler J, von Segesser LK, Kappenberger L, Vassalli G (2004) Helper-dependent adenovirus vectors devoid of all viral genes cause less myocardial inflammation compared with first-generation adenovirus vectors. Basic Res Cardiol 99:247–256
77. Yoshimitsu M, Higuchi K, Dawood F, Rasaiah VI, Ayach B, Chen M, Liu P, Medin JA (2006) Correction of cardiac abnormalities in Fabry mice by direct intraventricular injection of a recombinant lentiviral vector that engineers expression of alpha-galactosidase A. Circ J 70:1503–1508
78. Rosen AB, Kelly DJ, Schuldt AJ, Lu J, Potapova IA, Doronin SV, Robichaud KJ, Robinson RB, Rosen MR, Brink PR, Gaudette GR, Cohen IS (2007) Finding fluorescent needles in the cardiac haystack: tracking human mesenchymal stem cells labeled with quantum dots for quantitative in vivo 3-D fluorescence analysis. Stem Cells 25:2128–2138
79. Furuta A, Miyoshi S, Itabashi Y, Shimizu T, Kira S, Hayakawa K, Nishiyama N, Tanimoto K, Hagiwara Y, Satoh T, Fukuda K, Okano T, Ogawa S (2006) Pulsatile cardiac tissue grafts using a novel three-dimensional cell sheet manipulation technique functionally integrates with the host heart, in vivo. Circ Res 98:705–712

80. Feinberg AW, Feigel A, Shevkoplyas SS, Sheehy S, Whitesides GM, Parker KK (2007) Muscular thin films for building actuators and powering devices. Science 317:1366–1370
81. Younes A, Lyashkov AE, Graham D, Sheydina A, Volkova MV, Mitsak M, Vinogradova TM, Lukyanenko YO, Li Y, Ruknudin AM, Boheler KR, van Eyk J, Lakatta EG (2008) Ca(2+)-stimulated basal adenylyl cyclase activity localization in membrane lipid microdomains of cardiac sinoatrial nodal pacemaker cells. J Biol Chem 283:14461–14468
82. DiFrancesco D, Ferroni A, Mazzanti M, Tromba C (1986) Properties of the hyperpolarizing-activated current (i_f) in cells isolated from the rabbit sino-atrial node. J Physiol (London) 377:61–88
83. Rosen MR, Brink PR, Cohen IS, Robinson RB (2004) Genes, stem cells and biological pacemakers. Cardiovasc Res 64:12–23

Tachyarrhythmia Therapies: Approaches to Atrial Fibrillation and Postinfarction Ventricular Arrhythmias

J. Kevin Donahue and Kenneth R. Laurita

Abstract Cardiac tachyarrhythmias are the leading cause of mortality, and a considerable cause of morbidity in the developed world. Conventional therapies, including antiarrhythmic drugs, implantable devices, and ablation, have been inadequate in solving this problem. Biological therapies are proposed as next-generation solutions for arrhythmias. The underlying idea is that biological therapies can attack the root cause of the arrhythmia, with the possibility of long-term palliation or even cure. In this chapter, we review general concepts in gene and cell therapy, and the application of these concepts to atrial fibrillation and post-myocardial infarction ventricular tachyarrhythmias.

Keywords Gene therapy • Cell therapy • Atrial fibrillation • Ventricular tachycardia • Ventricular fibrillation • Myocardial infarction

1 Introduction

Cardiac arrhythmias are responsible for extensive morbidity and mortality in the developed world. Cardiac arrest accounts for more than 300,000 deaths, and atrial fibrillation (AF) burdens two million to five million people in the USA [1–3]. Similar numbers are found in most other industrialized nations. Compounding the problem is the lack of curative therapies for these arrhythmias.

Until the early 1990s, antiarrhythmic drug therapy was considered the standard of care for treating most arrhythmias. Since that time, however, numerous studies of antiarrhythmic drugs have shown that they increase mortality [4–6]. The results of these trials left a void in available strategies for treatment of cardiac arrhythmias, filled incompletely by ablation and implantable cardiac devices. Radiofrequency ablation

J.K. Donahue (✉)
Heart and Vascular Research Center, MetroHealth Hospital,
Case Western Reserve University, Cleveland, OH
e-mail: kdonahue@metrohealth.org

I.S. Cohen and G.R. Gaudette (eds.), *Regenerating the Heart*, Stem Cell Biology and Regenerative Medicine, DOI 10.1007/978-1-61779-021-8_19,
© Springer Science+Business Media, LLC 2011

is a proven technology for focal arrhythmias [e.g. atrioventricular (AV) nodal reentry tachycardia or atrial flutter], but only an experimental approach for more common arrhythmias such as AF or infarct-related ventricular tachycardia (VT) [7, 8]. For most life-threatening arrhythmias, implantable devices are the only option. Pacemakers prevent bradycardia, and implantable defibrillators convert malignant ventricular tachyarrhythmias back to sinus rhythm, but problems with implantable devices include significant expense, potential complications from the invasive procedures, and in the case of defibrillators, pain related to the treatment strategy [9, 10].

Recently, interest has been expressed in the use of gene and cell therapies for cardiac arrhythmias. In this chapter, we will discuss general concepts relevant to development and implementation of these novel therapies and application of the therapies to AF and VT.

2 General Concepts in Gene and Cell Therapy

2.1 Gene Therapy

Basic elements common to all gene therapy approaches include a gene transfer vector and a delivery method. Other considerations, including the therapeutic gene (transgene), target, and genetic control elements, are less generalizable and must be individualized to the specific application.

2.1.1 Vectors

Gene transfer vectors are vehicles for transport of the genetic material (transgene) into the target cells. Gene delivery vectors can be divided into viral and nonviral types. Nonviral vectors are DNA plasmids with or without complexing agents to increase probability of cellular uptake (calcium phosphate, liposomes, proteins, etc.). The initial gene transfer studies used isolated plasmid DNA to show proof-of-concept that genes could be taken up and expressed by tissues, but these early studies also demonstrated the inefficiency of DNA vectors; only a negligible percentage of cells expressed reporter genes after DNA transfection [11]. Little sustained interest has been generated in the use of plasmid vectors for cardiac electrophysiological applications, because the efficiency of gene uptake is extremely poor, inflammatory reactions occur after plasmid delivery, and the ability to sustain gene expression after plasmid-mediated gene transfer has not been demonstrated.

The increased efficiency of viral vectors allowed them to quickly supplant plasmid-based vectors as the gene transfer vehicles of choice. Viral vectors are essentially wild-type viruses that have been genetically modified to prevent virus reproduction or disease and to insert the transgene. Adenoviruses, adeno-associated viruses (AAVs), and lentiviruses have been the most widely used and most successful vectors for myocardial applications. All of these vectors can efficiently transduce cardiac myocytes.

Other viruses described for successful myocardial gene transfer include herpesviruses, Semliki Forest viruses, and Coxsackie viruses. In general, these less commonly used vectors are capable of myocardial gene transfer, but they have not been shown to do so with any degree of efficiency. The available literature suggests that host immune responses and limited duration of expression are problems for all of these vectors.

Adenoviruses

Recombinant adenoviruses are the most commonly used gene transfer vectors for myocardial research. The adenovirus is a 74-nm protein-encapsidated virus with a 35-kb double-stranded DNA genome [12]. Virus binding is mediated by an initial attachment to the Coxsackie adenovirus receptor followed by a secondary interaction with α_V integrins, both of which are widely expressed [13, 14]. Most of the published work used so-called early-generation vectors that have deletions of the E1 and E3 genes. Deletion of E1 is sufficient to render the virus incapable of replication, and the E3 deletion was performed primarily to increase the space for insertion of the transgene. With these deletions, early-generation vectors could accommodate up to approximately 10 kb of genetic material. Problems with early-generation adenovirus vectors include broad biodistribution due to the ubiquitous expression of the virus receptors, inflammatory reactions to the vector causing hepatotoxicity, and limited transgene expression due at least in part to the inflammatory reaction.

Fully deleted adenoviral vectors solve some of the problems found with the early-generation vectors. These viruses, also called helper-dependent, gutted, or gutless, have had all adenoviral genes deleted. Production of these vectors requires co-infection with helper viruses or plasmids that carry the relevant adenoviral genes, causing the problem that all gutted adenoviruses are contaminated at low levels with these helper viruses. The gutted viruses have capacity for over 30 kb of genetic material, and they appear capable of sustained gene expression when tissue-specific promoters are used to drive the transgene [15]. Problems with the initial immune response to the adenoviral particle and the broad biodistribution remain. In addition, the gutted vectors are considerably more difficult to produce than their earlier-generation counterparts.

Adeno-Associated Viruses

AAVs are a phylogenetically distinct class of viruses, unrelated to adenoviruses. The origin of the name (and the confusion) is that AAVs were originally isolated from cultures of adenovirus-infected cells; thus, they were "associated" with adenovirus infection [16]. AAV is a small icosahedral virus with a diameter of approximately 20 nm, considerably smaller than adenovirus [17]. In the last 10 years, over 100 AAV serotypes have been described. Tropism varies significantly with serotype, due in part to the different receptors utilized by the different viruses. Serotype 2 was used for the early myocardial gene transfer work [18], but recent reports have shown that serotypes 1, 6, 8, and 9 more effectively target the myocardium in rodents than does serotype 2 [19–22]. The best vector for large-mammal myocardial application is still an open question. AAV-2 binding was initially linked to heparan sulfate proteoglycans [23]. In addition, AAVs 2, 3, 8, and 9 attach to the 37/67-kDa laminin receptor [24], and AAVs 1 and 6 attach to α2,3 and α2,6 N-linked sialic acids [25]. The chief advantage of AAV vectors is the possibility of long-term (potentially permanent) gene expression. Disadvantages include difficulties in production and a relatively slow onset of gene expression, possibly due to cytoplasmic trafficking, vector uncoating, and conversion of the single-stranded genome into double-stranded DNA [26, 27].

Lentiviruses

Retrovirus vectors have been used for a number of gene transfer applications. A principal limitation of these vectors is the need for active cell cycling for integration and expression of the transgene, making these vectors inappropriate for use in end-differentiated tissues such as myocardium. An exception to this limitation is the human immunodeficiency (HIV)-based vectors, also called lentiviral vectors, which have been shown to efficiently transfer genes to postmitotic cells, including cardiac myocytes. HIV is an approximately 100–120 nm diameter enveloped virus with a 9.7-kb positive, single-stranded RNA genome [28]. When compared with similar concentrations of adenoviral vectors, "advanced-generation" lentiviral vectors transduce cardiac myocytes at roughly the same efficiency in vitro and by intramyocardial injection methods in vivo [29]. A major advantage of lentiviral vectors is long-term gene expression, possibly related to integration of the transgene into the host genome. A disadvantage of lentiviral vectors is that current production methods are not capable of concentrating the virus to levels generally required for most reported myocardial gene delivery methods, other than intramyocardial injection. The major safety concerns for lentiviral vectors include possibilities of insertional mutagenesis from genomic integration and of wild-type reversion. Modern production techniques for lentiviral vectors have several built-in safeguards to virtually eliminate the possibility of wild-type reversion: the vector is devoid of any wild-type HIV genetic material other than the long terminal repeat segments, and the cell lines used to produce these vectors contain modified and isolated wild-type HIV genes to further reduce the possibility of wild-type reversion [30].

2.1.2 Gene Delivery Methods

Ventricular Delivery

Previously reported myocardial delivery methods include intramyocardial injection [31], coronary catheterization [32], pericardial delivery [33], ventricular cavity infusion during aortic cross-clamping [34, 35], and perfusion during cardiopulmonary bypass [36]. Each of these methods is limited by either efficacy or tolerability. A viable technique to transfer genes to all cardiac ventricular myocytes continues to elude investigators in the field.

Intramyocardial injection is the most frequently reported method for ventricular delivery, primarily because the method is easy to implement. In essence, the method involves sticking a needle into the middle of the myocardium and infusing virus. The principal factors affecting gene transfer efficacy include the infusion rate and the vector concentration. Logically, one might expect injectate volume to be a factor, but in practice, injection volumes greater than 0.1 ml tend to just flow backwards along the needle track and out of the tissue. The principal limitations for intramyocardial injection are the tissue damage and resulting inflammation caused by delivery, and the extremely limited spread of vector beyond the needle track. After injection, gene transfer is rarely seen when looking more than a few millimeters away from the site of injection [37]. As such, focal arrhythmias might be cured using mapping and injection approaches similar to the mapping and ablation techniques currently in use. More common atrial or ventricular arrhythmias with diffuse mechanisms could not be efficiently treated with direct injection.

Intracoronary perfusion appears to be the best option for widespread ventricular gene delivery. Unfortunately, coronary vascular dynamics and the vascular endothelial barrier also make this the least efficient delivery option. A series of investigations have shown that the best circumstances for intracoronary gene transfer included maximal local vasodilation, exposure to endothelial permeability-enhancing agents (inflammatory agents, vascular endothelial growth factor, phosphodiesterase 5 inhibitors, nitric oxide or cyclic GMP donors, etc.), and perfusion with the highest tolerable virus concentration for the longest possible time [38–40]. With attention to these details, evidence of gene transfer can be seen in approximately half of cells in the target with antegrade perfusion of the target artery and in more than 80% of cells with simultaneous antegrade and retrograde perfusion of the artery and vein pair (Fig. 1a) [41].

Atrial Delivery

Gene delivery to the atria is considerably more complicated than ventricular delivery. Size, geometry, and tissue thickness limit the utility of myocardial injection methods. The absence of a dedicated atrial vasculature limits intracoronary perfusion methods. To date, the only reported widespread atrial gene transfer method is epicardial gene painting [42]. We showed that complete transmural gene transfer could be achieved by painting a solution containing the vector, a polymerization compound, and dilute protease onto the atrial epicardium (Fig. 1b). The polymerization compound (pluronic F127) caused the vector to stick to the tissue, increasing contact time and probability of gene transfer. The inclusion of trypsin in the mixture allowed transmural penetration of the vector. In safety testing, we found that effective concentrations of trypsin were sufficiently dilute that atrial structure and tensile strength were preserved. The specificity of the technique was twofold: the control provided by direct application with a paint brush limited spread beyond the target area, and the ventricular tissue

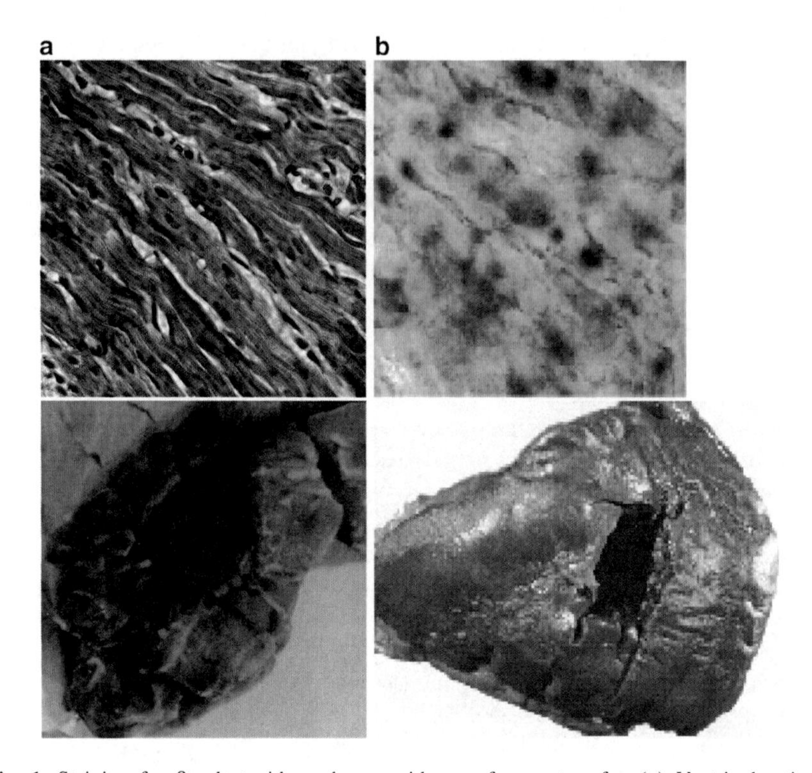

Fig. 1 Staining for β-galactosidase shows evidence of gene transfer. (**a**) Ventricular tissues stained after simultaneous perfusion of the left anterior descending coronary artery and great cardiac vein with adenovirus encoding the β-galactosidase gene. *Blue coloration* indicates transgene expression. Dense but highly localized gene transfer is possible with this delivery method. (**b**) Atrial gene transfer after painting the epicardium with the β-galactosidase-encoding adenovirus. Homogeneous, transmural gene transfer occurs with coadministration of 0.5% trypsin. This method transduces all cells in areas accessible from the epicardium. (With permission from (**a**) Sasano et al. [41] and (**b**) Kikuchi et al. [42])

appeared impervious to gene transfer by this method (even when the solution was directly applied to the ventricular epicardium), suggesting that some intrinsic difference between atrial and ventricular tissue allowed atrial gene transfer but not ventricular gene transfer. With this method, we achieved complete, transmural atrial gene transfer without any evidence of ventricular gene transfer.

2.1.3 Current Problems in the Field of Gene Therapy

In the ideal world, a gene therapeutic would provide one-time curative therapy for an underlying arrhythmia. The action would be localized to the target, minimizing side effects, and the effect would eliminate the problem at the subcellular level. Obviously, we are a long way from such ideal therapy. Current problems in the field of gene therapy include the inefficiency of the gene transfer process, the inability to control the duration and level of gene expression, and the immune response generated by the gene transfer vector. Research to solve these problems is underway in numerous laboratories, and solutions are becoming evident.

Advances in vector design have the potential to solve many of the current problems. Ultimately, the answer to these problems may be found in ongoing research to retarget adenoviruses and AAVs [43–45]. The goal of this work is to direct the virus vector toward proteins that occur specifically and in high abundance on the surface of target cells. More specific virus binding could improve delivery efficiency and reduce both nontarget binding and inflammation. Ultimately, the ability to solve these vector issues may determine the success of gene therapy as a field.

2.2 *Cell Therapy*

The general procedure for myocardial cell therapy is to obtain viable cells from a donor, purify and expand the desired cell type in culture, and then deliver the cells to the target area of injury. Ideally, these cells would regenerate viable myocytes by direct or indirect (paracrine) mechanisms and increase blood supply, resulting in improved hemodynamic function and, importantly, reduce arrhythmia risk compared with no treatment. The optimal cell type for administration has yet to be determined but many cell types have been tested, including skeletal myoblasts (SKMBs) from skeletal muscle, mesenchymal stem cells (MSCs) and hematopoietic stem cells derived from the bone marrow, cardiac stem cells (CSCs) from cardiac tissue, and embryonic stem cells (ESCs) [46]. A distinguishing feature of the various cell types is their ability to differentiate into any cell type (pluripotent), into a limited number of cell types (multipotent), or into only one cell type (unipotent). It is likely that additional cell types will be identified and tested in the future, such as induced pluripotent cells [47]. Donor cells can be obtained from the recipient (autologous) and administered at a later date, which is unlikely to elicit an immune response; however, this may impose significant limitations as an acute therapy. Allogeneic transplantation has the therapeutic advantage of being banked and administered to anyone at any time. ESC therapy is inherently an allogeneic procedure and may always impose some risk. In contrast, bone-marrow-derived MSCs are immuno-privileged [48] and, thus, can be obtained from an allogeneic source. A recent clinical trial has shown that intravenous allogeneic human MSC therapy in patients with myocardial infarction (MI) is safe [49].

2.2.1 Cell Delivery Methods

Cells can be delivered by direction injection into the myocardium from the epicardial surface during a surgical procedure or from the endocardial surface via percutaneous access. The advantage of this method is that cells can be placed exactly at the site of injury, irrespective of target perfusion status or cell "homing" ability. Alternatively, some cell types can be delivered intravenously. MSCs and CSCs have the ability to migrate through the endothelium and accumulate at the site of acute injury, and thus can be delivered intravenously. In contrast, SKMBs do not have this ability and need to be directly injected at the site of injury. Finally, it may also be possible to deliver "intact tissue," which is cells already incorporated into an extracellular matrix [50].

2.2.2 Current Problems in the Field of Cell Therapy

One of the main challenges of cell therapy is that to replace the entire scar with viable cells, billions of cells are required, which is significantly more than the surviving number of cells from any currently reported study [51]. To improve survival, Fischer et al. [52] engineered cardiac progenitor cells to overexpress Pim-1 kinase and showed that improved cellular engraftment, persistence, and organ function continued at least 32 weeks after delivery relative to animals with control cell injections. Even if all cells survive, this amount may still be insufficient to completely replace the function of scarred tissue; thus, other, paracrine-mediated mechanisms will be needed.

Finally, the mechanism of benefit (direct or paracrine) is not well understood. For example, even though numerous studies have shown evidence of cell surface markers and protein expression (for the most part) and some functional characteristics akin to cardiac myocytes, it is unclear if there are enough of these cells to elicit a clinically significant effect. In light of recent clinical trials [49], one might ask if mechanisms really matter as long as patients are better off. However, to further improve cell therapy such that it becomes a new standard of care, a better understanding of the underlying mechanisms responsible for significant clinical benefit are needed.

2.3 *Atrial Fibrillation*

2.3.1 Disease Mechanisms

For over a century, there has been debate about the mechanism for AF. A number of possibilities have been proposed, each with accompanying animal or human data. Engelmann [53] was first to propose that AF resulted from a single focus that was firing so rapidly that uniform electrical conduction away from the focus was not possible (so-called fibrillatory conduction). Lewis et al. [54] proposed similarly that AF originated in a single site, but they considered it to be a single reentrant circuit rather than a point focus. Mines [55] and Garrey [56] described multiple reentrant circuits as a source for fibrillation, claiming that each reentrant circuit captured localized atria and thus prevented uniform conduction throughout the tissue. Moe et al. [57] later modified the multiple reentrant circuit hypothesis of Mines and Garrey with a proposal of "multiple wavelets" that were reentrant in nature but not fixed in position. Moe et al. proposed that individual reentry circuits formed and dissipated over time and that continuous atrial activation was perpetuated by the continuous formation of new wavelets. The number and distribution of these wavelets was a probabilistic function of the conduction and refractory properties of the tissue. Most recently, Davidenko et al. [58] introduced the concept of spiral wave reentry with formation and dissipation of rotors in the tissue. The fundamental distinction between leading circle reentry of Mines, Garrey, and Moe et al. and spiral wave reentry is that the core of the reentrant circuit is continuously invaded and inexcitable in the leading circle mechanism, and it is excitable but not excited owing to the effects of curvature on electrical activation in spiral wave theory. Both theoretical mechanisms are susceptible to disruption from alterations in conduction or refractory properties of the tissue.

Thus, between the competing theories for fibrillation, we have a combination of single or multiple point sources activating by reentry or triggered activity, covering all known arrhythmia mechanisms. Each of these mechanistic hypotheses is supported by human studies. In all likelihood, AF is probably a final common pathway of fibrillatory conduction that is possible from either single or multiple sources and the exact mechanism is dependent on the tissue substrate. For healthy atria, the substrate for multiple-site, intra-atrial reentry is generally not present, so AF is most probably from a single focus (e.g. a pulmonary vein), whereas the diseased atrial substrate is capable of supporting multiple site reentry, and AF is most likely sustained (if not initiated) with intra-atrial conduction.

With the exception of whatever mechanism is operational inside the pulmonary veins, one consistent conclusion of recent AF literature is that reentrant mechanisms sustain AF. Atienza et al. [59] published the most recent human data that support this conclusion. They mapped the dominant frequency of atrial activation and evaluated the effect of adenosine infusion. The logic behind their investigation was that adenosine-induced shortening of action potential duration (APD) in the tissue responsible for maintaining AF would either speed up the dominant frequency if AF came from reentry or slow down the dominant frequency if AF came from triggered activity. They found that dominant frequency consistently increased in both paroxysmal and persistent AF, suggesting reentry as a mechanism for both.

Sustained reentry is facilitated by electrical and mechanical remodeling that occur with both the underlying heart disease and the presence of AF. Human data for remodeling in AF come largely from studies using tissue taken from cardiac surgery patients, comparing chronic AF patients with sinus rhythm patients. These data obviously cannot be considered the "normal" AF situation since all patients had an indication for cardiac surgery. With that weakness acknowledged, these reports contain the best available data in humans. Compared with the patients in sinus rhythm, those with chronic AF had evidence of electrical remodeling, including altered action potential morphology, decreased APD to 90% repolarization (APD90), shortened effective refractory period, and a loss of normal rate adaptive behavior in APD and effective refractory period [60, 61]. The molecular and functional perturbations associated with APD changes included a decrease in the transient outward current (I_{to}) with reduction in KCND3 (Kv4.3) messenger RNA (mRNA) and protein [62, 63], decreased ultrarapid component of the delayed rectifier current

(I_{Kur}) with reduced KCNA5 (Kv1.5) protein [62], increased inward rectifier current (I_{K1}) with increased KCNJ1 at the mRNA level [64], and constitutive activation of the acetylcholine-activated potassium current ($I_{K,Ach}$) without change in protein level but associated with increased protein kinase Cε activity [64–66]. In humans, the rapid component of the inward rectifier current (I_{Kr}) has generally been found unchanged, as has the corresponding KCNH2 mRNA [63, 66].

The L-type calcium current ($I_{Ca,L}$) was considerably reduced in AF patients [67], but unlike other current reductions the changes in $I_{Ca,L}$ appeared to occur at the post-translational level. mRNA and protein levels were unaltered in AF [68]. Increased phosphatase activity and impaired src kinase activity were associated with the reduction in $I_{Ca,L}$, suggesting a reduction in calcium channel phosphorylation as a mechanism [68, 69]. The actual phosphorylation status of the calcium channel has not been reported to confirm this hypothesis.

Overall, the human studies have shown consistent electrical remodeling, predominately in ionic currents active during repolarization. Considerable data from the goat burst pacing AF model and the dog atrial and ventricular tachypacing models corroborate the human cardiac surgery findings [70–77]. This remodeling causes important effects on stability and persistence of AF.

2.3.2 Gene and Cell Therapy for AF

Rate Control Therapy

The initial proof-of-concept study for antiarrhythmic gene therapy targeted AV nodal conduction [78]. Our approach was designed to test the hypothesis that transduction of the AV node with a gene encoding an inhibitory G protein α subunit (Gαi2) would suppress AV nodal function via an adenylate cyclase/cyclic AMP mechanism. By using adenovirus vectors for highly efficient (albeit short-term, self-limited) gene transfer and delivery by perfusion of the AV nodal artery, overexpression of wild-type Gαi2 slowed AV nodal conduction and prolonged the effective refractory period of the AV node. The ventricular response rate to acutely induced AF was 20% lower with Gαi2 overexpression compared with controls (Fig. 2a). This result set the stage to assess AV nodal manipulation in the setting of chronic AF. In a follow-up study, we induced persistent AF by repetitive, high-rate, burst atrial pacing, and then transduced pigs via AV nodal artery perfusion with adenovirus carrying wild-type Gαi2, a constitutively active Gαi2 mutant (Gαi2-Q205L), or a control gene [79]. Wild-type Gαi2 only produced a significant effect during sedation, but the Gαi2-Q205L-infected pigs showed a continuous 20% heart rate reduction over time. This improvement in ventricular rate was sufficient to reverse the tachycardia-induced heart failure inherent to that animal model. Furthermore, the gene transfer result was more effective than the AV nodal slowing drugs (esmolol, diltiazem, and digoxin) in this model (Fig. 2b). We subsequently found similar results with gene transfer of a G protein (GEM) known to inhibit L-type calcium channel function [80].

Bunch et al. [81] investigated a cellular approach for rate control of AF. They used cell therapy to slow AV nodal conduction by increasing collagen deposition in

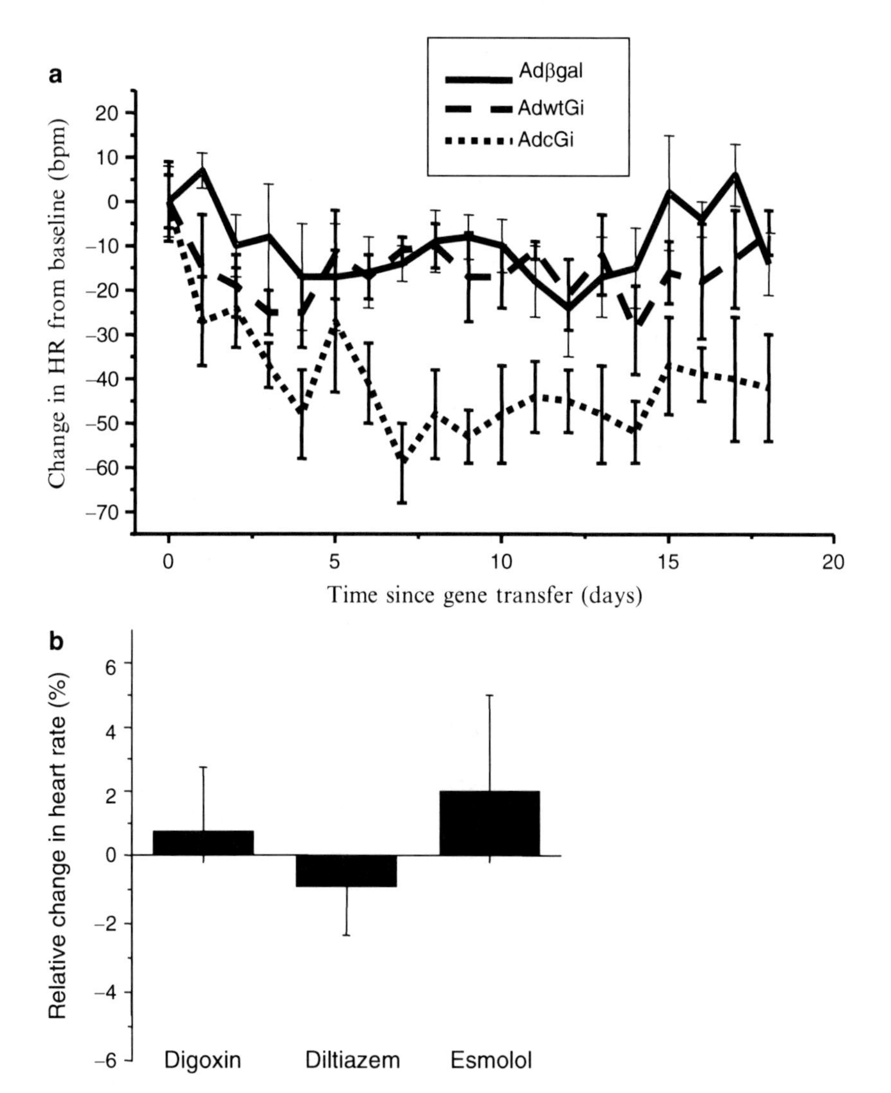

Fig. 2 Atrioventricular nodal conduction in atrial fibrillation. (**a**) Gene transfer with adenovirus encoding wild-type Gαi2 (*AdcwtGi*), a constitutively active mutation of Gαi2 (Gαi2-Q205L) (*AdcGi*), or an electrically inert reporter gene (β-galactosidase) (*Adβgal*) shows reduction in heart rate during atrial fibrillation with the constitutively active Gαi2-Q205L. Animals were evaluated while alert and awake. Gene transfer effects in the wild-type Gαi2 animals were not apparent in awake animals, but the effects became apparent when the animals were sedated. (**b**) In contrast to the gene transfer effects, no atrioventricular nodal suppression was possible in sedated animals with typical atrioventricular nodal suppressing drugs. The lack of effect has been attributed to the very robust baseline atrioventricular nodal function in these young pigs. *HR* heart rate. (With permission from Bauer et al. [79])

the perinodal area, thus forming conduction barriers. Fibroblasts derived from dogs' skin biopsies were cultured and injected along the AV nodal conduction pathways using an electroanatomic mapping system. In total, 80 ml of fibroblast-containing solution was injected into 320 sites around the AV node for each dog. They found that fibroblast injections prolonged the AH intervals during sinus rhythm and atrial pacing after 4 weeks. The ventricular rate was slower in the fibroblast group during acutely induced AF, but this result did reach not statistical significance. The use of cells to modify the AV node has fascinating possibilities, particularly if the functional result could be turned toward rejuvenation of AV nodal function in cases of AV block rather than suppression in cases of AF.

Rhythm Control Therapy

So far the only report of a gene transfer strategy to eliminate AF is our investigation with a dominant negative mutation of the I_{Kr} channel α subunit, KCNH2-G628S [82]. Previous work on this mutation by Craig January's laboratory had shown that the mutation traffics normally, and it blocks the current by obstructing the channel pore. We hypothesized that atrial expression of this mutation would block I_{Kr} and thereby prolong atrial APD and increase the wavelength of reentry. We tested the hypothesis using an adenovirus vector encoding KCNH2-G628S (AdG628S) and the afore-mentioned epicardial gene painting method. All animals had an atrial pacemaker placed for burst pacing to induce AF. The pacemaker bursts had a 2-s on/off duty cycle, and we evaluated the rhythm during the 2-s off periods. Gene transfer and pacemaker implantation were performed at the same time, and then we checked rhythm on a daily basis (Fig. 3). We found that animals in all groups had a roughly 50% incidence of AF over the first 3 days. After that time point, the percentage of control animals in AF gradually increased until all were in AF by day 10.

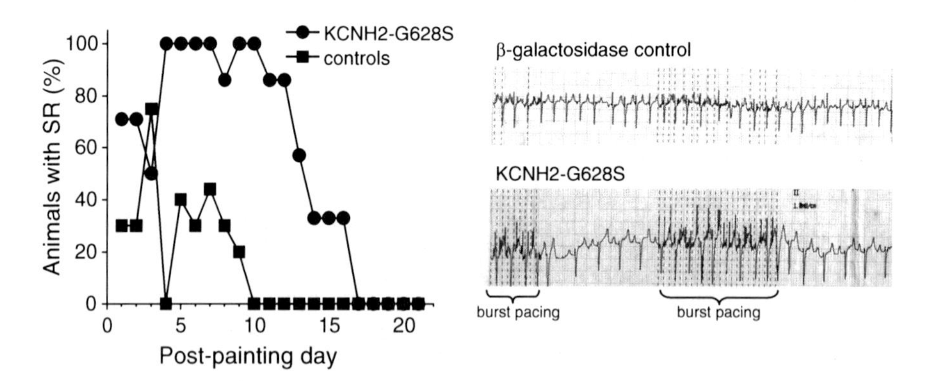

Fig. 3 Rhythm on daily telemetry as a function of time since gene transfer. The *left panel* shows that control animals ($n = 10$) continually progress toward persistent atrial fibrillation after initiation of burst atrial pacing, whereas the G628S animals ($n = 7$) show an abrupt increase in percentage with sinus rhythm (*SR*) 3 days after gene transfer, correlating with onset of gene expression. G628S animals maintained sinus rhythm until days 11–18, correlating with the time when gene expression is lost. The *right panel* shows examples of telemetry recordings from animals 8 days after gene transfer. (With permission from Amit et al. [82])

In contrast, the number of AdG628S animals in AF dropped to zero on day 4. With exception of one animal who had AF on day 9, all of the AdG628S animals were in sinus rhythm between days 4 and 10. After day 10, the AdG628S animals progressively deteriorated into AF, and by day 17 all animals were in AF. We evaluated animals with invasive electrophysiology study 7 and 21 days after gene transfer. We found that the atrial APD was prolonged on day 7 but not on day 21 in the animals receiving AdG628S relative to controls. Tissue analysis showed that transgene expression correlated with phenotypic effect, suggesting that gene transfer vectors capable of long-term expression may give longer-lasting results. With this delivery method, the ventricular electrophysiological properties were not affected.

Burton et al. have investigated an alternative method for prolonging atrial repolarization with focal gene transfer of KCNE2-Q9E [83]. KCNE2 is an auxiliary subunit for the I_{Kr} current, and the Q9E mutation increases the sensitivity of the current to erythromycin-mediated block. They used the myocardial direct injection method to deliver naked DNA plasmids containing the mutation into a focal spot in porcine right atria and found dose-dependent prolongation in monophasic APD with clarithromycin exposure. This graded effect from the drug exposure was a novel modification of current strategies for antiarrhythmic gene therapy. So far, the published results with KCNH2-G628S and KCNE2-Q9E have only been in sinus rhythm, so efficacy against AF remains unknown.

2.4 *Post-MI Ventricular Arrhythmias*

2.4.1 Disease Mechanisms

Most of what has been learned about ventricular arrhythmias in the postinfarcted human heart is from a limited number of animal models that attempted to reproduce the natural disease process. The mechanisms of arrhythmias after infarction are mostly due to reentrant excitation originating from the infarct zone [84, 85]. Impulse conduction through the infarct scar and border zone is slow, with evidence of irregular excitation patterns, anisotropic conduction, and functional conduction block [85, 86]. Non-reentrant ectopic activity originating from the infarct border zone may also play an important role in the formation of ventricular arrhythmias in the postinfarcted heart [87]. These seminal studies make it clear that the mechanisms of arrhythmias after infarction are not merely reentrant excitation around a region of anatomical block. Importantly, significant mechanical and electrical remodeling combine to create a potent substrate for arrhythmias.

Mechanical Remodeling

Mechanical ventricular remodeling after infarction is a complicated and dynamic process that consists of expanding scar tissue and pressure/volume overload that significantly affects infarcted and noninfarcted tissue. The structural changes can include apoptosis, fibrosis, edema, and dilation with wall thinning. At the cellular level, cells become hypertrophic [88] and sarcomere/myofilament proteins are altered [89, 90]. Mechanical remodeling begins immediately and continues indefinitely, potentially contributing to the electrical substrate for fatal arrhythmias.

Electrical Remodeling

In addition to mechanical remodeling, significant electrical remodeling occurs in the postinfarcted heart at sites near and distal to the infarct. For a more detailed description of these changes, refer to Pinto et al. [91]. Several days after infarction, the most notable action potential changes in the vicinity of the infarct, including the scar and border zones, are a decrease in the rate of depolarization during the upstroke and a triangulation of the action potential leading to a shorter APD. Reduction of the amplitude and rate of depolarization can be attributed to a decrease in sodium current and altered sodium channel kinetics. Altered sodium channel kinetics can slow recovery such that it is difficult to initiate an action potential, even though membrane potential has returned to resting levels (post-repolarization refractoriness). These changes can lead to abnormal impulse conduction, including slow conduction and functional conduction block, meaning block only appears under certain circumstances (e.g. rapid impulses). Side-to-side gap junction conductance is reduced in the border zone of 5-day-old canine infarcts, which may be due to a cellular lateralization of connexin 43 protein [92]. These changes may also contribute to abnormal impulse conduction in the peri-infarct zone.

In general, less is known about electrical remodeling several months after infarction, which is problematic because most patients fall into this category. After several weeks to months, the cells in the border zone exhibit reduced upstrokes and prolonged APD. Even though normal sodium-dependent action potentials can be observed after infarct healing, conduction is slow and discontinuous, possibly due to fibrosis and abnormalities in gap junction distribution and function [93]. In a rat healed infarct model, we found that APD90 from border zone tissue tended to be longer (approximately 12.5 ms) than APD90 measured from normal hearts [94]. Ohara et al. [95] reported border zone APDs that were not very different compared with those from sites remote from the border zone, which may be prolonged (compared with normal) as well. Not all results are the same. In felines with 2-month-old MI, the L-type calcium current is reduced and may account for a reduction in phase 2 of the action potential near the scar [91]. Other, more exotic changes may also play a role. For example, longer APDs from the border zone in rats 5 weeks after infarction may be accentuated via stretch-activated channels [96]. Moreover, non-uniform stretch that may occur in the border zone can elicit spontaneous calcium release and may even cause triggered activity [97].

2.4.2 Gene Therapy for Postinfarction Ventricular Arrhythmias

Different strategies to modify the substrate that supports postinfarction ventricular arrhythmias include improvement of conduction or prolongation of the refractory period. The overall strategy has been to attack the elements that control reentrant wavelength to make the tachycardia unsustainable. We looked at this issue in a porcine model of postinfarction VT [98]. In that model, pigs underwent infarction by 2.5-h balloon occlusion of the mid-left anterior descending coronary artery. As the infarct scar healed, approximately 90% of animals became inducible for sustained monomorphic VT with programmed electrical stimulation. We looked at the effects of KCNH2-G628S gene transfer, to test the hypothesis that I_{Kr} block by this dominant negative hERG mutant would extend the refractory period in the VT circuit and eliminate VT [99]. Active-treatment animals were compared with controls receiving either no gene transfer or the nonelectrically active gene *Escherichia coli* β-galactosidase. All of the animals in the study had moderate-sized anterior septal infarcts and reliably inducible VT. One week after gene transfer, the animals receiving KCNH2-G628S no longer had any ventricular arrhythmias, but the control animals continued to have inducible sustained monomorphic VT (Fig. 4). Since the area of gene transfer was small and localized, there was no global QT prolongation and no evidence of proarrhythmia with therapy.

Lau et al. [100] investigated the relationship between cellular excitability and ventricular arrhythmias in a canine coronary occlusion model. This dog model has been extensively investigated [101–104]. Consistent findings include VT inducible 3–7 days after infarction and slow conduction velocity in the epicardial infarct border VT sites. Cells in this area show reduced resting membrane potential and sodium current. Lau et al. investigated border-zone gene transfer of SCN4b, a skeletal muscle isoform of the sodium channel that has greater activity in reduced membrane potential conditions than the cardiac isoform. They found that SCN4b gene transfer increased sodium current in the still relatively depolarized cells of the infarct border, and that this improvement resulted in increased conduction velocity and decreased VT inducibility.

2.4.3 Cell Therapy for Postinfarction Ventricular Arrhythmias

Skeletal Myoblasts

Even though numerous studies have shown that cell therapy is associated with a modest improvement in cardiac function, much less is known about the electrophysiological consequence (good or bad). Unlike the antiarrhythmic effects of gene therapy described earlier, unfortunately the most common electrophysiological perception of cell therapy is proarrhythmia [105]. Our laboratory and other experimental and clinical studies have associated SKMBs with abnormal conduction and proarrhythmia [106–109]. This has been specifically attributed to the lack of connexin expression in vivo associated with SKMBs. Abraham et al. [108] co-cultured human SKMBs

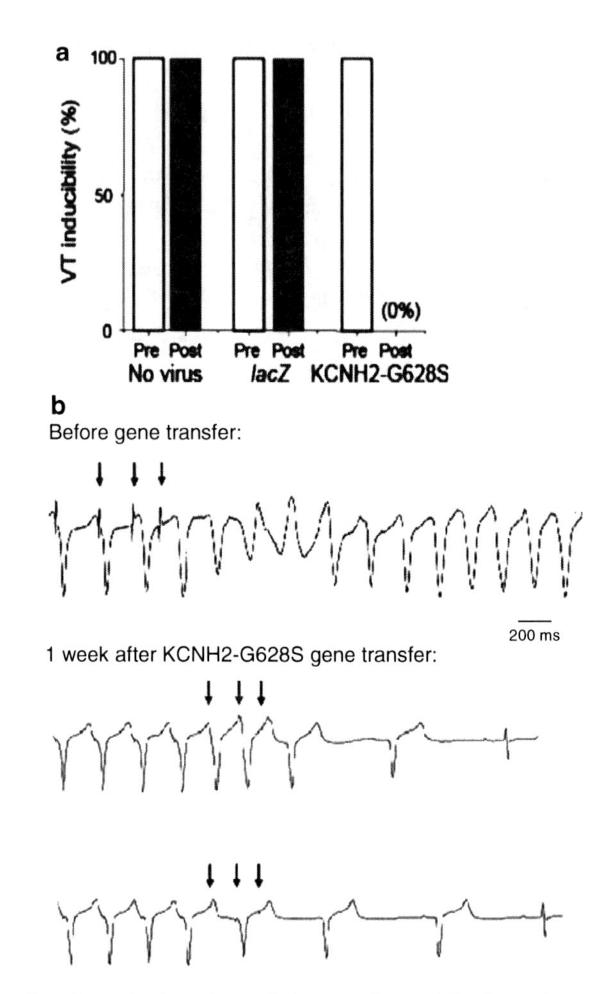

Fig. 4 Elimination of ventricular tachycardia in a porcine model of healed myocardial infarction (MI) after gene transfer of a dominant negative mutant of the KCNH2 potassium channel, encoding the α-subunit of I_{Kr}. (**a**) Summary data showing reliably inducible ventricular tachycardia before and 1 week after gene transfer in five animals receiving no virus and five animals receiving adenovirus encoding β-galactosidase. In contrast, no ventricular arrhythmias could be induced in any animal 1 week after KCNH2-G628S gene transfer. (**b**) Example of ventricular tachycardia induced before gene transfer in one animal, and that same animal after gene transfer when paced with triple extra-stimuli down to refractoriness. (With permission from Sasano et al. [99])

with rat ventricular myocytes in a monolayer, and using optical mapping techniques, they measured impulse propagation and arrhythmia inducibility. When co-cultures contained more than 30% myoblasts, abnormal impulse propagation and reentrant excitation were observed. We and others have demonstrated that SKMBs injected directly into normal or border zone tissue can form cell clusters of surviving skeletal

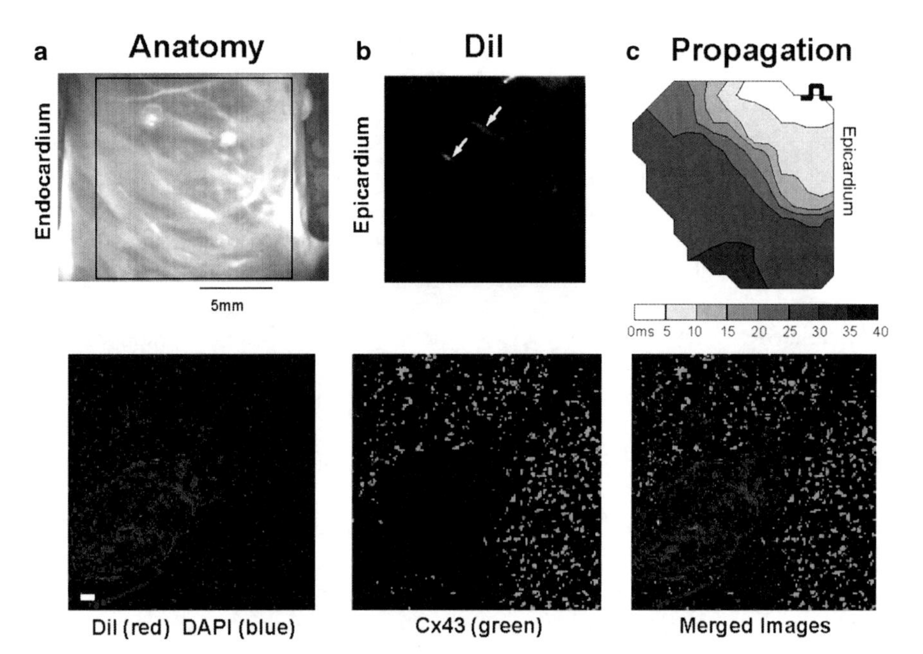

Fig. 5 (**a**) The wedge preparation under ambient room light. The *large box* shows the area that was optically mapped. (**b**) The location of dioctadecyl-3,3,3′,3′-tetramethylindocarbocyanine perchlorate (*DiI*) fluorescent cells (*orange*) from the same wedge as in (**a**). (**c**) The corresponding activation map while pacing from the epicardium (*pacing symbol*). Abnormal conduction (crowding of isochrone lines) is apparent where DiI fluorescence is most intense (**b**, *arrows*). (**d**) Confocal images taken from a wedge preparation with skeletal myoblast (SKMB) injections. Tissue sections were stained with 4′,6-diamidino-2-phenylindole (DAPI; *blue*), and SKMBs were stained with DiI (*red*) before transplantation. DiI fluorescence was colocalized to DAPI fluorescence indicating the presence of viable transplanted cells. Connexin 43 (*Cx43*) staining (center) was not present in DiI-positive cells (merged images). *Scale bars* 40 μm. (With permission from Fouts et al. [107])

myotubes that do not express connexin protein and, importantly, can cause abnormal impulse propagation, requirements for the initiation of reentrant excitation [107]. Shown in Fig. 5 is abnormal transmural impulse propagation caused by SKMBs injected into normal canine myocardium (25 million cells/ml in 20–40 epicardial injections). Figure 5a shows a left ventricular wedge preparation 4–5 weeks after cells had been injected, and the location from which measurements were made (black box). 1,1′-Dioctadecyl-3,3,3′,3′-tetramethylindocarbocyanine perchlorate (DiI)-labeled SKMBs (Fig. 5b, orange) were most obvious near the epicardial surface and at "needle tracks" orientated oblique to the epicardial surface (arrows). In this example impulse propagation (Fig. 5c) was abnormally slow (i.e. closely spaced isochrone lines) at the location of intense SKMB engraftment (arrows). Confocal imaging indicated the presence of viable transplanted cells (Fig. 5d; 4′,6-diamidino-2-phenylindole, blue; DiI, red) without connexin 43 expression (Fig. 5d; green). This finding is consistent with our study in rats with SKMB therapy, where impulse conduction velocity in the MI border zone (0.21 ± 0.05 m/s) was slower compared with control MI (0.33 ± 0.05 m/s; $p=0.02$) [106]. Similarly, Fernandes et al. [110] showed increased arrhythmia inducibility with SKMB therapy. Thus, it is very unlikely that SKMB therapy will reduce the risk of sudden cardiovascular death associated with MI by any mechanisms directly related to the cells. Therefore, these results raise serious concerns regarding the safety of unaltered SKMBs.

Several groups have reported that increasing the expression of gap junction proteins in SKMBs may be beneficial. Overexpression of connexin 43 can improve the formation and function of gap junctions [111, 112]. Abraham et al. [108] showed in SKMB and cardiac myocyte co-cultures that increasing the expression of connexin 43 reduced the occurrence of arrhythmias. Similarly, Roell et al. [113] showed in whole hearts with scar that injection of SKMBs engineered to over-express connexin protein restored some impulse conduction.

Mesenchymal Stem Cells

There are a significant number of reports from experimental models showing that MSCs can improve hemodynamic function and, possibly, differentiate into cardiac myocytes when delivered by infusion or direct injection [114]. MSCs offer many advantages for developing clinical cellular therapeutics: easy isolation and expansion, compatibility with different delivery methods and formulations, and they are hypoimmunogenic [48]. Importantly, MSCs are rarely associated with proarrhythmia or uncontrolled cell proliferation (e.g. teratomas). MSCs have also been approved for clinical trials and seem to show some electrical benefit [49]. Heubach et al. [115] described the electrophysiological properties of isolated, undifferentiated human MSCs. These cells revealed a distinct pattern of ion channel subunit expression (e.g. Kv4.2, Kv4.3, MaxiK, HCN2, α1C of the L-type calcium channel), suggesting that MSCs may contribute to electrophysiological function. In contrast, others have shown that isolated bone-marrow-derived c-kit(+) cells co-cultured on rat neonatal myocytes expressed Na^+ and Ca^{2+} channel protein, but no function was evident [116].

Presently, there are very few reported investigations of the electrical connectivity between MSCs and host cardiomyocytes. Potapova et al. [117] reported that MSCs can couple with native cardiomyocytes both in vitro and in vivo; however, the percentage of coupling pairs was very low. Recently, it was reported that adult human MSCs are able to conduct action potentials between two fields of cardiomyocytes divided by an experimental conduction block [118, 119]. In contrast, Chang et al. [120] showed conduction slowing and increased susceptibility to reentrant arrhythmias in co-cultures of human MSCs with neonatal rat ventricular myocytes, despite the presence of functional gap junctions. How most MSCs function in vivo is not well understood and a challenge to determine.

In previous studies, we have shown that when MSCs are administered at the time of MI by intravenous injection, they are retained in the MI border zone and enhance electrical viability [106, 121]. Shown in Fig. 6 is arrhythmia vulnerability as measured by an arrhythmia inducibility score, where a higher number indicates a higher

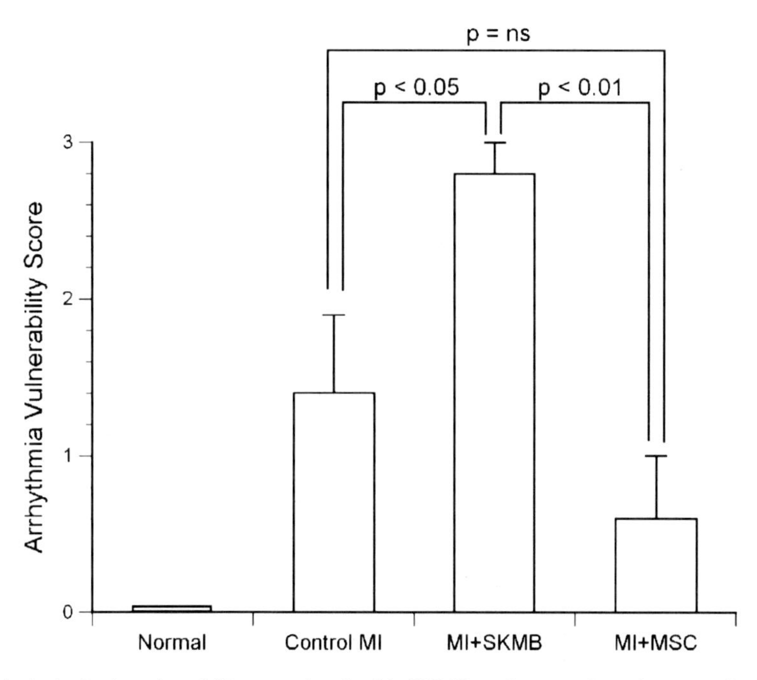

Fig. 6 Arrhythmia vulnerability associated with SKMB and mesenchymal stem cell (MSC) therapy. Arrhythmia vulnerability was determined by an arrhythmia inducibility score where the larger number indicates higher arrhythmia inducibility. SKMB therapy ($n = 6$) was associated with increased vulnerability to ventricular tachycardia compared with normal hearts ($n = 5$), control MI ($n = 7$), and MI + MSCs ($n = 7$). MSC therapy was associated with a decrease in arrhythmia vulnerability compared with control MI. *ns* not significant. (With permission from Mills et al. [106])

arrhythmia vulnerability [106]. During programmed stimulation, VT was induced in 57% of hearts with MI alone (control MI) compared with 0% in normal hearts. Importantly, arrhythmia inducibility in MI plus SKMB group (100%) was significantly greater than the in control MI group. Interestingly, the MI plus MSC group tended to be less vulnerable to arrhythmias than the control MI group; however, this difference did not reach statistical significance. These data suggest that SKMB therapy for MI increases vulnerability to ventricular arrhythmias, whereas MSC therapy may reduce arrhythmia vulnerability. Fernandes et al. [110] reported similar results in an almost identical experimental model.

Why might arrhythmia vulnerability decrease when MSCs are used? Shown in Fig. 7 are representative examples of electrical activation isochrone maps and optical action potentials, recorded from normal (Fig. 7a), control MI (Fig. 7b), MI plus SKMB (Fig. 7c), and MI plus MSC (Fig. 7d) treated Langendorff perfused rat hearts. The images on the left show the location from which action potentials were recorded (on the right) corresponding to noninfarcted tissue (site 1), border zone (sites 2 and 3), and infarct zone (site 4). The activation map and action potentials from a normal heart (Fig. 7a) demonstrate uniform impulse propagation without evidence of block (thick black lines), and uniform amplitude and normal action potential morphology.

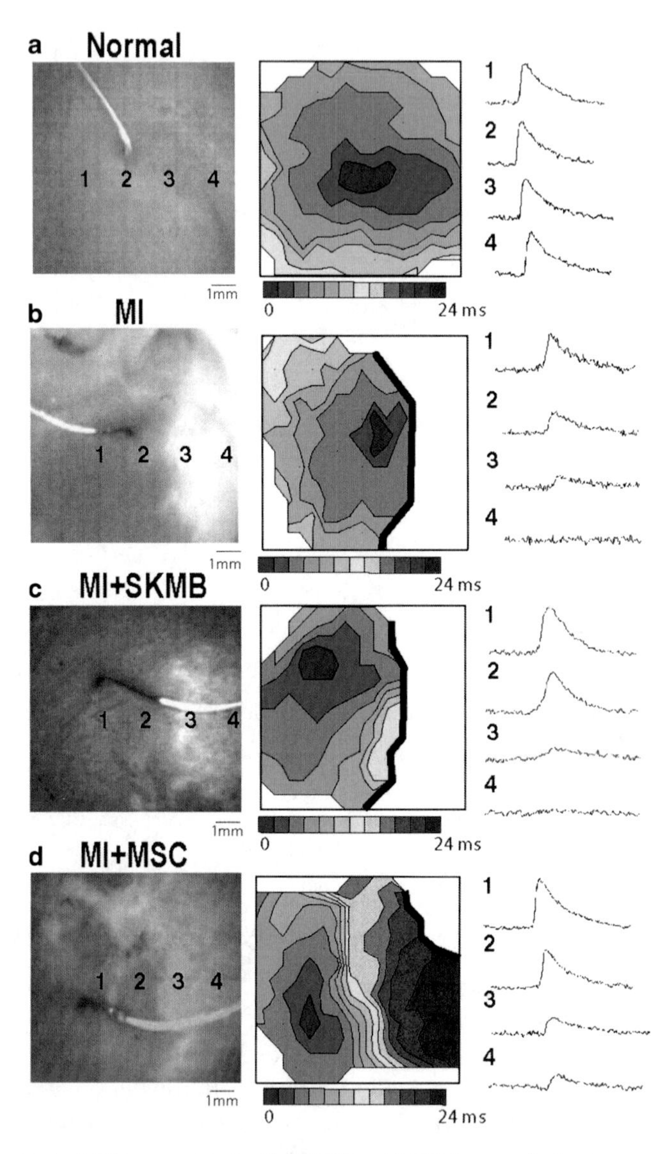

Fig. 7 Electrical viability associated with SKMB and MSC therapy. Activation maps of impulse propagation and unnormalized optically recorded action potentials obtained from normal (**a**), control MI (**b**), MI + MSC (**c**), and MI + SKMB (**d**) hearts. The images on the *left* correspond directly to the area mapped (approximately 8 mm × 8 mm) in each heart, and the signals beside each activation map were recorded from noninfarcted (site 1), border zone (sites 2 and 3), and infarcted (site 4) tissue. Action potentials are plotted as absolute florescence. For all activation maps, each *contour line* represents 2 ms. The *thick black lines* indicate impulse block. (With permission from Mills et al. [106])

It is important to note that we have previously shown that optical action potential amplitude can be used to assess the amount of electrophysiologically viable tissue [94]. Shown in Fig. 7b is a control MI heart. The activation map indicates regions of impulse block (thick black line) corresponding to the location of the border zone (between sites 2 and 3). Action potentials recorded from the border zone (2 and 3) and infarct (site 4) demonstrated an abrupt decrease in amplitude and a slower depolarization phase. For hearts treated with SKMBs (Fig. 7c), impulse propagation, block, and unnormalized action potential amplitudes are similar to those for control MI. However, for hearts treated with MSCs, the results (Fig. 7d) are different. The activation map reveals impulse propagation beyond the border zone (sites 3 and 4) toward the center of the infarct and a smaller region of conduction block compared with MI plus SKMBs and control MI. Action potentials recorded from the border zone (sites 2 and 3) were smaller than normal, but not as small as those recorded from the border zone in control MIs (Fig. 7b). Enhanced electrical viability was observed only in hearts that received MSC therapy. In addition, we observed the presence of MSCs expressing connexins 40, 43, and 45. In preliminary studies we have shown that MSCs exert their effect by, largely, an electrotonic influence. If so, MSCs may enhance electrophysiological viability by forming a passive electrical "bridge" between surviving myocytes, as fibroblasts can [122]. Moreover, the natural expression of connexin proteins may suppress (via electrical loading) the tendency for abnormal impulse formation at the border zone [123]. Can MSCs restore impulse conduction in vivo by actively generating action potentials? Despite convincing reports, it is not clear that MSCs can differentiate into a large number of cells with a cardiac myocyte phenotype [124]. It may be possible to enhance a cardiac linage by deriving MSCs from spheroid bodies, as demonstrated by Potapova et al. [50].

Cardiac Stem Cells

Although it is possible for MSCs to transdifferentiate into cardiac myocytes, it is unclear if this happens at a level required to achieve clinical significance [124]. In contrast, CSCs have a much more robust ability to produce several cardiac cells types [125]. CSCs are shed from cardiac tissue samples (i.e. explants) and express the cell surface markers c-kit and Sca-1. In suspension, these cells form clusters of cells (10–100 cells), called cardiospheres. Cardiospheres express a higher level of c-kit and are composed of clonally derived cells that partially differentiated toward a cardiac lineage. Rota et al. [125] showed that when cardiac progenitor cells were either injected or stimulated by growth factors, a significant portion of infarct scar was replaced by newly formed myocardium. When co-cultured with rat neonatal ventricular myocytes in vitro, CDCs exhibited calcium transients and action potentials that were in sync with neighboring myocytes. In addition, these CDCs expressed I_{Na}, I_{K1}, and $I_{Ca,L}$ (when transduced with a Ca-β-subunit). When they were directly injected into the infarct border zone of mice, hemodynamic function improved and several administered cells were identified as differentiated myocytes.

The electrophysiological benefit of CSCs in vivo is still unknown. Like MSCs, CSCs can migrate and be retained at the site of acute injury. Dawn et al. [126] showed that intravenous injection of CSCs after reperfusion limited infarct size and improved left ventricular function. Similarly, using intracoronary delivery of cardiosphere-derived cells in pig, Johnston et al. [127] showed a significant reduction of infarct size (from 19.2 to 14.2%) compared with no effect of a placebo (from 17.7 to 15.3%). Importantly, no evidence of enhanced arrhythmia risk or tumor formation was observed. Finally, it has been shown that bone-marrow-derived MSCs may participate in the expression of CSCs [128].

ESCs and Induced Pluripotent Cells

ESCs have the potential to differentiate into cardiac-like myocytes [129], but they have been associated with arrhythmic potentials in vitro and the formation of teratomas when undifferentiated cells are used [123, 130]. In addition, since ESCs are allogeneic, they may pose some inherent immunological risk. Nevertheless, it has been demonstrated that ESCs are more able to restore cardiac function than bone-marrow-derived cells [131]. In addition, ESCs isolated from spontaneously beating embryoid bodies express functional Na^{+}, Ca^{2+}, and K^{+} channels [132]. He et al. [133] showed that cardiac myocytes isolated from embryoid bodies are able to generate action potentials with embryonic, nodal, atrial, and ventricular action potential characteristics. Furthermore, Satin et al. [134] showed that impulse conduction in multicellular spontaneously contracting embryoid bodies is slowed with tetrodotoxin, suggesting the presence of functioning sodium channels. Finally, several groups have shown that cardiac myocytes derived from ESC express key calcium regulatory proteins and the ability for excitation–contraction coupling [135].

Despite these promising results, few studies have characterized the ion channel phenotype in vivo. In addition to appropriate ion channel expression and function, another important phenotype of cardiac myocytes is the ability to electrically integrate in vivo. ESCs have been shown to express connexin proteins [136]. Importantly, Kehat et al. [129] have shown that human ESCs can integrate with host myocardium in vitro and in vivo, improving function of infarcted rat hearts. Finally, it may also be possible to dedifferentiate cells (induced pluripotent cells) to restore embryonic-type potential [47]. Currently, the electrophysiological benefits of these cells are completely unknown.

Timing of Cell Therapy Administration

In many patients, stem cell therapy can only be administered long after a previous ischemic event, when the infarct has already healed. In such a setting, the chemical signals present during acute ischemia have long since abated. In that setting, it is not clear that stem cells will engraft into the scar and show benefits similar to delivery during acute ischemia. Hypoxic stress increases the production of several paracrine factors including stromal-cell-derived factor 1 (SDF-1) [137]. SDF-1 is a chemokine secreted by ischemic cardiomyocytes that leads to an increase in vascular density and improvement in cardiac function [138]. We have shown that by reestablishing the expression of SDF-1 several weeks after the initial ischemic event, MSCs can engraft at the scar border zone [139].

Paracrine Factors and Effects

Chemokines, cytokines, and growth factors act in a paracrine fashion and promote stem cell homing, survival, and engraftment, neovascularization, attenuation of left ventricular remodeling, and enhanced regeneration [137]. We have shown that MSCs engineered to overexpress SDF-1 can recruit endogenous stem cells (possibly cardiac) that express its conjugate receptor (CXCR4), and are associated with enhance electrical viability in the border zone [121]. Other paracrine and growth factors may also be important [125]. Some examples are the vascular endothelial growth factor and the pretreatment of MSCs with cardiomyogenic growth factors [140, 141]. More recently, the important role of GATA-4 following MI has been recognized. The GATA proteins are a family of zinc-finger-containing transcription factors. Bayes-Genis et al. [142] demonstrated that expanded MSCs express cardiac markers in vitro, in the absence of myocyte induction media. GATA-4 is crucial for the viability and hypertrophic response of cardiomyocytes following myocardial injury [143]. It has been demonstrated that MSCs exposed to cardiac fibroblasts overexpressing a chimeric protein encoding GATA-4 and a cell-penetrating protein VP22 increased the expression of troponin I and Nkx 2.5 [144].

What electrophysiological changes can be attributed to paracrine effects? This is difficult to ascertain; however, one obvious indirect electrical effect is that if paracrine factors reduce the overall size of infarcted tissue, then the pathway for reentrant excitation will be less and so will the chance of sudden cardiovascular death. Are there any paracrine factors that have a direct effect on cellular electrophysiology? Yes, SDF-1 may recruit endogenous CSCs, which is a good example of a beneficial paracrine effect [121]. However, Pedrotty et al. [145] have shown that cardiac fibroblast paracrine factors can cause a reduction of conduction velocity and a prolongation of APD in rat neonate monolayers. This effect was not due to cell apoptosis or fibroblast proliferation or a reduction in gap junction proteins, but rather could be attributed to the reduction of ionic currents responsible for depolarization and repolarization. Despite the important role paracrine effects have, it is difficult to discount the direct effects of administered cells.

3 Summary

So, where are we now? Currently, the tools for gene and cell transfer are rudimentary, but effective. In large mammals with "human-size" hearts, we can transfer genes in high density to very localized areas by intramyocardial injection or intracoronary perfusion. We can also achieve localized cell delivery (albeit with considerable cell loss). We can also transfer genes in high density to the atrial free wall using the epicardial painting method, and we can transfer genes and cells to the AV nodal region by AV nodal arterial perfusion or intramyocardial injection. With these tools, we can affect cardiac arrhythmias, as demonstrated in the examples described in this chapter. Ongoing research in delivery and vector systems should increase the possibility of biological therapies for the treatment of AF and postinfarction ventricular arrhythmias.

References

1. Lloyd-Jones D, Adams R, Carnethon M et al (2009) Heart disease and stroke statistics – 2009 update: a report from the American Heart Association Statistics Committee and Stroke Statistics Subcommittee. Circulation 119:480–486
2. Miyasaka Y, Barnes ME, Gersh BJ et al (2006) Secular trends in incidence of atrial fibrillation in Olmsted County, Minnesota, 1980 to 2000, and implications on the projections for future prevalence. Circulation 114:119–125
3. Go AS, Hylek EM, Phillips KA et al (2001) Prevalence of diagnosed atrial fibrillation in adults: national implications for rhythm management and stroke prevention: the An Ticoagulation and Risk Factors in Atrial Fibrillation (ATRIA) Study. JAMA 285:2370–2375
4. Echt D, Liebson P, Mitchell L et al (1991) Mortality and morbidity in patients receiving encainide, flecainide, or placebo. N Engl J Med 324:781–788
5. Waldo A, Camm A, deRuyter H et al (1996) Effect of *d*-sotalol on mortality in patients with left ventricular dysfunction after recent and remote myocardial infarction. Lancet 348:7–12
6. MacMahon S, Collins R, Peto R et al (1988) Effect of prophylactic lidocaine in suspected acute myocardial infarction: an overview of results from the randomized, controlled trials. JAMA 260:1910–1916
7. Cappato R, Calkins H, Chen SA et al (2010) Updated worldwide survey on the methods, efficacy and safety of catheter ablation for human atrial fibrillation. Circ Arrhythm Electrophysiol 3:32–38
8. Stevenson WG, Wilber DJ, Natale A et al (2008) Irrigated radiofrequency catheter ablation guided by electroanatomic mapping for recurrent ventricular tachycardia after myocardial infarction: the multicenter thermocool ventricular tachycardia ablation trial. Circulation 118:2773–2782
9. Parsonnet V, Cheema A (2003) The nature and frequency of postimplant surgical interventions: a realistic appraisal. Pacing Clin Electrophysiol 26:2308–2312
10. Gould PA, Gula LJ, Champagne J et al (2008) Outcome of advisory implantable cardioverter-defibrillator replacement: one-year follow-up. Heart Rhythm 5:1675–1681
11. Lin H, Parmacek M, Morle G et al (1990) Expression of recombinant genes in the myocardium in vivo after direct injection of DNA. Circulation 82:2217–2221
12. Shenk T (1996) Adenoviridae: the viruses and their replication. In: Fields BN, Knipe DM, Howley PM (eds) Fields virology, Lippincott-Raven, Philadelphia, PA.
13. Bergelson J, Cunningham JA, Droguett G et al (1997) Isolation of a common receptor for coxsackie B viruses and adenoviruses 2 and 5. Science 275:1320–1323
14. Wickham TJ, Mathias P, Cheresh DA et al (1993) Integrins $\alpha_v\beta_3$ and $\alpha_v\beta_5$ promote adenovirus internalization but not virus attachment. Cell 73:309–319
15. Chen HH, Mack LM, Kelly R et al (1997) Persistence in muscle of an adenoviral vector that lacks all viral genes. Proc Natl Acad Sci USA 94:1645–1650
16. Brandon F, McLean I, Jr. (1962) Adenoviruses. Adv Virus Res 13:157–193
17. Hoggan M, Blacklow N, Rowe W (1966) Studies of small DNA viruses found in various adenovirus preparations: physical, biological, and immunological characteristics. Proc Natl Acad Sci USA 55:1467–1474
18. Svensson E, Marshall D, Woodard K et al (1999) Efficient and stable transduction of cardiomyocytes after intramyocardial or intracoronary perfusion with recombinant adeno-associated vectors. Circulation 99:201–205
19. Du L, Kido M, Lee D et al (2004) Differential myocardial gene delivery by recombinant serotype-specific adeno-associated viral vectors. Mol Ther 10:604–608
20. Gregorevic P, Blankinship M, Allen J et al (2004) Systemic delivery of genes to striated muscles using adeno-associated viral vectors. Nat Med 10:828–834
21. Wang Z, Zhu T, Qiao C et al (2005) Adeno-associated virus serotype 8 efficiently delivers genes to muscle and heart. Nat Biotechnol 23:321–328

22. Inagaki K, Fuess S, Storm T et al (2006) Robust systemic transduction of AAV9 vectors in mice: efficient global cardiac gene transfer superior to that of AAV8. Mol Ther 14:45–53
23. Summerford C, Samulski R (1998) Membrane-associated heparan sulfate proteoglycan is a receptor for adeno-associated virus type 2 virions. J Virol 72:1438–1445
24. Akache B, Grimm D, Pandey K et al (2006) The 37/67-kilodalton laminin receptor is a receptor for adeno-associated virus serotypes 8, 2, 3, and 9. J Virol 80:9831–9836
25. Wu Z, Miller E, Agbandje-McKenna M et al (2006) Alpha2,3 and alpha2,6 N-linked sialic acids facilitate efficient binding and transduction by adeno-associated virus types 1 and 6. J Virol 80:9093–9103
26. Ferrari F, Samulski T, Shenk T et al (1996) Second-strand synthesis is a rate-limiting step for efficient transduction by recombinant adeno-associated virus vectors. J Virol 70:3227–3234
27. Hauck B, Zhao W, High K et al (2004) Intracellular viral processing, not single-stranded DNA accumulation, is crucial for recombinant adeno-associated virus transduction. J Virol 78:13678–13686
28. Cleghorn F, Reitz M, Popovic M et al (2005) Human immunodeficiency viruses. In: Mandell G, Bennett J, Dolin R (eds) Principles and practice of infectious diseases, Elsevier, Philadelphia, PA
29. Bonci D, Cittadini A, Latronico MV et al (2003) "Advanced" generation lentiviruses as efficient vectors for cardiomyocyte gene transduction in vitro and in vivo. Gene Ther 10:630–636
30. Cockrell AS, Kafri T (2007) Gene delivery by lentivirus vectors. Mol Biotechnol 36:184–204
31. Guzman RJ, Lemarchand P, Crystal R et al (1993) Efficient gene transfer into myocardium by direct injection of adenovirus vectors. Circ Res 73:1202–1207
32. Muhlhauser J, Jones M, Yamada I et al (1996) Safety and efficacy of in vivo gene transfer into the porcine heart with replication-deficient, recombinant adenovirus vectors. Gene Ther 3:145–153
33. Lamping K, Rios CD, Chun JA et al (1997) Intrapericardial administration of adenovirus for gene transfer. Am J Physiol 272:H310–H317
34. Hajjar R, Schmidt U, Matsui T et al (1998) Modulation of ventricular function through gene transfer in vivo. Proc Natl Acad Sci USA 95:5251–5256
35. Maurice J, Hata J, Shah A et al (1999) Enhancement of cardiac function after adenoviral-mediated in vivo intracoronary beta2-adrenergic receptor gene delivery. J Clin Invest 104:21–29
36. Bridges C, Burkman J, Malekan R et al (2002) Global cardiac-specific transgene expression using cardiopulmonary bypass with cardiac isolation. Ann Thorac Surg 73:1939–1946
37. Kass-Eisler A, Falck-Pedersen E, Alvira M et al (1993) Quantitative determination of adenovirus-mediated gene delivery to rat cardiac myocytes in vitro and in vivo. Proc Natl Acad Sci USA 90:11498–11502
38. Donahue JK, Kikkawa K, Johns DC et al (1997) Ultrarapid, highly efficient viral gene transfer to the heart. Proc Natl Acad Sci USA 94:4664–4668
39. Donahue JK, Kikkawa K, Thomas AD et al (1998) Acceleration of widespread adenoviral gene transfer to intact rabbit hearts by coronary perfusion with low calcium and serotonin. Gene Ther 5:630–634
40. Nagata K, Marban E, Lawrence J et al (2001) Phosphodiesterase inhibitor-mediated potentiation of adenovirus delivery to myocardium. J Mol Cell Cardiol 33:575–580
41. Sasano T, Kikuchi K, Feng N et al (2007) Targeted high-efficiency, homogeneous myocardial gene transfer. J Mol Cell Cardiol 42:954–961
42. Kikuchi K, McDonald AD, Sasano T et al (2005) Targeted modification of atrial electrophysiology by homogeneous transmural atrial gene transfer. Circulation 111:264–270
43. Warrington K, Gorbatyuk O, Harrison J et al (2004) Adeno-associated virus type 2 VP2 capsid protein is nonessential and can tolerate large peptide insertions at its N terminus. J Virol 78:6595–6609

44. Everts M, Curiel D (2004) Transductional targeting of adenoviral cancer gene therapy. Curr Gene Ther 4:337–346
45. Beck C, Uramoto H, Boren J et al (2004) Tissue-specific targeting for cardiovascular gene transfer. Potential vectors and future challenges. Curr Gene Ther 4:457–467
46. Dimmeler S, Zeiher AM, Schneider MD (2005) Unchain my heart: the scientific foundations of cardiac repair. J Clin Invest 115:572–583
47. Takahashi K, Tanabe K, Ohnuki M et al (2007) Induction of pluripotent stem cells from adult human fibroblasts by defined factors. Cell 131:861–872
48. Ryan JM, Barry FP, Murphy JM et al (2005) Mesenchymal stem cells avoid allogeneic rejection. J Inflamm (Lond) 2:8
49. Hare JM, Traverse JH, Henry TD et al (2009) A randomized, double-blind, placebo-controlled, dose-escalation study of intravenous adult human mesenchymal stem cells (prochymal) after acute myocardial infarction. J Am Coll Cardiol 54:2277–2286
50. Potapova IA, Doronin SV, Kelly DJ et al (2008) Enhanced recovery of mechanical function in the canine heart by seeding an extracellular matrix patch with mesenchymal stem cells committed to a cardiac lineage. Am J Physiol Heart Circ Physiol 295:H2257–H2263
51. Blank AC, van Veen TA, Jonsson MK et al (2009) Rewiring the heart: stem cell therapy to restore normal cardiac excitability and conduction. Curr Stem Cell Res Ther 4:23–33
52. Fischer KM, Cottage CT, Wu W et al (2009) Enhancement of myocardial regeneration through genetic engineering of cardiac progenitor cells expressing Pim-1 kinase. Circulation 120:2077–2087
53. Engelmann T (1896) Ueber den Einfluss der Systole auf die motorische Leitung in der Herzkammer, mit Bemerkungen zur Theorie allorhythmischer Herzstoerungen. Pflugers Arch Gesamte Physiol Menschen Tiere 62:543–566
54. Lewis T, Drury A, Iliescu C (1921) A demonstration of circus movement in clinical fibrillation of the auricles. Heart 8:361–369
55. Mines G (1913) On dynamic equilibrium in the heart. J Physiol 46:349–383
56. Garrey W (1914) The nature of fibrillary contraction of the heart: its relation to tissue mass and form. Am J Physiol 33:397–414
57. Moe G, Rheinboldt W, Abildskov J (1964) A computer model of atrial fibrillation. Am Heart J 67:200–220
58. Davidenko JM, Pertsov AV, Salomonsz R et al (1992) Stationary and drifting spiral waves of excitation in isolated cardiac muscle. Nature 355:349–351
59. Atienza F, Almendral J, Moreno J et al (2006) Activation of inward rectifier potassium channels accelerates atrial fibrillation in humans: evidence for a reentrant mechanism. Circulation 114:2434–2442
60. Boutjdir M, Le Heuzey J, Lavergne T et al (1986) Inhomogeneity of cellular refractoriness in human atrium: factor of arrhythmia? Pacing Clin Electrophysiol 9:1095–1100
61. Franz M, Karasik P, Li C et al (1997) Electrical remodeling of the human atrium: similar effects in patients with chronic atrial fibrillation and atrial flutter. J Am Coll Cardiol 30:1785–1792
62. Van Wagoner D, Pond A, McCarthy P et al (1997) Outward K+ current densities and Kv1.5 expression are reduced in chronic human atrial fibrillation. Circ Res 80:772–781
63. Brundel B, Van Gelder I, Henning R et al (2001) Alterations in potassium channel gene expression in atria of patients with persistent and paroxysmal atrial fibrillation: differential regulation of protein and mRNA levels for K+ channels. J Am Coll Cardiol 37:926–932
64. Dobrev D, Graf E, Wettwer E et al (2001) Molecular basis of downregulation of G-protein-coupled inward rectifying K(+) current (I(K,ACh) in chronic human atrial fibrillation: decrease in GIRK4 mRNA correlates with reduced I(K,ACh) and muscarinic receptor-mediated shortening of action potentials. Circulation 104:2551–2557
65. Dobrev D, Friedrich A, Voigt N et al (2005) The G protein-gated potassium current I(K,ACh) is constitutively active in patients with chronic atrial fibrillation. Circulation 112:3697–3706
66. Voigt N, Friedrich A, Bock M et al (2007) Differential phosphorylation-dependent regulation of constitutively active and muscarinic receptor-activated IK,ACh channels in patients with chronic atrial fibrillation. Cardiovasc Res 74:426–437

67. Van Wagoner D, Pond A, Lamorgese M et al (1999) Atrial L-type Ca2+ currents and human atrial fibrillation. Circ Res 85:428–436
68. Christ T, Boknik P, Wohrl S et al (2004) L-type Ca2+ current downregulation in chronic human atrial fibrillation is associated with increased activity of protein phosphatases. Circulation 110:2651–2657
69. Greiser M, Halaszovich C, Frechen D et al (2007) Pharmacological evidence for altered src kinase regulation of I (Ca,L) in patients with chronic atrial fibrillation. Naunyn Schmiedebergs Arch Pharmacol **375**:383–392
70. Wijffels M, Kirchhof C, dorland R et al (1995) Atrial fibrillation begets atrial fibrillation. A study in awake chronically instrumented goats. Circulation 92:1954–1968
71. Ausma J, Wijffels M, Thone F et al (1997) Structural changes of atrial myocardium due to sustained atrial fibrillation in the goat. Circulation 96:3157–3163
72. van der Velden H, van Kempen M, Wijffels M et al (1998) Altered pattern of connexin40 distribution in persistent atrial fibrillation in the goat. J Cardiovasc Electrophysiol 9:596–607
73. Allessie M, Ausma J, Schotten U (2002) Electrical, contractile and structural remodeling during atrial fibrillation. Cardiovasc Res 54:230–246
74. Thijssen V, van der Velden H, van Ankeren E et al (2002) Analysis of altered gene expression during sustained atrial fibrillation in the goat. Cardiovasc Res 54:427–437
75. Li D, Fareh S, Leung T et al (1999) Promotion of atrial fibrillation by heart failure in dogs: atrial remodeling of a different sort. Circulation 100:87–95
76. Schram G, Pourrier M, Melnyk P et al (2002) Differential distribution of cardiac ion channel expression as a basis for regional specialization in electrical function. Circ Res 90:939–950
77. Gaborit N, Steenman M, Lamirault G et al (2005) Human atrial ion channel and transporter subunit gene-expression remodeling associated with valvular heart disease and atrial fibrillation. Circulation 112:471–481
78. Donahue J, Heldman A, Fraser H et al (2000) Focal modification of electrical conduction in the heart by viral gene transfer. Nat Med 6:1395–1398
79. Bauer A, McDonald AD, Nasir K et al (2004) Inhibitory G protein overexpression provides physiologically relevant heart rate control in persistent atrial fibrillation. Circulation 110:3115–3120
80. Murata M, Cingolani E, McDonald A et al (2004) Creation of a genetic calcium channel blocker by targeted Gem gene transfer in the heart. Circ Res 95:398–405
81. Bunch T, Mahapatra S, Bruce G et al (2006) Impact of transforming growth factor-beta1 on atrioventricular node conduction modificaiton by injected autologous fibroblasts in the canine heart. Circulation 113:2485–2494
82. Amit G, Kikuchi K, Greener ID et al (2010) Selective molecular potassium channel blockade prevents atrial fibrillation. Circulation 121:2263–2270.
83. Burton D, Song C, Fishbein I et al (2003) The incorporation of an ion channel gene mutation associated with the long QT syndrome (Q9E-hMiRP1) in a plasmid vector for site-specific arrhythmia gene therapy: in vitro and in vivo feasibility studies. Hum Gene Ther 10:907–922
84. Wit AL, Allessie MA, Bonke FI et al (1982) Electrophysiologic mapping to determine the mechanism of experimental ventricular tachycardia initiated by premature impulses. Experimental approach and initial results demonstrating reentrant excitation. Am J Cardiol 49:166–185
85. El-Sherif N, Scherlag BJ, Lazzara R et al (1977) Re-entrant ventricular arrhythmias in the late myocardial infarction period. 1. Conduction characteristics in the infarction zone. Circulation 55:686–702
86. Dillon SM, Allessie MA, Ursell PC et al (1988) Influences of anisotropic tissue structure on reentrant circuits in the epicardial border zone of subacute canine infarcts. Circ Res 63:182–206
87. Hirose M, Stuyvers BD, Dun W et al (2008) Function of Ca(2+) release channels in Purkinje cells that survive in the infarcted canine heart: a mechanism for triggered Purkinje ectopy. Circ Arrhythm Electrophysiol 1:387–395

88. Yuan F, Pinto JM, Li Q et al (1999) Characteristics of I(K) and its response to quinidine in experimental healed myocardial infarction. J Cardiovasc Electrophysiol 10:844–854
89. Neagoe C, Kulke M, del Monte F et al (2002) Titin isoform switch in ischemic human heart disease. Circulation 106:1333–1341
90. Walker LA, Walker JS, Ambler SK et al (2010) Stage-specific changes in myofilament protein phosphorylation following myocardial infarction in mice. J Mol Cell Cardiol 48:1180–1186.
91. Pinto J, Yuan F, Wasserlauf B et al (1997) Regional gradation of L-type calcium currents in the feline heart with a healed myocardial infarct. J Cardiovasc Electrophysiol 8:548–560
92. Peters N, Coromilas J, Severs N et al (1997) Disturbed connexin43 gap junction distribution correlates with the location of reentrant circuits in the epicardial border zone of healing canine infarcts that cause ventricular tachycardia. Circulation 95:988–996
93. Smith J, Green C, Peters N et al (1991) Altered patterns of gap junction distribution in ischemic heart disease. An immunohistochemical study of human myocardium using laser scanning confocal microscopy. Am J Pathol 139:801–821
94. Mills WR, Mal N, Forudi F et al (2006) Optical mapping of late myocardial infarction in rats. Am J Physiol Heart Circ Physiol 290:H1298–H1306
95. Ohara T, Ohara K, Cao JM et al (2001) Increased wave break during ventricular fibrillation in the epicardial border zone of hearts with healed myocardial infarction. Circulation 103:1465–1472
96. Kiseleva I, Kamkin A, Wagner KD et al (2000) Mechanoelectric feedback after left ventricular infarction in rats. Cardiovasc Res 45:370–378
97. Miura M, Boyden PA, ter Keurs HE (1998) Ca2+ waves during triggered propagated contractions in intact trabeculae. Am J Physiol 274:H266–H276
98. Sasano T, Kelemen K, Greener ID et al. (2009) Ventricular tachycardia from the healed myocardial infarction scar: validation of an animal model and utility of gene therapy. Heart Rhythm 6:S91–S97
99. Sasano T, McDonald AD, Kikuchi K et al. (2006) Molecular ablation of ventricular tachycardia after myocardial infarction. Nat Med 12:1256–1258
100. Lau DH, Clausen C, Sosunov EA et al. (2009) Epicardial border zone overexpression of skeletal muscle sodium channel SkM1 normalizes activation, preserves conduction, and suppresses ventricular arrhythmia: an in silico, in vivo, in vitro study. Circulation 119:19–27
101. Gardner PI, Ursell PC, Fenoglio JJ, Jr. et al. (1985) Electrophysiologic and anatomic basis for fractionated electrograms recorded from healed myocardial infarcts. Circulation 72:596–611
102. Ursell PC, Gardner PI, Albala A et al. (1985) Structural and electrophysiological changes in the epicardial border zone of canine myocardial infarcts during infarct healing. Circ Res 56:436–451
103. Lue W, Boyden P (1992) Abnormal electrical properties of myocytes from chronically infarcted canine heart: alterations in Vmax and the transient outward current. Circulation 85:1175–1188
104. Pu J, Boyden P (1997) Alterations of Na+ currents in myocytes from epicardial border zone of the infarcted heart. A possible ionic mechanism for reduced excitability and postrepolarization refractoriness. Circ Res 81:110–119
105. Dudley SC, Jr. (2005) Beware of cells bearing gifts: cell replacement therapy and arrhythmic risk. Circ Res 97:99–101
106. Mills WR, Mal N, Kiedrowski MJ et al. (2007) Stem cell therapy enhances electrical viability in myocardial infarction. J Mol Cell Cardiol 42:304–314
107. Fouts K, Fernandes B, Mal N et al. (2006) Electrophysiological consequence of skeletal myoblast transplantation in normal and infarcted canine myocardium. Heart Rhythm 3:452–461
108. Abraham MR, Henrikson CA, Tung L et al. (2005) Antiarrhythmic engineering of skeletal myoblasts for cardiac transplantation. Circ Res 97:159–167
109. Menasche P, Hagege AA, Vilquin JT et al. (2003) Autologous skeletal myoblast transplantation for severe postinfarction left ventricular dysfunction. J Am Coll Cardiol 41:1078–1083

110. Fernandes S, Amirault JC, Lande G et al. (2006) Autologous myoblast transplantation after myocardial infarction increases the inducibility of ventricular arrhythmias. Cardiovasc Res 69:348–358
111. Suzuki K, Brand NJ, Allen S et al. (2001) Overexpression of connexin 43 in skeletal myoblasts: relevance to cell transplantation to the heart. J Thorac Cardiovasc Surg 122:759–766
112. Stagg MA, Coppen SR, Suzuki K et al. (2006) Evaluation of frequency, type, and function of gap junctions between skeletal myoblasts overexpressing connexin43 and cardiomyocytes: relevance to cell transplantation. FASEB J 20:744–746
113. Roell W, Lewalter T, Sasse P et al. (2007) Engraftment of connexin 43-expressing cells prevents post-infarct arrhythmia. Nature 450:819–824
114. Amado LC, Saliaris AP, Schuleri KH et al. (2005) Cardiac repair with intramyocardial injection of allogeneic mesenchymal stem cells after myocardial infarction. Proc Natl Acad Sci USA 102:11474–11479
115. Heubach JF, Graf EM, Leutheuser J et al. (2004) Electrophysiological properties of human mesenchymal stem cells. J Physiol 554:659–672
116. Lagostena L, Avitabile D, De FE et al. (2005) Electrophysiological properties of mouse bone marrow c-kit+ cells co-cultured onto neonatal cardiac myocytes. Cardiovasc Res 66:482–492
117. Potapova I, Plotnikov A, Lu Z et al. (2004) Human mesenchymal stem cells as a gene delivery system to create cardiac pacemakers. Circ Res 94:952–959
118. Beeres SL, Atsma DE, van der Laarse A et al. (2005) Human adult bone marrow mesenchymal stem cells repair experimental conduction block in rat cardiomyocyte cultures. J Am Coll Cardiol 46:1943–1952
119. Pijnappels DA, Schalij MJ, van Tuyn J et al. (2006) Progressive increase in conduction velocity across human mesenchymal stem cells is mediated by enhanced electrical coupling. Cardiovasc Res 72:282–291
120. Chang MG, Tung L, Sekar RB et al. (2006) Proarrhythmic potential of mesenchymal stem cell transplantation revealed in an in vitro coculture model. Circulation 113:1832–1841
121. Unzek S, Zhang M, Mal N et al. (2007) SDF-1 recruits cardiac stem cell-like cells that depolarize in vivo. Cell Transplant 16:879–886
122. Gaudesius G, Miragoli M, Thomas SP et al. (2003) Coupling of cardiac electrical activity over extended distances by fibroblasts of cardiac origin. Circ Res 93:421–428
123. Zhang YM, Hartzell C, Narlow M et al. (2002) Stem cell-derived cardiomyocytes demonstrate arrhythmic potential. Circulation 106:1294–1299
124. Rota M, Kajstura J, Hosoda T et al. (2007) Bone marrow cells adopt the cardiomyogenic fate in vivo. Proc Natl Acad Sci USA 104:17783–17788
125. Rota M, Padin-Iruegas ME, Misao Y et al. (2008) Local activation or implantation of cardiac progenitor cells rescues scarred infarcted myocardium improving cardiac function. Circ Res 103:107–116
126. Dawn B, Stein AB, Urbanek K et al. (2005) Cardiac stem cells delivered intravascularly traverse the vessel barrier, regenerate infarcted myocardium, and improve cardiac function. Proc Natl Acad Sci USA 102:3766–3771
127. Johnston PV, Sasano T, Mills K et al. (2009) Engraftment, differentiation, and functional benefits of autologous cardiosphere-derived cells in porcine ischemic cardiomyopathy. Circulation 120:1075–1083
128. Barile L, Cerisoli F, Frati G et al (2009) Bone marrow-derived cells can acquire cardiac stem cells properties in damaged heart. J Cell Mol Med Blackwell Publishing Ltd 1582–4934, http://dx.doi.org/10.1111/j.1582-4934.2009.00968.x, DOI 10.1111/j.1582-4934.2009.00968.x
129. Kehat I, Khimovich L, Caspi O et al. (2004) Electromechanical integration of cardiomyocytes derived from human embryonic stem cells. Nat Biotechnol 22:1282–1289
130. Nussbaum J, Minami E, Laflamme MA et al. (2007) Transplantation of undifferentiated murine embryonic stem cells in the heart: teratoma formation and immune response. FASEB J 21:1345–1357
131. Kolossov E, Bostani T, Roell W et al. (2006) Engraftment of engineered ES cell-derived cardiomyocytes but not BM cells restores contractile function to the infarcted myocardium. J Exp Med 203:2315–2327

132. Doevendans PA, Kubalak SW, An RH et al. (2000) Differentiation of cardiomyocytes in floating embryoid bodies is comparable to fetal cardiomyocytes. J Mol Cell Cardiol 32:839–851
133. He JQ, Ma Y, Lee Y et al. (2003) Human embryonic stem cells develop into multiple types of cardiac myocytes: action potential characterization. Circ Res 93:32–39
134. Satin J, Kehat I, Caspi O et al. (2004) Mechanism of spontaneous excitability in human embryonic stem cell derived cardiomyocytes. J Physiol 559:479–496
135. Itzhaki I, Schiller J, Beyar R et al. (2006) Calcium handling in embryonic stem cell-derived cardiac myocytes: of mice and men. Ann N Y Acad Sci 1080:207–215
136. Menard C, Hagege AA, Agbulut O et al. (2005) Transplantation of cardiac-committed mouse embryonic stem cells to infarcted sheep myocardium: a preclinical study. Lancet 366:1005–1012
137. Gnecchi M, Zhang Z, Ni A et al. (2008) Paracrine mechanisms in adult stem cell signaling and therapy. Circ Res 103:1204–1219
138. Zhang M, Mal N, Kiedrowski M et al. (2007) SDF-1 expression by mesenchymal stem cells results in trophic support of cardiac myocytes after myocardial infarction. FASEB J 21:3197–3207
139. Askari AT, Unzek S, Popovic ZB et al. (2003) Effect of stromal-cell-derived factor 1 on stem-cell homing and tissue regeneration in ischaemic cardiomyopathy. Lancet 362:697–703
140. Hakuno D, Fukuda K, Makino S et al. (2002) Bone marrow-derived regenerated cardiomyocytes (CMG Cells) express functional adrenergic and muscarinic receptors. Circulation 105:380–386
141. Bartunek J, Croissant JD, Wijns W et al. (2007) Pretreatment of adult bone marrow mesenchymal stem cells with cardiomyogenic growth factors and repair of the chronically infarcted myocardium. Am J Physiol Heart Circ Physiol 292:H1095–H1104
142. Bayes-Genis A, Roura S, Soler-Botija C et al. (2005) Identification of cardiomyogenic lineage markers in untreated human bone marrow-derived mesenchymal stem cells. Transplant Proc 37:4077–4079
143. Bisping E, Ikeda S, Kong SW et al. (2006) Gata4 is required for maintenance of postnatal cardiac function and protection from pressure overload-induced heart failure. Proc Natl Acad Sci USA 103:14471–14476
144. Bian J, Popovic ZB, Benejam C et al. (2007) Effect of cell-based intercellular delivery of transcription factor GATA4 on ischemic cardiomyopathy. Circ Res 100:1626–1633
145. Pedrotty DM, Klinger RY, Kirkton RD et al. (2009) Cardiac fibroblast paracrine factors alter impulse conduction and ion channel expression of neonatal rat cardiomyocytes. Cardiovasc Res 83:688–697

Long-Term Prospects for Arrhythmia Treatment: Advantages and Limitations of Gene and Cell Therapies

Michael R. Rosen

Abstract Gene and cell therapies of cardiac arrhythmias are novel approaches to long-standing clinical problems. They have been developed in recognition of the fact that the many drug and device treatments available for arrhythmias, all have shortcomings. In this chapter the current status of therapies for bradyarrhythmias and tachyarrhythmias is reviewed, the prospects for gene and cell therapies are considered, and the steps required in moving toward clinical application are presented.

Keywords Stem cells • Viral gene therapy • Ventricular tachycardia and fibrillation • Atrial fibrillation • Bradyarrhythmias

1 Introduction

Including a section on "electrical regeneration" in a book on regenerating the heart creates an interesting diversion: interesting in the sense that many investigators who focus on myocardial repair and regeneration anticipate that replacing diseased cells and tissue with healthy cells and tissue would incorporate normal sinus rhythm or a normal pacemaker-induced rhythm as an obvious benefit. And if we consider cardiac transplantation as an extreme variant on myocardial regeneration – in the sense of using a new heart to regenerate the function of the old one – the normalization of rhythm is the rule, and not the exception [1]. We must consider that in addressing *electrical regeneration* as an entity, we are focusing on one particular function of a nodal or specialized conducting fiber or myocyte: we are stating our intent to fix a specific electrical dysfunction while not necessarily doing anything

M.R. Rosen (✉)
Department of Pharmacology, Department of Pediatrics,
Center for Molecular Therapeutics, Columbia University,
New York, NY
e-mail: mrr1@columbia.edu

I.S. Cohen and G.R. Gaudette (eds.), *Regenerating the Heart*, Stem Cell Biology and Regenerative Medicine, DOI 10.1007/978-1-61779-021-8_20,
© Springer Science+Business Media, LLC 2011

else to that cell. This can play out in a variety of ways. If the intent is to use gene therapy to regenerate normal function in the setting of a genetic anomaly (as in congenital long QT syndrome), then such an approach could be completely curative. However, if the intent is to normalize conduction and prevent reentry in the setting of a myocardial infarction, then a myocyte that has poor contractile function will likely continue to contract poorly even though it begins to conduct normally.

The preceding chapters in this section of the book considered mechanisms for electrical regeneration, integration of stem cells into a myocardial syncytium, and the potential for treating bradyarrhythmias and tachyarrhythmias. The approaches incorporate gene therapy, cell therapy, and hybrids of both. As such they carry the potential benefits and risks of each individual intervention as well as of the combined interventions. In other words, combined gene/cell therapy can and should be looked at as harboring safety issues of each type of approach, with the combination perhaps cancelling, perhaps augmenting the risks of either alone. Additionally, augmentation might represent a simple sum of risks or a synergy of risks.

So, as is so often the case with new therapeutic ventures, the situation is complex. In the following sections I shall try to dissect the promise and the concerns regarding gene/cell therapy of arrhythmias within the framework of our needs, the available treatments, and how we might go forward. I shall deliberately avoid discussing ion channelopathies [2, 3] and other genetic anomalies [4] as in those settings the cause is a single genetic lesion or a concordance of such lesions and their modifiers. "Cures" when they come should and will focus on (1) identifying the gene or genes responsible, (2) depending on the nature of the lesion(s), means to increase or decrease functionality in afflicted individuals, and (3) means to prevent passage of the lesion(s) from generation to generation. Thus far, combinations of drugs, devices, and counseling have been at the forefront here and gene therapy is a potential but not yet a practical reality.

2 What Is Available and What Do We Need?

Whether we consider bradyarrhythmias or tachyarrhythmias, we do so with the understanding that the state of the art uses drugs and/or devices to prevent/treat them [5–10] and may require cardiac transplants in settings where advanced heart failure is an overriding concern [8]. With regard to devices, the implantation of electrical pacemakers for single chamber or biventricular pacing, left ventricular assist devices, internal cardioverter/defibrillators, and the use of catheter techniques for ablation have moved to the center of our therapeutic arsenal over the last 50 years [5–7]. Drugs, whether antiarrhythmic or targeted at "upstream" modulators of electrical activity, still find a variety of uses [9, 10], but the toxicity of many antiarrhythmics has been documented and a variety of concerns have been voiced. One result has been that antiarrhythmic drug discovery, so active in the 1960s–1980s has become almost dormant.

Most arrhythmias that require treatment are tachycardias occurring in the setting of acquired heart disease. Those most in need of therapy are (1) atrial fibrillation, because of its frequency (5% of the population over age 65 in the USA), morbidity, and association with embolism and stroke [11, 12], and (2) ventricular tachycardia/ ventricular fibrillation, because of their role in arrhythmic sudden death (250,000 occurrences per year in patients with high coronary risk profiles in the USA) [13].

2.1 Atrial Fibrillation

The prevention and therapy of atrial fibrillation are complex because of the variety of causes as well as the progression of the arrhythmia from paroxysmal to persistent to permanent [14]. One of the knottiest problems is atrial fibrillation resulting from triggered foci in the pulmonary veins [15, 16]. This variant afflicts a younger age group (40s and up) such that attendant disability has not only the implications and progression of that in older patients but the added psychological and economic burden of afflicting individuals in the active workforce. Drug therapy is often of limited use here and the treatment now actively employed is catheter ablation of the foci within the pulmonary veins or isolation of the electrical output into the atrium by placing ablation scars in the atria [17, 18]. Although these approaches have met with some success, pulmonary vein stenosis and recurrence of atrial fibrillation after an arrhythmia-free interval of varying duration have been two major concerns [19].

The drug therapy of atrial fibrillation has involved two schools of thought. One focuses on rate control, in which the chronicity of the arrhythmia is accepted and the achievement of sufficient atrioventricular block to ensure a physiologically acceptable heart rate is the target [20]. Here the traditional role of digitalis to induce atrioventricular block has largely been supplanted by the use of β-adrenergic blockers and/or calcium channel blockers alone or in combination with one another or with digitalis [20]. The second approach centers on rhythm control, and in this case efforts are made to return the patient to sinus rhythm and to maintain that rhythm [20]. Here the approaches are cardioversion (usually electrical) to terminate the fibrillation and the subsequent administration of drugs that slow conduction and/or prolong repolarization and refractoriness [21]. Nonantiarrhythmic drugs that interfere with angiotensin II actions on the cardiac substrate reduce the frequency of recurrences of paroxysmal atrial fibrillation [22, 23] as do anti-inflammatory agents [22]. Additional modalities for maintaining sinus rhythm include the use of surgical or catheter maze procedures [24].

Most studies comparing the benefits and risks of rate control with those of rhythm control have seen long-term outcomes sufficiently similar that neither approach is generally favored over the other [20]. Most patients with atrial fibrillation that is frequently recurrent, persistent, or chronic require anticoagulation, which reduces the risk of thromboembolus but which adds to the complexity of care and carries with it the risk of bleeding.

In sum, if we ask whether medical arts and sciences have impacted favorably on the management of atrial fibrillation, the answer is clearly yes. Atrial fibrillation is recognized as a disease of multiple causality, responsive to a variety of treatment approaches, and one that occurs as a chronic disease in the population. Most cases of atrial fibrillation are not permanently cured, and until prevention and cure become the rule new therapies will be sought.

2.2 Ventricular Tachycardia/Ventricular Fibrillation

Ventricular tachycardia and fibrillation have long caught the imagination of the public because of their devastating impact on individual lives. These arrhythmias can lead to catastrophic collapse and death in the setting of acute myocardial infarction or in the postinfarction period. The post-World War II era was one of optimism regarding the enlightened use of antiarrhythmic drugs to prevent/treat these arrhythmias, and the Vaughan Williams criteria for antiarrhythmic drug classification [21] provided a framework for mechanistic understanding of how prevention/treatment might be achieved. But cold reality arrived in the CAST [25], SWORD [26], and subsequent trials that demonstrated the proarrhythmic risks negated the potential for antiarrhythmic gain [9, 10]. The development of internal cardioversion and defibrillation provided electronic means for terminating the arrhythmias that had more consistency, more effectiveness, and less toxicity than drugs, as well as means for monitoring a patient's rhythm status through memory loops in the devices. Moreover, the utility of these devices has been supplemented by concomitant administration of amiodarone, which in some studies has decreased the number of shocks delivered to patients over time [27].

In sum, there is an effective therapy here, but one that is cumbersome, often psychologically impacts on the patient, and carries the risk of failure of the device and of inappropriate sensing and delivery of unneeded shocks. So in this arena too, alternative therapies are needed.

2.3 Bradyarrhythmias

The bradyarrhythmia of most general concern is complete heart block, whether congenital or acquired. The latter is usually a result of ischemic heart disease. Although catecholamine therapy had been attempted as an intervention, it was not until the era of transvenous cardiac pacing commenced that a widely applicable and consistently beneficial treatment became available [5]. There is no question that electronic pacing is life-saving in this setting: but it is far from perfect [28, 29]. The risks of device insertion and maintenance and of lead fracture and the need for extraction are real. Also problematic are the needs for constant monitoring and maintenance, replacement of battery packs, and possibilities of infection and of interference from other devices. Finally, with insertion of the pacemaker lead into the most usual site, the right ventricular endocardial apex, there occurs an asymmetry of ventricular contraction, a suboptimal cardiac output (although far better than that in the setting of complete heart block) and over the years, an increasing tendency to cardiac failure [28, 29]. Although there is perhaps less immediacy for new therapies in this setting than in ventricular tachycardia/fibrillation, it is clear that safer and more long-lasting therapies than the electronic device are desirable.

3 What Are the Prospects for Gene/Cell Therapy?

Regardless of which arrhythmia and regardless of the effectiveness of therapy to date, it is clear that longer-lasting, more effective, and less toxic therapies are desirable. My own prejudice is to agree with those individuals who say that if one can repair/regenerate abnormal myocardium, then the arrhythmia should cease to be an issue. However, this does not augur as clear-cut a direction as it might appear. It is not clear-cut because not everyone at risk for or experiencing an arrhythmia need be subject to a procedure targeted at regenerating parts or all of his/her heart. And so there is a place for thinking specifically about preventing/treating the arrhythmia and not intervening in the totality of the myocardium. It is in this spectrum of cardiac disease that I believe gene and cell therapies have the most to offer.

In querying "what are the prospects for gene/cell therapy," one need only read this chapter and chapters "Regenerating Blood Vessels," "Regenerating Heart Valves," and "Tissue Engineering Strategies for Cardiac Regeneration." These enumerate aspects of the mechanistic factors to be considered and progress to date in research relating to cell therapy and to specific arrhythmias. Depending on whether the focus is on complete heart block, atrial fibrillation, or ventricular tachycardia/ fibrillation, there is a varied risk/benefit ratio to be considered. On the one hand, we can say that we are adequately treating large numbers of patients using existing approaches. On the other hand, we can say that "better" is possible and cure would be ideal. The downside is, what do we know about gene therapy or cell therapy or the combination? Not only what do we know about efficacy, but what do we know about safety?

Viral vectors are the major means for administering gene therapy, and there have been and remain important questions regarding both the genes and the vectors. The early interventions used to treat otherwise incurable conditions manifested various toxicities, including neoplasia [30, 31]. More recent viral vectors are replication-deficient and genes can be incorporated into the cell either episomally or genomically. Viral gene therapies are being actively used in trials on human subjects, often in the setting of neurological disease. Yet questions persist regarding whether in long-term settings the intervention retains an oncogenic potential, an infectious potential, and/or other as yet undefined toxicities.

A variety of nonviral means for loading genes have been described (reviewed in [32]). The extent to which such loading is episomal versus genomic appears to vary with the type of method used. One issue common to viral vectors or nonviral loading centers on where in the genome the gene being administered is integrated. Is it at a site where it achieves only its anticipated function? Or is it at a site at which it can give rise to unintended consequences ranging from loss of function to active oncogenesis? Methods are being developed to ensure appropriate integration, many centering on the zinc finger nucleases [33–35]. However, this question remains an open one which will require more study.

We have the greatest body of information about the safety of cell therapy when the cell is an adult human mesenchymal stem cell (hMSC) or another bone-marrow-derived cell [36, 37]. We know this because of the long history of bone marrow transplant and the small but still intriguing body of information regarding administration of bone-marrow-derived cells autologously or allogeneically to patients with myocardial infarction or congestive failure. Of great concern has been the possibility of arrhythmogenesis. Yet this has not happened with hMSCs and has only been reported with skeletal myoblasts [38]. This complication seems to reflect the failure of skeletal myoblasts to couple with myocytes to provide continuity of impulse propagation in the myocardial syncytium [39].

What are the issues one might consider when implanting bone-marrow-derived stem cells in the heart? The answer reflects the use to which they are being put. If the issue is myocardial repair/regeneration, the question of safety has long been supplanted by concern over cell loss due to rejection or migration and the modus operandi for the repair that occurs [36, 37]. Cell loss due to rejection has not been of concern with autologous transplants and – in general – with allogeneic hMSC transplants because of the apparent "immunoprivilege" of the hMSCs [36, 40, 41]. The mechanisms for this are still under active study. As for migration, we know that the cells are lost to other sites in the body, but the mechanisms for loss and the sites favored by the cells are at issue because of the inadequacy and expense of methods for tracking them.

When stem cells are given for the purpose of repair/regeneration, the expectation is that myocytes will be formed. Much of the literature indicates that myocyte development in this setting results from a paracrine effect of the stem cells to recruit endogenous cardiac precursors rather than from their own differentiation [36, 37]. Another approach is to engineer stem cells such that they tend to grow in a cardiogenic lineage [42]. The expectation is that when implanted in the heart, they will continue in this lineage. However, when hMSCs are used in the treatment of an arrhythmia, as in fabricating a biological pacemaker (see chapter "Regenerating Heart Valves"), they are not expected to differentiate into another cell type. Rather, they are expected to remain undifferentiated, to cease multiplying, to couple to adjacent myocytes, and to transmit an ion current to the adjacent myocyte [28, 29]. Their function to generate this ion current is the result of their being loaded with the gene of interest using either a viral vector or electroporation [43].

Although this strategy has worked well for several weeks, persisting until cell loss presumably due to migration occurs [43], it carries with it the problem inherent in the loading of the hMSCs with the gene of interest. Whether the vector is viral or whether electroporation is used, we have to ask where the genetic material is incorporated and what the potential downside is, whether oncogenic or otherwise. One thought has been to encapsulate hMSCs carrying pacemaker genes or to load them onto a biomaterial that will prevent their migration [44]. But we need to learn not only whether a biomaterial will hold hMSCs in place, prevent their differentiation, and permit them to function, but also whether the cells have an oncogenic potential.

It might be argued that implanting and testing encapsulated cells would be the proper way to proceed with preclinical evaluation. Although the FDA has no specific guidelines here, if we use combinations of medications as examples, it is more likely that testing each component of a complex system is the safe way to proceed. Via this strategy one understands both the function and the limits of function of each component, as well as the deleterious potential of each component. One might also argue that hMSCs that are encapsulated have less oncogenic potential than those implanted in an untethered setting, but it is extremely likely that if untethered stem cells are shown to be oncogenic, one would not want to implant them even if they are encapsulated. What would happen if even a few cells were to escape or the capsule were to fracture?

And so the issues are many and varied.

4 When Will Clinical Application Happen?

If the issues are many and varied, how should we proceed? This depends on the requisites of regulatory agencies, the therapeutic potential of the intervention, and the nature of the disease being treated, in other words, the risk/benefit ratio. We can argue at the outset that none of the arrhythmias under consideration puts the patient in the situation comparable to having a cancer that is refractory to known therapies. In the latter setting, or in its cardiac equivalent, the individual in end-stage congestive failure with no availability of a transplant and only transient relief from a left ventricular assist device, consideration of a trial therapy which will provide knowledge and may or may not benefit the patient is valid. Ultimately, it will be the patient's choice. Here, regulatory guidelines can permit phase 1/2 trials without the full range of preclinical testing that would be required in less pressing medical settings.

Since we have somewhat effective therapies for many of the arrhythmias described, what must be done? Clearly, one thing that needs to occur is animal studies demonstrating efficacy. In general in academic laboratories, the major focus is on efficacy linked to understanding of the mechanism. The use of an intervention that tests the mechanism and is shown through that mechanism to be therapeutic carries with it the affirmation of knowledge as well as its extension to a potentially beneficial outcome (see chapters "Regenerating Heart Valves" and "Tissue Engineering Strategies for Cardiac Regeneration"). Using our own work as an example, understanding that fragmented or slow conduction is associated with reentry in postinfarction ventricular tachycardia led us to test a gene therapy using the skeletal muscle sodium channel [45], which the literature showed was largely functional at low membrane potentials. This was important as in the depolarized milieu of an infarct border zone the normal cardiac sodium channel is largely inactivated and conduction is slow, setting up the conditions for reentry. Modeling studies and cell culture experiments suggested overexpressing the skeletal muscle sodium channel would be effective here [46, 47], and subsequent studies in a canine infarct showed speeding of activation [48] and reduction in occurrence of inducible ventricular tachycardia [46].

Studies in large-animal models of myocardial infarction, of atrial fibrillation, or of heart block are useful for testing efficacy or lack thereof, and also for testing proarrhythmia. FDA recommendations for studies of cell therapies recommend the use of dogs, sheep, or pig models to help understand some aspects of efficacy/activity of administered cell products, but even more importantly the safety of the technologies used to deliver cells to the heart [49]. Mapping systems, delivery catheters, dose, rate and volume of injection, and site of injection of constructs are all more realistically tested for human administration in such large-animal models than in small animals [49]. Studies in small-animal models are recommended – usually in immunocompromised rats or mice – to provide information on long-term cell survival and differentiation in the myocardium as well as on safety issues such as the oncogenic potential already mentioned [49].

It is only after a thorough investigation in small and large animals for the reasons stated above, as well as complete documentation of the properties of the gene and/or cell therapy to be administered and the technology for its administration (see [49] for details) that clinical testing can be considered. The steps recommended are clearly mapped by the FDA, but only as they refer to myocardial regeneration and repair. There are as yet no firm recommendations for antiarrhythmic gene/cell therapy interventions.

5 How and in Whom Do We Test?

In developing a gene and/or cell therapy, clearly one goal is to improve the therapeutic armamentarium for man. The question arises: how and in whom do we test? FDA recommendations speak of populations with myocardial disease, specifically refractory angina and ischemia, acute ischemia and infarction, and heart failure [49]. A conversation has to occur with regard to interventions focused on arrhythmias. When such a conversation has been held and recommendations have been provided, what happens next? Any clinical trial that is performed likely will be a combined phase 1/phase 2 trial, in which patients with the arrhythmia of interest are provided with a state-of-the-art therapy plus the therapy to be tested. Hence, the patient with refractory ventricular tachycardia/fibrillation will likely be given an implantable cardioverter/defibrillator plus the experimental therapy. The patient with atrial fibrillation and complete heart block will likely be given a right ventricular apical electronic pacemaker as well as the biological pacemaker. For the implantable cardioverter/defibrillator patient, the likely therapeutic measure will be a surrogate: the reduction of shocks delivered. For the pacemaker patient, it will be the extent to which the biological pacemaker functions to drive the heart in an autonomically responsive fashion without having to be rescued by the electronic unit [50].

But as much as the therapeutic benefit will be monitored, the primary purpose of an initial clinical trial is to evaluate safety. Safety will in part be a short-term consideration and in part a long-term one. For a stem cell intervention, will the cells stay in place and be functional without generating proarrhythmia? For an encapsulated construct, will the capsule stay in place and/or be fractured or extruded, and if it is extruded, to where will it embolize? For a gene therapy or a stem cell loaded via a viral vector or electroporation, how long will function persist and will there be oncogenesis?

None of these issues is trivial. Each must be addressed as completely as possible in the animal studies that precede consideration of human intervention. As is usually the case, a negative outcome, such as increased oncogenesis, in the animal trial will forestall human investigation. But even when all preclinical work is permissive and "all systems are go," experience tells us that apparently safe procedures still can go awry. As an example, we need only recall the postmarketing observations demonstrating that administering terfenadine, an antihistamine that blocks the delayed rectifier current, to persons with allergies resulted in lethal arrhythmias in a small subset [51].

It is at this juncture that risk/benefit encounters individual beliefs and/or prejudices with regard to proceeding in human trials. In such a setting, the persons least appropriate for making decisions are those with a vested interest who may be too aggressive or too cautious. We are best served by evidence-based approaches evaluated by disinterested third parties who weigh the risks and the benefits. The final part of the equation is provided by the prospective study subject who needs the information provided to him/her to understand that the phase 1/2 trial is a study which may/may not provide certain benefits, transient or long term, but participation in which is designed first to understand matters of safety even as data on outcome are collected. The conundrum for those who design studies, those who participate in them, those who review them, and those who publish them was recently summarized in the following comments relating to a specific clinical trial: "… it is clear that even with both formal guidelines and common sense, clinical research will continue to provide challenging ethical dilemmas where researchers, ethical review boards and journal editors will disagree [52]." To that, I can only add, "Amen."

Acknowledgment Supported by USPHS-NHLBI grant HL-094410 and NYSTEM grant CO24344.

References

1. Liem LB, DiBiase A, Schroeder JS. Arrhythmias and clinical electrophysiology of the transplanted human heart. Semin Thorac Cardiovasc Surg. 1990:2:271–8.
2. Moss AJ, Schwartz PJ. 25th Anniversary of the International Long-QT Syndrome Registry: an ongoing quest to uncover the secrets of long-QT syndrome. Circulation. 2005:111:1199–201.
3. Antzelevitch C, Brugada P, Borggrefe M, Brugada J, Brugada R, Corrado D, Gussak I, LeMarec H, Nademanee K, Perez Riera AR, Shimizu W, Schulze-Bahr E, Tan H, Wilde A.

Brugada syndrome: report of the second consensus conference: endorsed by the Heart Rhythm Society and the European Heart Rhythm Association. Circulation. 2005:111:659–70.

4. Hayashi M, Denjoy I, Extramiana F, Maltret A, Buisson NR, Lupoglazoff JM, Klug D, Hayashi M, Takatsuki S, Villain E, Kamblock J, Messali A, Guicheney P, Lunardi J, Leenhardt A. Incidence and risk factors of arrhythmic events in catecholaminergic polymorphic ventricular tachycardia. Circulation. 2009:119:2426–34.
5. Gregoratos G, Abrams J, Epstein AE, Freedman RA, Hayes DL, Hlatky MA, Kerber RE, Naccarelli GV, Schoenfeld MH, Silka MJ, Winters SL, Gibbons RJ, Antman EM, Alpert JS, Gregoratos G, Hiratzka LF, Faxon DP, Jacobs AK, Fuster V, Smith SC Jr. ACC/AHA/NASPE 2002 guideline update for implantation of cardiac pacemakers and antiarrhythmia devices: summary article: a report of the American College of Cardiology/American Heart Association Task Force on Practice Guidelines (ACC/AHA/NASPE Committee to Update the 1998 Pacemaker Guidelines). Circulation. 2002:106:2145–61.
6. Passman R, Kadish A. Sudden death prevention with implantable devices. Circulation. 2007:116:561–71.
7. Hall MCS, Todd DM. Modern management of arrhythmias. Postgrad Med J. 2006:82:117–25.
8. Boyle A. Current status of cardiac transplantation and mechanical circulatory support. Curr Heart Fail Rep. 2009:6:28–33.
9. Members of the Sicilian Gambit. The search for novel antiarrhythmic strategies. Eur Heart J. 1998:19:1178–96.
10. Members of the Sicilian Gambit. New approaches to antiarrhythmic therapy. Emerging therapeutic applications of the cell biology of cardiac arrhythmias. Cardiovasc Res. 2001:52:345–60.
11. Allessie MA, Boyden PA, Camm AJ, Kleber AG, Lab MJ, Legato MJ, Rosen MR, Schwartz PJ, Spooner PM, Van Wagoner DR, Waldo AL. Pathophysiology and prevention of atrial fibrillation. Circulation. 2001:103:769–77.
12. Sohara H, Amitani S, Kurose M, Miyahara K. Atrial fibrillation activates platelets and coagulation in a time-dependent manner: a study in patients with paroxysmal atrial fibrillation. J Am Coll Cardiol. 1997:29:106–12.
13. Huikuri HV, Castellanos A, Myerburg RJ. Sudden death due to cardiac arrhythmias. N Engl J Med. 2001:345:1473–82.
14. Gallagher MG, Camm AJ. Classification of atrial fibrillation. Pacing Clin Electrophysiol. 1997:20:1603–5.
15. Hirose M, Laurita KR. Calcium-mediated triggered activity is an underlying cellular mechanism of ectopy originating from the pulmonary vein in dogs. Am J Physiol. 2007:292:H1861–7.
16. Wit AL, Boyden PA. Triggered activity and atrial fibrillation. Heart Rhythm. 2007:4(Suppl):S17–23.
17. Haissaguerre M, Jais P, Shah DC, Takahashi A, Hocini M, Quiniou G, Garrigue S, Le Mouroux A, Le Metayer P, Clementy J. Spontaneous initiation of atrial fibrillation by ectopic beats originating in the pulmonary veins. N Engl J Med. 1998:339:659–66.
18. O'Neill MD, Jais P, Hocini M, Sacher F, Klein GJ, Clementy J, Haissaguerre M. Catheter ablation for atrial fibrillation. Circulation. 2007:116:1515–23.
19. Kantachuvessiri A. Pulmonary veins: preferred site for catheter ablation of atrial fibrillation. Heart Lung. 2002:31:271–8.
20. Fuster V, Ryden LE, Cannom DS, Crijns HJ, Curtis AB, Ellenbogen KA, Halperin JL, Le Heuzey JY, Kay GN, Lowe JE, Olsson SB, Prystowsky EN, Tamargo JL, Wann S, Smith SC Jr, Jacobs AK, Adams CD, Anderson JL, Antman EM, Halperin JL, Hunt SA, Nishimura R, Ornato JP, Page RL, Riegel B, Priori SG, Blanc JJ, Budaj A, Camm AJ, Dean V, Deckers JW, Despres C, Dickstein K, Lekakis J, McGregor K, Metra M, Morais J, Osterspey A, Tamargo JL, Zamorano JL. ACC/AHA/ESC 2006 guidelines for the management of patients with atrial fibrillation: full text: a report of the American College of Cardiology/American Heart Association Task Force on practice guidelines and the European Society of Cardiology Committee for Practice Guidelines. Circulation. 2006:114:e257–354.
21. Vaughan Williams EM. The relevance of cellular to clinical electrophysiology in classifying antiarrhythmic actions. J Cardiovasc Pharmacol. 1992:20(Suppl 2):S1–7.

22. Guglin M, Garcia M, Yarnoz MJ, Curtis AB. Non-antiarrhythmic medications for atrial fibrillation: from bench to clinical practice. J Interv Card Electrophysiol. 2008:22:119–28.
23. Vermes E, Tardif JC, Bourassa MG, Racine N, Levesque S, White M, Guerra PG, Ducharme A. Enalapril decreases the incidence of atrial fibrillation in patients with left ventricular dysfunction. Circulation. 2003:107:2926–31.
24. Schaff HV, Dearani JA, Daly RC, Orszulak TA, Danielson GK. Cox-Maze procedure for atrial fibrillation: Mayo Clinic experience. Semin Thorac Cardiovasc Surg. 2000:12:30–7.
25. Akhtar M, Breithardt G, Camm AJ, Coumel P, Janse MJ, Lazzara R, Myerburg RJ, Schwartz PJ, Waldo AL, Wellens HJ. CAST and beyond. Implications of the Cardiac Arrhythmia Suppression Trial. Circulation. 1990:81:1123–7.
26. Pratt CM, Camm AJ, Cooper W, Friedman PL, MacNeil DJ, Moulton KM, Pitt B, Schwartz PJ, Veltri EP, Waldo AL. Mortality in the Survival with ORal D-sotalol (SWORD) trial: why did patients die? Am J Cardiol. 1998:81:869–76.
27. Gilman JK, Jalal S, Naccarelli GV. Predicting and preventing sudden death from cardiac causes. Circulation. 1994:90:1083–92.
28. Rosen MR, Brink PR, Cohen IS, Robinson RB. Genes, stem cells and biological pacemakers. Cardiovasc Res. 2004:64:12–23.
29. Rosen MR, Brink PR, Cohen IS, Robinson RB. Cardiac pacing: from biological to electronic… to biological? Circ Arrhythm Electrophysiol. 2008:1:54–61.
30. Puck JM. Severe combined immunodeficiency: new advances in diagnosis and treatment. Immunol Res. 2007:38:64–7.
31. Bekeredjian R, Shohet RV. Cardiovascular gene therapy: angiogenesis and beyond. Am J Med Sci. 2004:327:139–48.
32. Al-Dosari MS, Gao X. Nonviral gene delivery: principle, limitations, and recent progress. AAPS J. 2009:11:671–81.
33. Porteus MH, Carroll D. Gene targeting using zinc finger nucleases. Nat Biotechnol. 2005:23:967–77.
34. Kim HJ, Lee HJ, Kim H, Cho SW, Kim JS. Targeted genome editing in human cells with zinc finger nucleases constructed via modular assembly. Genome Res. 2009:19:1279–88.
35. Cornu TI, Thibodeau-Beganny S, Guhl E, Alwin S, Eichtinger M, Joung JK, Cathomen T. DNA-binding specificity is a major determinant of the activity and toxicity of zinc-finger nucleases. Mol Ther. 2008:16:352–8.
36. Zimmett JM, Hare JM. Emerging role for bone marrow derived mesenchymal stem cells in myocardial regenerative therapy. Basic Res Cardiol. 2005:100:471–81.
37. Rosen MR. Are stem cells drugs? The regulation of stem cell research and development. Circulation. 2006:114:1992–2000.
38. Menasche P. Myoblast transplantation: feasibility, safety and efficacy. Ann Med. 2002:34:314–5.
39. Stagg MA, Coppen SR, Suzuki K, Varela-Carver A, Lee J, Brand NJ, Fukushima S, Yacoub MH, Terracciano CMN. Evaluation of frequency, type, and function of gap junctions between skeletal myoblasts overexpressing connexin43 and cardiomyocytes: relevance to cell transplantation. FASEB J. 2006:20:744–6.
40. Groh ME, Maitra B, Szekely E, Koc ON. Human mesenchymal stem cells require monocyte-mediated activation to suppress alloreactive T cells. Exp Hematol. 2005:33:928–34.
41. Di Nicola M, Carlo-Stella C, Magni M, Milanesi M, Longoni PD, Matteucci P, Grisanti S, Gianni AM. Human bone marrow stromal cells suppress T-lymphocyte proliferation induced by cellular or nonspecific mitogenic stimuli. Blood. 2002:99:3838–43.
42. Fischer KM, Cottage CT, Wu W, Din S, Gude NA, Avitabile D, Quijada P, Collins BL, Fransioli J, Sussman MA. Enhancement of myocardial regeneration through genetic engineering of cardiac progenitor cells expressing Pim-1 kinase. Circulation. 2009:120:2077–87.
43. Plotnikov AP, Shlapakova I, Szabolcs MJ, Danilo P Jr, Lorell BH, Potapova IA, Lu Z, Rosen AB, Mathias RT, Brink PR, Robinson RB, Cohen IS, Rosen MR. Xenografted adult human mesenchymal stem cells provide a platform for sustained biological pacemaker function in canine heart. Circulation. 2007:116:706–13.

44. Feinberg AW, Feigel A, Shevkoplyas SS, Sheehy S, Whitesides GM, Parker KK. Muscular thin films for building actuators and powering devices. Science. 2007:317:1366–70.
45. Kallen RG, Sheng Z, Yang Y, Chen L, Rogart RB, Barchi RL. Primary structure and expression of a sodium channel characteristic of denervated and immature rat skeletal muscle. Neuron. 1990:4:233–42.
46. Lau DH, Clausen C, Sosunov EA, Shlapakova IN, Anyukhovsky EP, Danilo P Jr, Rosen TS, Kelly CW, Duffy HS, Szabolcs MJ, Chen M, Robinson RB, Lu J, Kumari S, Cohen IS, Rosen MR. Epicardial border zone overexpression of skeletal muscle sodium channel, SkM1, normalizes activation, preserves conduction and suppresses ventricular arrhythmia: an in silico, in vivo, in vitro study. Circulation. 2009:119:19–27.
47. Protas L, Dun W, Jia Z, Lu J, Bucchi A, Kumari S, Chen M, Cohen IS, Rosen MR, Entcheva E, Robinson RB. Expression of skeletal but not cardiac Na^+ channel isoform preserves normal conduction in a depolarized cardiac syncytium. Cardiovasc Res. 2009:81:528–35.
48. Coronel R, Lau DH, Sosunov EA, Janse MJ, Danilo P, Anyukhovsky EP, Wilms-Schopman FJG, Opthof T, Shlapakova IN, Ozgen N, Prestia K, Kryukova Y, Cohen IS, Robinson RB, Rosen MR. Cardiac expression of skeletal muscle sodium channels increases longitudinal conduction velocity in the canine one week myocardial infarction. Heart Rhythm. 2010:7(8):1104–10.
49. http://www.fda.gov/downloads/BiologicsBloodVaccines/GuidanceComplianceRegulatoryInformation/Guidances/CellularandGeneTherapy/UCM164345.pdf.
50. Bucchi A, Plotnikov AN, Shlapakova I, Danilo P Jr, Kryukova Y, Qu J, Lu Z, Liu H, Pan Z, Potapova I, Knight BK, Girouard S, Cohen IS, Brink PR, Robinson RB, Rosen MR. Wild-Type and mutant HCN channels in a tandem biological-electronic cardiac pacemaker. Circulation. 2006:114:992–9.
51. Rosen MR. Of oocytes and runny noses. Circulation. 1996:94:607–9.
52. Lund LH, Ekman I. Individual rights and autonomy in clinical research. Eur J Heart Fail. 2010:12:311–2.

Part III
Regenerating Cardiac Tissues

Regenerating Blood Vessels

Tracy A. Gwyther and Marsha W. Rolle

Abstract Maintaining coronary artery blood flow and myocardial perfusion is critically important for the survival and function of cardiac muscle. Stem cells can contribute to the generation and maintenance of vascular networks that support the myocardium either directly, as a source of new vascular cells, or indirectly, by secreting soluble factors to promote angiogenesis. In addition to generating microvascular networks, stem cells are increasingly being used to create transplantable replacement arteries for coronary artery bypass grafting, the standard of care for surgical cardiac revascularization. This chapter reviews current progress in blood vessel tissue engineering, sources of stem cells for vascular graft synthesis, and opportunities and challenges associated with using stem cells to create blood vessel substitutes for clinical applications.

Keywords Blood vessels • Coronary artery bypass grafting • Media • Intima • Adventitia • Scaffold • Fibrin gels • Smooth muscle α-actin • Multipotent adult progenitor cells • Human mesenchymal stem cells • SM22α

1 Clinical Need for Blood Vessel Substitutes

In spite of advances in myocardial regeneration therapies, the current standard of care for patients who suffer a myocardial infarction is to restore blood flow to the heart by removing or bypassing the blockage that is occluding the coronary artery. Approximately 450,000 coronary artery bypass grafting (CABG) procedures are performed each year in the USA alone [1]. The standard of surgical care in CABG is to use the patient's own vessels, most commonly the internal mammary artery or saphenous vein, as the conduit for the bypass procedure [2, 3]. However, in a subset

M.W. Rolle (✉)
Department of Biomedical Engineering, Worcester Polytechnic Institute, Worcester, MA
e-mail: mrolle@wpi.edu

I.S. Cohen and G.R. Gaudette (eds.), *Regenerating the Heart*, Stem Cell Biology and Regenerative Medicine, DOI 10.1007/978-1-61779-021-8_21,
© Springer Science+Business Media, LLC 2011

of patients, such as those undergoing reoperation, sufficient autologous arterial or venous graft material may not be available. Synthetic grafts have been used widely for peripheral vascular surgery to replace large vessels, but they are unsuitable for applications such as CABG in which smaller vessels are needed (approximate inner diameter less than 6 mm), as synthetic grafts of this size occlude owing to thrombosis [4, 5].

The lack of vascular graft alternatives for patients, coupled with the growing incidence of coronary artery disease, has led to increased interest in tissue engineering as a means of creating blood vessel substitutes. The potential advantages of tissue-engineered over synthetic grafts are their ability to remodel, integrate, and grow with the patient's own tissues, which is particularly important for pediatric cardiovascular reconstruction applications.

2 Vascular Tissue Engineering – Emulating Blood Vessel Structure

Since the first published report of tissue-engineered blood vessel construction in the late 1980s [6], engineers have looked to native blood vessel structure and function as a model for vascular graft synthesis. Normal blood vessels consist of a medial layer, or *media*, composed of layers of contractile smooth muscle cells (SMCs) surrounded by collagen and elastin. A thicker, elastin-rich layer at the luminal edge of the media is the internal elastic lamina. On the luminal surface of the internal elastic lamina is a basal lamina, to which a confluent monolayer of endothelial cells attach to form a structure known as the *intima*. Intimal endothelial cells form the blood-contacting inner lining of all blood vessels, and function to maintain homeostasis, prevent thrombosis, and regulate the contractility and phenotype of medial SMCs. Finally, the outside of the blood vessel is surrounded by a connective tissue layer known as the *adventitia*. Rich in collagen and fibroblasts, the adventitia provides structural support to the vessel and imparts tensile strength. Ideally, tissue-engineered vascular grafts will possess (1) sufficient tensile strength to withstand arterial pressure while maintaining adequate compliance to maintain normal blood flow, (2) contractile function to allow vasoconstriction and dilation in response to physiological signals and (3) a stable nonthrombogenic blood-contacting luminal surface.

Vascular tissue engineering uses cells, scaffolds, or combinations of cells and scaffolds to create living blood vessel equivalents (reviewed in [7, 8]). Most strategies developed and tested to date have involved seeding adult or neonatal somatic cells harvested from blood vessels onto scaffolds made from synthetic or natural polymer materials. In a landmark study, Weinberg and Bell [6] fabricated the first tissue-engineered blood vessel from concentric layers of vascular SMCs and fibroblasts polymerized in collagen gels to form artificial medial and adventitial layers, respectively, followed by luminal seeding with vascular endothelial cells. These constructs exhibited normal cell morphology and function, but had very low strength (less than 10 mmHg burst pressure), even when reinforced with a Dacron mesh (approximately 40–70 mmHg) [6]. Regardless, this pioneering work demonstrated for the first time the possibility of constructing living blood vessel equivalents from vascular cells and scaffold materials.

Subsequently, a variety of different strategies for blood vessel tissue engineering were explored; many of which have been tested in vivo. Both synthetic polymer [9–13] and natural biopolymer [6, 7, 14–17] materials have been developed and evaluated as scaffolds for vascular tissue engineering. Synthetic polymer materials are more readily processed with a variety of manufacturing techniques to tailor the porosity, fiber architecture, and geometry of the scaffolds, and their degradation rates and material properties may be tuned by controlling the chemical composition of the polymer. For tissues such as blood vessels which bear a mechanical load, it is critically important to balance strength with degradation rates that match the rate of cellular growth and tissue remodeling. If the cells are unable to create a matrix that provides sufficient structural integrity to the tissue as it degrades, the results can be catastrophic [12]. In a study by Opitz et al. [12], segments of ovine aorta were surgically replaced by engineered tissue composed of vascular SMCs seeded onto a biodegradable scaffold. Although the tissue appeared to have successfully integrated with the host aorta by 12 weeks after transplantation, at 6 months the authors noted dramatic distension in the tissue-engineered aortic segments, which they attributed to insufficient elastic fiber generation in the newly formed vascular tissue.

In one of the earliest reports using this approach, Shinoka et al. [13] designed a biodegradable polymer graft for venous reconstructive surgery composed of a polyglycolic acid (PGA) mesh seeded with cells (myofibroblasts and endothelial cells) isolated from carotid artery or jugular vein harvested from juvenile sheep. In subsequent studies, the scaffold material was changed to a blend of ε-polycaprolactone (PCL) and polylactic acid (PLA), reinforced with PGA or PLA mesh, for improved strength and compliance, with gradual complete degradation to allow tissue remodeling by host cells [18]. Grafts created by seeding these scaffolds with autologous bone marrow cells have been used successfully as vascular conduits in pediatric patients [19, 20].

A variety of natural materials have also been explored for vascular tissue engineering, including fibrin [15, 21], collagen [6, 17], fibrin/collagen mixtures [22], hyaluronan [23], silk fibroin [24], and decellularized extracellular matrix (ECM) [25, 26]. Cells suspended in collagen and fibrin gels can be injection-molded to form tubular constructs which are then remodeled by the cells. Although these constructs lack strength after short-term static culture [6, 17, 22], extended culture times, addition of soluble factors (e.g. sodium ascorbate [27], transforming growth factor β_1, (TGF-β_1) [28], and insulin [28]), and mechanical conditioning [11, 29] have been shown to strengthen these constructs.

Alternatively, vascular grafts have been created using cells as the starting material, without the use of a preformed scaffold for cell seeding. These "scaffold-less" tissue engineering approaches include rolling cultured cell sheets [30, 31], organ printing [32, 33], and fusion of cell clusters [34]. Most notably, L'Heureux et al. [30] demonstrated that sheets of cultured cells could be harvested and rolled around a mandrel to form a multilayered vascular construct consisting of cells and cell-derived ECM. Autologous vascular grafts produced with this method exhibit tensile strength comparable to that of human saphenous veins [35], although graft fabrication and maturation requires 2–3 months [36]. However, vascular grafts created with this method have already shown clinical promise as arteriovenous fistulas [37] and it currently stands alone as the only strategy to yield a vascular graft used clinically as a small-diameter artery replacement [37, 38].

3 Cell Sources for Vascular Tissue Engineering

The source of cells used to generate tissue-engineered blood vessels is also an important design parameter. Most experimental studies have used either autologous or syngeneic cells or have implanted xenogeneic or allogeneic cells into immunocompromised animals. Vascular cells (endothelial cells, SMCs, fibroblasts) harvested from autologous blood vessels are an obvious choice and have been the most widely used cell source to date in experimental vascular tissue engineering studies. However, to obtain autologous SMCs requires a tissue biopsy from a patient's vein or artery, which causes pain and morbidity at the donor site. Furthermore, patients who are candidates for CABG with tissue-engineered vascular grafts are typically older and have more cardiovascular disease risk factors than the young, healthy donors from which cells are obtained for most experimental studies. SMCs from these patients were shown to have a limited replicative life span, and impaired ability to synthesize collagen and elastin, which worsens with increasing patient age [39]. Tissue-engineered blood vessels comprising SMCs from patients 67 and 74 years of age seeded onto biodegradable PGA scaffolds disintegrated during in vitro conditioning [39]. Interestingly, cell proliferation could be extended for several passages in these cells by overexpressing human telomerase reverse transcriptase [39]. This approach raised safety concerns owing to the use of genetic manipulation of these cells and the potential for cell cycle dysregulation and malignancy, although it was not shown to increase the likelihood of neoplastic transformation [40].

Unlike vascular SMCs, fibroblasts and endothelial cells are more easily obtained in clinically useful quantities from autologous tissues. Fibroblasts and venule endothelial cells obtained from a single dermal biopsy provide sufficient numbers of cells to create patient-matched cell-sheet-based vascular grafts, regardless of patient age, cardiovascular disease status, and other risk factors [35–37]. Multiple sources of adult endothelial cells have been explored for vascular graft seeding, including microvascular endothelial cells from autologous adipose tissue [41]. However, a patient-matched, expandable source of vascular SMCs has been elusive.

4 Sources of Stem Cells for Vascular Tissue Engineering

The challenges associated with obtaining sufficient numbers of patient-matched cells for vascular tissue engineering have led to exploration of alternative cell sources, including stem cells. The observation that circulating cells of bone marrow origin appeared to contribute to new blood vessel formation in injured tissues suggested that adult progenitor cells may be a source of vascular cells for tissue engineering applications. Over the past decade, cells from bone marrow [9, 42–47], adipose tissue [48, 49], muscle [10, 50], and, more recently, hair follicles [51] have been isolated and seeded onto scaffolds to create vascular grafts. Applications of these cells have included isolation of (1) vascular cell populations selected from stem cell precursors using genetic approaches, (2) vascular cells (SMCs or endothelial cells) generated by directed differentiation of stem and progenitor cells, and (3) freshly isolated, undifferentiated cells.

Many stem cell sources have been shown to differentiate down a SMC lineage when cultured in the proper conditions, including hair follicle, bone marrow, circulating blood, skeletal muscle, embryonic stem cells, and adipose tissue. Cells isolated from the bone marrow, including freshly purified mononuclear cells [18], multipotent adult progenitor cells [44], and adherent mesenchymal cells [9, 43], are the predominant stem cell sources that have been studied to date. Bone marrow-derived stem cells have been cultured and used to differentiate into SMCs, and have also been used fresh, without ex vivo culture, which is advantageous for clinical use. Additionally, these cells have proven to be a valuable model cell type for animal studies because they can be isolated from many species.

Most strategies for selection and differentiation rely on evidence of expression of genes that are characteristic of differentiated, contractile SMCs. The most common early marker used to identify SMCs is smooth muscle α-actin (SMA); however, it is also expressed in myofibroblasts and therefore is not specific. More stringent criteria include expression of another early contractile protein, calponin, which is restricted to SMCs, transcription factors such as SM22α, and proteins expressed at later stages of differentiation, including smooth muscle myosin heavy chain. Most studies rely on a combination of gene expression, contractile protein synthesis, and functional contraction assays as evidence of SMC differentiation from progenitor cells.

4.1 Stem Cell Differentiation into Smooth Muscle Cells

Several differentiation protocols have been utilized to direct stem cells to express SMC proteins. Most of these methods include the use of growth factors, mechanical stimulation, culture on purified ECM molecules, or combinations of these factors. The most widely used medium supplement that has been shown to stimulate SMC protein expression is TGF-β_1. TGF-β_1 increases SMC protein marker expression when added to cultures in combination with platelet-derived growth factor [44], cyclic loading [43, 46], or bone morphogenetic protein 4 [48]. In addition to soluble factors, chemical cues from the culture surface may play a role in SMC protein expression [43]. Fibronectin [44] and fibronectin/gelatin [52] have been used in combination with growth factors to culture bone marrow-derived progenitor cells to stimulate SMC differentiation. Finally, culture duration is a critical parameter in differentiating stem cells toward SMC-like cells. Certain treatments require at least 6–14 days [43, 44, 48, 53] before there are notable increases in smooth muscle protein expression.

One of the remaining challenges of applying the results of in vitro stem cell differentiation studies is that they are primarily performed on tissue culture plastic (TCP). The hope is that the same differentiation protocol that increases SMC markers on TCP will then translate to 3D vascular tissue and have the same if not an increased effect. However, this is not always the case. When mesenchymal stem cells (MSCs) were embedded in fibrin gels and cultured statically for 7 days, their expression of SMA decreased compared with culture on TCP [46]. With mechanical stimulation introduced to the culture, the expression of SMA increased significantly compared with the static controls, but that effect was diminished when it was examined next to the expression of SMA on TCP [46].

4.2 Isolation of Vascular Smooth Muscle Cells from Stem Cells by Genetic Selection

Although growth factor addition is the most common method used to generate vascular cells from stem cells, genetic manipulation and promoter sorting is another alternative. Kashiwakura et al. [54] demonstrated this concept by transfecting bone marrow-derived cells with an SM22α promoter–green fluorescent protein (GFP) reporter construct and using neomycin selection to isolate the cells expressing this SMC-specific transcription factor. This method was found to be inefficient at purifying SMCs (1–3% of adherent cells were GFP-positive after transfection) and only a fraction of the cells isolated with the SM22α promoter selection stained positively for other SMC proteins [54]. A similar approach has also been reported for isolation of contractile SMCs by promoter sorting with a SMA promoter driving enhanced GFP expression [45]. SMCs derived from stem cells from both ovine bone marrow [45] and ovine hair follicle [51] have been isolated with flow cytometry based on fluorescent expression of the SMA-GFP construct. The cells isolated with this method exhibited several characteristics of SMCs, including gene expression of SM22α, calponin, caldesmon, and smoothelin, positive immunostaining for SMA and calponin, and fibrin gel contraction [45, 51].

4.3 *Undifferentiated Stem Cells in Vascular Tissue Engineering*

Using chemical factors to differentiate stem cells toward SMCs is an appealing approach owing to the ease of use and lack of genetic manipulation; however, there has yet to be a differentiation protocol that yields a homogeneous population of differentiated SMCs. In addition, extensive ex vivo culture times may limit widespread clinical utility of tissue-engineered vascular grafts. Alternatively, several studies have suggested that the successful application of bone marrow-derived stem cells is not dependent upon their differentiation into SMCs. Vascular grafts comprising degradable scaffolds seeded with undifferentiated mononuclear cells from autologous bone marrow a few hours prior to use have been successfully implanted in pediatric patients for reconstructive vascular surgery [19, 20]. In an experimental animal model, undifferentiated human bone marrow-derived MSCs seeded on PLA sheets were rolled into tubular structures and exhibited a lower instance of thrombus formation than the same PLA scaffolds without MSCs, and it was determined that compared with SMCs, MSCs had a lower percentage of platelet adhesion [9].

5 Summary

Over the past decade, stem cells have been explored as a new source of cells for seeding tissue-engineered vascular grafts. Potential advantages of using stem cells over differentiated vascular cells such as SMCs are the relative ease of harvesting, abundance, and proliferative capacity. It has been shown that for some applications it may not be necessary to differentiate stem cells, thereby eliminating the need for extensive cell culture and graft fabrication time, and simplifying the process of vascular graft construction. Early clinical results with autologous bone marrow are promising in pediatric patients, although successful fabrication and transplantation of vascular grafts generated from adult human stem cells has yet to be achieved.

References

1. Lloyd-Jones D, Adams R, Carnethon M, De Simone G, Ferguson TB, Flegal K, et al. Heart disease and stroke statistics – 2009 update: a report from the American Heart Association Statistics Committee and Stroke Statistics Subcommittee. Circulation. 2009;119:480–6.
2. Goldman S, Zadina K, Moritz T, Ovitt T, Sethi G, Copeland JG, et al. Long-term patency of saphenous vein and left internal mammary artery grafts after coronary artery bypass surgery: results from a Department of Veterans Affairs Cooperative Study. J Am Coll Cardiol. 2004;44:2149–56.
3. Khot UN, Friedman DT, Pettersson G, Smedira NG, Li J, Ellis SG. Radial artery bypass grafts have an increased occurrence of angiographically severe stenosis and occlusion compared with left internal mammary arteries and saphenous vein grafts. Circulation. 2004;109:2086–91.

4. Kakisis JD, Liapis CD, Breuer C, Sumpio BE. Artificial blood vessel: the Holy Grail of peripheral vascular surgery. J Vasc Surg. 2005;41:349–54.
5. Nerem RM. Tissue engineering of the vascular system. Vox Sang. 2004;87 Suppl 2:158–60.
6. Weinberg CB, Bell E. A blood vessel model constructed from collagen and cultured vascular cells. Science. 1986;231:397–400.
7. Stegemann JP, Kaszuba SN, Rowe SL. Review: advances in vascular tissue engineering using protein-based biomaterials. Tissue Eng. 2007;13:2601–13.
8. Isenberg B, Williams C, Tranquillo R. Small-diameter artificial arteries engineered in vitro. Circ Res. 2006;98:25–35.
9. Hashi CK, Zhu Y, Yang GY, Young WL, Hsiao BS, Wang K, et al. Antithrombogenic property of bone marrow mesenchymal stem cells in nanofibrous vascular grafts. Proc Natl Acad Sci U S A. 2007;104:11915–20.
10. Nieponice A, Soletti L, Guan J, Deasy BM, Huard J, Wagner WR, et al. Development of a tissue-engineered vascular graft combining a biodegradable scaffold, muscle-derived stem cells and a rotational vacuum seeding technique. Biomaterials. 2008;29:825–33.
11. Niklason LE, Gao J, Abbott WM, Hirschi KK, Houser S, Marini R, et al. Functional arteries grown in vitro. Science. 1999;284:489–93.
12. Opitz F, Schenke-Layland K, Cohnert TU, Starcher B, Halbhuber KJ, Martin DP, et al. Tissue engineering of aortic tissue: dire consequence of suboptimal elastic fiber synthesis in vivo. Cardiovasc Res. 2004;63:719–30.
13. Shinoka T, Shum-Tim D, Ma P, Tanel R, Isogai N, Langer R, et al. Creation of viable pulmonary artery autografts through tissue engineering. J Thorac Cardiovasc Surg. 1998;115:536–45; discussion 45–6.
14. Grassl ED, Oegema TR, Tranquillo RT. A fibrin-based arterial media equivalent. J Biomed Mater Res A. 2003;66:550–61.
15. Swartz DD, Russell JA, Andreadis ST. Engineering of fibrin-based functional and implantable small-diameter blood vessels. Am J Physiol Heart Circ Physiol. 2005;288:H1451–60.
16. Zavan B, Vindigni V, Lepidi S, Iacopetti I, Avruscio G, Abatangelo G, et al. Neoarteries grown in vivo using a tissue-engineered hyaluronan-based scaffold. FASEB J. 2008;22:2853–61.
17. Seliktar D, Black RA, Vito RP, Nerem RM. Dynamic mechanical conditioning of collagen-gel blood vessel constructs induces remodeling in vitro. Ann Biomed Eng. 2000;28:351–62.
18. Matsumura G, Hibino N, Ikada Y, Kurosawa H, Shin'oka T. Successful application of tissue engineered vascular autografts: clinical experience. Biomaterials. 2003;24:2303–8.
19. Hibino N, McGillicuddy E, Matsumura G, Ichihara Y, Naito Y, Breuer C, et al. Late-term results of tissue-engineered vascular grafts in humans. J Thorac Cardiovasc Surg. 2010;139:431–6, 436.e1–2.
20. Shinoka T, Breuer C. Tissue-engineered blood vessels in pediatric cardiac surgery. Yale J Biol Med. 2008;81:161–6.
21. Grassl ED, Oegema TR, Tranquillo RT. Fibrin as an alternative biopolymer to type-I collagen for the fabrication of a media equivalent. J Biomed Mater Res. 2002;60:607–12.
22. Rowe SL, Stegemann JP. Interpenetrating collagen-fibrin composite matrices with varying protein contents and ratios. Biomacromolecules. 2006;7:2942–8.
23. Arrigoni C, Camozzi D, Imberti B, Mantero S, Remuzzi A. The effect of sodium ascorbate on the mechanical properties of hyaluronan-based vascular constructs. Biomaterials. 2006;27:623–30.
24. Lovett M, Cannizzaro C, Daheron L, Messmer B, Vunjak-Novakovic G, Kaplan D. Silk fibroin microtubes for blood vessel engineering. Biomaterials. 2007;28:5271–9.
25. Schaner P, Martin N, Tulenko T, Shapiro I, Tarola N, Leichter R, et al. Decellularized vein as a potential scaffold for vascular tissue engineering. J Vasc Surg. 2004;40:146–53.
26. Dahl S, Koh J, Prabhakar V, Niklason L. Decellularized native and engineered arterial scaffolds for transplantation. Cell Transplant. 2003;12:659–66.
27. Ahlfors JE, Billiar KL. Biomechanical and biochemical characteristics of a human fibroblast-produced and remodeled matrix. Biomaterials. 2007;28:2183–91.
28. Long JL, Tranquillo RT. Elastic fiber production in cardiovascular tissue-equivalents. Matrix Biol. 2003;22:339–50.

29. Isenberg BC, Tranquillo RT. Long-term cyclic distention enhances the mechanical properties of collagen-based media-equivalents. Ann Biomed Eng. 2003;31:937–49.
30. L'Heureux N, Paquet S, Labbe R, Germain L, Auger FA. A completely biological tissue-engineered human blood vessel. FASEB J. 1998;12:47–56.
31. Gauvin R, Ahsan T, Larouche D, Lévesque P, Dubé J, Auger F, et al. A novel single-step self-assembly approach for the fabrication of tissue-engineered vascular constructs. Tissue Eng Part A. 2010;16:1737–47.
32. Mironov V, Visconti R, Kasyanov V, Forgacs G, Drake C, Markwald R. Organ printing: tissue spheroids as building blocks. Biomaterials. 2009;30:2164–74.
33. Norotte C, Marga F, Niklason L, Forgacs G. Scaffold-free vascular tissue engineering using bioprinting. Biomaterials. 2009;30:5910–7.
34. Kelm J, Lorber V, Snedeker J, Schmidt D, Broggini-Tenzer A, Weisstanner M, et al. A novel concept for scaffold-free vessel tissue engineering: self-assembly of microtissue building blocks. J Biotechnol. 2010;148(1):46–55.
35. Konig G, McAllister T, Dusserre N, Garrido S, Iyican C, Marini A, et al. Mechanical properties of completely autologous human tissue engineered blood vessels compared to human saphenous vein and mammary artery. Biomaterials. 2009;30:1542–50.
36. L'Heureux N, Dusserre N, Konig G, Victor B, Keire P, Wight TN, et al. Human tissue-engineered blood vessels for adult arterial revascularization. Nat Med. 2006;12:361–5.
37. McAllister T, Maruszewski M, Garrido S, Wystrychowski W, Dusserre N, Marini A, et al. Effectiveness of haemodialysis access with an autologous tissue-engineered vascular graft: a multicentre cohort study. Lancet. 2009;373:1440–6.
38. L'Heureux N, McAllister T, de la Fuente L. Tissue-engineered blood vessel for adult arterial revascularization. N Engl J Med. 2007;357:1451–3.
39. Poh M, Boyer M, Solan A, Dahl SL, Pedrotty D, Banik SS, et al. Blood vessels engineered from human cells. Lancet. 2005;365:2122–4.
40. Klinger R, Blum J, Hearn B, Lebow B, Niklason L. Relevance and safety of telomerase for human tissue engineering. Proc Natl Acad Sci U S A. 2006;103:2500–5.
41. Hewett P. Vascular endothelial cells from human micro- and macrovessels: isolation, characterisation and culture. Methods Mol Biol. 2009;467:95–111.
42. Cho S, Kim I, Lim S, Kim D, Kang S, Kim S, et al. Smooth muscle-like tissues engineered with bone marrow stromal cells. Biomaterials. 2004;25:2979–86.
43. Gong Z, Niklason LE. Small-diameter human vessel wall engineered from bone marrow-derived mesenchymal stem cells (hMSCs). FASEB J. 2008;22:1635–48.
44. Ross JJ, Hong Z, Willenbring B, Zeng L, Isenberg B, Lee EH, et al. Cytokine-induced differentiation of multipotent adult progenitor cells into functional smooth muscle cells. J Clin Invest. 2006;116:3139–49.
45. Liu JY, Swartz DD, Peng HF, Gugino SF, Russell JA, Andreadis ST. Functional tissue-engineered blood vessels from bone marrow progenitor cells. Cardiovasc Res. 2007;75:618–28.
46. O'Cearbhaill E, Murphy M, Barry F, McHugh P, Barron V. Behavior of human mesenchymal stem cells in fibrin-based vascular tissue engineering constructs. Ann Biomed Eng. 2010;38:649–57.
47. Hibino N, Shin'oka T, Matsumura G, Ikada Y, Kurosawa H. The tissue-engineered vascular graft using bone marrow without culture. J Thorac Cardiovasc Surg. 2005;129:1064–70.
48. Wang C, Cen L, Yin S, Liu Q, Liu W, Cao Y, et al. A small diameter elastic blood vessel wall prepared under pulsatile conditions from polyglycolic acid mesh and smooth muscle cells differentiated from adipose-derived stem cells. Biomaterials. 2010;31:621–30.
49. Wang C, Yin S, Cen L, Liu Q, Liu W, Cao Y, et al. Differentiation of adipose-derived stem cells into contractile smooth muscle cells induced by transforming growth factor-beta1 and bone morphogenetic protein-4. Tissue Eng Part A. 2010;16:1201–13.
50. Soletti L, Hong Y, Guan J, Stankus J, El-Kurdi M, Wagner W, et al. A bilayered elastomeric scaffold for tissue engineering of small diameter vascular grafts. Acta Biomater. 2010;6:110–22.
51. Liu JY, Peng HF, Andreadis ST. Contractile smooth muscle cells derived from hair-follicle stem cells. Cardiovasc Res. 2008;79:24–33.

52. Roh J, Brennan M, Lopez-Soler R, Fong P, Goyal A, Dardik A, et al. Construction of an autologous tissue-engineered venous conduit from bone marrow-derived vascular cells: optimization of cell harvest and seeding techniques. J Pediatr Surg. 2007;42:198–202.
53. Gong Z, Calkins G, Cheng E, Krause D, Niklason L. Influence of culture medium on smooth muscle cell differentiation from human bone marrow-derived mesenchymal stem cells. Tissue Eng Part A. 2009;15:319–30.
54. Kashiwakura Y, Katoh Y, Tamayose K, Konishi H, Takaya N, Yuhara S, et al. Isolation of bone marrow stromal cell-derived smooth muscle cells by a human SM22alpha promoter: in vitro differentiation of putative smooth muscle progenitor cells of bone marrow. Circulation. 2003;107:2078–81.

Regenerating Heart Valves

Benedikt Weber and Simon P. Hoerstrup

Abstract Valvular heart disease is a significant cause of morbidity and mortality worldwide. Current options for surgical heart valve replacement are associated with several major disadvantages as clinically available valve prostheses represent nonviable structures and lack the potential to grow, repair, and remodel. Heart valve tissue engineering represents a promising scientific concept to overcome these limitations, aiming at the fabrication of living autologous heart valves with a thromboresistant surface and a viable interstitium with repair and remodeling capabilities. Following the in vitro tissue engineering concept, autologous cells are harvested and seeded onto three-dimensional matrices followed by biomimetic conditioning enabling the development of neo-heart valve tissue. Here, we review the concept of both in vitro and in vivo heart valve tissue engineering, focusing in particular on different synthetic scaffold materials and available cell sources for the fabrication of living autologous heart valve substitutes.

Keywords Heart valves • Tissue engineering • Scaffolds • Polymeric starter matrices • Aliphatic polyesters • Polyhydroxyalkanoates • Heart valve bioreactors • Synthetic scaffolds • Bone-marrow-derived cells • Endothelial progenitor cells • Umbilical-cord-derived cells • Chorionic-villi-derived cells • Amniotic-fluid-derived cells • Pluripotent stem cells

Abbreviations

α-SMA	α-Smooth muscle actin
BMSC	Bone-marrow-derived stem cell
ECM	Extracellular matrix

S.P. Hoerstrup (✉)
Swiss Center for Regenerative Medicine, Department of Surgical Research and Center for Clinical Research, University of Hospital of Zurich, Zurich, Switzerland
e-mail: simon_philipp.hoerstrup@usz.ch

I.S. Cohen and G.R. Gaudette (eds.), *Regenerating the Heart*, Stem Cell Biology and Regenerative Medicine, DOI 10.1007/978-1-61779-021-8_22,
© Springer Science+Business Media, LLC 2011

EPC	Endothelial progenitor cell
MSC	Marrow stromal cell
P4HB	Poly(4-hydroxybutyrate)
PGA	Polyglycolic acid
PHA	Polyhydroxyalkanoate
PLA	Polylactic acid
VEGF	Vascular endothelial growth factor
WMF	Myofibroblast derived from Wharton's jelly

1 Introduction

1.1 Valvular Heart Disease

Valvular heart disease represents a major cause of global disease load, with a high incidence in the developing world owing to rheumatic disease and an increasing number of patients in the developed world affected by degenerative valve disease [1–3]. In return, the progress in medical treatment of cardiovascular diseases has also been significant; particularly tissue substitution has shown that functional replacements of tissue and organs could be lifesaving. Also, the replacement of diseased and insufficient heart valves has markedly improved the life expectancy of patients with severe valve disease receiving optimum medical therapy [4]. Worldwide, approximately 290,000 heart valve replacements are performed annually and the number of patients requiring heart valve replacement is estimated to triple by the year 2050 [5, 6]. Despite these remarkable advances in treatment, valvular heart disease still remains a significant cause of morbidity and mortality worldwide [7].

1.2 Heart Valve Replacements

The most common treatment for end-stage valvular disease is surgical replacement of the dysfunctional valve with either a mechanical or a bioprosthetic substitute. Today's valve replacement surgery is efficacious, and currently available heart valve substitutes display excellent structural durability (reviewed by Ruel and Lachance [8]). However, the state-of-the-art valve prostheses in clinical use have substantial limitations. These include the lack of growth, repair, and remodeling capabilities once they are implanted into the body. Additionally, mechanical valve substitutes are inherently susceptible to thromboembolic events due to high shear stress, nonphysiological flow profiles, and blood damage. Lifelong anticoagulation therapy is indispensable in these patients, and results in a substantial risk of spontaneous hemorrhage and embolism (reviewed by Yoganathan et al. [9] and Dasi et al. [10]). Bioprotheses, taken either from animal origin (xenografts) or from human donors (homografts), are more prone to structural valve degeneration, and the associated need for repeat reoperations makes them less suitable for middle-aged and younger patients [11]. This is also reflected by the current clinical guidelines that recommend the use of bioprosthetic valves in patients aged 65 years and older [12]. Theoretically, cryopreserved donor heart valves are the valvular replacements closest to natural valves, with low thrombogenicity and infectious risk [13]. However, the lack of availability, reflecting worldwide organ scarcity, represents a seemingly unsolvable obstacle with regard to a widespread implementation of this valve replacement concept [14].

Overall, these considerable shortcomings indicate that current options for heart valve replacements are suboptimal for a substantial group of patients, and that the ideal cardiac valve replacement yet has to be developed [6]. The native heart valve is composed of living, dynamic tissue capable of continuous remodeling to adapt to its constantly alternating hemodynamic environment [15]. None of the currently available valvular replacements is capable of fully restoring the native valvular function owing to the lack of these adaptive capacities. Ultimately, this will also affect the integrity of physiological cardiac function. Tissue engineering of heart valves represents an emerging field of research with the potential to overcome these limitations by creating a living autologous valve replacement that prevents an immune response, clotting activation, and valvular degeneration on the one hand, and allows for growth, remodeling, and repair throughout the patient's lifetime on the other hand.

1.3 Architecture of Native Heart Valves

The ultimate rationale of tissue engineering is the fabrication of living neotissues identical or at least very similar to native human valvular structures. All four human heart valves share several microstructural similarities. However, the trileaflet aortic valve best illustrates the essential features, and serves as a paradigm for microstructural and cellular adaption to functional requirements. This semilunar heart valve facilitates the blood flow from the left ventricle to the aorta and peripheral blood vessels through a complex opening and closing mechanism in an unrelenting, physiologically demanding environment over a wide range of hemodynamic conditions [4, 16, 17]. The aortic valve leaflets are covered by a continuous layer of endothelial cells, enabling smooth blood flow during opening and closing of the valve and regulating immune and inflammatory reactions (reviewed by Simmons [18]). The load-bearing part of adult aortic valve leaflets has a layered architecture with endothelial coverings, enabling the remarkable changes in shape and dimension. The heart valve leaflet tissue itself consists of three layers: the ventricularis, the fibrosa, and the spongiosa. These individual layers of valve leaflets have different mechanical characteristics, which are mainly due to their disparities in microstructural composition. In particular, the elastic and collagen fibers have a preferential arrangement and orientation within specific valvular connective tissue layers (reviewed by Misfeld and Sievers [19]).

The ventricularis, which is closest to the inflow surface, predominantly consists of radially aligned elastin fibers. The inner spongiosa layer is largely composed of loosely arranged collagen and copious amounts of glycosaminoglycans. The fibrosa or arterial layer, which is situated at the outflow surface, contains coarse bundles of circumferential collagen fibers that form the macroscopic folds parallel to the free edge of the leaflets. These collagen fibers represent the strongest part of the valve leaflet and are mainly responsible for bearing diastolic stress. All of these collagen bundles blend into the aortic wall, thereby transferring the gross load of the leaflets to the wall of the aortic root. Thus, the fibrosa is considered to be the major load-bearing layer of the valvular leaflet, enabling a stress increase while preventing a prolapse of the leaflets [20–22]. In contrast, the elastin in the ventricularis restores the contracted configuration of the leaflets in a systolic pressure environment. During valve opening, elastin stretches by extension of collagen crimp and corrugations. When the valve is closing, the elastin becomes unfolded and the main load shifts from elastin to collagen. In addition, the hydrophilic glycosaminoglycans in the loose spongiosa layer may serve as a shock and shear absorber by readily absorbing water and decreasing the pressure difference across the valve [23–25].

Between the extracellular components reside valvular interstitial cells that can be subdivided into two major cellular phenotypes: smooth muscle cells, arranged in bundles or just as single cells [26, 27], and fibroblasts maintaining the extracellular matrix (ECM). Approximately 60% of the fibroblasts are myofibroblasts [28], cells that have phenotypic features of both fibroblasts and smooth muscle cells depending on their biological and mechanical microenvironment [29, 30].

2 Strategies in Heart Valve Tissue Engineering

Langer and Vacanti [31] defined the term "tissue engineering" as an interdisciplinary field, applying the principles and methods of engineering to the development of biological substitutes that can restore, maintain, or improve tissue formation. According to this predefinition, two principal strategies have been developed to generate living autologous heart valve replacements. One requires an in vitro phase generating the valvular substitute ex vivo [31]. This traditional tissue engineering paradigm comprises the isolation and expansion of cells from the patient, subsequent seeding onto an appropriate scaffold material, in vitro tissue formation, and finally, implantation into the patient from whom the cells were taken. This paradigm, further referred to as the in vitro tissue engineering approach, is being employed as the principal approach for heart valve tissue engineering and is aimed at full development of the tissue substitute in vitro. The second approach of in situ heart valve tissue engineering circumvents the in vitro tissue culture phase by straight implantation of natural tissue-derived heart valve matrices, aiming at potential cell ingrowth and remodeling in vivo [32] (Fig. 1).

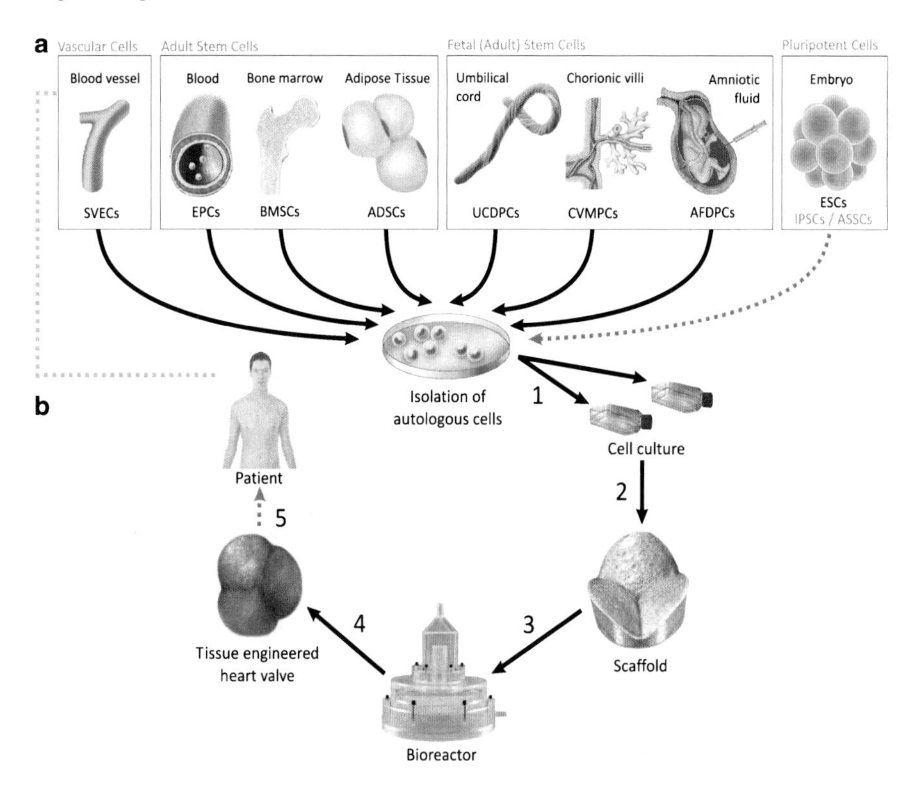

Fig. 1 (**a**) Available cell sources for heart valve tissue engineering. Standard vascular endothelial cells (*SVECs*), endothelial progenitor cells (*EPCs*), bone-marrow-derived mesenchymal stem cells (*BMSCs*), adipose-derived stem cells (*ADSCs*), umbilical-cord-derived progenitor cells (*UCDPCs*), chorionic-villi-derived mesenchymal progenitor cells (*CVMPC*s), amniotic-fluid-derived progenitor cells (*AFDPCs*), embryonic stem cells (*ESCs*), adult spermatogonial stem cells (*ASSCs*), and induced pluripotent stem cells (*IPSCs*). (**b**) Concept of in vitro heart valve tissue engineering. Autologous cells are harvested from the patient and expanded in vitro (*1*). When sufficient numbers are reached, cells are seeded onto a biodegradable scaffold (*2*). Constructs are positioned in a bioreactor (*3*) and conditioned. When tissue formation is sufficient, tissue-engineered heart valves (*4*) are ready for implantation (*5*)

3 Concept of In Vitro Heart Valve Tissue Engineering

3.1 Principal Strategy

According to the approach of in vitro tissue engineering, the successful fabrication of autologous living heart valve replacements similar to the native prototype is supported by three main elements: First, autologous cells that resemble their native counterparts in phenotype and functionality are isolated and expanded using standard cell culture techniques. Second, these cells are seeded onto a temporary biodegradable supporter matrix fabricated in the shape of a trileaflet heart valve, termed the "*scaffold*," which promotes tissue strength until the ECM produced guarantees functionality on its own. Third, to promote tissue formation and maturation, the seeded scaffolds are exposed to stimulation transmitted via a culture medium (biological stimuli) or via "conditioning" of the tissue in a bioreactor (mechanical stimuli). This aims at adequate cellular differentiation, proliferation, and ECM production to form a living tissue model, called the "*construct*." This construct is subsequently implanted orthotopically as a valve replacement, and further in vivo remodeling is intended to recapitulate physiological valvular architecture and function [15, 33–36].

The attempt to develop a scaffold for heart valve tissue engineering has proceeded along two fronts: a biological matrix material and a fully synthetic scaffold [37]. Regardless of the material of the scaffold matrix, the design of a scaffold capable of supporting cellular growth and of withstanding the unrelenting cardiovascular environment, while forming a tight seal during closure, is critical to the success of the tissue-engineered construct. In addition to meeting all the standard design criteria of traditional tissue valves, in which durability and biocompatibility are effectively passive attributes of the underlying materials, and selecting the best scaffold material, it requires consideration of the active behavior of the cells in the regulation of tissue growth, remodeling, and homeostasis to fully lay the foundation for clinical application. Taken as a whole, the major goal is the in vitro creation of a living autologous tissue-engineered heart valve with structural differentiation, anatomically appropriate and high-quality ECM, viable valvular interstitial cells available to respond to varying physiological needs and to repair structural injury by remodeling ECM, and the capacity to grow with the patient (reviewed by Brody and Pandit [38] and Sacks et al. [39]). In the following sections, we provide a comprehensive overview of the two key bioengineering aspects of in vitro heart valve tissue engineering: the development of an optimal biomimetic and biodegradable scaffold material, and the use of bioreactors targeting the guidance of valvular tissue formation in vitro.

3.2 Scaffold Materials

In the in vitro heart valve tissue engineering approach, isolated and expanded cells are seeded onto appropriate scaffolds as starter matrices for the valvular fabrication process. The matrices must be able to support cell growth and cell-to-cell interaction guiding tissue formation into a functional organ with organotypic ECM. The surfaces of these starter vehicles must be biocompatible, allowing cellular ingrowth and the formation of antithrombogenic cell linings, and biodegradable, providing an optimized degradation rate for cellular expansion (reviewed by Brody and Pandit [38]). These specific requirements entailed the development of various approaches to identify the optimal scaffold material, including the creation of synthetic [34–36] and biological [40] scaffold materials. These can be further subdivided into native tissue-derived ECM scaffolds [41], polymeric scaffolds [42–46], biological/

polymeric hybrid scaffolds [47, 48], and collagen or fibrin gel scaffolds [49–52]. Although significant advances have been made in all these approaches, the polymeric scaffolds have to date received most attention as to in vitro heart valve tissue engineering.

3.2.1 Decellularized Tissue-Derived Starter Matrices

Donor heart valves (homografts) or animal-derived heart valves (xenografts) are among the most obvious choices for scaffold materials. They are fixed and depleted of cellular antigens, which makes them less immunogenic, and/or thus eligible to be used as a scaffold material in tissue engineering. The removal of cellular components results in a template composed of ECM proteins that serve as an intrinsic medium for subsequent cell attachment. Nevertheless, they still possess a nativelike geometry and architecture with biomechanical and hemodynamic properties similar to those of their native counterpart (reviewed by Brody and Pandit [38], Schmidt et al. [34–36], and Mol et al. [53]).

Various decellularization techniques have been extensively investigated to minimize the residual immunologic potential of biological matrices. Although removal of all cellular components is indispensible, the decellularization treatment should avoid any harm or alteration of the ECM properties. This preservation of matrix integrity, as well as the efficiency of cell removal, is highly dependent on the method used for decellularization [54]. Several different decellularization methods for heart valve scaffold fabrication have been reported, including trypsin/EDTA [54–56], freeze drying [57], osmotic gradients [58], nonenzymatic detergent treatment [54, 59–62], and multistep enzymatic procedures [63]. The use of nonenzymatic detergent-based techniques has been shown to result in much more efficient cell removal, while preserving the overall matrix integrity of the scaffold, when compared with other, more aggressive acellularization methods such as trypsin/EDTA [54, 62, 64, 65]. To avoid this impairment of the matrix integrity and function due to tissue-derived protease activation, the use of suitable protease inhibitors has been recommended [61]. Accessorily, nuclease digestion steps should be embedded into the decellularization procedure to remove any residual RNA or DNA within the scaffold matrix [59–61].

After seeding of acellular matrix scaffolds with endothelial cells, and subsequent in vitro culturing, the constructs revealed a confluent endothelial cell lining [57, 59, 62, 66–68]. In 2003, Schenke-Layland et al. [68] even demonstrated that complete decellularization of porcine pulmonary heart valves with preservation of ECM structure and subsequent reseeding is feasible. Also, in vivo implantation of decellularized valves repopulated with autologous endothelial cells and myofibroblasts as orthotopic pulmonary valve substitutes in a sheep model showed promising results [56]. In addition, Kim et al. [69] demonstrated partial recellularization and endothelialization of decellularized scaffolds repopulated with bone-marrow-derived cells 3 weeks after implantation as aortic valve replacements in dogs.

Analysis of explanted valve substitutes also revealed that originally seeded cells were still present on the engineered valve constructs, implying that seeded cells contribute to the regenerative process of implanted tissue.

In spite of this proof of principle and great strides toward a clinically applicable xenogenic scaffold, decellularized biological matrices, as an obvious source of scaffold materials for heart valve replacement therapy, have mainly been used in the in situ tissue engineering approach. To some extent this might be due to their major shortcomings, especially when compared with synthetic materials. When using xenografts, the risk of zoonoses, in terms of human diseases caused by animal-derived infectious agents, is an important aspect that has to be considered [70, 71]. In particular, the identification of porcine endogenous retroviruses [72–79] and prionic diseases [80, 81] has given rise to widespread concern. Even if deemed to be minimal [82], the hypothetical risk of infection in the course of xenotransplantation constitutes a substantial limitation of a possible large-scale clinical application of the concept [83]. When the matrix material is from homogenic in origin, the limited availability of donor valves and the associated ethical concerns represent a considerable shortcoming. Moreover, the lack of evidence of growth and remodeling capacities of valve replacements when using biological scaffolds seems to be a further drawback, as this represents an indispensable prerequisite for a future clinical implementation into the pediatric field [84, 85]. These drawbacks and uncertainties raise a common concern associated with the use of decellularized starter matrices, also fuelling the search for synthetic scaffold alternatives.

3.2.2 Polymeric Starter Matrices

The use of polymeric scaffold materials for heart valve tissue engineering has already been broadly demonstrated. The ideal scaffold matrix has to be at least 90% porous [86], and comprises an interconnected pore network, as this is essential for cell growth, nutrient supply, and removal of metabolic waste products. Besides being biodegradable, biocompatible, and reproducible, the scaffold material should also display a cell-favorable surface chemistry and match the biomechanical properties of the native heart valve tissue [34–36]. In addition, the rate of matrix degradation should be controllable and commensurate with the rate of novel tissue formation to provide sufficient mechanical stability of the construct over time [15, 87, 88].

Several synthetic biodegradable polymers have been investigated as potential starter matrices for heart valve tissue engineering. Polyglactin, polyglycolic acid (PGA), polylactic acid (PLA), polyhydroxyalkanoates (PHAs), poly(vinyl alcohol), polycaprolactone, poly(L-lactic acid), poly(4-hydroxybutyrate) (P4HB), and a copolymer of PGA and PLA [poly(glycolic)-*co*-lactic acid] were the materials of choice (reviewed by Brody and Pandit [38]). These are all biodegradable scaffold materials that vary in their manufacturing possibilities and degradation rates.

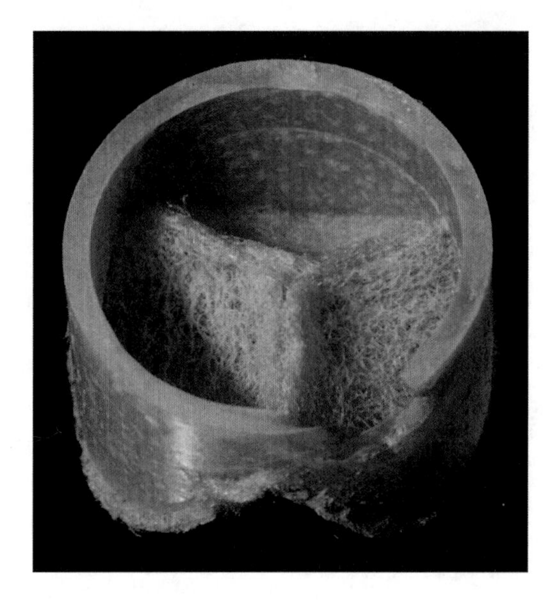

Fig. 2 Stented heart valve based on a nonwoven polyglycolic acid mesh coated with 1% poly (4-hydroxybutyrate) (from Mol et al. [116, 156])

Aliphatic Polyesters

Polyglactin, PGA, and PLA are part of the aliphatic polyester family, and degrade by cleavage of the polymer chains owing to hydrolysis of their ester bonds. The resulting monomer is either excreted via urinal secretion or enters the tricarboxylic acid cycle [86]. To fabricate single heart valve leaflets, the creation of scaffolds was initially based on combinations of aliphatic polyesters, including polyglactin nonwoven PGA meshes with layers of poly(glycolic-*co*-lactic acid) and nonwoven PGA meshes. The major limitations of aliphatic polyesters are their thickness, initial stiffness, and nonpliability, making the fabrication of trileaflet heart valves a difficult process (Fig. 2).

Polyhydroxyalkanoates

The PHA family is composed of polyesters built up from hydroxyacids that are produced as intracellular granules by various bacteria [89]. Polyhydroxyoctanoates as well as P4HB have been used to create trileaflet heart valve conduits [90–93]. These materials possess thermoplastic properties and can be molded into any desired shape using stereolithography [94, 95]. The general drawback of PHAs is their slow degradation. Combinations of aliphatic polyesters and PHAs have also been tested as alternative composite materials [96–98]. Particularly, the use of PGA coated with P4HB, combining the thermoplastic properties of P4HB and the high porosity of PGA, for the fabrication of complete trileaflet heart valves revealed promising results in a rapidly growing sheep model [34, 96, 97, 99, 100].

3.2.3 Biological/Polymeric Hybrid Starter Matrices

In recent years a further strategy for the fabrication of scaffold materials in tissue engineering emerged using biological/polymeric composite materials as starter matrices [101]. These hybrid scaffolds can also constitute complex three-dimensional structures such as heart valves, e.g., fabricated from decellularized porcine aortic valves and dip-coated with a bioresorbable polymer [47, 48, 102]. Furthermore, scaffolds equipped with molecular cues mimicking certain functional or structural aspects of extracellular microenvironments have been developed in recent years [103, 104]. In particular, gene delivery from the scaffold represents a versatile approach to manipulate the local environment for directing cell function and to promote tissue formation [105].

3.2.4 Collagen-Based Scaffolds

In the effort to provide heart valve replacements consisting solely of autologous tissue, biodegradability turned out to be an indispensable prerequisite for scaffold materials. Collagen is one of these biological materials that display bioresorbable properties and can theoretically be obtained from the patient. Thus, it has been investigated as a possible scaffold material for heart valve tissue engineering by using collagen (type I) foams [106, 107], gels or sheets [49, 87, 108–112], sponges [113], and even as fiber-based scaffolds [114, 115]. However, besides a low degradation rate, collagen has the major disadvantage of very low availability in humans as it is difficult to obtain from the patient. Therefore, most collagen scaffolds are based on animal-derived collagens, which potentially involve common shortcomings of animal-derived tissues such as the risk of transmission of zoonoses.

3.2.5 Fibrin-Based Scaffolds

Fibrin is a further biological material that also possesses controllable bioresorbable properties. Besides its potential to serve as an autologous biodegradable scaffold material, fibrin gel can also be used as a cell carrier to seed cells into a fiber-based [116] or porous synthetic [117] scaffold. As the patient's blood serves as a source for fibrin gel production, no toxic degradation or inflammatory response is expected upon reimplantation [13]. With use of injection-molding of the cell–gel mixture followed by enzymatic polymerization of fibrinogen, three-dimensional structures can be obtained. Degradation is controlled by adding aprotinin, a proteinase inhibitor that slows down or even stops fibrinolysis [118–120]. Immobilization of growth factors in specific areas has also been shown as a possible option to control degradation of fibrin scaffolds [121].

Besides these seemingly favorable features, the use of fibrin as a scaffold material has several disadvantages, including its tendency to shrink, its poor initial mechanical properties, and importantly its reduced diffusion and washout capacity into the surrounding medium compared with other porous matrices [118, 120]. Although measures have been developed to limit these shortcomings, such as poly(L-lysine) chemical fixation, preventing shrinkage and improving mechanical properties, further investigations are needed to determine its effective value as a scaffold design for heart valve tissue engineering.

3.2.6 From Animal to Human

In 1995, the first successful replacement of a single pulmonary valve leaflet with an in vitro tissue-engineered substitute, based on a synthetic scaffold, was demonstrated in a lamb model [122, 123]. In the following years, complete trileaflet valve replacements based on synthetic scaffolds were introduced [91, 98, 124, 125]. After the development of mechanical conditioning as a substantial improvement of the in vitro culture phase, the fabricated valves showed adequate functionality and remodeling into nativelike heart valves up to 5 months in animal studies [96]. A 2-year follow-up study of pulmonary artery substitutes in a lamb model even revealed clear evidence for growth of implanted tissue-engineered structures [126]. However, the actual proof of a growth capacity of valvular structures fabricated on the basis of synthetic scaffolds is still missing. In the first clinical experiences, synthetic-scaffold-based tissue-engineered patches were successfully used for pulmonary artery reconstruction in humans [127, 128]. Besides these auspicious in vivo results, biodegradable synthetic polymers offer a number of advantages over biological scaffold materials, including greater control over mechanical properties and fabrication, better reproducibility and degradation rates, advanced geometrical design specifications, and lower immunogenicity. Even though several challenges, such as improvement of the elastic network formation and matrix degradation control, remain to be addressed by further investigations prior to a definite clinical translation of the concept, synthetic-scaffold-based heart valve replacements hold great promise for the future.

3.3 Heart Valve Tissue Formation In Vitro

A major challenge in tissue engineering of functional valvular substitutes is to mimic the biomechanical properties and dominant tissue structures of native heart valves to provide valve replacements that hold promise for long-term graft functionality after implantation. In native tissues, these biomechanical properties are predominantly based on the composition, structural organization, and quality of the ECM. In an attempt to create adequate ECM production, cellular differentiation, and proliferation, the seeded scaffolds are exposed to stimulation transmitted via a culture medium (biochemical stimuli) or via "mechanical conditioning" of the tissue (mechanical stimuli) prior to in vivo implantation. This goes along with strict isolation of the in vitro system ensuring long-term maintenance of sterile culture conditions [96, 129]. Besides ECM formation, the collagen cross-link concentration has also been shown to play a critical role in the biomechanical behavior of heart valves [130].

Biochemical conditioning, in terms of using biological stimuli such as growth factors to promote tissue growth and regeneration during in vitro development, represents a common conditioning approach in tissue engineering. In heart valve tissue engineering, different growth factors, such as hepatocyte growth factor [131], transforming growth factor β_1 [50, 132], basic fibroblast growth factor [133], epidermal growth factor [56], vascular endothelial growth factor (VEGF) [133], and platelet-derived growth factor [132], have been employed to render preferred cellular differentiation within the engineered constructs. However, precaution is recommended when considering a possible clinical implementation of this concept, as several pathways of the growth factors used have not been fully elucidated yet.

Mechanical stimulation during the in vitro culture phase has also been widely utilized in heart valve tissue engineering for improving tissue architecture and biomechanical properties [134–137]. Various mechanical conditioning approaches have been investigated with regard to their influence on ECM development and biomechanical tissue properties using model systems of fast-degrading scaffolds seeded with human adult vascular-derived cells. As a considerable number of these vascular-derived cells exhibit a myofibroblast-like phenotype, they generate a long-lasting tensile activity in terms of continuous isometric tension on their environment [138, 139]. Allowing these myofibroblasts to exert their tensile activity by application of a static mechanical constraint has been proven to be an asset for tissue component alignment during tissue organization [140]. In particular, the beneficial effects of continuous long-term dynamic straining [141, 142] and intermittent dynamic straining [143] have been shown to further optimize tissue formation in various ways. In return, large straining magnitudes displayed a negative effect on the mechanical properties of tissues [141]. In addition, cyclic flexure, stretch, and flow have been demonstrated to exhibit both independent and coupled stimulatory effects on cell and ECM development in tissue-engineered heart valves [141, 144–146]. By utilizing this system, Engelmayr et al. [147, 148] demonstrated the profound role of the coupled effects of two physiological mechanical stimuli – cyclic flexure and laminar flow – that are highly relevant to heart valve tissue formation in vitro using a novel flex–stretch–flow bioreactor system (reviewed by Sacks et al. [39]). Recently, Balguid et al. [149] presented another highly promising conditioning approach using hypoxia. By culturing tissue-engineered heart valve constructs under hypoxic culture conditions (7% O_2), they were able to generate valvular substitutes displaying near-native mechanical properties [149] (Fig. 3).

3.4 Heart Valve Bioreactors

A bioreactor is a biomimetic system that exposes the developing tissue to mechanical conditioning, primarily through cyclic flow and pressure changes (reviewed by Breuer et al. [37], Mertsching and Hansmann [150], Mol et al. [53], Ruel and

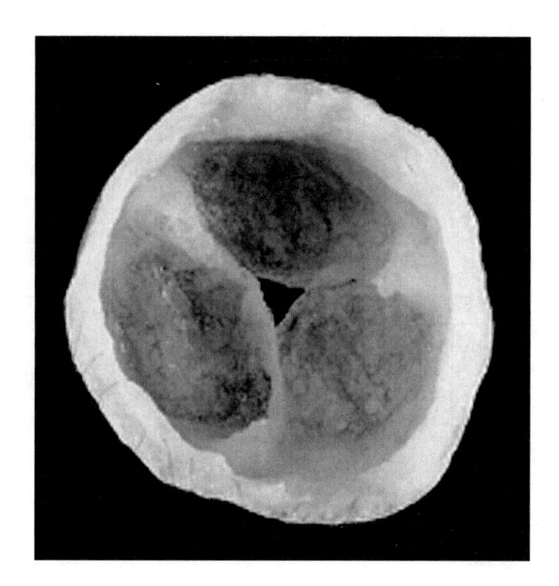

Fig. 3 Autologous tissue-engineered heart valve before implantation in an animal model (from Hoerstrup et al. [96])

Lachance [8], and Sacks et al. [39]). By mimicking the physiological environment of native heart valves in vitro, the bioreactor enhances neotissue development, improves cell attachment, changes cellular orientation, and increases production of ECM proteins, such as collagen and elastin [151]. Importantly, bioreactors can also be used to modulate the biomechanical properties of these neotissues, which is of particular importance in the development of tissue-engineered heart valves [129, 135, 136, 144, 146, 152, 153]. Different bioreactors have been developed using flow [95, 129, 153–155], strain [156–158], and cyclic stretching [51, 159] as their main mechanical cues to engineer living blood vessels and heart valves in vitro. For the engineering of heart valves, a diastolic pulse duplicator system has been developed to mimic only the diastolic phase of the cardiac cycle – the phase in which the leaflet tissue is primarily exposed to cyclic straining [116, 156]. This pulsatile conditioning approach provides physiological pressure and flow to the developing valvular structure and promotes both the development of mechanical strength and the modulation of cellular function [129, 152, 153]. In particular, the application of a pressure difference across the growing leaflets inducing local dynamic strains is of critical importance. Overall, this conditioning strategy, which solely mimics the straining phase of the cardiac cycle, offers a highly innovative bioreactor design that has been used by several groups for heart valve in vitro conditioning [34–36, 99, 129, 160–162]. By rendering human tissue-engineered heart valves that may even sustain the systemic circulation, the diastolic pulse duplicator regime represents a major step toward a tissue-engineered aortic human heart valve replacement [100]. Recent investigations have also pursued the idea of monitoring and simultaneously controlling the applied deformations during tissue culture in terms of a feedback control loop design [39, 163]. In consideration of this

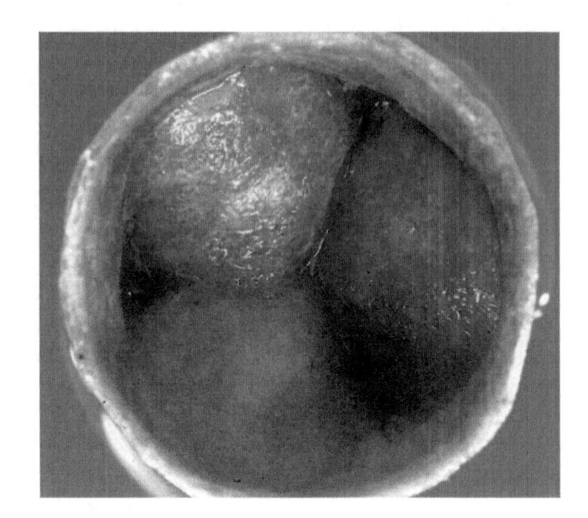

Fig. 4 BMSC-based tissue-engineered heart valve after 14 days in the pulse duplicator in vitro system (bioreactor) (from Hoerstrup et al. [99])

attempt, Kortsmit et al. [164, 165] demonstrated the value of volumetric deformation measurement and the inverse experimental–numerical estimation method as to the realization of a real-time quality control system next to a controllable culture environment. On the whole, the optimal conditioning protocol depends on various parameters, such as the sensitivity of the cell phenotype to mechanical cues, the scaffold design used, the transfer of mechanical cues from the scaffold to the cells, as well as the type and degree of mechanical exposure (Fig. 4).

4 Concept of In Vivo Heart Valve Tissue Engineering

4.1 Principal Strategy

The in vivo tissue engineering approach, also referred to as "guided tissue regeneration," relies on the natural regenerative potential of the body. Its principal strategy aims at the implantation of a nonthrombogenic, nonhemolytic, and nonimmunogenic functional scaffold that is capable of autologous host cell reseeding and subsequent remodeling in vivo. It is sometimes deemed to have the potential to yield a more clinically attractive alternative in comparison with the in vitro approach considering a possible off-the-shelf availability. By circumventing the extensive ex vivo processing of valve substitutes of the in vitro tissue engineering concept, it would certainly represent a quick, cheap, and on-demand approach. In particular, the absence of the complex preseeding procedure minimizes and simplifies the fabrication process, and after implantation this approach holds potential to reduce immunogenic reactions (reviewed by Vesely [166], Sievers [167], and Schleicher et al. [168]).

4.2 *Decellularized Scaffold Materials*

Various studies have focused on decellularized xenografts or homografts as these are likely highly attractive scaffold materials, in particular for guided tissue regeneration of heart valves. Several groups have demonstrated the feasibility of in situ repopulation of decellularized scaffolds in animal models [55, 169–177]. However, the success of these attempts was mainly linked to the actual mode of decellularization and if either xenografts or homografts were used [64]. When using SynerGraft™ as the decellularization method, leaflet explants showed up to 80% cellular repopulation and a distribution of α-smooth muscle actin (α-SMA)-positive cells comparable to that of natural valves [170, 171, 173]. Conversely, when trypsin/EDTA digestion was used for acellularization, allogenic ovine valve conduits did not reveal any interstitial valve tissue reconstitution, owing to severe calcification. Only xenogenic porcine valves implanted into the pulmonary position in sheep showed histologic reconstitution of valve tissue [55, 176]. A comparison study of different decellularization approaches indicated that a novel detergent-based technology, using Triton X-100 combined with sodium deoxycholate, yields the most favorable scaffold for cellular repopulation [64]. Consequently, several in vivo investigations further indicated that besides cellular repopulation also adaptive matrix remodeling and even growth can be found in implanted acellularized valve constructs [169, 172].

However, the transferability of these promising results from animal models to the human organism is still disputed, even though several groups have reported promising results in adult valve recipients [173, 178, 179] – particularly, when considering the clinical results of Simon et al. [180] that revealed early failure of porcine SynerGraft™-treated heart valves in children. In their clinical trial, implanted heart valve substitutes elicited a strong inflammatory response, followed by structural failure and rapid degeneration of the graft within a few weeks to months [180]. These results, clearly demonstrating the ineffectiveness of the SynerGraft™ technology, also suggest that decellularization of xenogenic heart valves, according to this technology, should be avoided. Subsequent investigations focusing on the molecular mechanism of this residual immunogenicity of decellularized grafts indicated that decellularization of porcine xenografts, even when fully depleted of cells and cellular components, did not eliminate thrombogenicity and inflammatory stimulation response. In spite of missing attraction of lymphocytes and monocytes in completely decellularized tissue, the recruitment of granulocytes was still evident [181, 182]. Bastian et al. [183] suggested that this inflammatory response is predominantly due to the loss of specific ECM proteins during the decellularization process, which usually prevents these excessive immunologic phenomena. Several investigations striving for a substantial decrease of the susceptibility of decellularized matrices to an immunologic response, i.e., by using polymer impregnation, offered several improvements, also with regard to a possible future clinical establishment [102, 184–187]. However, some of them may have exceeded the scope of in vivo tissue engineering, i.e., by using ex vivo endothelialization of matrices.

By contrast, the use of decellularized homografts seems to be beneficial over the use of xenogenic decellularized scaffold materials owing to their lower thrombogenicity and reduced risk of infection [188, 189]. Even the clinical use of SynerGraft™-treated homografts as pulmonary valve substitutes in adults showed astonishingly good results [170, 171, 190, 191]. Yet, actual clinically manifest benefits of decellularized substitutes when compared with conventional homografts, and thus conclusive reasons for preferable clinical use, are missing [192, 193]. Altogether, the clinical applicability of decellularized xenogenic as well as homogenic heart valve replacements with spontaneous cellular repopulation in vivo remains a controversial field within tissue engineering.

4.3 Synthetic Scaffold Materials

The use of synthetic scaffold materials constitutes another approach to in vivo heart valve tissue engineering, having several advantages when compared with decellularized scaffolds. Besides their diminished immunogenic potential, they also allow for a better definition of expectant chemical and physical properties. Although the implantation of a plain scaffold was unsuccessful, rendering thrombus formation and inappropriate cellular repopulation [194], scaffolds compounded with a collagen microsponge matrix showed promising results in several animal studies [195–198].

A further concept, also utilizing synthetic matrix materials, is the seeding of freshly isolated autologous cells onto scaffold materials prior to implantation. Several different cell sources and surgical procedures have been investigated, using this principle of same-day harvesting of desired cell types via a single biopsy and subsequent reimplantation after seeding onto a synthetic scaffold [127, 199–203]. However, it seems as if this is an approach that is again located at the interface between in vitro and in vivo heart valve tissue engineering considering the use of ex vivo cell seeding technologies.

4.4 The "Biovalve" Concept

First implemented by De Visscher et al. [204], a novel subtle way of repopulating scaffold materials for heart valve replacement surgery is the use of the peritoneal cavity as a "natural bioreactor." With use of this approach, scaffolds are preimplanted into the abdominal cavity, thereby inducing an intraperitoneal foreign body reaction and an in situ cellularization that is driven by a host-mediated immune response [204, 205]. Also a subcutaneous implantation of scaffold prototypes into rabbits showed respectable results [206, 207].

At any rate, the concept of in situ autologous tissue engineering with the patient as an autologous "bioreactor" is doubtlessly appealing. However, it remains exceedingly questionable whether this approach renders correct cell types and if these cells, which are primarily due to an encapsulation reaction, are capable of withstanding the unrelenting hemodynamic forces of a native heart valve position [167, 208].

4.5 Significance of the In Vivo Concept

The guided tissue regeneration approach in heart valve tissue engineering seems to be highly seductive with regard to avoiding the entire in vitro conditioning phase. It may have also shown some success in animal studies [209]; however, there is currently no striking scientific evidence that heart valve scaffolds will be sufficiently recellularized when implanted into humans [166]. In return, several authors described this concept as "highly speculative" or "highly unlikely" [210]. Even though endothelial monolayers have been found to develop on vascular grafts implanted in animals, studies have also proven that vascular grafts implanted in humans do not spontaneously form an endothelial lining [180]. As the endothelial lining is essential for providing a thromboresistant surface and a precondition for the formation of a proper interstitial cell compartment, the in situ approach remains highly questionable and will have to justify its position as a feasible approach for heart valve tissue engineering in the future.

5 Cell Sources for Heart Valve Tissue Engineering

In tissue engineering of implantable cardiovascular structures, such as heart valves, the in vitro formation of a strong, well-structured, and viable tissue is indispensable for successful in vivo function.

Several technologies have been introduced to meet these demands of designing a heart valve replacement with optimum functional durability in vivo. Considering the progress in the cardiovascular tissue engineering field, these related interdisciplinary technologies have already reached an advanced state. Synthetic scaffolds with adjustable biodegradation properties are available and can be fabricated into three-dimensional biomimetic matrices [38, 94, 100, 211]. The advancements in the field of in vitro cell culturing allow for optimal control of tissue growth conditions. However, the cell source used for engineering these cardiovascular structures is the least controlled factor, yet being the most important for the quality of the viable part of the replacement. The quality of these cells depends on the individual tissue characteristics of their origin and thus varies between individuals [212, 213]. Therefore, the choice of the right cell source constitutes the key to the long-term success of heart valve regeneration [15].

Besides mechanical stability, regenerative potential, expansion capacity, and substantial cell growth, the feasibility to develop a cell phenotype that matches the native counterpart is deemed to be an important issue. Native heart valves, for example, have a specific three-layered composition of cellular elements. The imitation of this complex three-dimensional microstructure of native cell phenotypes is expected to have a major impact on the long-term functionality of cardiovascular replacements in general [108]. Accessorily, the capability of rapid ECM development is of critical importance to create functional replacements with biomechanical properties similar to those of their native counterparts [212, 214]. Consequently, the choice of cells, which are responsible for production of ECM, is considered a key factor. In addition to fulfilling these roles of native valvular cell elements, the cells seeded onto scaffolds must also be available in sufficient quantities, free of contamination, and free of pathogens of any type [38].

In most heart valve tissue engineering approaches, cells were harvested from donor tissues, e.g., from peripheral arteries, and mixed vascular cell populations were obtained for seeding [33, 215]. From these, two cell lines can be isolated: endothelial cells forming a confluent endothelial cell lining with antithrombogenic features, and myofibroblast/fibroblast-like cells responsible for the ECM development [123, 213, 216]. With regard to clinical applications, several human cell sources have been investigated throughout the last few years, including cardiovascular-derived, bone-marrow-derived, and blood-derived cells (reviewed by Schmidt and Hoerstrup [217] and Ugurlucan et al. [218]). However, in the new concept of pediatric cardiovascular tissue engineering that primarily aims at the fabrication of living, autologous, growing replacements for the repair of congenital malformations, cells are ideally obtained during pregnancy to have the tissue-engineered replacements, such as heart valves, available at or shortly after birth. Several recent studies have addressed this issue of finding the appropriate prenatal cell source for this novel approach. In particular, cells derived from umbilical cord, amniotic fluid, and chorionic villi are among the candidates with most potential being investigated.

5.1 Cardiovascular-Derived Cells

In most cardiovascular tissue engineering approaches cells are harvested from vascular donor tissues. Many groups started their work in heart valve tissue engineering using animal-derived cells as they are easy to obtain and directly implementable in animal studies [96, 98, 123, 213, 219]. A common difficulty with tissue-engineered heart valves fabricated from animal-derived cells was the existing differences compared with their human equivalents. In particular, the disparities concerning the cell size and responsiveness to mechanical conditioning were conspicuous [53].

With regard to future clinical application, several human vascular cell sources, including mammary artery, radial artery, and saphenous veins, have also been investigated [23, 94, 100, 211, 212, 217]. Pure endothelial and myofibroblast cell lines are obtained from mixed vascular cell populations, e.g., by labeling the cells with low-density lipoprotein markers, and subsequent fluorescence-activated cell sorting. Alternatively, biopsies are digested with collagenase to detach endothelial cells from the luminal layer. The rest of the tissue is minced into small pieces, attached to culture dishes, and cultivated until primary myofibroblasts grow out [123, 216, 220].

Interestingly, the comparison of cells derived from saphenous veins with those of aortic origin revealed significant disparities concerning collagen formation and mechanical properties. Cells derived from saphenous veins showed increased proliferation in monolayer culture and increased ECM formation when cultivated on three-dimensional structures compared with artery-derived myofibroblasts [212]. These results suggest that saphenous vein cells, which can be easily obtained by minor surgical interventions, represent an attractive cell source for cardiovascular tissue engineering, especially when considering these beneficial properties. However, the importance of using venous in place of arterial myofibroblasts for tissue engineering has not been fully clarified [221], most notably because several groups reported the successful generation of heart valves, pulmonary conduits, or patch material by using venous as well as arterial cell lines [56, 91, 96, 98, 128, 222]. In a recent investigation, Schaefermeier et al. [94, 211] compared the morphology and marker gene expression of endothelial cells derived from the pulmonary heart valve with those of cells harvested from the saphenous vein and umbilical cord with regard to their eligibility for heart valve tissue engineering. Aside from marginal differences concerning the collagen and α-SMA content, no significant disparities were found between different cell origins, implying that endothelial cells derived from umbilical cord and saphenous veins have similar applicability for tissue engineering of valvular substitutes [94, 211]. Another interesting aspect was addressed by Rogers et al. [132] based on the fact that the phenotype of vascular smooth muscle cells, a multiply used cell source in tissue engineering, differs from the native valvular interstitial cell phenotype in the constitution of ECM components. By using a specific growth factor combination, of transforming growth factor β_1 with epidermal growth factor and platelet-derived growth factor, they rendered vascular smooth muscle cells that produced ECM constituents similar to those found in native aortic valves [132]. In contrast to the vascular cell origin, the age of the cell donor of vascular cell populations did not reveal any significant influence on the ECM protein production or mechanical properties of the tissue-engineered constructs [34–36]. Therefore, heart valve tissue engineering using vascular-derived cells would also be a practical option for elderly patient populations.

Overall, the utilization of vascular-derived cells is also associated with certain fundamental shortcomings. Cell harvesting before seeding necessitates the sacrifice of intact vascular structures of the donor organism, which represents a considerable limitation – particularly for pediatric applications. In addition, cardiovascular risk factors and comorbidities such as atherosclerosis and diabetes mellitus cause endothelial dysfunction (reviewed by De Vriese et al. [223] and Bonetti et al. [224]) and thus may influence the consistency of endothelial and subendothelial cell layers. To avoid the sacrifice of healthy human vascular tissue and additional surgical procedures, other potential cell sources have been investigated.

5.2 Bone-Marrow-Derived Cells

With regard to future clinical application of the cardiovascular tissue engineering concept, bone-marrow-derived stem cells (BMSCs) represent an attractive alternative heterogeneous cell source [99, 225, 226]. Aside from continuous progenitor cells of different hematopoietic lineages, the bone marrow is also a source of cells that meet criteria for stem cells of nonhematopoietic tissue. These multipotent stemlike cells are referred to either as "mesenchymal stem cells," because of their ability to differentiate into cells that can roughly be defined as mesenchymal, or as "marrow stromal cells" (MSCs), because they appear to arise from the complex array of supporting structures found in the marrow [227]. The capability of BMSCs, including hematopoietic stem cells, to form solid organ tissue cells, in particular functional cardiomyocytes and vascular structures, has already been demonstrated by several groups [218, 228, 229].

Human BMSCs were successfully used for the in vitro fabrication of living trileaflet heart valves [99, 225], and also showed satisfactory in vivo function as pulmonary valve substitutes for at least 8 months [125]. Flow-cytometric characterization of isolated BMSCs prior to seeding confirmed a myofibroblast-like phenotype and showed tissue formation on bioabsorbable scaffolds with ECM production comparable to that of vascular-cell-derived tissue-engineered constructs [225]. Histology of the tissue-engineered valve leaflets revealed viable tissue organized in a layered fashion with ECM proteins characteristic of heart valve tissue such as collagen I and collagen III, as well as glycosaminoglycans. Also mechanical properties comparable to those of native tissue were demonstrated. The cells showed the expression of α-SMA and vimentin, whereas the remaining expression pattern indicated an absence of myeloid and hematopoietic cell differentiation [99, 225]. A similar staining pattern, indicating myofibroblast-like cell populations, was also reported in native semilunar valve tissue [30, 230] and populations of ovine BMSCs [226]. Despite the layered fashion in histology, the typical three-layered structural composition of native valve leaflets was not achieved in tissue-engineered heart valve leaflets using this cell type [99]. Engelmayr et al. [147, 148] were able to optimize the in vitro generation of tissue-engineered heart valves fabricated from BMSCs by using a novel bioreactor system. They demonstrated that cyclic flexure and laminar flow can synergistically accelerate BMSC-mediated tissue formation, providing a basis for the rational design of in vitro conditioning regimens for BMSC-seeded tissue-engineered heart valves.

Taken as a whole, the usage of BMSCs in cardiovascular tissue engineering may offer several advantages, especially when compared with vascular-derived cells: (1) BMSCs can be obtained without major surgical interventions, thus representing an easy-to-access cell source in a clinical scenario, (2) they exhibit the potential to differentiate into multiple cell lineages under biochemical [231] and mechanical [232] stimulation, (3) they demonstrate unique immunological characteristics allowing persistence in allogenic settings [233], and (4) they show an extensive in vitro proliferation capacity. Most notably, the easy accessibility of BMSCs, by a simple puncture of the iliac crest avoiding a major surgical intervention, seems indispensable with regard to a future routine clinical realization of the concept. However, in spite of all advantages Xin et al. [234] recently revealed a possible limitation of the concept of BMSCs by comparing the in vitro properties of BMSCs for construction of tissue-engineered heart valves of different age groups. They were able to show that the biological properties of cells in terms of proliferation, differentiation, VEGF release, and migratory abilities were significantly impaired in elderly patients. This seems to be highly relevant when considering the increasing number of older patients with heart valve disease, and illustrates the demand for development of new strategies to modify older BMSCs such as culture medium enrichment and antisenescence gene transfection.

5.3 Endothelial Progenitor Cells

It has been shown that the presence of endothelium on cardiovascular surfaces significantly reduces the risk for both coagulation and inflammatory complications (reviewed by Tanaka et al. [235]). Aside from these common endothelial properties, recent studies have also indicated a specific role of valvular endothelial cells in regulating the mechanical properties of heart valve cusps, suggesting a fundamental importance of endothelial cells for optimal valve function [236]. Therefore, to improve the functional capacities, the tissue-engineered heart valve constructs are usually covered with a layer of autologous human endothelial cells as an antithrombogenic lining [237]. Fully matured endothelial cells have been isolated from different vascular donor sources and have demonstrated promising results in heart valve tissue engineering [91, 96, 122, 123]. However, the expansion rate of these harvested endothelial cells is comparatively slow and the proliferation capabilities are limited, requiring a large number of endothelial cells to be harvested for therapeutic use [238]. In addition, tissue-specific phenotype expression in endothelial cells varies tremendously from one tissue type to another [239], which might interfere with the appropriate function of endothelial cells. Beyond that, the harvest of endothelial cells from donor vessels requires an invasive procedure, which is usually associated with a substantial risk for the donor. Therefore, finding a source of rapidly proliferating and ready-to-use endothelial cells, lacking a tissue-specific phenotypic expression, is of critical importance. Endothelial progenitor cells (EPCs), first discovered in human peripheral blood by Asahara et al. [240], have been explored as possible sources of endothelial cells. These blood-derived EPCs constitute a highly attractive alternative cell source of endothelial cells, in particular for pediatric patients, as they can be isolated from peripheral blood, bone marrow, as well as from umbilical cord blood (reviewed by Kim and von Recum [241, 242]). The latter can be obtained during pregnancy using the well-established method of ultrasound-guided cordocentesis, providing an elegant method of prenatal EPC harvest.

In general, EPCs represent a rare heterogeneous population of mononuclear blood cells. Controversy exists with respect to their phenotype, identification, and origin. Commonly, there is consensus that EPCs can derive from bone marrow. However, several other possible sources of EPCs have been discussed critically, including hematopoietic stem cells, myeloid cells, other circulating progenitor cells, and circulating mature endothelial cells shed off vessel walls (reviewed by Urbich and Dimmeler [243, 244]). Since their discovery, the therapeutic application and potency of EPCs for cardiovascular regeneration have been the subjects of intense experimental and clinical investigation. In particular, EPCs have been shown to exhibit the potential to differentiate into mature endothelial cells and to show distinct regenerative features (reviewed by Kawamoto and Losordo [245]). They have been successfully used for the repair of injured vessels, neovascularization, or regeneration of ischemic tissue in both preclinical models [246–248] and, more recently, early-phase clinical trials (reviewed by Martin-Rendon [249] and Pearson [250]). Accessorily, the ability of EPCs to promote revascularization has also been used for coating of synthetic vascular grafts [251], endothelialization of decellularized grafts in animal models [252], and seeding of hybrid grafts [253].

Importantly, Schmidt et al. [254–256] demonstrated the feasibility of using EPCs derived from human umbilical cord blood for the generation of constant neoendothelial phenotypes in tissue-engineered cardiovascular replacements. Recently, they also showed the fabrication of biologically active living heart valve leaflets using prenatally available progenitor cells derived from human umbilical cord as the only cell source [84]. Matrix stromal cells and EPCs derived from umbilical cord blood were seeded on biodegradable scaffolds and cultured in a biomimetic system. The leaflets showed mature layered tissue formation with functional neoendothelia and ECM production comparable with that of native tissues, which convincingly demonstrates the feasibility of using prenatal umbilical-cord-derived progenitor cells including EPCs for in vitro heart valve leaflet fabrication. This promising strategy of in vitro fabrication of autologous antithrombogenic surfaces on tissue-engineered heart valve replacements was also confirmed by Fang et al. [257]. EPCs, isolated from peripheral blood, showed stable phenotypes when they were co-cultured with nonendothelial cells as well as when exposed to mechanical stimuli. The ECM production of undifferentiated EPCs was shown to be insufficient, even though their differentiation into endothelial cells on biodegradable scaffolds was observed [258].

Overall, EPCs represent an auspicious cell source for endothelialization of engineered cardiovascular replacements, such as heart valves. Despite these pioneering studies showing great promise, the application of EPCs in tissue engineering is still in its infancy [241]. Since EPCs are easily accessible, current research increasingly aims at their transdifferentiation into a mesenchymal, myofibroblast-like phenotype in an effort to provide new strategies to guide tissue formation in engineered cardiac valves [259] and to ultimately enable blood as a single cell source for heart valve tissue engineering.

5.4 Umbilical-Cord-Derived Cells

To provide tissue-engineered constructs for congenital heart defects including heart valve disease, alternative cell sources have been investigated, with particular attention to preserving the intact vascular donor structure of newborn patients.

With regard to this approach in the pediatric population, the umbilical cord may serve as an optimal perinatal autologous cell source for tissue engineering of cardiovascular constructs.

In general, the human umbilical cord as a crucial part of the embryonic circulation is composed of two arteries and one vein embedded in mucoid embryonic connective tissue, called Wharton's jelly [260, 261]. Isolated cells of the umbilical cord represent a mixed cell population derived from these tissues [221, 262]. Interestingly, Kobayashi et al. [263] demonstrated that all three of these cell types exhibit myofibroblast-like characteristics by coexpressing α-SMA and vimentin. Additionally, Wang et al. [261] identified mesenchymal cells within Wharton's jelly expressing significant amounts of MSC markers, indicating the existence of mesenchymal progenitor cells. Ultimately, the presence of these MSCs, termed "umbilical cord matrix stromal cells," including their potential of multilineage differentiation and specific immune properties, has been confirmed by several groups, suggesting that umbilical cord matrix stromal cells represent a promising cell source for mesenchymal-cell-based therapies comprising tissue engineering [264–268]. Taken as a whole, the umbilical cord contains several cell sources that can be utilized for heart valve tissue engineering: (1) standard vascular endothelial cells, including endothelial cells derived from human umbilical cord vein and endothelial cells derived from human umbilical cord artery, (2) myofibroblasts derived from human umbilical cord, and (3) EPCs derived from human umbilical cord blood.

Presently, only sporadic experience exits with cells derived from umbilical cord for cardiovascular tissue engineering. In 1996, Sipehia et al. [269] first described the use of endothelial cells derived from human umbilical cord vein for creating a cell monolayer on artificial vascular prostheses. In 2002, myofibroblasts derived from human umbilical cord were established as a new promising cell source for cardiovascular tissue engineering and were used for the in vitro fabrication of pulmonary conduits using a biomimetic culture environment. The morphologic and mechanical features of the engineered constructs approximated the native human pulmonary artery, and the human umbilical cord cells demonstrated excellent growth properties in culture [262, 270]. A comparative study of vascular myofibroblast cells isolated from umbilical cord artery, umbilical cord vein, whole umbilical cord, and saphenous vein segments revealed similar cell growth, morphology, and tissue formation with regard to the cardiovascular tissue engineering approach, implying a comparable applicability of all cell sources investigated [221].

In 2004, Koike et al. [271] created a network of long-lasting blood vessels with human endothelial cells derived from the umbilical cord vein in a three-dimensional fibronectin-type I collagen gel connected to the mouse circulatory system and demonstrated its in vivo function up to 1 year. Schmidt et al. [254] used differentiated EPCs derived from human umbilical cord blood seeded on vascular scaffolds for the formation of vascular neotissue in both a biomimetic and a static in vitro environment. In 2005, Schmidt et al. [255] demonstrated the successful in vitro generation of living autologous cardiovascular replacements (patches) based on myofibroblasts derived from Wharton's jelly (WMFs) and EPCs derived from umbilical cord blood.

Recently, they were able to optimize the usage of umbilical-cord-derived cells for cardio vascular tissue engineering by generating functional tissue-engineered blood vessels [256] and by ultimately fabricating biologically active living heart valve leaflets in vitro [84]. Sodian et al. [162] demonstrated the use of cryopreserved human umbilical cord cells for the in vitro fabrication of tissue-engineered heart valves. Importantly, WMFs of the engineered valve leaflets exhibited phenotypic profiles of a fibroblast–myofibroblast lineage, indicating substantial similarity to native valvular interstitial cells [30]. Additionally, WMFs revealed excellent ECM production, in amounts comparable to that of native tissue and similar to that of other cells that have been successfully used for heart valve tissue engineering applications [96, 270], thus making umbilical cord cells, as WMFs combined with umbilical-cord-derived EPCs, a highly attractive sole cell source for pediatric heart valve tissue engineering [84].

Overall, the major advantage of umbilical cord cells, as an autologous cell source, is deemed to be the avoidance of invasive harvesting of intact vascular structures from pediatric patients and the availability of approximately 25-cm vascular tissue sections, which allows the isolation of a large number of juvenile, fast-growing cells for the generation of a sufficient cell number for scaffold seeding in a short period of time [272]. Additional advantages include the possibility to preserve the postnatal cords by standard cell and tissue banking technology to obtain an autologous cell pool for the patient's lifetime [270, 273], and the feasibility of isolating EPCs from the cord vessels [274] to create an antithrombogenic endothelial layer on the tissue-engineered constructs, which may be crucial for their long-term function. Moreover, the presence of mesenchymal progenitor cells in Wharton's jelly of human umbilical cords with multilineage potential [261, 266] and the possibility to obtain these cells prenatally using ultrasound-guided sampling technology make this cell source even more attractive [217].

However, despite these beneficial properties of umbilical cord cells and the achievements of recent investigations, the fabrication of autologous pediatric cardiovascular tissue, such as heart valves, from umbilical cord is still at a very early stage of development and a number of issues remain to be investigated.

5.5 Chorionic-Villi-Derived Cells

The human placenta, particularly its chorionic villi, provides extraembryonically situated fetal mesenchymal cells, including progenitor cells that are routinely obtained for prenatal genetic diagnostics by biopsy (reviewed by Pappa and Anagnou [275]). These cells might present a further attractive cell source for pediatric tissue engineering applications as indicated by recent investigations [276]. Schmidt et al. [85] first demonstrated the feasibility of this approach in heart valve tissue engineering using cells derived from chorionic villi for the fabrication of viable heart valve leaflets in vitro and showed promising results.

5.6 Amniotic-Fluid-Derived Cells

Congenital heart valve defects are often detectable prior to delivery by routine ultrasound examination. Therefore, the ideal pediatric tissue engineering paradigm would comprise a prenatal cell harvest providing time for the in vitro fabrication of an autologous living implant that is ready to use directly or shortly after birth to prevent secondary damage to the immature heart. Besides umbilical cord tissue and chorionic villi, amniotic fluid represents an attractive fetal cell source for this concept as it enables easy prenatal access to fetal progenitor cells from all three germ layers in a low-risk procedure (reviewed by Miki and Strom [277], Toda et al. [278], and Parolini et al. [279]). Several studies have shown promising results based on cells derived from human amniotic fluid with regard to noncardiovascular tissue engineering [280, 281].

In 2007, Schmidt et al. demonstrated the feasibility of generating living autologous heart valve leaflets in vitro, using human amniotic fluid as a single cell source. Cell populations required for the fabrication of heart valves, namely, mesenchymal-like progenitor cells and endothelium-like progenitor cells, have been successfully differentiated and expanded [34–36]. To expand the versatility of these cells also for adult application, cryopreserved cells derived from amniotic-fluid were investigated as a potential life-long available cell source, once again showing successful fabrication of viable heart valve leaflets in vitro [161]. Despite the futuristic strategy of this novel class of replacements potentially leading to innovative therapies for both pediatric and elderly patients, several concerns possibly related to this cell source, such as oncogenicity, long-term durability, and adequate growth behavior, have to be addressed in further in vivo investigations.

5.7 Adipose-Tissue-Derived Cells

Human adipose tissue has been shown to contain MSCs that can differentiate into different phenotypes in vitro, depending on the presence of lineage-specific induction factors [282, 283]. Owing to the high availability of adipose tissue – as it can be obtained in large quantities with minimal discomfort – adipose-derived stem cells (ADSCs) have been considered a potential alternative stem cell source to MSCs [284]. Several studies demonstrated that ADSCs show characteristics of EPCs, express endothelial specific markers in vitro when cultured with VEGF, and have the potential to differentiate into endothelial cells in vivo [285–287]. The first investigations of ADSCs, focusing on their use as a potential reservoir of autologous stem cells for cell-based therapies, suggest that fat appears to be a viable cell source for cardiovascular tissue engineering [288, 289]. The potential advantages of ADSCs when compared with vascular-derived cells lie in the ease of harvest and high availability, preventing sacrifice of donor vascular structures. However, presently, the long time required for differentiation may offset this advantage of ADSCs for their use in cardiovascular tissue engineering.

5.8 *Pluripotent Stem Cells*

The use of human embryonic stem cells for cell-based therapies is ethically problematic and will not readily be accessible as an autologous cell source. The availability of induced pluripotent stem cells, which are almost indistinguishable from embryonic stem cells, now represents significant progress toward the development of pluripotent patient-derived autologous cells for tissue engineering and other cell-based therapies [290–292]. Recent studies demonstrated the feasibility to differentiate murine [293] and human [294] induced pluripotent stem cells into functional cardiomyocytes.

In 2006, Guan et al. [295] first demonstrated the pluripotent character of adult mouse testis stem cells, suggesting a further potential cell source for cardiovascular tissue engineering applications. These adult spermatogonial stem cells represent germline stem cells found in the adult testis, which have also been converted into multipotent embryonic-stem-cell-like structures in vitro [296]. Even though these experiments are at a very preliminary stage, when considering these advances, it is only a question of time until the translation of these concepts to heart valve tissue engineering approaches occurs.

6 Future Perspectives

Heart valve tissue engineering is a promising approach for living functional autologous replacements that holds exciting potential for tremendously improving treatment of valvular heart disease. Particularly pediatric patients would benefit most from growing replacement materials for the repair of congenital heart defects. Since its inception in the early 1990s, significant advances have been made, and for the first time, the ultimate goal of the transition from experimental models to clinical reality is within our grasp. However, before clinical application of the heart valve tissue engineering concept will be routine, numerous steps have to be surmounted in the laboratory. Primary among these is our limited knowledge of the biochemical and immunological characteristics of the cells and tissues undergoing in vitro growth. Even having the first indications as to the influence of age and in vitro conditions, we still have to define the most favorable cell sources for in vitro seeding und ultimate in vivo survival. In addition, the disposition of implanted structures to typical biomaterial tissue interactions with medical devices, such as calcification, thrombosis, and excessive inflammatory response has to be elucidated. Another important consideration concerns the definition of the ideal scaffold material for heart valve tissue engineering, providing a template for directing new tissue growth and organization, as well as for regulating cellular adhesion, migration, and differentiation. Importantly, as ECM is produced and organized, the ideal scaffold would degrade until the matrix material has been completely replaced by functionally integrated neotissue. With regard to the inevitable concerns as to the potential transmission of microbiological hazards by tissues of xenogenic origin, it seems as if synthetic scaffold materials will make the grade as standard scaffolds for future clinical applications. These biodegradable synthetic polymers may offer several advantages over biological materials, including greater control over degradation rates, mechanical properties, fabrication, and reproducibility.

A further key consideration is the fact that currently available valve replacements have predictable behavior in terms of durability and biocompatibility, whereas tissue-engineered substitutes, potentially relying on in vivo remodeling, might display substantial variability among different individuals, owing to the heterogeneity of the physiological tissue remodeling potential. This leads directly to the assumption that the implementation of tissue-engineered constructs, such as heart valves, in the clinical field has to be accompanied by the development of consistent clinical guidelines specifying the inclusion criteria for certain patient populations according to safety, efficacy, and quality of the engineered products. Therefore, the demonstration of a long-term benefit and efficacy, as well as safety of implanted constructs in specific patient populations in preclinical studies will be of critical importance for a possible translation of the concept into clinics. In the course of this attempt to understand, monitor, and potentially control individual differences of in vivo tissue remodeling capacities, the identification of biomarkers as independent predictors of implant outcomes increasingly becomes a point of contention. This is closely related to the need for substantial advances of our understanding of heart valve cell biology. A thorough understanding of molecular mechanisms as well as embryonic and fetal heart valve development may also permit control of heart valve morphogenesis both in vitro and in vivo.

Owing to the demographic development, the prevalence of heart valve disease shows an increasing trend with increasing age. More and more elderly patients will suffer from valvular heart disease and will thus require replacement therapy. Therefore, the development of tissue-engineered heart valves based on a stented design might be an important step toward improving quality of life and life expectancy of these patients. Ideally, when tissue engineering technologies make it to the clinical scenario in the future, the schedule for this patient population, primarily diagnosed with degenerative valve disease, may be (1) cell isolation from designated sources (i.e., by a bone marrow puncture under local anesthesia), (2) differentiation and expansion of cells and engineering of a heart valve prostheses in vitro, and (3) following yet to be defined quality criteria (biological, histological, biosafety, etc.), reimplantation of living autologous valve replacements into patients after a time period of at most 6–8 weeks.

References

1. Iung B, Vahanian A (2006) Valvular heart diseases in elderly people. Lancet 368(9540):969–71
2. Otto CM, Lind BK, Kitzman DW et al (1999) Association of aortic-valve sclerosis with cardiovascular mortality and morbidity in the elderly. N Engl J Med 341(3):142–7
3. Supino PG, Borer JS, Preibisz J (2006) The epidemiology of valvular heart disease: a growing public health problem. Heart Fail Clin 2:379–93
4. Yacoub MH, Cohn LH (2004) Novel approaches to cardiac valve repair: from structure to function: part I. Circulation 109:942–50

5. Mikos AG, Herring SW, Ochareon P et al (2006) Engineering complex tissues. Tissue Eng 12(12):3307–39
6. Yacoub MH, Takkenberg JJ (2005) Will heart valve tissue engineering change the world? Nat Clin Pract Cardiovasc Med 2:60–1
7. Nkomo VT, Gardin JM, Skelton TN et al (2006) Burden of valvular heart diseases: a population-based study. Lancet 368(9540):1005–11
8. Ruel J, Lachance G (2009) A new bioreactor for the development of tissue-engineered heart valves. Ann Biomed Eng 37(4):674–81
9. Yoganathan AP, He Z et al (2004) Fluid mechanics of heart valves. Annu Rev Biomed Eng 6:331–62
10. Dasi LP, Simon HA, Sucosky P et al (2009) Fluid mechanics of artificial heart valves. Clin Exp Pharmacol Physiol 36(2):225–37
11. Zilla P, Brink J, Human P et al (2008) Prosthetic heart valves: catering for the few. Biomaterials 29:385–406
12. Bonow RO, Carabello BA, Kanu C et al (2006) ACC/AHA 2006 guidelines for the management of patients with valvular heart disease: a report of the American College of Cardiology/American Heart Association Task Force on Practice Guidelines. Circulation 114(5):e84–231
13. Lee KY, Mooney DJ (2001) Hydrogels for tissue engineering. Chem Rev 101(7):1869–79
14. Senthilnathan V, Treasure T, Grunkemeier G et al (1999) Heart valves: which is the best choice? Cardiovasc Surg 7(4):393–7
15. Schoen FJ (2008) Evolving concepts of cardiac valve dynamics: the continuum of development, functional structure, pathobiology, and tissue engineering. Circulation 118:1864–80
16. Schoen JF (1997) Aortic valve structure–function correlations: role of elastic fibers no longer a stretch of imagination. J Heart Valve Dis 6:1–6
17. Yacoub MH, Kilner PJ, Birks EJ, Misfeld M (1999) The aortic outflow and root: a tale of dynamism and crosstalk. Ann Thorac Surg 68(3 Suppl):S37–43
18. Simmons CA (2009) Aortic valve mechanics: an emerging role for the endothelium. J Am Coll Cardiol 53:1456–8
19. Misfeld M, Sievers HH (2007) Heart valve macro- and microstructure. Philos Trans R Soc Lond B Biol Sci 362(1484):1421–36
20. Butcher JT, Simmons CA, Warnock JN (2008) Mechanobiology of the aortic heart valve. Heart Valve Dis 17(1):62–73
21. Peskin CS, McQueen DM (1994) Mechanical equilibrium determines the fractal fiber architecture of aortic heart valve leaflets. Am J Physiol 266(1 Pt 2):H319–28
22. Thubrikar MJ, Aouad J, Nolan SP (1986) Comparison of the in vivo and in vitro mechanical properties of aortic valve leaflets. J Thorac Cardiovasc Surg 92:29–36
23. Schoen FJ, Levy RJ (1999) Tissue heart valves: current challenges and future research perspectives. Biomed Mater Res 47(4):439–65
24. Scott M, Vesely I (1995) Aortic valve cusp microstructure: the role of elastin. Ann Thorac Surg 60(2 Suppl):S391–4
25. Scott MJ, Vesely I (1996) Morphology of porcine aortic valve cusp elastin. J Heart Valve Dis 5(5):464–71
26. Bairati A Jr, De Biasi S, Pilotto F (1978) Smooth muscle cells in the cusps of the aortic valve of pigs. Experientia 34(12):1636–8
27. Bairati A, DeBiasi S (1981) Presence of a smooth muscle system in aortic valve leaflets. Anat Embryol (Berl) 161(3):329–40
28. Roy A, Brand NJ, Yacoub MH (2000) Molecular characterization of interstitial cells isolated from human heart valves. J Heart Valve Dis 9(3):459–64; discussion 464–5
29. Della Rocca F, Sartore S, Guidolin D et al (2000) Cell composition of the human pulmonary valve: a comparative study with the aortic valve – the VESALIO Project. Ann Thorac Surg 70(5):1594–600
30. Messier RH Jr, Bass BL, Aly HM et al (1994) Dual structural and functional phenotypes of the porcine aortic valve interstitial population: characteristics of the leaflet myofibroblast. J Surg Res 57(1):1–21

31. Langer R, Vacanti JP (1993) Tissue engineering. Science 260:920–6
32. Matheny RG, Hutchison ML, Dryden PE et al (2000) Porcine small intestine submucosa as a pulmonary valve leaflet substitute. J Heart Valve Dis 9:769–75
33. Mol A, Bouten CV, Baaijens FP et al (2004) Review article: tissue engineering of semilunar heart valves: current status and future developments. J Heart Valve Dis 13(2):272–80
34. Schmidt D, Stock UA, Hoerstrup SP (2007a) Tissue engineering of heart valves using decellularized xenogeneic or polymeric starter matrices. Philos Trans R Soc Lond B Biol Sci 362(1484):1505–12
35. Schmidt D, Mol A, Kelm JM et al (2007b) In vitro heart valve tissue engineering. Methods Mol Med 140:319–30
36. Schmidt D, Achermann J, Odermatt B et al (2007c) Prenatally fabricated autologous human living heart valves based on amniotic fluid derived progenitor cells as single cell source. Circulation 116:I64–70
37. Breuer CK, Mettler BA, Anthony T et al (2004) Application of tissue-engineering principles toward the development of a semilunar heart valve substitute. Tissue Eng 10(11–12):1725–36
38. Brody S, Pandit A (2007) Approaches to heart valve tissue engineering scaffold design. J Biomed Mater Res B Appl Biomater 83(1):16–43
39. Sacks MS, Schoen FJ, Mayer JE (2009) Bioengineering challenges for heart valve tissue engineering. Annu Rev Biomed Eng 11:289–313
40. Lichtenberg A, Cebotari S, Tudorache I et al (2007) Biological scaffolds for heart valve tissue engineering. Methods Mol Med 140:309–17
41. Sales VL, Engelmayr GC Jr, Johnson JA Jr et al (2007) Protein precoating of elastomeric tissue-engineering scaffolds increased cellularity, enhanced extracellular matrix protein production, and differentially regulated the phenotypes of circulating endothelial progenitor cells. Circulation 116(11 Suppl):I55–63
42. Affonso da Costa FD, Dohmen PM, Lopes SV et al (2004) Comparison of cryopreserved homografts and decellularized porcine heterografts implanted in sheep. Artif Organs 28:366–70
43. Allen BS, El-Zein C, Cuneo B et al (2002) Pericardial tissue valves and Gore-Tex conduits as an alternative for right ventricular outflow tract replacement in children. Ann Thorac Surg 74:771–77
44. Bielefeld MR, Bishop DA, Campbell DN et al (2001) Reoperative homograft right ventricular outflow tract reconstruction. Ann Thorac Surg 71:482–7; discussion 7–8
45. Carr-White GS, Glennan S, Edwards S et al (1999) Pulmonary autograft versus aortic homograft for rereplacement of the aortic valve: results from a subset of a prospective randomized trial. Circulation 100(19 Suppl):II103–6
46. Grauss RW, Hazekamp MG, van Vliet S et al (2003) Decellularization of rat aortic valve allografts reduces leaflet destruction and extracellular matrix remodeling. J Thorac Cardiovasc Surg 126:2003–10
47. Hong H, Dong GN, Shi WJ et al (2008) Fabrication of biomatrix/polymer hybrid scaffold for heart valve tissue engineering in vitro. ASAIO J 54(6):627–32
48. Hong H, Dong N, Shi J et al (2009) Fabrication of a novel hybrid heart valve leaflet for tissue engineering: an in vitro study. Artif Organs 33(7):554–8
49. Neidert MR, Tranquillo RT (2006) Tissue-engineered valves with commissural alignment. Tissue Eng 12(4):891–903
50. Robinson PS, Johnson SL, Evans MC et al (2008) Functional tissue-engineered valves from cell-remodeled fibrin with commissural alignment of cell-produced collagen. Tissue Eng Part A 14:83–95
51. Syedain ZH, Weinberg JS, Tranquillo RT (2008) Cyclic distension of fibrin-based tissue constructs: evidence of adaptation during growth of engineered connective tissue. Proc Natl Acad Sci 105(18):6537–42
52. Williams C, Johnson SL, Robinson PS et al (2006) Cell sourcing and culture conditions for fibrin-based valve constructs. Tissue Eng 12:1489–502
53. Mol A, Smits AI, Bouten CV et al (2009) Tissue engineering of heart valves: advances and current challenges. Expert Rev Med Devices 6(3):259–75

54. Kasimir MT, Rieder E, Seebacher G et al (2003) Comparison of different decellularization procedures of porcine heart valves. Int J Artif Organs 26(5):421–7
55. Leyh RG, Wilhelmi M, Rebe P et al (2003) In vivo repopulation of xenogeneic and allogeneic acellular valve matrix conduits in the pulmonary circulation. Ann Thorac Surg 75(5):1457–63; discussion 1463
56. Steinhoff G, Stock U, Karim N et al (2000) Tissue engineering of pulmonary heart valves on allogenic acellular matrix conduits: in vivo restoration of valve tissue. Circulation 102:50–5
57. Curtil A, Pegg DE, Wilson A (1997) Repopulation of freeze-dried porcine valves with human fibroblasts and endothelial cells. J Heart Valve Dis 6(3):296–306
58. Wilson GJ, Courtman DW, Klement P et al (1995) Acellular matrix: a biomaterials approach for coronary artery bypass and heart valve replacement. Ann Thorac Surg 1995 60:353–8
59. Bader A, Schilling T, Teebken OE et al (1998) Tissue engineering of heart valves – human endothelial cell seeding of detergent acellularized porcine valves. Eur J Cardiothorac Surg 14(3):279–84
60. Bertiplaglia B, Ortolani F, Petrelli L et al (2003) Cell characterization of porcine aortic valve and decellularized leaflets repopulated with aortic valve interstitial cells: the VESALIO project. Ann Thorac Surg 75:1274–82
61. Booth C, Korossis SA, Wilcox HE et al (2002) Tissue engineering of cardiac valve prostheses I: development and histological characterization of an acellular porcine scaffold. J Heart Valve Dis 11(4):457–62
62. Kim WG, Park JK, Lee WY (2002) Tissue-engineered heart valve leaflets: an effective method of obtaining acellularized valve xenografts. Int J Artif Organs 25(8):791–7
63. Zeltinger J, Landeen LK, Alexander HG et al (2001) Development and characterization of tissue-engineered aortic valves. Tissue Eng 7:9–22
64. Rieder E, Kasimir MT, Silberhumer G et al (2004) Decellularization protocols of porcine heart valves differ importantly in efficiency of cell removal and susceptibility of the matrix to re-cellularization with human vascular cells. J Thorac Cardiovasc Surg 127(2):399–405
65. Tudorache I, Cebotari S, Sturz G et al (2007) Tissue engineering of heart valves: biomechanical and morphological properties of decellularized heart valves. J Heart Valve Dis 16:567–73
66. Cushing MC, Jaeggli MP, Masters KS et al (2005) Serum deprivation improves seeding and repopulation of acellular matrices with valvular interstitial cells. J Biomed Mater Res A 75(1):232–41
67. Knight RL, Booth C, Wilcox HE et al (2005) Tissue engineering of cardiac valves: re-seeding of acellular porcine aortic valve matrices with human mesenchymal progenitor cells. J Heart Valve Dis 14(6):806–13
68. Schenke-Layland K, Opitz F, Gross M (2003) Complete dynamic repopulation of decellularized heart valves by application of defined physical signals – an in vitro study. Cardiovasc Res 60:497–509
69. Kim SS, Lim SH, Hong YS et al (2006) Tissue engineering of heart valves in vivo using bone marrow-derived cells. Artif Organs 30(7):554–7
70. Takeuchi Y (2000) Risk of zoonosis in xenotransplantation. Transplant Proc 32:2698–700
71. Weiss RA, Magre S, Takeuchi Y (2000) Infection hazards of xenotransplantation. J Infect 40:21–5
72. Martin U, Kiessig V, Blusch JH et al (1998) Expression of pig endogenous retrovirus by primary porcine endothelial cells and infection of human cells. Lancet 352:692–4
73. Martin U, Winkler ME, Id M et al (2000) Productive infection of primary human endothelial cells by pig endogenous retrovirus (PERV). Xenotransplantation 7(2):138–42
74. Moza AK, Mertsching H, Herden T et al (2001) Heart valves from pigs and the porcine endogenous retrovirus: experimental and clinical data to assess the probability of porcine endogenous retrovirus infection in human subjects. J Thorac Cardiovasc Surg 121:697–701
75. Patience C, Takeuchi Y, Weiss RA (1997) Infection of human cells by an endogenous retrovirus of pigs. Nat Med 3(3):282–6
76. Patience C, Switzer WM, Takeuchi Y et al (2001) Multiple groups of novel retroviral genomes in pigs and related species. J Virol 75:2771–5

77. Prabha S, Verghese S (2008) Existence of proviral porcine endogenous retrovirus in fresh and decellularised porcine tissues. Indian J Med Microbiol 26(3):228–32
78. Specke V, Rubant S, Denner J (2001) Productive infection of human primary cells and cell lines with porcine endogenous retroviruses. Virology 285:177–80
79. Wilson CA, Wong S, Muller J et al (1998) Type C retrovirus released from porcine primary peripheral blood mononuclear cells infects human cells. J Virol 72:3082–7
80. Knight R, Brazier M, Collins SJ (2004) Human prion diseases: cause, clinical and diagnostic aspects. Contrib Microbiol 11:72–97
81. Knight R, Collins S (2001) Human prion diseases: cause, clinical and diagnostic aspects. Contrib Microbiol 7:68–92
82. Kallenbach K, Leyh RG, Lefik E et al (2004) Guided tissue regeneration: porcine matrix does not transmit PERV. Biomaterials 25(17):3613–20
83. Walles T, Lichtenberg A, Puschmann C (2003) In vivo model for cross-species porcine endogenous retrovirus transmission using tissue engineered pulmonary arteries. Eur J Cardiothorac Surg 24:358–63
84. Schmidt D, Mol A, Odermatt B et al (2006) Engineering of biologically active living heart valve leaflets using human umbilical cord-derived progenitor cells. Tissue Eng 12(11):3223–32
85. Schmidt D, Mol A, Breymann C et al (2006) Living autologous heart valves engineered from human prenatally harvested progenitors. Circulation 114(1 Suppl):I125–31
86. Agrawal CM, Ray RB (2001) Biodegradable polymeric scaffolds for musculoskeletal tissue engineering. J Biomed Mater Res 55(2):141–50
87. Hutmacher DW, Goh JC, Teoh SH (2001) An introduction to biodegradable materials for tissue engineering applications. Ann Acad Med Singapore 30(2):183–91
88. Hutmacher DW, Schantz T, Zein I et al (2001) Mechanical properties and cell cultural response of polycaprolactone scaffolds designed and fabricated via fused deposition modeling. J Biomed Mater Res 55(2):203–16
89. Kessler B, Witholt B (2001) Factors involved in the regulatory network of polyhydroxyalkanoate metabolism. J Biotechnol 86(2):97–104
90. Kidane AG, Burriesci G, Cornejo P et al (2009) Current developments and future prospects for heart valve replacement therapy. Biomed Mater Res B Appl Biomater 88(1):290–303
91. Sodian R, Hoerstrup SP, Sperling JS et al (2000) Early in vivo experience with tissue-engineered trileaflet heart valves. Circulation 102:22–9
92. Sodian R, Hoerstrup SP, Sperling JS et al (2000) Evaluation of biodegradable, three-dimensional matrices for tissue engineering of heart valves. ASAIO J 46:107–10
93. Sodian R, Sperling JS, Martin DP et al (2000) Fabrication of a trileaflet heart valve scaffold from a polyhydroxyalkanoate biopolyester for use in tissue engineering. Tissue Eng 6:183–8
94. Schaefermeier PK, Szymanski D, Weiss F et al (2009) Design and fabrication of three-dimensional scaffolds for tissue engineering of human heart valves. Eur Surg Res 42(1):49–53
95. Sodian R, Loebe M, Hein A et al (2002) Application of stereolithography for scaffold fabrication for tissue engineered heart valves. ASAIO J 48:12–6
96. Hoerstrup SP, Sodian R, Daebritz S et al (2000) Functional living trileaflet heart valves grown in vitro. Circulation 102(19 Suppl 3):III44–9
97. Rabkin E, Hoerstrup SP, Aikawa M et al (2002) Evolution of cell phenotype and extracellular matrix in tissue-engineered heart valves during in-vitro maturation and in-vivo remodeling. J Heart Valve Dis 11(3):308–14
98. Stock UA, Nagashima M, Khalil PN (2000) Tissue-engineered valved conduits in the pulmonary circulation. J Thorac Cardiovasc Surg 119:732–40
99. Hoerstrup SP, Kadner A, Melnitchouk S et al (2002) Tissue engineering of functional trileaflet heart valves from human marrow stromal cells. Circulation 106(12 Suppl 1):I143–50
100. Mol A, Rutten MC, Driessen NJ et al (2006) Autologous human tissue-engineered heart valves: prospects for systemic application. Circulation 114(1 Suppl):I152–8
101. Grabow N, Schmohl K, Khosravi A et al (2004) Mechanical and structural properties of a novel hybrid heart valve scaffold for tissue engineering. Artif Organs 28(11):971–9

102. Stamm C, Khosravi A, Grabow N et al (2004) Biomatrix/polymer composite material for heart valve tissue engineering. Ann Thorac Surg 78:2084–92
103. Lutolf MP, Hubbell JA (2005) Synthetic biomaterials as instructive extracellular microenvironments for morphogenesis in tissue engineering. Nat Biotechnol 23(1):47–55
104. Tabata Y (2009) Biomaterial technology for tissue engineering applications. J R Soc Interface 6:311–24
105. De Laporte L, Shea LD (2007) Matrices and scaffolds for DNA delivery in tissue engineering. Adv Drug Deliv Rev 59(4–5):292–307
106. Rothenburger M, Volker W, Vischer JP et al (2002) Tissue engineering of heart valves: formation of a three-dimensional tissue using porcine heart valve cells. ASAIO J 48(6):586–91
107. Rothenburger M, Völker W, Vischer P et al (2002) Ultrastructure of proteoglycans in tissue-engineered cardiovascular structures. Tissue Eng 8(6):1049–56
108. Butcher JT, Nerem RM (2004) Porcine aortic valve interstitial cells in three-dimensional culture: comparison of phenotype with aortic smooth muscle cells. J Heart Valve Dis 13(3):478–85; discussion 485–6
109. Butcher JT, Nerem RM (2006) Valvular endothelial cells regulate the phenotype of interstitial cells in co-culture: effects of steady shear stress. Tissue Eng 12(4):905–15
110. Flanagan TC, Wilkins B, Black A et al (2006) A collagen-glycosaminoglycan co-culture model for heart valve tissue engineering applications. Biomaterials 27(10):2233–46
111. Taylor PM, Sachlos E, Dreger SA et al (2006) Interaction of human valve interstitial cells with collagen matrices manufactured using rapid prototyping. Biomaterials 27:2733–7
112. Tedder ME, Liao J, Weed B et al (2009) Stabilized collagen scaffolds for heart valve tissue engineering. Tissue Eng Part A 15:1257–68
113. Taylor PM, Allen SP, Dreger SA (2002) Human cardiac valve interstitial cells in collagen sponge: a biological three-dimensional matrix for tissue engineering. J Heart Valve Dis 11:298–306
114. Rothenburger M, Vischer P, Völker W et al (2001) In vitro modelling of tissue using isolated vascular cells on a synthetic collagen matrix as a substitute for heart valves. Thorac Cardiovasc Surg 49(4):204–9
115. Shi Y, Ramamurthi A, Vesely I (2002) Towards tissue engineering of a composite aortic valve. Biomed Sci Instrum 38:35–40
116. Mol A, van Lieshout MI, Dam-de Veen CG et al (2005) Fibrin as a cell carrier in cardiovascular tissue engineering applications. Biomaterials 26(16):3113–21
117. Ameer GA, Mahmood TA, Langer R (2002) A biodegradable composite scaffold for cell transplantation. J Orthop Res 20(1):16–9
118. Jockenhoevel S, Chalabi K, Sachweh JS et al (2001) Tissue engineering: complete autologous valve conduit – a new moulding technique. Thorac Cardiovasc Surg 49(5):287–90
119. Jockenhoevel S, Zund G, Hoerstrup SP et al (2001) Fibrin gel – advantages of a new scaffold in cardiovascular tissue engineering. Eur J Cardiothorac Surg 19(4):424–30
120. Ye Q, Zünd G, Benedikt P et al (2000) Fibrin gel as a three dimensional matrix in cardiovascular tissue engineering. Eur J Cardiothorac Surg 17:587–91
121. Schense JC, Hubbell JA (1999) Cross-linking exogenous bifunctional peptides into fibrin gels with factor XIIIa. Bioconjug Chem 10(1):75–81
122. Shin'oka T, Breuer CK, Tanel RE (1995) Tissue engineering heart valves: valve leaflet replacement study in a lamb model. Ann Thorac Surg 60(6):513–6
123. Shin'oka T, Ma PX, Shum-Tim D et al (1996) Tissue-engineered heart valves. Autologous valve leaflet replacement study in a lamb model. Circulation 94:II164–8
124. Kim WG, Cho SK, Kang MC et al (2001) Tissue-engineered heart valve leaflets: an animal study. Int J Artif Organs 24(9):642–8
125. Sutherland FW, Perry TE, Yu Y et al (2005) From stem cells to viable autologous semilunar heart valve. Circulation 111:2783–91
126. Hoerstrup SP, Cummings Mrcs I, Lachat M et al (2006) Functional growth in tissue-engineered living, vascular grafts: follow-up at 100 weeks in a large animal model. Circulation 114 (1 Suppl):I159–66

127. Matsumura G, Hibino N, Ikada Y et al (2003) Successful application of tissue engineered vascular autografts: clinical experience. Biomaterials 24(13):2303–8
128. Shin'oka T, Imai Y, Ikada Y (2001) Transplantation of a tissue-engineered pulmonary artery. N Engl J Med 344:532–3
129. Hoerstrup SP, Sodian R, Sperling JS et al (2000) New pulsatile bioreactor for in vitro formation of tissue engineered heart valves. Tissue Eng 6(1):75–9
130. Balguid A, Rubbens MP, Mol A et al (2007) The role of collagen cross-links in biomechanical behavior of human aortic heart valve leaflets – relevance for tissue engineering. Tissue Eng 13(7):1501–110
131. Huang SD, Liu XH, Bai CG et al (2007) Synergistic effect of fibronectin and hepatocyte growth factor on stable cell-matrix adhesion, re-endothelialization, and reconstitution in developing tissue-engineered heart valves. Heart Vessels 22(2):116–22
132. Rogers KA, Boughner D, Appleton CT et al (2009) Vascular smooth muscle cells as a valvular interstitial cell surrogate in heart valve tissue engineering. Tissue Eng Part A 15(12):3889–97
133. Stock UA, Vacanti JP, Mayer Jr JE (2002) Tissue engineering of heart valves – current aspects. Thorac Cardiovasc Surg 50:184–93
134. Isenberg BC, Tranquillo RT (2003) Long-term cyclic distention enhances the mechanical properties of collagen-based media-equivalents. Ann Biomed Eng 31(8):937–49
135. Mendelson K, Schoen FJ (2006) Heart valve tissue engineering: concepts, approaches, progress, and challenges. Ann Biomed Eng 34(12):1799–819
136. Mol A, Bouten CV, Zünd G et al (2003) The relevance of large strains in functional tissue engineering of heart valves. Thorac Cardiovasc Surg 51(2):78–83
137. Seliktar D, Nerem RM, Galis ZS et al (2003) Mechanical strain-stimulated remodeling of tissue-engineered blood vessel constructs. Tissue Eng 9(4):657–66
138. Hinz B, Gabbiani G (2003) Mechanisms of force generation and transmission by myofibroblasts. Curr Opin Biotechnol 14(5):538–46
139. Parizi M, Howard EW, Tomasek JJ (2000) Regulation of LPA-promoted myofibroblast contraction: role of Rho, myosin light chain kinase, and myosin light chain phosphatase. Exp Cell Res 254(2):210–20
140. Grenier G, Rémy-Zolghadri M, Larouche D et al (2005) Tissue reorganization in response to mechanical load increases functionality. Tissue Eng 11(1–2):90–100
141. Boerboom RA, Rubbens MP, Driessen NJ et al (2008) Effect of strain magnitude on the tissue properties of engineered cardiovascular constructs. Ann Biomed Eng 36(2):244–53
142. Rubbens MP, Mol A, van Marion MH et al (2009) Straining mode-dependent collagen remodeling in engineered cardiovascular tissue. Tissue Eng Part A 15(4):841–9
143. Rubbens MP, Mol A, Boerboom RA et al (2009) Intermittent straining accelerates the development of tissue properties in engineered heart valve tissue. Tissue Eng Part A 15(5):999–1008
144. Engelmayr GC Jr, Hildebrand DK, Sutherland FW et al (2003) A novel bioreactor for the dynamic flexural stimulation of tissue engineered heart valve biomaterials. Biomaterials 24(14):2523–32
145. Engelmayr GC Jr, Rabkin E, Sutherland FW et al (2005) The independent role of cyclic flexure in the early in vitro development of an engineered heart valve tissue. Biomaterials 26(2):175–87
146. Jockenhoevel S, Zund G, Hoerstrup SP et al (2002) Cardiovascular tissue engineering: a new laminar flow chamber for in vitro improvement of mechanical tissue properties. ASAIO J 48(1):8–11
147. Engelmayr GC Jr, Sales VL, Mayer JE Jr et al (2006) Cyclic flexure and laminar flow synergistically accelerate mesenchymal stem cell-mediated engineered tissue formation: implications for engineered heart valve tissues. Biomaterials 27(36):6083–95
148. Engelmayr GC Jr, Soletti L, Vigmostad SC et al (2008) A novel flex-stretch-flow bioreactor for the study of engineered heart valve tissue mechanobiology. Ann Biomed Eng 36(5):700–12
149. Balguid A, Mol A, van Vlimmeren MA et al (2009) Hypoxia induces near-native mechanical properties in engineered heart valve tissue. Circulation 119(2):290–7

150. Mertsching H, Hansmann J (2009) Bioreactor technology in cardiovascular tissue engineering. Adv Biochem 112:29–37
151. Stock UA, Wiederschain D, Kilroy SM et al (2001) Dynamics of extracellular matrix production and turnover in tissue engineered cardiovascular structures. J Cell Biochem 81:220–8
152. Sodian R, Lemke T, Loebe M et al (2001) New pulsatile bioreactor for fabrication of tissue-engineered patches. J Biomed Mater Res 58:401–5
153. Sodian R, Lemke T, Fritsche C et al (2002) Tissue-engineering bioreactors: a new combined cell-seeding and perfusion system for vascular tissue engineering. Tissue Eng 8:863–70
154. Narita Y, Hata K, Kagami H et al (2004) Novel pulse duplicating bioreactor system for tissue-engineered vascular construct. Tissue Eng 10(7–8):1224–33
155. Williams C, Wick TM (2004) Perfusion bioreactor for small diameter tissue-engineered arteries. Tissue Eng 10:930–41
156. Mol A, Driessen NJ, Rutten MC et al (2005) Tissue engineering of human heart valve leaflets: a novel bioreactor for a strain-based conditioning approach. Ann Biomed Eng 33(12):1778–88
157. Niklason LE, Gao J, Abbott WM et al (1999) Functional arteries grown in vitro. Science 284(5413):489–93
158. Stegemann JP, Nerem RM (2003) Phenotype modulation in vascular tissue engineering using biochemical and mechanical stimulation. Ann Biomed Eng 31:391–402
159. Syedain ZH, Tranquillo RT (2009) Controlled cyclic stretch bioreactor for tissue-engineered heart valves. Biomaterials 30(25):4078–84
160. Lee DJ, Steen J, Jordan JE et al (2009) Endothelialization of heart valve matrix using a computer-assisted pulsatile bioreactor. Tissue Eng Part A 15(4):807–14
161. Schmidt D, Achermann J, Odermatt B (2008) Cryopreserved amniotic fluid-derived cells: a lifelong autologous fetal stem cell source for heart valve tissue engineering. J Heart Valve Dis 17(4):446–55
162. Sodian R, Lueders C, Kraemer L et al (2006) Tissue engineering of autologous human heart valves using cryopreserved vascular umbilical cord cells. Ann Thorac Surg 81(6):2207–16
163. Hildebrand DK, Wu ZJ, Mayer JE Jr et al (2004) Design and hydrodynamic evaluation of a novel pulsatile bioreactor for biologically active heart valves. Ann Biomed Eng 32(8):1039–49
164. Kortsmit J, Driessen NJ, Rutten MC et al (2009) Nondestructive and noninvasive assessment of mechanical properties in heart valve tissue engineering. Tissue Eng Part A 15(4):797–806
165. Kortsmit J, Driessen NJ, Rutten MC et al (2009) Real time, non-invasive assessment of leaflet deformation in heart valve tissue engineering. Ann Biomed Eng 37(3):532–41
166. Vesely I (2005) Heart valve tissue engineering. Circ Res 97:743–55
167. Sievers HH (2007) In vivo tissue engineering an autologous semilunar biovalve: can we get what we want? J Thorac Cardiovasc Surg 134:20–2
168. Schleicher M, Wendel HP, Fritze O et al (2009) In vivo tissue engineering of heart valves: evolution of a novel concept. Regen Med 4(4):613–9
169. Dohmen PM, da Costa F, Holinski S et al (2006) Is there a possibility for a glutaraldehyde-free porcine heart valve to grow? Eur Surg Res 38(1):54–61
170. Elkins RC, Dawson PE, Goldstein S et al (2001) Decellularized human valve allografts. Ann Thorac Surg 71(5 Suppl):S428–32
171. Elkins RC, Goldstein S, Hewitt CW et al (2001) Recellularization of heart valve grafts by a process of adaptive remodeling. Semin Thorac Cardiovasc Surg 13(4 Suppl 1):87–92
172. Erdbrügger W, Konertz W, Dohmen PM et al (2006) Decellularized xenogenic heart valves reveal remodeling and growth potential in vivo. Tissue Eng 12(8):2059–68
173. Goldstein S, Clarke DR, Walsh SP et al (2000) Transpecies heart valve transplant: advanced studies of a bioengineered xeno-autograft. Ann Thorac Surg 70(6):1962–9
174. Iwai S, Torikai K, Coppin CM et al (2007) Minimally immunogenic decellularized porcine valve provides in situ recellularization as a stentless bioprosthetic valve. J Artif Organs 10(1):29–35
175. Kim WG, Huh JH (2004) Time related histopathologic changes of acellularized xenogenic pulmonary valved conduits. ASAIO J 50(6):601–5

176. Leyh RG, Wilhelmi M, Walles T et al (2003) Acellularized porcine heart valve scaffolds for heart valve tissue engineering and the risk of cross-species transmission of porcine endogenous retrovirus. J Thorac Cardiovasc Surg 126(4):1000–4
177. Takagi K, Fukunaga S, Nishi A et al (2006) In vivo recellularization of plain decellularized xenografts with specific cell characterization in the systemic circulation: histological and immunohistochemical study. Artif Organs 30:233–41
178. Dohmen PM, Konertz W (2005) Results with decellularized xenografts. Circ Res 97(8):743–55
179. Dohmen PM, Hauptmann S, Terytze A et al (2007) In-vivo repopularization of a tissue-engineered heart valve in a human subject. J Heart Valve Dis 16(4):447–9
180. Simon P, Kasimir MT, Seebacher G (2003) Early failure of the tissue engineered porcine heart valve SYNERGRAFT in pediatric patients. Eur J Cardiothorac Surg 23:1002–6
181. Kasimir MT, Rieder E, Seebacher G et al (2006) Decellularization does not eliminate thrombogenicity and inflammatory stimulation in tissue-engineered porcine heart valves. J Heart Valve Dis 15(2):278–86
182. Rieder E, Nigisch A, Dekan B et al (2006) Granulocyte-based immune response against decellularized or glutaraldehyde cross-linked vascular tissue. Biomaterials 27(33):5634–42
183. Bastian F, Stelzmüller ME, Kratochwill K et al (2008) IgG deposition and activation of the classical complement pathway involvement in the activation of human granulocytes by decellularized porcine heart valve tissue. Biomaterials 29(12):1824–32
184. Lichtenberg A, Cebotari S, Tudorache I et al (2006) Flow-dependent re-endothelialization of tissue-engineered heart valves. J Heart Valve Dis 15(2):287–93; discussion 293–4
185. Lichtenberg A, Tudorache I, Cebotari S et al (2006) In vitro re-endothelialization of detergent decellularized heart valves under simulated physiological dynamic conditions. Biomaterials 27(23):4221–9
186. Stamm C, Steinhoff G (2006) When less is more: go slowly when repopulating a decellularized valve in vivo! J Thorac Cardiovasc Surg 131:843–52
187. Stamm C, Kleine HD, Choi YH et al (2007) Intramyocardial delivery of CD133+ bone marrow cells and coronary artery bypass grafting for chronic ischemic heart disease: safety and efficacy studies. J Thorac Cardiovasc Surg 133:717–25
188. Rieder E, Seebacher G, Kasimir MT et al (2005) Tissue engineering of heart valves: decellularized porcine and human valve scaffolds differ importantly in residual potential to attract monocytic cells. Circulation 111(21):2712–4
189. Sayk F, Bos I, Schubert U et al (2005) Histopathologic findings in a novel decellularized pulmonary homograft: an autopsy study. Ann Thorac Surg 79(5):1755–8
190. Miller DV, Edwards WD, Zehr KJ (2006) Endothelial and smooth muscle cell populations in a decellularized cryopreserved aortic homograft (SynerGraft) 2 years after implantation. J Thorac Cardiovasc Surg 132(1):175–6
191. Zehr KJ, Yagubyan M, Connolly HM et al (2005) Aortic root replacement with a novel decellularized cryopreserved aortic homograft: postoperative immunoreactivity and early results. J Thorac Cardiovasc Surg 130(4):1010–5
192. Bechtel JF, Müller-Steinhardt M, Schmidtke C et al (2003) Evaluation of the decellularized pulmonary valve homograft (SynerGraft). J Heart Valve Dis 12(6):734–9
193. Bechtel JF, Stierle U, Sievers HH (2008) Fifty-two months' mean follow up of decellularized SynerGraft-treated pulmonary valve allografts. J Heart Valve Dis 17(1):98–104
194. Shin'oka T, Shum-Tim D, Ma PX et al (1998) Creation of viable pulmonary artery autografts through tissue engineering. J Thorac Cardiovasc Surg 115(3):536–45
195. Iwai S, Sawa Y, Ichikawa H et al (2004) Biodegradable polymer with collagen microsponge serves as a new bioengineered cardiovascular prosthesis. J Thorac Cardiovasc Surg 128(3):472–9
196. Iwai S, Sawa Y, Taketani S et al (2005) Novel tissue-engineered biodegradable material for re-construction of vascular wall. Ann Thorac Surg 80(5):1828
197. Torikai K, Ichikawa H, Hirakawa K et al (2008) A self-renewing, tissue-engineered vascular graft for arterial reconstruction. J Thorac Cardiovasc Surg 136:37–45

198. Yokota T, Ichikawa H, Matsumiya G et al (2008) In situ tissue regeneration using a novel tissue-engineered, small-caliber vascular graft without cell seeding. J Thorac Cardiovasc Surg 2008 136:900–7
199. Brennan MP, Dardik A, Hibino N et al (2008) Tissue-engineered vascular grafts demonstrate evidence of growth and development when implanted in a juvenile animal model. Ann Surg 248(3):370–7
200. Hibino N, Shin'oka T, Matsumura G et al (2005) The tissue-engineered vascular graft using bone marrow without culture. J Thorac Cardiovasc Surg 129(5):1064–70
201. Matsumura G, Miyagawa-Tomita S, Shin'oka T et al (2003) First evidence that bone marrow cells contribute to the construction of tissue-engineered vascular autografts in vivo. Circulation 108(14):1729–34
202. Matsumura G, Ishihara Y, Miyagawa-Tomita S et al (2006) Evaluation of tissue-engineered vascular autografts. Tissue Eng 12(11):3075–83
203. Shin'oka T, Matsumura G, Hibino N et al (2005) Midterm clinical result of tissue-engineered vascular autografts seeded with autologous bone marrow cells. J Thorac Cardiovasc Surg 129(6):1330–8
204. De Visscher G, Vranken I, Lebacq A et al (2007) In vivo cellularization of a cross-linked matrix by intraperitoneal implantation: a new tool in heart valve tissue engineering. Eur Heart J 28(11):1389–96
205. De Visscher G, Blockx H, Meuris B et al (2008) Functional and biomechanical evaluation of a completely recellularized stentless pulmonary bioprosthesis in sheep. J Thorac Cardiovasc Surg 135(2):395–404
206. Hayashida K, Kanda K, Yaku H et al (2007) Development of an in vivo tissue-engineered, autologous heart valve (the biovalve): preparation of a prototype model. J Thorac Cardiovasc Surg 134(1):152–9
207. Nakayama Y, Yamanami M, Yahata Y et al (2009) Preparation of a completely autologous trileaflet valve-shaped construct by in-body tissue architecture technology. J Biomed Mater Res B Appl Biomater 91(2):813–8
208. Vranken I, De Visscher G, Lebacq A et al (2008) The recruitment of primitive Lin(−) Sca-1(+), CD34(+), c-kit(+) and CD271(+) cells during the early intraperitoneal foreign body reaction. Biomaterials 29:797–808
209. Ruiz CE, Iemura M, Medie S et al (2005) Transcatheter placement of a low-profile biodegradable pulmonary valve made of small intestinal submucosa: a long-term study in a swine model. J Thorac Cardiovasc Surg 130(2):477–84
210. Stock UA, Schenke-Layland K (2006) Performance of decellularized xenogeneic tissue in heart valve replacement. Biomaterials 27:1–2
211. Schaefermeier PK, Cabeza N, Besser JC et al (2009) Potential cell sources for tissue engineering of heart valves in comparison with human pulmonary valve cells. ASAIO J 55(1):86–92
212. Schnell AM, Hoerstrup SP, Zund G et al (2001) Optimal cell source for cardiovascular tissue engineering: venous vs. aortic human myofibroblasts. Thorac Cardiovasc Surg 49(4):221–5
213. Shin'oka T, Shum-Tim D, Ma PX (1997) Tissue-engineered heart valve leaflets: does cell origin affect outcome? Circulation 96:II102–7
214. Zund G, Hoerstrup SP, Schoeberlein A et al (1998) Tissue engineering: a new approach in cardiovascular surgery: seeding of human fibroblasts followed by human endothelial cells on resorbable mesh. Eur J Cardiothorac Surg 13:160–4
215. Mol A, Hoerstrup SP (2004) Heart valve tissue engineering – where do we stand? Int J Cardiol 95(1 Suppl):S57–8
216. Hoerstrup SP, Zund G, Schoeberlein A et al (1998) Fluorescence activated cell sorting: a reliable method in tissue engineering of a bioprosthetic heart valve. Ann Thorac Surg 66(5):1653–7
217. Schmidt D, Hoerstrup SP (2007) Tissue engineered heart valves based on human cells. Swiss Med Wkly 155:80–5
218. Ugurlucan M, Yerebakan C, Furlani D et al (2009) Cell sources for cardiovascular tissue regeneration and engineering. Thorac Cardiovasc Surg 57:63–73

219. Breuer CK, Shin'oka T, Tanel RE et al (1996) Tissue engineering lamb heart valve leaflets. Biotechnol Bioeng 50(5):562–7
220. Shin'oka T (2002) Tissue engineered heart valves: autologous cell seeding on biodegradable polymer scaffold. Artif Organs 26:402–6
221. Kadner A, Zund G, Maurus C et al (2004) Human umbilical cord cells for cardiovascular tissue engineering: a comparative study. Eur J Cardiothorac Surg 25(4):635–41
222. Shum-Tim D, Stock U, Hrkach J (1999) Tissue engineering of autologous aorta using a new biodegradable polymer. Ann Thorac Surg 68:2298–304
223. De Vriese AS, Verbeuren TJ, Van de Voorde J et al (2000) Endothelial dysfunction in diabetes. Br J Pharmacol 130(5):963–74
224. Bonetti PO, Lerman LO, Lerman A (2003) Endothelial dysfunction: a marker of atherosclerotic risk. Arterioscler Thromb Vasc Biol 23(2):168–75
225. Kadner A, Hoerstrup SP, Zund G et al (2002) A new source for cardiovascular tissue engineering: human bone marrow stromal cells. Eur J Cardiothorac Surg 21(6):1055–60
226. Perry TE, Kaushal S, Sutherland FW et al (2003) Thoracic Surgery Directors Association Award. Bone marrow as a cell source for tissue engineering heart valves. Ann Thorac Surg 75(3):761–7
227. Prockop DJ (1997) Marrow stromal cells as stem cells for nonhematopoietic tissues. Science 276(5309):71–4
228. Orlic D, Kajstura J, Chimenti S et al (2001) Bone marrow cells regenerate infarcted myocardium. Nature 410(6829):701–5
229. Sata M, Saiura A, Kunisato A et al (2002) Hematopoietic stem cells differentiate into vascular cells that participate in the pathogenesis of atherosclerosis. Nat Med 8(4):403–9
230. Taylor PM, Allen SP, Yacoub MH (2000) Phenotypic and functional characterization of interstitial cells from human heart valves, pericardium and skin. J Heart Valve Dis 9:150–8
231. Oswald J, Boxberger S, Jorgensen B et al (2004) Mesenchymal stem cells can be differentiated into endothelial cells in vitro. Stem Cells 22(3):377–84
232. Huang CY, Hagar KL, Frost LE et al (2004) Effects of cyclic compressive loading on chondrogenesis of rabbit bone-marrow derived mesenchymal stem cells. Stem Cells 22(3):313–23
233. Liechty KW, MacKenzie TC, Shaaban AF et al (2000) Human mesenchymal stem cells engraft and demonstrate site-specific differentiation after in utero transplantation in sheep. Nat Med 6(11):1282–6
234. Xin Y, Wang YM, Zhang H et al (2009) Aging adversely impacts biological properties of human bone marrow-derived mesenchymal stem cells: implications for tissue engineering heart valve construction. Artif Organs 34(3):215–22
235. Tanaka KA, Key NS, Levy JH (2009) Blood coagulation: hemostasis and thrombin regulation. Anesth Analg 108:1433–46
236. El-Hamamsy I, Balachandran K, Yacoub MH et al (2009) Endothelium-dependent regulation of the mechanical properties of aortic valve cusps. J Am Coll Cardiol 53(16):1448–55
237. Kasimir MT, Weigel G, Sharma J et al (2005) The decellularized porcine heart valve matrix in tissue engineering: platelet adhesion and activation. Thromb Haemost 94(3):469–70
238. Alsberg E, von Recum HA, Mahoney MJ (2006) Environmental cues to guide stem cell fate decision for tissue engineering applications. Expert Opin Biol Ther 6:847–66
239. Aird WC (2007) Phenotypic heterogeneity of the endothelium: II. Representative vascular beds. Circ Res 100(2):174–90
240. Asahara T, Murohara T, Sullivan A et al (1997) Isolation of putative progenitor endothelial cells for angiogenesis. Science 275(5302):964–7
241. Kim S, von Recum HA (2008) Endothelial stem cells and precursors for tissue engineering: cell source, differentiation, selection, and application. Tissue Eng Part B Rev 14(1):133–47
242. Kim S, von Recum HA (2009) Endothelial progenitor populations in differentiating embryonic stem cells I: identification and differentiation kinetics. Tissue Eng Part A 15(12):3709–18
243. Urbich C, Dimmeler S (2004) Endothelial progenitor cells: characterization and role in vascular biology. Circ Res 95(4):343–53
244. Urbich C, Dimmeler S (2004) Endothelial progenitor cells functional characterization. Trends Cardiovasc Med 14:318–22

245. Kawamoto A, Losordo DW (2008) Endothelial progenitor cells for cardiovascular regeneration. Trends Cardiovasc Med 18(1):33–7
246. Hofmann M, Wollert KC, Meyer GP et al (2005) Monitoring of bone marrow cell homing into the infarcted human myocardium. Circulation 111:2198–202
247. Iwasaki H, Kawamoto A, Ishikawa M et al (2006) Dose-dependent contribution of CD34-positive cell transplantation to concurrent vasculogenesis and cardiomyogenesis for functional regenerative recovery after myocardial infarction. Circulation 113:1311–25
248. Kocher AA, Schuster MD, Szabolcs MJ et al (2001) Neovascularization of ischemic myocardium by human bone marrow-derived angioblasts prevents cardiomyocyte apoptosis, reduces remodeling and improves cardiac function. Nat Med 7:430–6
249. Martin-Rendon E, Brunskill S, Dorée C et al (2008) Stem cell treatment for acute myocardial infarction. Cochrane Database Syst Rev 8(4):CD006536
250. Pearson JD (2009) Endothelial progenitor cells – hype or hope? J Thromb Haemost 7(2):255–62
251. Shirota T, He H, Yasui H (2003) Human endothelial progenitor cell-seeded hybrid graft: proliferative and antithrombogenic potentials in vitro and fabrication processing. Tissue Eng 9:127–36
252. Kaushal S, Amiel GE, Guleserian KJ et al (2001) Functional small-diameter neovessels created using endothelial progenitor cells expanded ex vivo. Nat Med 7(9):1035–40
253. Shirota T, Yasui H, Shimokawa H et al (2003) Fabrication of endothelial progenitor cell (EPC)-seeded intravascular stent devices and in vitro endothelialization on hybrid vascular tissue. Biomaterials 24:2295–302
254. Schmidt D, Breymann C, Weber A et al (2004) Umbilical cord blood derived endothelial progenitor cells for tissue engineering of vascular grafts. Ann Thorac Surg 78(6):2094–8
255. Schmidt D, Mol A, Neuenschwander S et al (2005) Living patches engineered from human umbilical cord derived fibroblasts and endothelial progenitor cells. Eur J Cardiothorac Surg 27(5):795–800
256. Schmidt D, Asmis LM, Odermatt B (2006) Engineered living blood vessels: functional endothelia generated from human umbilical cord-derived progenitors. Ann Thorac Surg 82(4):1465–71
257. Fang NT, Xie SZ, Wang SM et al (2007) Construction of tissue-engineered heart valves by using decellularized scaffolds and endothelial progenitor cells. Chin Med J (Engl) 120(8):696–702
258. Dvorin EL, Wylie-Sears J, Kaushal S et al (2003) Quantitative evaluation of endothelial progenitors and cardiac valve endothelial cells: proliferation and differentiation on polyglycolic acid/poly-4-hydroxybutyrate scaffold in response to vascular endothelial growth factor and transforming growth factor beta1. Tissue Eng 9(3):487–93
259. Sales VL, Engelmayr GC Jr, Mettler BA et al (2006) Transforming growth factor-beta1 modulates extracellular matrix production, proliferation, and apoptosis of endothelial progenitor cells in tissue-engineering scaffolds. Circulation 114(1 Suppl):I193–9
260. Ferguson VL, Dodson RB (2009) Bioengineering aspects of the umbilical cord. Eur J Obstet Gynecol Reprod Biol 144(1):108–13
261. Wang HS, Hung SC, Peng ST et al (2004) Mesenchymal stem cells in Wharton's Jelly of the human umbilical cord. Stem Cells 22:1330–37
262. Kadner A, Hoerstrup SP, Tracy J et al (2002) Human umbilical cord cells: a new cell source for cardiovascular tissue engineering. Ann Thorac Surg 74(4):S1422–8
263. Kobayashi K, Kubota T, Aso T (1998) Study on myofibroblast differentiation in the stromal cells of Wharton's jelly: expression and localization of alpha-smooth muscle actin. Early Hum Dev 51:223–33
264. Kogler G, Sensken S, Airey JA et al (2004) A new human somatic stem cell from placental cord blood with intrinsic pluripotent differentiation potential. J Exp Med 200:123–35
265. Lee OK, Kuo TK, Chen WM et al (2004) Isolation of multipotent mesenchymal stem cells from umbilical cord blood. Blood 103:1669–75
266. Sarugaser R, Lickorish D, Baksh D et al (2005) Human umbilical cord perivascular (HUCPV) cells: a source of mesenchymal progenitors. Stem Cells 23(2):220–9

267. Weiss ML, Medicetty S, Bledsoe AR et al (2006) Human umbilical cord matrix stem cells: preliminary characterization and effect of transplantation in a rodent model of Parkinson's disease. Stem Cells 24:781–92
268. Weiss ML, Anderson C, Medicetty S et al (2008) Immune properties of human umbilical cord Wharton's jelly-derived cells. Stem Cells 26:2865–74
269. Sipehia R, Martucci G, Lipscombe J (1996) Transplantation of human endothelial cell monolayer on artificial vascular prosthesis: the effect of growth-support surface chemistry, cell seeding density, ECM protein coating, and growth factors. Artif Cells Blood Substit Immobil Biotechnol 24:51–63
270. Hoerstrup SP, Kadner A, Breymann CI et al (2002) Living, autologous pulmonary artery conduits tissue engineered from human umbilical cord cells. Ann Thorac Surg 74:46–52
271. Koike N, Fukumura D, Gralla O et al (2004) Tissue engineering: creation of long-lasting blood vessels. Nature 428(6979):138–9
272. Breymann C, Schmidt D, Hoerstrup SP (2006) Umbilical cord cells as a source of cardiovascular tissue engineering. Stem Cell Rev 2(2):87–92
273. Armson BA, Maternal/Fetal Medicine Committee, Society of Obstetricians and Gynaecologists of Canada (2005) Umbilical cord blood banking: implications for perinatal care providers. J Obstet Gynaecol Can 27(3):263–90
274. Ruhil S, Kumar V, Rathee P (2009) Umbilical cord stem cell: an overview. Curr Pharm Biotechnol 10(3):327–34
275. Pappa KI, Anagnou NP (2009) Novel sources of fetal stem cells: where do they fit on the developmental continuum? Regen Med 4(3):423–33
276. Zhang X, Mitsuru A, Igura K et al (2006) Mesenchymal progenitor cells derived from chorionic villi of human placenta for cartilage tissue engineering. Biochem Biophys Res Commun 340:944–52
277. Miki T, Strom SC (2006) Amnion-derived pluripotent/multipotent stem cells. Stem Cell Rev 2(2):133–42
278. Toda A, Okabe M, Yoshida T, Nikaido T (2007) The potential of amniotic membrane/amnion-derived cells for regeneration of various tissues. J Pharmacol Sci 105(3):215–28
279. Parolini O, Soncini M, Evangelista M et al (2009) Amniotic membrane and amniotic fluid-derived cells: potential tools for regenerative medicine? Regen Med 4(2):275–91
280. Kaviani A, Guleserian K, Perry TE et al (2003) Fetal tissue engineering from amniotic fluid. J Am Coll Surg 196(4):592–7
281. Kunisaki SM, Armant M, Kao GS et al (2007) Tissue engineering from human mesenchymal amniocytes: a prelude to clinical trials. J Pediatr Surg 42(6):974–9; discussion 979–80
282. Pansky A, Roitzheim B, Tobiasch E (2007) Differentiation potential of adult human mesenchymal stem cells. Clin Lab 53(1–2):81–4
283. Tuan RS, Boland G, Tuli R (2003) Adult mesenchymal stem cells and cell-based tissue engineering. Arthritis Res Ther 5:32–45
284. Zuk PA, Zhu M, Mizuno H et al (2001) Multilineage cells from human adipose tissue: implications for cell-based therapies. Tissue Eng 7:211–28
285. Cao Y, Sun Z, Liao L et al (2005) Human adipose tissue-derived stem cells differentiate into endothelial cells in vitro and improve postnatal neovascularization in vivo. Biochem Biophys Res Commun 332(2):370–9
286. Miranville A, Heeschen C, Sengenès C et al (2004) Improvement of postnatal neovascularization by human adipose tissue-derived stem cells. Circulation 110(3):349–55
287. Planat-Benard V, Silvestre JS, Cousin B et al (2004) Plasticity of human adipose lineage cells toward endothelial cells: physiological and therapeutic perspectives. Circulation 109(5):656–63
288. Colazzo F, Sarathchandra P, Chester AH et al (2009) An evaluation of adipose-derived stem cells for heart valve tissue engineering. Paper presented at the 5th biennial meeting of the Society of Heart Valve Disease, Berlin, 28–30 June 2009
289. DiMuzio P, Tulenko T (2007) Tissue engineering applications to vascular bypass graft development: the use of adipose-derived stem cells. J Vasc Surg 45(Suppl A):A99–103

290. Okita K, Ichisaka T, Yamanaka S (2007) Generation of germline-competent induced pluripotent stem cells. Nature 448(7151):313–7
291. Takahashi K, Yamanaka S (2006) Induction of pluripotent stem cells from mouse embryonic and adult fibroblast cultures by defined factors. Cell 126(4):663–76
292. Wernig M, Meissner A, Foreman R (2007) In vitro reprogramming of fibroblasts into a pluripotent ES-cell-like state. Nature 448:318–24
293. Mauritz C, Schwanke K, Reppel M et al (2008) Generation of functional murine cardiac myocytes from induced pluripotent stem cells. Circulation 118(5):507–17
294. Zhang J, Wilson GF, Soerens AG (2009) Functional cardiomyocytes derived from human induced pluripotent stem cells. Circ Res 104:e30–41
295. Guan K, Nayernia K, Maier LS et al (2006) Pluripotency of spermatogonial stem cells from adult mouse testis. Nature 440(7088):1199–203
296. Seandel M, James D, Shmelkov SV et al (2007) Generation of functional multipotent adult stem cells from GPR125+ germline progenitors. Nature 449(7160):346–50

Tissue Engineering Strategies for Cardiac Regeneration

Amandine F.G. Godier-Furnémont, Yi Duan, Robert Maidhof, and Gordana Vunjak-Novakovic

Abstract Once injured, cardiac muscle does not regenerate. Massive and irreversible loss of cardiomyocytes due to myocardial infarction remains the main cause of heart failure. Cardiac tissue engineering has a potential to reestablish the structure and function of injured myocardium, by (1) injection of cardiogenic cells, (2) transplantation of functional cardiac tissue constructs, or (3) mobilization of endogenous repair cells. Irrespective of the therapeutic strategy, the regeneration depends on both the biological potential of the repair cells and the environment (matrix, signals) to which the cells are subjected. In general, the biological potential of the cells – the actual "tissue engineers" – needs to be mobilized by providing highly controllable "biomimetic" environments designed to enhance cell survival, differentiation, and electromechanical coupling. For cardiac regeneration, some of the key requirements include the establishment of cardiac tissue matrix, electromechanical cell coupling, robust contractile function, and functional vascularization. Engineered tissue constructs can also serve as high-fidelity models to study cardiac development and disease. We review the potential and challenges of cardiac tissue engineering to provide for the development of therapies for heart regeneration.

Keywords Tissue scaffolds • Bioreactors • Electrical stimulation • Engineered heart tissue • Vascularization • Cell sheets • Cell stacking

1 Developmental Principles, Engineering Designs

The goal of tissue engineering is to grow tissue grafts in vitro (by using cells, specialized scaffolds, and bioreactors) or induce tissue regeneration in situ (by providing the necessary environments for exogenous or host cells). Engineered tissues can also

G. Vunjak-Novakovic (✉)
Department of Biomedical Engineering, Columbia University,
New York, NY
e-mail: gv2131@columbia.edu

I.S. Cohen and G.R. Gaudette (eds.), *Regenerating the Heart*, Stem Cell Biology and Regenerative Medicine, DOI 10.1007/978-1-61779-021-8_23,
© Springer Science+Business Media, LLC 2011

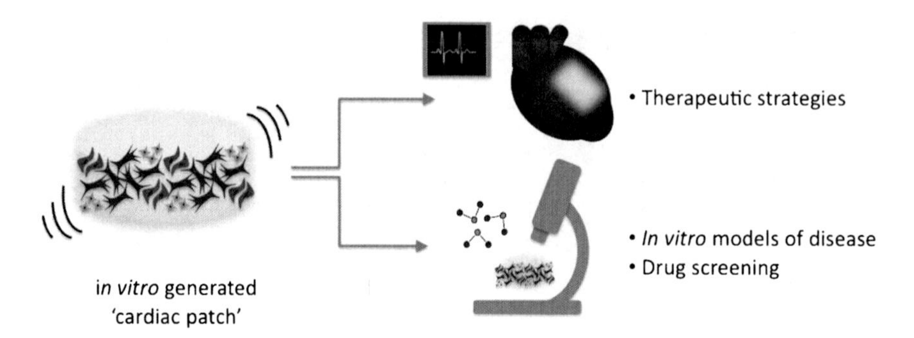

Fig. 1 Applications of cardiac tissue engineering. An engineered cardiac patch could be used to regenerate cardiac tissue lost to myocardial infarction, as a model of disease, and as a platform for screening of drugs and biologicals

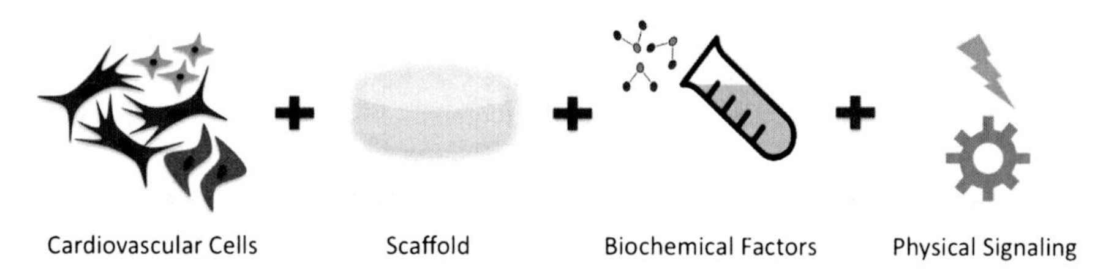

Fig. 2 Tissue engineering paradigm. Cells (the actual "tissue engineers") are directed to form functional tissue structures by biomaterial scaffolds (providing a structural and informational template for tissue formation) and cascades of biochemical and physical signals (provided in vitro by a bioreactor and in vivo by the tissue regeneration milieu)

be used for study of disease and drug screening (Fig. 1). Engineered tissues should be customized to the needs of a specific patient and clinical situation, and provide some immediate functionality upon implantation. To create a functioning tissue one needs to unlock the full potential of stem cells, by mobilizing and controlling factors involved in tissue development and regeneration. Cellular, biochemical, and biophysical mechanisms collectively regulate native tissue development, at various spatial and temporal scales. In engineered tissues, some of these developmental events need to be recapitulated to yield functional tissues. A standard tissue engineering paradigm involves a coordinated use of cells, biomaterial scaffolds, and the in vitro and in vivo environments providing biochemical and physical regulatory signals (Fig. 2). Our attempts to mimic the necessary conditions for tissue regeneration are limited by how much is known about the combinations and timing of regulatory factors involved in cell differentiation. It is thus important to first understand how uncommitted stem cells respond to the external and internal signals and stressors.

Since the first-ever tissue culture study performed by Harrison in 1907 [1], mammalian cells have been cultured in two-dimensional (2D) settings, resulting in our current knowledge of molecular regulation of stem cells. More recently, several groups have used these 2D cultures to interrogate geometric control of cell life and death in culture [2], and the control of cell differentiation by substrate stiffness [3, 4]. These studies go beyond the traditional focus on molecular regulatory factors and open a new era of utilization of physical factors mediating cell function, with direct implication for tissue engineering and regenerative medicine. However, there is a growing notion that cell confinement to 2D culture is an inherently unnatural situation owing to the attachment of one side of the cell to the substrate, and limited contact with other cells and the matrix. Therefore, many current approaches involve cell culture in three-dimensional (3D) environments.

The biomechanical signals cells sense in vivo are associated with cellular deformation (due to compressibility or shear), mechanical stress (in response to pressure or shear force), and the deformation of the cell nucleus (elongation in response to tension). Among these factors, the change in cell shape appears to be the most obvious indicator of physical effects on cell function. Variation in cellular function, including gene expression leading to matrix synthesis and expression of surface markers, is often associated with the changes in cellular morphology. A series of groundbreaking studies demonstrated that the cell shape and cytoskeletal tension can be effectively manipulated by varying the substrate stiffness [5, 6]. These studies showed that cells prefer the mechanical environment they normally experience in vivo, to the extent that substrate stiffness alone can direct cell differentiation [5]. For example, embryonic cardiomyocytes beat best on a substrate with a stiffness similar to that of native heart tissue, which is optimal for transmitting contractile work to the matrix, and only poorly on stiffer substrates matching the stiffness of scar tissue formed following cardiac infarction [6].

In general, cell behavior is more natural when culture is performed in 3D environments. For many years, 3D scaffolds have been designed to provide a biocompatible structural template for cell attachment that would be permissive for the exchange of nutrients, metabolites, and cytokines and allow cell seeding. Numerous studies (reviewed in [7–9]) have demonstrated that the cell phenotype depends on the entire context of the cellular microenvironment. For engineering cardiac muscle, structural requirements include the optimization of scaffold pores (to provide the right balance between pore size determining cell migration and pore curvature determining cell attachment) and hierarchical structure (orientation, anisotropy, channels for vascular conduits). New "cell-instructive" materials are being utilized to mimic the native matrix and actively interact with the cells. These new scaffolds are functional at multiple length scales and timescales: *molecular* (by incorporation of integrin-binding ligands and regulation of availability of growth factors), *cellular* (directed migration, mediation of cell–cell contacts and stiffness as a differentiation factor), and *tissue* levels (establishment of interfaces, structural and mechanical anisotropy). The enormous variation of cell/tissue properties has led to the development of "designer scaffolds" [10] (Fig. 3).

One specialized scaffold for cardiac tissue engineering is the elastomer poly(glycerol sebacate) (PGS) [11]. The pores, stiffness, and channel geometry in this scaffold have been designed to enable engineering of vascularized cardiac tissue. The structural and mechanical properties of native cardiac tissue have

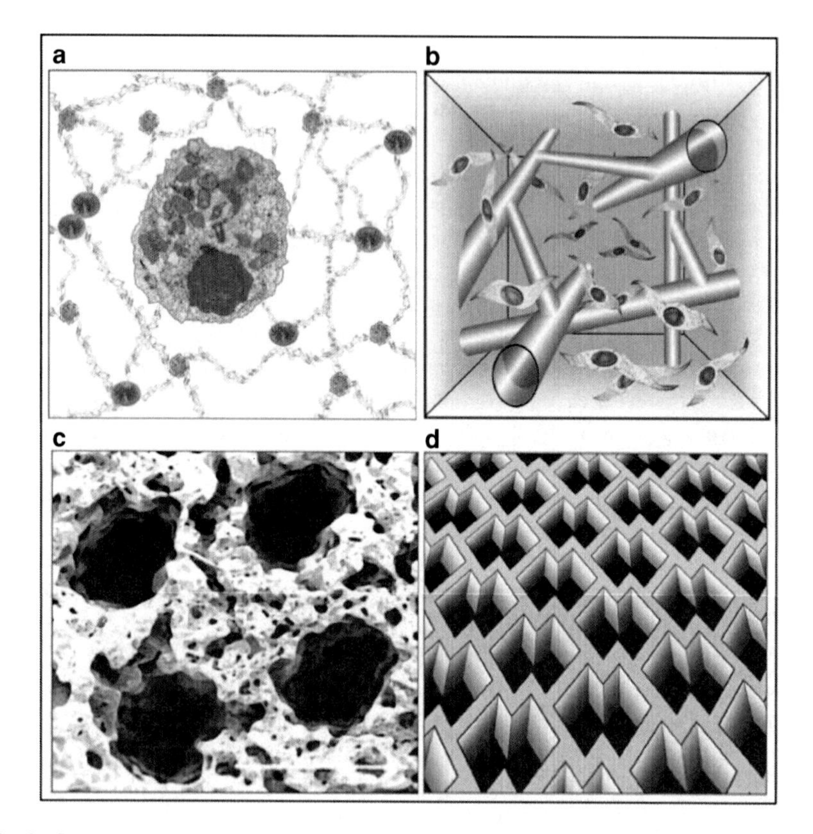

Fig. 3 Cardiac tissue engineering scaffolds. Scaffolds used to provide structural and mechanical signals to engineered cardiac tissue include (**a**) hydrogels with tunable molecular, mechanical, and degradation properties, (**b**) hydrogels modified by laser light to form channels for cell migration, (**c**) a soft, highly porous channeled elastomer [poly(glycerol sebacate) (PGS)] scaffold, and (**d**) accordion-like elastomer scaffolds with structural and mechanical anisotropy. (Reproduced with permission from [10])

guided the scaffold design. Another "designer scaffold" was engineered to mimic the anisotropic structure and biomechanics of cardiac muscle [3]. The scaffold (also composed of PGS) was processed by microfabrication into an accordion-like honeycomb. When cultured with neonatal cardiomyocytes, the scaffold induced cell alignment and coupling, and resulted in direction-dependent contractile behavior, a situation much closer to native heart tissue properties than can be achieved with isotropic scaffolds.

Advanced methods to improve control over the 3D cellular microenvironment are being developed. The conventional culture setting (e.g., well plates) provides environmental control only through periodic exchange of culture medium, and fundamentally lacks the capability for coordination of molecular and physical regulatory signals. These conditions are far from the in vivo situation, where cells reside in a precisely controlled environment, and are subjected to spatial and temporal gradients of multiple factors. To overcome these limitations, bioreactors can be used to provide tightly controlled, dynamic culture settings that expose cells to various biochemical and biophysical conditions with gradients in space and time.

Mechanical forces play similar roles in native and engineered tissues. During development, mechanical manipulation of the embryo causes abnormal axis formation [14], and abnormal blood flow impairs the formation of heart chambers and valves [15]. In tissue engineering, physical regulation during cultivation of mechanically active tissues has emerged as a new paradigm of "functional tissue engineering" [16].

For cells seeded in collagen scaffolds fixed at both ends, mechanical forces are generated inside the gel, as cells are rearranging, aligning, and actively pulling collagen fibrils along the scaffold axis [17, 18]. Tensile forces could also be generated by external mechanical tension. When applied to cardiac tissue engineering, this concept induced rapid formation of intensively interconnected, longitudinally oriented cardiac muscle bundles with morphological features resembling those of adult native tissue [19]. The cellular tension can also be induced via excitation–contraction coupling by applying cardiac-like electrical stimulation to cultured cells on scaffolds using bioreactors fitted with conducting electrodes [20]. Similar to the application of mechanical stretch, the electrically stimulated cells conduct electrical pacing signals over macroscopic distances, and beat synchronously at the frequency of stimulation. The application of mechanical and electrical stimulation enhanced cellular, structural, and functional properties of engineered myocardium.

1.1 Summary

Tissue engineering has resulted in a new generation of culture systems of high biological fidelity that are finding applications in fundamental biological research, engineering of functional tissue grafts, and studies of disease. These engineering designs are guided by a "biomimetic" approach, which attempts to recapitulate in vitro some of the important aspects of the native cellular milieu associated with tissue development and regeneration [10].

2 Current Concepts in Cardiac Tissue Engineering

We review here six fundamentally different landmark approaches to building functional tissue grafts.

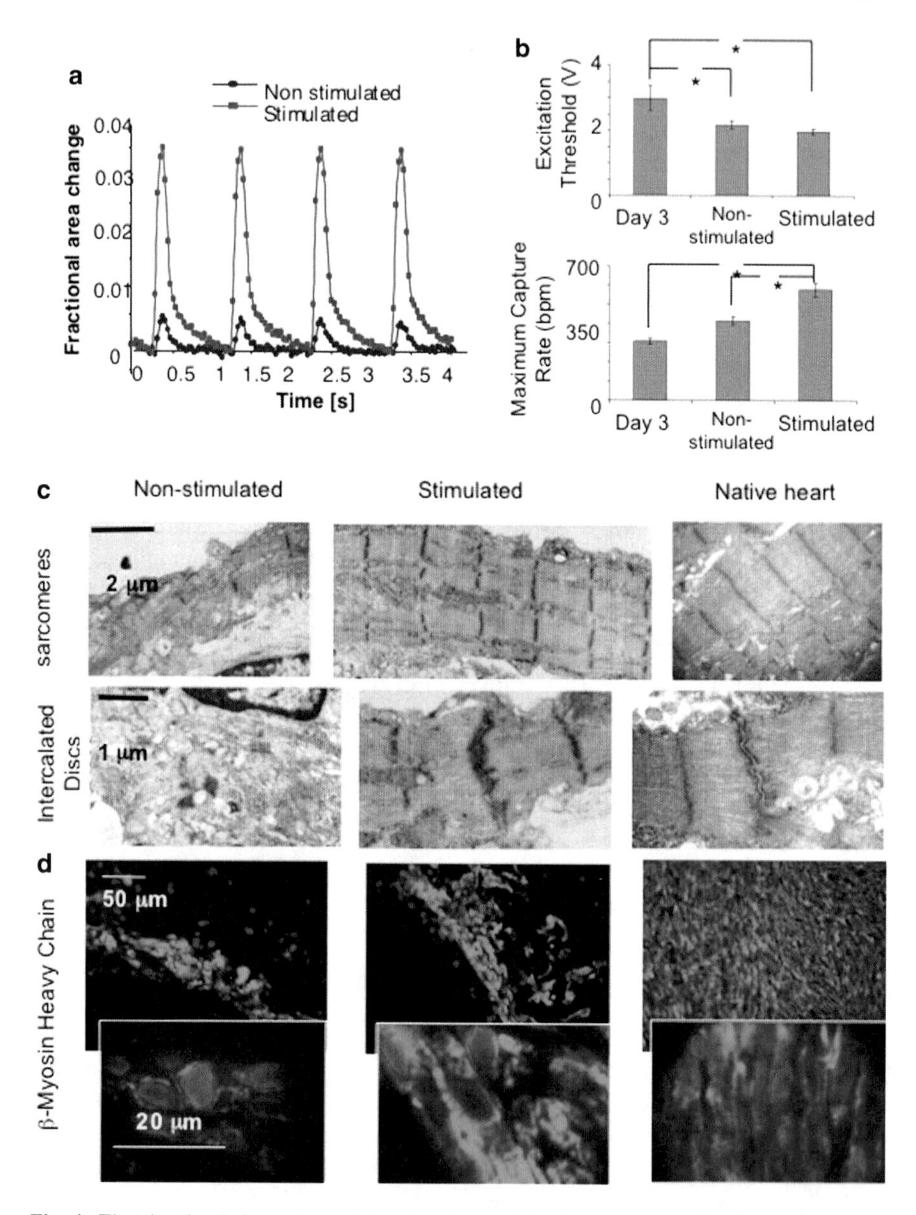

Fig. 4 Electric stimulation enhanced functional assembly of cardiac tissue in vitro. Neonatal rat heart cells were cultured for a total of 8 days (3 days of preculture and 5 days with or without electrical stimulation) and compared with the parent heart tissue. (**a**) Contraction amplitude. (**b**) Excitation threshold and maximum capture rate. (**c**) Ultrastructure (sarcomeres and intercalated discs). (**d**) β-Myosin heavy chain. (Reproduced with permission from [12])

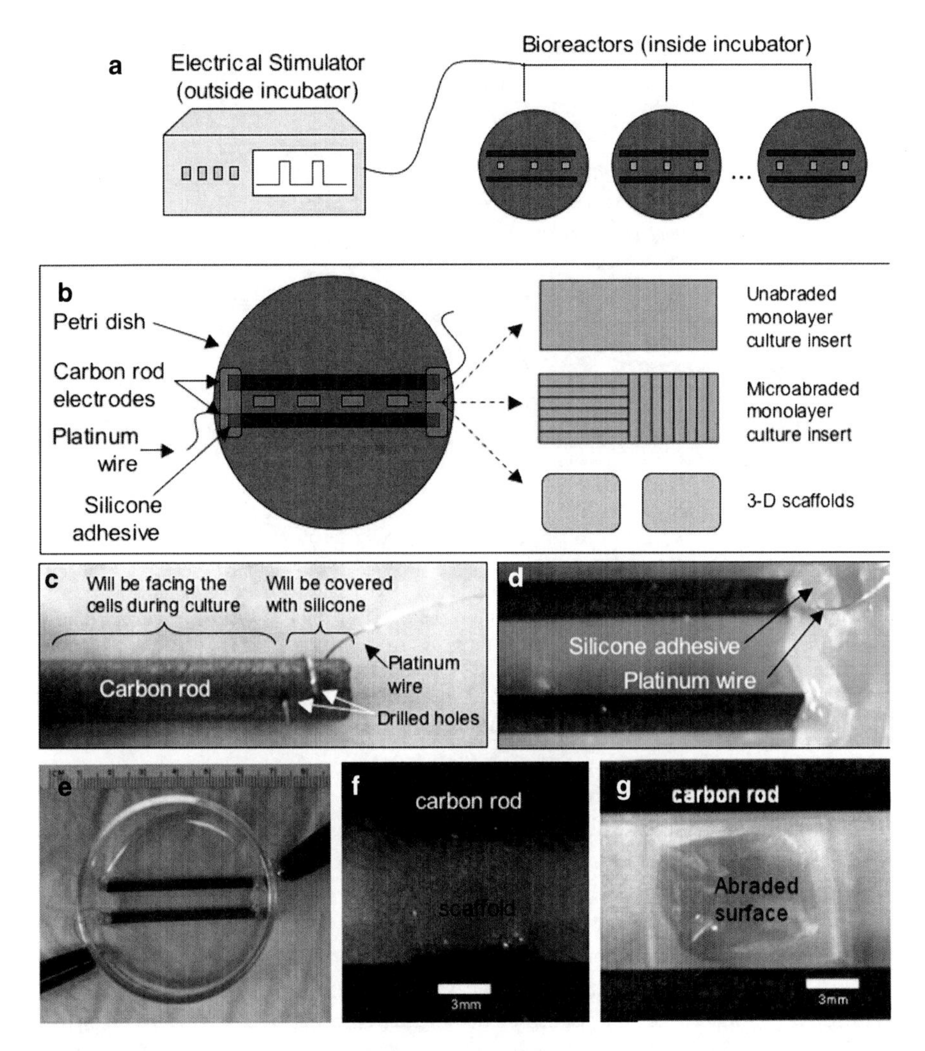

Fig. 5 Cardiac tissue engineering with electric stimulation. (**a**) Experimental setup. (**b**) Electrical stimulation chamber (modified Petri dish with carbon rod electrodes). An unabraded monolayer culture insert, a microabraded monolayer culture insert, or three-dimensional scaffolds may be placed between the electrodes. (**c**) Photograph of an assembled electrical stimulation chamber. (**d**) Close-up view of a scaffold positioned between electrodes. (**e**) Close-up view of abraded monolayer culture insert placed between electrodes. (Reproduced with permission from [12])

2.1 *Engineering Cardiac-Specific Matrix*

The anisotropic properties of native myocardium play a major role in the transduction of mechanical and electrical signals across cardiomyocytes. Muscle fibers are surrounded by the extracellular matrix (ECM), which allows for sustained

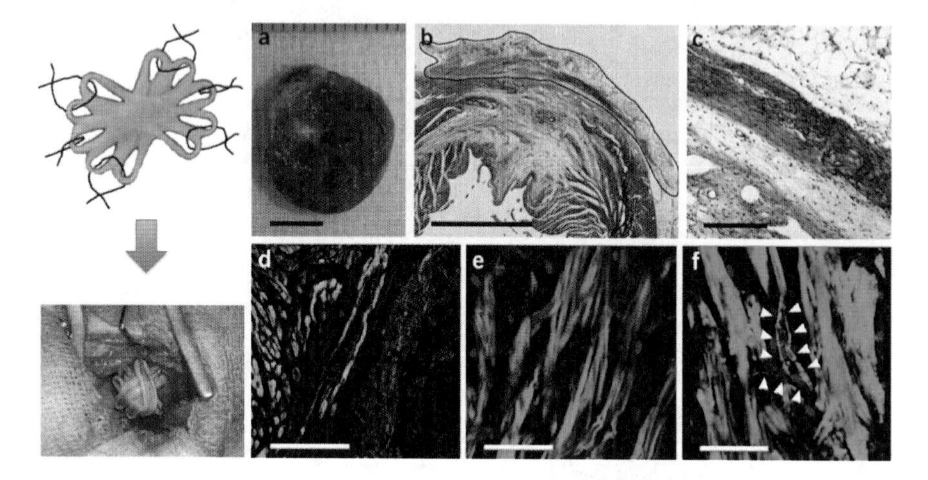

Fig. 6 Engineered heart tissue (EHT) improved function of infarcted rat hearts. EHTs were made from neonatal heart cells, collagen, and Matrigel, reconstituted in circular molds, subjected to mechanical strain, and implanted for 4 weeks: (**a**) EHTs remained pale but attached to the infarct scar; (**b**) thick cardiac muscle on top of the infarct scar (hematoxylin and eosin staining); (**c**) engrafted EHTs formed compact and oriented heart muscle spanning the transmural infarct; (**d**) engrafted EHTs (nuclei *blue*, actin *green*); (**e**) sarcomeric organization (actin *green*, nuclei *blue*); (**f**) EHT grafts were vascularized in vivo (actin, *green*; nuclei *blue*; arrows indicate a putative vessel in a reconstituted confocal image). Scale bars (**a**, **b**) 5 mm, (**c**, **d**) 500 μm, (**e**, **f**) 50 μm. (Reproduced with permission from [13])

alignment of the cardiomyocytes within highly ordered anisotropic sheets of cardiomyocyte fibers. The fiber alignment contributes to the direction-dependent electromechanical cell coupling that has not been reproduced in vitro using conventional biomaterials. The Freed group recently reported the development of a 3D PGS scaffold that successfully reproduced cardiac anisotropy on a 3D scale. This accordion-like honeycomb scaffold has mechanical properties comparable to those of native myocardium (and tunable via the curing time of PGS) [3]. Unlike traditional biomaterials that generally do not provide the necessary load-bearing capability, suturing ability, and compliance with native tissue, the PGS scaffold induced alignment of cardiomyocytes and supported their contractile function. This study indicated that the control of cell alignment throughout a 3D construct may be necessary to achieve nativelike properties of engineered tissue.

2.2 *Native Tissue for Cardiac Tissue Engineering*

Although tissue engineering relies on taking lessons from nature to inform the design of novel biomaterials and the selection of appropriate regulatory signals, several groups have sought to use biological tissues as scaffolds for cardiac tissue engineering. The Badylak group pioneered decellularization of biological tissues, and has used ECM extracts to develop gels for in vivo use. In general, the cellular content is removed from biological tissue, leaving intact the underlying ECM in its original form. They have used the ECM components from a variety of tissues and demonstrated their superiority over synthetic biomaterials, in terms of decreased immune response to the implanted tissue. Additionally, the pockets of regenerated tissue formed throughout the infarct bed evidence that the engrafted ECM promoted the recruitment, survival and differentiation of endogenous cells. Further investigation of ECM-derived scaffolds may elucidate how such signaling may be utilized to induce differentiation of stem cells in vitro, for use as an implanted viable cardiac tissue [21].

In 2007, Taylor's group reported the repopulation of whole decellularized rat myocardium with cardiovascular cell populations and restoration of 2% of adult rat heart function [22]. Although their approach is unlikely to reach the clinical setting, the study demonstrated that native myocardium is a compelling scaffold for engineering myocardium as it retains the native ECM components, with the appropriate long-range architecture, anisotropy, and mechanical properties characteristic of the native tissue. Recently the Badylak group reported on derivation of ECM from a porcine heart, demonstrating preservation of ECM components [21] and, like the Taylor group, cited advantages of using the embedded vascular conduits for cell delivery to repopulate the tissue, and as a method for vascularizing engineered tissue in the scaffold.

2.3 Electrical Stimulation of Cardiac Constructs

In 2005, the Vunjak-Novakovic group reported that electrical stimulation of neonatal rat ventricular cardiomyocytes seeded with Matrigel onto collagen sponges significantly enhanced functional tissue assembly, as assessed by increased cell alignment and improved cardiomyocyte ultrastructure that showed a level of maturation toward native adult rat myocardium and formation of functional gap junctions [20]. Various drug studies showed that functional tissue assembly is dependent on electromechanical coupling, functional gap junctions, and cytoskeletal ultrastructure. Moreover, analysis of the stimulated engineered tissues revealed sarcomeric density (by volume) that was indistinguishable from that of native rat tissue – an important indicator of functional cell assembly. Constructs were cultured with three drugs independently, each of which plays an inhibitory role in one of the above-mentioned parameters, and cultured with or without electrical stimulation to deconvolute effects of stimulation on each parameter. When verapamil, an L-type Ca^{2+} inhibitor, was introduced, electrical stimulation was able to recover function of treated constructs, sustaining levels of connexin 43 (Cx-43), although with diminished functional characteristics. Palmitoleic acid and LY294002, which block gap junctions and the phosphatidylinositol three-kinase pathway (implicated in cytoskeletal arrangement), respectively, had effects that were incapable of recovery with electrical stimulation. These drug studies demonstrate that electrical stimulation can be used to override communication effects in engineered tissues, and overall the study demonstrates that suprathreshold stimulation was critical for tissue excitability and ultrastructural organization.

Further studies have explored variations in electrical signaling [12, 23]. The original study modeled the signal based on physiological parameters (2-ms pulses, 1 Hz) and used a square wave with an amplitude of 5 V/cm [20]. In their 2008 study, Chiu et al. reported culture of cardiac organoids, with different ratios of endothelial cells, cardiac fibroblasts, and cardiomyocytes, subjected to monophasic (5 V/cm, 1 Hz, 2 ms) or biphasic (2.5 V/cm, 1 Hz, 1 ms) electrical stimulation, or no stimulation [24]. Biphasic stimulation enriched cardiomyocyte populations in the organoids, and electrical excitability of the cells, as evidenced by increased cell elongation, density, Cx-43 expression, and expression of α-actin. Electrical stimulation, along with biochemical signaling, markedly contributed to the biomimetic nature of culture platforms for cardiac tissue engineering.

2.4 Mechanical Stimulation of Engineered Heart Tissue

Cardiac tissue function is intrinsically related to electrical and mechanical signals. In the developing heart, it is the evolution of strains, stresses, and originating currents that instructs the cells down the myogenic path. These mechanical signals continue throughout the life of the cardiomyocyte, sustaining the cells' ability to indefinitely contract, and guiding the formation of heart tubes and their subsequent assembly into a functional heart. The study of cardiomyocytes under stretch was first conducted by Komuro and Yazaki in 1991 [25], who used deformable silicone membranes to apply stretch to the cells. In 2000, Eschenhagen's laboratory reported the use of chronic stretch of engineered heart tissue [26], and through subsequent incarnations of their bioreactor systems, they developed a differentiated cardiac construct and the functional in vivo benefits of the engrafted engineered heart tissues [13, 19]. They demonstrated that cell alignment and hypertrophy enhanced contractile function, and increased the RNA/DNA and protein/cell ratio, actin expression, and mitochondrial density.

In 2002, this group compared engineered heart tissues undergoing phasic stretch with native newborn, 6-day-old, and adult rat heart tissue [19]. The engineered constructs, after 14 days in culture, were able to "surpass the differentiation state of their source tissue," neonatal rat myocardium, resembling instead adult myocardium. This was evident through extensive interconnectivity, elongation, and multinucleation of the cardiomyocytes within the tissue, high sarcomeric density, and the presence of adherens, desmosomes and gap junctions. It was proposed that the engineered heart tissue would readily integrate with the host myocardium, as the engineered heart tissues exhibited extensive capillary density and nonmyocte cells, which may facilitate integration. The force of contraction surpassed the forces previously reported, but remained approximately one tenth of that in adult rat myocardium – consistent with the compaction density of engineered heart tissues that was one tenth that of native myocardium.

In 2006, Zimmermann and Eschenhagen demonstrated that engineered heart tissues significantly improve systolic and diastolic function in rats with infarcted hearts [13]. The culture conditions were further optimized from their 2002 work by culture in hyperoxic conditions, under auxotonic load and with insulin-supplemented media. At 28 days after implantation, the engrafted constructs formed a thick (approximately 440 μm) layer of cardiac muscle in the infarct, and demonstrated de novo vascularization of the graft with both the host and the engrafted cells forming the conduits. The engrafted engineered heart tissues demonstrated undelayed signal conduction and attenuated the remodeling process by inducing systolic wall thickening and improving fractional area shortening as compared with both sham-operated controls and rats implanted with noncontractile grafts. This study clearly demonstrated that engineered heart tissue could suspend further deterioration of cardiac function, and perhaps most interestingly, it was found that animals with the worst function at the baseline showed significant improvement, and represented the only group in which engineered heart tissue engraftment led to improved ejection function.

2.5 *Methods for Vascularization*

Functional vascularization – with the establishment of blood supply – remains a major unsolved problem of cardiac tissue engineering, and tissue engineering in general. Several different approaches are currently under investigation, ranging from the engineering of prevascularized tissues with capability for connection to the blood supply of the host, to the induction of vascularization by host cells using bioactive materials (Fig. 7).

In 2009 the Cohen group described a study whereby vascularization of a cardiac graft was induced by implantation on the rat omentum for 7 days, prior to implantation over the rat's infarct [27]. The incorporation of prosurvival and angiogenic factors [Matrigel, insulin-like growth factor, stromal-cell derived factor 1 (SDF-1), vascular endothelial growth factor (VEGF)] capable of sustained release in the scaffold further facilitated the vascularization of the patch. These "mixture-supplemented" cardiac patches, implanted 7 days after myocardial infarction, integrated with host vasculature and formed thick scar by day 28, demonstrating significantly more vessel density as compared with controls. Rats implanted with mixture-supplemented cardiac patches, and prevascularized by omentum, attenuated left ventricular dilation and were electromechanically coupling with the host. Omentum-mediated vascularization of the patch led to accelerated anastomosis with the host and further facilitated integration with the underlying infarct.

The Vunjak-Novakovic and Radisic groups recently described methods for perfusion-based culture of cardiac tissues (Fig. 8) [29]. On the basis of their previous work measuring oxygen gradients in static cultures [30], they hypothesized that perfusion culture of cardiac constructs would enhance viability, distribution, and density of cells, and that supplementation of media with perfluorocarbon

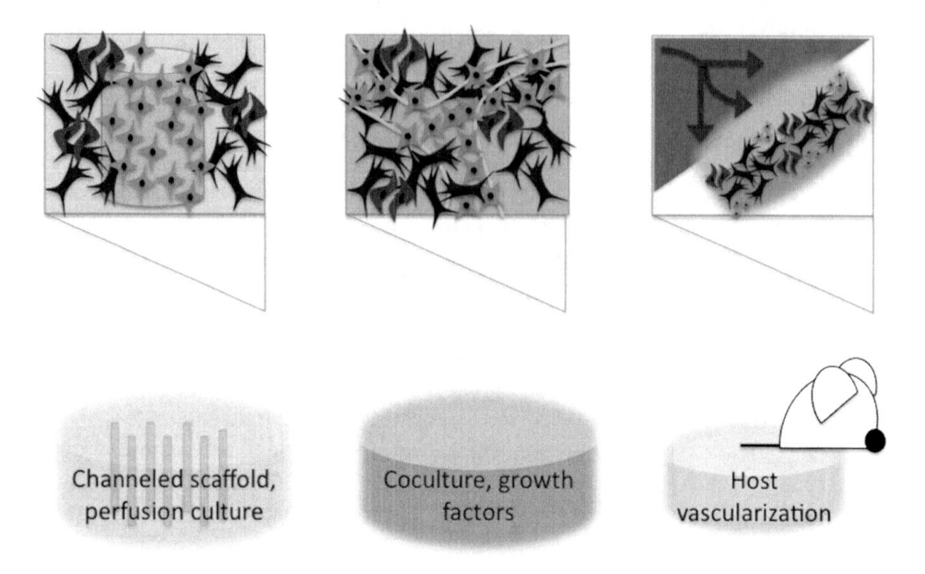

Fig. 7 Methods for vascularization. *Left*: implantation of prevascularized tissue engineered in vitro, with connection to the host vasculature. *Center*: implantation of tissue constructs containing vasculogenic cells and factors. *Right*: vascularization in situ

(an oxygen carrier to mimic hemoglobin) would further improve construct quality. They described assembly of perfusion bioreactors and outlined fundamental equations to use in calculating appropriate flow rates for both fibrous and porous scaffolds [29]. Further, they addressed the issue of nonphysiologic shear stress that cardiac cells are exposed to (earlier demonstrated to activate the p38 pathway, inducing apoptosis; [27]) by developing a channeled scaffold, whereby medium perfusion through capillary-like structures shields cardiomyocytes from shear, while allowing a thick compact cardiac tissue with sufficient oxygen and nutrients to be supplied via diffusion from the aqueous phase through the tissue's depth. Cardiac constructs cultured in the presence of perfluorocarbon and in channeled scaffolds exhibited superior contractile behavior, with higher cell density, DNA, and troponin I and Cx-43 expression. Thus, perfusion-based culture appears to provide a way of overcoming current size limitations due to diffusion constraints, for generating thick, compact cardiac tissues.

In 2009, the Gepstein group demonstrated the importance of a heterogeneous cell population for the successful formation and maintenance of functional vessels in cardiac grafts [31]. They cultured human embryonic stem cell (hESC)-derived cardiomyocytes with or without supporting endothelial cells and fibroblasts. They demonstrated a significantly greater number of vessels positive for smooth muscle α-actin, and greater size of vessels, in the triculture constructs. When these scaffolds were placed in an in vivo model, they formed functional vessels, as assessed by fluorescent microsphere or lectin injection into the left ventricular cavity. The microspheres and lectin were observed in human CD31+ vessels and demonstrate connected and perfused host–graft vasculature. Triculture grafts were associated

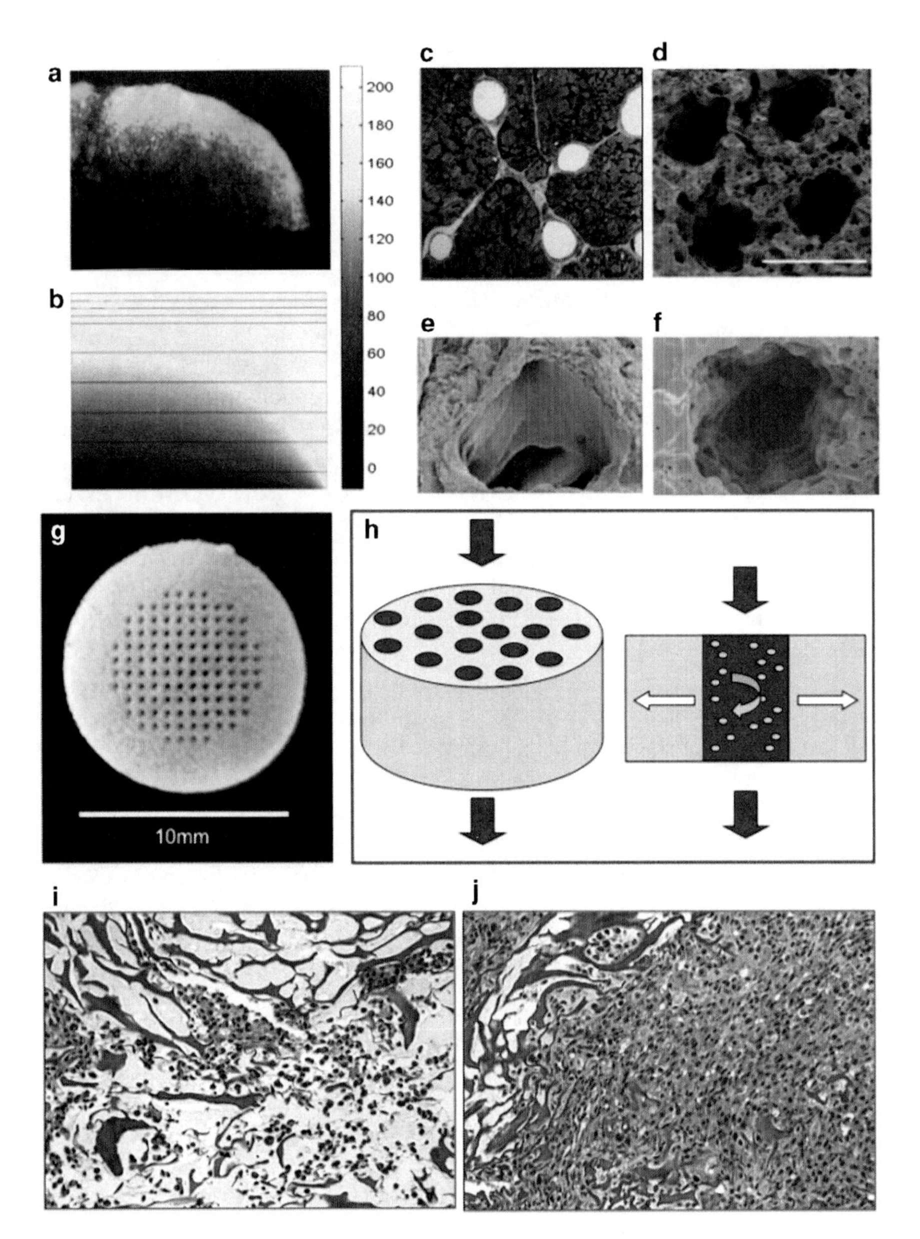

Fig. 8 Perfusion culture of cardiac cells on channeled scaffolds. (**a**) Live–dead staining of engineered cardiac tissue cultivated under static conditions. (**b**) Mathematical model of oxygen concentration, indicating that when oxygen is supplied by diffusion alone the interior of the construct remains hypoxic. (**c**) Capillary network in native myocardium. (**d**) Elastomer (PGS) scaffold with an array of channels. (**e**) One capillary in the native myocardium. (**f**) One channel in the PGS scaffold. (**g**) Macroscopic view of the scaffold. (**h**) Oxygen supply to a channeled engineered tissue (to mimic capillary flow) using culture medium with an oxygen carrier (to mimic hemoglobin). Hematoxylin and eosin staining of a collagen scaffold cultured statically (**i**) or with perfusion (interstitial flow) of culture medium (**j**). (Reproduced with permission from [28])

with more complex embedded vascular structures, and demonstrated the dependence of cardiomyocyte function and survival, as well as graft integration with the host, on the presence of these supportive cells.

2.6 Cell Sheet Engineering and Cell Stacking

The Okano group pioneered the development of contractile cell sheets for restoring function to the injured hearts [32–34]. In 2002 came their first report [34] utilizing neonatal rat ventricular myocytes, in which four layers were sequentially cultured and stacked to form a multilayer sheet. To prevent disruption of intracellular junctions, cell sheets were cultured on temperature-sensitive dishes that allowed them to be removed as a single monolayer. Undelayed coupling and signal propagation were observed between the cell sheets. When implanted subcutaneously in rats, cell sheets exhibited neovascularization and increased presence of myogenic structures (sarcomeres, desmosomes, gap junctions). Subsequently, they demonstrated bidirectional communication [32] and functional integration between cell sheets and the host myocardium [35]. Functional benefits were also reported [36] in terms of improved ejection fraction and attenuated left ventricular remodeling.

In 2009, the Okano group reported the generation of vascularized multilayered cell sheets [37]. Using a plunger device, the group sequentially stacked monolayer myoblast sheets while seeding human umbilical vein endothelial cells (HUVECs) between the layers. In vitro, they observed formation of partial capillary-like structures interconnected between sheets, and a subcutaneous model revealed neovascularization of the muscle graft. Although this study was carried out with skeletal myoblasts, we readily anticipate the application of this method for cardiovascular applications. Cell sheet technology remains a dominating approach to cell therapy as the Okano group has consistently shown effective therapeutic outcomes using these several-hundred-micron-thick grafts, evading diffusion concerns and maintaining intracellular communication throughout the graft.

2.7 Summary

Although each of the studies described above utilized a different approach – be it with cell source, scaffold, biochemical or physical signaling, and functional assessments – they highlight the critical questions that remain in the field of cardiac tissue engineering. These include finding a suitable cell source that can demonstrate the ability to engineer functional cardiac tissues that have already been achieved using largely primary cell cultures. Future characterization of cell–matrix interactions, and how to select appropriate biomaterials, culture techniques, and biophysical stimuli appropriate for the study all remain challenges. Lastly, the challenge of vascularizing cardiac constructs, facilitating rapid integration and connection with host vasculature, and achieving electromechanical coupling, all remain pivotal areas of interest in the field. In the following sections, we will outline these challenges and discuss how past studies inform the current work that is being done to meet these goals.

3 Challenge 1: Directing Cell Differentiation

Tissue-engineered cardiac regeneration is largely driven by the idea that physiologically representative environments for directing cell differentiation and function need to include more than biochemical factors alone. The success of such cell-instructive environments is dependent on providing factors that match the physiological requirements for the cells of choice. Each of these factors must be carefully orchestrated to optimize the function of engineered constructs for cardiovascular repair. We focus here on the spectrum of cells that may be used for cardiovascular repair (each reviewed in detail in earlier chapters) and the ways to optimize their function by culture in a controlled microenvironment.

3.1 Biophysical Stimulation of Differentiating Cells

One major focus of cardiac tissue engineering is the creation of functional muscle grafts. For the purpose of regenerative medicine, several myogenic cell sources exist. Embryonic stem cell (ESC)-derived and induced pluripotent stem cell (iPSC)-derived cardiomyocytes continue to offer the greatest possibility for obtaining sufficient numbers of cells necessary for human therapy. However, several groups (as discussed in other chapters) are working on utilizing other myocyte progenitors (resident cardiac stem cells, circulating progenitors) and gene therapy approaches to induce cell cycle reentry and allow for expansion of terminally differentiated cell populations. Importantly, the features of ESC- and iPSC-derived cardiomyocytes are not yet fully understood. The ability of cardiovascular cells to undergo continued proliferation following differentiation has been reported for both the in vitro and the in vivo settings [38], and may allow for engraftment of smaller cell populations that would proliferate over time in vivo.

Cardiomyocytes are characterized by arrested mitosis, but DNA replication and nuclear division may still occur, resulting in the polyploid and/or polynucleated myocytes found in the heart [39]. The Murry group [38] reported the unique ability of cardiomyocytes derived from hESCs to undergo mitosis following differentiation, in comparison with the mitotic arrest that occurs in rodent embryonic, fetal, or neonatal cardiomyocytes. Whether stem-cell-derived cardiomyocytes will undergo hypertrophic growth remains to be determined. Such cell characteristics lead to the next question: How can we define cell maturity, develop systems to mature the cells, and determine the level of cell maturity that is optimal for graft survival?

The genotypic and phenotypic stability of stem-cell-derived cardiovascular cell populations in vivo is likely to be critical for the efficacy and safety of cell therapy for the heart. Tissue engineering strategies have been developed to direct cell differentiation and to further mature cell populations through the combined use of biomaterials and biochemical and biophysical stimuli.

3.2 The Argument for Mixed Cell Populations

Early reports on cardiac constructs engineered in vitro argued for the use of purified cell populations. These studies utilized plating separation, magnetic sorting, or flow-activated cell sorting to enrich populations for cardiomyocytes, or multipotent progenitor cells [40]. However, in native heart tissue, cardiomyocytes coexist with the cardiac fibroblasts and endothelial cells (each representing a third of the total cell population by number) and small fractions of pericytes, macrophages, and smooth muscle cells. Early reports characterized critical interactions between cardiomyocytes, endothelial cells, and cardiac fibroblasts [41, 42]. Two subsequent studies [31, 43] supported the hypothesis that multicellular cardiovascular populations stabilize and enhance cardiomyocyte phenotype in engineered constructs and promote vascularization and support graft survival and in vivo function. These studies suggest that the ESCs and iPSCs should be differentiated into all three cardiovascular lineages to enable the formation of functional tissue constructs.

Recent studies have also shown the importance of the timing of cell interactions in co-cultures of cardiac myocytes, fibroblasts, and endothelial cells. For example, the Radisic group [43a] reported that preculture of scaffolds with fibroblasts enhanced the quality of engineered muscle, owing to the matrix deposition and scaffold remodeling by fibroblasts prior to cardiomyocyte culture. The Zimmermann group determined the developmental origin of fibroblasts as concurrent with cardiomyocyte differentiation [44], hinting at the necessary roles of fibroblasts in the formation of engineered heart tissue. Each of these studies contributes to the emerging theory that multiple cell types are necessary for the formation of a robust engineered tissue, and that cardiomyocyte function can be enhanced by the presence of stabilizing endothelial, smooth muscle, and fibroblast cells.

3.3 The Argument for Vascularization

Although cardiac tissue engineering typically focuses on the use of biomaterial, cells, and signaling to create a beating myocardium, various other approaches have found success as engineered constructs to deliver cells or molecules to induce revascularization of the heart. Mesenchymal stem cells have been extensively used in preclinical and clinical trials of heart repair [45–48], with some beneficial effects that could be attributed to vascularization rather than myogenesis. As a result of implantation of a vascularized patch, deterioration of heart function following myocardial infarction has been delayed, presumably owing to the increased ventricular wall thickening and neovascularization within the infarct [40, 49, 50].

3.4 Summary

Various studies have investigated the role of the scaffold, biochemical signals, and physical stimuli in the behavior and differentiation of cells. These signaling paradigms are employed as a means of imitating natural phenomena. Optimal conditions for generating functional bona fide cardiomyocytes and engineered heart tissues remain to be determined, but the future of directing stem cell differentiation may depend on the ability to integrate all of these signaling paradigms, to achieve a true biomimetic environment for tissue regeneration.

4 Challenge 2: Scaffold Designs for Cell Culture and Delivery

Biomaterials are generally designed to interact with biological systems [51]. The scaffold should be gradually replaced by the newly formed tissue, by safely degrading at a rate similar to that of the new tissue formation, and eventually being removed from the body without toxic by-products [52]. Biomaterials for cardiac tissue engineering are available in many different forms [53], including porous, fibrous, and hydrogel scaffolds – like the different tissue structures they are made to resemble [54]. For example, porous scaffolds provide macroscopic voids for the migration of cells, whereas fibrous scaffolds can be fabricated to mimic the hierarchy of the native ECM and control cellular alignment [55].

Cells respond to the entire context of their environment, with the ECM being one of the key regulators of cell behavior. For "normal" function, cardiac cells need a "normal" matrix, with cardiac-like molecular composition, structure, and mechanical properties. In a typical tissue engineering approach, culture begins with cell seeding into a 3D scaffold. Immediately, the cell microenvironment is influenced by the physical and chemical cues of the scaffold, and the cells alter their shape and function in response to these cues. Careful selection of a biomaterial, including molecular composition, hierarchical structure, and mechanical properties, can thus be used to direct development of engineered cardiac tissue.

Two broad classes of cardiac tissue engineering scaffolds are natural materials (derived from the native ECM) and synthetic materials (made by chemical processing). Frequently used natural materials include collagen [20, 56], fibrin [57], and hyaluronic acid [58]. Natural materials readily provide signals to cells by surface receptor interactions and uptake and degradation of the matrix molecules. A drawback, however, is that natural scaffolds may be difficult to process without disrupting potentially important hierarchical structures. Hydrogels formed from natural materials are suitable for a range of tissue engineering applications, but these materials generally have poor mechanical properties and may also elicit an immune response. Frequently used synthetic materials include polyesters such as poly(lactic acid) and poly(glycolic acid) [59], polylactones such as poly(ε-caprolactone) [60], polyurethanes [61], and polysebasic acid [11, 62]. Synthetic biomaterials have a wide range of properties, can be customized with respect to mechanics, chemistry, and structure, but may be limited in functional cellular interactions. To overcome the lack of interaction, these materials can be modified to incorporate adhesion peptides or designed to release biological molecules.

In a series of studies by Eschenhagen and Zimmermann, cardiac constructs were made using a mixture of liquid collagen I, Matrigel, and growth supplements to encapsulate heart cells [19, 56, 63]. The cell-hydrogel solution was reconstituted in circular molds to form ring-shaped constructs 15 mm in diameter and 1–4 mm thick, and subjected to mechanical loading (10% stretch, 2 Hz). After a short culture period, these constructs developed into cardiac organoids, and showed contractile as well as electrophysiological properties of working myocardium. Further evolution of this model includes the evaluation of different cell populations and the effort to replace certain components of their protocol to adhere to clinical requirements [43].

Another natural biomaterial that has been successfully applied to cardiac tissue engineering is hyaluronic acid. Although hyaluronic acid macromolecules maintain the biomechanical properties of cardiac tissue, their fragments are highly bioactive and involved in many regulatory events, including cellular function and development, tumor progression, angiogenesis, inflammation, wound healing, and regeneration. Although hyaluronic acid has not received as much attention as other biomaterials for cardiac tissue engineering, it has been shown to support a differentiated cell phenotype in adult cardiomyocytes [58].

An entire decellularized organ can serve as a natural biomaterial scaffold. As mentioned in the last section, the study by Taylor's group showed that rat hearts were decellularized by perfusion through the coronary blood vessels with detergents, leaving the preserved ECM, vascular architecture, acellular valves, and intact vascular geometry [22]. The decellularized hearts were then repopulated with cardiac cells and maintained in culture for up to 28 days by coronary perfusion. After 4 days, macroscopic contractions could be observed and by day 8 these recellularized hearts could generate pump function equivalent to about 2% of adult heart function. Although this approach offers promise in terms of engineering tissues with relevant architecture and geometry, it remains to be seen if it can be scaled up to engineer functional hearts for implantation.

PGS is a tough and biodegradable elastomer with excellent biocompatibility and nontoxic degradation products [64]. The advantage of using a synthetic biomaterial is that it can easily be fabricated to have mechanical and microstructural properties matching those of a native tissue matrix. In recent studies, PGS has been used to engineer patches of cardiac tissue (approximately 10 mm diameter, 1–2 mm thick) by cell culture in bioreactors [29]. The PGS-based patches can then be surgically sutured to the heart.

For cardiac tissue engineering, the final stiffness of the patch, as well as its active, force-generating functional properties, will play an important role in healing following implantation. Human mesenchymal stem cells were shown to exhibit lineage specification depending on the substrate mechanics, which correlated with the in vivo matrix elasticity (8–17 kPa for muscle) in the absence of any additional inducing factors [5]. The stem cells are thought to "pull" on the matrix and generate signals based on this force. Also, the differentiation of a skeletal myoblast could be directly influenced by the stiffness of the substrate [65]. Myotubes differentiated optimally on materials with tissuelike stiffness [66].

An important property of the native heart architecture is that myocytes are highly aligned in a specific pattern that varies through the tissue, leading to anisotropic structural and mechanical properties. Several studies have aimed to recapitulate cardiac cell alignment in vitro by using aligned biomaterials. In one approach, patterned hyaluronic acid substrates were generated [58] to serve as inductive templates for the assembly of cardiac organoids that elongated and aligned along the pattern direction. In another approach, microfabrication techniques were used to create an accordion-like honeycomb microstructure in PGS scaffolds, which yielded porous, elastomeric 3D scaffolds with controllable stiffness and anisotropy [3]. These scaffolds provided, for the first time, a structurally and mechanically anisotropic scaffold matching native heart tissue that promoted cell alignment and function. In future clinical applications, such aligned cell-scaffold constructs may play a major role in guiding the regeneration of damaged myocardium.

Biomaterials could also aid the cell delivery applications that do not include the in vitro culture step. In early cell therapy studies, cells such as skeletal myoblasts [67] or bone-marrow-derived stem cells [68] were injected into the heart wall of infarct patients, resulting in modest gains in cardiac function. The key limitation of these cell therapies was the massive cell loss following injection. A biocompatible hydrogel polymerizable in situ could potentially be used as a cell delivery vehicle to improve cell retention, survival, and function in the ischemic myocardium. In one study [50], fibrin glue – a hydrogel widely used in surgery, cell culture, and tissue engineering – was used to deliver human mesenchymal stem cells to the ventricles of infarcted nude rats. Ninety minutes after injection, the addition of fibrin glue resulted in a substantial increase in local cardiac retention over that achieved using saline. However, cell retention and engraftment after longer periods of time, and the restoration of cardiac functionality, remain to be evaluated.

4.1 Summary

The choice of biomaterial for cardiac tissue engineering directly influences the phenotype of seeded cells and the resulting development and functionality of the engineered tissue. Scaffold materials provide a logistical template for cell attachment and tissue formation, but the mechanical and chemical properties of the material can also direct cells to differentiate, generate ECM and other signaling proteins, and organize into complex structures. Therefore, biomaterial selection is a key tool for engineering functional tissues.

5 Challenge 3: In Vitro Models of Development and Disease

In many ways, engineered 3D tissues span the gap between traditional 2D cultures and whole-animal systems. By mimicking the features of the in vivo environment and taking advantage of the same tools used to study cells in 2D culture, 3D tissue models provide unique insights into the cell behavior in developing tissues, and in response to drugs and pathological conditions.

Studies in 2D cell culture have produced important conceptual advances. Nevertheless, cells grown on flat substrates can differ considerably in their morphology, cell–cell and cell–matrix interactions, and gene expression pattern from those growing in 3D environments [69–73]. On the other hand, animal models are difficult to control, costly, and in many cases quite different from the human system. In particular, cardiac physiological function is quite different between the small-animal models and humans, and the results obtained in these animal studies are not necessarily predictive of the human situation. Therefore, the potential of cardiac tissue engineering based on human stem cells extends to the use of human tissue equivalents for medical research.

One simple model system for study of cardiac development and disease is provided by 3D cultivation of cardiac cells in the form of well-differentiated spheroids [74]. The embryoid body is another spheroid system used in vitro to differentiate ESCs into cardiomyocytes [75]. These systems can provide an efficient way to probe the function of candidate genes and proteins before proceeding to laborious gene targeting approaches in animals. Tissue models allow screening of multiple genes and proteins, cell survival factors, biomaterials, inflammatory cytokines, other cells, and small molecules.

Experiments can be designed to control and vary any number of parameters to generate quantitative data for system analysis. For example, an in vitro injection system was developed to serve as a test bed to identify the optimal cell type to drive robust cardiac myogenesis [76]. This system can be engineered to create disease models such as myocardial infarction with an artificial scar tissue layer to screen the effect of matrix proteolytic enzymes. Novel micropatterned cell-pair assays enable studies of cellular interactions at different scales by controlling cell morphology and position on the substrate. These assays can enhance our understanding of heterocellular interactions in pathological cardiac states, and facilitate the rational design of cardiac cell therapies, in settings involving multiple cell types and matrix molecules [77]. Model tissues can also be assayed after altering levels of differentiation factors and nutrients in combinations and temporal sequences [78].

Tissue models can also provide high-throughput genomic and therapeutic screening. For example, it was shown in a culture of chick embryonic heart aggregates that the effects of a clinically important antitumor agent were due to a depolarizing action of the cardiac membrane [74]. More recently, hESC-derived cardiomyocytes assessed with a combination of single cell electrophysiology and microelectrode array mapping were used for electrophysiological drug screening [79].

Using human 3D tissue surrogates for in vivo-like drug testing might help reduce the high failure rate in the development of drugs. It might also be possible to use 3D models for individualized therapy and analysis of the pathogenesis of disease. iPSCs, produced by reprogramming somatic cells using retroviral overexpression of the stem cell factors Oct4, Sox2, c-myc, and Klf4 [80–83], have a particularly promising future. These cells appear to have characteristics similar to those of ESCs, can be maintained for several months in culture, have the capacity to differentiate into all three germ layers, and can give rise to functional cardiomyocytes [84]. The development of physiological assays for contractility, conduction, and mechanical work with patient-specific heart progenitors may allow the generation of heart muscle that harbors specific genetic backgrounds, facilitating functional analysis of genetic variations in human populations. In a recent study, the Eschenhagen group developed a screening platform in a 24-well plate format, containing miniature cardiac organoids (0.2–1.3 mm in diameter, 8 mm long) consisting of neonatal rat cardiomyocytes in fibrinogen/Matrigel [84a]. Such platforms are finding an increasing use in drug screening and study of disease.

5.1 Summary

Three-dimensional tissue models can thus provide in vitro test beds to dissect biological and pathological mechanisms predictive of in vivo outcomes. Challenges remain to design a "valid" 3D model that integrates biological, structural, and mechanical factors with ECM that can reflect in vivo physiological condition. These factors are fundamental to any approach toward developing a clinically viable cardiac tissue. Overall a hierarchy of approaches will likely be necessary to connect the organ–tissue function with underpinning cellular events.

6 Challenge 4: Vascularization and Blood Perfusion

A major limitation of all of tissue engineering is the lack of methods to produce thick tissues with the ability to provide vascular supply. The human left ventricle is typically 1–1.5 cm thick, yet diffusion can only supply nutrients to a depth of approximately 100 μm. When an engineered tissue construct is implanted, the seeded cells will have a limited capacity to take up nutrients (and most critically oxygen) and to clear metabolic products, leading to intensive apoptosis and necrosis. Several strategies have been proposed to induce vascularization of engineered tissues: (1) optimization of scaffold material and design to favor cell infiltration, (2) the incorporation of growth factors into the scaffold to induce angiogenesis, (3) preseeding of the implant with mature or precursor endothelial cells, and (4) scaffold vascularization with the aid of host coronary circulation in vivo (Fig. 7).

To overcome the limitations of nutrient transport, the Kofidis group established pulsatile flow around cardiomyocytes encapsulated in fibrin glue around a rat artery [85]. Dvir et al. developed a perfusion bioreactor that employs a distributing mesh upstream from the construct to provide homogeneous fluid flow and maximum exposure to perfusing medium, resulting in somewhat increased cell viability because of the convective supply of oxygen [86, 87]. The Vunjak-Novakovic group has implemented the in vivo-like mechanisms of convective–diffusive oxygen transport by perfusion of culture medium through the cultured construct [12, 29, 30, 88, 89]. Cells were cultured in a porous scaffold with an array of channels perfused with medium containing an oxygen carrier. Both the perfusion and the supplementation of an oxygen carrier improved the viability and differentiation of cultured cells and enabled the formation of compact layers of viable cells around the channels. However, perfusion alone cannot substitute for vascularization. It is debatable what extent of vascularization needs to be provided before implantation (channels only, channels lined with vascular cells, or primitive blood vessels perfusable by blood [13, 31]).

Microelectromechanical systems (MEMS) technology has been used in cardiovascular tissue engineering as it allows for the creation of complex structures for cell seeding, such as polymer wafers containing branched vascular-like spaces [90]. The MEMS process entails the use of a microetched wafer as a master mold for pattern transfer to a biocompatible polymer material. Such patterns can be designed to mimic specific vascular structures. Recent studies have demonstrated the feasibility of creating ordered branched arrays of channels lined with living endothelial cells [90–92]. Topologically complex 3D microfluidic channel systems can be fabricated in poly(dimethylsiloxane) [93, 94], with the aim to support large and viable tissue constructs both in vitro and in vivo.

During the past decade, gene therapy has targeted revascularization within the failing myocardium by introducing genes encoding angiogenic factors. Among them, VEGF and basic fibroblast growth factor (bFGF) have been most widely used. VEGF plays a major role in the early development of blood-cell progenitors [95], and bFGF is a potent inducer of endothelial cell proliferation and blood-vessel growth, in vitro and in vivo. When VEGF and bFGF are integrated into a scaffold, local angiogenesis has been observed [96]. Strategies to tailor the synthesis and composition of scaffolds to obtain controlled release of these factors have been explored. A highly porous, biodegradable elastomeric poly(ester urethane)urea scaffold was created for controlled release of bFGF [97]. The incorporation of heparin into the scaffold contributed to the favorable release pattern of bFGF. The release of SDF-1 from a vasculogenic cardiac patch attracted human CD34+ cells to a great extent [27]. However, we still do not have the ability for precise long-term control of contributing factors.

Endothelial cells have been extensively used to form vascular networks in vitro. Aortic endothelial cells seeded in polyglycolide matrices and implanted in rat resulted in heterogeneous vascular structures [98]. Perivascular cells (such as pericytes and smooth muscle cells) helped remodel and mature the primitive vascular networks, and stabilize neovasculature toward the formation of long-lasting blood vessels [99, 100]. Autologous endothelial cells may pose problems because patients with obstructive vascular disease may have dysfunctional endothelial cells, precluding their use for tissue engineering applications. The timing is also an issue, as a vascular graft seeded with autologous cells would need to be created for surgery. Stem cells are therefore gaining increased interest, as they can contribute to angiogenesis both directly, by participating in new vessel formation, by stabilizing immature vessels [101, 102], and by producing supportive ECM, and indirectly, by secreting angiogenic and antiapoptotic factors [31, 43, 103]. Levenberg et al. derived endothelial cells from hESCs and showed that when cultured on Matrigel, these cells formed tubelike structures, and when transplanted into SCID mice, they appeared to form microvessels containing mouse blood cells [104].

Endothelial cells co-cultured with cardiomyocytes were able to establish capillary-like structures in a 3D scaffold that were stable in vitro [105]. Preformed endothelial cell networks reduced apoptosis and necrosis of cardiomyocytes and promoted their synchronous contractions. Cardiac tissue patches created by co-culture of HUVEC, mouse embryonic fibroblasts, and mesenchymal stem cells resulted in a significantly higher survival rate, more myocardial-like mechanical properties, and more vessel structures compared with cardiomyocytes only and cardiomyocytes plus HUVEC patches [106]. The great advantage of heterotypic cell culture is its self-sustainability, obviating the exogenous supply of angiogenic stimuli. Regarding the cross talk between endothelial cells and cardiomyocytes, it has been suggested [105] that the proangiogenic factor VEGF might enhance the Cx-43 expression in cardiomyocytes. The in vitro prevascularization might accelerate the establishment of a vascular supply following implantation. However, it remains to be determined whether microcapillary structures are able to establish efficient connections with the host microvasculature.

An alternative to the in vitro establishment of a microcapillary network is to culture the mixture of cells in the scaffold for a short time (hours to a few days), followed by implantation. Angiogenesis from arteriovenous loop and flow-through vascular pedicle models has been shown to result in the creation of a newly formed "flap" of tissue. Neonatal cardiac myocytes cultured in vivo in silicone chambers, close to an intact vascular pedicle, were found to generate a spontaneous beating cardiac muscle tissue after 3 weeks [107]. This approach takes advantage of the in vivo environment to orchestrate the cellular interaction for the establishment of a functional vasculature, but has the drawback of two separate surgeries (one to implant the construct at the arteriovenous loop and one to implant the construct to the infarcted heart).

Combination approaches have also emerged. A combination of gene therapy (biodegradable polymer–DNA nanoparticles) and cell therapy (stem cells) was found to stimulate angiogenesis through secretion of angiogenic and antiapoptotic factors (e.g., VEGF) [108]. Neonatal cardiac cells with a mixture of prosurvival and angiogenic factors on an alginate scaffold were implanted into rat omentum for 7 days to enhance vascularization, and then explanted and implanted in the infarcted heart site. It was shown that this approach improved cardiac function and prevented further dilation of the chamber and ventricular dysfunction [27].

6.1 Summary

Vascularization remains one of the main obstacles that need to be overcome before thick tissue-engineered constructs can be applied clinically. Current strategies have shown promising results; however, none of them will be sufficient to sustain engineered constructs that are larger than several millimeters after implantation. Furthermore, apart from histology, functional tests need to be conducted to assess the vessel perfusion, stability, and the construct viability.

7 Challenge 5: Functional Evaluation

Engineered tissue constructs enable researchers to interrogate the capacity of cell populations to function as a tissue. The ability of cardiovascular cells to electromechanically couple into a functional syncytium can be evaluated in vitro at the subcellular, cellular, and tissue levels. Readouts for graft survival are also being developed through novel animal models and imaging modalities.

When evaluating cell function in engineered cardiac constructs, we are interested in understanding the effects of the environment (scaffold, biochemical factors, physical signaling) on the differentiation, maturation, or functionality of cells. This involves looking at functionality of junctions, cell alignment, and cytoskeletal arrangement. Specifically for cardiomyocytes, development and localization of connexins, the proteins that make up gap junctions, can be used as a predictor of effective cell–cell communication. Additionally, understanding calcium handling (intracellular and extracellular) is possible through calcium imaging, as well as by visualizing sarcomeric densities within the cells.

At the subcellular level, we can use transmission electron microscopy to investigate the contractile apparatus, the arrangement and density of sarcomeres and mitochondria, the elongation of nuclei, and sarcomere composition (M and Z lines, I, A, and H bands, which are made up of actin and myosin filaments). Proper ultrastructural organization of sarcomeres is pivotal in cardiomyocyte function, and an indicator of cell maturation. The gap junction protein Cx-43 is essential for ventricular conduction and remains a hallmark of mature cardiac tissue. Cx-43-expressing cells in the infarcted heart allow for ventricular protection against arrhythmias through enhanced cell coupling [109]. Such reports have been corroborated by in vitro tissue engineering studies, where enhanced conduction is accompanied by increased Cx-43 density. The presence of other cell junctions including desmosomes, which anchor the cell cytoskeleton and retain cell–cell connections during contraction, is an indicator of tissue assembly. Along with conduction mapping, cell junction imaging reveals the level of electromechanical coupling in cells.

Understanding the effects of environmental conditions on cell differentiation can be increased through transcriptional profiling and the use of molecular tags to track cell differentiation in real time. Using microarrays, the Wu group investigated hESCs throughout their differentiation and identified transcriptional changes that occur between the undifferentiated state and the stage at which they resemble fetal heart cells [110]. Taking advantage of high-throughput screening to track gene expression and identify what proteins are being translated and are contributing to contractile function, matrix adhesion, or intracellular junctions, we can begin to identify the signals that will lead to formation of cardiac tissues that are optimal for implantation. Additionally, molecular markers that enable real-time and long-term tracking of cell fate in vivo can easily be extended to in vitro tissue engineering applications, to investigate how different biochemical signals and scaffolds of physical stimuli activate biological processes.

Function of engineered heart tissue should be rigorously assessed for electrical homogeneity and mechanical integrity. Electrical conduction can be visualized using voltage-sensitive dyes to map signal propagation across the tissue. These readouts show the localization, density, and distribution of cardiomyocytes within the 3D structure, and the magnitude of action potential and conduction velocity. Integration can be assessed by visualizing signal propagation through the host–graft interface. In addition, voltage imaging can aid in the realm of disease models, where it may be possible to visualize the origins of abnormal conduction through simultaneous drug treatment and imaging.

7.1 Designing In Vivo Studies: Considerations for Timing

Two time-dependent variables must be considered for in vivo implantation of cardiac grafts: (1) cell function within the engineered construct (which depends on culture conditions and duration) and (2) the timing of implantation following myocardial injury (which depends on the progression of remodeling in the injured heart).

Early studies of engineered heart tissue used neonatal rat cardiomyocytes as the cell source. As we move into more translational approaches, myogenic stem cell sources, such as ESCs and iPSCs, are being used [106, 111, 112]. It has not yet been determined what level of maturity is optimal for in vivo survival and performance of the cells. ESC- and iPSC-derived cardiomyocytes pass through maturation processes similar to those seen during the development. With maturation, the cell morphology, conduction profile, and functional apparatus develop and change, as does the ability of the cell to connect with other cells, survive, and function in vivo (i.e., subjected to hypoxia, strain, electrical signals, and biological factors).

In one study, hESC-derived cardiomyocytes resembling fetal cardiomyocytes were identified as optimal for in vivo implantation [110]. In another study, hESC-derived cardiomyocytes were characterized with respect to the effects of construct vascularization on cell maturity [79]. Future studies should investigate if a mature engineered tissue couples with the host, which secreted factors might play a role in host vasculature recruitment, and how the maturation level of cardiomyocytes influences their ability to survive.

There is also an emerging recognition that timing for engraftment of engineered tissues must be optimized. For the heart, this involves careful consideration of the remodeling process following injury (myocardial infarction) – both the dynamic changes in muscle wall composition and stiffness and the presence of inflammatory signals and cells [113, 114]. The interactions between engrafted cells and host cells have not been studied in sufficient detail, and will remain a determining factor of the success of cell-based therapies.

Tissue-engineered constructs are cell-protective to some extent, preventing anoikis and providing local matrix mechanical properties that are optimal for cell function – as compared with the cells injected into a fibrotic scar. With appropriate construct vascularization and inclusion of prosurvival cocktails, the cells are further shielded from the ischemic infarct. With a greater understanding of the interaction between dynamic populations of inflammatory cells and engrafted cells, it may be possible to harness the inflammatory response in favor of construct survival and remodeling of the heart.

7.2 How Does One Evaluate the Success of Engrafted Patches?

Whether a cardiac implant is a functional cardiac patch or a biomaterial with incorporated bioactive components, the main output from animal studies has traditionally been the global improvement in cardiac function. The parameters of interest include ejection fraction, diastolic and systolic volumes, wall thickness, and dilation. Many studies over the past few years have shown that cell therapy may delay or halt functional deterioration rather than reverse the decline in cardiac function. Positive functional results, however, do not always correlate with macroscopic assays, such as wall thickness and dilation, and hint at the need for other assays: biochemical, imaging, and biomechanical. The role of biomechanics arises when we compare cell sheet implantation (a few hundred microns) within hydrogel grafts (millimeter scale). Whether or not a graft may alter the functional behavior of the heart simply as a result of its size has not yet been extensively studied. Additional instructive assays for engineered constructs may include graft integration and cell migration, survival and cross talk of engrafted cells in the infarct, and making use of technologies such as single-photon-emission computed tomography to evaluate functionality of vessels in the infarct.

7.3 Summary

Linking biochemical assays to functional imaging will be crucial in furthering our understanding of the biological networks that govern the function of engineered tissues. In vitro functional assays will be the determining output for in vitro models of disease, and serve as a predictor for construct engineering for implantation. Continued development of noninvasive assays, from the molecular to the macroscopic level, will drive translation of tissue engineering strategies to the clinic, as we begin to understand how to optimize these constructs.

Acknowledgements The work described in this article has been supported by the NIH (HL076485, HL089913, and EB002520 to G.V.N.), an NIH graduate fellowship (to R.M., T 32 HL087745), and an NSF graduate fellowship (to A.F.G.G.-F.).

References

1. Harrison, R. (1907) Observations on the living developing nerve fiber. *Anat Rec,* 1, 116–8.
2. Chen, H., Feyereisen, M., Long, X. P. & Fitzgerald, G. (1993) Stability, bonding, and geometric structure of Ti8C12, Ti8N12, V8C12, and Zr8C12. *Phys Rev Lett,* 71, 1732–35.
3. Engelmayr, G. C., Jr., Cheng, M., Bettinger, C. J., Borenstein, J. T., Langer, R. & Freed, L. E. (2008) Accordion-like honeycombs for tissue engineering of cardiac anisotropy. *Nat Mater,* 7, 1003–10.
4. Discher, D. E., Mooney, D. J. & Zandstra, P. W. (2009) Growth factors, matrices, and forces combine and control stem cells. *Science,* 324, 1673–7.
5. Engler, A. J., Sen, S., Sweeney, H. L. & Discher, D. E. (2006) Matrix elasticity directs stem cell lineage specification. *Cell,* 126, 677–89.
6. Engler, A. J., Carag-Krieger, C., Johnson, C. P., Raab, M., Tang, H. Y., Speicher, D. W., Sanger, J. W., Sanger, J. M. & Discher, D. E. (2008) Embryonic cardiomyocytes beat best on a matrix with heart-like elasticity: scar-like rigidity inhibits beating. *J Cell Sci,* 121, 3794–802.
7. Langer, R. & Tirrell, D. A. (2004) Designing materials for biology and medicine. *Nature,* 428, 487–92.
8. Lutolf, M. P. & Hubbell, J. A. (2005) Synthetic biomaterials as instructive extracellular microenvironments for morphogenesis in tissue engineering. *Nat Biotechnol,* 23, 47–55.
9. Tibbitt, M. W. & Anseth, K. S. (2009) Hydrogels as extracellular matrix mimics for 3D cell culture. *Biotechnol Bioeng,* 103, 655–63.
10. Freytes, D. O., Wan, L. Q. & Vunjak-Novakovic G. (2009) Geometry and force control of cell function. *J Cell Biochem*, 108, 1047–58.
11. Radisic, M., Park, H., Chen, F., Salazar-Lazzaro, J. E., Wang, Y., Dennis, R., Langer, R., Freed, L. E. & Vunjak-Novakovic, G. (2006) Biomimetic approach to cardiac tissue engineering: oxygen carriers and channeled scaffolds. *Tissue Eng,* 12, 2077–91.
12. Tandon, N., Cannizzaro, C., Chao, P. H., Maidhof, R., Marsano, A., Au, H. T., Radisic, M. & Vunjak-Novakovic, G. (2009) Electrical stimulation systems for cardiac tissue engineering. *Nat Protoc,* 4, 155–73.
13. Zimmermann, W. H., Melnychenko, I., Wasmeier, G., Didie, M., Naito, H., Nixdorff, U., Hess, A., Budinsky, L., Brune, K., Michaelis, B., Dhein, S., Schwoerer, A., Ehmke, H. & Eschenhagen, T. (2006) Engineered heart tissue grafts improve systolic and diastolic function in infarcted rat hearts. *Nat Med,* 12, 452–8.
14. Belousov, L. V. & Ermakov, A. S. (2001) [Artificially applied tensions normalize development of relaxed *Xenopus laevis* embryos]. *Ontogenez,* 32, 288–94.
15. Hove, J. R., Koster, R. W., Forouhar, A. S., Acevedo-Bolton, G., Fraser, S. E. & Gharib, M. (2003) Intracardiac fluid forces are an essential epigenetic factor for embryonic cardiogenesis. *Nature,* 421, 172–7.
16. Butler, D. L., Goldstein, S. A. & Guilak, F. (2000) Functional tissue engineering: the role of biomechanics. *J Biomech Eng,* 122, 570–5.

17. Vandenburgh, H., Del Tatto, M., Shansky, J., Lemaire, J., Chang, A., Payumo, F., Lee, P., Goodyear, A. & Raven, L. (1996) Tissue-engineered skeletal muscle organoids for reversible gene therapy. *Hum Gene Ther,* 7, 2195–200.
18. Vandenburgh, H., Shansky, J., Benesch-Lee, F., Skelly, K., Spinazzola, J. M., Saponjian, Y. & Tseng, B. S. (2009) Automated drug screening with contractile muscle tissue engineered from dystrophic myoblasts. *FASEB J,* 23, 3325–34.
19. Zimmermann, W. H., Schneiderbanger, K., Schubert, P., Didie, M., Munzel, F., Heubach, J. F., Kostin, S., Neuhuber, W. L. & Eschenhagen, T. (2002) Tissue engineering of a differentiated cardiac muscle construct. *Circ Res,* 90, 223–30.
20. Radisic, M., Park, H., Shing, H., Consi, T., Schoen, F. J., Langer, R., Freed, L. E. & Vunjak-Novakovic, G. (2004) Functional assembly of engineered myocardium by electrical stimulation of cardiac myocytes cultured on scaffolds. *Proc Natl Acad Sci USA,* 101, 18129–34.
21. Robinson, K. A., Li, J., Mathison, M., Redkar, A., Cui, J., Chronos, N. A., Matheny, R. G. & Badylak, S. F. (2005) Extracellular matrix scaffold for cardiac repair. *Circulation,* 112, I135–43.
22. Ott, H. C., Matthiesen, T. S., Goh, S. K., Black, L. D., Kren, S. M., Netoff, T. I. & Taylor, D. A. (2008) Perfusion-decellularized matrix: using nature's platform to engineer a bioartificial heart. *Nat Med,* 14, 213–21.
23. Park, H., Bhalla, R., Saigal, R., Radisic, M., Watson, N., Langer, R. & Vunjak-Novakovic, G. (2008) Effects of electrical stimulation in C2C12 muscle constructs. *J Tissue Eng Regen Med,* 2, 279–87.
24. Chiu, L. L., Iyer, R. K., King, J. P. & Radisic, M. (2008) Biphasic electrical field stimulation aids in tissue engineering of multicell-type cardiac organoids. *Tissue Eng Part A*. DOI: 10.1089/ten.tea.2007.0244.
25. Komuro, I., Katoh, Y., Kaida, T., Shibazaki, Y., Kurabayashi, M., Hoh, E., Takaku, F. & Yazaki, Y. (1991) Mechanical loading stimulates cell hypertrophy and specific gene expression in cultured rat cardiac myocytes. Possible role of protein kinase C activation. *J Biol Chem,* 266, 1265–8.
26. Fink, C., Ergun, S., Kralisch, D., Remmers, U., Weil, J. & Eschenhagen, T. (2000) Chronic stretch of engineered heart tissue induces hypertrophy and functional improvement. *FASEB J,* 14, 669–79.
27. Dvir, T., Kedem, A., Ruvinov, E., Levy, O., Freeman, I., Landa, N., Holbova, R., Feinberg, M. S., Dror, S., Etzion, Y., Leor, J. & Cohen, S. (2009) Prevascularization of cardiac patch on the omentum improves its therapeutic outcome. *Proc Natl Acad Sci USA,* 106, 14990–5.
28. Grayson, W. L., Martens, T. P., Eng, G. M., Radisic, M. & Vunjak-Novakovic, G. (2009) Biomimetic approach to tissue engineering. *Semin Cell Dev Biol,* 20, 665–73.
29. Radisic, M., Marsano, A., Maidhof, R., Wang, Y. & Vunjak-Novakovic, G. (2008) Cardiac tissue engineering using perfusion bioreactor systems. *Nat Protoc,* 3, 719–38.
30. Radisic, M., Malda, J., Epping, E., Geng, W., Langer, R. & Vunjak-Novakovic, G. (2006) Oxygen gradients correlate with cell density and cell viability in engineered cardiac tissue. *Biotechnol Bioeng,* 93, 332–43.
31. Caspi, O., Lesman, A., Basevitch, Y., Gepstein, A., Arbel, G., Habib, I. H., Gepstein, L. & Levenberg, S. (2007) Tissue engineering of vascularized cardiac muscle from human embryonic stem cells. *Circ Res,* 100, 263–72.
32. Furuta, A., Miyoshi, S., Itabashi, Y., Shimizu, T., Kira, S., Hayakawa, K., Nishiyama, N., Tanimoto, K., Hagiwara, Y., Satoh, T., Fukuda, K., Okano, T. & Ogawa, S. (2006) Pulsatile cardiac tissue grafts using a novel three-dimensional cell sheet manipulation technique functionally integrates with the host heart, in vivo. *Circ Res,* 98, 705–12.
33. Shimizu, T., Sekine, H., Yang, J., Isoi, Y., Yamato, M., Kikuchi, A., Kobayashi, E. & Okano, T. (2006) Polysurgery of cell sheet grafts overcomes diffusion limits to produce thick, vascularized myocardial tissues. *FASEB J,* 20, 708–10.
34. Shimizu, T., Yamato, M., Isoi, Y., Akutsu, T., Setomaru, T., Abe, K., Kikuchi, A., Umezu, M. & Okano, T. (2002) Fabrication of pulsatile cardiac tissue grafts using a novel 3-dimensional cell sheet manipulation technique and temperature-responsive cell culture surfaces. *Circ Res,* 90, e40–8.

35. Sekine, H., Shimizu, T., Hobo, K., Sekiya, S., Yang, J., Yamato, M., Kurosawa, H., Kobayashi, E. & Okano, T. (2008) Endothelial cell coculture within tissue-engineered cardiomyocyte sheets enhances neovascularization and improves cardiac function of ischemic hearts. *Circulation,* 118, S145–52.
36. Miyagawa, S., Sawa, Y., Sakakida, S., Taketani, S., Kondoh, H., Memon, I. A., Imanishi, Y., Shimizu, T., Okano, T. & Matsuda, H. (2005) Tissue cardiomyoplasty using bioengineered contractile cardiomyocyte sheets to repair damaged myocardium: their integration with recipient myocardium. *Transplantation,* 80, 1586–95.
37. Sasagawa, T., Shimizu, T., Sekiya, S., Haraguchi, Y., Yamato, M., Sawa, Y. & Okano, T. (2009) Design of prevascularized three-dimensional cell-dense tissues using a cell sheet stacking manipulation technology. *Biomaterials,* 31, 1646–54.
38. Laflamme, M. A., Gold, J., Xu, C., Hassanipour, M., Rosler, E., Police, S., Muskheli, V. & Murry, C. E. (2005) Formation of human myocardium in the rat heart from human embryonic stem cells. *Am J Pathol,* 167, 663–71.
39. Murry, C. E. & Lee, R. T. (2009) Development biology. Turnover after the fallout. *Science,* 324, 47–8.
40. Martens, T. P., See, F., Schuster, M. D., Sondermeijer, H. P., Hefti, M. M., Zannettino, A., Gronthos, S., Seki, T. & Itescu, S. (2006) Mesenchymal lineage precursor cells induce vascular network formation in ischemic myocardium. *Nat Clin Pract Cardiovasc Med,* 1, S18–22.
41. Banerjee, I., Yekkala, K., Borg, T. K. & Baudino, T. A. (2006) Dynamic interactions between myocytes, fibroblasts, and extracellular matrix. *Ann N Y Acad Sci,* 1080, 76–84.
42. Nag, A. C., Cheng, M. & Healy, C. J. (1980) Studies of adult amphibian heart cells in vitro: DNA synthesis and mitosis. *Tissue Cell,* 12, 125–39.
43. Naito, H., Melnychenko, I., Didie, M., Schneiderbanger, K., Schubert, P., Rosenkranz, S., Eschenhagen, T. & Zimmermann, W. H. (2006) Optimizing engineered heart tissue for therapeutic applications as surrogate heart muscle. *Circulation,* 114, I72–8.
43a. Radisic, M., Park, H., Martens, T. P., Salazar-Lazaro, J. E., Geng, W., Wang, Y., Langer, R., Freed, L. E. & Vunjak-Novakovic, G. (2008) Pre-treatment of synthetic elastomeric scaffolds by cardiac fibroblasts improves engineered heart tissue. *J Biomed Mater Res A,* 86, 713–24.
44. Christalla, P., Didie, M., Eschenhagen, T., Tozakidou, M., Ehmke, H. & Zimmermann, W.-H. (2009) Abstract 2279: Fibroblasts are essential for the generation of embryonic stem cell derived bioengineered myocardium. *Circulation,* 120, S615-a.
45. Hare, J. M., Traverse, J. H., Henry, T. D., Dib, N., Strumpf, R. K., Schulman, S. P., Gerstenblith, G., Demaria, A. N., Denktas, A. E., Gammon, R. S., Hermiller, J. B., Jr., Reisman, M. A., Schaer, G. L. & Sherman, W. (2009) A randomized, double-blind, placebo-controlled, dose-escalation study of intravenous adult human mesenchymal stem cells (prochymal) after acute myocardial infarction. *J Am Coll Cardiol,* 54, 2277–86.
46. Perin, E. C., Silva, G. V., Assad, J. A. R., Vela, D., Buja, L. M., Sousa, A. L. S., Litovsky, S., Lin, J., Vaughn, W. K., Coulter, S., Fernandes, M. R. & Willerson, J. T. (2008) Comparison of intracoronary and transendocardial delivery of allogeneic mesenchymal cells in a canine model of acute myocardial infarction. *J Mol Cell Cardiol,* 44, 486–95.
47. Dimmeler, S., Zeiher, A. M. & Schneider, M. D. (2005) Unchain my heart: the scientific foundations of cardiac repair. *J Clin Invest,* 115, 572–83.
48. Murry, C. E., Field, L. J. & Menasche, P. (2005) Cell-based cardiac repair: reflections at the 10-year point. *Circulation,* 112, 3174–83.
49. Chen, C. H., Wei, H. J., Lin, W. W., Chiu, I., Hwang, S. M., Wang, C. C., Lee, W. Y., Chang, Y. & Sung, H. W. (2008) Porous tissue grafts sandwiched with multilayered mesenchymal stromal cell sheets induce tissue regeneration for cardiac repair. *Cardiovasc Res,* 80, 88–95.
50. Martens, T. P., Godier, A. F., Parks, J. J., Wan, L. Q., Koeckert, M. S., Eng, G. M., Hudson, B. I., Sherman, W. & Vunjak-Novakovic, G. (2009) Percutaneous cell delivery into the heart using hydrogels polymerizing in situ. *Cell Transplant,* 18, 297–304.
51. Ratner, B., Hoffman, A., Schoen, F. & Lemons, J. (2004) *Biomaterials Science*, Oxford: Academic.

52. Wu, K., Liu, Y. L., Cui, B. & Han, Z. (2006) Application of stem cells for cardiovascular grafts tissue engineering. *Transpl Immunol,* 16, 1–7.
53. Dawson, E., Mapili, G., Erickson, K., Taqvi, S. & Roy, K. (2008) Biomaterials for stem cell differentiation. *Adv Drug Deliv Rev,* 60, 215–28.
54. Nair, L. S. & Laurencin, C. T. (2006) Polymers as biomaterials for tissue engineering and controlled drug delivery. *Adv Biochem Eng Biotechnol,* 102, 47–90.
55. Nerurkar, N. L., Elliott, D. M. & Mauck, R. L. (2007) Mechanics of oriented electrospun nanofibrous scaffolds for annulus fibrosus tissue engineering. *J Orthop Res,* 25, 1018–28.
56. Zimmermann, W. H., Fink, C., Kralisch, D., Remmers, U., Weil, J. & Eschenhagen, T. (2000) Three-dimensional engineered heart tissue from neonatal rat cardiac myocytes. *Biotechnol Bioeng,* 68, 106–14.
57. Christman, K. L., Fok, H. H., Sievers, R. E., Fang, Q. & Lee, R. J. (2004) Fibrin glue alone and skeletal myoblasts in a fibrin scaffold preserve cardiac function after myocardial infarction. *Tissue Eng,* 10, 403–9.
58. Khademhosseini, A., Eng, G., Yeh, J., Kucharczyk, P. A., Langer, R., Vunjak-Novakovic, G. & Radisic, M. (2007) Microfluidic patterning for fabrication of contractile cardiac organoids. *Biomed Microdevices,* 9, 149–57.
59. Bursac, N., Papadaki, M., Cohen, R. J., Schoen, F. J., Eisenberg, S. R., Carrier, R., Vunjak-Novakovic, G. & Freed, L. E. (1999) Cardiac muscle tissue engineering: toward an in vitro model for electrophysiological studies. *Am J Physiol,* 277, H433–44.
60. Ishii, O., Shin, M., Sueda, T. & Vacanti, J. P. (2005) In vitro tissue engineering of a cardiac graft using a degradable scaffold with an extracellular matrix-like topography. *J Thorac Cardiovasc Surg,* 130, 1358–63.
61. Mcdevitt, T. C., Woodhouse, K. A., Hauschka, S. D., Murry, C. E. & Stayton, P. S. (2003) Spatially organized layers of cardiomyocytes on biodegradable polyurethane films for myocardial repair. *J Biomed Mater Res A,* 66, 586–95.
62. Gao, J., Crapo, P. M. & Wang, Y. (2006) Macroporous elastomeric scaffolds with extensive micropores for soft tissue engineering. *Tissue Eng,* 12, 917–25.
63. Zimmermann, W. H., Melnychenko, I. & Eschenhagen, T. (2004) Engineered heart tissue for regeneration of diseased hearts. *Biomaterials,* 25, 1639–47.
64. Wang, Y., Ameer, G. A., Sheppard, B. J. & Langer, R. (2002) A tough biodegradable elastomer. *Nat Biotechnol,* 20, 602–6.
65. Engler, A. J., Griffin, M. A., Sen, S., Bonnemann, C. G., Sweeney, H. L. & Discher, D. E. (2004) Myotubes differentiate optimally on substrates with tissue-like stiffness: pathological implications for soft or stiff microenvironments. *J Cell Biol,* 166, 877–87.
66. Levy-Mishali, M., Zoldan, J. & Levenberg, S. (2009) Effect of scaffold stiffness on myoblast differentiation. *Tissue Eng Part A,* 15, 935–44.
67. Sherman, W. (2007) Myocyte replacement therapy: skeletal myoblasts. *Cell Transplant,* 16, 971–5.
68. Schachinger, V., Assmus, B., Britten, M. B., Honold, J., Lehmann, R., Teupe, C., Abolmaali, N. D., Vogl, T. J., Hofmann, W. K., Martin, H., Dimmeler, S. & Zeiher, A. M. (2004) Transplantation of progenitor cells and regeneration enhancement in acute myocardial infarction: final one-year results of the TOPCARE-AMI Trial. *J Am Coll Cardiol,* 44, 1690–9.
69. Birgersdotter, A., Sandberg, R. & Ernberg, I. (2005) Gene expression perturbation in vitro – a growing case for three-dimensional (3D) culture systems. *Semin Cancer Biol,* 15, 405–12.
70. Cukierman, E., Pankov, R. & Yamada, K. M. (2002) Cell interactions with three-dimensional matrices. *Curr Opin Cell Biol,* 14, 633–9.
71. Griffith, L. G. & Swartz, M. A. (2006) Capturing complex 3D tissue physiology in vitro. *Nat Rev Mol Cell Biol,* 7, 211–24.
72. Nelson, C. M. & Bissell, M. J. (2006) Of extracellular matrix, scaffolds, and signaling: tissue architecture regulates development, homeostasis, and cancer. *Annu Rev Cell Dev Biol,* 22, 287–309.

73. Yamada, K. M. & Cukierman, E. (2007) Modeling tissue morphogenesis and cancer in 3D. *Cell,* 130, 601–10.
74. Mitrius, J. C. & Vogel, S. M. (1990) Doxorubicin-induced automaticity in cultured chick heart cell aggregates. *Cancer Res,* 50, 4209–15.
75. Kehat, I., Khimovich, L., Caspi, O., Gepstein, A., Shofti, R., Arbel, G., Huber, I., Satin, J., Itskovitz-Eldor, J. & Gepstein, L. (2004) Electromechanical integration of cardiomyocytes derived from human embryonic stem cells. *Nat Biotechnol,* 22, 1282–9.
76. Song, H., Yoon, C., Kattman, S. J., Dengler, J., Masse, S., Thavaratnam, T., Gewarges, M., Nanthakumar, K., Rubart, M., Keller, G. M., Radisic, M. & Zandstra, P. W. (2009) Regenerative medicine special feature: interrogating functional integration between injected pluripotent stem cell-derived cells and surrogate cardiac tissue. *Proc Natl Acad Sci U S A,* 107, 3329–34.
77. Pedrotty, D. M., Klinger, R. Y., Badie, N., Hinds, S., Kardashian, A. & Bursac, N. (2008) Structural coupling of cardiomyocytes and noncardiomyocytes: quantitative comparisons using a novel micropatterned cell pair assay. *Am J Physiol Heart Circ Physiol,* 295, H390–400.
78. Mousses, S., Kallioniemi, A., Kauraniemi, P., Elkahloun, A. & Kallioniemi, O. P. (2002) Clinical and functional target validation using tissue and cell microarrays. *Curr Opin Chem Biol,* 6, 97–101.
79. Caspi, O., Itzhaki, I., Arbel, G., Kehat, I., Gepstien, A., Huber, I., Satin, J. & Gepstein, L. (2009) In vitro electrophysiological drug testing using human embryonic stem cell derived cardiomyocytes. *Stem Cells Dev,* 18, 161–72.
80. Takahashi, K. & Yamanaka, S. (2006) Induction of pluripotent stem cells from mouse embryonic and adult fibroblast cultures by defined factors. *Cell,* 126, 663–76.
81. Takahashi, K., Tanabe, K., Ohnuki, M., Narita, M., Ichisaka, T., Tomoda, K. & Yamanaka, S. (2007) Induction of pluripotent stem cells from adult human fibroblasts by defined factors. *Cell,* 131, 861–72.
82. Yu, J., Vodyanik, M. A., Smuga-Otto, K., Antosiewicz-Bourget, J., Frane, J. L., Tian, S., Nie, J., Jonsdottir, G. A., Ruotti, V., Stewart, R., Slukvin, Ii & Thomson, J. A. (2007) Induced pluripotent stem cell lines derived from human somatic cells. *Science,* 318, 1917–20.
83. Nakagawa, M., Koyanagi, M., Tanabe, K., Takahashi, K., Ichisaka, T., Aoi, T., Okita, K., Mochiduki, Y., Takizawa, N. & Yamanaka, S. (2008) Generation of induced pluripotent stem cells without Myc from mouse and human fibroblasts. *Nat Biotechnol,* 26, 101–6.
84. Zhang, J., Wilson, G. F., Soerens, A. G., Koonce, C. H., Yu, J., Palecek, S. P., Thomson, J. A. & Kamp, T. J. (2009) Functional cardiomyocytes derived from human induced pluripotent stem cells. *Circ Res,* 104, e30–41.
84a. Hansen, A., Eder, A., Bönstrup, M., Flato, M., Mewe, M., Schaaf, S., Aksehirlioglu, B., Schwörer, A., Uebeler, J. & Eschenhagen, T. (2010) Development of a drug screening platform based on engineered heart tissue. *Circ Res,* 107, 35–44.
85. Kofidis, T., Lenz, A., Boublik, J., Akhyari, P., Wachsmann, B., Mueller-Stahl, K., Hofmann, M. & Haverich, A. (2003) Pulsatile perfusion and cardiomyocyte viability in a solid three-dimensional matrix. *Biomaterials,* 24, 5009–14.
86. Dvir, T., Benishti, N., Shachar, M. & Cohen, S. (2006) A novel perfusion bioreactor providing a homogenous milieu for tissue regeneration. *Tissue Eng,* 12, 2843–52.
87. Dvir, T., Levy, O., Shachar, M., Granot, Y. & Cohen, S. (2007) Activation of the ERK1/2 cascade via pulsatile interstitial fluid flow promotes cardiac tissue assembly. *Tissue Eng,* 13, 2185–93.
88. Radisic, M., Euloth, M., Yang, L., Langer, R., Freed, L. E. & Vunjak-Novakovic, G. (2003) High-density seeding of myocyte cells for cardiac tissue engineering. *Biotechnol Bioeng,* 82, 403–14.
89. Radisic, M., Yang, L., Boublik, J., Cohen, R. J., Langer, R., Freed, L. E. & Vunjak-Novakovic, G. (2004) Medium perfusion enables engineering of compact and contractile cardiac tissue. *Am J Physiol Heart Circ Physiol,* 286, H507–16.
90. Ochoa, E. R. & Vacanti, J. P. (2002) An overview of the pathology and approaches to tissue engineering. *Ann N Y Acad Sci,* 979, 10–26; discussion 35–8.

91. Kaihara, S., Borenstein, J., Koka, R., Lalan, S., Ochoa, E. R., Ravens, M., Pien, H., Cunningham, B. & Vacanti, J. P. (2000) Silicon micromachining to tissue engineer branched vascular channels for liver fabrication. *Tissue Eng,* 6, 105–17.
92. Frame, M. D. & Sarelius, I. H. (1995) A system for culture of endothelial cells in 20–50-microns branching tubes. *Microcirculation,* 2, 377–85.
93. Anderson, J. R., Chiu, D. T., Jackman, R. J., Cherniavskaya, O., Mcdonald, J. C., Wu, H., Whitesides, S. H. & Whitesides, G. M. (2000) Fabrication of topologically complex three-dimensional microfluidic systems in PDMS by rapid prototyping. *Anal Chem,* 72, 3158–64.
94. Chiu, D. T., Jeon, N. L., Huang, S., Kane, R. S., Wargo, C. J., Choi, I. S., Ingber, D. E. & Whitesides, G. M. (2000) Patterned deposition of cells and proteins onto surfaces by using three-dimensional microfluidic systems. *Proc Natl Acad Sci USA,* 97, 2408–13.
95. Losordo, D. W. & Dimmeler, S. (2004) Therapeutic angiogenesis and vasculogenesis for ischemic disease. Part I: angiogenic cytokines. *Circulation,* 109, 2487–91.
96. Perets, A., Baruch, Y., Weisbuch, F., Shoshany, G., Neufeld, G. & Cohen, S. (2003) Enhancing the vascularization of three-dimensional porous alginate scaffolds by incorporating controlled release basic fibroblast growth factor microspheres. *J Biomed Mater Res A,* 65, 489–97.
97. Guan, J., Stankus, J. J. & Wagner, W. R. (2007) Biodegradable elastomeric scaffolds with basic fibroblast growth factor release. *J Control Release,* 120, 70–8.
98. Holder, W. D., Jr., Gruber, H. E., Moore, A. L., Culberson, C. R., Anderson, W., Burg, K. J. & Mooney, D. J. (1998) Cellular ingrowth and thickness changes in poly-L-lactide and polyglycolide matrices implanted subcutaneously in the rat. *J Biomed Mater Res,* 41, 412–21.
99. Park, H. J., Yoo, J. J., Kershen, R. T., Moreland, R. & Atala, A. (1999) Reconstitution of human corporal smooth muscle and endothelial cells in vivo. *J Urol,* 162, 1106–9.
100. Atala, A., Bauer, S. B., Soker, S., Yoo, J. J. & Retik, A. B. (2006) Tissue-engineered autologous bladders for patients needing cystoplasty. *Lancet,* 367, 1241–6.
101. Koike, N., Fukumura, D., Gralla, O., Au, P., Schechner, J. S. & Jain, R. K. (2004) Tissue engineering: creation of long-lasting blood vessels. *Nature,* 428, 138–9.
102. Loffredo, F. & Lee, R. T. (2008) Therapeutic vasculogenesis: it takes two. *Circ Res,* 103, 128–30.
103. Levenberg, S., Rouwkema, J., Macdonald, M., Garfein, E. S., Kohane, D. S., Darland, D. C., Marini, R., Van Blitterswijk, C. A., Mulligan, R. C., D'amore, P. A. & Langer, R. (2005) Engineering vascularized skeletal muscle tissue. *Nat Biotechnol,* 23, 879–84.
104. Levenberg, S., Golub, J. S., Amit, M., Itskovitz-Eldor, J. & Langer, R. (2002) Endothelial cells derived from human embryonic stem cells. *Proc Natl Acad Sci U S A,* 99, 4391–6.
105. Narmoneva, D. A., Vukmirovic, R., Davis, M. E., Kamm, R. D. & Lee, R. T. (2004) Endothelial cells promote cardiac myocyte survival and spatial reorganization: implications for cardiac regeneration. *Circulation,* 110, 962–8.
106. Stevens, K. R., Kreutziger, K. L., Dupras, S. K., Korte, F. S., Regnier, M., Muskheli, V., Nourse, M. B., Bendixen, K., Reinecke, H. & Murry, C. E. (2009) Physiological function and transplantation of scaffold-free and vascularized human cardiac muscle tissue. *Proc Natl Acad Sci USA,* 106, 16568–73.
107. Morritt, A. N., Bortolotto, S. K., Dilley, R. J., Han, X., Kompa, A. R., Mccombe, D., Wright, C. E., Itescu, S., Angus, J. A. & Morrison, W. A. (2007) Cardiac tissue engineering in an in vivo vascularized chamber. *Circulation,* 115, 353–60.
108. Yang, F., Cho, S. W., Son, S. M., Bogatyrev, S. R., Singh, D., Green, J. J., Mei, Y., Park, S., Bhang, S. H., Kim, B. S., Langer, R. & Anderson, D. G. (2009) Regenerative medicine special feature: genetic engineering of human stem cells for enhanced angiogenesis using biodegradable polymeric nanoparticles. *Proc Natl Acad Sci USA,* 107, 3317–22.
109. Roell, W., Lewalter, T., Sasse, P., Tallini, Y. N., Choi, B. R., Breitbach, M., Doran, R., Becher, U. M., Hwang, S. M., Bostani, T., Von Maltzahn, J., Hofmann, A., Reining, S.,

Eiberger, B., Gabris, B., Pfeifer, A., Welz, A., Willecke, K., Salama, G., Schrickel, J. W., Kotlikoff, M. I. & Fleischmann, B. K. (2007) Engraftment of connexin 43-expressing cells prevents post-infarct arrhythmia. *Nature,* 450, 819–24.

110. Cao, F., Wagner, R. A., Wilson, K. D., Xie, X., Fu, J. D., Drukker, M., Lee, A., Li, R. A., Gambhir, S. S., Weissman, I. L., Robbins, R. C. & Wu, J. C. (2008) Transcriptional and functional profiling of human embryonic stem cell-derived cardiomyocytes. *PLoS One,* 3, e3474.
111. Lesman, A., Habib, M., Caspi, O., Gepstein, A., Arbel, G., Levenberg, S. & Gepstein, L. (2009) Transplantation of a tissue-engineered human vascularized cardiac muscle. *Tissue Eng Part A,* 16, 115–25.
112. Laflamme, M. A., Chen, K. Y., Naumova, A. V., Muskheli, V., Fugate, J. A., Dupras, S. K., Reinecke, H., Xu, C., Hassanipour, M., Police, S., O'sullivan, C., Collins, L., Chen, Y., Minami, E., Gill, E. A., Ueno, S., Yuan, C., Gold, J. & Murry, C. E. (2007) Cardiomyocytes derived from human embryonic stem cells in pro-survival factors enhance function of infarcted rat hearts. *Nat Biotechnol,* 25, 1015–24.
113. Souders, C. A., Bowers, S. L. & Baudino, T. A. (2009) Cardiac fibroblast: the renaissance cell. *Circ Res,* 105, 1164–76.
114. Frangogiannis, N. G. (2008) The immune system and cardiac repair. *Pharmacol Res,* 58, 88–111.

Part IV
Technical Issues for Stem Cell Therapy in the Heart

Methods of Cell Delivery for Cardiac Repair

Sarah Fernandes and Hans Reinecke

Abstract Cellular cardiomyoplasty describes the procedure of cell delivery to the injured heart with the goal to either reverse or at least halt the detrimental effects of myocardial injury. Several strategies for cell delivery to the injured myocardium are available. Direct intramyocardial injection may be performed using a transepicardial, surgical, transendocardial, or transvenous approach. Other strategies include the delivery of cells through coronary arteries, coronary veins, or peripheral veins. In this chapter, we review the successes and the shortcomings of the various procedures that have been tried in animal models of cardiac injury and clinically in patient trials.

Keywords Intramyocardial cell delivery • Transvascular cell delivery • Intracoronary cell delivery • Intravenous cell delivery • Cellular cardiomyoplasty • TOPCARE • REPAIR-AMI • BOOST

1 Introduction

For about 15 years now, cellular cardiomyoplasty has been explored for the repair of the injured myocardium. Pioneering studies in animal models of cardiac injury used a variety of cell types, e.g., primary cardiomyocytes, skeletal muscle cells, smooth muscle cells, and bone-marrow-derived cells, in conjunction with various modes of delivery. Maximum cell engraftment is a prerequisite for the success of cellular cardiomyoplasty. Therefore, the optimal delivery method needs to deliver a sufficient number of cells to the site of myocardial injury while maximizing cell survival and retention within the impaired cardiac muscle. Indeed, cell homing

H. Reinecke (✉)
Center for Cardiovascular Biology, University of Washington Medicine at South Lake Union, Seattle, WA
e-mail: hreineck@u.washington.edu

I.S. Cohen and G.R. Gaudette (eds.), *Regenerating the Heart*, Stem Cell Biology and Regenerative Medicine, DOI 10.1007/978-1-61779-021-8_24,
© Springer Science+Business Media, LLC 2011

and/or cell retention is affected not only by the route of delivery but also by the timing of injection and the vehicle used for cell delivery. The cardiac disease (i.e., acute myocardial infarction with or without reperfusion or chronic ischemic coronary disease) and the type of cell to be injected will determine the cell delivery strategy. Direct transepicardial cell delivery (i.e., injection into the ventricular wall with direct visualization through thoracotomy) was the first and still is the main method used in both animal and clinical studies. However, less invasive cell delivery methods such as the transvascular route (e.g., intracoronary or intravenous cell delivery) or transendocardial cell delivery using a vascular approach have been explored more recently. In this chapter, we discuss the advantages and disadvantages of the various cell delivery methods.

2 Cell Delivery in Experimental Animal Studies

Major methods of cell delivery (e.g., direct injection into the ventricular wall and transvascular infusion) have been extensively tested in various small- and large-animal models and many of these studies demonstrated an improvement of heart function with cell delivery, thus justifying the beginning of clinical studies. However, to date only few experimental studies have evaluated the effects of different cell delivery methods side-by-side [1–11] (Table 1). These studies are important as they contribute to our understanding of cell fate after implantation, a prerequisite to improve the outcome of cell therapy.

2.1 Intramyocardial Cell Delivery

Intramyocardial cell delivery (direct injection into the left ventricular wall) is the most effective means to deliver a therapeutic agent to the infarcted heart. This approach is applicable to both patients with acute coronary disease and those with ischemic cardiomyopathy (Table 2). Of advantage, cells with a larger diameter can be injected without the risk of embolization. In fact, several studies showed that intramyocardial injection maximizes cell retention compared with transvascular cell delivery when analyzed immediately after cell injection [4, 6, 8–10]. However, it appears that the beneficial effect of intramyocardial injection on cardiac cell retention may be transient. Indeed, 50% of cells escape from the heart within 1-h after intramyocardial injection owing to immediate leakage from the needle track after injection [12, 13], and only 15% and less than 5% of the injected cells are still present at 1 and 6 weeks after injection, respectively [6, 14, 15]. This progressive cell loss within days after injection resulted in an intracardiac cell number similar to that with intracoronary or intravenous injection within 2 weeks after cell delivery [2, 8]. When performing intramyocardial injections, the repeated needle puncture, exaggerated or too forceful injections, and excessive injection volumes can induce

Table 1 Potential route of cell delivery: advantages and disadvantages

	Advantages	Disadvantages
Transvascular cell delivery	Suitable for acute myocardial infarction	Not suitable for an unperfused region (chronic ischemic cardiomyopathy)
Intravenous (peripheral vein)	non invasive	Poor cell homing/low cardiac retention
	Does not require an operating room or a catheterization laboratory	May increase likelihood that cells locate in organs other than the heart
Retrograde infusion via coronary vein	non invasive	Requires a catheterization laboratory
	No concern for embolization	Low cardiac retention
		Potential for perforation of the coronary sinus
Intracoronary infusion	Does not require major surgery	Requires a catheterization laboratory
	Can be performed at the same time as other intracoronary procedures (angioplasty, stenting)	Low cardiac retention Potential embolization hazard
Direct injection into the ventricular wall	Suitable for chronic myocardial infarction	
Transendocardial catheter delivery	Can target the area of interest using electromechanical guidance	Requires a catheterization laboratory for electromechanical guidance
	Does not require major surgery	Potential cardiac perforation or hemorrhage
	A large number of cells can be implanted in the myocardium without embolism risk	
Transepicardial (direct intramyocardial injection)	Can be associated with coronary artery bypass	Invasive, requires surgery
	Area of interest can be visualized	
	A large number of cells can be implanted directly in the myocardium	

ecchymosis, leading to cardiac tissue inflammation and to accumulation of CD45+ cells [3, 16]. Such inflammatory responses specific to intramyocardial injection can be responsible for cell death over time and may present a limitation of the use of direct intramyocardial injection as the injection volume and the number of injected cells cannot be increased without exacerbating the inflammation. Another concern regarding intramyocardial injection pertains to the particular engraftment pattern of the injected cells. After intramyocardial injection, most of the cells stay at the injection site without migrating, thus creating islet-like clusters within the myocardium (Fig. 1, unpublished results). This cell-cluster-like distribution pattern may be

Table 2 Preclinical studies

Reference	Animal model/time between injury and cell injection	Cell type/cell number	Mode of injection	Follow-up after injection	Outcomes
Li et al. [10]	Rat Sham 3 days after infarct 4 weeks after infarct	BMC 3×10^6	• Intraarterial • Intravenous • Intramyocardial	2 h 24 h 48 h	Greater cardiac cell retention in the chronic model compared with the acute model Intravenous injection leads to nearly complete trapping of the injected cells in the lungs Intraarterial injection leads to the smallest cell retention in the heart Intramyocardial injection facilitates cell targeting ($P < 0.05$ vs. intravenous and intraaortic injection)
Fukushima et al. [3]	Rat 3 weeks after infarct	SkM 5×10^6	• Retrograde intracardiac vein injection • Intramyocardial	84 days	Functional improvement is similar in both groups Retrograde intracoronary injection: attenuated early-phase arrhythmogenicity but late-phase arrhythmogenicity was similar, regardless of the delivery route
Hale et al. [6]	Rat Immediately after infarct	MSC 2×10^6	• Intravenous • Intramyocardial	1 week	Intramyocardial: 15% of cells were retained in the heart Intravenous: no MSCs were detected in the heart, no evidence of homing Cell were distributed to other organs (the number of MSCs in liver, spleen, and lung was similar with both intravenous and intramyocardial injection)
Freyman et al. [2]	Pig Immediately after infarct	MSC 5×10^7	• Intravenous • Intracoronary • Endocardiac	14 ± 3 days	At 14 days, the number of engrafted cells within the infarct zone corresponds to the order intracoronary infusion > endocardiac injection > intravenous infusion There is less systemic delivery to the lungs following endocardiac injection vs. intravenous infusion or vs. intracoronary infusion After intracoronary infusion 1/2 of pigs had decreased blood flow distal to the infusion site

George et al. [4]	Pig Injection 1 week after infarct	BMC 1×10^7	• Transvenous intramyocardic • Intracoronary	1 h	Transvenous intramyocardic approach is safe and feasible Transvenous intramyocardic approach may provide an advantage over intracoronary infusion in retention of the cellular product. However, the study was limited by a small size sample ($n=2$ per group)
Hou et al. [9]	Pig Immediately after infarct	hPB-MNC 1×10^7	• Retrograde coronary venous • Intracoronary • Intramyocardial	1 h	For each modality, most of the cells are not retained in the heart Cardiac retention: intramyocardial > intracoronary > retrograde coronary venous Pulmonary retention: intracoronary > retrograde coronary venous > intramyocardial
Hayashi et al. [8]	Rat Immediately after infarct	BMC $2–20 \times 10^7$	• Intravenous • 10 times intravenous • Intramyocardial	1, 3, 7, and 14 days	Cell therapy is more effective when cells are delivered intramyocardially rather than intravenously Cell survival: intramyocardial > intravenous > 10 times intravenous at 3 days Blood flow: intramyocardial = 10 times intravenous > intravenous at 14 days Left ventricular ejection fraction is higher in intramyocardial delivery and 10 times intravenous delivery than in intravenous delivery
Barbash et al. [21]	Rat Sham 2 days after infarct 2 weeks after infarct	BMC 4×10^6	• Intravenous • Intraventricular	4 h 1 week	Intravenous delivery leads to cell entrapment in the lung With intraventricular injection cells can migrate to and colonize the ischemic myocardium
Perin et al. [15]	Dogs 1 week after infarct	MSC 100×10^6	• Transendocardial • Intracoronary	2 weeks	Transendocardial injection induces a higher retention rate than intracoronary injection Functional beneficial effects were observed only with transendocardial injection

BMCs bone marrow cell, *SkM* skeletal myoblast, *MSC* mesenchymal stem cell, *hPB-MNC* human peripheral blood mononuclear cell

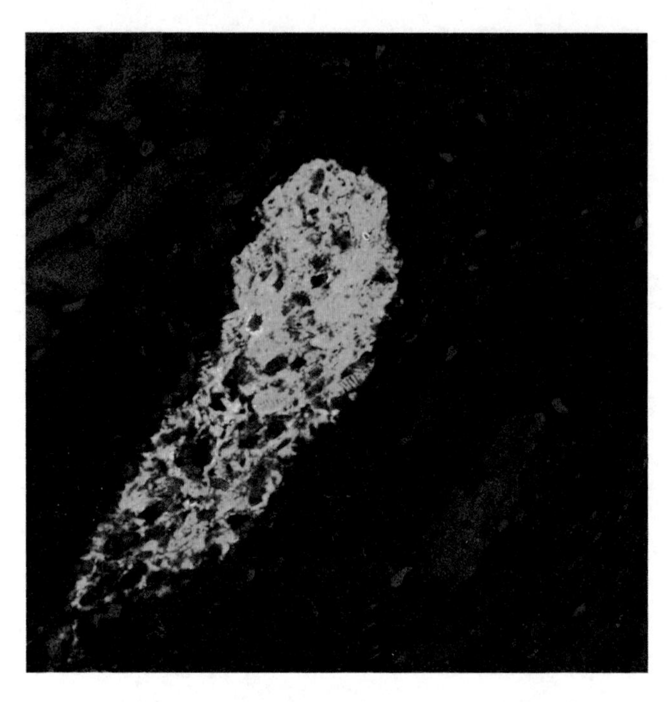

Fig. 1 Clusterlike distribution pattern after direct intramyocardial injection. Three months after direct intramyocardial injection into the infarcted rat heart, human embryonic stem cell derived cardiomyocytes (*green fluorescence* corresponds to β-myosin heavy chain) are still present in the rat myocardium (*red fluorescence* corresponds to cardiac troponin T). Human cardiomyocytes did not migrate after injection but formed cell islets that were separated from host myocardium by scar tissue. *Blue fluorescence* Hoechst 33342 nuclear counterstain (Fernandes S and Murry CE, unpublished results)

prone to induce arrhythmia. Indeed, a comparative study showed that intramyocardial injection leads to greater arrhythmic risk than retrograde venous injection [3].

2.2 *Transvascular Cell Delivery*

2.2.1 Intracoronary Cell Delivery

Intracoronary infusion of a cell suspension allows the delivery of larger volumes and higher cell numbers targeting the area of interest in the heart. Coronary infusion can be performed with a catheter introduced into the coronary artery, commonly performed during a reperfusion procedure after acute myocardial infarction. In addition, this delivery mode is less traumatic and invasive compared with direct intramyocardial injection. Four different studies performed in swine or dog models of acute myocardial infarction showed that intracoronary injection is feasible. However, there were mixed results with regard to cell retention within the cardiac tissue [2, 4, 9, 15]. In three of the four studies cardiac retention after intracoronary injection was lower than that after intramyocardial injection [4, 9, 15]. The main adverse side effect of intracoronary injection was a decreased coronary blood flow during the injection in a pig model [2]. Conversely, Vulliet et al. observed ST-segment changes on EKG during the intracoronary delivery characteristic of acute myocardial ischemia. Histologic evaluation performed at 7 days after cell injection show macroscopic and microscopic evidence of myocardial infarction: sections of myocardium showed several scattered regions of dense fibroplasia accompanied by macrophage infiltrates only in areas where the injected cells were observed [17]. In a clinical study Kang et al. observed a mild increase in creatine phosphokinase-MB after cell infusion [18], confirming the potential risk of micro-infarction following intracoronary cell injection. Dependent on their size, injected cells may cause capillary plugging leading to decreased distal coronary blood flow and reduced tissue perfusion [2]. This phenomenon (called "no reflow" by interventional cardiologists) may further increase infarct size and expansion of an already infarcted patient, worsening scar thinning, and long-term cardiac complications, potentially related to adverse ventricular remodeling [19, 20]. In summary, coronary plugging after intracoronary cell delivery is a serious potential side effect and may limit the number of cells that can be injected.

2.2.2 Intravenous Cell Delivery

Intravenous delivery of stem cells is an attractive and noninvasive strategy that allows repeated administration of large numbers of cells. Once delivered, the cells migrate through the systemic circulation, settle in the infarcted myocardium, and receive local signals directing differentiation [21]. However, mixed results from clinical trials using this particular route of delivery stimulated the investigation of cell distribution [22]. Several preclinical studies have shown that intravenous injection results in lower cell retention in the myocardium compared with other delivery routes [2, 4, 6, 8, 10, 21], independently of the animal size (rat or pig) or the type of injury (acute myocardial infarction or remote infarct injury). Because normal coronary blood flow corresponds to only approximately 3% of the cardiac output [23], once injected in the systemic circulation, only a very small fraction of the infused cells can reach the myocardial tissue or even the infarcted region. Retrograde coronary venous injection (e.g., infusion of cells through the venous coronary system) did not improve myocardial retention, compared with intramyocardial injection or even intracoronary injection [9]. In fact, immediately after intravenous injection or retrograde coronary venous injection most of the cells are retained in lung, liver, spleen, and kidney rather than the heart [2, 9, 10, 21]. However, using intramyocardial injection of cardiomyocytes derived from human embryonic stem cells, our group recently demonstrated that 4 weeks after cell injection, cells are spontaneously washed out of nontargeted organs, suggesting innocuity of such cell dispersion [24]. Another limitation of transvascular delivery is that the therapeutic cells need to be capable of transendothelial migration, a process unnatural for some cell types that have been used in cardiac cell therapy, e.g., skeletal myoblasts. However, most of the studies cited here were performed with cell types that display some plasticity [e.g., bone marrow cells, mesenchymal stem cells (MSCs), mononuclear peripheral blood cells, and endothelial cells] and these may be more likely to cross the endothelial barrier.

2.2.3 Impact of Cell Delivery on Myocardial Function

Most studies evaluating the mode of delivery focused on cell biodistribution and more particularly on cardiac retention within a few days after injection. However, some studies correlated the cardiac cell retention with the long-term functional effects of cell transplantation. Interestingly, when evaluating intravenous injection and direct intramyocardial injection side-by-side by injecting the same number of cells in a canine model of acute myocardial infarction model, Perin et al. demonstrated that a beneficial effect is observed only with intramyocardial cell injection (transendocardial cell delivery with electrical mapping) but not with intracoronary cell delivery [15]. Hayashi et al. suggested that to obtain the same beneficial effect, the number of intravenously injected cells needed to be ten times higher than the number of cells injected via direct intramyocardial injection [8]. In contrast, Fukushima et al. observed the same beneficial effect using retrograde intravenous cell delivery or direct intramyocardial cell delivery [3]. However, contrasting results may be explained by the use of different animal models (rat vs. dog), the timing of cell delivery (immediately vs. 1 week after myocardial infarction), and the time point of functional evaluation (2 weeks vs. 3 months after cell injection) and thus little insight can be gained as to the best cell delivery route.

3 Cell Delivery in Humans

3.1 Transvascular Cell Delivery

3.1.1 Intravenous Cell Delivery via Peripheral Veins

Intravenous infusion of therapeutic cells via peripheral veins represents a convenient and minimally invasive route of delivery. Indeed, Kocher et al. [25] showed that human CD34+ cells delivered via the tail vein in a rat model of mycocardial infarction [permanent ligation of the left anterior descending (LAD) coronary artery] resulted in the infiltration of the infarct zone within 48 h of LAD ligation. Histologic examination at 2 weeks after infarction revealed that injection of CD34+ cells was accompanied by a significant increase in infarct zone microvascularity, cellularity, and numbers of factor VIII+ angioblasts and capillaries, and was also accompanied by reduction in matrix deposition and fibrosis [25]. In our laboratory,

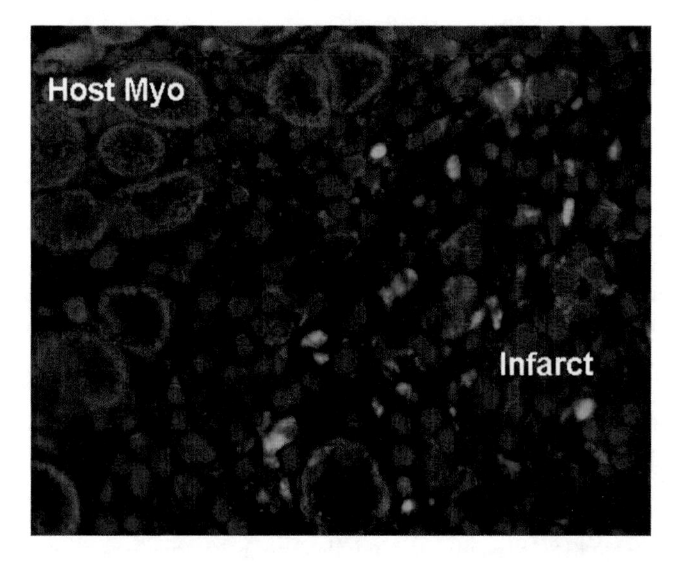

Fig. 2 Homing of monocytes to the infarct. Mouse monocytes/macrophages were derived from green fluorescent protein (GFP)-positive bone marrow (GFP mice) and injected into the tail vein of recipient mice that were infarcted 2 days before. GFP+ cells (*green fluorescence*) were readily detected in the infarct area at 3 days after peripheral injection. *Myo* myocardium, *red fluorescence* rhodamine-phalloidin counterstaining identifies myocardium and smooth muscle, *blue fluorescence* Hoechst 33342 nuclear counterstain (Reinecke H, Minami E, and Murry CE, unpublished results)

we followed the hypothesis that bone-marrow-derived monocytes could be used as gene delivery vehicles that would home to a site of injury. In these experiments, we injected green fluorescent protein (GFP)-tagged monocytes into the tail vein of infarcted mice. Immunohistologic analysis revealed the presence of GFP+ monocytes/macrophages in the area of infarct (Fig. 2, unpublished results).

Gao et al. studied the in vivo distribution of rat bone-marrow-derived MSCs, labeled with indium oxine, and infused into syngeneic rats via intraartery, intravenous, and intraperitoneal cavity infusions [26]. In addition, for intraartery and intravenous infusions, a vasodilator, sodium nitroprusside, was administered prior to the cell infusion and examined for its effect on MSC circulation. After 48 h, radioactivity in excised organs, including liver, lungs, kidneys, spleen, and long bones, was measured in a gamma well counter and expressed as a percentage of injected doses. After both intraartery and intravenous infusion, radioactivity associated with MSCs was detected primarily in the lungs and then secondarily in the liver and other organs. When sodium nitroprusside was used, more labeled MSCs cleared the lungs, resulting in a larger proportion detected in the liver. These results indicate multiple homing sites for injected MSCs and that the distribution of MSCs can be influenced by administration of a vasodilator [26].

In summary, despite the principal feasibility of the intravenous delivery route, a major drawback is the poor recovery of the injected cells at the site of injury as most cells are trapped in the microvasculature of lungs, liver, and other organs.

A further complication is represented by the fact that peripherally infused cells only home shortly (within days) after the injury occurred when homing signals are upregulated. Hence, this delivery route appears not useful in treating chronic myocardial ischemia.

3.1.2 Intracoronary Cell Delivery

Intracoronary infusion of cell populations has been widely used clinically and is the most common delivery route after acute myocardial infarction [27–38]. Cell delivery via this route is relatively easy to perform and a practically unlimited number of cells can be infused; however, cell size is limiting. As described previously, Vulliet et al. showed a potential complication of injecting MSCs or cells of similar size into the coronary circulation of healthy dogs. Although differences between canine and human coronary circulation exist, and different cell types and sizes have been used for selected cytotherapeutic applications, this potential complication should be considered before MSCs are injected into the arterial circulation of patients [17].

Strauer et al. performed the first feasibility study of human percutaneous intracoronary cell delivery in 2002 [27]. They analyzed ten patients who were treated by intracoronary transplantation of autologous, mononuclear bone marrow cells in addition to standard therapy after myocardial infarction. The cells were delivered via a balloon catheter placed into the infarct-related artery during balloon dilatation (percutaneous transluminal coronary angioplasty). Another ten patients with acute myocardial infarction were treated by standard therapy alone. At 3 months' follow-up, the infarct region had decreased significantly within the cell therapy group. Likewise, infarction wall movement velocity increased significantly only in the cell therapy group. In addition, the cell therapy group showed significant improvement in stroke volume index, left ventricular end-systolic volume and contractility (ratio of systolic pressure and end-systolic volume) and myocardial perfusion of the infarct region. It was concluded that selective intracoronary transplantation of autologous mononuclear bone marrow cells is safe and effective under clinical conditions [27].

The technique is similar to that used for coronary angioplasty, which involves over-the-wire positioning of an angioplasty balloon in one of the coronary arteries. The coronary blood flow is then stopped for a few minutes while the stem cells are infused under pressure. This maximizes contact between the infused cells and the microcirculation of the infarct-related artery, allowing for transendothelial migration to perivascular spaces. The latter implies that cells used for intracoronary infusion must be capable of transendothelial migration. Therefore, this delivery technique appears suitable only in the setting of acute ischemia after adhesion molecules and cytokine signaling are temporarily upregulated. In addition, cell preparations that are viscous, e.g., by suspending the cells in a hydrogel or such, may not be suitable for intracoronary infusion, owing to the risk of microvascular obstruction and myocardial ischemia.

Since Strauer's pioneering feasibility study, many others have followed with mixed results [39]. For example, the TOPCARE [36, 40] and REPAIR-AMI

[37, 38] trials reported sustained improvements of myocardial performance, whereas the BOOST [32, 33] and ASTAMI [35] trials reported only transient or no benefits, respectively. All trials used bone marrow mononuclear cells and intracoronary delivery. However, it was suggested that differences in cell isolation protocols may have impacted the different outcomes [41].

In summary, although intracoronary infusion is widely used, this method of delivery lacks a strong experimental background with regard to safety and efficacy. Attractive are the ease of use, the potentially broad distribution of cells within the myocardium, and the targeting of a specific coronary branch, although overall cell retention is poor. Major disadvantages are the limitation of cell size and the quick washout of cells. The latter, depending on the cell type used, may be a concern for adverse effects such as tumorigenesis.

3.2 *Intramyocardial Cell Delivery*

As described earlier, intramyocardial delivery of therapeutic cells was largely pioneered in animal models of cardiac injury and became the foundation for the first trials in humans with the direct transepicardial injection of skeletal myoblasts by Menasché's group [42] and bone marrow cells by Stamm et al. [43]. Catheter-based methods were soon developed, and in 2003 Perin et al. reported transendocardial injections of autologous mononuclear bone marrow cells in patients with end-stage ischemic heart disease [44, 45].

Intramyocardial cell delivery is achieved via three possible routes: epicardium, endocardium, or coronary vein. The cell preparation is introduced into the myocardium under pressure using a hollow needle. This delivery route is the method of choice in patients with chronic occlusion of coronary arteries and where homing signals are expected to be weak or absent, such as in chronic congestive heart failure. The advantages include the accurate targeting of the desired areas, ability to deliver larger cell types, and greater cell retention (less washout) at the injection site that can be further enhanced by suspending the cells in a gel-like matrix. Limitations include the restriction of the target area (e.g., the septum cannot be reached), the limited injection volume (volumes up to 8 mL have been reported, [46], although cells can be highly concentrated), the risk of myocardial perforation, and, in the case of direct intramyocardial injection, the highly invasive nature of this procedure (sternotomy) and its associated increased surgical morbidity. However, in a planned open-heart procedure, the concomitant delivery of cell therapy with this technique can be easily justified.

3.2.1 Transepicardial Cell Delivery

Transepicardial cell delivery has been pioneered and used extensively in animal models of heart disease. As mentioned earlier, the direct intramyocardial injection of a cell suspension was first performed by Menasché in 2001 during open-surgical revascularization surgery [42]. In this groundbreaking trial, the area of infarction was visualized on the left-ventricular posterior wall and 33 autologous myoblast cell suspensions for a total of 800 million cells, suspended in 5 mL albumin, were injected in and around the white spots of necrosis with a customized right-angled 27-G needle designed specifically to allow the creation of subepicardial channels in a blister-like pattern. From Menasché's report it becomes evident that direct visualization and targeting of the infarcted areas are relatively easy to achieve. In this case, the cell concentration in the suspension was 160,000 cells/mL, in our experience the upper limit that can be used. This delivery technique for autologous skeletal myoblasts was used in other trials [46–51]; however, the Myoblast Autologous Grafting in Ischemic Cardiomyopathy (MAGIC) trial failed to improve echocardiographic heart function in patients with depressed left ventricular function. In addition, there was an increased number of early postoperative arrhythmic events after myoblast transplantation [49]. Nonetheless, cell delivery by direct intramyocardial injection is still used clinically with other cell types, such as bone-marrow-derived cells [52].

3.2.2 Transendocardial Cell Delivery

In catheter-based transendocardial cell delivery the injection site is approached via the femoral artery, then crossing the aortic valve into the left ventricular chamber. Four catheter systems have been developed for this approach: Helix (BioCardia, South San Francisco, CA, USA), MyoCath (Bioheart, Sunrise, FL, USA), Myostar (Biologics Delivery Systems, Cordis, Diamond Bar, CA, USA), and Stiletto (Boston Scientific, Natick, MA, USA). Transendocardial cell injection can be performed during routine left-sided heart catheterization procedures.

The current state-of-the-art technology for patients with severe chronic ischemia is the use of an electromagnetic mapping system (NOGA, Biologics Delivery Systems, Cordis) incorporated into the Myostar injection catheter. The technique uses magnetic fields generated by a magnetic pad positioned beneath the patient. The magnetic fields intersect with a location sensor built into the mapping catheter. A baseline three-dimensional endocardial map is created by the NOGA system that can be color-coded to identify regions of viable and ischemic myocardium [53]. This allows for precise targeting of desired areas of cell injection. The system was extensively tested in both preclinical and clinical studies [44, 45, 54]. Perin et al. pioneered cell delivery using this technique clinically [44, 45]. In this study, 21 patients were enrolled suffering from severe chronic heart failure, refractory angina pectoris, and having no other option for standard revascularization therapies. Bone marrow mononuclear cells were harvested, isolated, washed, and resuspended in saline for injection by a NOGA-guided Myostar catheter (15 injections of 0.2 mL). Electromechanical mapping was used to identify viable myocardium for treatment. At 4 months, there was improvement in ejection fraction from a baseline of 20–29% and a reduction in end-systolic volume in the treated patients. On electromechanical mapping, segmental analysis revealed a significant mechanical improvement of stem-cell-treated segments [44, 45]. Other studies in both animals and humans have confirmed the excellent safety profile of this technique [55–57].

Recently, Perin et al. evaluated a novel integrated platform in which a magnetic navigation system is used to remotely guide electromechanical mapping of the left ventricle and transendocardial cell injections [58]. Integrating NOGA electromechanical mapping (NOGA EMM) with the stereotactic magnetic navigation system allows remote navigation inside the left ventricular cavity. This new catheter system allows for transendocardial injections in remote areas of the left ventricle that are not easily reached with manually directed catheters. Perin et al. used pigs to deliver mesenchymal precursor cells to targeted myocardial segments and determined success and failure rates by myocardial contrast staining. The success rate for transendocardial cell injections was approximately 95% and no epicardial hemorrhage or injury (from inadvertent perforation) was observed [58]. It can be expected that further studies will reveal the safety profile of this system for clinical use.

3.2.3 Comparison of Direct Intramyocardial Injection and Transendocardial Cell Delivery

It is commonly believed that approximately equal numbers of cells can be delivered by either intramyocardial or transendocardial injection; however, few data are available comparing the two techniques directly. In an interesting study, Grossman et al. compared the fate of materials administered via a percutaneous endomyocardial catheter or via surgical epicardial injection in pigs [59]. In this study, the authors used multiple neutron-activated microsphere species as tracers to evaluate the acute retention of agents injected directly into the myocardium and to compare epicardial delivery with the percutaneous endocardial delivery. In the pigs, myocardial injections were performed using a percutaneous endomyocardial catheter and an epicardial needle via an open chest. Injections into two or three myocardial wall sites were performed in each animal using 3.5 mm, 27–28-G needles with various injectate volumes. The pigs were killed immediately and analyzed. With direct intramyocardial administration, a significant fraction of injectate was not retained locally (mean 15% retention), whereas catheter-based needle endomyocardial injection was associated with equivalent or superior injectate retention (mean 43% retention). Proportionately, more injectate was retained at lower volumes. Loss of cells may occur through a combination of channel leakage, venous return, and lymphatic return [59].

3.2.4 Intravenous Cell Delivery via Coronary Veins

The technique to deliver cells via the coronary sinus was pioneered by Thompson et al. in 2003 [57]. In this study, Yorkshire swine bone-marrow-derived adherent cells were tagged with GFP and resuspended in a collagen hydrogel for injection. An intravascular ultrasound (IVUS)-guided catheter system (TransAccess) was inserted into the coronary sinus and advanced to the anterior interventricular coronary vein. This composite catheter system (TransAccess, Medtronic Vascular, Santa Rosa, CA, USA) incorporates a phased-array ultrasound tip for guidance and a sheathed, extendable nitinol needle for transvascular myocardial access. A micro-infusion (IntraLume) catheter was advanced through the needle, deep into remote myocardium, and the autologous cell-hydrogel suspension was injected into the normal heart of recipient Yorkshire swine. The authors reported widespread intramyocardial access to the anterior, lateral, septal, apical, and inferior walls from the anterior interventricular coronary vein. No death, cardiac tamponade, ventricular arrhythmia, or other procedural complications were reported. Importantly, histologic analysis of the GFP signal demonstrated delivery of cells to the targeted sites [57]. The same group then pioneered the use of this delivery technique in chronically infarcted swine (5 weeks after myocardial infarction) [60]. Here, the IVUS-guided TransAccess catheter was used to deliver an enriched, predominantly lineage-negative mononuclear autologous bone-marrow-derived cell subpopulation through the coronary veins directly into infarct and peri-infarct myocardium. Two months after transplant, histologic evaluation revealed robust, viable cell grafts in all treated animals. In addition, cell therapy improved overall left ventricular systolic function by recruiting previously hypokinetic or akinetic myocardium [60]. Interestingly, in a small study George et al. compared intracoronary and transvenous cell delivery in infarcted pig hearts [4]. They found that delivery of GFP-tagged bone marrow mononuclear cells via the transvenous route resulted in better cell retention in the targeted area compared with intracoronary delivery [4].

To date, only one small phase I clinical trial, the POZNAN study, has been documented. Siminiak et al. assessed the feasibility and safety of autologous skeletal myoblast transplantation performed via a percutaneous trans-coronary-venous approach in patients with postinfarction left ventricular dysfunction [61]. Ten patients with heart failure and presence of an akinetic or a dyskinetic postinfarction injury with no viable myocardium were included in the study. Skeletal myoblasts were obtained from a biopsy specimen and expanded in culture. Skeletal myoblast transplantations were performed uneventfully in nine patients using the TransAccess catheter system under fluoroscopic and IVUS guidance. The TransAccess delivery system is unique among the intramyocardial devices in approaching the myocardium through the epicardial surface. To achieve this, a support catheter is positioned in specific branches of the cardiac venous system, by way of the femoral vein. In five patients the anterior interventricular vein and in four patients the middle cardiac vein were used to access the myocardium. Two to four intramyocardial injections 1.5–4.5 cm deep were performed in each patient, delivering up to 100 million cells in 0.4–2.5 mL of saline. During a 6-month follow-up, New York Heart Association class improved in all patients and ejection fraction increased by 3–8% in six of nine patients. Despite the promising results suggesting feasibility and procedural safety of cell delivery performed via the trans-coronary-venous approach using the TransAccess catheter system, no further studies have been reported. Limitations of this approach include the technical challenge of targeting a specific myocardial area because of the intrinsic variability and tortuosity of the coronary venous system.

3.2.5 Cell Delivery Mediated by Transmyocardial Laser Revascularization

Transmyocardial laser revascularization (TMR) is used as a biomechanical trigger to enhance the angiogenic response. Patel and Sherman hypothesized that cells have an inadequate microvascular environment in order to survive once implanted into scar tissue [62]. They used TMR to create a microvascular environment as a pretreatment before MSC implantation in a porcine infarct model. The authors found increased cell retention when the cells were injected into the border zone of a laser channel, suggesting the microenvironment created by the laser may be important for stem cell retention in ischemic tissue. On the basis of this study, Reyes et al. used TMR in combination with intramyocardial injection of concentrated autologous bone-marrow-derived stem cells [63]. Fourteen patients with diffuse coronary artery disease and medically refractory class III/IV angina who were not candidates for conventional therapies were treated. Bone marrow cell injection and TMR were performed using a Phoenix hand piece (Cardiogenesis, Irvine, CA, USA) which consists of a 1-mm flexible optical fiber connected to a 20-W pulsed laser. Three retractable needles with multiple side holes were incorporated into the hand piece, allowing injection of the cell suspension into the border zone around each laser channel (an average of 20 laser channels were created in each patient). Following creation of each laser channel, the three retractable needles were deployed and 1 mL cell suspension containing approximately 80 million cells was injected intramyocardially. At 7 months' follow-up, average angina class was significantly improved and there was no death during the follow-up. The authors concluded that delivery of stem cells combined with TMR in a single device is efficient, safe, and effective for treating otherwise unmanageable angina [63].

3.3 *Strategies to Improve Cell Delivery*

3.3.1 Adjunctive Treatment to Enhance Cell Delivery

Most small-animal experiments used direct intramyocardial (transepicardial) injection of the cells suspended in physiologic vehicle or mixed with some kind of extracellular matrix, e.g., fibrin glue, Matrigel, or collagen. The latter addresses two major problems associated with cellular cardiomyoplasty: (1) cell retention after delivery and (2) cell death. Numerous studies have shown that cell death due to apoptosis and necrosis can eliminate up to 99% of the engrafted cells [14, 64, 65]. Efforts in our laboratory have addressed this problem by using preconditioning of to-be-transplanted cells with heat shock and suspension in a prosurvival cocktail before injection, thus inhibiting apoptotic pathways and activating survival pathways [24]. The problem of low cell retention is mostly addressed by suspending the cells in a gel-like matrix with the goal to prevent leakage after direct intramyocardial injection and to prevent a form of programmed cell death, which is induced by anchorage-dependent cells detaching from the surrounding extracellular matrix (anoikis). Christman et al. evaluated the beneficial effect of adding a semirigid scaffold (fibrin) to the cell therapy product [66]. In a rat ischemia reperfusion model, adjunction of a fibrin scaffold to the cell therapy product did not modify the number of myoblasts present in the heart tissue 24 h after cell intramyocardial injections. However, long-term survival and beneficial effects on heart function were significantly higher in the fibrin scaffold group, suggesting that fibrin scaffolds may enhance cell transplant survival but not cell retention in infarcted myocardium.

4 Conclusion and Outlook

To date, myocardial repair remains the "holy grail" of regenerative cardiology. On the basis of a large number of studies in experimental animal models as well as clinical studies, cell-based therapies have emerged as a promising strategy and several techniques have been developed for their delivery to the heart: surgical injection, intracoronary infusion, retrograde venous infusion, transendocardial injection, and peripheral infusion. Each technique has been tested extensively in both animals and humans and they can be applied safely. Thus, the delivery side of cell-based cardiac repair appears well developed, with several options available. However, critical questions remain with regard to the cell type that should be used and the respective mechanism of action and how to improve cell survival after delivery. So far, clinical studies have entailed use of skeletal myoblasts and bone-marrow-derived cells. In the case of skeletal myoblasts, the randomized placebo-controlled MAGIC trial failed to show that myoblast injections increased ejection fraction beyond that seen in controls [49]. However, the observation that the highest dose of myoblasts resulted in a significant anti-remodeling effect compared with the placebo group is encouraging for cell grafting per se. In the case of bone-marrow-derived cells, surgical injections of the mononuclear fraction combined with coronary artery bypass surgery has not shown a substantial benefit, but positive results have been reported with intraoperative epicardial injections of CD133+ bone marrow progenitor cells [67]. Interestingly, a meta-analysis of 18 eligible studies (with a total of 999 patients) in which adult bone marrow mononuclear cells, bone marrow MSCs, or bone-marrow-derived circulating progenitor cells were transplanted showed a modest but significant improvement of 3.66% in left ventricular ejection fraction in the cell-treated group [22]. Many questions remain as to the mechanism of action and the long-term benefit and safety of these different cell populations. To find the optimal bone-marrow-derived cell type, it is mandatory to compare specific bone marrow subsets directly in large randomized controlled trials.

Future cell-based therapies may also involve the use of either human embryonic stem cell derived cardiomyocytes or induced pluripotent stem cell derived cardiomyocytes. The latter hold great potential in the emerging field of personalized regenerative medicine. Here, patient-derived cells (e.g., skin fibroblasts) may be reprogrammed to the pluripotent state and then redifferentiated into any desired cell type, including cardiomyocytes.

Cell-based therapies for the cure of heart disease offer tremendous potential. What is needed is cautious and careful investigation of the various cell types, delivery routes, and timing of delivery to achieve optimal cardiac repair.

References

1. Baklanov DV, Moodie KM, McCarthy FE, et al (2006) Comparison of transendocardial and retrograde coronary venous intramyocardial catheter delivery systems in healthy and infarcted pigs. Catheter Cardiovasc Interv 68:416–423
2. Freyman T, Polin G, Osman H, et al (2006) A quantitative, randomized study evaluating three methods of mesenchymal stem cell delivery following myocardial infarction. Eur Heart J 27:1114–1122
3. Fukushima S, Coppen SR, Lee J, et al (2008) Choice of cell-delivery route for skeletal myoblast transplantation for treating post-infarction chronic heart failure in rat. PLoS One 3:e3071
4. George JC, Goldberg J, Joseph M, et al (2008) Transvenous intramyocardial cellular delivery increases retention in comparison to intracoronary delivery in a porcine model of acute myocardial infarction. J Interv Cardiol 21:424–431
5. Grogaard HK, Sigurjonsson OE, Brekke M, et al (2007) Cardiac accumulation of bone marrow mononuclear progenitor cells after intracoronary or intravenous injection in pigs subjected to acute myocardial infarction with subsequent reperfusion. Cardiovasc Revasc Med 8:21–27
6. Hale SL, Dai W, Dow JS, et al (2008) Mesenchymal stem cell administration at coronary artery reperfusion in the rat by two delivery routes: a quantitative assessment. Life Sci 83:511–515
7. Hamdi H, Furuta A, Bellamy V, et al (2009) Cell delivery: intramyocardial injections or epicardial deposition? A head-to-head comparison. Ann Thorac Surg 87:1196–1203
8. Hayashi M, Li TS, Ito H, et al (2004) Comparison of intramyocardial and intravenous routes of delivering bone marrow cells for the treatment of ischemic heart disease: an experimental study. Cell Transplant 13:639–647
9. Hou D, Youssef EA, Brinton TJ, et al (2005) Radiolabeled cell distribution after intramyocardial, intracoronary, and interstitial retrograde coronary venous delivery: implications for current clinical trials. Circulation 112:I150–I156
10. Li SH, Lai TY, Sun Z, et al (2009) Tracking cardiac engraftment and distribution of implanted bone marrow cells: comparing intra-aortic, intravenous, and intramyocardial delivery. J Thorac Cardiovasc Surg 137:1225.e1–1233.e1
11. Sun Z, Wu J, Fujii H, et al (2008) Human angiogenic cell precursors restore function in the infarcted rat heart: a comparison of cell delivery routes. Eur J Heart Fail 10:525–533
12. Anderl JN, Robey TE, Stayton PS, et al (2009) Retention and biodistribution of microspheres injected into ischemic myocardium. J Biomed Mater Res A 88:704–710
13. Yasuda T, Weisel RD, Kiani C, et al (2005) Quantitative analysis of survival of transplanted smooth muscle cells with real-time polymerase chain reaction. J Thorac Cardiovasc Surg 129:904–911
14. Muller-Ehmsen J, Krausgrill B, Burst V, et al (2006) Effective engraftment but poor mid-term persistence of mononuclear and mesenchymal bone marrow cells in acute and chronic rat myocardial infarction. J Mol Cell Cardiol 41:876–884
15. Perin EC, Silva GV, Assad JA, et al (2008) Comparison of intracoronary and transendocardial delivery of allogeneic mesenchymal cells in a canine model of acute myocardial infarction. J Mol Cell Cardiol 44:486–495
16. Ben-Dor I, Fuchs S, and Kornowski R (2006) Potential hazards and technical considerations associated with myocardial cell transplantation protocols for ischemic myocardial syndrome. J Am Coll Cardiol 48:1519–1526

17. Vulliet PR, Greeley M, Halloran SM, et al (2004) Intra-coronary arterial injection of mesenchymal stromal cells and microinfarction in dogs. Lancet 363:783–784
18. Kang HJ, Kim HS, Zhang SY, et al (2004) Effects of intracoronary infusion of peripheral blood stem-cells mobilised with granulocyte-colony stimulating factor on left ventricular systolic function and restenosis after coronary stenting in myocardial infarction: the MAGIC cell randomised clinical trial. Lancet 363:751–756
19. Morishima I, Sone T, Okumura K, et al (2000) Angiographic no-reflow phenomenon as a predictor of adverse long-term outcome in patients treated with percutaneous transluminal coronary angioplasty for first acute myocardial infarction. J Am Coll Cardiol 36:1202–1209
20. Reffelmann T, Hale SL, Dow JS, et al (2003) No-reflow phenomenon persists long-term after ischemia/reperfusion in the rat and predicts infarct expansion. Circulation 108:2911–2917
21. Barbash IM, Chouraqui P, Baron J, et al (2003) Systemic delivery of bone marrow-derived mesenchymal stem cells to the infarcted myocardium: feasibility, cell migration, and body distribution. Circulation 108:863–868
22. Abdel-Latif A, Bolli R, Tleyjeh IM, et al (2007) Adult bone marrow-derived cells for cardiac repair: a systematic review and meta-analysis. Arch Intern Med 167:989–997
23. Aicher A, Brenner W, Zuhayra M, et al (2003) Assessment of the tissue distribution of transplanted human endothelial progenitor cells by radioactive labeling. Circulation 107:2134–2139
24. Laflamme MA, Chen KY, Naumova AV, et al (2007) Cardiomyocytes derived from human embryonic stem cells in pro-survival factors enhance function of infarcted rat hearts. Nat Biotechnol 25:1015–1024
25. Kocher AA, Schuster MD, Szabolcs MJ, et al (2001) Neovascularization of ischemic myocardium by human bone-marrow-derived angioblasts prevents cardiomyocyte apoptosis, reduces remodeling and improves cardiac function. Nat Med 7:430–436
26. Gao J, Dennis JE, Muzic RF, et al (2001) The dynamic in vivo distribution of bone marrow-derived mesenchymal stem cells after infusion. Cells Tissues Organs 169:12–20
27. Strauer BE, Brehm M, Zeus T, et al (2002) Repair of infarcted myocardium by autologous intracoronary mononuclear bone marrow cell transplantation in humans. Circulation 106:1913–1918
28. Chen SL, Fang WW, Ye F, et al (2004) Effect on left ventricular function of intracoronary transplantation of autologous bone marrow mesenchymal stem cell in patients with acute myocardial infarction. Am J Cardiol 94:92–95
29. Fernandez-Aviles F, San Roman JA, Garcia-Frade J, et al (2004) Experimental and clinical regenerative capability of human bone marrow cells after myocardial infarction. Circ Res 95:742–748
30. Janssens S, Dubois C, Bogaert J, et al (2006) Autologous bone marrow-derived stem-cell transfer in patients with ST-segment elevation myocardial infarction: double-blind, randomised controlled trial. Lancet 367:113–121
31. Kang HJ, Lee HY, Na SH, et al (2006) Differential effect of intracoronary infusion of mobilized peripheral blood stem cells by granulocyte colony-stimulating factor on left ventricular function and remodeling in patients with acute myocardial infarction versus old myocardial infarction: the MAGIC Cell-3-DES randomized, controlled trial. Circulation 114:I145–I151
32. Meyer GP, Wollert KC, Lotz J, et al (2006) Intracoronary bone marrow cell transfer after myocardial infarction: eighteen months' follow-up data from the randomized, controlled BOOST (BOne marrOw transfer to enhance ST-elevation infarct regeneration) trial. Circulation 113:1287–1294
33. Wollert KC, Meyer GP, Lotz J, et al (2004) Intracoronary autologous bone-marrow cell transfer after myocardial infarction: the BOOST randomised controlled clinical trial. Lancet 364:141–148
34. Zohlnhofer D, Ott I, Mehilli J, et al (2006) Stem cell mobilization by granulocyte colony-stimulating factor in patients with acute myocardial infarction: a randomized controlled trial. JAMA 295:1003–1010
35. Lunde K, Solheim S, Aakhus S, et al (2006) Intracoronary injection of mononuclear bone marrow cells in acute myocardial infarction. N Engl J Med 355:1199–1209

36. Schachinger V, Assmus B, Britten MB, et al (2004) Transplantation of progenitor cells and regeneration enhancement in acute myocardial infarction: final one-year results of the TOPCARE-AMI trial. J Am Coll Cardiol 44:1690–1699
37. Schachinger V, Erbs S, Elsasser A, et al (2006) Intracoronary bone marrow-derived progenitor cells in acute myocardial infarction. N Engl J Med 355:1210–1221
38. Schachinger V, Erbs S, Elsasser A, et al (2006) Improved clinical outcome after intracoronary administration of bone-marrow-derived progenitor cells in acute myocardial infarction: final 1-year results of the REPAIR-AMI trial. Eur Heart J 27:2775–2783
39. Rosenzweig A (2006) Cardiac cell therapy – mixed results from mixed cells. N Engl J Med 355:1274–1277
40. Assmus B, Honold J, Schachinger V, et al (2006) Transcoronary transplantation of progenitor cells after myocardial infarction. N Engl J Med 355:1222–1232
41. Seeger FH, Zeiher AM, and Dimmeler S (2007) Cell-enhancement strategies for the treatment of ischemic heart disease. Nat Clin Pract Cardiovasc Med 4 Suppl 1:S110–S113
42. Menasche P, Hagege AA, Scorsin M, et al (2001) Myoblast transplantation for heart failure. Lancet 357:279–280
43. Stamm C, Westphal B, Kleine HD, et al (2003) Autologous bone-marrow stem-cell transplantation for myocardial regeneration. Lancet 361:45–46
44. Perin EC, Dohmann HF, Borojevic R, et al (2003) Transendocardial, autologous bone marrow cell transplantation for severe, chronic ischemic heart failure. Circulation 107:2294–2302
45. Perin EC, Geng YJ, and Willerson JT (2003) Adult stem cell therapy in perspective. Circulation 107:935–938
46. Hagege AA, Marolleau JP, Vilquin JT, et al (2006) Skeletal myoblast transplantation in ischemic heart failure: long-term follow-up of the first phase I cohort of patients. Circulation 114:I108–I113
47. Dib N, Michler RE, Pagani FD, et al (2005) Safety and feasibility of autologous myoblast transplantation in patients with ischemic cardiomyopathy: four-year follow-up. Circulation 112:1748–1755
48. Herreros J, Prosper F, Perez A, et al (2003) Autologous intramyocardial injection of cultured skeletal muscle-derived stem cells in patients with non-acute myocardial infarction. Eur Heart J 24:2012–2020
49. Menasche P, Alfieri O, Janssens S, et al (2008) The Myoblast Autologous Grafting in Ischemic Cardiomyopathy (MAGIC) trial: first randomized placebo-controlled study of myoblast transplantation. Circulation 117:1189–1200
50. Pagani FD, DerSimonian H, Zawadzka A, et al (2003) Autologous skeletal myoblasts transplanted to ischemia-damaged myocardium in humans. Histological analysis of cell survival and differentiation. J Am Coll Cardiol 41:879–888
51. Siminiak T, Kalawski R, Fiszer D, et al (2004) Autologous skeletal myoblast transplantation for the treatment of postinfarction myocardial injury: phase I clinical study with 12 months of follow-up. Am Heart J 148:531–537
52. van Ramshorst J, Bax JJ, Beeres SL, et al (2009) Intramyocardial bone marrow cell injection for chronic myocardial ischemia: a randomized controlled trial. JAMA 301:1997–2004
53. Opie SR and Dib N (2006) Surgical and catheter delivery of autologous myoblasts in patients with congestive heart failure. Nat Clin Pract Cardiovasc Med 3 Suppl 1:S42–S45
54. Vale PR, Losordo DW, Milliken CE, et al (2000) Left ventricular electromechanical mapping to assess efficacy of phVEGF(165) gene transfer for therapeutic angiogenesis in chronic myocardial ischemia. Circulation 102:965–974
55. Kawamoto A, Gwon HC, Iwaguro H, et al (2001) Therapeutic potential of ex vivo expanded endothelial progenitor cells for myocardial ischemia. Circulation 103:634–637
56. Smits PC, van Geuns RJ, Poldermans D, et al (2003) Catheter-based intramyocardial injection of autologous skeletal myoblasts as a primary treatment of ischemic heart failure: clinical experience with six-month follow-up. J Am Coll Cardiol 42:2063–2069
57. Thompson CA, Nasseri BA, Makower J, et al (2003) Percutaneous transvenous cellular cardiomyoplasty. A novel nonsurgical approach for myocardial cell transplantation. J Am Coll Cardiol 41:1964–1971

58. Perin EC, Munger T, Pandey A, et al (2007) First experience with remote LV mapping and injection utilizing a novel integrated NOGA XP/stereotaxis platform. EuroInterv 3:142–148
59. Grossman PM, Han Z, Palasis M, et al (2002) Incomplete retention after direct myocardial injection. Catheter Cardiovasc Interv 55:392–397
60. Thompson CA, Reddy VK, Srinivasan A, et al (2005) Left ventricular functional recovery with percutaneous, transvascular direct myocardial delivery of bone marrow-derived cells. J Heart Lung Transplant 24:1385–1392
61. Siminiak T, Fiszer D, Jerzykowska O, et al (2005) Percutaneous trans-coronary-venous transplantation of autologous skeletal myoblasts in the treatment of post-infarction myocardial contractility impairment: the POZNAN trial. Eur Heart J 26:1188–1195
62. Patel AN and Sherman W (2007) Cardiac stem cell therapy from bench to bedside. Cell Transplant 16:875–878
63. Reyes G, Allen KB, Aguado B, et al (2009) Bone marrow laser revascularisation for treating refractory angina due to diffuse coronary heart disease. Eur J Cardiothorac Surg 36:192–194
64. Muller-Ehmsen J, Whittaker P, Kloner RA, et al (2002) Survival and development of neonatal rat cardiomyocytes transplanted into adult myocardium. J Mol Cell Cardiol 34:107–116
65. Zhang M, Methot D, Poppa V, et al (2001) Cardiomyocyte grafting for cardiac repair: graft cell death and anti-death strategies. J Mol Cell Cardiol 33:907–921
66. Christman KL, Vardanian AJ, Fang Q, et al (2004) Injectable fibrin scaffold improves cell transplant survival, reduces infarct expansion, and induces neovasculature formation in ischemic myocardium. J Am Coll Cardiol 44:654–660
67. Bartunek J, Vanderheyden M, Vandekerckhove B, et al (2005) Intracoronary injection of CD133-positive enriched bone marrow progenitor cells promotes cardiac recovery after recent myocardial infarction: feasibility and safety. Circulation 112:I178–I183

Tracking of Stem Cells In Vivo

Yingli Fu and Dara L. Kraitchman

Abstract Clinical and basic studies of stem-cell-based therapies have shown promising results for cardiovascular diseases. Despite a rapid transition from animal studies to clinical trials, the mechanisms of action by which stem cells improve heart function remain largely unknown. To optimize stem cell therapies in patients, a method to noninvasively monitor stem cell delivery and to evaluate cell survival, biodistribution, and the fate of implanted cells in the same subject over time would be desirable. Many different methods have been adapted from histopathological cell labeling techniques to enable tracking of stem cells with noninvasive imaging. This chapter focuses on the most promising stem cell labeling techniques that can be combined with clinically available imaging modalities for the evaluation of cardiac function.

Keywords Cell tracking • Radiotracer labeling • Magnetic resonance contrast cell labeling • Quantum dots • Fluorescence imaging • Positron emission tomography/single photoemission computed tomography reporter genes • Ultrasonic reporter gene imaging • Multimodality imaging

1 Introduction

Meta-analyses of recent cardiovascular clinical trials have shown that stem cell therapy is safe and could be of benefit to patients [1–4]. However, taken individually, the clinical trials to date have demonstrated variable degrees of improvement in cardiac function after stem cell administration [5–10]. Speculation abounds as to

D.L. Kraitchman (✉)
Russell H. Morgan Department of Radiology and Radiological Science, School of Medicine, The Johns Hopkins University, Baltimore, MD
e-mail: dkraitc1@jhmi.edu

I.S. Cohen and G.R. Gaudette (eds.), *Regenerating the Heart*, Stem Cell Biology and Regenerative Medicine, DOI 10.1007/978-1-61779-021-8_25,
© Springer Science+Business Media, LLC 2011

whether differences in the stem cell product, cell therapy handling, administration, and timing could be related to positive or negative clinical outcomes. However, all clinical trials suffer from the inability to determine whether the therapeutic product was delivered to the heart and the degree of stem cell engraftment. Visualization of transplanted stem cells, which relies on appropriate stem cell labeling using clinically available imaging modalities, is vital to explaining the conflicting outcomes from current clinical trials. The ability to monitor and track stem cells could provide further insights into the unresolved critical issues, including (1) the optimal cell delivery route; (2) biodistribution; (3) cell retention in the target region; (4) survival of transplanted cells; and (5) the fate and mechanisms of action. Ideally, this acquired information could be used to refine cardiac stem cell therapeutic approaches.

In this chapter, we will describe the available stem cell labeling methods and some recent advances in noninvasive imaging of stem cell therapy with an emphasis on techniques that show promise for clinical applications to cardiovascular diseases. The advantages and disadvantages of stem cell labeling techniques will be discussed. Finally, the potential directions for future development in stem cell labeling and imaging will be discussed.

2 Stem Cell Labeling Methods

The ideal method for tracking stem cells requires the labeling agent be biocompatible, have low toxicity, be highly specific to the target cells, yet enable the detection of a single cell or a few cells. In this section, we will highlight the current stem cell labeling strategies in conjunction with imaging modalities. In general, stem cell labeling methods fall onto two categories: (1) direct cell labeling and (2) reporter gene methods.

2.1 Direct Stem Cell Labeling

Direct labeling is the most straightforward and widely used method to label stem cells. Typically, a label is incorporated into the cells prior to cell implantation. The simplest method is direct incubation of cells with the markers. Similar to dyes that are used for cell tracking with histopathology [11], superparamagnetic iron oxides (SPIOs), fluorescent probes, and radionuclides can be incubated with cells for magnetic resonance imaging (MRI), optical imaging, or radionuclide imaging, respectively. Because there is usually no background radioactivity, a high sensitivity to radiolabeled cells can be achieved. Thus, whole body biodistribution of the stem cells can be readily studied using radiolabeled stem cells. MRI direct labeling may not be as sensitive, but the excellent soft-tissue detail and lack of ionizing radiation using this technique can be advantageous. Although clinical optical imaging systems are not commercially available, they may eventually offer a low-cost alternative without ionizing radiation for cell tracking.

2.1.1 Radiotracer Direct Cell Labeling

Direct cell labeling with radioactive isotopes, such as indium-111 oxyquinoline (oxine), indium-111 tropolone, technicium-99m, and ^{18}F-fluoro-2-deoxy-D-glucose (FDG), has been widely used due to the lack of a native background signal which imparts an extremely high sensitivity (10^{-11} to 10^{-12} mol/L) for detection. Clinical tomographic imaging with positron emission tomography (PET) or single photoemission computed tomography (SPECT) can also provide spatial resolutions in the 1–2-mm range. With the introduction of combined SPECT-computed tomography (CT) and PET–CT clinical scanners, high-resolution anatomical information can be obtained for localization of radioactive hot spots.

One radiotracer, indium-111 oxine, is especially attractive for stem cell labeling for several reasons. Historically, indium-111 oxine has been used clinically to label lymphocytes for several decades [12]. Direct cellular labeling with indium-111 oxine occurs as the radiotracer passively diffuses into the cell, dissociates from the oxine moiety, and becomes bound in the cytoplasm [13]. Thus, reuptake of the radiotracer by other nonphagocytic cells should be limited. In addition, the relatively long half-life of approximately 2.8 days allows for serial imaging on the order of 5–7 days after labeling. For stem cell labeling, the first studies were performed by Gao et al. and entailed incubation of mesenchymal stem cells (MSCs) with indium-111 oxine to follow MSC distribution after intravenous injection [14].

Subsequently, several radiotracers, such as FDG, indium-111 tropolone, and technicium-99m, have been used to label stem cells [15], determine the best delivery route for engraftment [16], study biodistribution/homing (Fig. 1) [17–21], optimize delivery dosing [22], and compare the engraftment of different cell types in animal models of cardiovascular disease [16].

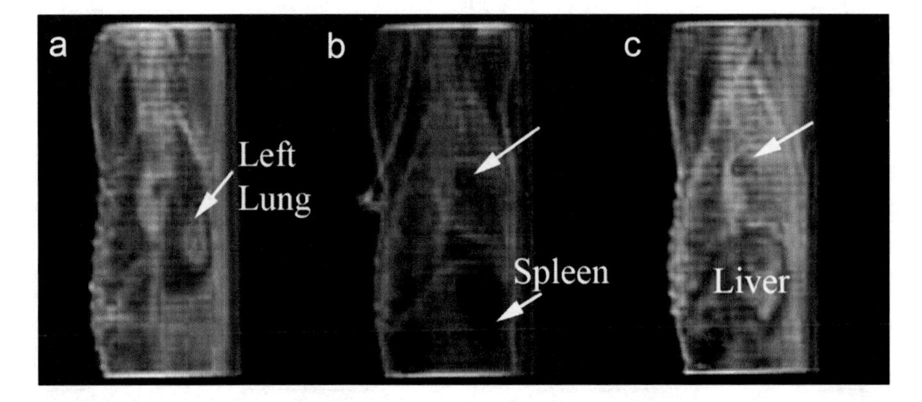

Fig. 1 Coronal fused single photoemission computed tomography (SPECT) (*color*) and computed tomography (CT) (*gray*) image of a dog with myocardial infarction (MI) showing predominant lung uptake 1 h with increased uptake to the dependent left lung after intravenous injection of mesenchymal stem cells (MSCs) labeled with indium-111 oxine (**a**). Redistribution of MSCs labeled with indium-111 oxine to predominantly the spleen (**b**) and the liver (**c**) occurs at 24 h after intravenous injection. (Adapted from [20, 21] with permission)

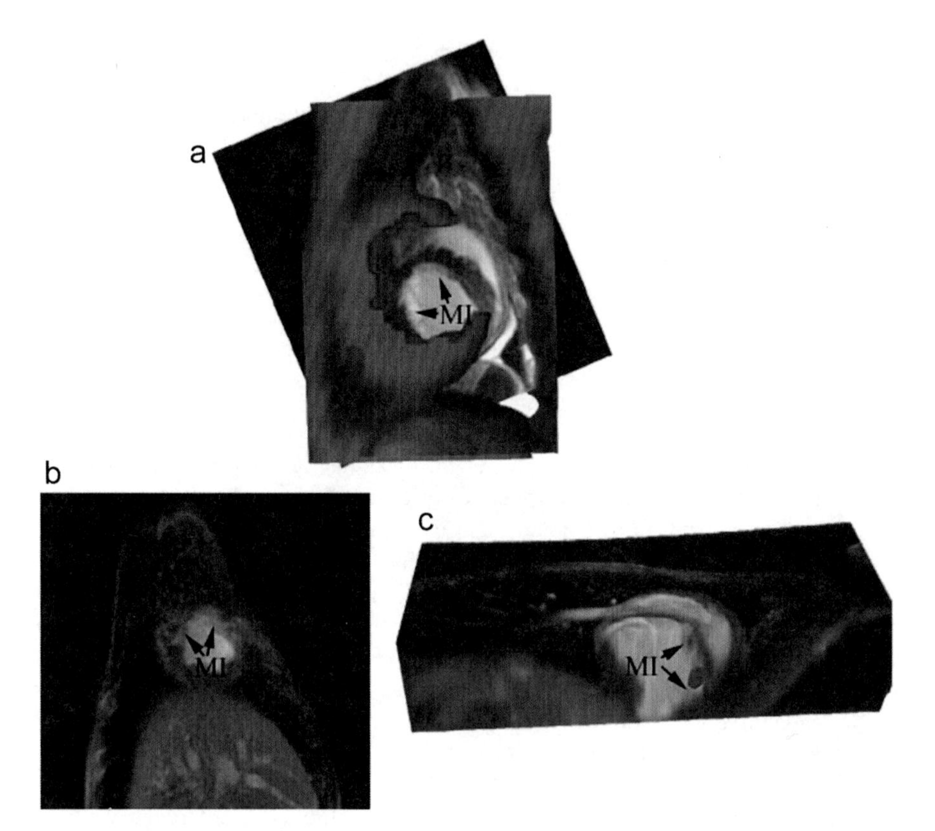

Fig. 2 Registration of SPECT/CT with cardiac magnetic resonance imaging (MRI) in a dog with a reperfused myocardial infarction receiving MSCs labeled with indium-111 oxine and superparamagnetic iron oxides (SPIOs). (**a**) Short-axis view of alignment of CT (*gold*) with MRI (*gray scale*) and SPECT (*red*) showing focal uptake in the septal region of the MI in a representative dog. (**b**) Focal uptake on SPECT (*red*) in another animal demonstrating localization of the MSCs to the infarcted myocardium in the short-axis (**b**) and long-axis (**c**) views. SPECT, owing to the higher sensitivity, was able to detect approximately 8,000 labeled MSCs/g of tissue, whereas MRI was unable to detect the SPIO-labeled MSCs. (Reprinted from [21] with permission)

Most radiotracer studies have been performed in murine models owing to the high cost of radiotracers. These studies have shown that only a small percentage, e.g., less than 5%, of endothelial or hematopoietic progenitor cells are retained in the heart [17, 18]. Thus, a large dose of therapeutic is delivered to nontarget organs. This was subsequently confirmed in large-animal studies using MSCs labeled with indium-111 oxine delivered intravenously where most cells were initially trapped in the lungs – presumably the pulmonary capillaries owing to the large cell size [19–21]. Redistribution of the MSCs over time to injured myocardium as well as nontarget organs, such as the liver, was also demonstrated. In addition, the enhanced sensitivity of radiotracer direct labeling over MRI to detect approximately 8,000 cells/g of tissue was demonstrated in a canine model of acute reperfused myocardial infarction (Fig. 2) [21].

The first clinical cardiovascular study with radiolabeled stem cells demonstrated higher retention of FDG-labeled CD34-positive bone marrow cells in the infarcted myocardium than unselected bone marrow cells at 70 min after intracoronary delivery [23]. In another clinical study, involving 20 myocardial infarction patients, PET imaging indicated that 0.2–3.3% of intracoronarily infused FDG-labeled hematopoietic stem cells accumulated within the infarcted myocardium at 2 h after administration [24]. Another, more recent study showed that peripheral blood proangiogenic cells labeled with indium-111 oxine showed the highest retention after intracoronary infusion in acute myocardial infarction versus more chronic stages [25].

Major concerns of direct radiotracer labeling include the potential radiation damage to the cells and leakage of radiotracers over the time course. The binding of some radiotracers, such as copper-64 pyruvaldehyde bis(N^4-methylthiosemicarbazone), is often reversible, which may lead to label loss even from viable cells [26]. Labeling of stem cells with high levels of radioactivity can significantly interfere with proliferation and function. However, careful titration of the radioactivity and incubation time can minimize these effects [27]. The cytotoxic effects of Indium-111 on human stem cells may be more time-dependent than dose-dependent. Therefore, assessment of stem cell viability immediately after radioactive labeling may fail to detect cytotoxic effects of this radiopharmaceutical on stem cell integrity [28]. However, the greatest disadvantage of direct labeling with radionuclides is the radioactive decay, leading to limited imaging time intervals, which may be as short as hours with short-lived radiotracers, such as FDG.

2.1.2 Magnetic Resonance Contrast Direct Cell Labeling

Unlike radiotracers, where one directly counts the emissions from the radiotracer, most MRI contrast agents are not imaged directly, but rather their effect on water is imaged. The paramagnetic contrast agents, such as gadolinium and manganese, primarily exert their effect on T_1 (spin–lattice) relaxation and, thereby, cause a hyperintense signal intensity on T_1-weighted MRI. Gadolinium is chelated for clinical use since free gadolinium is highly toxic. Attempts with direct cellular labeling with FDA-approved gadolinium-based contrast agents, which results in endosomal localization of the contrast agent, have not provided sufficient contrast because of poor access of the tissue water to the contrast agent. Some effort has been made to design gadolinium-based contrast agents for cellular labeling that would achieve sufficient signal contrast. However, much research will be required to demonstrate the stability and biocompatibility of these agents as dead labeled cells are removed from the body.

SPIO nanoparticles are another class of contrast agents that have become the predominate cellular label used for MRI. Initially developed for liver imaging [29], SPIOs in small quantities are able to cause large changes in T_2 (spin–spin) relaxation, which were seen as hypointensities on T_2^*-weighted MRI. Fortunately, an endosomal location of SPIOs causes only minor reductions in the ability of these agents to affect T_2 relaxation. More recently, several groups have developed imaging sequences that create hyperintense signal intensities from SPIO-labeled cells [30–32].

Simple incubation of stem cells with SPIOs has been less successful at loading the cells with sufficient quantities for detection using MRI [33]. Whereas macrophages and other cells of phagocytic lineages will readily take up SPIOs [29], most stem cells will require long incubation times to achieve a sufficient uptake, if any, of SPIOs. Thus, for direct labeling of stem cells with SPIOs, a transfection agent is often used to facilitate uptake, which has been coined "magnetofection" [34, 35]. The SPIOs are allowed to complex with a cationic transfection agent prior to addition to the stem cell growth medium. Stable labeling of stem cells with the SPIOs contained in endosomes is usually achieved within 24–48 h with intracellular iron concentrations of 10–20 pg/cell [34]. Although most studies have shown little effect of SPIO labeling at these levels on cellular function, one study showed that MSCs failed to differentiate in vitro down a chondrocytic lineage after iron labeling [36]. Another study demonstrated that stem cell migration was reduced after SPIO labeling in a dose-dependent manner [37]. However, these effects may be related not only to the SPIO, but also to the transfection agent used [38].

Because formulations of protamine sulfate (as a transfection agent) and SPIOs that are FDA-approved have been developed for direct cell labeling [39], the regulatory hurdles for clinical approval of these techniques are markedly lessened. However, because of the long incubation times required for direct labeling with SPIOs using transfection agents, their use in an acute setting, such as myocardial infarction, will be limited without the ability to create prelabeled products similar to the blood bank.

For cardiovascular applications, the first study demonstrating MRI-tracking of SPIO-labeled bone-marrow-derived MSCs using the magnetofection method was performed in an acute reperfused myocardial infarction in swine (Fig. 3) [40]. This study demonstrated that transmyocardial delivery resulted in unsuccessful delivery to the myocardium in about 30% of the injections [40]. Subsequently, several

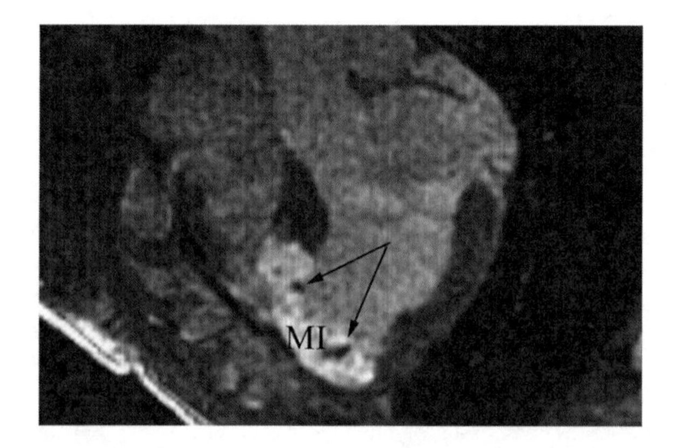

Fig. 3 Delay contrast-enhanced magnetic resonance image (1.5 T) of the heart in a swine with acute reperfused MI showing the detection of SPIO-labeled MSCs as hypointensities at 24 h after transendocardial delivery. The MSCs were labeled using the magnetofection method. (Reprinted from [40] with permission)

groups developed specialized MRI-compatible devices to deliver SPIO-labeled stem cells using interventional MRI [20, 41–45]. Delayed contrast-enhanced MRI was used to target injections to the peri-infarcted regions in swine and canine acute myocardial infarction models [20, 42, 43, 46–48]. Combining MRI tracking with MRI measurement of global and regional myocardial function has enabled the evaluation of efficacy of stem cells therapies in preclinical models [20, 49, 50]. However, to date, MRI in cardiovascular clinical trials has been limited to defining global measures of ventricular function and infarct size [51, 52].

A more rapid method to label stem cells with SPIOs was developed on the basis of the electroporation techniques that were developed to incorporate DNA into cells. In general, electroporation was poor at translocation of DNA into the nucleus. However, for cellular labeling, a cytoplasmic location such as this is preferred. Initial attempts at "magnetoelectroporation" by Daldrup-Link et al. [33] were unsuccessful, but probably reflect the need to fine-tune the voltage pulses to the particular cell lineage. MSCs have been labeled with SPIOs using magnetoelectroporation with no detectable effects on cellular metabolism, viability, or differentiation capacity [53]. In addition, bone-marrow-derived stem cells were labeled with SPIO using magnetoelectroporation, and tracked by MRI with a positive contrast sequence [32] in a rabbit model of peripheral arterial disease [42]. Because no transfection agents are required, magnetoelectroporation may represent a more viable method of SPIO labeling for cardiovascular clinical trials.

Direct cell labeling with fluorinated nanoparticles has also been investigated [54]. The gyromagnetic ratio of ^{19}F nuclei is comparable to that of a proton. Therefore, it is a sensitive nucleus for MRI detection. In addition, ^{19}F atoms are naturally absent from the body. Thus, like radionuclides, hot spot imaging is also possible with fluorinated cells [55–57]. Nonetheless, special imaging hardware/coils and pulse sequences are needed that have limited the widespread use of such techniques.

2.1.3 Optical Imaging Direct Cell Labeling

A new class of fluorescence probes, quantum dots (QDs), have recently been adopted for in vitro as well as in vivo stem cell labeling and bioimaging [58–60]. QDs are multifunctional inorganic fluorescent semiconductor nanoparticles. Compared with organic fluorescent dyes, QDs provide unique optical advantages, including narrower band emission and broader band excitation with a high quantum yield, exceptional photostability, and resistance to chemical and metabolic degradation [61]. These properties make QDs suitable for cell labeling and multicolor optical imaging. The narrow emission wavelength of QDs is also very attractive for distinguishing native autofluorescence that occurs in the heart from the optical labeling agent. Furthermore, the narrow wavelength of a particular QD lends the technique to multicolor labeling with QDs.

MSCs and embryonic stem cells have been effectively labeled with QDs using passive incubation and a peptide-mediated uptake (QTracker) [58, 59, 62]. In vitro studies have demonstrated biocompatibility at low concentrations and functional integration of QD-labeled MSCs in a cardiac myocyte co-culture system [59]. These studies have also shown efficient labeling without altering stem cell viability, proliferation, and differentiation capacity [58, 63]. In addition, QDs did not transfer to adjacent stem cells owing to their large size (approximately 10 nm) relative to the size of gap junctions (approximately 1 nm) of the cell [62]. In vivo, QD-labeled stem cells could be detected with multiplex optical imaging up to 2 weeks after injection in mice with a sensitivity to approximately 1×10^5 cells [58]. For cardiovascular applications, human MSCs have been labeled with QDs and transplanted into the hearts of dogs and rats. QD-labeled human MSCs can be unambiguously imaged in vivo and in postmortem histological sections at least 8 weeks after delivery (Fig. 4) [62]. However, the effects of QDs on long-term stem cell differentiation and proliferation remain largely unknown. Moreover, the nonspecific binding of QDs to multiple cells and the potential toxicity concerns related to the cadmium base of QDs may hamper their translation to clinical applications.

In addition, all visible-light-based optical imaging techniques are inherently limited by light absorption and scattering in tissues. Presently, only near-infrared (NIR) (700–1,000 nm) fluorophores show potential for clinical applications. Intracoronary delivery of bone-marrow-derived MSCs labeled with IR-786 (a NIR fluorophore) were successfully tracked in vivo 90 min after injection in a swine model of myocardial infarction [64]. However, the practical penetration depth of NIR fluorophores is limited to the 4–10-cm range, which presents a formidable challenge in their applications to the heart and other deep tissue in large animals or humans.

In summary, the ease of stem cell labeling using direct labeling techniques also creates the primary disadvantage of direct labeling techniques. Since the label is not firmly bound to the stem cell, the relationship between the label and the stem cell location after delivery may no longer coincide. For some radiotracers, this may be related to the high efflux rate that occurs within the first hours after labeling [65]. Also, because many cells fail to survive the early transplantation period, cellular debris, including the label, may remain behind, whereas no viable stem cells remain [66–68]. However, the latter appears to be more problematic in poorly revascularized tissue [69, 70]. Finally, if the stem cell is successful at replicating to replace missing tissue, then the label will be rapidly diluted, leading to an inability to detect the remaining viable cells [71].

3 Reporter Gene Labeling Methods

As an alternative to direct cell labeling, reporter gene imaging relies on infecting a cell with a gene that expresses a receptor, a protein, or an enzyme that can be detected after administration of a reporter probe. Cells are transfected with the

Fig. 4 (continued) staining (green) shows the differentiation of QD-hMSCs along an endothelial lineage. The insert shows a magnified view of QD-positive cells with colocalized PECAM-1 expression. Scale bars represent 20 mm in (a, b), 50 μm in (c) (10 μm in the insert), and 20 μm in (d) (5 μm in the insert). (Adapted from [62] with permission)

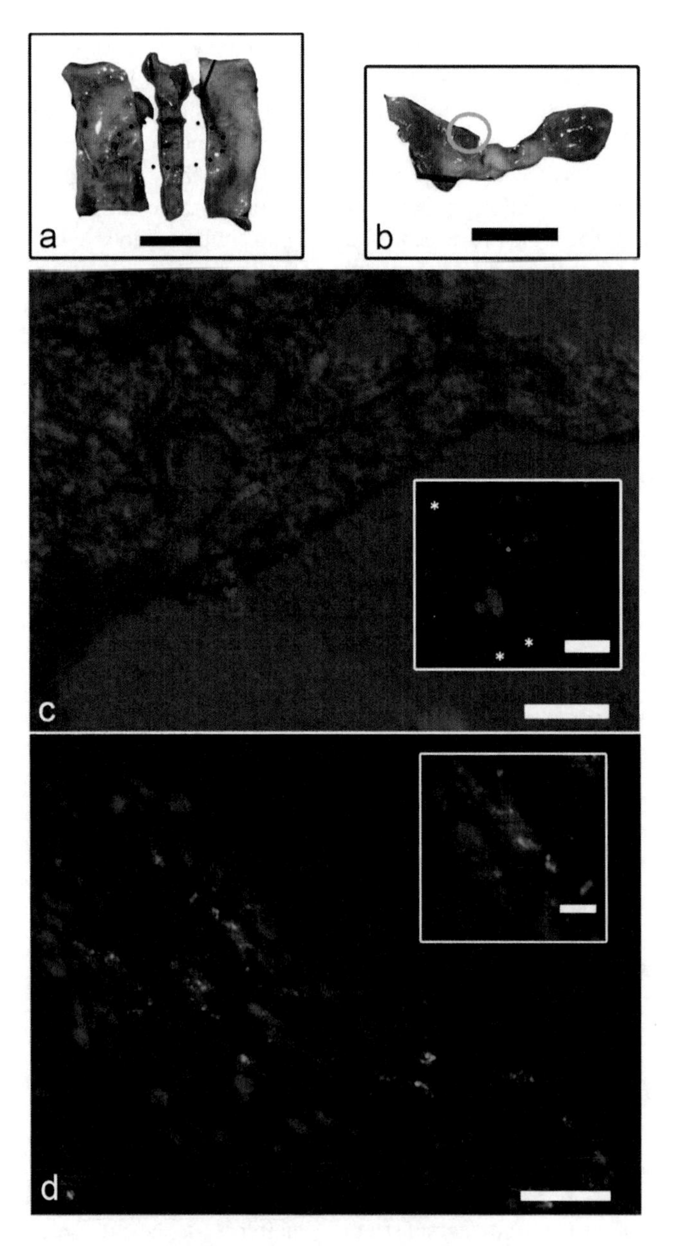

Fig. 4 Quantum dot (QD)-labeled human MSCs (QD-hMSCs) delivered to the heart of a canine on an extracellular matrix scaffold were histologically identifiable 8 weeks after injection. (**a**) Fixed tissue from one animal illustrating the region of interest (*blue line*) and the border of the scaffold (*black dotted ellipse*). (**b**) Transmural view of the section from epicardium (*top*) to endocardium (*bottom*). The *green circle* shows the location of identified QDs 8 weeks after delivery. (**c**) Composite image of phase-contrast and fluorescence image of the tissue section showing the presence of QD-hMSCs. The *insert* shows QD-hMSCs amid endogenous tissue (*asterisks*) (*blue* nuclei stained with Hoechst 33342). (**d**) Fluorescence image of the tissue demonstrating the detectability of QD-hMSCs (*red*) 8 weeks after delivery. Platelet/endothelial cell adhesion molecule 1 (PECAM-1)

reporter gene DNA construct exogenously using a nonviral vector (i.e., plasmid), a retroviral vector, or a viral vector (e.g., lentivirus, adenovirus). The advantage of this approach over direct cell labeling is that (1) if there is long-term expression of the gene, there is no concern about image signal dilution with cell proliferation, (2) stem cell expression of the reporter gene can be potentially controlled by appropriate selection of specific promoters, and (3) only viable cells express reporter gene products. Thus, the presence of the imaging signal reliably indicates survival of the implanted cells.

Reporter genes have been developed for optical imaging (fluorescence imaging and bioluminescence imaging, BLI), radionuclide imaging, and MRI. Detailed examples will be given next according to the imaging modalities.

3.1 Fluorescence Imaging with Green Fluorescent Protein

Cells transfected to express the green fluorescent protein (GFP) gene generate fluorescence when excited with an appropriate light source. In general, this green fluorescence can be detected only microscopically. However, cell labeling with fluorescent reporter genes suffers from problems similar to those with direct optical probe labeling. Only NIR fluorophores are useful for in vivo stem cell tracking because of the high photon absorbance and tissue scattering in the visible range. Owing to limited tissue penetration depth, in vivo detection of GFP-labeled stem cells is formidable. Therefore, fluorescent reporter gene labeling is used for postmortem histological validation of stem cells or cell sorting in conjunction with other, better imaging visible labels. For example, with use of a double-fusion construct expressing enhanced GFP and firefly (*Photinus pyralis*) luciferase (fluc) reporter genes, bone marrow mononuclear homing to the ischemic heart could be detected in mice [72]. Although tremendous improvement has been made in optical imaging techniques, their applications are primarily limited to small-animal studies.

3.2 Bioluminescent Reporter Gene Imaging

BLI has emerged as a unique tool for localizing infused stem cells within living subjects [73]. The most widely used luminescent protein is fluc, a 61-kDa monomeric protein, which reacts with its substrate, D-luciferin, in the presence of oxygen, Mg^{2+}, and ATP to release green light with a peak wavelength at 562 nm [74]. The dynamic homing and engraftment property of hematopoietic stem cells transfected with fluc reporter genes could be examined in vivo over a long period of time using BLI [75]. A recent study has shown that rat embryonic cardiomyoblasts (H9c2) transfected with plasmids (pCMV-Fluc-SV40-neo) expressing fluc under the regulation of a cytomegalovirus promoter can be tracked with BLI in the heart up to 6 days [76]. Thus, BLI is a useful method for longitudinal monitoring of cell survival, comparing cell typesm and delivery routes in animals [77–81]. However, this technique is restricted to small animals because the low-energy photons emitted are subject to attenuation within deep tissues. Thus, BLI reporter gene labeling is not applicable to large-animal studies nor clinical patient trials.

3.3 PET/SPECT Reporter Genes

Several reporter gene/probe combinations have been developed specifically for radionuclide imaging. Initially, SPECT and PET reporter genes were developed as a noninvasive method to monitor gene therapy. One of the earliest cardiac applications used the herpes simplex virus type 1 mutant thymidine kinase (HSV1-sr39tk) to monitor transgene expression of vascular endothelial growth factor, a potential angiogenic factor, using PET imaging [82]. The enzyme thymidine kinase will phosphorylate a variety of substrates, such as pyrimidine analogs and acycloguanosines, in addition to the native substrate thymidine. The thymidine kinase reaction results in the production of monophosphates, which are in turn converted to diphosphates and/or triphosphates by cellular enzymes. The latter are trapped within the cells. Only the cells expressing thymidine kinase are able to metabolize the reporter probe and thus these cells will retain the reporter probe. This ability of reporter gene imaging to only detect viable cells is one of the major advantages of this technique.

Although thymidine kinase reporter genes have dominated the field, several other reporter genes, including the Na^+/I^- symporter (NIS) [83, 84] for SPECT or PET imaging and the receptor-based dopamine type 2 receptor (D2R) for PET imaging [85, 86], have been used in cardiovascular applications.

NIS is a transporter-based PET/SPECT reporter gene that has been used for cardiac imaging [67, 68, 87, 88]. NIS is an endogenous membrane glycoprotein in a limited number of organs, i.e., thyroid, gastric mucosa, and salivary and lactating mammary glands, for transport of sodium and iodine. One of the major limitations with any reporter gene method is an immune response elicited toward the foreign protein/enzyme or viral vector. Thus, immunogenicity with NIS is reduced relative to other reporter gene imaging methods. Because NIS is not expressed in the heart, there is a high sensitivity to cells expressing the reporter gene. NIS can be used in combination with iodine-123 and technicium-99m pertechnetate for SPECT imaging or iodine-124 for PET imaging. However, cells that are transfected with the NIS gene do not retain the reporter probe to the same extent as occurs natively in the thyroid gland, where thyroperoxidase is present to effectively trap the iodine. Thereby, efflux of radioactive iodine in transfected cells has been a major stumbling block with this reporter system [89]. To address this issue, coexpression of the enzyme thyroperoxidase with the NIS gene may be required to improve the radioactive iodine retention in transfected cells [90].

D2R is a receptor-based reporter gene that was initially developed for cell tracking in the brain because the reporter probe, ^{18}F-fluoroethylspiperone, can cross the intact blood–brain barrier and bind to D2R with high affinity [91, 92]. A potential problem in both the brain and the heart is that endogenous D2R expression could provide sufficient background signal to make interpretation of PET images difficult. Furthermore, activation of endogenous D2R leads to increases in the levels of cyclic adenosine monophosphate, which may have untoward side effects, when the reporter probe is administered. To increase imaging sensitivity and minimize these physiological side effects, a mutant form of the D2R reporter gene, D2R80a, has been developed and fused with HSV1-sr39tk via an internal ribosomal entry site based bicistronic adenoviral vector in cardiomyocytes (H9c2) [93]. Longitudinal PET imaging demonstrated a good correlation between the cardiac expressions of the two PET reporter genes after intramyocardial delivery [93].

Among the PET reporter probes for imaging HSV1-*tk* variants, 9-[3-fluoro-1-hydroxy-2-(propoxymethyl)]guanine (^{18}F-FHBG) and ^{124}I-labeled 2′-fluoro-2′deoxy-15-iodo-1-β-D-arabinofuranosyluracil (^{124}I-FIAU) are the most widely used and have been applied in preclinical studies for evaluation of gene therapy in the myocardium [84, 94, 95] as well as in clinical studies [96]. Enzyme-based PET reporter gene labeling has the advantage of signal amplification, because each enzyme can potentially cleave more than a single reporter probe. Thus, low levels of reporter gene expression and, therefore, a small number of transplanted cells can potentially be detected. Also, there may be rate-limited transport of the reporter probe into the cells [97]. Improvement of the HSV1-*sr39tk* reporter gene, the truncated thymidine kinase (ttk), which accumulates preferentially in the cytoplasm rather than the nucleus, has been shown to be less cytotoxic and provided an improved signal [98, 99].

3.4 MRI Reporter Genes

The first MRI reporter gene that was demonstrated involved direct expression of a creatine kinase isozyme, an enzyme that catalyzes ATP to ADP and generates phosphocreatine, which can be detected by ^{31}P magnetic resonance spectroscopy [100]. Several other candidate MRI reporter genes, including those for iron transport and storage proteins, such as ferritin, transferrin, and transferrin receptor, have been reported [101–104]. Overexpression of transgenetic human ferritin receptor and ferritin heavy chain subunit has been induced in neural stem cell [104] and embryonic stem cell [103] lines. MRI contrast was generated through upregulation of transferrin receptor that led to increased cellular iron storage in the ferritin-bound format. In vivo cell graft monitoring was successfully achieved using T_2^* -weighted MRI. MRI reporter genes that rely on metals (e.g., gadolinium and iron) can induce long-term background contrast in the surrounding cellular environment – potentially long after the cell containing the reporter gene has died [105–107]. Recently, chemical exchange saturation transfer (CEST)-based reporter genes have emerged as the most promising MRI reporter gene method. Based on the production of an artificial lysine residue, this CEST-based reporter gene can be used for imaging multiple cell types simultaneously by applying different excitation frequencies akin to QD imaging [105, 106]. Presently, this technique has been used in neuroimaging applications, but may be suitable for stem cell tracking in cardiac applications.

3.5 Ultrasonic Reporter Gene Imaging

Echocardiography or cardiac ultrasound imaging is the most commonly used noninvasive imaging modality to evaluate cardiac function and anatomy. However, to use echocardiographic contrast agents to view stem cells, the stem cells must be transformed to express unique cell surface markers and the cells must be accessible to the ultrasound contrast agent, which is essentially limited to the vascular space. Recently, Kuliszewski et al. genetically engineered endothelial progenitor cells to express H-2Kk on the cell surface [108]. Matrigel plugs containing the modified endothelial progenitor cells were implanted subcutaneously in syngeneic rats. Contrast-enhanced ultrasound imaging demonstrated retained targeted microbubbles indicative of the development of a vascular network in the Matrigel plugs. Unfortunately, expression of the surface marker on the endothelial progenitor cells declined markedly after 1 week. Currently, the delivery of the ultrasonic reporter probe via the vascular system limits this technique to tracking of stem cells that differentiate down an endothelial lineage.

3.6 Hybrid Reporter Gene for Multimodality Imaging

To take advantage of the strength of each labeling technique and imaging modality, multimodality reporter genes have been developed. Among those, a novel triple-fusion reporter gene consisting of *fluc*, monomeric red fluorescence protein (*mrfp*), and HSV-truncated thymidine kinase (HSV1-*ttk*) has been used for cardiac imaging of mouse embryonic stem cells [77] and for cancer research of tumor xenografts in mice [109]. Cao et al. used the triple-fusion reporter gene to study the kinetics of embryonic stem cell survival, proliferation, and migration after intramyocardial delivery in rats for up to 4 weeks using BLI and PET imaging [77]. The ability of reporter gene imaging to not be diluted by rapid cell division as occurs with embryonic stem cells was demonstrated in this study. Not only was the potential for undifferentiated embryonic stem cells to cause teratomas demonstrated, but the administration of ganciclovir to turn on a suicide gene to kill transfected embryonic stem cells that had gone awry was also demonstrated. However, for the most part, the triple-fusion reporter gene has been used to validate in vivo imaging results with postmortem histopathology of fluorescence. Recently, a hybrid reporter gene imaging was used in a large-animal model of myocardial infarction with delivery using electromechanical mapping (Fig. 5) [110], which is a current delivery route in patient trials. This study again highlights the poor retention of MSCs in the heart after direct transendocardial delivery as well as the potential for cell migration to nontarget organs [110]. A nonpermanent transfection technique was used in this study. Whereas serial imaging will only be possible for up to a few weeks using this transfection technique, the associated concerns about permanent expression or viral transfection of other endogenous cells are greatly reduced.

In summary, reporter-gene-based imaging offers unique capabilities for noninvasive and longitudinal measurements of viable engrafted cells. Concerns about the

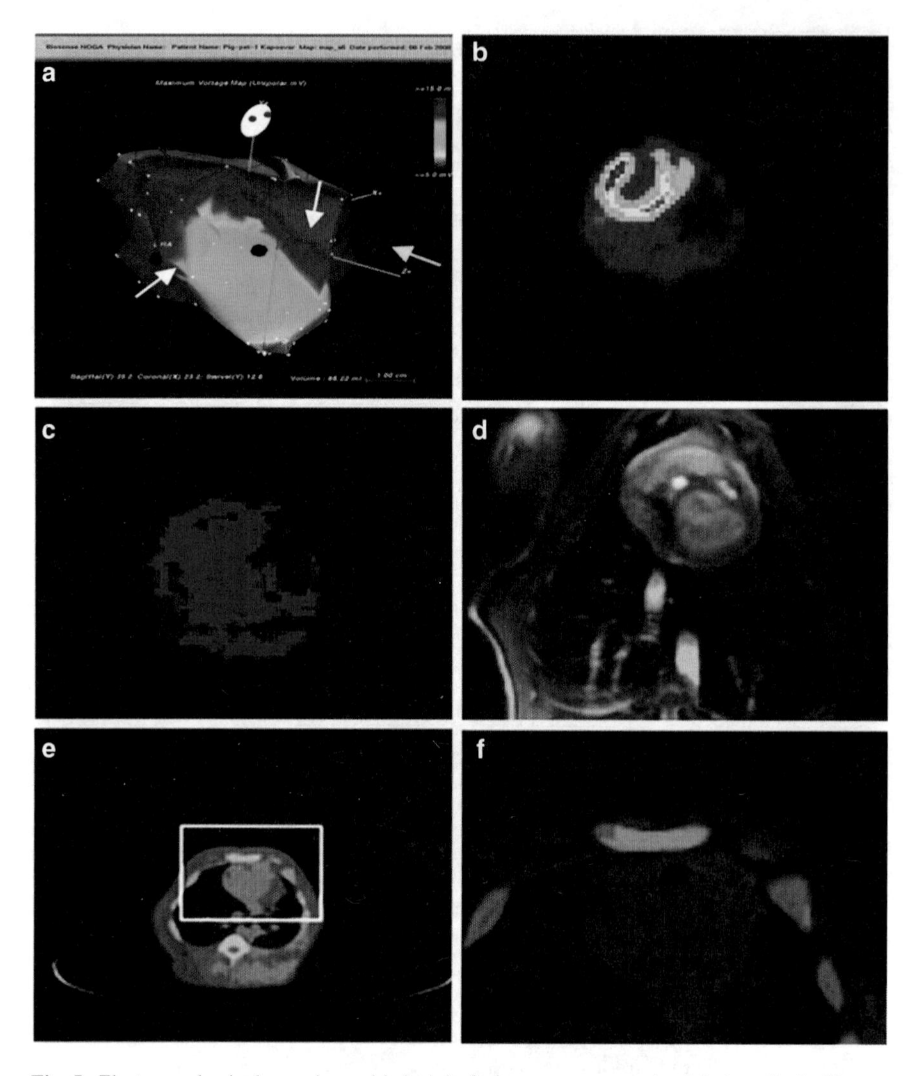

Fig. 5 Electromechanical mapping guided triple-fusion reporter gene lentiviral renilla luciferase, red fluorescent protein and herpes simplex truncated thymidine kinase (LV-RL-RFP-tTk)-labeled MSC (LV-RL-RFP-tTk-MSC) delivery in a swine MI model. (**a**) Endocardial mapping of a pig heart 16 days after MI. LV-RL-RFP-tTk-MSCs were intramyocardially injected into the border zone of the infarction (*white arrows*), and unlabeled MSCs were delivered into noninfarcted posterior wall (*yellow arrow*). (**b**) ^{13}N-ammonia positron emission tomography (PET) with transmission scan of the pig heart showing perfusion defect in the anterior wall and apex 16 days after MI. (**c**) ^{18}F-9-[3-fluoro-1-hydroxy-2-(propoxymethyl)]guanine (^{18}F-FHBG) PET image of the pig heart demonstrating the location of injected LV-RL-RFP-tTk-MSCs in the two injection sites 8 h after injection. Unlabeled MSCs were not detectable. (**d**) Registration of ^{18}F-FHBG PET (hot scale) with MRI (*gray scale*) demonstrating tracer uptake only in LV-RL-RFP-tTk-MSC injection sites. (**e**) ^{18}F-FHBG PET/CT image for localization of LV-RL-RFP-tTk-MSCs injected in the anterior wall. (**f**) Magnified region of interest of ^{18}F-FHBG PET/CT. (Reprinted from [110] with permission)

possible biohazard risks associated with expression of nonhuman genes in conjunction with the delivery of the immunogenicity of nonhuman substrates remain the significant hurdles for clinical translations. Although there are a few published human studies using reporter gene systems for in vivo PET imaging [111–114], none of these studies have progressed to stem cells or cardiovascular patient applications.

4 Tissue-Engineered Cellular Delivery

One of the major pitfalls of cellular therapy is that only a small percentage of transplanted cells are retained in the myocardium regardless of the delivery route. Although some of the cells may die owing to immunodestruction, the failure of many cells to survive may be related to poor nutrient supply, loss of cell–cell contact, and limited survival signals [11, 115]. Various groups have engineered scaffolds that may be able to better support cellular grafts in the heart [116]. A recent study has demonstrated that stem cell retention in myocardium could be increased by coadministering a Matrigel scaffold with the cells [117]. Recently, Ott et al. have shown the potential for whole-heart reconstitution using a decellularized scaffold [118].

Cellular labeling within scaffolds has been performed to a very limited extent for cardiovascular applications. Terrovitis et al. showed SPIO-labeled MSCs in collagen scaffolds survived 4 weeks in culture [119].

In a slightly different twist, our group has been labeling the scaffold rather than the cells directly. We have modified the formulation of an alginate microencapsulation method to include the incorporation of contrast agents to make them visible by several imaging methods. Microencapsulation was originally developed to provide a porous barrier for allogeneic and xenogenic cellular therapies that would exclude immunoglobulins and immune cellular mediators yet allow the free diffusion of nutrients and waste products. Because the contrast agent is added to the alginate layer surrounding the cell rather than directly labeling the cell, higher concentrations of contrast agents can be achieved without causing cellular toxicity. In addition, several formulations of imaging-visible alginate microcapsules have been studied. Importantly, the addition of the contrast agent does not appear to alter the pore size or mechanical characteristics of the microcapsule.

The first proof-of-principle of imaging-visible microcapsules was demonstrated in encapsulated islets for treatment of type I diabetes mellitus [120]. SPIOs were added to the alginate microcapsules to create magnetocapsules, which could be visualized as both hyperintensities and hypointensities by MRI depending on the imaging parameters (Fig. 6) [120]. Subsequently, the addition of radiopaque contrast agents to the alginate microcapsule was developed to create the first method to image stem cells with X-ray imaging techniques (Fig. 7) [121, 122]. Theretofore, X-ray-visible stem cell tracking methods were not available because direct labeling of cells with radiopaque contrast agents was highly cytotoxic. For stem cell therapies, we recently developed novel perfluorinated microcapsules that can be detected in a rabbit peripheral arterial disease model using ^{19}F MRI, X-ray, CT, and ultrasound [123].

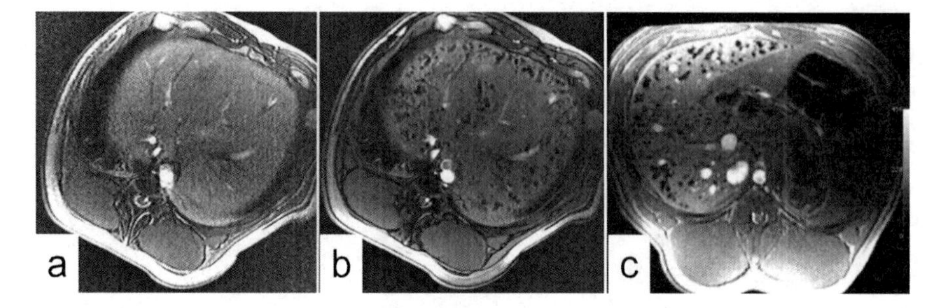

Fig. 6 In vivo MRI of SPIO-labeled microcapsules (magnetocapsules) before (**a**) and 5 min (**b**) and 3 weeks (**c**) after intraportal infusion of magnetocapsules in a swine. With use of inversion recovery with on-resonant water suppression sequence for generating positive contrast, magnetocapsules can be seen as a hypointense signal. (Adapted from [120]) with permission)

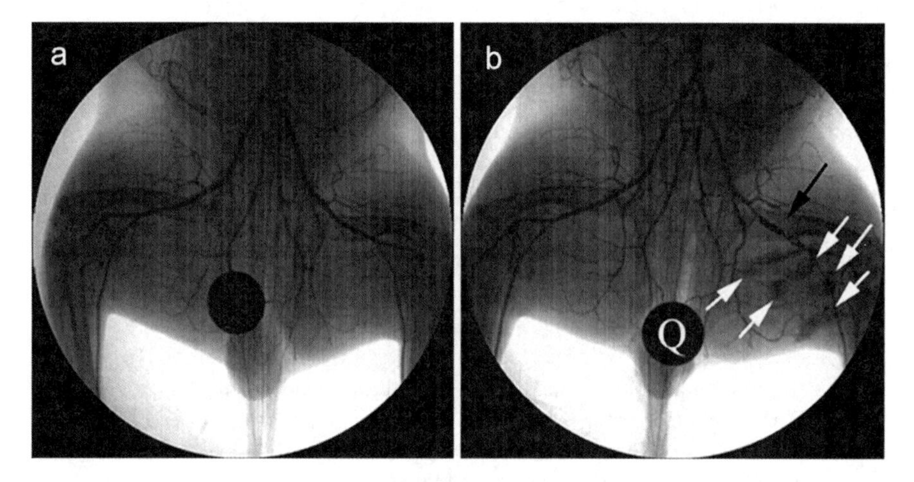

Fig. 7 X-ray angiogram of the rabbit peripheral hind limb before intervention (**a**) and after femoral artery occlusion using a platinum coil (*black arrow*) (**b**). X-ray-visible MSCs containing microcapsules appear as radiopacities in the medial thigh of the rabbit after intramuscular delivery (**b**). *Q* quarter for reference of size and opacity. (Adapted from [121] with permission)

The addition of perfluorooctylbromide to the alginate matrix provided an ideal environment for enhancing the viability of the encapsulated MSCs. Microencapsulation can also be used in concert with reporter gene imaging to assess cell viability [124]. However, the size of the microcapsules is currently 300–500 μm, which may be problematic for intraarterial or direct myocardial delivery.

As a general rule, microencapsulation will fix cells to the location where they are delivered and prevent cellular migration. Furthermore, microencapsulated stem cells are not able to participate directly in tissue regeneration, but rather enhance tissue mechanical properties or participate by releasing cytokines, e.g., paracrine effect [125–128], that recruit cells for active incorporation in tissue regeneration.

Research is needed to further elucidate the therapeutic efficacy of microencapsulated stem cell transplantation in a relevant animal model before clinical application.

5 Clinical Outlook

Successful stem cell labeling and tracking strategies require the use of agents that are sensitive and specific to the target cells, and the ability to image both anatomy and function with high sensitivity and spatial resolution. From an imaging sensitivity point of view, fluorescence imaging and BLI are the most sensitive imaging modalities, with detectabilities of 10^{-9} to 10^{-12} mol/L and 10^{-15} to 10^{-17} mol/L, respectively [129]. Yet both imaging techniques are not suitable for clinical applications because of limited tissue penetration and low-energy-photon attenuation. Radionuclide labeling, although highly sensitive (10^{-8} to 10^{-9} mol/L), is hindered by radioactive decay, which limits temporal tracking of stem cells. MRI is a noninvasive three-dimensional imaging technique that provides excellent soft tissue contrast without ionization radiation. Although MRI is well accepted for cardiac function assessment, the sensitivity for cellular detection remains quite low (10^{-3} to 10^{-5} mol/L). Reporter gene labeling/imaging offers a means for long-term viable cell fate assessment. However, significant hurdles for clinical translations will remain until safety concerns are addressed. Hybrid imaging including reporter gene labeling may play a significant role in stem cell tracking in the myocardium with evaluation of myocardial function recovery using the combination of optical imaging, radionuclide imaging, and MRI techniques. As the field of stem cell labeling and imaging gains momentum, smart labeling strategies will enable us to detect trace amounts of stem cell targets noninvasively. The recent development of ultrasonic and X-ray methods for labeling stem cells is especially attractive owing to widespread use of echocardiography and X-ray angiography in the diagnosis and treatment of cardiovascular disease.

References

1. Abdel-Latif A, Bolli R, Tleyjeh IM, et al. (2007) Adult bone marrow-derived cells for cardiac repair: a systematic review and meta-analysis. Arch Intern Med 167: 989–997
2. Lipinski MJ, Biondi-Zoccai GG, Abbate A, et al. (2007) Impact of intracoronary cell therapy on left ventricular function in the setting of acute myocardial infarction: a collaborative systematic review and meta-analysis of controlled clinical trials. J Am Coll Cardiol 50: 1761–1767
3. Singh S, Arora R, Handa K, et al. (2009) Stem cells improve left ventricular function in acute myocardial infarction. Clin Cardiol 32: 176–180
4. Zhang SN, Sun AJ, Ge JB, et al. (2009) Intracoronary autologous bone marrow stem cells transfer for patients with acute myocardial infarction: a meta-analysis of randomised controlled trials. Int J Cardiol 136: 178–185

5. Bartunek J, Vanderheyden M, Vandekerckhove B, et al. (2005) Intracoronary injection of CD133-positive enriched bone marrow progenitor cells promotes cardiac recovery after recent myocardial infarction: feasibility and safety. Circulation 112: I178–I183
6. Fernandez-Aviles F, San Roman JA, Garcia-Frade J, et al. (2004) Experimental and clinical regenerative capability of human bone marrow cells after myocardial infarction. Circ Res 95: 742–748
7. Lunde K, Solheim S, Aakhus S, et al. (2005) Autologous stem cell transplantation in acute myocardial infarction: the ASTAMI randomized controlled trial. Intracoronary transplantation of autologous mononuclear bone marrow cells, study design and safety aspects. Scand Cardiovasc J 39: 150–158
8. Perin EC, Dohmann HF, Borojevic R, et al. (2004) Improved exercise capacity and ischemia 6 and 12 months after transendocardial injection of autologous bone marrow mononuclear cells for ischemic cardiomyopathy. Circulation 110: I213–I218
9. Schachinger V, Assmus B, Britten MB, et al. (2004) Transplantation of progenitor cells and regeneration enhancement in acute myocardial infarction: final one-year results of the TOPCARE-AMI trial. J Am Coll Cardiol 44: 1690–1699
10. Wollert KC, Meyer GP, Lotz J, et al. (2004) Intracoronary autologous bone-marrow cell transfer after myocardial infarction: the BOOST randomised controlled clinical trial. Lancet 364: 141–148
11. Zhang M, Methot D, Poppa V, et al. (2001) Cardiomyocyte grafting for cardiac repair: graft cell death and anti-death strategies. J Mol Cell Cardiol 33: 907–921
12. Lavender JP, Goldman JM, Arnot RN, et al. (1977) Kinetics of indium-III labelled lymphocytes in normal subjects and patients with Hodgkin's disease. Br Med J 2: 797–799
13. Thakur ML, Segal AW, Louis L, et al. (1977) Indium-111-labeled cellular blood components: mechanism of labeling and intracellular location in human neutrophils. J Nucl Med 18: 1022–1026
14. Gao J, Dennis JE, Muzic RF, et al. (2001) The dynamic in vivo distribution of bone marrow-derived mesenchymal stem cells after infusion. Cell Tissues Organs 169: 12–20
15. Zhou R, Thomas DH, Qiao H, et al. (2005) In vivo detection of stem cells grafted in infarcted rat myocardium. J Nucl Med 46: 816–822
16. Hou D, Youssef EA, Brinton TJ, et al. (2005) Radiolabeled cell distribution after intramyocardial, intracoronary, and interstitial retrograde coronary venous delivery: implications for current clinical trials. Circulation 112: I150–I156
17. Aicher A, Brenner W, Zuhayra M, et al. (2003) Assessment of the tissue distribution of transplanted human endothelial progenitor cells by radioactive labeling. Circulation 107: 2134–2139
18. Brenner W, Aicher A, Eckey T, et al. (2004) 111In-labeled CD34+ hematopoietic progenitor cells in a rat myocardial infarction model. J Nucl Med 45: 512–518
19. Chin BB, Nakamoto Y, Bulte JW, et al. (2003) 111In oxine labelled mesenchymal stem cell SPECT after intravenous administration in myocardial infarction. Nucl Med Commun 24: 1149–1154
20. Kraitchman DL, Mahmood A, Soto AV, et al. (2005) Targeted magnetic resonance imaging fluoroscopic delivery of magnetically-labeled mesenchymal stem cells improves myocardial function without altering infarction size. Circulation 112: U175
21. Kraitchman DL, Tatsumi M, Gilson WD, et al. (2005) Dynamic imaging of allogeneic mesenchymal stem cells trafficking to myocardial infarction. Circulation 112: 1451–1461
22. Doyle B, Kemp BJ, Chareonthaitawee P, et al. (2007) Dynamic tracking during intracoronary injection of 18F-FDG-labeled progenitor cell therapy for acute myocardial infarction. J Nucl Med 48: 1708–1714
23. Hofmann M, Wollert KC, Meyer GP, et al. (2005) Monitoring of bone marrow cell homing into the infarcted human myocardium. Circulation 111: 2198–2202
24. Kang HJ, Lee HY, Na SH, et al. (2006) Differential effect of intracoronary infusion of mobilized peripheral blood stem cells by granulocyte colony-stimulating factor on left ventricular function and remodeling in patients with acute myocardial infarction versus old myocardial infarction: the MAGIC Cell-3-DES randomized, controlled trial. Circulation 114: I145–I151

25. Schachinger V, Aicher A, Dobert N, et al. (2008) Pilot trial on determinants of progenitor cell recruitment to the infarcted human myocardium. Circulation 118: 1425–1432
26. Adonai N, Nguyen KN, Walsh J, et al. (2002) Ex vivo cell labeling with 64Cu-pyruvaldehyde-bis(N^4-methylthiosemicarbazone) for imaging cell trafficking in mice with positron-emission tomography. Proc Natl Acad Sci USA 99: 3030–3035
27. Jin Y, Kong H, Stodilka RZ, et al. (2005) Determining the minimum number of detectable cardiac-transplanted 111In-tropolone-labelled bone-marrow-derived mesenchymal stem cells by SPECT. Phys Med Biol 50: 4445–4455
28. Gholamrezanezhad A, Mirpour S, Ardekani JM, et al. (2009) Cytotoxicity of 111In-oxine on mesenchymal stem cells: a time-dependent adverse effect. Nucl Med Commun 30: 210–216
29. Stark DD, Weissleder R, Elizondo G, et al. (1988) Superparamagnetic iron oxide: clinical application as a contrast agent for MR imaging of the liver. Radiology 168: 297–301
30. Cunningham CH, Arai T, Yang PC, et al. (2005) Positive contrast magnetic resonance imaging of cells labeled with magnetic nanoparticles. Magn Reson Med 53: 999–1005
31. Mani V, Saebo KC, Itskovich V, et al. (2006) Gradient echo acquisition for superparamagnetic particles with positive contrast (GRASP): sequence characterization in membrane and glass superparamagnetic iron oxide phantoms at 1.5T and 3T. Magn Reson Med 55: 126–135
32. Stuber M, Gilson WD, Schär M, et al. (2007) Positive contrast visualization of iron oxide-labeled stem cells using inversion recovery with ON-resonant water suppression (IRON). Magn Reson Med 58: 1072–1077
33. Daldrup-Link HE, Meier R, Rudelius M, et al. (2005) In vivo tracking of genetically engineered, anti-HER2/neu directed natural killer cells to HER2/neu positive mammary tumors with magnetic resonance imaging. Eur Radiol 15: 4–13
34. Frank JA, Miller BR, Arbab AS, et al. (2003) Clinically applicable labeling of mammalian and stem cells by combining superparamagnetic iron oxides and transfection agents. Radiology 228: 480–487
35. Frank JA, Zywicke H, Jordan EK, et al. (2002) Magnetic intracellular labeling of mammalian cells by combining (FDA-approved) superparamagnetic iron oxide MR contrast agents and commonly used transfection agents. Acad Radiol 9: S484–S487
36. Kostura L, Kraitchman DL, Mackay AM, et al. (2004) Feridex labeling of mesenchymal stem cells inhibits chondrogenesis but not adipogenesis or osteogenesis. NMR Biomed 17: 513–517
37. Schafer R, Kehlbach R, Muller M, et al. (2009) Labeling of human mesenchymal stromal cells with superparamagnetic iron oxide leads to a decrease in migration capacity and colony formation ability. Cytotherapy 11: 68–78
38. Arbab AS, Yocum GT, Rad AM, et al. (2005) Labeling of cells with ferumoxides-protamine sulfate complexes does not inhibit function or differentiation capacity of hematopoietic or mesenchymal stem cells. NMR Biomed 18: 553–559
39. Arbab AS, Yocum GT, Kalish H, et al. (2004) Efficient magnetic cell labeling with protamine sulfate complexed to ferumoxides for cellular MRI. Blood 104: 1217–1223
40. Kraitchman DL, Heldman AW, Atalar E, et al. (2003) In vivo magnetic resonance imaging of mesenchymal stem cells in myocardial infarction. Circulation 107: 2290–2293
41. Hill JM, Dick AJ, Raman VK, et al. (2003) Serial cardiac magnetic resonance imaging of injected mesenchymal stem cells. Circulation 108: 1009–1014
42. Kraitchman DL and Bulte JW (2008) Imaging of stem cells using MRI. Basic Res Cardiol 103: 105–113
43. Kraitchman DL, Gilson WD, and Lorenz CH (2008) Stem cell therapy: MRI guidance and monitoring. J Magn Reson Imaging 27: 299–310
44. Rickers C, Gallegos R, Seethamraju RT, et al. (2004) Applications of magnetic resonance imaging for cardiac stem cell therapy. J Interv Cardiol 17: 37–46
45. Saeed M, Saloner D, Weber O, et al. (2005) MRI in guiding and assessing intramyocardial therapy. Eur Radiol 15: 851–863
46. Dick AJ, Guttman MA, Raman VK, et al. (2003) Magnetic resonance fluoroscopy allows targeted delivery of mesenchymal stem cells to infarct borders in swine. Circulation 108: 2899–2904

47. Kraitchman DL (2007) Non-invasive imaging and labelling techniques in stem cell therapy. In Rebuilding the infarcted heart, K.C. Wollert, and L.J. Field, eds. (London, UK Informa UK Ltd), pp. 135–149
48. Rickers C, Kraitchman D, Fischer G, et al. (2005) Cardiovascular interventional MR imaging: a new road for therapy and repair in the heart. Magn Reson Imaging Clin N Am 13: 465–479
49. Amado LC, Salrais AP, Schuleri KH, et al. (2005) Cardiac repair with intramyocardial injection of allogeneic mesenchymal stem cells after myocardial infarction. Proc Natl Acad Sci USA 102: 11474–11479
50. Schuleri KH, Amado LC, Boyle AJ, et al. (2008) Early improvement in cardiac tissue perfusion due to mesenchymal stem cells. Am J Physiol Heart Circ Physiol 294: H2002–H2011
51. Britten MB, Abolmaali ND, Assmus B, et al. (2003) Infarct remodeling after intracoronary progenitor cell treatment in patients with acute myocardial infarction (TOPCARE-AMI): mechanistic insights from serial contrast-enhanced magnetic resonance imaging. Circulation 108: 2212–2218
52. Janssens S, Dubois C, Bogaert J, et al. (2006) Autologous bone marrow-derived stem-cell transfer in patients with ST-segment elevation myocardial infarction: double-blind, randomised controlled trial. Lancet 367: 113–121
53. Walczak P, Kedziorek D, Gilad AA, et al. (2005) Instant MR labeling of stem cells using magnetoelectroporation. Magn Reson Med 54: 769–774
54. Partlow KC, Chen J, Brant JA, et al. (2007) 19F magnetic resonance imaging for stem/progenitor cell tracking with multiple unique perfluorocarbon nanobeacons. FASEB J 21: 1647–1654
55. Ahrens ET, Flores R, Xu H, et al. (2005) In vivo imaging platform for tracking immunotherapeutic cells. Nat Biotechnol 23: 983–987
56. Bulte JW (2005) Hot spot MRI emerges from the background. Nat Biotechnol 23: 945–946
57. Srinivas M, Turner MS, Janjic JM, et al. (2009) In vivo cytometry of antigen-specific t cells using (19)F MRI. Magn Reson Med 63: 747–753
58. Lin S, Xie X, Patel MR, et al. (2007) Quantum dot imaging for embryonic stem cells. BMC Biotechnol 7: 67
59. Muller-Borer BJ, Collins MC, Gunst PR, et al. (2007) Quantum dot labeling of mesenchymal stem cells. J Nanobiotechnol 5: 9
60. Slotkin JR, Chakrabarti L, Dai HN, et al. (2007) In vivo quantum dot labeling of mammalian stem and progenitor cells. Dev Dyn 236: 3393–3401
61. Medintz IL, Uyeda HT, Goldman ER, et al. (2005) Quantum dot bioconjugates for imaging, labelling and sensing. Nat Mater 4: 435–446
62. Rosen AB, Kelly DJ, Schuldt AJ, et al. (2007) Finding fluorescent needles in the cardiac haystack: tracking human mesenchymal stem cells labeled with quantum dots for quantitative in vivo three-dimensional fluorescence analysis. Stem Cells 25: 2128–2138
63. Shah BS, Clark PA, Moioli EK, et al. (2007) Labeling of mesenchymal stem cells by bioconjugated quantum dots. Nano Lett 7: 3071–3079
64. Hoshino K, Ly HQ, Frangioni JV, et al. (2007) In vivo tracking in cardiac stem cell-based therapy. Prog Cardiovasc Dis 49: 414–420
65. Ma B, Hankenson KD, Dennis JE, et al. (2005) A simple method for stem cell labeling with fluorine 18. Nucl Med Biol 32: 701–705
66. Amsalem Y, Mardor Y, Feinberg MS, et al. (2007) Iron-oxide labeling and outcome of transplanted mesenchymal stem cells in the infarcted myocardium. Circulation 116: I38–I45
67. Terrovitis J, Kwok KF, Lautamaki R, et al. (2008) Ectopic expression of the sodium-iodide symporter enables imaging of transplanted cardiac stem cells in vivo by single-photon emission computed tomography or positron emission tomography. J Am Coll Cardiol 52: 1652–1660
68. Terrovitis J, Stuber M, Youssef A, et al. (2008) Magnetic resonance imaging overestimates ferumoxide-labeled stem cell survival after transplantation in the heart. Circulation 117: 1555–1562
69. Ebert SN, Taylor DG, Nguyen HL, et al. (2007) Noninvasive tracking of cardiac embryonic stem cells in vivo using magnetic resonance imaging techniques. Stem Cells 25: 2936–2944

70. Stuckey DJ, Carr CA, Martin-Rendon E, et al. (2006) Iron particles for noninvasive monitoring of bone marrow stromal cell engraftment into, and isolation of viable engrafted donor cells from, the heart. Stem Cells 24: 1968–1975
71. Walczak P, Kedziorek DA, Gilad AA, et al. (2007) Applicability and limitations of MR tracking of neural stem cells with asymmetric cell division and rapid turnover: the case of the shiverer dysmyelinated mouse brain. Magn Reson Med 58: 261–269
72. Sheikh AY, Lin SA, Cao F, et al. (2007) Molecular imaging of bone marrow mononuclear cell homing and engraftment in ischemic myocardium. Stem Cells 25: 2677–2684
73. Lin Y, Molter J, Lee Z, et al. (2008) Bioluminescence imaging of hematopoietic stem cell repopulation in murine models. Methods Mol Biol 430: 295–306
74. DeLuca M and McElroy WD (1974) Kinetics of the firefly luciferase catalyzed reactions. Biochemistry 13: 921–925
75. Wang X, Rosol M, Ge S, et al. (2003) Dynamic tracking of human hematopoietic stem cell engraftment using in vivo bioluminescence imaging. Blood 102: 3478–3482
76. Chen IY, Greve JM, Gheysens O, et al. (2009) Comparison of optical bioluminescence reporter gene and superparamagnetic iron oxide MR contrast agent as cell markers for non-invasive imaging of cardiac cell transplantation. Mol Imaging Biol 11: 178–187
77. Cao F, Lin S, Xie X, et al. (2006) In vivo visualization of embryonic stem cell survival, proliferation, and migration after cardiac delivery. Circulation 113: 1005–1014
78. Min JJ, Ahn Y, Moon S, et al. (2006) In vivo bioluminescence imaging of cord blood derived mesenchymal stem cell transplantation into rat myocardium. Ann Nucl Med 20: 165–170
79. van der Bogt KE, Sheikh AY, Schrepfer S, et al. (2008) Comparison of different adult stem cell types for treatment of myocardial ischemia. Circulation 118: S121–S129
80. Wilson K, Yu J, Lee A, et al. (2008) In vitro and in vivo bioluminescence reporter gene imaging of human embryonic stem cells. J Vis Exp 14: 740
81. Wu JC, Chen IY, Sundaresan G, et al. (2003) Molecular imaging of cardiac cell transplantation in living animals using optical bioluminescence and positron emission tomography. Circulation 108: 1302–1305
82. Wu JC, Chen IY, Wang Y, et al. (2004) Molecular imaging of the kinetics of vascular endothelial growth factor gene expression in ischemic myocardium. Circulation 110: 685–691
83. Lee Z, Dennis JE, and Gerson SL (2008) Imaging stem cell implant for cellular-based therapies. Exp Biol Med (Maywood) 233: 930–940
84. Miyagawa M, Anton M, Wagner B, et al. (2005) Non-invasive imaging of cardiac transgene expression with PET: comparison of the human sodium/iodide symporter gene and HSV1-tk as the reporter gene. Eur J Nucl Med Mol Imaging 32: 1108–1114
85. Sun X, Annala AJ, Yaghoubi SS, et al. (2001) Quantitative imaging of gene induction in living animals. Gene Ther 8: 1572–1579
86. Yaghoubi SS, Wu L, Liang Q, et al. (2001) Direct correlation between positron emission tomographic images of two reporter genes delivered by two distinct adenoviral vectors. Gene Ther 8: 1072–1080
87. Kang JH, Lee DS, Paeng JC, et al. (2005) Development of a sodium/iodide symporter (NIS)-transgenic mouse for imaging of cardiomyocyte-specific reporter gene expression. J Nucl Med 46: 479–483
88. Ricci D, Mennander AA, Pham LD, et al. (2008) Non-invasive radioiodine imaging for accurate quantitation of NIS reporter gene expression in transplanted hearts. Eur J Cardiothorac Surg 33: 32–39
89. Shin JH, Chung JK, Kang JH, et al. (2004) Feasibility of sodium/iodide symporter gene as a new imaging reporter gene: comparison with HSV1-tk. Eur J Nucl Med Mol Imaging 31: 425–432
90. Huang M, Batra RK, Kogai T, et al. (2001) Ectopic expression of the thyroperoxidase gene augments radioiodide uptake and retention mediated by the sodium iodide symporter in non-small cell lung cancer. Cancer Gene Ther 8: 612–618
91. Barrio JR, Satyamurthy N, Huang SC, et al. (1989) 3-(2′-[18F]fluoroethyl)spiperone: in vivo biochemical and kinetic characterization in rodents, nonhuman primates, and humans. J Cereb Blood Flow Metab 9: 830–839

92. MacLaren DC, Gambhir SS, Satyamurthy N, et al. (1999) Repetitive, non-invasive imaging of the dopamine D2 receptor as a reporter gene in living animals. Gene Ther 6: 785–791
93. Chen IY, Wu JC, Min JJ, et al. (2004) Micro-positron emission tomography imaging of cardiac gene expression in rats using bicistronic adenoviral vector-mediated gene delivery. Circulation 109: 1415–1420; Epub 2004 Mar, 1418
94. Inubushi M, Wu JC, Gambhir SS, et al. (2003) Positron-emission tomography reporter gene expression imaging in rat myocardium. Circulation 107: 326–332
95. Wu JC, Inubushi M, Sundaresan G, et al. (2002) Positron emission tomography imaging of cardiac reporter gene expression in living rats. Circulation 106: 180–183
96. Penuelas I, Mazzolini G, Boan JF, et al. (2005) Positron emission tomography imaging of adenoviral-mediated transgene expression in liver cancer patients. Gastroenterology 128: 1787–1795
97. Luker GD, Sharma V, Pica CM, et al. (2002) Noninvasive imaging of protein–protein interactions in living animals. Proc Natl Acad Sci USA 99: 6961–6966
98. Gambhir SS, Bauer E, Black ME, et al. (2000) A mutant herpes simplex virus type 1 thymidine kinase reporter gene shows improved sensitivity for imaging reporter gene expression with positron emission tomography. Proc Natl Acad Sci USA 97: 2785–2790
99. Ray P, De A, Min JJ, et al. (2004) Imaging tri-fusion multimodality reporter gene expression in living subjects. Cancer Res 64: 1323–1330
100. Koretsky AP, Brosnan MJ, Chen LH, et al. (1990) NMR detection of creatine kinase expressed in liver of transgenic mice: determination of free ADP levels. Proc Natl Acad Sci USA 87: 3112–3116
101. Deans AE, Wadghiri YZ, Bernas LM, et al. (2006) Cellular MRI contrast via coexpression of transferrin receptor and ferritin. Magn Reson Med 56: 51–59
102. Genove G, DeMarco U, Xu H, et al. (2005) A new transgene reporter for in vivo magnetic resonance imaging. Nat Med 11: 450–454
103. Liu J, Cheng EC, Long Jr RC, et al. (2009) Noninvasive monitoring of embryonic stem cells in vivo with MRI transgene reporter. Tissue Eng Part C Methods 15: 739–747
104. Pawelczyk E, Arbab AS, Pandit S, et al. (2006) Expression of transferrin receptor and ferritin following ferumoxides-protamine sulfate labeling of cells: implications for cellular magnetic resonance imaging. NMR Biomed 19: 581–592
105. Gilad AA, McMahon MT, Walczak P, et al. (2007) Artificial reporter gene providing MRI contrast based on proton exchange. Nat Biotechnol 25: 217–219
106. Gilad AA, Winnard PT, Jr., van Zijl PC, et al. (2007) Developing MR reporter genes: promises and pitfalls. NMR Biomed 20: 275–290
107. Gilad AA, Ziv K, McMahon MT, et al. (2008) MRI reporter genes. J Nucl Med 49: 1905–1908
108. Kuliszewski MA, Fujii H, Liao C, et al. (2009) Molecular imaging of endothelial progenitor cell engraftment using contrast-enhanced ultrasound and targeted microbubbles. Cardiovasc Res 83: 653–662; doi:10.1093/cvr/cvp218
109. Ray P, Wu AM, and Gambhir SS (2003) Optical bioluminescence and positron emission tomography imaging of a novel fusion reporter gene in tumor xenografts of living mice. Cancer Res 63: 1160–1165
110. Gyöngyösi M, Blanco J, Marian T, et al. (2008) Serial noninvasive in vivo positron emission tomographic tracking of percutaneously intramyocardially injected autologous porcine mesenchymal stem cells modified for transgene reporter gene expression. Circ Cardiovasc Imaging 1: 94–103
111. Jacobs A, Braunlich I, Graf R, et al. (2001) Quantitative kinetics of [124I]FIAU in cat and man. J Nucl Med 42: 467–475
112. Jacobs A, Voges J, Reszka R, et al. (2001) Positron-emission tomography of vector-mediated gene expression in gene therapy for gliomas. Lancet 358: 727–729
113. Ponomarev V, Doubrovin M, Shavrin A, et al. (2007) A human-derived reporter gene for noninvasive imaging in humans: mitochondrial thymidine kinase type 2. J Nucl Med 48: 819–826

114. Yaghoubi SS, Barrio JR, Namavari M, et al. (2005) Imaging progress of herpes simplex virus type 1 thymidine kinase suicide gene therapy in living subjects with positron emission tomography. Cancer Gene Ther 12: 329–339
115. Reinecke H, Zhang M, Bartosek T, et al. (1999) Survival, integration, and differentiation of cardiomyocyte grafts: a study in normal and injured rat hearts. Circulation 100: 193–202
116. Dar A, Shachar M, Leor J, et al. (2002) Optimization of cardiac cell seeding and distribution in 3D porous alginate scaffolds. Biotechnol Bioeng 80: 305–312
117. Willmann JK, Paulmurugan R, Rodriguez-Porcel M, et al. (2009) Imaging gene expression in human mesenchymal stem cells: from small to large animals. Radiology 252: 117–127
118. Ott HC, Matthiesen TS, Goh SK, et al. (2008) Perfusion-decellularized matrix: using nature's platform to engineer a bioartificial heart. Nat Med 14: 213–221
119. Terrovitis JV, Bulte JW, Sarvananthan S, et al. (2006) Magnetic resonance imaging of ferumoxide-labeled mesenchymal stem cells seeded on collagen scaffolds – relevance to tissue engineering. Tissue Eng 12: 2765–2775
120. Barnett BP, Arepally A, Karmarkar PV, et al. (2007) Magnetic resonance-guided, real-time targeted delivery and imaging of magnetocapsules immunoprotecting pancreatic islet cells. Nat Med 13: 986–991
121. Nahrendorf M, Sosnovik D, French B, et al. (2009) Multimodality cardiovascular molecular imaging – part II. Circ Cardiovasc Imaging 2: 56–70
122. Barnett BP, Kraitchman DL, Lauzon C, et al. (2006) Radiopaque alginate microcapsules for X-ray visualization and immunoprotection of cellular therapeutics. Mol Pharm 3: 531–538
123. Fu Y, Kedziorek D, Ouwerkerk R, et al. (2009) Multifunctional perfluorooctylbromide alginate microcapsules for monitoring of mesenchymal stem cell delivery using CT and MRI. J Cardiovasc Magn Reson 11: O7
124. Kraitchman DL and Bulte JW (2009) In vivo imaging of stem cells and beta cells using direct cell labeling and reporter gene methods. Arterioscler Thromb Vasc Biol 29: 1025–1030
125. Burchfield JS and Dimmeler S (2008) Role of paracrine factors in stem and progenitor cell mediated cardiac repair and tissue fibrosis. Fibrogenesis Tissue Repair 1: 4
126. Doyle B, Sorajja P, Hynes B, et al. (2008) Progenitor cell therapy in a porcine acute myocardial infarction model induces cardiac hypertrophy, mediated by paracrine secretion of cardiotrophic factors including TGFbeta1. Stem Cells Dev 17: 941–951
127. Kinnaird T, Stabile E, Burnett MS, et al. (2004) Local delivery of marrow-derived stromal cells augments collateral perfusion through paracrine mechanisms. Circulation 109: 1543–1549
128. Tang YL, Zhao Q, Qin X, et al. (2005) Paracrine action enhances the effects of autologous mesenchymal stem cell transplantation on vascular regeneration in rat model of myocardial infarction. Ann Thorac Surg 80: 229–236; discussion 236–237
129. Zhang SJ and Wu JC (2007) Comparison of imaging techniques for tracking cardiac stem cell therapy. J Nucl Med 48: 1916–1919

Assessing Regional Mechanical Function After Stem Cell Delivery

Jacques P. Guyette and Glenn R. Gaudette

Abstract One of the desired outcomes of cellular cardiomyoplasty is to restore contractile cellular mass to the injured heart for the improvement of cardiac function. Myocardial mechanical function is composed of both passive and active components, where passive function is derived primarily from the compliance of the myocardial tissue and active function is generated by contracting cardiomyocytes. Although materials and cell therapies have been shown to improve mechanical function in the myocardium after infarction, the preferred goal of regenerative therapy is to replace necrotic myocardium with contractile cells that will contribute active function. In this chapter, we discuss these concepts in further detail and provide background for how cardiac function is measured.

Keywords Magnetic resonance imaging • Echocardiography • Sonomicrometry • High-density mapping • Regional mechanical function • Global mechanical function • Ejection fraction • Stroke volume • Stroke work • Regional stroke work • Active contractile properties • Passive mechanical properties

1 Introduction

Cell therapy strategies for cardiac regeneration aim to overcome the chief limitation of current treatments for myocardial infarction and heart failure by replacing necrotic cardiomyocytes with viable cells that can restore normal cardiac mechanical function. A myocardial infarct can cause the loss of one billion cardiomyocytes that once provided contractile work necessary for sustaining healthy heart function [1].

G.R. Gaudette (✉)
Department of Biomedical Engineering, Worcester Polytechnic Institute, Worcester, MA, USA
e-mail: gaudette@wpi.edu

I.S. Cohen and G.R. Gaudette (eds.), *Regenerating the Heart*, Stem Cell Biology and Regenerative Medicine, DOI 10.1007/978-1-61779-021-8_26,
© Springer Science+Business Media, LLC 2011

This loss of contractile mass jeopardizes the heart's ability to perform as an electromechanical pump, decreasing systemic circulation, which can result in heart failure. Therefore, one of the higher-order goals of cell therapy for cardiac repair is to restore this lost contractile mass. Several different cell types that have been delivered to the heart in clinical and preclinical studies have been shown to regenerate cardiac mechanical function. Thus far, however, conclusive evidence of a correlation between regenerated striated cardiomyocytes and improvements in regional mechanical function has not been established. This leads to a fundamental question of whether the delivered cells increase active function by adding contractile elements, or whether they improve passive function by increasing the compliance within the region of the developed scar [2]. Recently, Wall et al. [3] used a computational model to demonstrate that the delivery of passive materials to the infarcted heart can result in improved mechanical function. With the goal of restoring contractile mass, however, the desired expectation of cell therapies is to contribute active mechanical function. In this chapter, we discuss different methods and metrics used to measure mechanical function in the heart. Although it is beyond the scope of this chapter to provide detailed explanations of all imaging and acquisition techniques, we aim to provide an understanding of some of the most commonly used techniques and assessment parameters. In addition, we evaluate the data suggesting that cell therapy can improve function through both active and passive mechanisms.

2 Mechanical Function

Myocardial infarction and heart failure result in cardiomyocyte cell death, thereby eliciting a remodeling cascade that not only reduces the number of contractile units, but also generates a scar that compromises both the integrity and the compliance of the myocardial wall. With respect to the heart as an electromechanical pump, healthy cardiomyocytes respond to cyclic electrical impulses to shorten or elongate their sarcomeres. Synchronous action by cardiomyocytes permits the contraction and relaxation of the myocardial walls, which act to pressurize atrial and ventricular cavities to pump blood for systemic circulation. Cardiomyocyte cell death reduces active contraction within a region of the heart owing to the decreased number of working sarcomeres, impairing the heart's function as a mechanical pump. Necrosis and scarring within the affected region changes the material properties of the myocardial wall, creating an area that may not comply with the dynamics of either surrounding contractile cardiomyocytes or pressures developed within heart cavities. The ultimate goal of cardiac regeneration for the treatment of myocardial infarction and heart failure is to restore mechanical function to the injured heart.

Consequently, the effectiveness of any novel therapy for the improvement of myocardial function must be based on the resulting changes it demonstrates on mechanical function. Currently, there are several techniques that can be applied for measuring global or regional mechanical cardiac function, and which have been used in either clinical or preclinical studies. In addition, there are several parameters that can be measured as indicators of cardiac function, but inevitably the technique chosen often limits the parameters that can be assessed. Unfortunately, resource availability can present logistical challenges in choosing an appropriate technique. Often, investigators are limited by equipment expense or availability. In addition, the need for technological expertise may limit the experimental approach or flexibility in choosing unfamiliar methods. Knowing the capabilities of different measurement techniques and the metrics that can be determined from acquired data should allow investigators to better interpret their findings regarding myocardial function.

3 Techniques for Measuring Mechanical Cardiac Function

Clinically, echocardiography and magnetic resonance imaging (MRI) are the most commonly used techniques to assess cardiac mechanical function in patients [4–15]. Both methods provide options that are noninvasive or minimally invasive, and have also been applied in experimental and preclinical studies. Although studies on smaller animals require appropriate equipment, the data acquisition and analysis methods are most similar to what clinicians use with human patients.

Echocardiography is a cardiac ultrasound technique, and employs a system to send and receive high-frequency sound waves to create either 2D or 3D images of the heart. The system requires the use of a probe to send sound across the chest cavity, which then reflects off internal tissue. The reflection signature is then received by a transducer and translated to visualize the structure and motion of the tissue. The probe used for echocardiography either can be placed externally on the outside of the chest for transthoracic signaling or inserted down the esophagus for a transesophageal approach. These applications can be used to examine both the shape and the velocity of the heart tissue in real time, allowing for the determination of wall movement, wall thickness, and blood velocity. Although echocardiography is a relatively inexpensive technique, it does have experimental limitations. The spatial resolution of echocardiography has been reported to be on the order of 5 mm, with a temporal resolution of 15 ms, which is more suited for evaluating global function (compared with regional function) for short-term cardiac-cycle analysis [16]. However, technological advances are continually improving upon these spatial and temporal constraints. Recent advances in signal analysis, such as speckle tracking, have improved the ability of echocardiography to assess regional function [17]. Echocardiography is also subject to processing limitations, such as "noise" resulting from interference and image variability depending on probe orientation [16, 18]. Although transthoracic echocardiography is a less invasive approach, some regions of the ventricle can be difficult to image [19]. However transesophageal echocardiography may provide better access for tissue imaging, it is a slightly more invasive procedure.

MRI is another noninvasive technique used to determine cardiac function, commonly used by radiologists to visualize detailed structure and function of internal body tissues. In this system, an MRI machine uses a radio-frequency transmitter to energize photons to a resonance frequency, creating an electromagnetic field. When a patient lies within the machine, the energized photons reorient the spin of aligned protons in water molecules within myocardial tissue. When the machine is turned off, the protons return to their natural spin state, generating a field which can be detected by a scanner to generate images of the heart. Tagged MRI can provide more regional information than shape and size data obtained with conventional MRI. This technique uses intersecting "presaturation" planes that form "tag lines" throughout the myocardial tissue, in which intersections serve as markers that move with the heart wall. The movements of the intersections can be tracked, allowing for the determination of 3D deformation and the enhancement of region-specific analysis. Despite these improvements, the data acquisition process for this technique can be time-consuming as usually only one plane can be imaged at a time. After the data have been acquired for one plane, the procedure is repeated for the next plane (usually orthogonal to the base–apex axis). As a result, the total time for data acquisition can be 35–40 min [20]. The spatial resolution of the technique (as defined as the distance between tags) is generally on the order of 5 mm. Improvements in image detail are limited, as better resolution comes at the risk of the MRI tags disappearing before the end of the diastolic/systolic interval being examined [11]. In addition to laborious data acquisition, data analysis is also complicated and time-consuming. Owing to the required equipment and expertise needed, MRI can be very expensive.

Another clinical technique gaining popularity involves the electromechanical mapping of the heart with magnetic fields, through the use of a NOGA system (originally developed as the NOGA cardiac navigational system by Biosense Webster in 1997). In this system, a catheter is inserted into the ventricular cavity through the ventricular wall. A magnetic pad lying under the patient is used to generate a low-level magnetic field. Through a triangulation algorithm, the 3D location of the catheter can be determined. The catheter can then be positioned to sample several different points on the endocardial surface of the ventricle. The point measurements are then used to create a map of the endocardial surface, detailing information regarding mechanical deformation. With a spatial resolution of 1 mm, this technique is very useful for determining the movement of the ventricular wall at multiple locations and defined regions. The inclusion of a sensing electrode in the tip of the catheter allows for determination of unipolar voltage in addition to spatial location.

In the animal laboratory setting, sonomicrometry is commonly used to determine regional function of the heart in vivo [10, 21–24]. Sonomicrometry is a uniaxial technique involving the use of ultrasonic transducers, in which one transducer sends a sound wave to another transducer that receives the signal. The distance between the two transducers is proportional to the time it takes for the sound to travel, thereby providing a continuous measure of the distance between the two transducers as the heart expands and contracts. The transducer crystals must be implanted or sutured into the heart wall and the tissue should be homogeneously transmissive to sound waves for the signal to pass correctly. Sonomicrometry also assumes a homogenous deformation between the transducers (approximately 7–20 mm) and can only determine function along this axis. One pair of transducers provides a measure of distance along one axis, whereas two pairs of transducers can be used along two orthogonal axes to measure area deformation. If three pairs of transducers are used at three orthogonal axes, the volume of the ventricle can be calculated making the assumption the ventricle is ellipsoidal in shape. In a regional setting, sonomicrometry can be used to determine changes in distance surrounding a specific area of the myocardium (e.g. an infarct region). This method is commonly used to determine ventricular volume, regional area, segment length, and wall thickening.

A major limitation of sonomicrometry, however, is that the implantation of the transducer crystals is more invasive and consequently more "fixed" in terms of region. To measure multiple regions would require either reimplanting the transducers at new locations or implanting more transducers, both of which result in more invasiveness and possible damage to the heart wall. As myocardial infarction usually results in a region (or regions) of the heart becoming dysfunctional, a technique to determine regional function over many regions (whole field) in the heart may be more helpful. To determine whole-field function, Prinzen et al. [25] used 40–60 paper markers (1.5-mm diameter) arranged on the epicardium to determine epicardial wall displacements. By tracking the markers, they were able to determine 2D wall displacements, and when using several markers, they could generate mechanical deformation data at multiple locations on the heart surface for whole-field analysis.

To improve the spatial resolution of the epicardial-marker technique, an innovative phase correlation algorithm called computer aided speckle interferometry (CASI) was developed [26]. In this system, silicon carbide particles (40 μm in diameter) are applied to the epicardium, which is then illuminated with white light to create a speckled contrast for particle tracking (Fig. 1). The phase-correlation algorithm allows for increased computational efficiency compared with conventional particle tracking or digital correlation techniques. This algorithm has been shown to determine uniaxial strain that is equivalent to what can be obtained with sonomicrometry in isolated rabbit hearts [26]. Unlike sonomicrometry, however, CASI is a whole-field technique that is able to determine 2D deformations with high spatial resolution.

Further improvements to the accuracy and precision of CASI were made by incorporating a subpixel algorithm, based on a sinc function method proposed by Foroosh and Zerubia [27]. The sinc function, or cardinal sine function, is used to interpolate band-limited discrete-time signals with respect to sampling rate and time by using uniformly spaced samples of each signal. This new method, referred to as high-density mapping (HDM), demonstrated an accuracy of 0.09 pixels, a precision of 0.04 pixels, and repeatability of measuring regional work loops in the beating heart of 0.04 mmHg (normalized to end-diastolic area) [28].

To determine regional function with HDM, the heart must be exposed to allow for a region to be imaged. A region of interest is then divided into subimages, with the displacement of each subimage being determined between consecutive images. Through this algorithm, displacement can be determined at hundreds of locations,

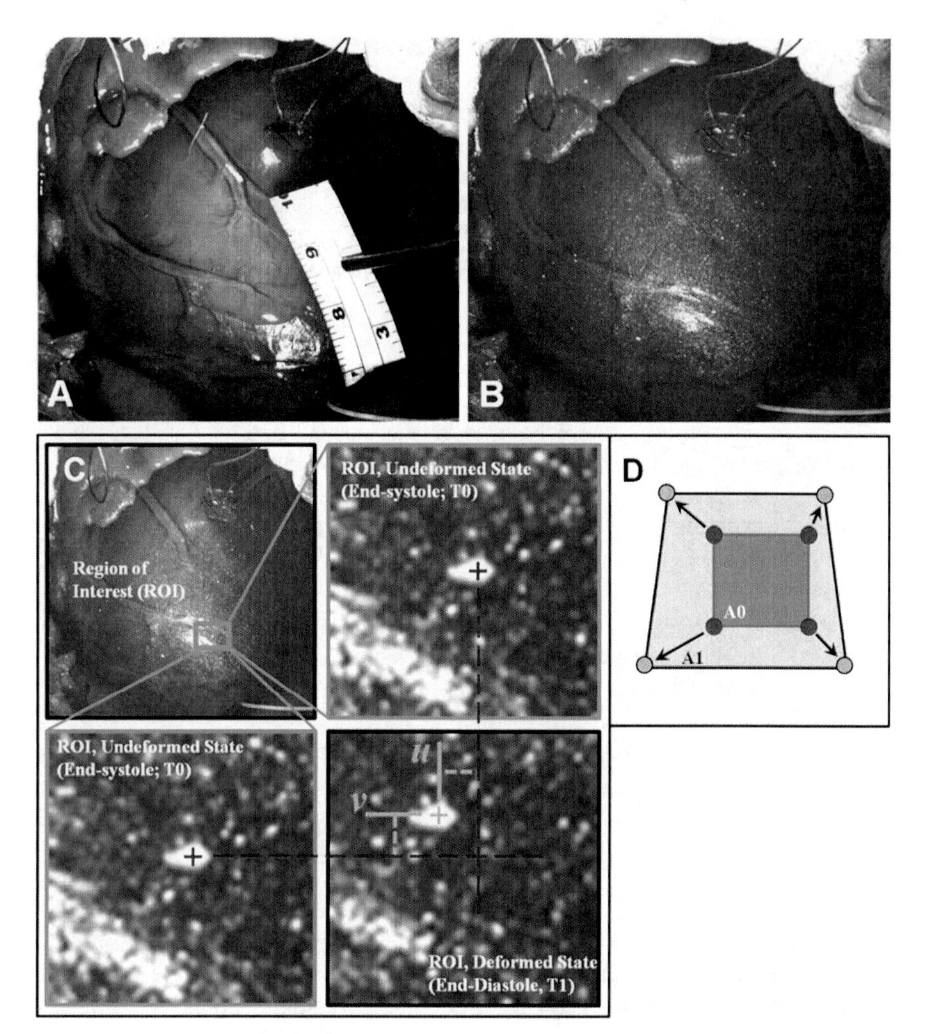

Fig. 1 Determination of regional cardiac function using high-density mapping (*HDM*). (**a**) Representative image of a porcine heart containing a region of interest (*ROI*). (**b**) Image containing a ROI, the same area as in (**a**), prepared with silicon carbide particles and retroreflective beads, creating speckled contrast for particle tracking. (**c**) An identified ROI outlined in *blue* in the *upper-left quadrant*. An enlargement of the ROI in the undeformed state at T_0 is shown in the *upper-right quadrant* and the *lower-left quadrant* (the same image is used for *u*, *v* reference points), with the location of a particle marked in *red*. An enlargement of the ROI in the deformed state at T_1, with the location of a particle marked in *yellow*, is shown in the *lower-right quadrant*, indicating the shift of speckle particles throughout the cardiac cycle. (**d**) Representation of how several areas within the ROI can be tracked through the cardiac cycle. As the heart expands during diastole, regions of normally functioning myocardium also expand. By application of this algorithm to several areas in the ROI, regional mechanical function can be determine in a whole-field manner, with high spatial resolution

providing a whole-field measurement. Regional stroke work can then be determined in very small regions on the basis of the change in area between four neighboring points. This method determines function in regions of less than 10 mm^2, as opposed to sonomicrometry, in which the regional areas are generally greater than 100 mm^2. This technique has been used to determine regional function in the isolated rabbit heart, the in vivo porcine heart, and the in vivo canine heart [29–33].

While the methods described herein provide data for both systolic and diastolic function, there are several limitations to them as well. Both MRI and echocardiography require expensive equipment to image the heart. Along with this equipment, a lot of time is invested to collect data using MRI. The spatial resolution of both methods is less than optimal. Sonomicrometry and optical methods require exposure of the heart, leading to an invasive procedure. In addition, optical methods generally require placement of nonsterile tracking markers, making them useful only during the terminal phases of an experiment.

4 Parameters for Assessing Mechanical Cardiac Function

Having discussed the clinical and experimental techniques used for analyzing heart structure and function, it is also necessary to review the relevant parameters that can be measured or calculated as functional indicators of cardiac mechanics. As the heart operates as an electromechanical pump, it functions to perpetuate and control circulation. Whereas the right ventricle pumps deoxygenated blood to the lungs (within relatively close proximity in the chest), the left ventricle pumps oxygenated blood throughout the body (accessing all internal and extremity tissues/organs). Consequently, the left ventricle performs more work than the right ventricle and has a thicker wall owing to increased muscularity. There are several measures that can be used to assess the function of the heart as a whole. The functional measures include stroke volume, stroke work, ejection fraction, tau (τ), maximum change in pressure over time (dP/dt_{max}), and minimum change in pressure as a function of time (dP/dt_{min}).

In describing these measures, it is useful to review the basic physiological dynamics that constitute a cardiac cycle. A cardiac cycle involves the complete succession of one phase of ventricular contraction (systole), followed by one phase of ventricular relaxation (diastole; Fig. 2). Ventricular systole begins as the mitral valve closes at the end of atrial contraction (Fig. 2, point A), and consists of two main stages known as isovolumic contraction and ejection. Isovolumic contraction is the brief period of time between the onset of systole and the opening of the aortic valve, during which the ventricular volume remains constant as the ventricular pressure increases sharply. As pressure continues to increase within the ventricle, the aortic valve opens (at approximately 80 mmHg in a healthy heart), marking the end of isovolumic contraction and the onset of ejection. Ventricular pressure continues to rise during rapid ejection, forcing blood from the ventricle to the aorta as myocardium contracts and ventricular volume decreases. As the pressure gradient, which drives blood flow, between the ventricle and the aorta becomes equalized,

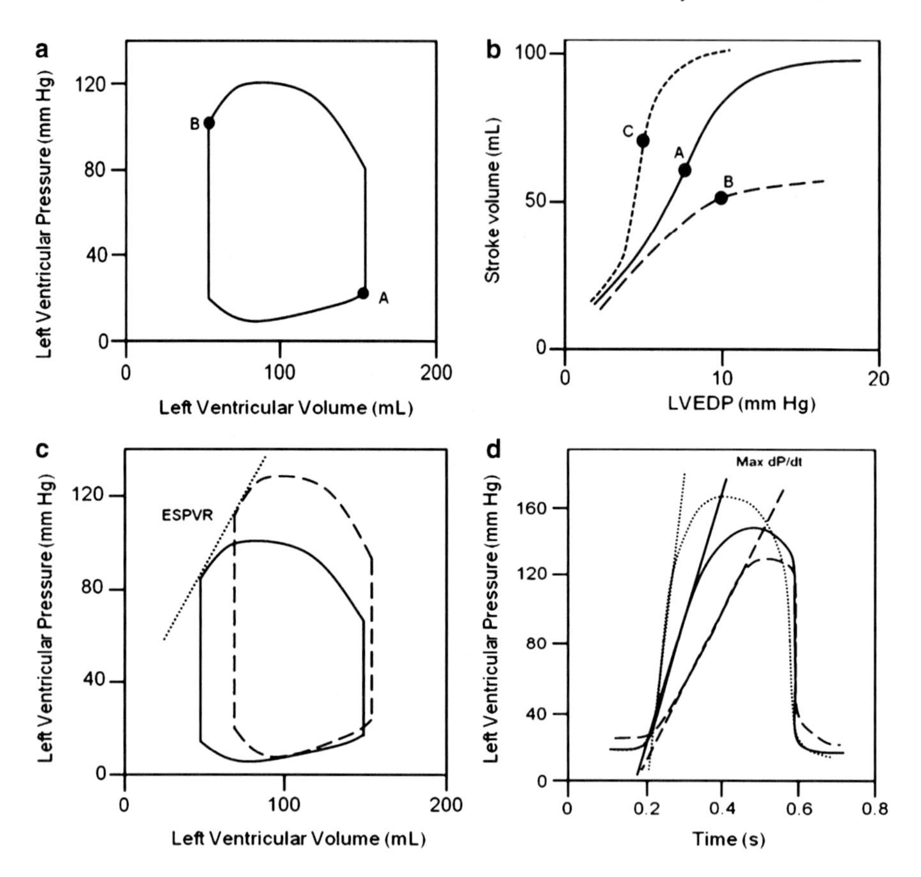

Fig. 2 Variables used in determining regional function in the heart. (**a**) Pressure–volume work loop of the left ventricle through one cardiac cycle. Point *A* indicates "end-diastole" and point *B* indicates "end-systole." Integrating for the area within the loop provides the amount of work done by the left ventricle over the course of one cardiac cycle. By substituting regional area for left ventricular volume, one constructs a regional stroke work loop, with regional stroke work being determined from the integral of left ventricular volume with respect to regional area. (**b**) Frank–Starling curves show the effects that preload can have on stroke volume, which is directly related to stretching during diastolic filling. *A* indicates a normal operating curve, whereas *B* indicates decreased contractility and *C* indicates increased contractility. (**c**) End-systolic pressure–volume relationship (*ESPVR*) is determined by plotting this point for several different-sized work loops. The size of the work loop can be altered by changing the preload on the ventricle. Preload-recruitable stroke work, the linear relationship between stroke work and end-diastolic volume, can also be determined with the information from this plot. (**d**) Maximum (or minimum) d*P*/d*t* (change in pressure with change in time) can be obtained by plotting the left ventricular pressure as a function of time. This relationship can be altered, providing a measure of the mechanical function in the left ventricle. *LVEDP* left ventricular end-diastolic pressure

pressure within the ventricle begins to decrease for reduced ejection. When the ventricle fully contracts to a minimum volume and the aortic pressure is greater than the ventricular pressure, the aortic valve closes to mark the end of the contraction phase (end-systole; Fig. 2, point B). End-systole also signifies the onset of the diastolic ventricular relaxation phase. Diastole begins with isovolumic relaxation, in which ventricular pressure decreases sharply as ventricular volume remains steady. During this period, the atria are filling with blood. The isovolumic relaxation stage ends as the mitral valve opens, signifying the start of diastolic filling. During the filling stage, ventricular volume rapidly increases, allowing the ventricular pressure to remain low for efficient blood flow from the higher-pressured atrium. When the ventricle is fully relaxed and the ventricular pressure exceeds atrial pressure, the mitral valve closes to mark the end of the relaxation phase (end-diastole; Fig. 2, point A), thus completing the cycle.

Another key concept of cardiac physiology is the effective preload and afterload on cardiac muscle. Preload can be described as the stretching of cardiomyocytes that make up the intact heart, effectively represented by the end-diastolic ventricular volume at the end of filling. Several conditions can either increase or decrease preload, thereby having a direct effect on end-diastolic volume and the amount of blood filling the ventricle. Afterload can be described as the amount of pressure that must be generated by the left ventricle to overcome aortic pressure, which is necessary for the ejection of blood into the aorta. Just as with preload, there are a number of conditions that can either increase or decrease afterload. Although preload and afterload do not directly effect each other, there is a well-known relationship, the Frank–Starling mechanism, in which the heart can change its force contraction in response to venous return of blood. Venous return affects ventricular filling, thereby affecting left ventricular end-diastolic pressure (LVEDP, or preload). Ventricular filling also affects end-diastolic volume (preload), and changes in sarcomere length affect force generation in the form of stroke volume. The Frank–Starling mechanism can be represented by plotting stroke volume as a function of LVEDP (Fig. 2). The dynamic relationship shows how increases in LVEDP can increase stroke volume and decreases in LVEDP can decrease stroke volume (Fig. 2b). This section will focus on measured parameters with reference to the cardiac cycle and physiologic phenomenon described (Fig. 2).

Stroke volume is a very practical volumetric measure, defined as the amount of blood that can be pumped into the aorta in one cardiac contraction (Eq. 1). Stroke volume is calculated by subtracting the end-systolic volume (the minimum ventricular volume) from the end-diastolic volume (the maximum ventricular volume):

$$\mathrm{SV} = \mathrm{EDV} - \mathrm{ESV}, \tag{1}$$

where SV is stroke volume, EDV is end-diastolic volume, and ESV is end-systolic volume.

Stroke volume is also considered to be a measure of mechanical function of the heart. Integrating ventricular volume (V) with respect to ventricular pressure (P) yields stroke work as shown in Eq. 2:

$$\mathrm{SW} = \int P\mathrm{d}V, \tag{2}$$

where SW is stroke work.

Stroke work is the amount of work performed by the ventricle as it ejects blood to the aorta in one cycle, or beat [34]. Stroke work can change with the presentation of cardiomyopathies (e.g. myocardial infarction), resulting from physiological changes in ventricular volume and/or pressure. To assess regional contractile function, regional area is substituted for ventricular volume in Eq. 2, resulting in regional stroke work.

Many clinical and experimental studies have also used ejection fraction as a measure of contractility. Ejection fraction is the ratio of the volume of blood ejected into the aorta in one cycle (i.e. end-diastolic volume minus end-systolic volume, which is equivalent to stroke volume) to the end-diastolic volume, as shown in Eq. 3:

$$\mathrm{EF} = \frac{\mathrm{EDV} - \mathrm{ESV}}{\mathrm{EDV}} = \frac{\mathrm{SV}}{\mathrm{EDV}}, \tag{3}$$

where EF is ejection fraction, EDV is end-diastolic volume, ESV is end-systolic volume, and SV is stroke volume.

Different from stroke volume, ejection fraction provides an indication of contractile efficiency by expressing the percentage of blood pumped out of the ventricle with respect to the maximum ventricular volume.

Parameters such as stroke volume, stroke work, and ejection fraction are relatively easy to determine in the clinic with commonly available equipment. This advantage allows the same variable to be measured at multiple testing sites and times. In addition, it offers a way to assess patients and compare the effects of potential treatment options. However, these parameters can be altered by ventricular aneurysms, valve incompetence, or changes in the overall ventricular size. In addition, changes in the passive properties of infarcted myocardium can lead to improved diastolic filling without improving the active contractile properties of the ventricle. Measurements of regional parameters that are synced with the ventricular pressure or EKG are better for determining improved contractile function.

Wall thickening, regional strain, and systolic contraction can be acquired clinically to provide information on the contractile status of the ventricular wall. It is important to determine if the wall is contracting in synchrony with the rest of the ventricle. A passive material in the heart that is loaded with intracavitary pressure will decrease in thickness, whereas a contractile material will increase in thickness (which leads to the increase in intracavitary pressure).

Animal models provide more opportunity for measuring independent variables that are not easy to control in the clinic (e.g. heart rate and afterload). Preload recruitable stroke work has been used to evaluate the global function of the heart. Preload recruitable stroke work, the linear relationship of the stroke work to end-diastolic volume, is independent of heart rate and afterload. Measurements such as end-systolic volume, end-systolic pressure–volume relationship, and change in ventricular pressure over time ($\mathrm{d}P/\mathrm{d}t$) can provide valuable information on the contractile status in the global heart, but can also be dependent on heart rate and aortic pressure (Fig. 2).

Although many techniques focus on systolic function, diastolic function also plays an important role in heart failure as stiffness and wall thickness can affect relaxation and filling. Diastolic function has commonly been assessed on the basis of a global measurement of the time constant of isovolumic pressure decay, tau (τ). The isovolumic relaxation constant was determined by Weiss et al. [35] by assessing pressure fall during isovolumic beats. They found pressure fall, dP/dt, was exponential during isovolumic relaxation and was characterized by a time constant of relaxation, τ. More recently, apical recoil or untwisting and the temporal change in this variable have been used to assess diastolic function of the ventricle [36]. As this can be measured at different locations within the ventricle, it can be used to define regional diastolic function.

5 Regional Cardiac Mechanics

Myocardial defects such as myocardial infarction generally occur within small regions of the heart. To study these effects and potential treatments, it is necessary to be able to determine the mechanical function in the defined regions. Clinically, these measurements can be taken with several techniques, including MRI, echocardiography, and NOGA. Tagged MRI is a method that can noninvasively measure the motions of artificial markers within the heart wall [37, 38]. As mentioned, MRI is advantageous because it can be used to determine 3D measurements of both stroke volume and stroke work. It can also be used for local strain measurements to determine function in regions of the myocardium that may be damaged by ischemia, hypertrophy, or infarction [39–42]. Dong et al. [43] have also used tagged MRI on canines to measure the angle displacement of the base of the heart relative to the apex. The displacement angles were indicative of the twisting of the heart due to an applied force in the circumferential direction, resulting in ventricular torsion. Torsion measurements determined during diastole could then be correlated to the global measure of the diastolic relaxation time constant, τ, during isovolumic relaxation [43].

Echocardiography is also used to determine regional changes in wall size, wall movements, and cavity size, leading to parameters including wall thickening, systolic contraction, strain, and strain rate. Most recently, echocardiography has been used to determine strain and strain rate in the myocardium. A new method of echocardiography called ultrasound speckle tracking imaging has been employed, which uses B-scan ultrasound to give the appearance of speckle patterns within the tissue. The speckle pattern can be tracked using methods similar to HDM (as described earlier). Motion analysis can then be done on the speckles to determine strain and ventricular torsion [44].

With use of the aforementioned methods, deformation can be determined in the beating heart, leading to the determination of engineering strain, defined in the uniaxial case as the change in length divided by the original undeformed length. The heart is exposed to 3D deformations, leading to strain along three axes and shear stress acting on orthogonal planes. There exists an orientation in which there is no shear strain. In this orientation, the maximum and minimum strain exists. The maximum strain is the principal strain in the two direction and the minimum strain is the principal strain in the two direction, which is orthogonal to the one direction.

The strain in the one direction is the expansion of the myocardium during systole. Since the myocardium is incompressible and hence the volume needs to be conserved, expansion has to be coupled with contraction in a different direction in the wall. Principal strain changes in both magnitude and angle resulting from cardiac events [45–47]. Using MRI to measure principal strain changes in mice with heart failure, Hankiewicz et. al. [46] found that the principal strain in the one direction decreased with advancement of heart failure compared with controls. In addition, they detected decreases in principal strains prior to dilation of the ventricle [46]. Cupps et al. [47] used MRI to measure the principal strain orientation in the normal human left ventricle. Their results suggested that principal strain in the maximum contraction direction occurs in the circumferential–longitudinal plane that aligns with the apex-to-base axis of the heart [47]. Their work shows that orienting displacement data along the apex-to-base cardiac axis will provide the maximum contractile strain for the heart.

6 Active and Passive Mechanical Function

Mechanical function in the heart consists of both active and passive components, both of which are potential targets for regenerative therapy. The active function is manifested by the contraction of cardiomyocytes, which is primarily responsible for the systolic phase of the cardiac cycle. Active function (i.e. systolic contraction) decreases during myocardial infarction and heart failure owing to the loss or dysfunction of cardiomyocytes. Just as important as active function, the passive function of the ventricle also plays an important role by providing a compliant structure. In the cardiac cycle, the compliance of the myocardium allows for maximum shortening during systolic contraction and maximum expansion during diastolic relaxation.

Myocardium is a complex tissue, comprising contractile cardiomyocytes that interact with each other and several other cell types (e.g. vascular smooth muscle cells, endothelial cells, and fibroblasts) in an extracellular matrix (ECM) composed of structural proteins (e.g. collagen, elastin, laminin, and fibronectin). Whereas cardiomyocytes are responsible for active function, the interactions of cells and ECM are responsible for passive function. Cell–ECM interactions provide the structural network for anchoring cardiomyocytes and translating the sarcomeric shortening of individual myocytes into organized myocardial contraction. The network also provides elasticity and integrity during diastole, allowing for maximum expansion while preventing overstretching or rupture.

Active and passive function can both be compromised during myocardial infarction. During a myocardial infarction, the heart undergoes a wound response that typically includes inflammation, cellular turnover, and ECM remodeling. Dead or afflicted cells (including cardiomyocytes) are replaced with new cellular components such as fibroblasts and vascular cells. The content and organization of the cardiac ECM is remodeled by both reparative and reactive fibrosis, causing the degradation of normal matrix and increasing the deposition of collagen [48]. A myocardial infarction leaves the affected area devoid of cardiomyocytes and, in general, the resulting scar is stiffer than healthy cardiac ECM. As the stress in the heart wall is inversely proportional to the wall thickness, a thin-walled scar will have higher wall stress than a thicker-walled scar. Therefore, there is greater concern that a thin-walled scar will rupture. For the non-thin-walled scar, not only does the infarct region lose contractile elements necessary for active function, but passive function is also impaired as the stiffer matrix is more difficult to deform during systolic contraction and diastolic expansion. For thin-walled scarred myocardium, an aneurysm can form owing to the intracavitary pressure causing increased stress in the ventricular wall, resulting in large deformation in this region. Considering the spared cardiomyocytes at the border zone of a myocardial infarction or in the case of a nontransmural myocardial infarction, cell–infarct interactions compromise systolic contraction as myocytes have a more difficult time shortening when anchored to a noncompliant substrate.

To repair the outcomes of a myocardial infarction and restore mechanical function, regenerative therapies can be targeted to improve active function, passive function, or both. Adding contractile elements to the infarct may increase active contraction. By delivery of exogenous cells (i.e. stem-cell-derived cardiomyocytes) or recruitment of endogenous cells to home and differentiate to cardiomyocytes, it may be possible to restore contractile function. Alternatively, improving the compliance of the infarct region may increase passive function. Whether the infarct is replaced surgically with a more compliant biomaterial or cells that improve cardiac tissue remodeling are delivered/recruited, it may be possible to decrease stiffness (or increase elasticity) for improved passive function. Implanting a more compliant biomaterial seeded with cardiomyocytes, or implanting a biomaterial that actively recruits cardiomyocytes, may combine efforts to improve both active and passive mechanical function.

It is difficult to distinguish the contractile properties from the passive properties in vivo. In normal myocardium, the increase in intracavitary pressure results from the contraction of myocytes. If one were to look at a region on the epicardial surface of the heart (Fig. 3), the area of this region would decrease with increased load (intracavitary pressure). However, if a material devoid of cells and contractile properties is implanted in the heart, it will expand under an increased cavitary load. This is similar to what happens when an elastic balloon is filled with water; the balloon expands as the pressure inside the balloon increases. If the balloon were composed of a stiff material, the balloon would still increase in size, however the increase would be less. When an acellular compliant scaffold, such as isolated porcine urinary bladder ECM, is implanted in the ventricular wall, the area of the scaffold will increase with intracavitary pressure (Fig. 3) [33]. If the scaffold is stiff, such as Dacron, the load is not enough to cause a significant change in the area of the implanted scaffold. Hence, changes in intracavitary pressure do not deform the scaffold, as seen in Fig. 3.

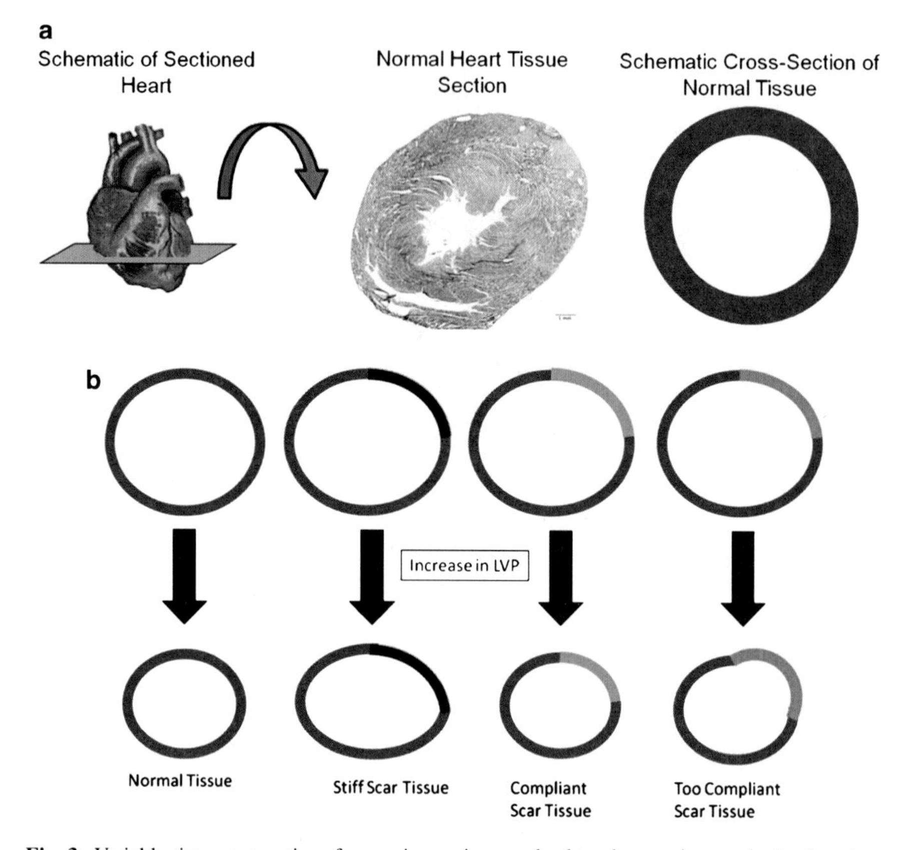

Fig. 3 Variable tissue properties of a passive region can lead to changes in ventricular function. For a region that is passive, that is, one that does not have any contractile elements, its deformation is dictated by the intracavitary pressure. Unlike normal myocardium, which contracts with increasing intracavitary pressure, passive materials will elongate (resulting in a thinner wall) under intracavitary loads. (**a**) The normal cross-sectional area of the left ventricle is shown as a *red circle*. (**b**) If the mechanical properties of a region of the heart are altered (e.g. myocardial infarct, scar, scaffold implantation), the shape of the cross section of the ventricle can be altered. As the pressure increases in the normal heart, the entire cross section of the ventricle contracts (*far left*). In the case of a stiff scar, the region cannot contract, leading to a region of the ventricle that does not change in shape (*second from left*). If the scar is compliant, it will deform with the ventricle, although it will not contract if it is passive (*second from right*). However, if the scar is thin and very compliant, it will balloon out as the pressure inside the ventricle increases (*far right*). *LVP* left ventricular pressure

If contractile cells begin to populate the implanted region, they must be able to deform the scaffold prior to performing useful mechanical work that contributes to pump function. If the material is expanding during contraction (Fig. 3), then the contractile cells in this region must generate enough force to balance the tensile forces due to the increased intracavitary load. If the region is stiff, as is often the case in collagen-populated infarcts, the contractile cells must be able to deform this stiff region to decrease systolic volume in the heart. For example, contracting embryoid bodies are able to deform attached cells, but they do not deform the cell culture dish. Thus, if myocytes were regenerated in a stiff infarcted region of the heart, they may not contribute to improving contraction in the heart. Thus, there appears to be a critical number of myocytes that need to be regenerated, which is dependent on the passive mechanical properties of the region, prior to contributing to the active function of the heart [49].

7 Regional Mechanical Function in the Dysfunctional Heart

Most animal models used to evaluate cell therapy aim to develop a well-defined region of dysfunction. Mechanical function within an ischemic or infarcted heart is generally heterogeneous, with function lowest toward the center of the infarct. For example, Fig. 4 demonstrates work loops from regions within the ischemic porcine heart. Mechanical function in the border zone, the region between infarcted tissue and normally perfused tissue, is very different from normally perfused tissue or tissue in the center of the ischemic zone. The work loops from the border zones (regions 2, 6, 11, 15, and 20 in Fig. 4) in general show very little work, as evident by the lack of any significant area within the work loop. Regions within the ischemic zone (regions 4, 8, 12, and 16 in Fig. 4) show reversed work loops, suggesting the

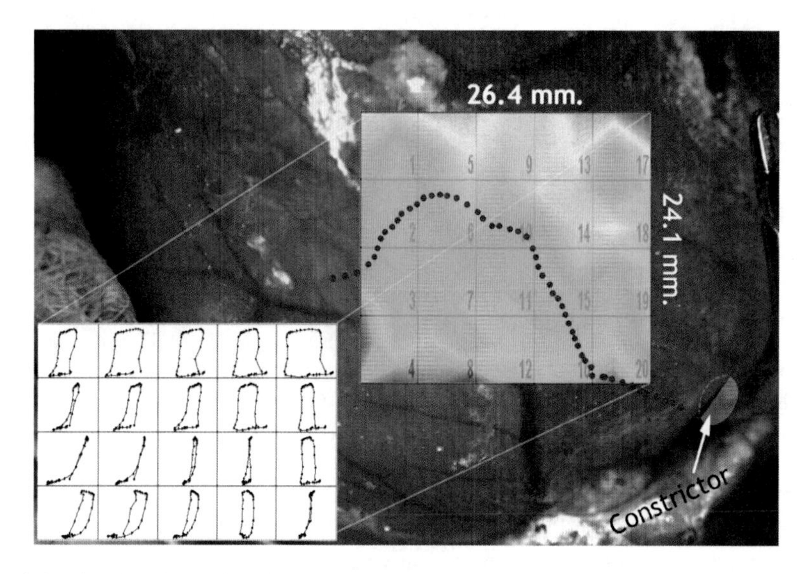

Fig. 4 Regional stroke work in the regionally ischemic porcine heart. The color map denotes the change in regional stroke work across the ischemic border (*orange* represents the highest function, whereas *blue* represents the lowest function). The *insert in the lower-left corner* shows some of the regional work loops determined with HDM in the region of interest. For clarity, only 20 of the 60 regional work loops computed in the ROI are shown. Note that the loops in *insert in the lower-left corner* are actually reversed in direction (clockwise), denoting negative regional work

contraction of these regions is out of sync with ventricular contraction. While regions of normally perfused tissue are contracting, resulting in increased intracavitary pressure, the ischemic regions are dyskinetic and expanding owing to the increased load on the tissue.

Similar to myocardial ischemia, various ranges of dysfunction can occur within the infarcted myocardium depending on the severity of the infarct, the time of recovery, and the location [50]. For more details on the mechanical properties of myocardial infarcts, see Holmes et al. [51].

To evaluate the effects of cell therapy in the dysfunctional heart, the region of cell delivery must first be identified. Delivery by intravenous or intracoronary methods can lead to dispersion of cells throughout the heart, making the location of cells difficult to identify. To clearly identify the delivery region, we have employed the use of a cell-seeded myocardial patch composed of an ECM [32]. When this patch is used to replace a full-thickness defect in the ventricular wall, a region for mechanical analysis is well defined. Using HDM (previously described) allowed for determination of function in regions approximately 5 mm×5 mm. Upon implantation of unseeded scaffold, there is no contractile function (regional stroke work equals zero). As the load inside the ventricle increase, the scaffold passively deforms, increasing in area. Such a response is expected of any compliant, passive material when it is exposed to an increase in intracavitary pressure. A stiff material, such as Dacron, is also a passive material and does not demonstrate any deformation when the heart contracts [49]. The deformation is limited owing to the stiff (noncompliant) nature of the material. Eight weeks after implantation of scaffolds seeded with human mesenchymal stem cells, regional stroke work is positive. This suggests that the scaffold region is contracting in sync with the rest of the heart because the scaffold region is contracting while the intracavitary pressure is increasing. Other parameters, including systolic shortening, can also provide a measure of contractility of a region.

8 Mechanical Function Post Stem Cell Delivery

A number of clinical trials have been initiated to assess the resultant function of various cell therapies delivered to the heart. The results from trials have been mixed, with some trials indicating improved left ventricular ejection fraction ranging from 2.5 to 6.0 percentage points with bone marrow cells at 4–6 months compared with controls [40–42] and others showing no effect from cell therapy (treatment effect of +0.6 and +1.0 percentage points on left ventricular ejection fraction) [43, 44]. Despite short-term improvements in function, it is unclear whether long-term enhancement of ejection fraction can be sustained as another study showed loss of improvements at 18 months after cell delivery [45]. To make these measurements in small animals such as mice and rats, echocardiography and MRI are frequently used. Both of these techniques are able to determine the global volume of the heart, allowing assessment of global function in the heart. However, the limited spatial resolution of the techniques decreases the ability to accurately assess regional function in the heart of small animals. In addition to echocardiography and MRI, sonomicrometry, conductance catheters [52], and digital imaging techniques have been used in large-animal models.

One major limitation facing the field is that the use of different parameters (e.g. ejection fraction, stroke work, developed pressure, and wall thickness) makes it difficult to compare different studies in a head-to-head fashion. In addition, many of these parameters are afterload- and/or preload-dependent, variables that can be different depending on the study. Therefore, it is difficult to compare the various cell types being used (as described throughout this book), which confounds the determination of an optimal cell type for cardiomyoplasty. Another major limitation that is often not considered is the use of human cells in animal models, whose hearts beat much faster than human hearts. It remains to be determined if human cells delivered in the human heart will engraft with similar efficiency and cause improvement of functional parameters similar to that of animal models. As mentioned previously, it is also difficult to separate the passive component and the active component of cardiac mechanical function. Moving forward, more studies are needed to correlate the improvement of mechanical function with the evidence of increased contractile mass. This correlation will help to distinguish whether the improvements can be attributed to regenerated, actively contracting cardiomyocytes, or whether the improvements are due to more optimal tissue compliance with the delivered cellular mass.

9 Summary

Although treatment for myocardial ischemia has made significant progress over the past few decades, myocardial infarction remains an elusive target. Current treatments for infarction need to address the problem at hand: the loss of contractile cells that provide mechanical function to the heart. The restoration of contractile myocytes to dysfunctional cardiac tissue can restore active function to regions of the heart. Although several options exist for evaluating the mechanical function in the heart, the limitations of each method must also be considered when interpreting results. Improved passive function of infarcted myocardium can increase cardiac function, but complete restoration of mechanical function in the heart must also include active properties.

References

1. Laflamme, M.A. and C.E. Murry, *Regenerating the heart.* Nat Biotechnol, 2005. **23**(7): p. 845–56.
2. Gaudette, G.R. and I.S. Cohen, *Cardiac regeneration: materials can improve the passive properties of myocardium, but cell therapy must do more.* Circulation, 2006. **114**(24): p. 2575–7.
3. Wall, S.T., et al., *Theoretical impact of the injection of material into the myocardium: a finite element model simulation.* Circulation, 2006. **114**(24): p. 2627–35.

4. Stoylen, A., et al., *Strain rate imaging by ulrasonography in the diagnosis of coronary artery disease.* J Am Soc Echocardiogr, 2000. **13**: p. 1053–64.
5. Spencer, K.T., et al., *The role of echocardiographic harmonic imaging and contrast enhancement for improvement of endocardial border delineation.* J Am Soc Echocardiogr, 2000. **13**(2): p. 131–8.
6. Voigt, J.U., et al., *Assessment of regional longitudinal myocardial strain rate derived from doppler myocardial imaging indexes in normal and infarcted myocardium.* J Am Soc Echocardiogr, 2000. **13**(6): p. 588–98.
7. Trambaiolo, P., et al., *New insights into regional systolic and diastolic left ventricular* function *with tissue Doppler echocardiography: from qualitative analysis to a quantitative approach.* J Am Soc Echocardiogr, 2001. **14**: p. 85–96.
8. Heimdal, A., et al., *Real-time strain rate imaging of the left ventricle by ultrasound.* J Am Soc Echocardiogr, 1998. **11**(11): p. 1013–9.
9. Derumeaux, G., et al., *Doppler tissue imaging quantitates regional wall motion during myocardial ischemia and reperfusion.* Circulation, 1998. **97**(19): p. 1970–7.
10. Urheim, S., et al., *Myocardial strain by Doppler echocardiography. Validation of a new method to quantify regional myocardial function.* Circulation, 2000. **102**(10): p. 1158–64.
11. Moulton, M.J., et al., *Spline surface interpolation for calculating 3-D ventricular strains from MRI tissue tagging.* Am J Physiol, 1996. **270**: p. H281–97.
12. Gotte, M.J., et al., *Recognition of infarct localization by specific changes in intramural myocardial mechanics.* Am Heart J, 1999. **138**(6 Pt 1): p. 1038–45.
13. McVeigh, E., *Regional myocardial function.* Cardiol Clin, 1998. **16**(2): p. 189–206.
14. Scott, C.H., et al., *Effect of dobutamine on regional left ventricular function measured by tagged magnetic resonance imaging in normal subjects.* Am J Cardiol, 1999. **83**(3): p. 412–7.
15. Bogaert, J., et al., *Remote myocardial dysfunction after acute anterior myocardial infarction: impact of left ventricular shape on regional function: a magnetic resonance myocardial tagging study.* J Am Coll Cardiol, 2000. **35**(6): p. 1525–34.
16. Stoylen, A., et al., *Strain rate imaging by ultrasonography in the diagnosis of coronary artery disease.* J Am Soc Echocardiogr, 2000. **13**(12): p. 1053–64.
17. Meunier, J. and M. Bertrand, *Ultrasonic texture motion analysis: theory and simulation.* IEEE Trans Med Imaging, 1995. **14**(2): p. 293–300.
18. Castro, P.L., et al., *Potential pitfalls of strain rate imaging: angle dependency.* Biomed Sci Instrum, 2000. **36**: p. 197–202.
19. Spencer, K.T., et al., *The role of echocardiographic harmonic imaging and contrast enhancement for improvement of endocardial border delineation.* J Am Soc Echocardiogr, 2000. **13**(2): p. 131–8.
20. Sayad, D.E., et al., *Dobutamine magnetic resonance imaging with myocardial tagging quantitatively predicts improvement in regional function after revascularization.* Am J Cardiol, 1998. **82**(9): p. 1149–51.
21. Cohen, M.V., et al., *Favorable remodeling enhances recovery of regional myocardial function in the weeks after infarction in ischemically preconditioned hearts.* Circulation, 2000. **102**(5): p. 579–83.
22. Atkins, B.Z., et al., *Myogenic cell transplantation improves in vivo regional performance in infarcted rabbit myocardium.* J Heart Lung Transplant, 1999. **18**(12): p. 1173–80.
23. von Degenfeld, G., W. Giehrl, and P. Boekstegers, *Targeting of dobutamine to ischemic myocardium without systemic effects by selective suction and pressure-regulated retroinfusion.* Cardiovasc Res, 1997. **35**(2): p. 233–40.
24. Bufkin, B.L., et al., *Preconditioning during simulated MIDCABG attenuates blood flow defects and neutrophil accumulation.* Ann Thorac Surg, 1998. **66**(3): p. 726–31.
25. Prinzen, F.W., et al., *Discrepancies between myocardial blood flow and fiber shortening in the ischemic border zone as assessed with video mapping of epicardial deformation.* Phlugers Arc (Eur J Physiol), 1989. **415**: p. 220–29.
26. Gaudette, G.R., et al., *Computer aided speckle interferometry: a technique for measuring deformation of the surface of the heart.* Ann Biomed Eng, 2001. **29**(9): p. 775–80.

27. Foroosh, H. and J. Zerubia, *Extension of phase correlation to subpixel registration.* IEEE Trans Image Process, 2002. **11**(3): p. 188–200.
28. Kelly, D.J., et al., *Accuracy and reproducibility of a subpixel extended phase correlation method to determine micron level displacements in the heart.* Med Eng Phys, 2007. **29**(1): p. 154–62.
29. Gaudette, G.R., et al., *Determination of regional area stroke work with high spatial resolution in the heart.* Cardiovasc Eng: Int J, 2002. **2**(4): p. 129–37.
30. Azeloglu, E.U., et al., *High resolution mechanical function in the intact porcine heart: mechanical effects of pacemaker location.* J Biomech, 2006. **39**(4): p. 717–25.
31. Kochupura, P.V., et al., *Tissue-engineered myocardial patch derived from extracellular matrix provides regional mechanical function.* Circulation, 2005. **112**(9 Suppl): p. I144–9.
32. Potapova, I.A., et al., *Enhanced recovery of mechanical function in the canine heart by seeding an extracellular matrix patch with mesenchymal stem cells committed to a cardiac lineage.* Am J Physiol Heart Circ Physiol, 2008. **295**(6): p. H2257–63.
33. Kelly, D.J., et al., *Increased myocyte content and mechanical function within a tissue-engineered myocardial patch following implantation.* Tissue Eng Part A, 2009. **15**(8): p. 2189–201.
34. Levy, M.N., A.J. Pappano, and R.M. Berne, *Cardiovascular physiology.* 9th ed. Mosby physiology monograph series. 2007, Philadelphia, PA: Mosby Elsevier. xiv, 269 p.
35. Weiss, J.L., J.W. Frederiksen, and M.L. Weisfeldt, *Hemodynamic determinants of the time-course of fall in canine left ventricular pressure.* J Clin Invest, 1976. **58**(3): p. 751–60.
36. Dong, S.J., et al., *Independent effects of preload, afterload, and contractility on left ventricular torsion.* Am J Physiol, 1999. **277**(3 Pt 2): p. H1053–60.
37. Axel, L. and L. Dougherty, *Heart wall motion: improved method of spatial modulation of magnetization for MR imaging.* Radiology, 1989. **172**(2): p. 349–50.
38. Axel, L. and L. Dougherty, *MR imaging of motion with spatial modulation of magnetization.* Radiology, 1989. **171**(3): p. 841–5.
39. Kraitchman, D., et al., *Myocardial perfusion and function in dogs with moderate coronary stenosis.* Magn Reson Med, 1996. **35**(5): p. 771–80.
40. Kuijpers, D., et al., *Dobutamine cardiovascular magnetic resonance for the detection of myocardial ischemia with the use of myocardial tagging.* Circulation, 2003. **107**(12): p. 1592–7.
41. Young, A., et al., *Three-dimensional left ventricular deformation in hypertrophic cardiomyopathy.* Circulation, 1994. **90**(2): p. 854–67.
42. Kramer, C., et al., *Regional differences in function within noninfarcted myocardium during left ventricular remodeling.* Circulation, 1993. **88**(3): p. 1279–88.
43. Dong, S.J., et al., *MRI assessment of LV relaxation by untwisting rate: a new isovolumic phase measure of tau.* Am J Physiol Heart Circ Physiol, 2001. **281**(5): p. H2002–9.
44. Notomi, Y., et al., *Measurement of ventricular torsion by two-dimensional ultrasound speckle tracking imaging.* J Am Coll Cardiol, 2005. **45**(12): p. 2034–41.
45. Hankiewicz, J. and E. Lewandowski, *Improved cardiac tagging resolution at ultra-high magnetic field elucidates transmural differences in principal strain in the mouse heart and reduced stretch in dilated cardiomyopathy.* J Cardiovasc Magn Reson, 2007. **9**(6): p. 883–90.
46. Hankiewicz, J., et al., *Principal strain changes precede ventricular wall thinning during transition to heart failure in a mouse model of dilated cardiomyopathy.* Am J Physiol Heart Circ Physiol, 2008. **294**(1): p. H330–6.
47. Cupps, B., et al., *Principal strain orientation in the normal human left ventricle.* Ann Thorac Surg, 2005. **79**(4): p. 1338–43.
48. Swynghedauw, B., *Molecular mechanisms of myocardial remodeling.* Physiol Rev, 1999. **79**(1): p. 215–62.
49. Kelly, D.J., et al., *Increased myocyte content and mechanical function within a tissue-engineered myocardial patch following implantation.* Tissue Eng Part A, 2009. **15**(8): p. 2189–201.
50. Fomovsky, G.M. and J.W. Holmes, *Evolution of scar structure, mechanics, and ventricular function after myocardial infarction in the rat.* Am J Physiol Heart Circ Physiol, 2010. **298**(1): p. H221–8.

51. Holmes, J.W., T.K. Borg, and J.W. Covell, *Structure and mechanics of healing myocardial infarcts.* Annu Rev Biomed Eng, 2005. **7**: p. 223–53.
52. Kim, B.O., et al., *Cell transplantation improves ventricular function after a myocardial infarction: a preclinical study of human unrestricted somatic stem cells in a porcine model.* Circulation, 2005. **112**(Suppl 9): p. I96–104.

Index

I.S. Cohen and G.R. Gaudette (eds.), *Regenerating the Heart*, Stem Cell Biology and Regenerative Medicine, DOI 10.1007/978-1-61779-021-8,
© Springer Science+Business Media, LLC 2011

C

LIBRARY-LRC
TEXAS HEART INSTITUTE